Fish Viruses and Bacteria

Pathobiology and Protection

Fish Viruses and Bacteria

Pathobiology and Protection

Edited by

Patrick T.K. Woo

University of Guelph, Canada

and

Rocco C. Cipriano

National Fish Health Research Laboratory, USA

CABI is a trading name of CAB International

CABI	CABI
Nosworthy Way	745 Atlantic Avenue
Wallingford	8th Floor
Oxfordshire OX10 8DE	Boston, MA 02111
UK	USA
Tel: +44 (0)1491 832111	Tel: +1 (617)682-9015
Fax: +44 (0)1491 833508	E-mail: cabi-nao@cabi.org
E-mail: info@cabi.org	
Website: www.cabi.org	

A catalogue record for this book is available from the British Library, London, UK.

Library of Congress Cataloging-in-Publication Data

Names: Woo, P. T. K., editor. | Cipriano, Rocco C., editor. | C.A.B.
    International, issuing body.
Title: Fish viruses and bacteria : pathobiology and protection / editors,
    Patrick T.K. Woo and Rocco C. Cipriano.
Description: Wallingford, Oxfordshire ; Boston, MA : CABI, [2017] | Includes
    bibliographical references and index.
Identifiers: LCCN 2016045664 (print) | LCCN 2016049737 (ebook) | ISBN
    9781780647784 (hardback : alk. paper) | ISBN 9781780647791 (pdf) | ISBN
    9781780647807 (ePub)
Subjects: LCSH: Fishes--Virus diseases. | Fishes--Microbiology. | Bacterial
    diseases in fishes. | MESH: Fish Diseases--virology | Fish
    Diseases--microbiology | Fish Diseases--prevention & control
Classification: LCC SH177.V57 F57 2017 (print) | LCC SH177.V57 (ebook) | NLM
    SH 171 | DDC 639.3--dc23
LC record available at https://lccn.loc.gov/2016045664

ISBN-13: 978 1 78064 778 4

Commissioning editor: Caroline Makepeace
Associate editor: Alexandra Lainsbury
Production editor: Shankari Wilford

Typeset by SPi, Pondicherry, India
Printed and bound in the UK by CPI Group (UK) Ltd, Croydon, CR0 4YY, UK

Contents

Contributors

Note: corresponding authors are indicated by an asterisk.

Maria Aamelfot, The Norwegian Veterinary Institute, Ullevålsveien 68, PO Box 750 Sentrum, N-0106 Oslo. E-mail: maria.aamelfot@vetinst.no

F. C. Thomas Allnutt, BrioBiotech LLC, PO Box 26, Glenelg, MD 21737, USA. E-mail: fct.allnutt@gmail.com

Ellen Ariel, College of Public Health, Medical and Veterinary Sciences, James Cook University, Townsville, QLD 4811, Australia. E-mail: ellen.ariel@jcu.edu.au

Kristen D. Arkush, formerly of Bodega Marine Laboratory, University of California-Davis, Bodega Bay, CA 94923, USA. Current address: 9489 Argonne Way, Forestville, CA 95436, USA. E-mail: kdarkush@gmail.com

Ruben Avendaño-Herrera, Laboratorio de Patología de Organismos Acuáticos y Biotecnología Acuícola, Facultad de Ciencias Biológicas, Universidad Andrés Bello, Viña del Mar, Chile and Interdisciplinary Center for Aquaculture Research (INCAR), Concepción, Chile. E-mail: reavendano@yahoo.com

Jerri Bartholomew,* Department of Microbiology, Oregon State University, Corvallis, OR 97331, USA. E-mail: bartholj@science.oregonstate.edu

Bryndis Bjornsdottir, Matís, Vinlandsleid 12, 113 Reykjavik, Iceland. E-mail: bryndis@matis.is

Rocco C. Cipriano, National Fish Health Research Laboratory, 11649 Leetown Road, Kearneysville, WV 25430, USA. E-mail: rcipriano@usgs.gov

Robert Davies,* Institute of Infection, Immunity and Inflammation, College of Medical, Veterinary and Life Sciences, University of Glasgow, 120 University Place, Glasgow, G12 8TA, UK. E-mail: Robert.Davies@glasgow.ac.uk

Arun K. Dhar,* BrioBiotech LLC, PO Box 26, Glenelg, MD 21737, USA; from January 2017, Aquaculture Pathology Laboratory Room 102, School of Animal and Comparative Biomedical Sciences, College of Agriculture and Life Sciences, The University of Arizona, 1117 E. Lowell Street, Tucson, Arizona 85721 USA. E-mail: arun_dhar@hotmail.com or adhar@email.arizona.edu

Peter Dixon, Cefas Weymouth Laboratory, Barrack Road, The Nothe, Weymouth, Dorset, DT1 1RY, UK. E-mail: pfdixon@waitrose.com

Diane G. Elliott,* US Geological Survey, Western Fisheries Research Center, 6505 NE 65th Street, Seattle, WA 98115, USA. E-mail: dgelliott@usgs.gov

Mohamed Faisal, Department of Pathobiology and Diagnostic Investigation, College of Veterinary Medicine, Michigan State University, East Lansing, MI 48824, USA. E-mail: faisal@cvm.msu.edu

Knut Falk,* The Norwegian Veterinary Institute, Ullevålsveien 68, PO Box 750 Sentrum, N-0106 Oslo. E-mail: knut.falk@vetinst.no

David T. Gauthier,* Department of Biological Sciences, Old Dominion University, Norfolk, VA 23529, USA. E-mail: dgauthie@odu.edu

Rodman G. Getchell,* C4-177 Veterinary Medicine Center, Department of Microbiology and Immunology, Cornell University, Ithaca, NY 14853, USA. E-mail: rgg4@cornell.edu

Christopher M. Good, The Conservation Fund's Freshwater Institute, Shepherdstown, WV 25443, USA. E-mail: c.good@freshwaterinstitute.org

Terrence E. Greenway, Thad Cochran National Warmwater Aquaculture Center, Mississippi State University, Stoneville, MS 38776, USA. E-mail: greenway@drec.msstate.edu

Matt J. Griffin,* Thad Cochran National Warmwater Aquaculture Center, Mississippi State University, Stoneville, MS 38776, USA. E-mail: griffin@cvm.msstate.edu

Geoffrey H. Groocock, Transit Animal Hospital, 6020 Transit Road, Depew, NY 14043, USA. E-mail: g.h.groocock@gmail.com

Bjarnheidur K. Gudmundsdottir,* Faculty of Medicine, University of Iceland, Vatnsmyrarvegur 16, 101 Reykjavik, Iceland. E-mail: bjarngud@hi.is

Larry A. Hanson,* Department of Basic Sciences, College of Veterinary Medicine, Mississippi State University, PO Box 6100, Mississippi, USA. E-mail: hanson@cvm.msstate.edu

John P. Hawke,* Department of Pathobiological Sciences, LSU School of Veterinary Medicine, Louisiana State University, Baton Rouge, LA 70803, USA. E-mail: jhawke1@lsu.edu

Paul Hick, Faculty of Veterinary Science, School of Life and Environmental Sciences, The University of Sydney, Camden, NSW 2570, Australia. E-mail: paul.hick@sydney.edu.au

Renate Johansen, PHARMAQ Analytiq, Thormøhlensgate 55, 5008 Bergen, Norway. E-mail: renate.johansen@zoetis.com

Marius Karlsen,* PHARMAQ AS, Harbitzalléen 2A, 0213 Oslo, Norway. E-mail: marius.karlsen@pharmaq.no

Hisae Kasai, Faculty of Fisheries Sciences, Hokkaido University, Minato, Hakodate, Hokkaido 041-8611 Japan. E-mail: hisae@fish.hokudai.ac.jp

Yasuhiko Kawato, National Research Institute of Aquaculture, Japan Fisheries Research and Education Agency, Nakatsuhamaura 422-1, Minami-Ise, Mie 516-0193, Japan. E-mail: ykawato@affrc.go.jp

Lester H. Khoo, Thad Cochran Warmwater Aquaculture Center, Mississippi State University, Stoneville, MS 38776, USA. E-mail: khoo@cvm.msstate.edu

Gael Kurath, Western Fisheries Research Center, U.S. Geological Survey, Seattle, WA 98115, USA. E-mail: gkurath@usgs.gov

Scott LaPatra, Clear Springs Foods, PO Box 712, Buhl, ID 83316, USA. E-mail: scott.lapatra@clearsprings.com

Jo-Ann C. Leong,* Hawaiʻi Institute of Marine Biology, School of Ocean and Earth Science and Technology, University of Hawaiʻi at Mānoa, Kāneʻohe, HI 96744, USA. E-mail: joannleo@hawaii.edu

Thomas P. Loch, Department of Pathobiology and Diagnostic Investigation, College of Veterinary Medicine, Michigan State University, East Lansing, MI 48824, USA. E-mail: lochthom@cvm.msu.edu

John S. Lumsden,* Department of Pathobiology, Ontario Veterinary College, University of Guelph, 50 Stone Road E., Guelph, ON N1G 2W1, Canada and (Adjunct Professor) Department of Pathobiology, St. George's University, True Blue, Grenada. E-mail: jsl@uoguelph.ca

Beatriz Magariños, Departamento de Microbiología y Parasitología, Facultad de Biología-CIBUS/ Instituto de Acuicultura, Universidade de Santiago de Compostela, 15782, Spain. E-mail: beatriz.magarinos@usc.es

David P. Marancik, Department of Pathobiology, St. George's University of Veterinary Medicine, PO Box 7, True Blue, St. George's, Grenada, West Indies. E-mail: dmaranci@sgu.edu

Kazuhiro Nakajima, National Research Institute of Aquaculture, Japan Fisheries Research and Education Agency, Nakatsuhamaura 422-1, Minami-Ise, Mie 516-0193, Japan. E-mail: kazuhiro@fra.affrc.go.jp

Michael Ormsby, Institute of Infection, Immunity and Inflammation, College of Medical, Veterinary and Life Sciences, University of Glasgow, 120 University Place, Glasgow, G12 8TA, UK. E-mail: M.ormsby.1@research.gla.ac.uk

Andrew Orry, Molsoft LLC, 11199 Sorrento Valley Road, S209 San Diego, CA 92121, USA. E-mail: andy@molsoft.com

Martha W. Rhodes, Department of Aquatic Health Sciences, Virginia Institute of Marine Science, The College of William and Mary, Gloucester Point, VA 23062, USA. E-mail: martha@vims.edu

Yoshihiro Sakoda, Graduate School of Veterinary Medicine, Hokkaido University, Sapporo, Hokkaido 060-0818 Japan. E-mail: y_sakoda@vetmed.hokudai.ac.jp

Motohiko Sano, Faculty of Marine Science, Tokyo University of Marine Science and Technology. Minato, Tokyo 108-8477, Japan. E-mail: msano00@kaiyodai.ac.jp

Natsumi Sano, Graduate School of Bioresources, Mie University. Tsu, Mie 514-8507, Japan. E-mail: natsumi5@bio.mie-u.ac.jp

Craig A. Shoemaker,* US Department of Agriculture Agricultural Research Service, Aquatic Animal Health Research Unit, 990 Wire Road, Auburn, AL 36832-4352, USA. E-mail: craig.shoemaker@ars.usda.gov

Esteban Soto,* Department of Medicine and Epidemiology, School of Veterinary Medicine, University of California-Davis, 2108 Tupper Hall, Davis, CA 95616-5270, USA. E-mail: sotomartinez@ucdavis.edu

David Stone,* Cefas Weymouth Laboratory, Barrack Road, The Nothe, Weymouth, Dorset, DT1 1RY, UK. E-mail: david.stone@cefas.co.uk

Kuttichantran Subramaniam, Department of Infectious Diseases and Pathology, College of Veterinary Medicine, University of Florida, Gainesville, FL 32611, USA. E-mail: kuttichantran@ufl.edu

Anna Toffan,* National Reference Laboratory for Fish Diseases, OIE Reference Centre for Viral Encephalopathy and Retinopathy, Istituto Zooprofilattico Sperimentale delle Venezie, Viale dell'Università 10, 35020 Legnaro (Padova), Italy. E-mail: atoffan@izsvenezie.it

Alicia E. Toranzo,* Departamento de Microbiología y Parasitología, Facultad de Biología-CIBUS/ Instituto de Acuicultura, Universidade de Santiago de Compostela, 15782, Spain. E-mail: alicia.estevez.toranzo@usc.es

Thomas Waltzek, Department of Infectious Diseases and Pathology, College of Veterinary Medicine, University of Florida, Gainesville, FL 32611, USA. E-mail: tbwaltzek@ufl.edu

Keith Way,* Cefas Weymouth Laboratory, Barrack Road, The Nothe, Weymouth, Dorset DT1 1RY, UK. E-mail: kman71@live.co.uk

Timothy J. Welch,* US Department of Agriculture Agricultural Research Service National Center for Cool and Cold Water Aquaculture, Kearneysville, WV 25430, USA. E-mail: tim.welch@ars.usda.gov

Richard Whittington,* Faculty of Veterinary Science, School of Life and Environmental Sciences, The University of Sydney, Camden, NSW 2570, Australia. E-mail: richard.whittington@sydney.edu.au

David J. Wise, Thad Cochran National Warmwater Aquaculture Center, Mississippi State University, Stoneville, MS 38776, USA. E-mail: dwise@drec.msstate.edu

Patrick T.K. Woo,* Department of Integrative Biology, University of Guelph, Guelph, ON, N1G 2W1, Canada. E-mail: pwoo@uoguelph.ca

De-Hai Xu, US Department of Agriculture Agricultural Research Service, Aquatic Animal Health Research Unit, 990 Wire Road, Auburn, AL 36832-4352, USA. E-mail: dehai.xu@ars.usda.gov

Mamoru Yoshimizu,* Faculty of Fisheries Sciences, Hokkaido University, Minato, Hakodate, Hokkaido 041-8611 Japan. E-mail: yosimizu@fish.hokudai.ac.jp

Preface

The main foci of *Fish Viruses and Bacteria: Pathobiology and Protection* (FVBPP) are on the pathobiology of and protective strategies against major viruses and bacteria that cause diseases and/or mortalities in economically important fish. The 25 chapters are written by scientists who have considerable expertise on the selected microbes, and the vast majority of the chapters are on notifiable microbes certified by the OIE (originally Office International des Epizooties, now the World Organisation for Animal Health). Contributors have made every effort to cite publications as recent as early 2016.

The pathogens and contributors for inclusion in FVBPP were selected by the editors. The selection of microbes/diseases was based on numerous criteria. These include the microbes/diseases that:

* were only briefly discussed in the predecessor volume published in 2011 – *Fish Diseases and Disorders, Volume 3: Viral, Bacterial and Fungal Infections*, 2nd edn, eds Woo, P.T.K. and Bruno, D.W. (e.g. Koi herpesvirus, *Weissella ceti*); or
* are relatively well-studied piscine pathogens (e.g. infectious haematopoietic necrosis, *Aeromonas* spp.) that may serve as disease models for other pathogens; or
* cause considerable financial hardships to specific sectors of the aquaculture industry (e.g. viral haemorrhagic septicaemia, *Vibrio* spp.); or
* have been introduced to new geographical regions through the transportation of infected fish (e.g. Koi herpesvirus in Europe) and have subsequently become significant threats to local fish populations; or
* are pathogenic to specific groups of fish (e.g. *Oncorhynchus masou* viral disease, oncogenic viruses to salmonids in Japan); or
* are highly adaptable and not host specific, and consequently have worldwide distributions (e.g. epizootic haematopoietic necrosis, *Streptococcus* spp.).

Each chapter is arranged to provide a brief description of the selected pathogens, its host(s), transmission, geographical distribution and impact(s) on fish production. The most current information is provided on detection and diagnosis of infection with discussions on clinical signs of the disease and additional details provided for external/internal lesions (macroscopic and microscopic). The focus in each chapter is on the pathophysiology of the disease, including its effects on osmoregulation, impacts on the host endocrine system, growth and reproduction.

Finally, the most current prevention and protective control strategies are presented. These include biological, physical and legislative approaches.

Many of these pathogens have been extensively studied; however, a few are less well studied, including the newly emergent pathogens (e.g. alphaviruses, *W. ceti*). In these cases, contributors have highlighted deficiencies in our knowledge and we hope that these treatises will spur additional research in 'neglected' areas.

FVBPP is directed at research scientists in the aquaculture industry and universities, fish health consultants, managers and supervisors of fish health laboratories, and veterinary specialists in commercial aquaria. The present volume is also appropriate for the training of fish health specialists, and for senior undergraduate/graduate and veterinary students who are conducting research on diseases of fish. In addition, FVBPP may also be a useful reference book for university courses on infectious diseases, general microbiology and the impacts of diseases on the aquaculture industry. A secondary audience includes pathologists who may wish to study the combined effects of microbial/parasitic infections on fish health, and environmental toxicologists and immunologists who are studying synergistic effects of pollutants and microbial infections. We expect this secondary audience to increase as it becomes increasingly evident that fish health may also serve as an indicator of ecosystem quality.

Patrick T.K. Woo and Rocco C. Cipriano

Previous titles by Patrick T.K. Woo

1 Infectious Pancreatic Necrosis Virus

Arun K. Dhar,[1,2]* Scott LaPatra,[3] Andrew Orry[4]
and F.C. Thomas Allnutt[1]

[1]BrioBiotech LLC, Glenelg, Maryland, USA; [2]Aquaculture Pathology Laboratory, School of Animal and Comparative Biomedical Sciences, The University of Arizona, Tucson, Arizona, USA; [3]Clear Springs Foods, Buhl, Idaho, USA; [4]Molsoft, San Diego, California, USA

1.1 Introduction

Infectious pancreatic necrosis virus (IPNV), the aetiological agent of infectious pancreatic necrosis (IPN), is a double-stranded RNA (dsRNA) virus in the family *Birnaviridae* (Leong *et al.*, 2000; ICTV, 2014). The four genera in this family include *Aquabirnavirus*, *Avibirnavirus*, *Blosnavirus* and *Entomobirnavirus* (Delmas *et al.*, 2005), and they infect vertebrates and invertebrates. *Aquabirnavirus* infects aquatic species (fish, molluscs and crustaceans) and has three species: IPNV, *Yellowtail ascites virus* and *Tellina virus*. IPNV, which infects salmonids, is the type species.

The IPNV genome consists of two dsRNAs, segments A and B (Fig. 1.1; Leong *et al.*, 2000). Segment A has ~ 3100 bp and contains two partially overlapping open reading frames (ORFs). The long ORF encodes a 106 kDa polyprotein (NH$_2$-pVP2-VP4-VP3-COOH) that is co-translationally cleaved by the VP4 (viral protein 4) protease (29 kDa) to generate pVP2 (62 kDa; the precursor of the major capsid protein VP2) and the 31 kDa VP3 (Petit *et al.*, 2000). The short ORF encodes VP5, a 17 kDa, arginine-rich, non-structural protein that is produced early in the replication cycle. VP5 is an anti-apoptosis protein similar to the Bcl-2 family of proto-oncogenes. VP5 is not required for IPNV replication *in vivo* and its absence does not alter virulence or persistence in the host (Santi *et al.*, 2005). Segment B has ~ 2900 bp and encodes the polypeptide VP1 (94 kDa) which is an RNA-dependent RNA polymerase. VP1 is found both within the mature virion as a free polypeptide with RNA-dependent RNA polymerase-associated activity

and as a genome-linked protein, VPg, via guanylylation of VP1 (Fig. 1.1 and Table 1.1).

Aquabirnaviruses have broad host ranges and differ in their optimal replication temperatures. They consist of four serogroups A, B, C and D (Dixon *et al.*, 2008), but most belong to serogroup A, which is divided into serotypes A1–A9. The A1 serotype contains most of the US isolates (reference strain West Buxton), serotypes A2–A5 are primarily European isolates (reference strains, Ab and Hecht) and serotypes A6–A9 include isolates from Canada (reference strains C1, C2, C3 and Jasper).

1.1.1 IPNV morphogenesis

Two types of particles (A and B) are produced during infection. After replication, dsRNA is assembled into the 66 nm diameter non-infectious particle A, in which the capsid is composed of both mature (VP2) and immature (pVP2) viral polypeptides. Proteolytic processing of the remaining pVP2 into VP2 compacts the capsid to the 60 nm diameter infectious particle, referred to as particle B (Villanueva *et al.*, 2004). The VP2 protein comprises the outer capsid, while the VP3 protein forms the inner layer of the mature virion. Additionally, VP3 remains associated with VP1 and VP4, as well as with the polymerase-associated genome.

1.1.2 IPNV tertiary structure

Virions are non-enveloped with a T13 lattice icosahedral morphology 60 nm in diameter, and have a

---

*Corresponding author e-mail: arun_dhar@hotmail.com or adhar@email.arizona.edu

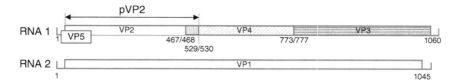

Fig. 1.1. Genome organization of infectious pancreatic necrosis virus (IPNV). The numbers at the bottom of the segments in the diagrams indicate the amino acid number(s). pVP2 is the precursor of the IPNV viral protein VP2. RNA 1 is segment A and RNA 2 is segment B of the viral dsRNA.

Table 1.1. Proteins encoded by infectious pancreatic necrosis virus (IPNV) and their functions.

Protein	Molecular mass	Function(s)	References
VP1	94 kDa	RNA-dependent-RNA polymerase; viral replication	Calvert et al., 1991; Leong et al., 2000; Graham et al., 2011
VP2	54 kDa	Major capsid protein, contains most antigenic determinants; has a structural function	Coulibaly et al., 2010
VP3	31 kDa	Minor capsid protein, interacts with the major capsid protein VP2 in capsid formation, associates with the dsRNA genome, recruits polymerase into capsids, contains some antigenic epitopes	Leong et al., 2000; Bahar et al., 2013
VP4	29 kDa	Protease involved in processing the polyprotein encoded in segment A of IPNV dsRNA	Feldman et al., 2006; Lee et al., 2007
VP5	17 kDa	Arginine-rich anti-apoptosis protein, similar to the Bcl-2 family of proto-oncogenes	Magyar and Dobos, 1994; Santi et al., 2005

buoyant density of 1.33 g/cm^3 in CsCl (Delmas et al., 2005). The viral capsid surface contains VP2 proteins, and the three-dimensional (3D) structures of these are known for IPNV and IBDV (Infectious bursal disease virus) (Fig. 1.2A). The IPNV VP2 capsid is made of 260 trimeric spikes that are projected radially and carry the antigenic domains as well as determinants for virulence and cellular adaptation. These are linked to VP3 in the interior of the virion (Fig. 1.2B). However, the spikes in IPNV are arranged differently from those in IBDV in that the amino acids controlling virulence and cell adaptation are located at the periphery in IPNV but in a central region for IBDV. The base of the spike contains an integrin-binding motif and is located in an exposed groove, which is conserved across all genera of birnaviruses (Coulibaly et al., 2010).

1.2 Geographical Distribution

IPN occurs worldwide among cultured and wild salmonid fishes. It was first detected in freshwater trout during the 1940s within Canada and during the 1950s within the USA (Wood et al., 1955). The virus

was first isolated in 1960 (Wolf et al., 1960). It was subsequently reported in Europe during the early 1970s and has also been reported in many other countries (e.g. Japan, Korea, Taiwan, China, Thailand, Laos, New Zealand, Australia, Turkey) that are involved either with importing salmonids or active in aquaculture. IPN outbreaks are often traced to importations and the subsequent distribution of infected ova/fingerlings (Munro and Midtlyng, 2011).

1.3 Economic Impacts of IPN

Historically, IPN is one of the top three causes of losses in the salmonid industry. This was reflected in a survey conducted by the Shetland Salmon Farmers Association in 2001 that showed an average loss of 20–30% with a cash value of 2 million pounds due to IPN (Ruane et al., 2007). From 1991 to 2002, IPN had an impact on salmon post-smolt survival in Norwegian epizootiological studies of from 6.4 to 12.0% (Munro and Midtlyng, 2011). In 1998, the economic losses were estimated to exceed €12 million (Munro and Midtlyng, 2011). Even today, IPN remains an important risk for

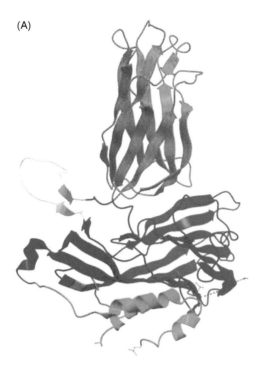

(A)

(B)

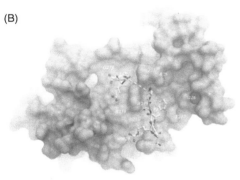

Fig. 1.2. (A) The crystal structure of infectious pancreatic necrosis virus (IPNV) viral protein VP2 showing the base domain (green), shell domain (blue) and variable P domain (red). The molecular graphic was prepared using the free ICM-Browser software (MolSoft LLC, San Diego, California, downloadable at http://www.molsoft.com/icm_browser.html). (B) The crystal structure of viral protein VP1 (blue surface with residue labels) and the VP3 protein C-terminal (white stick and ribbon). VP3 interacts with the finger domain of VP1. VP3 residues 227–231 and 236–238 are shown and missing residues that were not resolved are depicted by dashed lines in the ribbon. As in (A), the molecular graphic was prepared using the ICM-Browser software.

salmonid culture. For example, it was reported in 48 salmon farming facilities in Norway during 2014, although this was fewer than in previous years (Norwegian Veterinary Institute, 2015). IPN is most important in the first 6 months after sea transfer. The industry still reports significant losses due to mortalities and subsequent weakening of the surviving fish. A recent report on cumulative mortality in the first 6 months indicated an increase to 7.2% compared with a baseline mortality rate of 3.4%; this is more than doubling the cumulative mortality (Jensen and Kristoffersen, 2015). This same study showed that IPNV-infected cohorts challenged with other stressors showed increased levels of cumulative mortalities. For example, with pancreas disease (PD), mortality increased to 12.9%, whereas heart muscle and skeletal muscle inflammation (HSMI) increased to 16.6% when all other factors were normalized.

1.4 Diagnosis of the Infection

1.4.1 Clinical signs and viral transmission

IPNV and IPNV-like birnaviruses have been isolated from salmonids as well as from non-salmonid fishes (e.g. *Cyprinus carpio*, *Perca flavescens*, *Abramis brama* and *Esox lucius*), molluscs, crustaceans and pseudocoelomates (McAllister, 2007). External clinical signs include darkened colour, exophthalmia, abdominal distention, the presence of a mucoid pseudocast ('faecal cast') extruding from the vent, and haemorrhages on the body surface and at the bases of fins. Infected fish swim in a rotating manner along their longitudinal axis and death generally ensues within a few hours. Internal signs can include a pale liver and spleen, and an empty digestive tract filled with clean or milky mucus. Haemorrhages can occur in visceral organs (Munro and Midtlyng, 2011). IPN outbreaks characteristically consist of a sudden increase in fry and fingerling mortalities. The disease can also occur in post-smolts in the first few weeks after transfer to the sea (Jensen and Kristoffersen, 2015). Stress on the host plays a key role in enhancing viral replication, mutation and even reversion to virulence (Gadan *et al.*, 2013). The survivors of outbreaks often carry IPNV for their entire lives without clinical signs. These carriers serve as reservoirs that transmit the virus either horizontally through sheddings in faeces and urine, or vertically through contaminated reproductive products (Roberts and Pearson, 2005).

1.4.2 Viral detection

Clinical signs and pathology cannot be used to distinguish IPN from other viral diseases and the absence of clinical signs does not ensure that fish are free of IPNV. The tentative diagnosis of IPN is based on prior disease history of the farm and fish population, clinical signs and findings from gross necropsy. Confirmatory diagnosis involves isolation of the virus in cell culture followed by immunological or molecular confirmation. Serological or molecular techniques are especially useful for monitoring fish with and without clinical signs. Tissues suitable for virological examinations include the kidney, liver, spleen, the ovarian fluid from brood stock at spawning or whole alevins. The isolation of IPNV in cell culture is done using blue gill fry (BF-2), Chinook salmon embryo (CHSE-214) or rainbow trout gonad (RTG-2) cell lines (OIE, 2003). Identification of the virus from cell culture is done using neutralization assay, fluorescent antibody assay, enzyme-linked immunosorbent assay (ELISA), immunohistochemical staining using IPNV-specific antibody, or reverse-transcriptase-polymerase chain reaction (RT-PCR) (OIE, 2003; USFWS and AFS-FHS, 2007).

In recent years, SYBR Green and TaqMan-based real-time RT-PCR methods have been developed to detect IPNV (Bowers *et al.*, 2008; Orpetveit *et al.*, 2010). Real-time based methods are 100× more sensitive than conventional PCR, and detect the virus in subclinical animals (Orpetveit *et al.*, 2010). Using real-time RT-PCR, the IPNV load in pectoral fin clips was found to be as high as in the spleen and head kidney (Bowers *et al.*, 2008). Therefore, non-lethal tissue sampling coupled with real-time RT-PCR could be valuable tools for surveillance and monitoring wild and farmed fish, as well as minimizing the need to sacrifice brood stock at spawn.

1.5 Pathology

IPNV infection presents a variety of pathological changes. Pancreatic tissues undergo severe necrosis characterized with pyknosis (chromatin condensation), karyorrhexis (fragmentation of the nucleus) and cytoplasmic inclusion bodies (Fig. 1.3). The pylorus, pyloric caeca and anterior intestine also undergo extensive necrosis. Intestinal epithelial cells slough and combine with mucus to form thick, whitish exudates that may discharge from the vent. Degenerative changes also occur in the kidney, liver

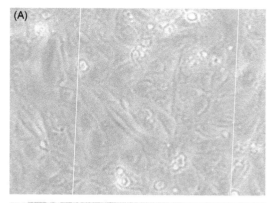

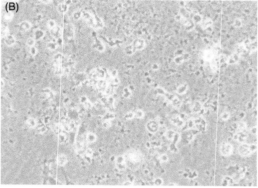

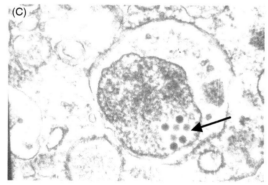

Fig. 1.3. (A) Cells of uninfected Chinook salmon embryos (CHSE-214); (B) cells infected by infectious pancreatic necrosis virus (IPNV) exhibiting a lytic type of cytopathic effect (magnification, 100×); and (C) a transmission electron microscope (TEM) image showing the IPN virus particles (indicated by an arrow, magnification, 27,500×) displaying a characteristic hexagonal profile in cytoplasmic vesicles.

and spleen. In persistently infected fish, IPNV is in macrophages within the haematopoietic tissue of the kidney, and can multiply in adherent leucocytes isolated from carrier fish (Johansen and Sommer,

1995). There are indications of reduced immune response in leucocytes isolated from carrier fish, and of increased *in vitro* viral replication after the stimulation of resting leucocytes with phytohaemagglutinin (Knott and Munro, 1986).

1.6 Pathophysiology

The susceptibility of fish to IPN infection and mortality depends on species, age or developmental stage, physiological condition of the host, virus strain, genetic background of the host, and environmental and management factors (Munro and Midtlyng, 2011). In cultured trout and salmon, infection varies from subclinical with little or no mortality to acutely virulent with high mortality. Although in trout and salmon the disease produces severe pancreatic necrosis, it also causes histological changes in the renal haematopoietic tissue, gut and liver. The liver is a key target (Ellis *et al.*, 2010), while the virus is also present in the islets of Langerhans and in the corpuscles of Stannius in the kidney (Fig. 1.4), which suggests that it could also affect metabolism. McKnight and Roberts (1976) reported the clinical sign of 'mucosal damage' with a description that fits what is currently referred to as acute enteritis caused by faecal casts. They postulated that this damage might be more lethal than necrosis of the pancreas. Cell necrosis of the digestive glands and of the mucosal gut epithelium is also thought to be responsible for the shedding of infective virus with faeces. Severe necrosis of the intestinal mucosa and pancreas may also cause anorexia that exacerbates conditions such as 'pinhead' fish and 'failing smolts', which are often observed among the survivors of an epizootic (Smail *et al.*, 1995). Roberts and Pearson (2005) also reported that in seawater, after a loss of 50% or more to IPN, many fish failed to grow, became chronically emaciated and were prone to sea louse infestation.

Subclinical infections may not affect the growth of Atlantic salmon (*Salmo salar*) parr or postsmolts. However, in laboratory studies, both feed intake and specific growth rates of healthy postsmolts were depressed after an immersion infection with IPNV (Damsgard *et al.*, 1998). Viral titres were determined in the kidney and pyloric caeca before and after the experimental infection, and no mortality occurred in the infected or in the control groups. In the infected fish, the titres in both the kidney and pyloric caeca increased significantly. Between 16 and 44 days after infection, the titre in the pyloric caeca decreased significantly from 10^6 to 10^3–10^4 plaque-forming units (pfu)/g. From approximately 20 days after infection, feed intake and specific growth rates were significantly lower in infected fish than in uninfected fish. The results indicated that IPNV-infected fish require relatively high viral titres in the kidney and pyloric caeca before reduced feeding is detectable.

IPNV induces programmed cell death as apoptosis markers have been found in hepatic, intestinal and pancreatic tissues that correspond to viral accumulation and pathological changes (Imatoh *et al.*, 2005; Santi *et al.*, 2005). It was hypothesized that apoptosis might limit rather than enhance the negative consequences of an IPNV infection.

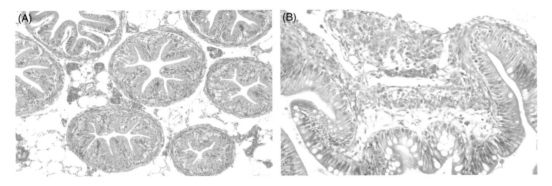

Fig. 1.4. Necrosis of pancreatic acinar cells that are located in the adipose tissue between the cylindrical pyloric caecae and sloughing of the intestinal mucosa from a fish infected with infectious pancreatic necrosis virus (IPNV) (H&E (haematoxylin and eosin) stained, magnification 400×).

Sadasiv (1995) found that viral clearance was minimal even in the presence of viral neutralizing antibodies. It was suggested that the virus may infect leucocytes, possibly persistently, and thereby subvert the neutralizing antibody response. Dual infections of rainbow trout (*Oncorhynchus mykiss*) with IPNV and IHNV (infectious haematopoietic necrosis virus) have been reported, but the potential effects of mixed infections on the immune system were not described (LaPatra *et al.*, 1993). Rainbow trout pre-exposed to IPNV and later challenged with viral haemorrhagic septicaemia virus (VHSV) had significant resistance to VHSV compared with fish that had not been previously exposed to IPNV (de Kinkelin *et al.*, 1992). The authors termed this phenomenon 'interference-mediated resistance' and suspected that it was due to the production of interferon.

In a similar manner, when Atlantic salmon carrying IPN were exposed to infectious salmon anaemia virus (ISAV), the mortality of non-IPNV infected post-smolts was consistently higher than that of fish that had been exposed to IPNV 3 weeks earlier (Johansen and Sommer, 2001). In contrast, when the fish were challenged with ISAV at 6 weeks post-IPNV infection, there was no difference in the mortality between IPNV carriers and non-carriers. These authors also reported a similar short-lived protection of subclinical IPNV infection against *Vibrio salmonicida* and attributed this to non-specific effects due to IPNV-induced interferon production. Additionally, no significant effects associated with intraperitoneal vaccination using a trivalent oil-adjuvanted bacterin were observed in the IPNV carrier group versus non-carrier controls. However, the IPNV carrier fish eventually had a moderate IPN outbreak with cumulative mortality in the unvaccinated carrier fish of 24% versus 7% in the bacterin-vaccinated carrier fish. In another study, no differences were seen in mortality between immunized IPNV carriers and non-carriers after experimental furunculosis or coldwater vibriosis challenge and in either group's humoral immune response to *Aeromonas salmonicida* (Johansen *et al.*, 2009). In addition, when IPNV carrier and non-carrier Atlantic salmon fry (mean weight 2–4 g) were immunized against enteric redmouth disease (ERM), there was no difference in protection after experimental challenge with *Yersinia ruckeri* (Bruno and Munro, 1989). These studies indicate that IPNV infection, even in small fish, had no detrimental effect on bacterial vaccine-induced protection.

1.7 Protective and Control Strategies

Because there is no therapy for IPN disease, avoidance is the best strategy. Epizootiological studies of IPNV transmission in salmon farms have shown that viral spread is unpredictable. Since non-clinical carriers serve as a source of infection through viral shedding in faeces and sexual products, intensive monitoring and biosecurity can reduce the prevalence of the virus. It is essential to obtain stock from pathogen-free sources and maintain strong biosecurity on a pathogen-free water supply whenever new fish are introduced. UV treatment of incoming water to the hatchery is an example of a suitable control measure. Treatment with disinfectants such as formalin (3% for 5 minutes), sodium hydroxide (pH 12.5 for 10 minutes), chlorine (30 ppm for 5 minutes) and iodine compounds is also capable of inactivating the virus (OIE, 2003).

1.7.1 Selection for improved IPN-resistant fish lines

Because significant numbers of fish survive IPN epizootics, it was postulated that breeding could enhance resistance. Ozaki and colleagues reported that quantitative trait loci (QTLs) could be correlated with improved resistance to IPN. A recent review on the current status of DNA marker-assisted breeding for improved disease resistance in commercially important fish is available (Ozaki *et al.*, 2012). The marker-assisted selection (MAS) of more resistant lines using genomic traits is a powerful tool for the development of IPNV-resistant salmonid lines (Moen *et al.*, 2009) and is also being expanded for other diseases (Houston *et al.*, 2008; Ozaki *et al.*, 2012). A recent report tied the epithelial cadherin gene (*cdh1*) with resistance to IPN (Moen *et al.*, 2015). Strains of IPN-resistant Atlantic salmon are marketed by companies such as Aquake, Trondheim, Norway (http://aquagen.no/en/products/salmon-eggs/product-documentation/resistance-against-ipn/); IPN resistance in these fish was linked to a single QTL that could prove useful for future efforts to develop resistant lines of fish using MAS (Moen *et al.*, 2009).

1.7.2 Available biologics

A number of IPN vaccines are available (Table 1.2), but there is a need to develop more cost-effective vaccines that can be delivered to all life stages. The

A.K. Dhar *et al.*

Table 1.2. Approved vaccines against infectious pancreatic necrosis (IPN).

Name	Content and/or vaccine against[a]	Type	Delivery	Company	Licensed in:
Alpha Ject® 1000	Whole IPNV	Inactivated virus	IP	Pharma AS, Norway	Chile
Alpha Ject® 2.2	IPN Furunculosis		IP	Pharma AS, Norway	UK
Alpha Ject® 4-1	IPN SRS Furunculosis Vibriosis		IP	Pharma AS, Norway	Chile
Alpha Ject® 5-1	IPN ISA SRS Furunculosis Vibriosis		IP	Pharma AS, Norway	Chile
Alpha Ject® 6-2	IPN Furunculosis Coldwater vibriosis Winter sore		IP	Pharma AS, Norway	Norway Faroe Islands
Alpha Ject® IPNV-Flevo 0.025	IPNV Flavobacteriosis		IP	Pharmaq AS, Norway	Chile
Alpha Ject® micro 2	IPN SRS		IP	Pharmaq AS, Norway	Chile
Alpha Ject® micro 3	IPN SRS Vibriosis		IP	Pharmaq AS, Norway	Chile
Alpha Ject® micro 7 ILA	IPN Vibriosis Furunculosis Coldwater vibriosis Winter sore ISA		IP	Pharmaq AS, Norway	Norway Faroe Islands
AquaVac® IPN Oral	VP2 and VP3 capsid proteins	Subunit	Oral	Merck Animal Health, New Jersey, USA	Canada, USA
Birnagen Forte	Whole IPNV	Inactivated virus	IP	Aqua Health Ltd., Novartis, Canada	Canada
IPNV	Whole IPNV	Inactivated virus	IP	Centrovet, Chile	Chile
Norvax® Minova 6 Norvax® Compact 6	VP2 capsid protein	Subunit	IP	Intervet International BV, The Netherlands Merck Animal Health	
SRS/IPNV/Vibrio	VP2 protein	Subunit	IP	Microtek International, Inc., British Columbia, Canada	Canada, Chile

[a]IPNV, infectious pancreatic necrosis virus; ISA, infectious salmon anaemia; SRS, salmon rickettsial septicaemia; VP2, VP3, viral proteins 2 and 3.

use of inactivated wild type virus to induce immunity was the earliest approach to fish viral vaccines and it is still a reliable standard by which other vaccines are evaluated. The Alpha Ject® micro 1 ISA (Pharma/Novartis) and Alpha Ject® 1000 vaccines (Table 1.2) are examples of such vaccines that target infectious salmon anaemia (ISA) and IPN, respectively (http://www.pharmaq.no/products/injectable/). Inactivated viral vaccines induce strong responses because they retain surface-exposed antigens and the inactivated genomic component.

Subunit vaccine based on major viral antigen(s) is another option for producing viral vaccines. The intrinsic ability of some viral structural proteins to self-assemble into particles that mimic the native virus in both size and processing by the host have led to the development of a class of subunit vaccines referred as virus-like particles (VLPs; Kushnir *et al.*, 2012). VLPs have been expressed in bacteria, yeast, transgenic plants and cell culture. A number of human vaccines (e.g. Gardasil vaccine®9, Human Papillomavirus 9-valent Vaccine, Recombinant) have been produced using this technology (Kushnir *et al.*, 2012). Recent efforts have used this approach to produce vaccines against IPNV. IPNV VLPs containing VP2 and VP3 proteins and measuring 60 nm in diameter have been produced in insect cells and *Trichoplusia ni* larvae using a baculovirus expression system. When Atlantic salmon post-smolts were intraperitoneally immunized with purified antigen and challenged via immersion, the cumulative mortalities 4 weeks post-challenge were lower (56%) than in control fish (77%) (Shivappa *et al.*, 2005).

Another IPN vaccine is based on IPNV VP2 protein alone (Allnutt *et al.*, 2007). The VP2-based subviral particles (SVPs) expressed in yeast were 22 nm in size compared with 60 nm for the native virus. SVPs induced a strong anti-IPNV antibody response in rainbow trout. The antigen was delivered via injection or through the diet, and the reduced IPNV load was 22- and 12-fold in injected and orally vaccinated fish, respectively (Allnutt *et al.*, 2007). To further explore the possibility of using the IPNV SVP to develop multivalent vaccine, a foreign epitope (human oncogene *c-myc*) was expressed on the SVP, and the chimeric SVPs induced antibody response to both IPNV and the *c-myc* epitope (Dhar *et al.*, 2010). Further research has led to the successful display of an ISAV haemagglutinin epitope on the surface of this IPNV SVP, and the chimeric SVPs, when injected into rainbow trout, induced antibody response against IPNV as well as ISAV (Dhar *et al.*, unpublished data). Three other vaccines based on the IPNV VP2 capsid protein are marketed, including IPNV (licensed in Chile and from Centrovet, Chile), Norvax (Intervet-International BV, The Netherlands), and SRS/IPNV/*Vibrio* (licensed in Canada and Chile and from Microtek International Inc., British Columbia, Canada; since December 2010 fully integrated into Zoetis Veterinary Medicine Research & Development and Zoetis Canada) (Gomez-Casado *et al.*, 2011). The Centrovet vaccine provides oral delivery of both an inactivated IPNV and a recombinant protein to provide flexibility

in delivery (http://www.centrovet.com/index.php/products/aqua/vaccines99). The Norvax vaccine is another recombinant protein vaccine that is delivered via intraperitoneal injection and only addresses IPN. The SRS/IPNV/Vibrio vaccine is a trivalent recombinant protein vaccine that is also delivered by intraperitoneal injection and provides the user with the convenience of addressing three different pathogens.

An experimental IPNV DNA vaccine (expressing the VP2 antigen), delivered via injection, provided almost 80% relative percent survival (RPS) upon challenge by an infectious homologous virus in 1–2 g rainbow trout fry (Cuesta *et al.*, 2010). Another DNA vaccine encapsulated in alginate and delivered in food pellets reduced or eliminated the IPNV titres in rainbow trout after waterborne virus challenge (Ballesteros *et al.*, 2015). In this study, the VP2 gene was cloned into a DNA vector, incorporated into alginate microspheres and delivered orally to rainbow trout using a pipette to assure uniform delivery of the vaccine (Ballesteros *et al.*, 2012). The alginate-bound DNA vaccine was also incorporated into food pellets to induce an immune response (Ballesteros *et al.*, 2014). Both IgM and IgT increased at 15 days postvaccination but were much higher at 30 days postvaccination. A cellular immune response was also monitored by looking at the T cell markers CD4 and CD8. Both markers were elevated at day 15 but returned to background levels by day 30. The RPS was 85.9 and 78.2% when fish were challenged with IPNV at days 15 and 30 postvaccination, respectively. Recently, another study reported a fusion protein of IPNV VP2–VP3 proteins expressed in *Escherichia coli* and delivered via injection-induced IgM production against IPNV; this provided an RPS of 83% in juvenile rainbow trout (Dadar *et al.*, 2015).

1.7.3 Subunit vaccines

Designer whole viral vaccines were produced using reverse genetics based on the Sp strain of IPNV (Munang'andu *et al.*, 2012). Avirulent and virulent motifs were added to the Sp strain, which was then inactivated for use as a vaccine. The inactivated virus was compared with DNA, subunit and nanoparticle subunit vaccines made against IPN using a cohabitation challenge system. The inactivated whole virus vaccine provided a similar antibody titre to the other vaccines but outperformed them in survival after viral challenge with 48–58% RPS while the VP2-fusion protein-, subunit- and DNA nanoparticle-based vaccines had values of 25.4–30.7, 22.8–34.2 and 16.7–27.2% RPS, respectively (Munang'andu et al., 2012).

The delivery of VP2 or VP3 antigens by a recombinant *Lactobacillus casei* was evaluated as a potential vaccine strategy (Liu et al., 2012). VP2 and VP3 were engineered either to be secreted by the *L. casei* or to be surface displayed. When these recombinant *L. casei* vaccines were orally delivered to rainbow trout, the VP2 secretory strain provided a much higher serum IgM titre than the other *L. casei* lines. On challenge with IPNV, the VP2-secreting *L. casei* was also more effective in reducing the viral load of the fish (~ 46-fold reduction compared with ~3 fold for the VP3-secreting strain).

Other ongoing research includes the improvement of the oral delivery of IPN antigens. For example, a recent study in Atlantic salmon showed that alginate-encapsulated IPNV antigens significantly improved the titre of IPNV-targeted antibodies and induction of immune-related genes when compared with the delivery of the same IPNV antigens in non-encapsulated form (Chen et al., 2014). The expense and inconvenience of injectable vaccines limit their usefulness in large-scale aquaculture beyond one vaccination cycle, so the use of oral vaccines to boost immune activation is attractive. The alginate-encapsulated vaccine was compared with the antigen alone and it was conclusively shown that protection of the antigen in the alginate was required for improved efficacy. The initial booster vaccination was done a year after the injection vaccination, when two oral doses were provided 7 weeks apart. Serum IgM was boosted after the first oral vaccination but IgT was not

upregulated. In contrast, after the second booster, both IgT and IgM were upregulated (Chen et al., 2014).

Some examples of experimental vaccines against IPNV are shown in Table 1.3.

1.8 Conclusions and Suggestions for Future Research

Aquabirnavirus is the largest and most diverse genus within the family *Birnaviridae*. IPNV is one of the most extensively studied and widely distributed viruses infecting marine and freshwater fishes. Since the initial report of IPN-associated disease outbreaks in the 1940s, large numbers of IPNV and IPNV-like viruses have been isolated worldwide from diseased and apparently healthy salmonids and non-salmonids, as well as invertebrates (ICTV, 2014). Due to the extensive diversity of viral species belonging to the genus *Aquabirnavirus*, it has been difficult to classify the virus to species level (Crane and Hyatt, 2011). It remains unknown whether phylogeny based on whole genome sequence data and structure-based analysis of the VP2 and VP3 proteins will help to delineate the aquabirnaviruses to species level.

IPN is economically important due to its lethality for salmonid fry in freshwater production, and in post-smolts after transfer to seawater. The development of improved biosecurity protocols, targeted vaccines and resistant brood stocks has been very beneficial to the control of IPN, but despite these efforts, the disease remains a serious challenge to salmonid farming worldwide. One of the major constraints in developing effective vaccines has been the lack of a repeatable infection model to evaluate vaccine efficacy. However, recently, a

Table 1.3. Experimental vaccines against *Infectious pancreatic necrosis virus* (IPNV).

Vaccine	Description	References
IPNV virus-like particle (VLP)	Both viral proteins VP2 and VP2/VP3-based VLPs have been produced and shown to induce a strong immune response	Martinez-Alonso et al., 2012
Live recombinant *Lactobacillus casei*-expressing VP2	Orally delivered recombinant bacteria expressing a secreted VP2 were most effective in reducing viral load	Liu et al., 2012
Plasmid expression of segment A of IPNV dsRNA (whole or in parts)	Protection only with all of the large open reading frame (ORF) polyprotein	Mikalsen et al., 2004
Recombinant inactivated whole virus vaccine (IWV)	Reverse genetic constructs of virulent and avirulent IPNV-based nanoparticles	Munang'andu et al., 2012

cohabitation challenge for IPNV has been developed in Atlantic salmon (Munang'andu et al., 2016). Further validation of this infection model, combined with non-invasive tissue sampling to determine viral titre, may enhance IPN management.

Considering the extensive diversity of IPNV and IPNV-like viruses, the efficacy of IPNV vaccines can perhaps be improved by employing a structure-based vaccine design approach. In a rapidly evolving *Norovirus* GII.4 infecting humans, it has been shown that chimeric VLP-containing epitopes from multiple strains incorporated into a single VLP background induce a broad blocking antibody response, not only against GII.4 VLPs from GII.4-1987 to GII.4-2012, but also against those strains that were not included in the chimeric VLP (Debbink et al., 2014). A similar approach could be applicable to developing IPNV vaccine that provides protection against a number of prevailing strains for salmonid aquaculture.

References

Allnutt, F.C., Bowers, R.M., Rowe, C.G., Vakharia, V.N., LaPatra, S.E. et al. (2007) Antigenicity of infectious pancreatic necrosis virus VP2 subviral particles expressed in yeast. *Vaccine* 25, 4880–4888.

Bahar, M.W., Sarin, L.P., Graham, S.C., Pang, J., Bamford, D.H. et al. (2013) Structure of a VP1-VP3 complex suggests how birnaviruses package the VP1 polymerase. *Journal of Virology* 87, 3229–3236.

Ballesteros, N.A., Saint-Jean, S.S.R., Encinas, P.A., Perez-Prieto, S.I. and Coll, J.M. (2012) Oral immunization of rainbow trout to infectious pancreatic necrosis virus (IPNV) induces different immune gene expression profiles in head kidney and pyloric ceca. *Fish and Shellfish Immunology* 33, 174–185.

Ballesteros, N.A., Rodriguez Saint-Jean, S. and Perez-Prieto, S.I. (2014) Food pellets as an effective delivery method for a DNA vaccine against infectious pancreatic necrosis virus in rainbow trout (*Oncorhynchus mykiss*, Walbaum). *Fish and Shellfish Immunology* 37, 220–228.

Ballesteros, N.A., Rodriguez Saint-Jean, S. and Perez-Prieto, S.I. (2015) Immune responses to oral pcDNA-VP2 vaccine in relation to infectious pancreatic necrosis virus carrier state in rainbow trout *Oncorhynchus mykiss*. *Veterinary Immunology and Immunopathology* 165, 127–137.

Bowers, R.M., Lapatra, S.E. and Dhar, A.K. (2008) Detection and quantitation of infectious pancreatic necrosis virus by real-time reverse transcriptase-polymerase chain reaction using lethal and non-lethal tissue sampling. *Journal of Virological Methods* 147, 226–234.

Bruno, D. and Munro, A. (1989) Immunity of Atlantic salmon, *Salmo salar* L., fry following vaccination against *Yersinia ruckeri*, and the influence of body weight and infectious pancreatic necrosis virus (IPNV) infection on the detection of carriers. *Aquaculture* 81, 205–211.

Calvert, J.G., Nagy, E., Soler, M. and Dobos, P. (1991) Characterization of the VPg-dsRNA linkage of infectious pancreatic necrosis virus. *Journal of General Virology* 72, 2563–2567.

Chen, L., Klaric, G., Wadsworth, S., Jayasinghe, S., Kuo, T.-Y. et al. (2014) Augmentation of the antibody response of Atlantic salmon by oral administration of alginate-encapsulated IPNV antigens. *PLoS ONE* 9(10): e109337.

Coulibaly, F., Chevalier, C., Delmas, B. and Rey, F.A. (2010) Crystal structure of an *Aquabirnavirus* particle: insights into antigenic diversity and virulence determinism. *Journal of Virology* 84, 1792–1799.

Crane, M. and Hyatt, A. (2011) Viruses of fish: an overview of significant pathogens. *Viruses* 3, 2025–2046.

Cuesta, A., Chaves-Pozo, E., de Las Heras, A.I., Saint-Jean, S.R., Pérez-Prieto, S. and Tafalla, C. (2010) An active DNA vaccine against infectious pancreatic necrosis virus (IPNV) with a different mode of action than fish rhabdovirus DNA vaccines. *Vaccine* 28, 3291–3300.

Dadar, M., Memari, H.R., Vakharia, V.N., Peyghan, R., Shapouri, M.S. et al. (2015) Protective and immunogenic effects of *Escherichia coli*-expressed infectious pancreatic necrosis virus (IPNV) VP2-VP3 fusion protein in rainbow trout. *Fish and Shellfish Immunology* 47, 390–396.

Damsgard, B., Mortensen, A. and Sommer, A.-I. (1998) Effects of infectious pancreatic necrosis virus (IPNV) on appetite and growth in Atlantic salmon, *Salmo salar* L. *Aquaculture* 163, 185–193.

de Kinkelin, P., Dorson, M. and Renault, T. (1992) Interferon and viral interference in viroses in salmonid fish. In: Kimura, T. (ed.) *Proceedings of the OJI Symposium on Salmonid Fish*. Hokkaido University Press, Sapporo, Japan, pp. 241–249.

Debbink, K., Lindesmith, L.C., Donaldson, E.F., Swanstrom, J. and Baric, R.S. (2014) Chimeric GII.4 norovirus virus-like-particle-based vaccines induce broadly blocking immune responses. *Journal of Virology* 88, 7256–7266.

Delmas, B., Kibenge, F., Leong, J., Mundt, E., Vakharia, V. et al. (2005) *Birnaviridae*. In: Fauquet, C., Mayo, M., Maniloff, J., Desselberger, U. and Ball, L. (eds) *Virus Taxonomy: Classification and Nomenclature of Viruses. Eighth Report of the International Committee on Taxonomy of Viruses*. Elsevier/Academic Press, San Diego, California and London, pp. 561–569.

Dhar, A.K., Bowers, R.M., Rowe, C.G. and Allnutt, F.C. (2010) Expression of a foreign epitope on infectious pancreatic necrosis virus VP2 capsid protein subviral particle (SVP) and immunogenicity in rainbow trout. *Antiviral Research* 85, 525–531.

Dixon, P.F., Ngoh, G.H., Stone, D.M., Chang, S.F., Way, K. et al. (2008) Proposal for a fourth aquabirnavirus serogroup. *Archives of Virology* 153, 1937–1941.

Ellis, A.E., Cavaco, A., Petrie, A., Lockhart, K., Snow, M. et al. (2010) Histology, immunocytochemistry and qRT-PCR analysis of Atlantic salmon, *Salmo salar* L., post-smolts following infection with infectious pancreatic necrosis virus (IPNV). *Journal of Fish Diseases* 33, 803–818.

Feldman, A.R., Lee, J., Delmas, B. and Paetzel, M. (2006) Crystal structure of a novel viral protease with a serine/lysine catalytic dyad mechanism. *Journal of Molecular Biology* 358, 1378–1389.

Gadan, K., Sandtrø, A., Marjara, I.S., Santi, N., Munang'andu, H.M. et al. (2013) Stress-induced reversion to virulence of infectious pancreatic necrosis virus in naive fry of Atlantic salmon (*Salmo salar* L.). *PloS ONE* 8(2): e54656.

Gomez-Casado, E., Estepa, A. and Coll, J.M. (2011) A comparative review on European-farmed finfish RNA viruses and their vaccines. *Vaccine* 29, 2657–2671.

Graham, S.C., Sarin, L.P., Bahar, M.W., Myers, R.A., Stuart, D.I. et al. (2011) The N-terminus of the RNA polymerase from infectious pancreatic necrosis virus is the determinant of genome attachment. *PLoS Pathogens* 7(6): e1002085.

Houston, R., Gheyas, A., Hamilton, A., Guy, D.R., Tinch, A.E. et al. (2008) Detection and confirmation of a major QTL affecting resistance to infectious pancreatic necrosis (IPN) in Atlantic salmon (*Salmo salar*). *Developmental Biology (Basel)* 132, 199–204.

ICTV (2014) *Virus Taxonomy: 2014 Release*. Available as *Virus Taxonomy: 2015 Release* at: http://www.ictvonline. org/virusTaxonomy.asp?src=NCBI&ictv_id= 19810087 (accessed 21 October 2016).

Imatoh, M., Hirayama, T. and Oshima, S. (2005) Frequent occurrence of apoptosis is not associated with pathogenic infectious pancreatic necrosis virus (IPNV) during persistent infection. *Fish and Shellfish Immunology* 18, 163–177.

Jensen, B.B. and Kristoffersen, A.B. (2015) Risk factors for outbreaks of infectious pancreatic necrosis (IPN) and associated mortality in Norwegian salmonid farming. *Diseases of Aquatic Organisms* 114, 177–187.

Johansen, L. and Sommer, A. (1995) Multiplication of infectious pancreatic necrosis virus (IPNV) in head kidney and blood leukocytes isolated from Atlantic salmon, *Salmo salar* (L.) *Journal of Fish Diseases* 18, 147–156.

Johansen, L.-H. and Sommer, A.-I. (2001) Infectious pancreatic necrosis virus infection in Atlantic salmon *Salmo salar* post-smolts affects the outcome of secondary infections with infectious salmon anemia virus or *Vibrio salmonicida*. *Diseases of Aquatic Organisms* 47, 109–117.

Johansen, L.-H., Eggset, G. and Sommer, A.-I. (2009) Experimental IPN virus infection of Atlantic salmon parr; recurrence of IPN and effects on secondary bacterial infections in post-smolts. *Aquaculture* 290, 9–14.

Knott, R.M. and Munro, A.L. (1986) The persistence of infectious pancreatic necrosis virus in Atlantic salmon. *Veterinary Immunology and Immunopathology* 12, 359–364.

Kushnir, N., Streatfield, S.J. and Yusibov, V. (2012) Virus-like particles as a highly efficient vaccine platform: diversity of targets and production systems and advances in clinical development. *Vaccine* 31, 58–83.

LaPatra, S., Lauda, K., Woolley, M. and Armstrong, R. (1993) Detection of a naturally occurring coinfection of IHNV and IPNV. *American Fisheries Society – Fish Health News* 21(1), 9–10.

Lee, J., Feldman, A.R., Delmas, B. and Paetzel, M. (2007) Crystal structure of the VP4 protease from infectious pancreatic necrosis virus reveals the acyl-enzyme complex for an intermolecular self-cleavage reaction. *The Journal of Biological Chemistry* 282, 24928–24937.

Leong, J., Brown, D., Dobos, P., Kibenge, F., Ludert, J. et al. (2000) Family *Birnaviridae*. In: Van Regenmortel, M., Bishop, D., Calisher, C., Carsten, E., Estes, M. et al. (eds) *Virus Taxonomy. Seventh Report of the International Committee for the Taxonomy of Viruses*. Academic Press, New York, pp. 481–490.

Liu, M., Zhao, L.-L., Ge, J.-W., Qiao, X.-Y., Li, Y.-J. et al. (2012) Immunogenicity of *Lactobacillus*-expressing VP2 and VP3 of the infectious pancreatic necrosis virus (IPNV) in rainbow trout. *Fish and Shellfish Immunology* 32(1), 196–203.

Magyar, G. and Dobos, P. (1994) Evidence for the detection of the infectious pancreatic necrosis virus polyprotein and the 17-kDa polypeptide in infected cells and of the NS protease in purified virus. *Virology* 204, 580–589.

Martinez-Alonso, S., Vakharia, V.N., Saint-Jean, S.R., Perez-Prieto, S. and Tafalla, C. (2012) Immune responses elicited in rainbow trout through the administration of infectious pancreatic necrosis virus-like particles. *Developmental and Comparative Immunology* 36, 378–384.

McAllister, P. (2007) 2.2.5 Infectious pancreatic necrosis. In: *AFS-FHS. Fish Health Section Blue Book: Suggested Procedures for the Detection and Identification of Certain Finfish and Shellfish Pathogens*, 2007 edn. American Fisheries Society-Fish Health Section, Bethesda, Maryland. Available at: http://www.afs-fhs. org/perch/resources/14069231202.2.5ipnv2007 ref2014.pdf (accessed 21 October 2016).

McKnight, I.J. and Roberts, R.J. (1976) The pathology of infectious pancreatic necrosis. I. The sequential histopathology of the naturally occurring condition. *British Veterinary Journal* 132, 76–85.

Mikalsen, A.B., Torgersen, J., Alestrom, P., Hellemann, A.L., Koppang, E.O. *et al.* (2004) Protection of Atlantic salmon *Salmo salar* against infectious pancreatic necrosis after DNA vaccination. *Diseases of Aquatic Organisms* 60, 11–20.

Moen, T., Baranski, M., Sonesson, A.K. and Kjoglum, S. (2009) Confirmation and fine-mapping of a major QTL for resistance to infectious pancreatic necrosis in Atlantic salmon (*Salmo salar*): population-level associations between markers and trait. *BMC Genomics* 10:368.

Moen, T., Torgersen, J., Santi, N., Davidson, W.S., Baranski, M. *et al.* (2015) Epithelial cadherin determines resistance to infectious pancreatic necrosis virus in Atlantic salmon. *Genetics* 200, 1313–1326.

Munang'andu, H.M., Fredriksen, B.N., Mutoloki, S., Brudeseth, B., Kuo, T.Y. *et al.* (2012) Comparison of vaccine efficacy for different antigen delivery systems for infectious pancreatic necrosis virus vaccines in Atlantic salmon (*Salmo salar* L.) in a cohabitation challenge model. *Vaccine* 30, 4007–4016.

Munang'andu, H.M., Santi, N., Fredriksen, B.N., Lokling, K.E. and Evensen, O. (2016) A systematic approach towards optimizing a cohabitation challenge model for infectious pancreatic necrosis virus in Atlantic salmon (*Salmo salar* L.). *PLoS ONE* 11(2): e0148467.

Munro, E.S. and Midtlyng, P.J. (2011) Infectious pancreatic necrosis and associated aquabirnaviruses. In: Woo, P. and Bruno, D. (eds) *Fish Diseases and Disorders: Viral, Bacterial and Fungal infections*, 2nd edn. CAB International, Wallingford, UK, pp. 1–65.

Norwegian Veterinary Institute (2015) *The Health Situation in Norwegian Aquaculture 2014*. Norwegian Veterinary Institute, Oslo.

OIE (2003) Chapter 2.1.8. Infectious pancreatic necrosis. In: *Manual of Diagnostic Tests for Aquatic Animals*, 4th edn,. World Organisation for Animal Health, Paris, pp. 142–151. Available at: http://www.oie.int/doc/ged/D6505.PDF (accessed 17 November 2016).

Orpetveit, I., Mikalsen, A.B., Sindre, H., Evensen, O., Dannevig, B.H. *et al.* (2010) Detection of infectious pancreatic necrosis virus in subclinically infected Atlantic salmon by virus isolation in cell culture or real-time reverse transcription polymerase chain reaction: influence of sample preservation and storage. *Journal of Veterinary Diagnostic Investigation* 22, 886–895.

Ozaki, A., Araki, K., Okamoto, H., Okauchi, M., Mushiake, K. *et al.* (2012) Progress of DNA marker-assisted breeding in maricultured fish. *Bulletin of Fish Research Agencies* 35, 31–37. Available at: https://www.fra. affrc.go.jp/bulletin/bull/bull35/35-5.pdf (accessed 21 October 2016).

Petit, S., Lejal, N., Huet, J.C. and Delmas, B. (2000) Active residues and viral substrate cleavage sites of the protease of the birnavirus infectious pancreatic necrosis virus. *Journal of Virology* 74, 2057–2066.

Roberts, R.J. and Pearson, M. (2005) Infectious pancreatic necrosis in Atlantic salmon, *Salmo salar* L. *Journal of Fish Diseases* 28, 383–390.

Ruane, N., Geoghegan, F. and Ó Cinneide, M. (2007) Infectious pancreatic necrosis virus and its impact on the Irish salmon aquacultue and wild fish sectors. Marine Environment and Health Series No. 30, Marine Institute, Oranmore Republic of Ireland.

Sadasiv, E.C. (1995) Immunological and pathological responses of salmonids to infectious pancreatic necrosis virus (IPNV). *Annual Review of Fish Diseases* 5, 209–223.

Santi, N., Sandtro, A., Sindre, H., Song, H., Hong, J.R. *et al.* (2005) Infectious pancreatic necrosis virus induces apoptosis *in vitro* and *in vivo* independent of VP5 expression. *Virology* 342, 13–25.

Shivappa, R.B., McAllister, P.E., Edwards, G.H., Santi, N., Evensen, O. *et al.* (2005) Development of a subunit vaccine for infectious pancreatic necrosis virus using a baculovirus insect/larvae system. *Developmental Biology (Basel)* 121, 165–174.

Smail, D., McFarlane, L., Bruno, D. and McVicar, A. (1995) The pathology of IPN-Sp and sub-type (Sh) in farmed Atlantic salmon, *Salmo salar* L., post-smolts in the Shetland Isles, Scotland. *Journal of Fish Diseases* 18, 631–638.

USFWS and AFS-FHS (2007) Section 2. USFWS [US Fish and Wildlife Service]/AFS-FHH Standard procedures for aquatic animal health inspections. In: *AFS-FHS. Fish Health Section Blue Book: Suggested Procedures for the Detection and Identification of Certain Finfish and Shellfish Pathogens*, 2007 edn. American Fisheries Society-Fish Health Section, Bethesda, Maryland.

Villanueva, R.A., Galaz, J.L., Valdes, J.A., Jashes, M.M. and Sandino, A.M. (2004) Genome assembly and particle maturation of the birnavirus infectious pancreatic necrosis virus. *Journal of Virology* 78, 13829–13838.

Wood, E.M., Snieszko, S.F. and Yasutake, W.T. (1955) Infectious pancreatic necrosis in brook trout. *American Medical Association Archives of Pathology* 60, 26–28.

Wolf, K., Snieszko, S.F., Dunbar, C.E. and Pyle, E. (1960) Virus nature of infectious pancreatic necrosis in trout. *Proceedings of Society for Experimental Biology and Medicine* 104, 105–108.

A.K. Dhar *et al.*

2 Infectious Haematopoietic Necrosis Virus

JO-ANN C. LEONG[1]* AND GAEL KURATH[2]

[1]Hawai'i Institute of Marine Biology, University of Hawai'i at Mānoa, Kāne'ohe, Hawai'i, USA; [2]Western Fisheries Research Center, US Geological Survey, Seattle, Washington, USA

2.1 Introduction

Infectious haematopoietic necrosis virus (IHNV, infectious hematopoietic necrosis virus) is a *Rhabdovirus* that causes significant disease in Pacific salmon (*Oncorhynchus* spp.), Atlantic salmon (*Salmo salar*), and rainbow and steelhead trout (*O. mykiss*). The disease that it causes, infectious haematopoietic necrosis (IHN), was first detected in cultured sockeye salmon (*O. nerka*) in the Pacific Northwest of North America and IHNV was first cultured in 1969 (see Bootland and Leong, 1999). IHNV is the type species and reference virus for the *Novirhabdovirus* genus of the family *Rhabdoviridae*. The viral genome is a linear, single-stranded RNA (~11,140 nucleotides in length) of negative sense with six genes that read from the 3′ end of the genome as N (nucleoprotein), P (phosphoprotein), M (matrix protein), G (glycoprotein), NV (non-virion protein) and L (RNA polymerase). The name *Novirhabdovirus* is derived from the unique *non-virion* gene present in this genus (Kurath and Leong, 1985; Leong and Kurath, 2011).

IHNV causes necrosis of the haematopoietic tissues, and consequently it was named infectious haematopoietic necrosis by Amend *et al.* (1969). This virus is waterborne and may transmit horizontally and vertically through virus associated with seminal and ovarian fluids (see Bootland and Leong, 1999). Convalescent rainbow trout fry often clear the virus, but some fish can harbour it for 46 days, (Drolet *et al.*, 1995). The virus persists in the kidneys of some survivors for a year after infection (Drolet *et al.*, 1995; Kim *et al.*, 1999).

Rainbow trout that survived the infection and were kept in virus-free water for 2 years had infectious virus in seminal and ovarian fluids at spawning (Amend, 1975). These fish are potential reservoirs. However, adult sockeye salmon collected in seawater and held to maturity in virus-free water had no detectable virus at spawn, while cohorts allowed to migrate naturally had prevalences of 90–100% (Amos *et al.*, 1989); this suggests the importance of horizontal transmission during river migration and that persistence of IHNV differs in different hosts. Recently, Müller *et al.* (2015) demonstrated the persistence of IHNV in the brains, but not in the kidneys, of sockeye salmon survivors. Despite the absence of disease and mortality among survivors, 4% of the fish had IHNV viral RNA in their brains at 9 months postexposure. This supports the hypothesis that a small percentage of infected fish become carriers. If the virus from the brain is infectious, the finding would have serious impacts on strategies for viral containment. Other potential reservoirs include virus adsorbed to sediment and virus detected, albeit rarely, in invertebrates or non-salmonid fish hosts.

IHNV is reportable to the World Organisation of Animal Health (OIE) and countries with confirmed or suspected cases include: Austria, Belgium, Bolivia, Canada, China, Croatia, Czech Republic, France, Germany, Iran, Italy, Japan, Korea (Republic of), The Netherlands, Poland, Russia, Slovenia, Spain, Switzerland and the USA (OIE, 2015; last updated 23 July 2015; and Cefas, 2011, last updated 31 January). The virus is originally endemic to western

*Corresponding author e-mail: joannleo@hawaii.edu

North America, where it has the largest diversity of host species and the longest history of disease impacts, infects both wild and cultured fishes, and is most genetically diverse. IHNV has also been introduced into Asia and Europe, where it has become established largely in rainbow trout farms. Global phylogenetic analysis of IHNV has defined five major genetic groups (genogroups): group U in North America and Russia; groups M and L in North America; group J in Asia; and group E in Europe (Kurath, 2012a).

Economic losses from IHNV can be a direct consequence of fish mortality, or an indirect effect related to regulations that restrict the movement of IHNV-infected fish or require that infected stocks be destroyed. Disease outbreaks have devastated both commercial aquaculture (e.g. rainbow trout and Atlantic salmon) and conservation/mitigation programmes for Pacific salmon and trout in western North America. Since the disease has spread to Europe and Asia, an example of the potential economic impact of IHNV on salmon and trout fisheries/aquaculture was recently provided by Fofana and Baulcomb (2012). Although IHNV has not been isolated in the UK, it has been estimated that the direct and indirect costs of a theoretical IHN outbreak over 10 years (1998–2008) would be 16.8 million British pounds (~$25.5 million US dollars). Direct costs would be due to culling, mortality and the disposal of dead fish. Indirect costs would include lost revenue from consumer responses, impact on reduced exports and the increased expense of implementing additional surveillance strategies. The OIE maintains a database of IHN outbreaks, and in 2012, there were detections of IHNV in the western USA, Germany, Italy, Poland, China, Japan, Korea and British Columbia in Canada.

2.2 Clinical Signs of Disease and Diagnosis

2.2.1 Clinical signs

IHNV infection causes serious disease in young salmonid fishes though the virus can infect salmonids at all ages. Typically, at the start of an epizootic, moribund fish become lethargic, with periods of sporadic whirling or hyperactivity; fry may have a dark coloration, distended abdomens, exophthalmia, pale gills and mucoid, opaque faecal casts (Fig. 2.1A). Petechial haemorrhages may occur at the base of the fins and vent and occasionally in the gills, mouth, eyes, skin and muscle (Figs 2.1B,C). Some fry may have a subdermal haemorrhage immediately behind the head. Older fish have fewer external clinical signs. Sockeye salmon smolts have gill and eye haemorrhages, clubbed and fused lamellae and cutaneous lesions, while 2-year-old kokanee salmon (landlocked sockeye salmon, O. nerka) have erratic swimming and haemorrhages near the base of the fins. Some fish succumb to IHN disease without visible signs (see Bootland and Leong, 1999).

The liver, spleen and kidney of infected fry are pale due to anaemia, there may be ascites and the stomach is filled with a milky fluid but without food. The intestine contains a watery, yellowish fluid and there may be petechial haemorrhages in the visceral mesenteries, adipose tissue, swim bladder, peritoneum, meninges and pericardium. Older fish may have empty stomachs, intestines filled with yellowish mucus and lesions in the musculature near the kidney (see Bootland and Leong, 1999).

2.2.2 Diagnosis

Preliminary diagnosis can be based on fish with clinical signs at a site where there is history of the disease. Histologically, the observation of necrosis of the granular cells of the alimentary tract is pathognomonic (Wolf, 1988). A preliminary diagnosis must be confirmed by specific identification. The most widely accepted diagnostic method is the isolation of the virus in cell culture (presumptive diagnosis) followed by identification using a serum neutralization test or PCR-based methods. Immunological and molecular methods are described in the (online) *Manual of Diagnostic Tests for Aquatic Animals* from the OIE (2015), the Canadian *Fish Health Protection Regulations: Manual of Compliance* (Department of Fisheries and Oceans, 1984; revised 2011), and the IHNV chapter (LaPatra, 2014) in the American Fisheries Society (AFS)-Fish Health Section (FHS) *FHS Blue Book: Suggested Procedures for the Detection and Identification of Certain Finfish and Shellfish Pathogens*, 2014 edn (AFS-FHS, 2014).

Many teleost cell lines are susceptible to IHNV infection, but those specified by the OIE Manual, the Canadian Manual of Compliance and the *FHS Blue Book* are the epithelioma papulosum cyprini (EPC) and/or fathead minnow (FHM) cell lines. The *FHS Blue Book* also recommends the Chinook

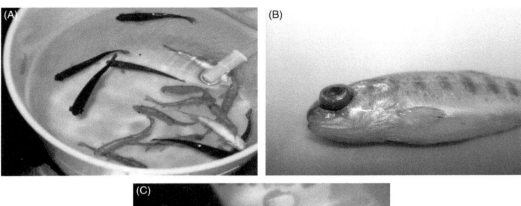

Fig. 2.1. (A) Juvenile rainbow trout at 7 days after immersion exposure to IHNV. The lighter fish show no evidence of infection and the darker fish exhibit the typical signs of infectious haematopoietic necrosis (IHN) disease with darkening coloration and pronounced exophthalmia. The dead fish to the right side of the tank exhibit petechial haemorrhages. (B) Juvenile rainbow trout with IHN disease showing pronounced exophthalmia with bleeding in the eye orbit. (C) Juvenile sockeye salmon fingerling with IHN disease showing petechial haemorrhages around the eye orbit and on the gills and fins.

salmon embryo cell line, CHSE-214. EPC cells are the most susceptible to IHNV (Lorenzen *et al.*, 1999). Other fish cell lines susceptible to IHNV cytopathogenicity are discussed extensively in Bootland and Leong (1999). The optimum temperature for growth is approximately 15°C (Mulcahy *et al.*, 1984); the higher range of 23–25°C does not support viral replication. IHNV is heat labile, and is inactivated within several hours at 32°C (Pietsch *et al.*, 1977).

Cell culture detection of IHNV can require 14 days. End-point titrations and plaque assays are typically used. The latter are more sensitive and are the standard for quantifying IHNV. Pretreatment of cell monolayers with polyethylene glycol improves the speed and sensitivity of plaque assays and produces larger plaques. The typical cytopathic effect (CPE) consists of grape-like clusters of rounded cells, with margination of the chromatin of the nuclear membrane. Typical IHNV plaques

consist of a cell sheet that retracts or piles up at the inner margins of the opening; the centre may contain granular debris. Cells should be examined for 14 days. If no CPE occurs, the supernatant may be passed on to fresh cells for a 'blind passage'. The absence of CPE indicates that the sample is virus negative (see Bootland and Leong, 1999)

The preferred tissues for isolating IHNV are the kidney and spleen; mucus and pectoral fin clippings have also been used as non-lethal samples. For testing brood stock, the ovarian fluid is preferred, because the virus is less frequently detected in the milt. The sampling of post-spawning females, storage of ovarian fluid or incubation of ovarian fluid cells enhances the sensitivity of viral detection. Milt samples should be centrifuged and, after the pellet is incubated in water, the water is assayed for virus (see Bootland and Leong, 1999, 2011).

Serological assays require either polyclonal or monoclonal antibodies and IHNV immunodetection kits are commercially available (Bio-X Diagnostics, Rochefort, France; see http://www.biox.com). Assays for the identification of IHNV include serum neutralization, indirect fluorescent antibody testing, direct alkaline phosphatase immunocyto-chemistry (APIC) and enzyme-linked immunosorb-ent assay (ELISA), as well as the Western blot, dot blot, staphylococcal co-agglutination and electron microscopy (see Bootland and Leong, 1999). The molecular methods include reverse transcriptase-dependent polymerase chain reaction (RT-PCR), real-time (quantitative) RT-PCR (qRT-PCR; see Bootland and Leong, 2011; Purcell *et al.*, 2013), multiplex RT-PCR (Liu *et al.*, 2008), loop-mediated isothermal amplification (Gunimaladevi *et al.*, 2005) and a molecular padlock probe (Millard *et al.*, 2006). With the exception of the latter, an initial RT step must be used to create cDNA from the IHNV viral RNA (mRNA, genome and anti-genome) followed by amplification using PCR. These methods and their relative sensitivities (if available) have been described by Bootland and Leong (2011) and their comparative sensitivities are given in Table 2.1. For research applications, modified qRT-PCR assays have been developed for the specific detection of positive- and negative-sense IHNV RNA (Purcell *et al.*, 2006a) and for genotype-specific detection of individual IHNV strains within mixed infections (Wargo *et al.*, 2010).

Salmonids infected with IHNV may mount a strong antibody response that persists for months (see Lorenzen and LaPatra, 1999). Monitoring the antibody response is not lethal and may be useful in surveillance of a population for previous expo-sure to IHNV (Bootland and Leong, 2011).

2.3 Pathology

2.3.1 Histopathology

Histopathological findings include degenerative necrosis in haematopoietic tissues, the posterior kidney, spleen, liver, pancreas and digestive tract. In the anterior kidney, the initial changes are small, lightly stained focal areas consisting of apparent macrophages and degenerating lymphoid cells. As the disease progresses, degenerative changes become noticeable throughout the kidney. Macrophages increase in number and may have a vacuolated cytoplasm and chromatin margination of the nuclei. Pyknotic and necrotic lymphoid cells

may be present. Necrosis may be so severe that the kidney tissue consists primarily of necrotic debris. Focal areas of cells in the spleen, pancreas, liver, adrenal cortex and intestine show nuclear poly-morphism and margination of the chromatin, with eventual necrosis. Extensive necrosis in all organs is accompanied by pyknosis, karyorrhexis and kary-olysis. A pathognomonic feature of IHN is degen-eration and necrosis of granular cells in the lamina propria, stratum compactum and stratum granulo-sum of the alimentary tract, and sloughing of intes-tinal mucosa may give rise to faecal casts. Smolts and yearlings tend to show less severe histopathol-ogy. The kidney, spleen, pancreas and liver may show necrosis, but there is only moderate slough-ing of the intestinal mucosa and no faecal casts. Fish have a normocytic aplastic anaemia and the blood of those that are affected has leucopenia with degenerating leucocytes and thrombocytes, a reduced haematocrit and osmolarity, and a slightly altered biochemical profile. Cellular debris (necro-biotic bodies) in blood smears or kidney imprints is pathognomonic for IHN (Wolf, 1988; in Bootland and Leong, 1999).

2.3.2 Disease progression

In rainbow trout, the IHN virus enters transiently through the gills, skin, fin bases and oral region to the oesophagus/cardiac stomach region before spreading to internal organs. The haematopoietic tissues of the kidney and spleen of young fish are most severely affected and are the first tissues to show extensive necrosis. Typically, within a day after immersion exposure, low titres of IHNV are detectable in the gills, skin and intestine of young rainbow trout before the infection spreads to the kidney (2–4 days) and subsequently becomes wide-spread (Bootland and Leong, 1999, 2011).

During infection, viral prevalence and titres peak within 5–14 days (Drolet *et al.*, 1994; Peñaranda *et al.*, 2009; Purcell *et al.*, 2009). Some rainbow trout have been shown to be still infected at 28 days but infectious virus was no longer detected after 54 days (Drolet *et al.*, 1994). Drolet *et al.* (1994) proposed that the infection progressed via two major routes: from the gills into the circulatory system and from the oral region into the gastro-intestinal (GI) tract and then into the circulatory system. Both routes induced systemic viraemia. These authors proposed that the initial infection of the kidney was not via the GI route but rather

Table 2.1. Sensitivity of different diagnostic methods for IHNV detection

Method	Reported sensitivity	Calculated no. of physical particles[a]	Reference
Infectivity in fish cell lines	1 infectious unit	1 TCID$_{50}$ units = ~ 500 1 PFU = 0.69 TCID$_{50}$ units = ~345	Durrin, 1980
Staphylococcal co-agglutination	1–6 × 10^4 PFU	3.45–20.7 × 10^6	Bootland and Leong, 1992
Western Blot			
Peroxidase labelled antibody	10 ng of virus[b]	7.67 × 10^7	Hsu and Leong, 1985
Iodine^{125} labelled antibody	2.5 ng of virus	1.53 × 10^7	Hsu and Leong, 1985
Alkaline Phosphatase labelled secondary antibody			
Polyclonal mouse anti IHNV N	1 PFU	345	Kim et al., 2001
Monoclonal mouse anti IHNV N	3 × 10^3 PFU	1.035 × 10^6	Kim et al., 2001
Enzyme-linked immunosorbent assay			
Polyclonal mouse anti-IHNV N	10 PFU	3450	Kim et al., 2001
Monoclonal mouse anti-IHNV N	10^3 PFU	3.45 × 10^6	Kim et al., 2001
Chicken and rabbit anti-IHNV	70 PFU	2.42 × 10^4	Medina et al., 1992
Immunoblot assay			
Peroxidase Labelled	0.85–4.0 ng virus	0.66–3.07 × 10^7	McAllister and Schill, 1986
Biotinylated antibody	10^2 PFU, 5.5 ng virus	3.45 × 10^4, 4.22 × 10^7	Schultz et al. 1989
RT-PCR, N gene, single round	4 × 10^2 PFU	1.38 × 10^5	Arakawa et al., 1990
RT-PCR, G gene, single round	2.2 × 10^3 G gene RNA copies[c]	5.9 × 10^2 viral gRNAs, 25 particles	Emmenegger et al., 2000, Purcell et al., 2013.
Immunomagnetic bead capture with monoclonal anti IHNV N, followed by RT-PCR	10^6 PFU	3.45 × 10^6	Kim et al., 2001
Dot blot RNA Hybridization (biotin or alkaline phosphatase label) if chemiluminescent reaction is used	20 pg viral RNA[d] 4 pg viral RNA	3.37 × 10^6 viral gRNA 4.22 × 10^4 viral gRNA	Gonzalez et al., 1997 Gonzalez et al., 1997
RT-qPCR, N gene	7 N gene RNA copies[e]	1.1 viral gRNA, 0.05 particles	Purcell et al., 2013

[a]The sensitivity of each method was used to determine the equivalent number of physical IHNV particles quantified using electron microscopy (Durrin, 1980).

[b]The weight of the IHNV particle was derived from a 21:1 ratio of virion protein to RNA and the estimate that the molecular weight of the viral genome is 3.57 × 10^6 (Kurath and Leong, unpublished observation). Thus, the molecular weight of a virion is 7.497 × 10^7 daltons of protein plus 3.57 × 10^6 daltons of RNA. The calculated weight of one virion is 1.30418 × 10^{-16} grams, and the calculated weight of one viral genome is 5.928 × 10^{-18} grams.

[c]The IHNV G gene conventional PCR assay detection was calibrated to viral RNA copy number based on data from Purcell et al., 2013 (M. Purcell and G. Kurath, unpublished), and includes both viral genomic (gRNA) and mRNA . Within this viral RNA copy number the ratio of gRNA:G mRNA was reported as 1:2.7, and the ratio of gRNA:PFU as 8000:1 (Purcell et al., 2006). These ratios were used to calculate the equivalent number of viral gRNA copies, PFU, and then physical particles.

[d]Gonzalez et al, 1997, used partially purified virus for their RNA extraction, consequently the actual virus RNA number might actually be lower.

[e]The universal IHNV N gene RT-qPCR assay detection limit was calibrated as described in footnote c, but was modified based on the reported molar ratio of IHNV G mRNA:N mRNA as approximately 1:2. Thus a ratio of gRNA: N mRNA of 1:5.4 was used to calculate the equivalent number of gRNA copies.

through the highly vascularized tissue lining the oral cavity. From there, the virus travelled through the blood to the kidney and haematopoietic tissues. Using Immunogold-labelled secondary antibody to detect the binding of an anti-N monoclonal antibody to infected cells, Helmick *et al.* (1995a,b) identified an early IHNV target area, the esophagus/cardiac stomach region (ECSR), particularly the cardiac mucus-secreting cardiac gland (MSSG). There was evidence of the attachment and internalization of IHNV in the ECSR mucosal epithelial cells in rainbow trout and coho salmon (*O. kisutch*) within 1 h postinfection. The MSSG of coho salmon showed a milder reaction, which supported previous findings that the virus replicated less efficiently in coho salmon. In Chinook and sockeye salmon fry there may be hepatic deposits of ceroid (Wood and Yasutake, 1956; Yasutake, 1970). In the final stages of the disease, necrosis is seen not only in the haematopoietic tissues of the kidney, but also in the glomeruli and kidney tubules.

Another study in rainbow trout used bioluminescence imaging in living fish to follow infection with recombinant IHNV strains expressing luciferase (Harmache *et al.*, 2006). In addition to identifying fin bases as a site of entry, there was continuing viral replication in fin tissues and internal organs. Some survivors maintained localized viral replication, mostly in the fins. More recently, the progression of infection with a high temperature-adapted IHNV strain was monitored in transparent zebrafish (*Danio rerio*) larvae (Ludwig *et al.*, 2011) infected by injection and held at 24°C. Macroscopic signs of infection slowed down and then arrested blood flow despite continuing heartbeat. This was followed by a loss of reactivity to touch and the fish were dead in 3–4 days. Using *in situ* hybridization in whole larvae, the first infected cells were detected at 6 h postinfection in the major blood vessels and the venous endothelium was a primary target of infection. This suggested that infection spread from damaged vessels to the underlying tissues. By transferring the larvae to 28°C (no viral replication), a critical threshold resulting in irreversible damage was reached in less than a day, before clinical signs appeared.

2.4 Pathophysiology

IHN is characterized by a severe depletion of the alkali reserve and an imbalance of blood electrolytes, resulting in decreased blood osmolality (Amend, 1973). These changes have been ascribed to the loss of renal function from the viral induced necrosis of kidney tissues. Packed cell volume, haemoglobin, red blood cell count and plasma bicarbonate were significantly depressed in 4 days (Amend and Smith, 1974). Plasma chloride, calcium, phosphorus, total protein and blood cell types did not change during the 9 days of study. An increase in lactate dehydrogenase B ($LDHB_4^{2\prime}$) isoenzyme levels was consistently associated with early development of the disease. Increased LDH was not observed in fish infected with infectious pancreatic necrosis virus (IPNV) or three bacterial pathogens.

Amend and Smith (1975) found reduced plasma bicarbonate, chloride, calcium, phosphorus, bilirubin and osmolality in moribund rainbow trout. When the fish showed signs of disease, plasma glucose and anterior kidney ascorbates were unchanged. Infected fish had reduced corpuscular counts, haemoglobin and packed cell volume, but mean corpuscular volume, mean corpuscular haemoglobin and mean corpuscular haemoglobin concentrations remained normal. The percentage of immature erythrocytes increased, but the percentage of leucocytes was unchanged. Neutrophils decreased, lymphocytes increased, but monocytes did not change. Plasma pH increased and the alpha 2 and 3 fractions of the serum proteins were altered. The alkali reserve diminished and acid:base and fluid balances were altered. Death probably resulted from severe electrolyte and fluid imbalances caused by renal failure.

2.5 Control Strategies

Control strategies rely primarily on avoidance of exposure to the virus through biosecurity policies and best practices for hygiene at culture facilities (Winton, 1991). The successful Alaskan sockeye salmon culture policy implemented in 1981 provides guidelines for IHNV control within a virus-endemic region based on three criteria: a virus-free water supply, rigorous disinfection; and the compartmentalization of incubating eggs and rearing juvenile fish (Meyers *et al.*, 2003). Thus, eggs are disinfected, typically using treatment with an iodophor (a disinfectant containing iodine complexed with a solubilizing agent), and then incubated in separate egg lots in virus-free water. Fry are also reared in virus-free water. Secure water sources (free of susceptible host fish) are used as long as possible throughout the rearing period. The recent success of a delayed exposure rearing strategy at a large steelhead hatchery illustrates the importance of water supply in minimizing viral transmission between free-ranging and hatchery fish (Breyta *et al.*, 2016). Disease outbreaks are controlled in areas where the disease is not endemic by culling, disinfection and quarantine (McDaniel *et al.*, 1994). Additional precautions generally practised within British Columbia salmon farms include the maintenance of single-year class and single-species sites, reduced fish movements between pens, the culling/accelerated harvest of smaller fish when infection occurs and fallowing between restockings (Saksida, 2006). The introduction of IHNV into new geographical areas has never been associated with the movement of killed fresh fish or frozen products and the risk is considered negligible for processed rainbow trout (LaPatra *et al.*, 2001a).

In cell cultures, IHNV replication is inhibited by methisoprinol (Siwicki *et al.*, 2002), and chloroquine inhibits IHNV *in vivo* by reducing viral binding and cell entry (Hasobe and Saneyoshi, 1985; De las Heras *et al.*, 2008). Similarly, the pretreatment of cultured cells with the antiviral agents amantadine or tributylamine also reduced IHNV binding and cell entry. When antisense phosphorodiamidate morpholino oligomers (PMOs) complementary to the 5′ end of IHNV genomic RNA were tested against IHNV in cell culture, inhibition was sequence and dose dependent. The compound was cross-linked to a membrane-penetrating peptide that enhanced entry of the MPO into cultured cells and live fish tissues (Alonso *et al.*, 2005). Anti-IHNV activity has also been reported in bacteria from aquatic environments (Myouga *et al.*, 1993). A peptide (46NW-64A), produced by *Pseudomonas fluorescens* biovar 1 completely inhibited the replication of IHNV (100% plaque reduction). Inclusion of the carotenoid, astaxanthin, in feed produced significant resistance to IHNV challenge in rainbow trout fry (Amar *et al.*, 2012). Bovine α_2-CN casein hydrolysates and total casein hydrolysates are potent antiviral agents against IHNV on fish cell monolayers (Rodríguez Saint-Jean *et al.*, 2012). When three month old brown trout (*Salmo trutta*) were fed casein at day 0 and day 3 postexposure to IHNV, the treated group showed nearly 50% protection against a virus dose that killed 93% of the control fish (Rodríguez Saint-Jean *et al.*, 2013).

Since McIntyre and Amend (1978) reported 30% heritability for resistance to IHNV in sockeye salmon, efforts have continued to develop IHNV-resistant strains of trout and salmon. The resistance can be species specific and confined to a particular viral genogroup. However, resistance to IHN disease does not mean resistance to infection. In terms of disease, sockeye salmon are resistant to M group virus but sensitive to U group virus, while rainbow trout have the converse phenotypes (Garver *et al.*, 2006). Rainbow trout hybrids from crosses with more resistant species, (e.g. coho salmon, or brook trout, *Salvelinus fontinalis*) are more resistant to IHNV (LaPatra *et al.*, 1993), whereas coho and chinook salmon hybrids are susceptible to IHNV-induced mortalities (Hedrick *et al.*, 1987).

The identification of the genes responsible for resistance has been the focus of selective breeding programmes for disease-resistant rainbow trout and Atlantic salmon. The majority of genetic studies on IHNV resistance have used rainbow trout backcrossed within families (Khoo *et al.*, 2004) or with less susceptible cutthroat trout (*O. clarkia*) (Palti *et al.*, 1999, 2001; Barroso *et al.*, 2008) or steelhead trout (Rodriguez *et al.*, 2004). Genetic linkage maps based on molecular markers (microsatellites and amplified fragment length polymorphisms (AFLPs)) and quantitative trait loci (QTLs) have been characterized, such studies but have not identified highly associated loci (Overturf *et al.*, 2010) or significant associations within very large blocks of linkage disequilibrium (Rodriguez *et al.*, 2004; Barroso *et al.*, 2008). Genome sequencing has identified 19 single-nucleotide polymorphism (SNP) markers in rainbow trout that were associated with IHNV resistance and can be used in a marker-assisted selection (MAS) programme (Campbell *et al.*, 2014). It is probable the resistance genes are related to MHC (major histocompatibility complex) haplotypes and T cell immunity, as has been reported by Yang *et al.* (2014). Interestingly, Verrier *et al.* (2012) found that cell lines derived from trout clonal lines resistant to viral haemorrhagic septicaemia virus (VHSV) were also resistant to rhabdovirus infection under *in vitro* conditions.

2.5.1 Vaccines

A DNA vaccine referred to as APEX-IHN (Novartis Animal Health Canada Inc.) was approved in 2005 by the Canadian Food Inspection Agency (CFIA) for farm-raised Atlantic salmon (CFIA, 2005). This was the first DNA vaccine approved for fish and one of only four DNA vaccines commercially available for animals. The vaccine was based on the demonstration that an intramuscular injection of plasmid DNA encoding the IHNV G gene under the control of a constitutive viral promoter induced protection in rainbow trout fry (Anderson *et al.*, 1996). Since then, numerous experimental studies in the USA have confirmed the efficacy of similar DNA vaccines and examined several practical aspects, such as vaccine dose, duration of protection, cross-species protection, routes of administration and vaccine safety (see Kurath, 2008). It was confirmed that the DNA vaccine must encode the IHNV G gene, not other IHNV genes, to induce protection and neutralizing antibodies in rainbow trout fry and in sockeye salmon. The vaccine was effective at extremely low doses (0.1–1 µg) in 1–2 g rainbow trout fry; larger fish required a higher dose. Vaccinated rainbow trout fry were protected against IHNV from different geographical locations, suggesting that it would be useful worldwide. The DNA vaccine induced rapid non-specific innate immunity (LaPatra *et al.*, 2001b; Lorenzen *et al.*, 2002) with the expression of Mx protein, an indicator of alpha/beta interferon induction (Kim *et al.*, 2000; Purcell *et al.*, 2006b), followed by long-term specific protection with relative percentage survival (RPS) > 90% for 3 months and RPS > 60% for 2 years (Kurath *et al.*, 2006). The IHNV-G DNA vaccine induced protection in rainbow trout (see Kurath, 2008), Atlantic salmon (Traxler *et al.*, 1999), Chinook salmon, sockeye salmon and kokanee salmon (*O. nerka*) (Garver *et al.*, 2005a). DNA vaccines containing G genes from either the M or U genogroups protected fish against intra- and cross-genogroup IHNV challenges (Peñaranda *et al.*, 2011).

The DNA vaccine is administered by intramuscular injection, which is labour intensive and difficult to implement with small fish. Equivalent high levels of efficacy were detected in rainbow trout fry when the vaccine was delivered by cutaneous particle bombardment using a gene gun. However, only partial protection was obtained by intraperitoneal injection, and several other routes of vaccination did not induce protection (Corbeil *et al.*, 2000). Recently, an oral DNA vaccine against IHNV was developed using alginate microspheres to encapsulate the DNA vaccine; using this, significant protection was observed in rainbow trout (Ballesteros *et al.*, 2015).

The IHNV DNA vaccine is safe and well tolerated by fish. No vaccine-specific pathological changes were observed in tissues for as long as 2 years after vaccination (Garver *et al.*, 2005b: Kurath *et al.*, 2006). The plasmid backbone of the vaccine contains the cytomegalovirus (CMV) immediate early promoter, which is derived from a human pathogen. As regulatory authorities may consider such a vaccine 'unsafe', Alonso *et al.* (2003) tested DNA vaccines containing the IHNV G gene linked to three different rainbow trout promoters. A vaccine with the interferon regulatory factor 1A promoter from rainbow trout provided protection equivalent to that of the pCMV-G vaccine (Alonso and Leong, 2012).

Extensive efforts have been made to develop more conventional vaccines against IHN, such as subunit vaccines and attenuated vaccines. Experimental IHNV vaccines reviewed in Winton (1997) and Bootland and Leong (1999) showed varying degrees of efficacy and have not resulted in a commercial product. Other vaccines, including a baculovirus-derived IHNV G protein vaccine produced in insect cells and a recombinant subunit vaccine produced on the surface of *Caulobacter crescentus* have shown some efficacy in experimental studies (Cain *et al.*, 1999; Simon *et al.*, 2001). A recent re-examination of killed IHNV vaccines demonstrated the critical importance of inactivating agents, and a β-propiolactone-inactivated whole virus vaccine induced both 7 day and 56 day protection (Anderson *et al.*, 2008). Hence, there is still potential for effective killed vaccines.

2.6 Conclusions and Suggestions for Future Research

IHNV is a global problem for wild and domestic salmonids, and its spread to salmonid farms is increasing. The virus has adapted to new hosts and biological environments as evidenced by the evolution of the trout-adapted M, J and E genogroups in North America, Japan and Europe, respectively. Effective control strategies are needed to prevent the spread of the virus to naive fish populations and IHNV-free geographic regions. Perhaps the spread of IHNV can be contained by developing a source of salmon or trout eggs that are certified as specific pathogen free (SPF). Briefly, eggs would be obtained from IHNV-free brood stock and the resultant progeny reared on SPF water. These viral-free fish can then produce eggs and fingerlings for sale as IHNV-free stock. Certainly, faster diagnostics and regular surveillance for IHNV can assess the viral status of both wild and domestic fish.

Sensitive detection assays for the virus are now available and thorough investigation of potential vectors and reservoirs for IHNV is possible. Genetic typing is used in molecular epizootiology and fish health management (Breyta *et al.*, 2016) and centralized databases of epizootiological and genetic information are publically available for virus isolated in North America or Europe (Johnstrup *et al.*, 2010; Emmenegger *et al.*, 2011; Kurath, 2012b). Although the first DNA vaccine has been approved for use in fish in Canada, other countries have not licensed this product. The efficacy of the vaccine is well established, but delivery by injection is often impractical, and finding an alternative method to deliver the vaccine remains a high priority (Munang'ando and Evensen, 2015).

References

AFS-FHS (2014) *FHS Blue Book: Suggested Procedures for the Detection and Identification of Certain Finfish and Shellfish Pathogens*, 2014 edn. American Fisheries Society-Fish Health Section, Bethesda, Maryland.

Alonso, M. and Leong, J.C. (2012) Licensed DNA vaccines against infectious hematopoietic necrosis virus (IHNV). *Recent Patents on DNA and Genes Sequences* 7, 62–65.

Alonso, M., Johnson, M., Simon, B. and Leong, J.C. (2003) A fish specific expression vector containing the interferon regulatory factor 1A (IRF1A) promoter for genetic immunization of fish. *Vaccine* 21, 1591–1600.

Alonso, M., Stein, D.A., Thomann, E., Moulton, H.M., Leong, J.C. *et al.* (2005) Inhibition of infectious haematopoietic necrosis virus in cell cultures with peptide-conjugated morpholino oligomers. *Journal of Fish Diseases* 28, 399–410.

Amar, E.C., Kiron, V., Akutsu, T., Satoh, S. and Watanabe, T. (2012) Resistance of rainbow trout, *Oncorhynchus mykiss*, to infectious hematopoietic necrosis virus (IHNV) experimental infection following ingestion of natural and synthetic carotenoids. *Aquaculture* 330–333, 148–155.

Amend, D.F. (1973) Pathophysiology of infectious hematopoietic necrosis virus disease in rainbow trout. PhD thesis, University of Washington, Seattle, Washington.

Amend, D.F. (1975) Detection and transmission of infectious hematopoietic necrosis virus in rainbow trout. *Journal of Wildlife Diseases* 11, 471–478.

Amend, D.F. and Smith, L. (1974) Pathophysiology of infectious hematopoietic necrosis virus disease in rainbow trout (*Salmo gairdneri*): early changes in blood and aspects of the immune response after injection of IHN virus. *Journal of the Fisheries Research Board of Canada* 31, 1371–1378.

Amend, D.F. and Smith, L. (1975) Pathophysiology of infectious hematopoietic necrosis virus disease in rainbow trout: hematological and blood chemical changes in moribund fish. *Infection and Immunity* 11, 171–179.

Amend, D.F., Yasutake, W.T. and Mead, R.W. (1969) A hematopoietic virus disease of rainbow trout and sockeye salmon. *Transactions of the American Fisheries Society* 98, 796–804.

Amos, K.H., Hopper, K.A. and Levander, L. (1989) Absence of infectious hematopoietic necrosis virus in adult sockeye salmon. *Journal of Aquatic Animal Health* 1, 281–283.

Anderson, E.D., Mourich, D.V., Fahrenkrug, S. and Leong, J.C. (1996) Genetic immunization of rainbow trout (*Oncorhynchus mykiss*) against infectious hematopoietic necrosis virus. *Molecular Marine Biology and Biotechnology* 5, 114–122.

Anderson, E.D., Clouthier, S., Shewmaker, W., Weighall, A. and LaPatra, S. (2008) Inactivated infectious haematopoietic necrosis virus (IHNV) vaccines. *Journal of Fish Diseases* 31, 729–745.

Arakawa, C.K., Deering, R.E., Higman, K.H., Oshima, K.H., O'Hara, P.J. and Winton, J. (1990) Polymerase chain reaction (PCR) amplification of a nucleoprotein gene sequence of infectious hematopoietic necrosis virus. *Diseases of Aquatic Organisms* 8, 165–170.

Ballesteros, N.A., Alonso, M., Rodríguez Saint-Jean, S. and Pérez-Prieto, S.I. (2015) An oral DNA vaccine against infectious haematopoietic necrosis virus (IHNV) encapsulated in alginate microspheres induces dose-dependent immune responses and significant protection in rainbow trout (*Oncorhynchus mykiss*). *Fish and Shellfish Immunology* 45, 877–888.

Barroso, R.M., Wheeler, P.A., LaPatra, S.E., Drew, R.E. and Thorgaard, G.H. (2008) QTL for IHNV resistance and growth identified in a rainbow (*Oncorhynchus mykiss*) × Yellowstone cutthroat (*Oncorhynchus clarki bouvieri*) trout cross. *Aquaculture* 277, 156–163.

Bootland, L.M. and Leong, J. (1992) Staphylococcal coagglutination – a rapid method of identifying infectious hematopoietic necrosis virus. *Applied and Environmental Microbiology* 58, 6–13.

Bootland, L.M. and Leong, J. (1999) Infectious hematopoietic necrosis virus. In: Woo, P.T.K. and Bruno, D.W. (eds) *Fish Diseases and Disorders. Volume 3: Viral Bacterial and Fungal Infections*, 2nd edn. CAB International, Wallingford, UK, pp. 57–121.

Bootland, L.M. and Leong, J. (2011) Infectious haematopoietic necrosis virus. In: Woo, P.T.K., Bruno, D.W. (eds) *Fish Diseases and Disorders. Volume 3: Viral, Bacterial, and Fungal Infections*, 2nd edn. CAB International. Wallingford, UK, pp. 66–109.

Breyta, R.B., Samson, C., Blair, M., Black, A. and Kurath, G. (2016) Successful mitigation of viral disease based on a delayed exposure rearing strategy at a large-scale steelhead trout conservation hatchery. *Aquaculture* 450, 213–224.

Cain, K.D., LaPatra, S.E., Shewmaker, B., Jones, J., Byrne, K.M. and Ristow, S.S. (1999) Immunogenicity of a recombinant infectious hematopoietic necrosis virus glycoprotein produced in insect cells. *Diseases of Aquatic Organisms* 36, 62–72.

Campbell, N.R., LaPatra, S.E., Overturf, K., Towner, R. and Narum, S.R. (2014) Association mapping of disease resistance traits in rainbow trout using restriction site associated DNA sequencing. *G3: Genes, Genomes, Genetics* 4, 2473–2481.

Cefas (2011) Non-OIE data for Infectious Haematopoietic Necrosis. Last updated 31 January 2014. Centre for Environment, Fisheries and Aquaculture Science, Lowestoft/Weymouth, UK. Available at: http://www.cefas.defra.gov.uk/idaad/disease.aspx?t=n&id=55 (accessed 24 October 2016).

CFIA (2005) Environmental assessment for licensing infectious haematopoietic necrosis virus vaccine, DNA vaccine in Canada. Canadian Food Inspection Agency, Ottawa.

Corbeil, S., Kurath, G. and LaPatra, S.E. (2000) Fish DNA vaccine against infectious hematopoietic necrosis virus: efficacy of various routes of immunization. *Fish and Shellfish Immunology* 10, 711–723.

De las Heras, A.I., Rodríguez Saint-Jean, S. and Pérez-Prieto, S.I. (2008) Salmonid fish viruses and cell interactions at early steps of the infective cycle. *Journal of Fish Diseases* 31, 535–546.

Department of Fisheries and Oceans (1984) *Fish Health Protection Regulations: Manual of Compliance*, rev. 2011. Fisheries and Marine Service Miscellaneous Special Publication 31, Fisheries Research Directorate Aquaculture and Resource Development Branch, Ottawa.

Drolet, B.S., Rohovec, J.S. and Leong, J.C. (1994) The route of entry and progression of infectious hematopoietic necrosis virus in *Oncorhynchus mykiss* (Walbaum): a sequential immunohistochemical study. *Journal of Fish Diseases* 17, 337–347.

Drolet, B.S., Chiou, P.P., Heidel, J. and Leong, J.C. (1995) Detection of truncated virus particles in a persistent RNA virus infection *in vivo*. *Journal of Virology* 69, 2140–2147.

Durrin, L. (1980) An electron micrographic study of IHNV and IPNV. Development of methods for determining particle-infectivity ratios and for rapid diagnosis. Master's thesis, Oregon State University, Corvallis, Oregon.

Emmenegger, E.J., Meyer, T.R., Burton, T. and Kurath, G. (2000) Genetic diversity and epidemiology of infectious hematopoietic necrosis virus in Alaska. *Diseases of Aquatic Organisms* 40, 163–176.

Emmenegger, E.J., Kentop, E., Thompson, T.M., Pittam, S., Ryan, A. *et al.* (2011) Development of an aquatic pathogen database (AquaPathogen X) and its utilization in tracking emerging fish virus pathogens in North America: *Journal of Fish Diseases* 34, 579–587.

Fofana, A. and Baulcomb, C. (2012) Counting the costs of farmed salmonids diseases. *Journal of Applied Aquaculture* 24, 118–136.

Garver, K.A., LaPatra, S.E. and Kurath, G. (2005a) Efficacy of an infectious hematopoietic necrosis (IHN) virus DNA vaccine in chinook *Oncorhynchus tshawytscha* and sockeye *O. nerka* salmon. *Diseases of Aquatic Organisms* 64, 13–22.

Garver, K.A., Conway, C.M., Elliott, D.G. and Kurath, G. (2005b) Analysis of DNA-vaccinated fish reveals viral antigen in muscle, kidney and thymus, and transient histopathologic changes. *Marine Biotechnology* 7, 540–553.

Garver, K.A., Batts, W.N. and Kurath, G. (2006) Virulence comparisons of infectious hematopoietic necrosis virus U and M genogroups in sockeye salmon and rainbow trout. *Journal of Aquatic Animal Health* 18, 232–243.

González, M.P., Sánchez, W., Ganga, M.A., López-Lastra, M., Jashés, M. and Sandino, A.M. (1997) Detection of the infectious hematopoietic necrosis virus directly from infected fish tissues by dot blot hybridization with a non-radioactive probe. *Journal of Virological Methods* 65, 273–279.

Gunimaladevi, I., Kono, T., LaPatra, S.E. and Sakai, M. (2005) A loop mediated isothermal amplification (LAMP) method for detection of infectious hematopoietic necrosis virus (IHNV) in rainbow trout (*Oncorhynchus mykiss*). *Archives of Virology* 150, 899–909.

Harmache, A., LeBerre, M., Droineau, S., Giovannini, M. and Brémont, M. (2006) Bioluminescence imaging of live infected salmonids reveals that the fin bases are the major portal of entry for *Novirhabdovirus*. *Journal of Virology* 80, 3655–3659.

Hasobe, M. and Saneyoshi, M. (1985) On the approach to the viral chemotherapy against infectious hematopoietic necrosis virus (IHNV). *Bulletin of the Japanese Society of Scientific Fisheries* 49, 157–163.

Hedrick, R.P., LaPatra, S.E., Fryer, J.L., McDowell, T. and Wingfield, W.H. (1987) Susceptibility of coho (*Oncorhynchus kisutch*) and chinook (*Oncorhynchus tshawytscha*) salmon hybrids to experimental infections with infectious hemato-poietic necrosis virus (IHNV). *Bulletin of the European Association of Fish Pathologists* 7, 97–100.

Helmick, C.M., Bailey, F.J., LaPatra, S. and Ristow, S. (1995a) Histological comparison of infectious hematopoietic necrosis virus challenged juvenile rainbow trout *Oncorhynchus mykiss* and coho salmon *O. kisutch* gill, esophagus/cardiac stomach region, small intestine and pyloric caeca. *Diseases of Aquatic Organisms* 23, 175–187.

Helmick, C.M., Bailey, F.J., LaPatra, S. and Ristow, S. (1995b) The esophagus/cardiac stomach region: site of attachment and internalization of infectious hematopoietic necrosis virus in challenged juvenile rainbow trout *Oncorhynchus mykiss* and coho salmon *O. kisutch. Diseases of Aquatic Organisms* 23, 189–199.

Hsu, Y.L. and Leong, J. (1985) A comparison of detection methods for infectious hematopoietic necrosis virus. *Journal of Fish Diseases* 8, 1–12.

Johnstrup, S.P., Schuetze, H., Kurath, G., Gray, T., Bang Jensen, B. and Olesen, N.J. (2010) An isolate and sequence database of infectious haematopoietic necrosis virus (IHNV). *Journal of Fish Diseases* 33, 469–471.

Khoo, S.K., Ozaki, A., Nakamura, F., Arakawa, T. and Ishimoto, S. (2004) Identification of a novel chromosomal region associated with IHNV resistance in rainbow trout. *Fish Pathology* 39, 95–102.

Kim, C.H., Dummer, D.M., Chiou, P.P. and Leong, J.C. (1999) Truncated particles produced in fish surviving infectious hematopoietic necrosis virus infection: mediators of persistence? *Journal of Virology* 73, 843–849.

Kim, C.H., Johnson, M.C., Drennan, J.D., Simon, B.E., Thomann, E. and Leong, J.C. (2000) DNA vaccines encoding viral glycoproteins induce nonspecific immunity and Mx protein synthesis in fish. *Journal of Virology* 74, 7048–7054.

Kim, S.J., Lee, K-Y., Oh, M.J. and Choi, T.J. (2001) Comparison of IHNV detection limits by IMS-RT-PCR, Western blot and ELISA. *Journal of Fisheries Science and Technology* 4, 32–38.

Kurath, G. (2008) Biotechnology and DNA vaccines for aquatic animals. *Revue Scientifique et Technique (Paris)* 27, 175–196.

Kurath, G. (2012a) Fish novirhabdoviruses. In: Dietzgen, R.G. and Kuzmin, I.V. (eds) *Rhabdoviruses: Molecular Taxonomy, Evolution, Genomics, Ecology, Host–vector Interactions, Cytopathology and Control.* Caister Academic Press, Wymondham, UK, pp. 89–116.

Kurath, G. (2012b) *An Online Database for IHN Virus in Pacific Salmonid Fish: MEAP-IHNV.* Fact Sheet 2012-3027, US Department of the Interior and US Geological Survey, Seattle, Washington. Available at: http://pubs.usgs.gov/fs/2012/3027/pdf/fs20123027.pdf (accessed 21 October 2016).

Kurath, G. and Leong, J.C. (1985) Characterization of infectious hematopoietic necrosis virus mRNA species reveals a nonvirion rhabdovirus protein. *Journal of Virology* 53, 462–468.

Kurath, G., Garver, K.A., Corbeil, S., Elliott, D.G., Anderson, E.D. and LaPatra, S.E. (2006) Protective immunity and lack of histopathological damage two years after DNA vaccination against infectious hematopoietic necrosis virus in trout. *Vaccine* 24, 345–354.

LaPatra, S.E. (2014) 2.2.4 Infectious hematopoietic necrosis. In: *AFS-FHS. Fish Health Section Blue Book: Suggested Procedures for the Detection and Identification of Certain Finfish and Shellfish Pathogens*, 2014 edn. American Fisheries Society-Fish Health Section, Bethesda, Maryland. Available at: http://www.afs-fhs.org/

perch/resources/14069231202.2.5ipnv2007ref2014.pdf (accessed 21 October 2010).

LaPatra, S.E., Parsons, J.E., Jones, G.R. and McRoberts, W.O. (1993) Early life stage survival and susceptibility of brook trout, coho salmon, rainbow trout, and their reciprocal hybrids to infectious hematopoietic necrosis virus. *Journal of Aquatic Animal Health* 5, 270–274.

LaPatra, S.E., Batts, W.N., Overturf, K., Jones, G.R., Shewmaker, W.D. and Winton, J.R. (2001a) Negligible risk associated with the movement of processed rainbow trout, *Oncorhynchus mykiss* (Walbaum), from an infectious hematopoietic necrosis virus (IHNV) endemic area. *Journal of Fish Diseases* 24, 399–408.

LaPatra, S.E., Corbeil, S., Jones, G.R., Shewmaker, W.D., Lorenzen, N. *et al.* (2001b) Protection of rainbow trout against infectious hematopoietic necrosis virus four days after specific or semi-specific DNA vaccination. *Vaccine* 19, 4011–4019.

Leong, J.C. and Kurath, G. (2011) Novirhabdoviruses. In: Tidona, C. and Darai, G. (eds) *The Springer Index of Viruses.* Springer, Heidelberg, Germany.

Liu, Z., Teng, Y., Liu, H., Jiang, Y., Xie, X. *et al.* (2008) Simultaneous detection of three fish rhabdoviruses using multiplex real-time quantitative RT-PCR assay. *Journal of Virological Methods* 149, 103–109.

Lorenzen, N. and LaPatra, S.E. (1999) Immunity to rhabdoviruses in rainbow trout: the antibody response. *Fish and Shellfish Immunology* 9, 345–360.

Lorenzen, E., Carstensen, B. and Olesen, N.J. (1999) Inter-laboratory comparison of cell line for susceptibility to three viruses: VHSV, IHNV and IPNV. *Diseases of Aquatic Organisms* 37, 81–88.

Lorenzen, N., Lorenzen, E., Einer-Jensen, K. and LaPatra, S.E. (2002) DNA vaccines as a tool for analysing the protective immune response against rhabdoviruses in rainbow trout. *Fish and Shellfish Immunology* 12, 439–453.

Ludwig, M., Palha, N., Torhy, C., Briolat, V., Colucci-Guyon, E. *et al.* (2011) Whole-body analysis of a viral infection: vascular endothelium is a primary target of infectious hematopoietic necrosis virus in zebrafish larvae. *PLoS Pathogens* 7(2): e1001269.

McAllister, P.E. and Schill, W.B. (1986) Immunoblot assay: a rapid and sensitive method for identification of salmonid fish viruses. *Journal of Wildlife Diseases* 22, 468–474.

McDaniel, T.R., Pratt, K.M., Meyers, T.R., Ellison, T.D., Follett, J.E. and Burke, J.A. (1994) *Alaska Sockeye Salmon Culture Manual.* Special Publication 6, Alaska Department of Fish and Game, Juneau, Alaska.

McIntyre, J.D. and Amend, D.F. (1978) Heritability of tolerance for infectious hematopoietic necrosis in sockeye salmon (*Oncorhynchus nerka*). *Transactions of the American Fisheries Society* 107, 305–308.

Medina, D.J., Chang, P.W., Bradley, T.M., Yeh, M-T. and Sadaasiv, E.C. (1992) Diagnosis of infectious hematopoietic necrosis virus in Atlantic salmon (*Salmo salar*) by enzyme-linked immunosorbent assay. *Diseases of Aquatic Organisms* 13, 147–150.

Meyers, T.R., Korn, D., Burton, T.M., Glass, K., Follett, J.E. et al. (2003) Infectious hematopoietic necrosis virus (IHNV) in Alaskan sockeye salmon culture from 1973 to 2000: annual virus prevalences and titers in broodstocks compared with juvenile losses. *Journal of Aquatic Animal Health* 15, 21–30.

Millard, P.J., Bickerstaff, L.E., LaPatra, S.E. and Kim, C.H. (2006) Detection of infectious haematopoietic necrosis virus and infectious salmon anaemia virus by molecular padlock amplification. *Journal of Fish Diseases* 29, 201–213.

Mulcahy, D., Pascho, R.J. and Jenes, C.K. (1984) Comparison of *in vitro* growth characteristics of ten isolates of infectious hematopoietic necrosis virus. *Journal of General Virology* 65, 2199–2207.

Müller, A., Sutherland, B.J.G., Koop, B.F., Johnson, S.C. and Garver, K.A. (2015) Infectious hematopoietic necrosis virus (IHNV) persistence in sockeye salmon: influence on brain transcriptome and subsequent response to the viral mimic poly(I:C). *BMC Genomics* 16:634.

Munang'andu, H.M. and Evensen, O. (2015) A review of intra and extracellular antigen delivery systems for virus vaccines of finfish. *Journal of Immunology Research* 2014: Article ID 960859.

Myouga, H., Yoshimizu, M., Ezura, Y. and Kimura, T. (1993) Anti-infectious hematopoietic necrosis virus (IHNV) substances produced by bacteria from aquatic environment. *Gyobyo Kenkyu [Fish Pathology]* 28, 9–13.

OIE (2015) Chapter 2.3.4. Infectious haematopoietic necrosis. In: *Manual of Diagnostic Tests for Aquatic Animals (2015)*. World Organisation for Animal Health, Paris. Updated 2016 version available at: http://www.oie.int/index.php?id=2439&L=0&htmfile=chapitre_ihn.htm (accessed 23 October 2016).

Overturf, K., LaPatra, S.E., Towner, R., Campbell, N.R. and Narum, S.R. (2010) Relationships between growth and disease resistance in rainbow trout, *Oncorhynchus mykiss* (Walbaum). *Journal of Fish Diseases* 33, 321–329.

Palti, Y., Parsons, J.E. and Thogaard, G.H. (1999) Identification of candidate DNA markers associated with IHN virus resistance in backcrosses of rainbow (*Oncorhynchus mykiss*) and cutthroat trout (*O. clarki*). *Aquaculture* 173, 81–94.

Palti, Y., Nichols, K.M., Waller, K.I., Parsons, J.E. and Thorgaard, G.H. (2001) Association between DNA polymorphisms tightly linked to MHC class II genes and IHN virus resistance in backcrosses of rainbow and cutthroat trout. *Aquaculture* 194, 283–289.

Peñaranda, M.M.D., Purcell, M.K. and Kurath, G. (2009) Differential virulence mechanisms of infectious hematopoietic necrosis virus in rainbow trout (*Oncorhynchus mykiss*) include host entry and viral replication kinetics. *Journal of General Virology* 90, 2172–2182.

Peñaranda, M.M.D., LaPatra, S.E. and Kurath, G. (2011) Genogroup specificity of DNA vaccines against infectious hematopoietic necrosis virus (IHNV) in rainbow trout (*Oncorhynchus mykiss*). *Fish and Shellfish Immunology* 31, 43–51.

Pietsch, J.P., Amend, D.F. and Miller, C.M. (1977) Survival of infectious hematopoietic necrosis virus held under various environmental conditions. *Journal of the Fisheries Research Board of Canada* 34, 1360–1364.

Purcell, M.K., Alexandra Hart, S., Kurath, G. and Winton, J.R. (2006a) Strand specific, real-time RT-PCR assays for quantification of genomic and positive-sense RNAs of the fish rhabdovirus infectious hematopoietic necrosis virus. *Journal of Virological Methods* 132, 18–24.

Purcell, M.K., Nichols, K.M., Winton, J.R., Kurath, G., Thorgaard, G.H. et al. (2006b) Comprehensive gene expression profiling following DNA vaccination of rainbow trout against infectious haematopoietic necrosis virus. *Molecular Immunology* 43, 2089–2106.

Purcell, M.K., Garver, K.A., Conway, C., Elliott, D.G. and Kurath, G. (2009) Infectious hematopoietic necrosis virus genogroup-specific virulence mechanisms in sockeye salmon (*Oncorhynchus nerka*) from Redfish Lake Idaho. *Journal of Fish Diseases* 32, 619–631.

Purcell, M.K., Thompson, R.L., Garver, K.A., Hawley, L.M., Batts, W.N. et al. (2013) Universal reverse-transcriptase real-time PCR for infectious hematopoietic necrosis virus (IHNV). *Diseases of Aquatic Organisms* 106, 103–115.

Rodriguez, M.F., LaPatra, S., Williams, S., Famula, T. and May, B. (2004) Genetic markers associated with resistance to infectious hematopoietic necrosis in rainbow and steelhead trout (*Oncorhynchus mykiss*) backcrosses. *Aquaculture* 241, 93–115.

Rodríguez Saint-Jean, S., Pérez Prieto, S.-L., Lopez-Exposito, I., Ramos, M., De Las Heras, A.I. and Recio, I. (2012) Antiviral activity of dairy proteins and hydrolysates on salmonid fish viruses. *International Dairy Journal* 23, 24–29.

Rodríguez Saint-Jean, S., De las Heras, A., Carrillo, W., Recio, I., Ortiz-Delgado, J.B. et al. (2013) Antiviral activity of casein and alpha-s2 casein hydrolysates against the infectious hematopoietic necrosis virus, a rhabdovirus from salmonid fish. *Journal of Fish Diseases* 36, 467–481.

Saksida, S.M. (2006) Infectious haematopoietic necrosis epidemic (2001 to 2003) in farmed Atlantic salmon *Salmo salar* in British Columbia. *Diseases of Aquatic Organisms* 72, 213–223.

Schultz, C.L., McAllister, P.E., Schill, W.B., Lidgerding, B.C. and Hetrick, F.M. (1989) Detection of infectious hematopoietic necrosis virus in cell culture fluid using immunoblot assay and biotinylated monoclonal antibody. *Diseases of Aquatic Organisms* 7, 31–31.

Simon, B., Nomellini, J., Chiou, P.P., Bingle, W., Thornton, J. et al. (2001) Recombinant vaccines against infectious hematopoietic necrosis virus: production by *Caulobacter crescentus* S-layer protein secretion system and evaluation in laboratory trials. *Diseases of Aquatic Organisms* 44, 17–27.

Siwicki, A.K., Pozet, F., Morand, M., Kazuń, B. and Trapkowska, S. (2002) *In vitro* effect of methisoprinol on salmonid rhabdoviruses replication. *Bulletin of the Veterinary Institute in Pulawy* 46, 53–58.

Traxler, G.S., Anderson, E.A., LaPatra, S.E., Richard, J., Shewmaker, B. and Kurath, G. (1999) Naked DNA vaccination of Atlantic salmon *Salmo salar* against IHNV. *Diseases of Aquatic Organisms* 38, 183–190.

Verrier, E.R., Langevin, C., Tohry, C., Houel, A., Ducrocq, V. et al. (2012) Genetic resistance to *Rhabdovirus* infection in teleost fish is paralleled to the derived cell resistance status. *PLoS ONE* 7(4): e33935.

Wargo, A.R., Garver, K.A. and Kurath, G. (2010) Virulence correlates with fitness *in vivo* for two M group genotypes of infectious hematopoietic necrosis virus (IHNV). *Virology* 404, 51–58.

Winton, J.R. (1991) Recent advances in detection and control of infectious hematopoietic necrosis virus in aquaculture. *Annual Review of Fish Diseases* 1, 83–93.

Winton, J.R. (1997) Immunization with viral antigens: infectious haematopoietic necrosis. In: Gudding, R., Lillehaug, A., Midtlyng, P.J. and Brown, F. (eds) *Fish Vaccinology*. Developments in Biological Standardization 90, Karger, Basel, Switzerland, pp. 211–220.

Wolf, K. (1988) Infectious hematopoietic necrosis. In: Wolf, K. (ed.) *Fish Viruses and Fish Viral Diseases*. Cornell University Press, Ithaca, New York, pp. 83–114.

Wood, E.M. and Yasutake, W.T. (1956) Histopathologic changes of a virus-like disease of sockeye salmon. *Transactions of the American Microscopical Society* 75, 85–90.

Yang, J., Liu, Z., Shi, H.-N., Zhang, J.-F., Want, J.-F. et al. (2014) Association between MHC II beta chain gene polymorphisms and resistance to infectious haematopoietic necrosis virus in rainbow trout (*Oncorhynchus mykiss*, Walbaum, 1792). *Aquaculture Research* 47, 570–578.

Yasutake, W.T. (1970) Comparative histopathology of epizootic salmonid virus diseases. In: Snieszko, S.F. (ed.) *A Symposium on Diseases of Fish and Shellfish*. American Fisheries Society Special Publication 5, Bethesda, Maryland, pp. 341–350.

3 Viral Haemorrhagic Septicaemia Virus

JOHN S. LUMSDEN*

Department of Pathobiology, Ontario Veterinary College, University of Guelph, Guelph, Ontario, Canada and Adjunct Professor, Department of Pathobiology, St. George's University, True Blue, Grenada

3.1 Introduction

Viral haemorrhagic septicaemia virus (VHSV), the aetiological agent of viral haemorrhagic septicaemia (VHS) is a member of the family *Rhabdoviridae* within the order *Mononegavirales*. These viruses have single-stranded, negative-sense (nega) RNA genomes with enveloped bullet- or cone-shaped nucleocapsids. The economically most important pathogenic rhabdoviruses are VHSV, infectious haematopoietic necrosis virus (IHNV; both in the genus *Novirhabdovirus*) and spring viraemia of carp virus (SVCV; new name *Carp sprivivirus*, genus *Sprivivirus*). Please refer to the International Committee on Taxonomy of Viruses for details (ICTV, 2015). VHSV and IHNV share the same six genes that read from the 3′ end to the 5′ end of the genome as N (nucleocapsid protein), P (phosphoprotein), M (matrix protein), G (glycoprotein), NV (non-virion) and L (RNA polymerase) (Einer-Jensen *et al.*, 2004). SVCV lacks the gene for NV and has other minor genomic differences (Hoffmann *et al.*, 2002). VSHV was comprehensively reviewed by Smail and Snow (2011) and LaPatra *et al.* (2016).

Serotyping was originally used to subgroup isolates (Olesen *et al.*, 1993); however, genotyping is currently done using the genes for G and N. Genotypes are broadly distributed geographically and their differentiation has proven immensely invaluable for understanding epizootiological spread and pathogenesis (Nishizawa *et al.*, 2002, Snow *et al.*, 2004, Brudaseth *et al.*, 2008). If the genotype and infected species is not specified, generalizations on VHSV pathogenesis, host range, virulence, etc. can be misleading.

Genotypes I and IV are the most relevant to this discussion, although all four genotypes I–IV occur in the northern Hemisphere. Genotypes I–III are European, with genotype I divided into freshwater rainbow trout (*Oncorhynchus mykiss*) isolates (Ia) and marine isolates from various species of wild fish (Ib). The Ia genotype is associated with the original descriptions of VHS in rainbow trout in which mortality approached 100%. Genotype IV is found in the Pacific Ocean (IVa) and North America, including the western Atlantic Ocean (IVb; Gagne *et al.*, 2007) and the Great Lakes (IVb; Lumsden *et al.*, 2007). A trend that has occurred in Europe and is very likely to have also occurred in North America is the movement of marine isolates into fresh water along with the expansion of host range (Einer-Jensen *et al.*, 2004; Snow *et al.*, 2004; Gagne *et al.*, 2007; Thompson *et al.*, 2011). Genotypes have an impact on the regulatory control of VHSV and the movement of fish because a diagnosis of genotype Ia outside Europe has greater implications than a geographic expansion of IVb in North America. The detection of any VHSV must be reported to the World Organisation for Animal Health (OIE, 2015).

One of the most remarkable features of VHSV is its broad host range; 28 species of fish are infected by genotype IVb in the Great Lakes (APHIS, 2009) and about 80 species are affected by all four genotypes (OIE, 2015). The ubiquitous phosphatidylserine and fibronectin, are receptors for VHSV in fish cells (Bearzotti *et al.*, 1999; see LaPatra *et al.*, 2016), which may account for its broad host range, though fibronectin is also a receptor for IHNV (Bearzotti *et al.*, 1999). Transmission is horizontal

---

*E-mail: jsl@uoguelph.ca

with virus excreted from the urine and reproductive fluids (Eaton *et al.*, 1991) and with uptake at least by gill epithelium (Brudeseth *et al.*, 2008). Even though VHSV is in the reproductive fluid and VHSV RNA is inside the gonads of fish infected with VHSV IVb (Al-Hussinee and Lumsden, 2011a), vertical transmission has not been documented.

VHS occurs at water temperatures between 4 and 15°C (McAllister, 1990), although the virus grows at higher temperatures in cell culture (Vo *et al.*, 2015), which emphasizes the importance of host factors in disease. Experimental infection using IVb gave bluegill (*Lepomis macrochirus*) mortality of 90% at 10°C, but 0% above 18°C (Goodwin and Merry, 2011). Temperature ranges and optima are similar for genotypes I and IV (Goodwin and Merry, 2011; Smail and Snow, 2011), and above 20°C mortalities due solely to VHSV are unlikely (Sano *et al.*, 2009). The results of surveillance performed at water temperatures above 20°C should be critically evaluated. In deep waters, thermoclines rarely reach the bottom, so the virus may still be detected into summer in demersal fish. Outside the host, the virus remains infective for variable periods. It survives for a year in filtered fresh water at 4°C (Hawley and Garver, 2008) and survival is longer at lower temperatures (Parry and Dixon, 1997). An average survival in raw seawater at 15°C is 4 days (Hawley and Garver, 2008).

3.2 Diagnosis

The clinical signs of and gross lesions in VHS-affected fish are numerous, but none are pathognomonic, and the disease should always be considered given the broad geographic area over which it is endemic. It should be borne in mind as a differential diagnosis if the temperature, geographic location and fish species are consistent with morbidity, and if the fish have haemorrhages, exophthalmia and enlarged abdomens with ascites. Infected fish may also have anaemia, a darkened colour, lethargy and abnormal swimming behaviour, and they may die rapidly (OIE, 2105). However, infected fish may also die without gross lesions; this is common for round goby (*Neogobius melanostomus*) in Lake Ontario (Groocock *et al.*, 2007), thus emphasizing the importance of histopathology and ancillary diagnostics. Similarly, clinically normal fish may be positive for VHSV, so emphasizing the need for either PCR or viral isolation to confirm detection.

Because VHS is an OIE reportable disease, that organization requires specific diagnostic techniques for confirmation (OIE, 2015). Viral isolation using BF-2 (bluegill fry), EPC (epithelioma papulosum cyprini) or other cell lines is used routinely, but it is labour intensive and expensive. Many laboratories use either standard reverse transcriptase (RT)-PCR or quantitative RT-PCR (qRT-PCR; Matejusova *et al.*, 2008, Hope *et al.*, 2010), which can detect all four genotypes of VHSV (Garver *et al.*, 2011). These techniques are more sensitive and rapid than viral isolation, and they are more easily standardized and allow high throughput (Hope *et al.*, 2010). ELISA and immunofluorescence (the indirect fluorescent antibody (IFA) test (IFAT); see OIE, 2015) and reverse transcription loop-mediated isothermal amplification (Soliman *et al.*, 2006) are also used.

When PCR is used, tissues should be saved to allow confirmatory testing, which includes viral isolation (OIE, 2015). Viral isolation also requires a minimum amount of tissue and/or pooling of fish to allow for a positive identification (AFS-FHS, 2014). Quantitative RT-PCR is ideal for a presumptive diagnosis or surveillance with banking of tissues in the case of a presumptive positive. A conclusive diagnosis of VHS requires viral isolation and serum neutralization followed by one of three confirmatory tests; IFAT, ELISA or RT-PCR (OIE, 2015).

3.3 Pathology

The pathology of VHSV has previously been reviewed (de Kinkelin *et al.*, 1979; Wolf 1988; Evensen *et al.*, 1994; Brudeseth *et al.*, 2002, 2005; OIE, 2015; LaPatra *et al.*, 2016). Gross lesions of many acute systemic diseases are often stereotypical, including haemorrhages, enlarged abdomen and exophthalmia, which are typical macroscopic lesions in VHSV-infected fish (OIE, 2015). Petechial haemorrhage in the dorsal musculature in rainbow trout infected with VHSV Ia, is very common (OIE, 2015). The lesions included in the present discussion are generalizations.

Mortality can occur without significant gross lesions, apart from pallor in round goby (Groocock *et al.*, 2007), or swim bladder haemorrhage in muskellunge (*Esox masquinongy*) (Elsayed *et al.*, 2006). The causes of many mortality events are multifactorial even when VHSV is detected in dead fish. Mortality in wild Pacific herring (*Clupea pallasii*) was associated with numerous pathogens, including VHSV (Marty *et al.*, 1998). In Conesus Lake (New York state), dead walleye (*Sander vitreus*) had lesions consistent with VHSV and the virus was isolated from the fish, but affected fish also had lesions typical of branchial columnaris (Al-Hussinee *et al.*, 2011b), a bacterial disease caused by *Flavobacterium columnare* (see Chapter 16).

Histological lesions vary, but are primarily necrotizing and affect almost all tissues. Patterns of the lesions can depend on the genotype and fish species. For example, VHSV Ia in rainbow trout will produce necrotizing lesions in the kidney and liver, but the spleen, heart, brain and other tissues may also be affected (de Kinkelin *et al.*, 1979; Wolf 1988; Evensen *et al.*, 1994). Turbot (*Scophthalmus maximus*) infected with VHSV genotype Ib (Brudaseth *et al.*, 2005) and some species affected with IVb from the Great Lakes had the most severe lesions in the heart, though the liver and haematopoietic tissues were also affected (Lumsden *et al.*, 2007; Al-Hussinee *et al.*, 2011b). The cardiac and vascular lesions in freshwater drum (*Aplodinotus grunniens*) were so spectacular that the likelihood of a foreign agent being responsible was readily apparent (see Figs 3.1 and 3.2). The infection produces a severe vasculitis in numerous tissues, but it can also be subtle (see Fig. 3.3). Freshwater drum had a

florid, often fibrinoid, vasculitis in the liver, kidney (Fig. 3.4), brain and spleen, and both this lesion and a tropism for endothelium have also been documented in rainbow trout infected with Ia and turbot infected with Ib (Brudaseth *et al.*, 2002, 2005). However, infected muskellunge had an equivocal lesion. Experimental reproduction of the disease by intraperitoneal injection in fathead minnow (*Pimphales promelas*), rainbow trout or walleye did not produce a notable vasculitis and heart lesions were minimal (Al-Hussinee *et al.*, 2010, 2011b; Grice *et al.*, 2016). Walleye that were bathed in formaldehyde, health checked

Fig. 3.2. Fibrinoid vasculitis in the liver of a freshwater drum (*Aplodinotus grunniens*) infected with viral haemorrhagic septicaemia virus (VHSV). The necrosis is centred on the venule, not the hepatocytes. H&E (haematoxylin and eosin) staining. Bar = 50 μm.

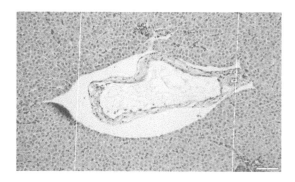

Fig. 3.3. Mild vasculitis in the liver of a freshwater drum (*Aplodinotus grunniens*) infected with viral haemorrhagic septicaemia virus (VHSV). There are a few necrotic cells within the wall of the vessel. H&E (haematoxylin and eosin) staining. Bar = 50 μm.

Fig. 3.1. Ventricle of a freshwater drum (*Aplodinotus grunniens*) infected with viral haemorrhagic septicaemia virus (VHSV). There is massive subacute necrosis and inflammation. The remnants of myocardial trabeculae are at bottom left. H&E (haematoxylin and eosin) staining. Bar = 50 μm.

J.S. Lumsden

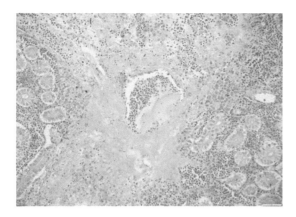

Fig. 3.4. Severe vasculitis in the kidney of a freshwater drum (*Aplodinotus grunniens*) infected with viral haemorrhagic septicaemia virus (VHSV). H&E (haematoxylin and eosin) staining. Bar = 50 μm.

and acclimatized to high quality 18°C well water developed VHS and columnaris shortly after injection with VHSV (Grice *et al.*, 2016).

Histopathology can be complemented by immunolocalization of the virus (Evensen *et al.*, 1994; Brudaseth *et al.*, 2005; Al-Hussinee *et al.*, 2010, 2011b; Grice *et al.*, 2016). Co-localization of the virus with the majority of lesions using immunohistochemistry (IHC) solidifies the association of the virus with pathogenesis (Al-Hussinee *et al.*, 2010, 2011b), but the sensitivity of this method is limited (Evensen *et al.*, 1994). Additionally, RNA probes can be used to localize VHSV (Al-Hussinee *et al.*, 2011a). The heart of freshwater drum with profound necrotizing myocardial lesions was surprisingly free of virus, at least as detected using IHC (Al-Hussinee *et al.*, 2011b); this is in contrast to turbot, where the virus was readily identified in affected hearts (Brudaseth *et al.*, 2005).

3.4 Pathophysiology, Pathogenesis and Virulence Factors

Aspects of the pathophysiology and pathogenesis of VHSV have been reviewed by Smail and Snow (2011). Briefly, genotype Ia is virulent in rainbow trout (Brudeseth *et al.*, 2002, 2008), while marine Ib isolates are virulent in turbot (Brudeseth *et al.*, 2005) but avirulent in rainbow trout (Brudaseth *et al.*, 2008). Many Ib isolates were avirulent in rainbow trout challenged by immersion, but their virulence varied when the virus was injected (Skall *et al.*, 2004). Genotype IVa was virulent in Pacific herring (*Clupea pallasi*) (Kocan *et al.*, 1997), but rainbow trout were resistant to IVa (Meyers and Winton, 1995) and to IVb (Al-Hussinee *et al.*, 2010; Kim and Faisal, 2010), although IVa showed better replication and cytopathic effect (CPE) in rainbow trout gill and splenic macrophage cells than IVb (Pham *et al.*, 2013).

Horizontal transmission of VHSV via the gill has been detailed (see Brudeseth *et al.*, 2008; Pham *et al.*, 2013; Al-Hussinee *et al.*, 2016). Viral replication (Ia) was detected early after infection in rainbow trout gill epithelium (Brudeseth *et al.*, 2002), whereas the endothelium of the kidney was the site of first detection of Ib in turbot (Brudeseth *et al.*, 2005) and the gill had virus only in a few fish, late in the infection. Several studies have correlated the capability of VHSV isolates to replicate in gill or skin epithelium with virulence. The avirulence of Ib in rainbow trout correlated with its inability to infect and translocate across the primary gill epithelial cells and to replicate in head kidney macrophages, in contrast to a rainbow trout-virulent Ia isolate (Brudeseth *et al.*, 2008). A IVa isolate with low virulence in rainbow trout replicated to a limited degree in excised fin tissue but not in excised gill tissue (Yamamoto *et al.*, 1992). Replication in excised fin tissue was correlated with resistance to waterborne infection in rainbow trout (Quillet *et al.*, 2001).

The macrophage plays a significant role in the virulence and pathogenesis of VHSV. The virus is detected in macrophages and melanomacrophages following natural and experimental infections by genogroups Ia (Evensen *et al.*, 1994; Brudeseth *et al.*, 2002), Ib (Brudeseth *et al.*, 2005) and IVb (Al-Hussinee *et al.*, 2010, 2011b); it has also replicated in primary cultures of macrophages *in vitro* (see Brudeseth *et al.*, 2008). Macrophages will phagocytose cells lysed by VHSV and, therefore, immunochemical detection of the virus inside macrophages may not indicate replication. Similarly, primary macrophage cultures of rainbow trout are heterogeneous and only a portion of the cells can support viral replication, even with a virulent Ia isolate (Tafalla *et al.*, 1998; Brudeseth *et al.*, 2008); hence, significant CPE may (Brudeseth *et al.*, 2008) or may not occur (Tafalla *et al.*, 1998). In the RTS11 cell line from rainbow trout splenic macrophages, replication of VHSV was aborted and

CPE did not occur for genotypes Ia, IVa and IVb isolates (Tafalla *et al.*, 2008; Pham *et al.*, 2013).

Much of this evidence was from experimental infections or *in vitro* studies; consequently, environmental factors that could predispose fish to mortality were excluded. Genotypes III and IVa are normally not highly virulent to salmonids; none the less, isolates from wild fish caused mortality in Atlantic salmon (*Salmo salar*) in net pens (Dale *et al.*, 2009; Garver *et al.*, 2013). The mechanism by which VHSV isolates adapt to new hosts is a critical question in the epizootiology of the virus. Five serial passages of a marine isolate in rainbow trout without intervening *in vitro* culture produced increased virulence, although there was no difference in the G gene sequence to explain the increased virulence (Snow and Cunningham, 2000). In a subsequent study, genotype Ib isolates that were >99.4% similar in the G amino acid sequence, but were of markedly different virulence in rainbow trout, were sequenced. Only four predicted amino acid substitutions were noted, one each in the N, G, NV and L proteins, which suggested that small genome mutations could induce large differences in virulence (Campbell *et al.*, 2009).

Selective apoptosis or excessive stimulation of autophagy in tissues, e.g. the heart, could help to explain extensive cell death without large numbers of viruses. The prevention or delay of apoptosis via the generation of anti-apoptotic factors is a common strategy of viruses for enhancing their replication (Skaletskaya *et al.*, 2001), but some viruses also stimulate apoptosis in late replication to promote viral spread (Hay and Kannourakis, 2002). Rainbow trout infected with VHSV had apoptotic cell death in the renal interstitium but not in the heart (Eleouet *et al.*, 2001). The use of NV gene deficient and NV gene knockout recombinant VHSV demonstrated that the NV gene product has an anti-apoptotic effect during early viral replication, and that wild-type VHSV infection triggered apoptosis by inducing caspases (Ammayappan and Vakharia, 2011). In contrast, the matrix protein of IHNV may induce apoptosis (Chiou *et al.*, 2000); such data for VHSV are lacking.

Autophagy is a conserved cellular process that involves autophagosome formation and degradation and the recycling of cellular constituents. The process also influences the immunity and survival of intracellular organisms (Deretic, 2011), including piscine viruses (Schiotz *et al.*, 2010; García-Valtanen *et al.*, 2014). Purified G proteins from SVCV and VHSV stimulated autophagosome formation, and VHSV-induced autophagy inhibited the replication of both viruses in zebrafish (*Danio rerio*) embryonic fibroblasts (García-Valtanen *et al.*, 2014). Autophagy is primarily a pro-survival mechanism but, when excessive, it can cause cell death (Ouyang *et al.*, 2012).

Immunosuppression due to environmental factors is commonly proposed to be a predisposing factor in VHS; however, VHSV itself can cause immunosuppression. Cell death by any mechanism, including viral lysis/necrosis in haematopoietic tissues, would contribute to immunosuppression. The NV protein has also been proposed to cause immunosuppression, resulting in the downregulation of many acquired and innate immune genes in the spleen and kidney (Chinchilla *et al.*, 2015). The virulence of VHSV may be partially due to the NV protein that downregulates the interferon (IFN) response and tumour necrosis factor alpha (TNFα)-mediated nuclear factor (NF)-kβ activation (Kim and Kim, 2012, 2013).

3.5 Protective and Control Strategies

Several disinfecting methods kill VHSV in cold fresh water (see Bovo *et al.*, 2005), including the use of UV irradiation, low and high pH and iodophors (disinfectants containing iodine complexed with a solubilizing agent); the latter is widely used for egg disinfection. The assessment of the risk factors, basic sanitation and the implementation of best practices markedly reduce the severity and scope of the disease; nevertheless, they must be in continual practice (see Gregory, 2008). Complete separation from the environment is the single best method to eliminate the transmission of VHSV. Fish in net pens or in land-based operations where surface water is used will always be at risk. In endemic areas, a risk-based approach is the most sensible and cost-effective method to predict outbreaks (Thrush and Peeler, 2013). Some details of appropriate control strategies are given in OIE (2015).

The selection of European rainbow trout with increased resistance to genotype Ia has been demonstrated by Dorson *et al.* (1995) and the resistance shows high heritability (Yanez *et al.*, 2014). The identification of extant strains of fish that are VHSV resistant would be an inexpensive management option in endemic areas when operations cannot be isolated from contaminated water. Extant strains of walleye vary in their susceptibility to experimental infection with genotype IVb (Grice *et al.*, 2016). A quantitative trait locus (QTL) was associated with the survival of rainbow trout and innate immunity to VHSV (Verrier *et al.*, 2013).

There are no commercial VHSV vaccines despite significant research (see Smail and Snow, 2011). This is partly due to the economic challenges of vaccinating small fish and also because VHSV in many locations requires eradication after confirmatory diagnosis (OIE, 2015). Numerous experimental DNA vaccines have been developed against rhaboviruses, including VHSV (Boudinot *et al.*, 1998; Lorenzen *et al.*, 2001; McLauclan *et al.*, 2003; Sommerset *et al.*, 2003; Chaves-Pozo *et al.*, 2010a; Martínez-López *et al.*, 2013; Sepúlveda and Lorenzen, 2016). The only commercially available DNA vaccine protects Atlantic salmon from IHNV in British Columbia, Canada; however, few of the fish at significant risk from VHSV compare with Atlantic salmon in value. The DNA experimental vaccines are based on VHSV glycoprotein G sequences, and immunity is directed at the protein responsible for cellular attachment (Bearzotti *et al.*, 1999). DNA vaccines can confer long-lasting high relative percentage survival with a single injection. The advantage of DNA vaccines over conventional products is that they stimulate not only an antibody response but also cell-mediated and innate immunity, though DNA vaccines to VHSV do not always provide sterile immunity (Chaves-Pozo *et al.*, 2010a; Sepúlveda and Lorenzen, 2016). Further, the neutralizing antibody response can be absent, either before it is seen to develop or after it has waned, yet protection is still provided (Lorenzen *et al.*, 1998; McLauclan *et al.*, 2003). Short-term non-specific protection against heterologous viruses is also afforded by VHSV DNA vaccines (Lorenzen *et al.*, 2002; Sommerset *et al.*, 2003) as the stimulation of innate immune effectors is vigorous (Byon *et al.*, 2005; Cuesta and Tafalla, 2009).

Immunity to rhabdoviruses has recently been reviewed by Purcell *et al.* (2012) and Pereiro *et al.* (2016), while this chapter emphasizes aspects of innate immunity to VHSV. In rainbow trout infected with VHSV and IHNV, neutralizing antibody, when present, is protective and depends on complement (see Lorenzen and LaPatra, 1999; Lorenzen *et al.*, 1999). Protection following intraperitoneal injection of a DNA vaccine induced IgT B cells (Martínez-López *et al.*, 2013). Rainbow trout had splenic lymphocytes expressing IgM, IgD and IgT, and clonal expansion of IgM and IgT occurred following systemic VHSV infection (Castro *et al.*, 2013). Research on cell-mediated immunity has made significant advances (see Laing and Hansen, 2011; Castro *et al.*, 2011); nevertheless, functional evidence for T cell-mediated viral resistance – apart from cell-mediated cytotoxicity (CMC) – is sparse. Infection with VHSV and DNA vaccination induced the clonal expansion of T cells and selected for CDR3 (complementarity determining region 3)-based clonotypic T cell diversity (Boudinot *et al.*, 2001, 2004). The upregulation of the T cell co-receptor CD3 (cluster of differentiation 3) and T cell activation in the skin of VHSV bath-infected rainbow trout was recently detected (Leal *et al.*, 2016). Leucocytes from VHSV-infected rainbow trout demonstrated CMC to VHSV-infected MHC (major histocompatibility complex) I-matched but not xenogenic cells (Utke *et al.*, 2007). Both CD8α (the α chain of CD8) and the natural killer (NK) cell enhancement factor were stimulated, although unexpectedly, the temporal response of the T cells preceded that of the NK cells. DNA vaccination with plasmids containing the VHSV genes for G and N stimulated CMC by peripheral blood leucocytes to VHSV-infected MHC I-matched cells, but only vaccination with the G gene also stimulated NK cells that killed xenogenic infected cells (Utke *et al.*, 2008).

The induction of interferon and production of interferon-stimulated genes (ISGs) constitute a critical antiviral mechanism in fish (see Zou and Secombes, 2011), but the IFNs are complex. There are two types of IFN, type I and type II (IFNγ), and also two groups of type I IFNs, and many have multiple genes. Their influences on immunity and on rhabdoviruses vary (see Zou and Secombes, 2011; Purcell *et al.*, 2012). Recombinant IFN I inhibits the replication of VHSV in rainbow trout ovary, and VHSV infection strongly stimulates an ovarian IFN response (Chaves-Pozo *et al.*, 2010b) and several chemokine genes (Chaves-Pozo *et al.*, 2010c). This vigorous response may explain the lack of vertical transmission of VHSV. Recently, intracellular IFNs were discovered in rainbow trout and their overexpression in RTG (rainbow trout gonad)-2 cells stimulated Mx protein, an indicator of alpha/beta interferon induction, and resulted in resistance to VHSV (Chang *et al.*, 2013).

Numerous ISGs influence the VHSV response (see Verrier *et al.*, 2011). The best characterized ISG in fish is Mx. Rainbow trout have three Mx isoforms that are differentially upregulated in tissues in response to poly(I:C) (polyinosinic–polycytidylic acid, a synthetic analogue of double-stranded RNA) or VHSV (Tafalla *et al.*, 2007). Mx protein was detected for 3 weeks at the site of injection following VHSV DNA vaccination of rainbow trout (Acosta *et al.*, 2005) along with the upregulation of transcription of the genes for Toll-like receptor 9, TNFα and interleukin 6 in the head kidney and spleen (Ortega-Villaizan *et al.*, 2009). The upregulation of Mx expression after vaccination with VHSV DNA was positively correlated with early protection (Boudinot *et al.*; 1998, McLauclan *et al.*, 2003). The VHSV-induced gene *vig-1* was highly expressed in rainbow trout lymphoid tissues following VHSV DNA immunization (Boudinot *et al.*, 1999). Vigs including *vig-1*, *vig-2*, *ISG15/vig-3* and chemokines were upregulated in VHSV-exposed leucocytes and were expressed in VHSV-infected rainbow trout (O'Farrell *et al.*, 2002). Protein kinase R was implicated in preventing poly(I:C)-dependent increases in the mRNA transcription of other ISGs, but did not control viral transcription when rainbow trout cells were infected with VHSV (Tafalla *et al.*, 2008). Overexpression of retinoic acid-inducible and mitochondrial antiviral signalling protein in salmonid cell lines resulted in protection from VHSV (Biacchesi *et al.*, 2009). Suppressive subtractive hybridization of rainbow trout leucocytes that were VHSV or mock infected revealed that VHSV infection stimulated the genes for the finTRIM proteins (van der Aa *et al.*, 2009), which are similar to the tripartite motif (TRIM) protein family, some of which are responsible for mammalian antiviral innate immunity. These and other gene families, e.g. the genes for the NOD (nucleotide-binding oligomerization domain)-like receptors, are likely to be associated with immunity to VHSV and await further characterization.

The production of nitric oxide (NO) was stimulated in turbot kidney macrophages following VHSV infection, and exogenous NO inhibited VHSV replication (Tafalla *et al.*, 1999). VHSV infection of turbot macrophages also reduced the production of NO induced by TNFα or macrophage-activating factor (Tafalla *et al.*, 2001). Other soluble factors associated with resistance to VHSV include antimicrobial peptides. There was differential expression of the genes for the antimicrobial proteins hepcidin-2 and nk-lysin in VHSV-susceptible and resistant families of turbot (Díaz-Rosales *et al.*, 2012). VHSV infection stimulated recruitment and increased the natural cytotoxic cell-like activity of leucocytes in gilthead sea bream (*Sparus aurata*), along with the production of reactive oxygen intermediates and myeloperoxidase (Esteban *et al.*, 2008).

The hepatic expression of serum amyloid A (SAA), an acute-phase reactant in fish, was stimulated by VHSV in rainbow trout (Rebl *et al.*, 2009; Castro *et al.*, 2014), and interleukin 6 (IL-6) was the likely stimulus for SAA (Castro *et al.*, 2014). The expression of galectin 9, (a chemoattractant protein with functions linked to immunity), was upregulated in response to VHSV (O'Farrell *et al.*, 2002), but evidence for fish lectins interacting with the virus is limited. Rainbow trout ladderlectin (but not intelectin), another lectin with a role in defence, bound to immobilized purified VHSV (Reid *et al.*, 2011), though the downstream effects are unknown because ladderlectin cannot fix complement.

3.6 Conclusions and Suggestions for Future Research

The future is relatively bright for technologies such as DNA vaccination despite the regulatory barriers that exist in many countries. The DNA vaccine (Apex-IHN, Elanco) for protection from IHNV that is commercially available in Canada is an example of a case where the benefits of the vaccine justify the high cost of the technology. Atlantic salmon are a high-value product and IHNV is a substantial disease threat to this species (LaPatra *et al.*, 2016). DNA vaccines for VHSV stimulate innate and acquired immunity and a prolonged duration of protection sufficient to protect later life stages, however, challenges for their use remain (see Holvold *et al.*, 2014). Their wider application for VHSV needs to be explored, particularly in high-value species like muskellunge released for stocking. Improvements in delivery, the enhanced development of immunity using molecular adjuvants, a marked reduction in cost and easing of the regulatory restrictions are also needed. Unfortunately, some of these factors are to likely remain significant barriers in the future. Continued advances in knowledge of teleost immunity are needed, particularly of the interplay between innate and acquired immune systems and of cell-mediated immunity (CMI).

Management of the spread and impacts of VHSV will rely on traditional approaches, including biosecurity, testing and either eradication or some method of restricting fish movement. Given the costs involved, some differentiation by regulatory authorities between the genotypes of VHSV would be welcome; this would need to occur at level of the OIE. For example, VHSV IVb is endemic in the Great Lakes and while occasional mortality events still occur, its impact is limited. Risk-based approaches based on the potential for exposure with annual or other testing is a practical approach that has been adopted by Canadian and other authorities. Thus, producers who wish to sell products internationally can establish a VHSV-free status within a VHSV endemic zone.

References

Acosta, F., Petrie, A., Lockhart, K., Lorenzen, N. and Ellis, A.E. (2005) Kinetics of Mx expression in rainbow trout (*Onchorhynchus mykiss* W) and Atlantic salmon (*Salmo salar* L) parr in response to VHSV-DNA vaccination. *Fish and Shellfish Immunology* 18, 81–89.

AFS-FHS (2014) *FHS Blue Book: Suggested Procedures for the Detection and Identification of Certain Finfish and Shellfish Pathogens*, 2014 edn. American Fisheries Society-Fish Health Section, Bethesda, Maryland.

Al-Hussinee, L. and Lumsden, J.S. (2011a) Detection of VHSV IVb within the gonads of Great Lakes fish using *in situ* hybridization. *Diseases of Aquatic Organisms* 95, 81–86.

Al-Hussinee, L., Huber, P., Russell, S., Lepage, V., Reid, A. *et al.* (2010) Viral haemorrhagic septicaemia virus IVb experimental infection of rainbow trout, *Oncorhynchus mykiss* (Walbaum), and fathead minnow, *Pimphales promelas* (Rafinesque). *Journal of Fish Diseases* 33, 347–360.

Al-Hussinee, L., Lord, S., Stevenson, R.M., Casey, R.N., Groocock, G.H. *et al.* (2011b) Immunohistochemistry and pathology of multiple Great Lakes fish from mortality events associated with viral hemorrhagic septicemia virus type IVb. *Diseases of Aquatic Organisms* 93, 117–127.

Al-Hussinee L., Tubbs L., Russell S., Pham J., Tafalla C. *et al.* (2016) Temporary protection of rainbow trout gill epithelial cells from infection with viral hemorrhagic septicemia virus IVb. *Journal of Fish Diseases* 38, 1099–1112.

Ammayappan, A. and Vakharia, V.N. (2011) Nonvirion protein of *Novirhabdovirus* suppresses apoptosis at the early stage of virus infection. *Journal of Virology* 85, 8393–8402.

APHIS (2009) *Viral Hemorrhagic Septicemia Virus IVb: U.S. Surveillance Report 2009*. US Department of Agriculture Animal and Plant Health Inspection Service, Veterinary Services, Riverdale, Maryland. Available at: https://www.aphis.usda.gov/animal_health/animal_dis_spec/aquaculture/downloads/vhs_surv_rpt.pdf (accessed 25 October 2016).

Bearzotti, M., Delmas, B., Lamoureux, A., Loustau, A.M., Chilmonczyk, S. *et al.* (1999) Fish rhabdovirus cell entry is mediated by fibronectin. *Journal of Virology* 73, 7703–7709.

Biacchesi S., LeBerre M., Lamoureux A., Louise Y., Lauret E. *et al.* (2009) MAVS plays a major role in induction of the fish innate immune response against RNA and DNA viruses. *Journal of Virology* 83, 7815–7827.

Boudinot, P., Blanco M., de Kinkelin P. and Benmansour A. (1998) Combined DNA immunization with the glycoprotein gene of the viral hemorrhagic septicemia virus and infectious hematopoietic necrosis virus induces a double-specific protective immunity and a nonspecific response in rainbow trout. *Virology* 249, 297–306.

Boudinot, P., Massin, P., Blanco, M., Riffault, S. and Benmansour A. (1999) *vig*-1, a new fish gene induced by the rhabdovirus glycoprotein, has a virus-induced homologue in humans and shares conserved motifs with the MoaA family. *Journal of Virology* 73, 1846–1852.

Boudinot, P., Boubekeur, S. and Benmansour, A. (2001) Rhabdovirus infection induces public and private T cell responses in teleost fish. *Journal of Immunology* 167, 6202–6209.

Boudinot, P., Bernard, D., Boubekeur, S., Thoulouze, M.I., Bremont, M. *et al.* (2004) The glycoprotein of a fish rhabdovirus profiles the virus-specific T-cell repertoire in rainbow trout. *Journal of General Virology* 85, 3099–3108.

Bovo, G., Hill, B., Husby, A., Håstein, T., Michel, C. *et al.* (2005) *Work Package 3 Report: Pathogen Survival Outside the Host, and Susceptibility to Disinfection*. Report QLK2-CT-2002-01546: Fish Egg Trade. VESO (Veterinary Science Opportunities, Oslo. Available at: http://www.eurl-fish.eu/-/media/Sites/EURL-FISH/english/activities/scientific%20reports/fisheggtrade-wp_3.ashx?la=da (accessed 25 October 2016).

Brudeseth, B.E., Castric, J. and Evensen, Ø. (2002) Studies on pathogenesis following single and double infection with viral hemorrhagic septicemia virus and infectious hematopoietic necrosis virus in rainbow trout (*Oncorhynchus mykiss*). *Veterinary Pathology* 39, 180–189.

Brudeseth, B., Raynard, R., King, J. and Evensen, Ø. (2005) Sequential pathology after experimental infection with marine viral hemorrhagic septicemia virus isolates of low and high virulence in turbot (*Scophthalmus maximus* L.). *Veterinary Pathology* 42, 9–18.

Brudeseth, B.E., Skall, H.F. and Evensen, O. (2008) Differences in virulence of marine and freshwater isolates of viral hemorrhagic septicemia virus *in vivo* correlate with *in vitro* ability to infect gill epithelial cells and macrophages of rainbow trout (*Oncorhynchus mykiss*). *Journal of Virology* 82, 10359–10365.

Byon, J.Y., Ohira, T., Hirono, I. and Aoki, T. (2005) Use of cDNA microarray to study immunity against viral hemorrhagic septicemia (VHS) in Japanese flounder (*Paralichthys olivaceus*) following DNA vaccination. *Fish and Shellfish Immunology* 18, 135–147.

Campbell, S., Collet, B., Einer-Jensen, K., Secombes, C.J. and Snow, M. (2009) Identifying potential virulence determinants in viral haemorrhagic septicemia virus (VHSV) for rainbow trout. *Diseases of Aquatic Organisms* 86, 205–212.

Castro, R., Bernard, D., Lefranc, M.P., Six, A., Benmansour, A. *et al*. (2011) T cell diversity and TcR repertoires in teleost fish. *Fish and Shellfish Immunology* 31, 644–654.

Castro, R., Jouneau, L., Pham, H., Bouchez, O., Guidicelli, V. *et al*. (2013) Teleost fish mount complex clonal IgM and IgT responses in spleen upon systemic viral infection. *PLoS Pathogens* 9(1): e1003098.

Castro, R., Abos, B., Pignatelli, J., von Gersdorff, J.L., Gonzalez Granja, A. *et al*. (2014) Early immune responses in rainbow trout liver upon viral hemorrhagic septicemia virus (VHSV) infection. *PLoS ONE* 9(10): e111084.

Chang, M.-X., Zou, J., Nie, P., Huang, B., Yu, Z. *et al*. (2013) Intracellular interferons in fish: a unique means to combat viral infection. *PLoS Pathogens* 9(11): e1003736.

Chaves-Pozo, E., Cuesta, A. and Tafalla, C. (2010a) Antiviral DNA vaccination in rainbow trout (*Oncorhynchus mykiss*) affects the immune response in the ovary and partially blocks its capacity to support viral replication in vitro. *Fish and Shellfish Immunology* 29, 579–586.

Chaves-Pozo, E., Zou, J., Secombes, C.J., Cuesta, A. and Tafalla, C. (2010b) The rainbow trout (*Oncorhynchus mykiss*) interferon response in the ovary. *Molecular Immunology* 47, 1757–1764.

Chaves-Pozo, E., Montero, J., Cuesta, A. and Tafalla, C. (2010c) Viral hemorrhagic septicemia and infectious pancreatic necrosis viruses replicate differently in rainbow trout gonad and induce different chemokine transcription profiles. *Developmental and Comparative Immunology* 34, 648–658.

Chinchilla, B., Encinas, P., Estepa, A., Coll, J.M. and Gomez-Casad, E. (2015) Transcriptome analysis of rainbow trout in response to non-virion (NV) protein of viral hemorrhagic septicemia virus (VHSV). *Applied Microbiology and Biotechnology* 99, 1827–1843.

Chiou, P., Kim, P., Carol, H., Ormonde, P. and Leong, J.C. (2000) Infectious hematopoietic necrosis virus matrix protein inhibits host-directed gene expression and induces morphological changes of apoptosis in cell cultures. *Journal of Virology* 74, 7619–7627.

Cuesta, A. and Tafalla, C. (2009) Transcription of immune genes upon challenge with viral hemorrhagic septicemia

virus (VHSV) in DNA vaccinated rainbow trout (*Oncorhynchus mykiss*). *Vaccine* 27, 280–289.

Dale, O.B., Orpetveit, I., Lyngstad, T.M., Kahns, S., Skall, H.F. *et al*. (2009) Outbreak of viral haemorrhagic septicaemia (VHS) in seawater-farmed rainbow trout in Norway caused by VHS virus genotype III. *Diseases of Aquatic Organisms* 85, 93–103.

de Kinkelin, P., Chilmonczyk, S., Dorson, M., le Berre, M. and Baudouy A.M. (1979) Some pathogenic facets of rhabdoviral infection of salmonid fish. In: Bachmann P.A. (ed.) *Munich Symposia of Microbiologia: Mechanisms of Viral Pathogenesis and Virulence.* WHO Collaborating Centre for Collection and Education of Data on Comparative Virology, Munich, Germany, pp. 357–375.

Deretic, V. (2011) Autophagy in immunity and cell-autonomous defense against intracellular microbes. *Immunological Reviews* 240, 92–104.

Díaz-Rosales, P., Romero, A., Balseiro, P., Dios, S. Novoa, B. *et al*. (2012) Microarray-based identification of differentially expressed genes in families of turbot (*Scophthalmus maximus*) after infection with viral haemorrhagic septicemia virus (VHSV). *Marine Biotechnology* 14, 515–529.

Dorson, M., Quillet, E., Hollebecq, M.G., Torhy, C. and Chevassus, B. (1995) Selection of rainbow trout resistant to viral haemorrhagic septicaemia virus and transmission of resistance by gynogenesis. *Veterinary Research* 26, 361–368.

Eaton, W., Hulett, J., Brunson, R. and True, K. (1991) The first isolation in North America of infectious, hematopoietic necrosis virus (IHNV) and viral hemorrhagic septicemia virus (VHSV) in coho salmon from the same watershed. *Journal of Aquatic Animal Health* 3, 114–117.

Einer-Jensen, K., Ahrens, P., Forsberg, R. and Lorenzen, N. (2004) Evolution of the fish rhabdovirus viral haemorrhagic septicaemia virus. *Journal of General Virology* 85, 1167–1179.

Eleouet, J.F., Druesne, N., Chilmonczyk, S. and Delmas, B. (2001) Comparative study of *in-situ* cell death induced by the viruses of viral hemorrhagic septicemia (VHS) and infectious pancreatic necrosis (IPN) in rainbow trout. *Journal of Comparative Pathology* 124, 300–307.

Elsayed, E., Faisal, M., Thomas, M., Whelan, G., Batts, W. *et al*. (2006) Isolation of viral haemorrhagic septicaemia virus from muskellunge, *Esox masquinongy* (Mitchill), in Lake St Clair, Michigan, USA reveals a new sublineage of the North American genotype. *Journal of Fish Diseases* 29, 611–619.

Esteban, M.A., Meseguer, K., Tafalla, C. and Cuesta, A. (2008) NK-like and oxidative burst activities are the main early cellular innate immune responses activated after virus inoculation in reservoir fish. *Fish and Shellfish Immunology* 25, 433–438.

Evensen, Ø., Meier, W., Wahli, T., Olesen, N., Vestergård Jørgensen, P. *et al*. (1994) Comparison of immunohistochemistry and virus cultivation for detection of viral haemorrhagic septicaemia virus in experimentally infected rainbow trout *Oncorhynchus mykiss*. *Diseases of Aquatic Organisms* 20, 101–109.

Gagne, N., Mackinnon, A.M., Boston, L., Souter, B., Cook-Versloot, M. *et al*. (2007) Isolation of viral haemorrhagic septicaemia virus from mummichog, stickleback, striped bass and brown trout in eastern Canada. *Journal of Fish Diseases* 30, 213–223.

García-Valtanen, P., Ortega-Villaizán, M.D., Martínez-López, A., Medina-Gali, R., Pérez, L. *et al*. (2014) Autophagy-inducing peptides from mammalian VSV and fish VHSV rhabdoviral G glycoproteins (G) as models for the development of new therapeutic molecules. *Autophagy* 10, 1666–1680.

Garver, K.A., Hawley, L.M., McClure, C.A., Schroeder, T., Aldous, S. *et al*. (2011) Development and validation of a reverse transcription quantitative PCR for universal detection of viral hemorrhagic septicemia virus. *Diseases of Aquatic Organisms* 95, 97–112.

Garver, K.A., Traxler, G.S., Hawley, L.M., Richard, J., Ross, J.P. *et al*. (2013) Molecular epidemiology of viral haemorrhagic septicaemia virus (VHSV) in British Columbia, Canada, reveals transmission from wild to farmed fish. *Diseases of Aquatic Organisms* 104, 93–104.

Goodwin, A.E. and Merry, G.E. (2011) Mortality and carrier status of bluegills exposed to viral hemorrhagic septicemia virus genotype IVb at different temperatures. *Journal of Aquatic Animal Health* 23, 85–91.

Gregory, A. (2008) A qualitative assessment of the risk of introduction of viral haemorrhagic septicaemia virus into the rainbow trout industry Scotland. Fisheries Research Services Internal Report No. 12/08. Fisheries Research Services, Marine Laboratory, Aberdeen, UK. Available at: http://www.gov.scot/uploads/documents/int1208.pdf (accessed 25 October 2016).

Grice, J., Reid, A., Peterson, A., Blackburn, K., Tubbs, L. *et al*. (2016) Walleye *Sander vitreus* Mitchill are relatively resistant to experimental infection with viral hemorrhagic septicemia virus IVb and extant strains vary in susceptibility. *Journal of Fish Diseases* 38, 859–872.

Groocock, G.H., Getchell, R.G., Wooster, G.A., Britt, K.L., Batts, W.N. *et al*. (2007) Detection of viral hemorrhagic septicemia in round gobies in New York State (USA) waters of Lake Ontario and the St. Lawrence River. *Diseases of Aquatic Organisms* 76, 187–192.

Hawley, L.M. and Garver K.A. (2008) Stability of viral hemorrhagic septicemia virus (VHSV) in freshwater and seawater at various temperatures. *Diseases of Aquatic Organisms* 82, 171–178.

Hay, S. and Kannourakis, G. (2002) A time to kill: viral manipulation of the cell death program. *Journal of General Virology* 83, 1547–1564.

Hoffmann, B., Schutze, H. and Mettenleiter, T.C. (2002) Determination of the complete genomic sequence and analysis of the gene products of the virus of spring viremia of carp, a fish rhabdovirus. *Virus Research* 84, 89–100.

Holvold, L.B., Myhr, A. and Dalmo, R.A. (2014) Strategies and hurdles using DNA vaccines to fish. *Veterinary Research* 45, 21.

Hope, K.M., Casey, R.N., Groocock, G.H., Getchell, R.G., Bowser, P.R. *et al*. (2010) Comparison of quantitative RT-PCR with cell culture to detect viral hemorrhagic septicemia virus (VHSV) IVb infections in the Great Lakes. *Journal of Aquatic Animal Health* 22, 50–61.

ICTV (2015) Virus Taxomomy: 2015 Release, EC47, London, UK, July 2015. International Committee on Taxonomy of Viruses. Available at: http://www.ictvonline.org/virusTaxonomy.asp (accessed 25 October 2016).

Kim, R. and Faisal, M. (2010) Comparative susceptibility of representative Great Lakes fish species to the North American viral hemorrhagic septicemia virus sublineage IVb. *Diseases of Aquatic Organisms* 91, 23–34.

Kim, M.S. and Kim, K.H. (2012) Effects of NV gene knock-out recombinant viral hemorrhagic septicemia virus (VHSV) on Mx gene expression in *Epithelioma papulosum cyprini* (EPC) cells and olive flounder (*Paralichthys olivaceus*). *Fish and Shellfish Immunology* 32, 459–463.

Kim, M.S. and Kim, K.H. (2013) The role of viral hemorrhagic septicemia virus (VHSV) NV gene in TNF-α and VHSV infection-mediated NF-kB activation. *Fish and Shellfish Immunology* 34, 1315–1319.

Kocan, R., Bradley, M., Elder, N., Meyers, T., Batts, W. *et al*. (1997) North American strain of viral hemorrhagic septicemia virus is highly pathogenic for laboratory-reared Pacific herring. *Journal of Aquatic Animal Health* 9, 279–290.

LaPatra, S., Al-Hussinee, L., Misk, E. and Lumsden J.S. (2016) Rhabdoviruses of fish. In: Kibenge, F. and Godoy, M. (eds) *Aquaculture Virology*. Elsevier/Academic Press, Amsterdam, pp. 266–298.

Laing, K.J. and Hansen, J.D. (2011) Fish T cells: recent advances through genomics. *Developmental and Comparative Immunology* 35, 1282–1295.

Leal, E., Granja, A.G., Zarza, C. and Tafalla, C. (2016) Distribution of T cells in rainbow trout (*Oncorhynchus mykiss*) skin and responsiveness to viral infection. *PLoS ONE* 11(1): e0147477.

Lorenzen, N. and LaPatra, S.E. (1999) Immunity to rhabdoviruses in rainbow trout: the antibody response. *Fish and Shellfish Immunology* 9, 345–360.

Lorenzen, N., Lorenzen E., Einer-Jensen, K., Heppell, J., Wu, T. *et al*. (1998) Protective immunity to VHS in rainbow trout (*Oncorhynchus mykiss* Walbaum) following DNA vaccination. *Fish and Shellfish Immunology* 8, 261–270.

Lorenzen, N., Olesen, N.J. and Koch, C. (1999) Immunity to VHS virus in rainbow trout. *Aquaculture* 172, 41–60. *Fish and Shellfish Immunology* 12, 439–453.

Lorenzen, N., Lorenzen, E. and Einer-Jensen, K. (2001) Immunity to viral haemorrhagic septicemia (VHS) following DNA vaccination of rainbow trout at an early lifestage. *Fish and Shellfish Immunology* 11, 585–591.

Lorenzen, N., Lorenzen, E., Einer-Jensen, K. and LaPatra, S.E. (2002) DNA vaccines as a tool for analyzing the protective immune response against rhabdoviruses in rainbow trout. *Fish and Shellfish Immunology* 12, 439–453.

Lumsden, J.S., Morrison, B., Yason, C., Russell, S., Young, K. *et al.* (2007) Mortality event in freshwater drum *Aplodinotus grunniens* from Lake Ontario, Canada, associated with viral haemorrhagic septicemia virus, type IV. *Diseases of Aquatic Organisms* 76, 99–111.

Martínez-López, A., García-Valtanen, P., Ortega-Villaizán, M.M., Chico, V., Medina-Gali, R.M. *et al.* (2013) Increasing versatility of the DNA vaccines through modification of the subcellular location of plasmid-encoded antigen expression in the in vivo transfected cells. *PLoS ONE* 8(10): e77426.

Marty, G.D., Freiberg, E.F., Meyers, T.R., Wilcock, J., Farver, T.B. *et al.* (1998) Viral hemorrhagic septicemia virus, *Ichthyophonus hoferi*, and other causes of morbidity in Pacific herring *Clupea pallasi* spawning in Prince William Sound, Alaska, USA. *Diseases of Aquatic Organisms* 32, 15–40.

Matejusova, I., McKay, P., McBeath, A.J., Collet, B. and Snow, M. (2008) Development of a sensitive and controlled real-time RT-PCR assay for viral haemorrhagic septicaemia virus (VHSV) in marine salmonid aquaculture. *Diseases of Aquatic Organisms* 80, 137–144.

McAllister, P.E. (1990) *Viral Hemorrhagic Septicemia of Fishes*. Fish Disease Leaflet 83, US Department of the Interior Fish and Wildlife Service, Washington, DC. Available at: http://digitalcommons.unl.edu/cgi/viewcontent.cgi?article=1141&context=usfwspubs (accessed 25 October 2016).

McLauclan, P.E., Collet, B., Ingerslev, E., Secombes, C.J., Lorenzen, N. *et al.* (2003) DNA vaccination against viral hemorrhagic septicemia (VHS) in rainbow trout: size, dose, route of injection and duration of protection–early protection correlates with Mx expression. *Fish and Shellfish Immunology* 15, 39–50.

Meyers, T.R. and Winton, J.R. (1995) Viral hemorrhagic septicemia virus in North America. *Annual Review of Fish Diseases* 5, 3–24.

Nishizawa, T., Iida, H., Takano, R., Isshiki, T., Nakajima, K. *et al.* (2002) Genetic relatedness among Japanese, American and European isolates of viral hemorrhagic septicemia virus (VHSV) based on partial G and P genes. *Diseases of Aquatic Organisms* 48, 143–148.

O'Farrell, C., Vaghefi, N., Cantonnet, M., Buteau, B., Boudinot, P. *et al.* (2002) Survey of transcript expression in rainbow trout leukocytes reveals a major contribution of interferon-responsive genes in the early response to rhabdovirus infection. *Journal of Virology* 76, 8040–8049.

OIE (2015) Viral haemorrhagic septicaemia, In: *Manual of Diagnostic Tests for Aquatic Animals*, 7th edn. World Organisation for Animal Health, Paris, pp. 374–396. Updated 2016 version (Chapter 2.3.10.) available at: http://www.oie.int/index.php?id=2439&L=0&htmfile=chapitre_vhs.htm (accessed 25 October 2016).

Olesen, N., Lorenzen, N. and Jørgensen, P. (1993) Serological differences among isolates of viral haemorrhagic septicaemia virus detected by neutralizing monoclonal and polyclonal antibodies. *Diseases of Aquatic Organisms* 16, 163–163.

Ortega-Villaizan, M., Chico, V., Falco, A., Perez, L., Coll, J.M. *et al.* (2009) The rainbow trout TLR9 gene, its role in the immune responses elicited by a plasma encoding the glycoprotein G of the viral haemorrhagic septicaemia rhabdovirus (VHSV). *Molecular Immunology* 46, 1710–1717.

Ouyang, L., Shi, Z., Zhao, S., Wang, F.T., Zhou, T.T. *et al.* (2012) Programmed cell death pathways in cancer: a review of apoptosis, autophagy and programmed necrosis. *Cell Proliferation* 45, 487–498.

Parry, L. and Dixon, P.F. (1997) Stability of viral haemorrhagic septicemia virus (VHSV) isolates in seawater. *Bulletin of the European Association of Fish Pathologists* 17, 31–36.

Pereiro, P., Figueras, A. and Novoa, B. (2016) Turbot (*Scophthalmus maximus*) vs. VHSV (viral hemorrhagic septicemia virus): a review. *Frontiers in Physiology* 7: 192.

Pham, J., Lumsden, J.S., Tafalla, C., Dixon, B. and Bols, N. (2013) Differential effects of viral hemorrhagic septicemia virus (VHSV) genotypes IVa and IVb on gill epithelial and spleen macrophage cell lines from rainbow trout (*Onchorhynchus mykiss*). *Fish and Shellfish Immunology* 34, 632–640.

Purcell, M.K., Laing, K.J. and Winton, J.R. (2012) Immunity to fish rhabdoviruses. *Viruses* 4, 140–166.

Quillet, E., Dorson, M., Aubard, G. and Torhy, C. (2001) *In vitro* viral hemorrhagic septicemia virus replication in excised fins of rainbow trout: correlation with resistance to waterborne challenge and genetic variation. *Diseases of Aquatic Organisms* 45, 171–182.

Rebl, A., Goldammer, T., Fischer, U., Köllner, B. and Sayfert, H.M. (2009) Characterization of two key molecules of teleost innate immunity from rainbow trout (*Oncorhynchus mykiss*): MyD88 and SAA. *Veterinary Immunology and Immunopathology* 131, 122–126.

Reid, A., Young, K. and Lumsden, J.S. (2011) Rainbow trout ladderlectin not intelectin binds VHSV IVb. *Diseases of Aquatic Organisms* 95, 137–143.

Sano, M., Ito, T., Matsuyama, T., Nakayasu, C. and Kurita, J. (2009) Effect of water temperature shifting on mortality of Japanese flounder *Paralichthys olivaceus* experimentally infected with viral hemorrhagic septicemia virus. *Aquaculture* 286, 254–258.

Schiotz, B.L., Roos, N., Rishovd, A.L. and Gjoen, T. (2010) Formation of autophagosomes and redistribution of LC3 upon *in vitro* infection with infectious salmon anemia virus. *Virus Research* 151, 104–107.

Sepúlveda, D. and Lorenzen, N. (2016) Can VHS bypass the protective immunity induced by DNA vaccination in rainbow trout? *PLoS ONE* 11(4): e0153306.

Skaletskaya, A., Bartle, L.M., Chittenden, T., McCormick, L., Mocarski, E.S. *et al.* (2001) A cytomegalovirus-encoded inhibitor of apoptosis that suppresses caspase-8 activation. *Proceedings of the National Academy of Sciences of the United States of America* 98, 7829–7834.

Skall, H.F., Slierendrecht, W.J., King, A. and Olesen, N.J. (2004) Experimental infection of rainbow trout *Oncorhynchus mykiss* with viral haemorrhagic septicemia virus isolates from European marine and farmed fishes. *Diseases of Aquatic Organisms* 58, 99–110.

Smail, D.A. and Snow, M. (2011) Viral hemorrhagic septicemia. In: Woo, P.T.K., Bruno, D.W. (eds) *Fish Diseases and Disorders. Volume 3: Viral Bacterial and Fungal Infections*, 2nd edn. CAB International, Wallingford, UK. pp. 123–146.

Soliman, H. and El-Matbouli, M. (2006) Reverse transcription loop-mediated isothermal amplification (RT-LAMP) for rapid detection of viral hemorrhagic septicaemia virus (VHS). *Veterinary Microbiology* 114, 205–213.

Sommerset, I., Lorenzen, E., Lorenzen, N., Bleie, H. and Nerland, A.H. (2003) A DNA vaccine directed against a rainbow trout rhabdovirus induces early protection against a nodavirus challenge in turbot. *Vaccine* 21, 4661–4667.

Snow, M. and Cunningham, C.O. (2000) Virulence and nucleotide sequence analysis of marine viral haemorrhagic septicaemia virus following *in vivo* passage in rainbow trout *Onchorhynchus mykiss*. *Diseases of Aquatic Organisms* 42, 17–26.

Snow, M., Bain, N., Black, J., Taupin, V., Cunningham, C.O. *et al.* (2004) Genetic population structure of marine viral haemorrhagic septicaemia virus (VHSV). *Diseases of Aquatic Organisms* 61, 11–21.

Tafalla, C., Figueras, A. and Novoa, B. (1998) *In vitro* interaction of viral haemorrhagic septicaemia virus and leukocytes from trout (*Oncorhynchus mykiss*) and turbot (*Scophthalmus maximus*). *Veterinary Immunology and Immunopathology* 62, 359–366.

Tafalla, C., Figueras, A. and Novoa, B. (1999) Role of nitric oxide on the replication of viral haemorrhagic septicemia virus (VHSV), a fish rhabdovirus. *Veterinary Immunology and Immunopathology* 72, 249–256.

Tafalla, C., Figueras, A. and Novoa, B. (2001) Viral hemorrhagic septicemia virus alters turbot *Scophthalmus maximus* macrophage nitric oxide production. *Diseases of Aquatic Organisms* 47, 101–107.

Tafalla, C., Chico, V., Pérez, L., Coll, J.M. and Estepa, A. (2007) *In vitro* and *in vivo* differential expression of rainbow trout (*Oncorhynchus mykiss*) Mx isoforms in response to viral haemorrhagic septicaemia virus (VHSV) G gene, poly I:C and VHSV. *Fish and Shellfish Immunology* 23, 210–221.

Tafalla, C, Sanchez, E., Lorenzen, N., DeWitte-Orr, S.J. and Bols, N.C. (2008) Effects of viral hemorrhagic septicemia virus (VHSV) on the rainbow trout (*Oncorhynchus mykiss*) monocyte cell line RTS-11. *Molecular Immunology* 45, 1439–1448.

Thompson, T.M., Batts, W.N., Faisal, M., Bowser, P., Casey, J.W. *et al.* (2011) Emergence of viral hemorrhagic septicemia virus in the North American Great Lakes region is associated with low viral genetic diversity. *Diseases of Aquatic Organisms* 96, 29–43.

Thrush, M.A. and Peeler, E.J. (2013) A model to approximate lake temperature from gridded daily air temperature records and its application in risk assessment for the establishment of fish diseases in the UK. *Transboundary and Emerging Diseases* 60, 460–471.

Utke, K., Bergmann, S., Lorenzen, N., Kollner, B., Ototake, M. *et al.* (2007) Cell-mediated cytotoxicity in rainbow trout, *Oncorhynchus mykiss*, infected with viral haemorrhagic septicemia virus. *Fish and Shellfish Immunology* 22, 182–196.

Utke, K., Kock, H., Schuetze, H., Bergmann, S.M., Lorenzen, N. *et al.* (2008) Cell-mediated immune response in rainbow trout after DNA immunization against the viral hemorrhagic septicemia virus. *Developmental and Comparative Immunology* 32, 239–252.

van der Aa, L.M., Levraud, J., Yahmi, M., Lauret, E., Briolat, V. *et al.* (2009) A large subset of TRIM genes highly diversified by duplication and positive selection in teleost fish. *BMC Biology* 7:7.

Verrier, E.R., Langevin, C., Benmansour, A. and Boudinot, P. (2011) Early antiviral response and virus-induced genes in fish. *Developmental and Comparative Immunology* 35, 1204–1214.

Verrier, E.R., Dorson, M., Mauger, S., Torhy, C., Ciobotaru, C. *et al.* (2013) Resistance to a rhabdovirus (VHSV) in rainbow trout: identification of a major QTL related to innate mechanisms. *PLoS ONE* 8(2): e55302.

Vo, N.T.K., Bender, A.W., Lee, L.E.J., Lumsden, J.S., Lorenzen, N. *et al.* (2015) Development of a walleye cell line and use to study the effects of temperature on infection by viral hemorrhagic septicemia virus group IVb. *Journal of Fish Diseases* 38, 121–136.

Wolf, K. (1988) *Fish Viruses and Fish Viral Diseases*. Comstock Publishing Associates, Cornell University Press, Ithaca, New York.

Yamamoto, T., Batts, W.N. and Winton, J.R. (1992) *In vitro* infection of salmonid epidermal tissues by infectious hematopoietic necrosis virus and viral hemorrhagic septicemia virus. *Journal of Aquatic Animal Health* 4, 231–239.

Yanez, J.M., Houston, R.D. and Newman, S. (2014) Genetics and genomics of disease resistance in salmonid species. *Frontiers in Genetics* 5, 415.

Zou, J. and Secombes, C.J. (2011) Teleost fish interferons and their role in immunity. *Developmental and Comparative Immunology* 35, 1376–1387.

4

Epizootic Haematopoietic Necrosis and European Catfish Virus

PAUL HICK[1], ELLEN ARIEL[2] AND RICHARD WHITTINGTON[1]*

[1]Faculty of Veterinary Science, University of Sydney, Camden, New South Wales, Australia; [2]College of Public Health, Medical and Veterinary Sciences, James Cook University, Townsville, Queensland, Australia

4.1 Introduction

Epizootic haematopoietic necrosis (EHN) disease is restricted to Australia and is caused by the ranavirus *Epizootic haematopoietic necrosis virus* (EHNV). This systemic disease causes high mortality among naturally infected wild redfin perch (*Perca fluviatilis*), has an impact on farmed rainbow trout (*Oncorhynchus mykiss*) and can threaten populations of native Australian fishes (Whittington *et al.*, 2010; Becker *et al.*, 2013). *European catfish virus* (ECV) is a closely related ranavirus that was originally isolated from sheatfish (*Silurus glanis*) in Germany (Ahne *et al.*, 1989). This virus causes high mortalities in sheatfish and black bullhead catfish (*Ameiurus melas*), and other economically important fishes may also be susceptible (Jensen *et al.*, 2009, 2011). Each of these viruses causes a similar disease with non-specific clinical signs and pathology characterized by widespread systemic necrosis that is most noticeable in haematopoietic tissues (Whittington *et al.*, 2010). Both diseases occur in wild fishes and in those in aquaculture. EHNV was the first finfish virus listed by the World Organisation for Animal Health (OIE) due to its virulence, lack of host specificity, geographic isolation and the existence of a reliable diagnostic test (OIE, 2015). Additionally, ranaviruses that infect amphibians are listed by the OIE because they have economic and ecological impacts (OIE, 2016a) and also the potential for interclass transmission (Brenes *et al.*, 2014). Experimental evidence indicates that a single fish species can be susceptible to multiple ranaviral species in addition to EHNV and

ECV (Jensen *et al.*, 2009). This chapter describes EHNV and ECV, which belong to a group of emerging pathogens – all of which are relevant to fish health – for which understanding of their epizootiology is incomplete (Gray *et al.*, 2009).

4.1.1 Classification

EHNV and ECV belong to the genus *Ranavirus* (Jancovich *et al.*, 2012) in the family *Iridoviridae*. Like all viruses in this family, they are large double-stranded DNA (dsDNA) viruses that undergo a complex nucleocytoplasmic replication cycle (Williams *et al.*, 2005). The family *Iridoviridae* includes numerous virus species that cause diseases of fish in aquaculture (Whittington *et al.*, 2010).

The five ranavirus lineages described by Jancovich *et al.* (2015) provide a framework for ranaviruses (Fig. 4.1). EHNV and ECV, together with the amphibian pathogen *Ambystoma tigrinum virus* (ATV) form a group in lineage 3. Each lineage contains pathogens of fish, and among these closely related viruses there are amphibian pathogens that also cause disease in fish (Waltzek *et al.*, 2014).

The genomes of EHNV and ECV contain approximately 127 kilobases (Jancovich *et al.*, 2010; Mavian *et al.*, 2012) and the dsDNA is circularly permuted and terminally redundant (Jancovich *et al.*, 2012). The guanine/cytosine (G/C) content is 54–55% and there is approximately 25% cytosine methylation (Chinchar *et al.*, 2011). The genome of EHNV has 88% nucleotide sequence identity with that of the ESV (European

---

*Corresponding author e-mail: richard.whittington@sydney.edu.au

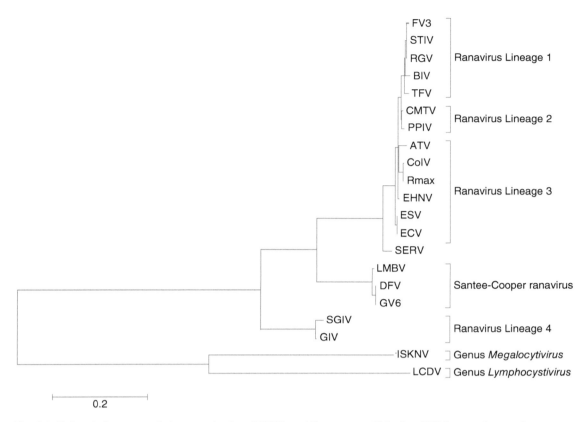

0.2

Fig. 4.1. *Epizootic haematopoietic necrosis virus* (EHNV) and *European catfish virus* (ECV) group in ranavirus Lineage 3 (Jancovich *et al.*, 2015). Molecular phylogenetics were calculated using 1127 base positions of the major capsid protein gene by the maximum likelihood method in Mega6 (Tamura *et al.*, 2013). The tree is drawn to scale, with branch lengths measured in the number of substitutions per site. Abbreviations (Lineage: virus name, Genbank reference): Lineage 1: FV3 (*Frog virus 3*, AY548484); STIV (soft-shelled turtle iridovirus, EU627010); RGV (Rana grylio iridovirus, JQ654586); BIV (*Bohle iridovirus*, AY187046); TFV (tiger frog virus, AF389451). Lineage 2: CMTV (Common midwife toad virus, JQ231222); PPIV (pike-perch iridovirus, FJ358610). Lineage 3: ATV (*Ambystoma tigrinum virus*, AY150217); CoIV (cod iridovirus, GU391284); Rmax (Ranavirus maxima, GU391285); EHNV (*Epizootic haematopoietic necrosis virus*, AY187045); ESV (European sheatfish virus, FJ358609); ECV (*European catfish virus*, FJ358608). SERV (short-finned eel ranavirus, FJ358612). Isolates of *Santee-Cooper ranavirus*: LMBV (largemouth bass ranavirus, FR682503); DFV (doctor fish virus, FR677324); GV6 (guppy virus 6, FR677325). Lineage 4: SGIV (Singapore grouper iridovirus, AY521625); GIV (grouper iridovirus, AY666015). Type species: genus *Megalocytivirus*, ISKNV (*Infectious spleen and kidney necrosis virus*, NC003494); genus *Lymphocystivirus*, LCDV (*Lymphocystis disease virus 1*, NC001824). Adapted from Jancovich *et al.* (2015).

sheatfish virus) isolate of ECV and the genome structure is very similar. These distinct but closely related species have a similar complement of approximately 100 genes; however, the function of many of the predicted proteins is not known (Grayfer *et al.*, 2012). Molecular evolutionary analyses suggest that all ranaviruses might have evolved following relatively recent host shifts from an ancestral fish virus (Jancovich *et al.*, 2010). On

this basis, new viruses may emerge or be detected and the diagnostic criteria for EHNV and ECV in the future may require consideration of whole genome sequence data.

4.1.2 Structure and replication

The capsids of EHNV and ECV are 150–160 nm in diameter with the icosahedral symmetry that is common to all iridovirids (Ahne *et al.*, 1989; Eaton *et al.*, 1991). The major capsid protein (MCP) makes up 40% of a complex virion composed of 36 proteins with structural and enzymatic activities (Chinchar *et al.*, 2011). An external lipid and glycoprotein membrane is not essential for infectivity, but appears to enhance it (Ariel *et al.*, 1995; Jancovich *et al.*, 2012). The unique nuclear and cytoplasmic replication strategy elucidated for the type species, *Frog virus 3* (FV3), has been reviewed (Chinchar *et al.*, 2011). There is a coordinated expression of early, delayed early and late genes (Teng *et al.*, 2008). Replication of the genome occurs initially in the nucleus followed by the generation of concatermeric copies of the genome and structural proteins in the cytoplasm in morphologically distinct virogenic stroma for the assembly of virions (Eaton *et al.*, 1991). EHNV is released by budding from the cell membrane, where an outer envelope is acquired, resulting in virions up to 200 nm in diameter. Alternatively, naked virions that can be released by cell lysis may aggregate in paracrystalline arrays in the cytoplasm (Eaton *et al.*, 1991).

4.1.3 Transmission

Further studies on the transmission of EHNV and ECV are required to explain recurrent outbreaks of disease. Vertical transmission has not been reported (Whittington *et al.*, 2010). Horizontal transmission of EHNV and ECV occurs via water or the ingestion of tissues from infected fish (Langdon, 1989; Jensen *et al.*, 2009; Gobbo *et al.*, 2010, Jensen *et al.*, 2011; Becker *et al.*, 2013; Leimbach *et al.*, 2014). Recurrence of EHN in rainbow trout was attributed to reinfection from wild redfin perch (Whittington *et al.*, 1996; Marsh *et al.*, 2002), and ESV may persist in apparently healthy fish (Ahne *et al.*, 1991).

EHNV has been spread from populations of farmed rainbow trout in which disease was active at low prevalence, but practically inapparent (Whittington *et al.*, 1999). Fish with subclinical infections are likely to be a source of the virus. EHNV was isolated from apparently healthy redfin perch that survived a natural disease outbreak, suggesting a possible reservoir, though the duration of infection is not known (Langdon and Humphrey, 1987). This suggestion is supported by the re-isolation of infectious virus from apparently healthy fish surviving experimental challenges (Langdon, 1989; Jensen *et al.*, 2009, 2011; Gobbo *et al.*, 2010;

Becker *et al.*, 2013; Leimbach *et al.*, 2014). These studies also provide evidence that hosts other than those that express disease may be a source of EHNV and ECV. There is no evidence of an amphibian reservoir, although there may be one because of the very broad host range and the capacity for inter-class transmission of ranaviruses (Chinchar and Waltzek, 2014).

The spread of EHNV via fomites or persistence outside a host is also possible because EHNV and similar ranaviruses are extremely stable in the environment under certain conditions. EHNV retained its infective titre for 97 days at 15°C and 300 days at 4°C. The virus was resistant to desiccation, retaining infectivity after drying for 113 days at 15°C (Langdon, 1989), and remained infective in frozen fish for at least 2 years (Whittington *et al.*, 1996). Infectivity of a related ranavirus, *Bohle iridovirus* (BIV) declined at 44°C and further declined at 52°C, but treatment at 58°C for 30 min was required for complete loss of infectivity (La Fauce *et al.*, 2012).

There is evidence that EHNV has spread by human activity, including the movement of infected rainbow trout fingerlings between farms (Langdon *et al.*, 1988; Whittington *et al.*, 1994, 1999). Disease outbreaks in redfin perch have occurred, with the progressive spread of EHNV in inland river systems such as the Murrumbidgee River in New South Wales and the Australian Capital Territory, and the Murray River in South Australia (Whittington *et al.*, 2010). This may be due to recreational fishing through illegal relocation of redfin perch, the use of frozen redfin perch as bait, or on fomites such as boats and fishing equipment. Birds may also spread EHNV by regurgitating ingesta within a few hours of feeding on EHNV-infected carcasses (Whittington *et al.*, 1996).

4.1.4 Geographic distribution

EHNV is found in Australia, while ECV is restricted to continental Europe (Whittington *et al.*, 2010). Within Australia, EHNV has a discontinuous distribution over time and throughout freshwater lakes, rivers and impoundments in south-eastern Australia. Its geographic range does not include the entire range of the known susceptible host species (Whittington *et al.*, 2010). Only a small number of rainbow trout farms are infected and these are restricted to two water catchment regions, the Shoalhaven River and the Murrumbidgee River

(Langdon *et al.*, 1988, Whittington *et al.*, 1999). Wild trout stocks, which are present in south-eastern Australia, have not been surveyed for the infection (Whittington *et al.*, 1999). Similarly, the farm-level prevalence of EHNV is unclear because not all farms have been tested. The virus has not been detected based on regular surveillance in salmonids in Tasmania and Western Australia.

The first reported outbreak of disease caused by ECV (sheatfish strain) occurred in sheatfish fry from south-east Europe that were reared in a warm-water aquaculture facility in Germany (Ahne *et al.*, 1989). Subsequently, the disease has been shown to occur in catfish in France, Italy, Hungary and Poland. Outbreaks have occurred in catfish reared in aquaculture facilities in Italy (Bovo *et al.*, 1993) and among brown bullheads (*A. nebulosus*) in Hungary (Juhasz *et al.*, 2013) and sheatfish in Poland (Borzym *et al.*, 2015). In France, catfish of all sizes in natural lakes were affected during summer months. The pathogen identified in France in 2007 closely resembled that which caused disease in wild catfish 15 years previously (Bigarre *et al.*, 2008).

EHNV or ECV could spread to new areas where susceptible hosts are present in permissive environments (Peeler *et al.*, 2009). The international trade and the movements of ornamental fish, reptiles and amphibians, and bait have been responsible for the distribution of some ranaviruses and have had serious impacts on populations of naive host species (Jancovich *et al.*, 2005; Whittington and Chong, 2007). EHNV is listed by the OIE (2016c) to reduce further impacts on aquaculture and wild fish reduced because of its broad host range and ability to adapt to environmental conditions. Surveillance and disease control zones are important in disease control. The international spread of EHNV has not been documented and a survey by Vesely *et al.* (2011) indicated that ranavirus was not detected in ornamental fish imported into Europe.

4.1.5 Impact of the pathogen on fish production

Recurrent outbreaks of disease occur in wild fish in natural environments and in aquaculture facilities (Ahne *et al.*, 1991; Whittington *et al.*, 2010). However, EHNV is an infrequent cause of mass mortality in wild redfin perch. All age classes may be affected, but in endemic areas mortality is limited to fingerling and juvenile fish (Langdon *et al.*, 1986; Langdon and Humphrey, 1987; Whittington

et al., 1996); this decline in the fish population has impacts on the recreational fishery (Whittington *et al.*, 2010). In contrast to mortality up to 95% in wild redfin perch, the impact of the disease on farmed rainbow trout is less than 5% of the farmed populations (Whittington *et al.*, 2010).

Disease caused by ECV also occurs infrequently, but can be recurrent at individual aquaculture facilities. Natural infections of sheatfish have caused 100% mortality in fingerlings and 10–30% mortality in older fish (Ahne *et al.*, 1991; Borzym *et al.*, 2015). Outbreaks in wild catfish caused high mortality of *A. melas* in French lakes, whereas other fishes were not affected (Bigarre *et al.*, 2008). The few documented cases of disease in reared catfish affected fish that included brood stocks (Bovo *et al.*, 1993; Juhasz *et al.*, 2013). Due to compliance with the European Union (EU) regulations and OIE guidelines on surveillance and reporting (OIE, 2010), EHNV is not found in Europe and the occasional outbreak of ECV disease is well documented.

Disease caused by natural EHNV or ECV infections is known only from a limited number of fish species, but many more species are susceptible in experimental infections (Langdon, 1989; Jensen *et al.*, 2009, 2011; Gobbo *et al.*, 2010; Becker *et al.*, 2013; Leimbach *et al.*, 2014). These pathogens therefore have the potential to have a serious impact on wild fish, ecosystems and emerging aquaculture industries. It is important to note that such experimental trials do not provide an estimate of the likelihood or impact of a natural disease outbreak, owing to the complexity of the host and environmental factors that can interact with the pathogen.

4.2 Diagnosis

Disease caused by EHNV or ECV is diagnosed using standard histological techniques with haematoxylin and eosin staining together with confirmation of the presence of the virus. The association of the virus with cellular pathology can be demonstrated using polyclonal antibodies developed against EHNV for immunohistochemical or immunofluorescent staining of fixed tissue sections (Reddacliff and Whittington, 1996; Bigarre *et al.*, 2008; Jensen *et al.*, 2009). These antibodies, which are available from the OIE Reference Laboratory for EHNV, cross-react with the immunodominant major capsid protein of all ranaviruses so that molecular characterization is required to determine the viral species (Hyatt *et al.*, 2000, OIE, 2016b). Electron microscopy can demonstrate the presence of iridovirids in tissue; however, EHNV and ECV are indistinguishable from each other and from other ranaviruses (Ahne *et al.*, 1998). Tests for the viruses can be applied to tissues from clinical cases. These include viral isolation in cell culture, antigen detection using immunological techniques such as enzyme-linked immunosorbent assays (ELISA) and, most commonly, the detection of specific nucleic acid sequence using polymerase chain reaction (PCR) (Whittington *et al.*, 2010; OIE, 2016b).

For the purpose of preventing spread of the pathogen, the OIE recognizes infection with EHNV rather than the occurrence of clinical disease or pathological changes. Protocols for sensitive laboratory assays with the highest level of validation are detailed in the OIE *Manual of Diagnostic Tests for Aquatic Animals* for the detection of subclinical infection (OIE, 2016b). Ideally, these tests should be undertaken in laboratories accredited according to international standards for quality control (ISO, 2005). The appropriate samples to test are the viscera; kidneys and the livers and spleens from individual fish are commonly pooled (Whittington and Steiner, 1993). Milt and ovarian fluid are not suitable samples. Tissues can be processed rapidly using the partially automated procedures of bead beating for tissue disruption (this is compatible with both viral isolation and PCR) and magnetic bead based nucleic acid purification for molecular tests (Rimmer *et al.*, 2012).

A statistically valid framework for calculating the sample size and selection method is required when undertaking surveillance for freedom from infection (OIE, 2010). Moribund fish are used for the diagnosis of disease, while biased or targeted sampling is recommended for surveillance. The latter is achieved on rainbow trout farms by freezing dead juvenile fish for later testing (Whittington *et al.*, 1999).

Viral isolation is the only method of identifying infectious virus. EHNV can be isolated between 15 and 22°C in several fish cell lines, including fathead minnow (FHM), rainbow trout gonad (RTG), bluegill fry (BF-2), and Chinook salmon embryo (CHSE-214) cells (Langdon *et al.*, 1986; Crane *et al.*, 2005; OIE, 2016b). Similarly, ECV and ESV can be isolated using BF-2 cells or epithelioma papulosum cyprinid (EPC) cells, FHM or channel catfish ovary (CCO) cells (Ahne *et al.*, 1989; Pozet *et al.*, 1992; Ariel *et al.*, 2009). Confirmation of the

identity of a virus that can be passaged and causes a cytopathic effect requires immunofluorescent staining (Bigarre *et al.*, 2008) or molecular characterization, typically of the MCP gene.

PCR assays most commonly target the MCP gene (Marsh *et al.*, 2002; Pallister *et al.*, 2007). A conventional PCR assay described in the OIE *Manual of Diagnostic Tests for Aquatic Animals* provides an amplicon that will distinguish EHNV, ECV and other ranaviruses using restriction enzyme digestion (Marsh *et al.*, 2002; OIE, 2016b). The real-time quantitative PCR (qPCR) assay described by Jaramillo *et al.* (2012) is convenient and supported by validation data on diagnostic sensitivity and specificity. Polyclonal antibodies that react with EHNV can be applied as an antigen capture ELISA on tissue homogenate samples to provide a lower cost approach to the identification of ranavirus antigen; the antibody reagents and controls can be supplied by the OIE reference laboratory for EHNV (Whittington and Steiner, 1993).

The MCP gene is conserved among ranaviruses, with the sequence identity of EHNV, ECV and BIV being >97.8% (Hossain *et al.*, 2008; Jancovich *et al.*, 2012). The identification of isolates to species level requires DNA sequencing, restriction enzyme digestion (Marsh *et al.*, 2002) or species-specific PCR assays (Pallister *et al.*, 2007). Nucleotide sequence analysis improves the diagnosis because isolates cluster in epidemiological events, thereby providing insights for disease control and prevention (Jancovich *et al.*, 2010). The nucleotide sequence of the DNA polymerase and neurofilament triplet H1-like protein genes can be used in addition to MCP (Holopainen *et al.*, 2009). However, in the future, whole genome sequencing is likely to become more accessible and replace candidate gene analysis for detailed epidemiological studies of ranaviruses (Epstein and Storfer, 2015).

4.2.1 Clinical signs

The clinical signs of disease caused by EHNV and ECV are non-specific and include loss of appetite, reduced activity, ataxia and multifocal haemorrhage, particularly in haematopoietic tissue (Whittington *et al.*, 2010). Outbreaks of EHN in wild redfin perch are typically described only as mass mortality (Langdon and Humphrey, 1987). Experimentally infected fish have a dark coloration and show erratic swimming preceding death (Langdon, 1989). Natural disease in rainbow trout

P. Hick *et al.*

may not be evident because of its low prevalence. In an experimental setting, infected rainbow trout had dark coloration, ataxia and reduced appetite (Reddacliff and Whittington, 1996).

Similarly, outbreaks of ranavirus in wild catfish were evident only as mortality (Bigarre *et al.*, 2008). Infected catfish in a culture facility had oedema, petechia of fin bases and viscera, ascites and gill pallor (Pozet *et al.*, 1992). Loss of appetite preceded gross signs of disease by several days. In experimental infections, moribund catfish assumed a vertical position with head above water and petechia of the skin and fin bases (Gobbo *et al.*, 2010). Disease in sheatfish was sometimes complicated by secondary bacterial infection (*Aeromonas hydrophila*) and there were skin ulcerations on dead and moribund fish (Ahne *et al.*, 1989, 1991). No obvious clinical signs of disease were noted in experimental infections, but diffuse subcutaneous haemorrhage in the region of the lower jaw as well as in the fins was found at necropsy (Leimbach *et al.*, 2014).

4.3 Pathology

Lesions are attributed to the vascular endothelial and haematopoietic tropism of the pathogen and are similar in all susceptible hosts. EHNV and ECV cause very similar pathology, with widespread systemic necrosis, especially in haematopoietic tissues (Whittington *et al.*, 2010). Grossly, there are petechial haemorrhages at the base of fins, excessive amounts of serosanguinous peritoneal fluid and swelling of kidney and spleen (Langdon and Humphrey, 1987; Reddacliff and Whittington, 1996). Often, the foci of necrosis are centred around small blood vessels. White to cream foci evident grossly in the liver of redfin perch correspond to necrotic foci. Microscopically, at the margins of necrotic foci in the liver, basophilic intracytoplasmic inclusion bodies may be found in hepatocytes (Fig. 4.2). These inclusion bodies are scarce and difficult to visualize in the kidney and spleen. Immunohistochemical stains reveal widespread localization of EHNV antigen in necrotic areas as well as in the endothelium and in individual circulating leucocytes within blood vessels (Fig. 4.3).

Microscopic lesions in experimentally infected sheatfish and catfish with isolates of ECV were characterized by necrosis of the haematopoietic tissues of the spleen and kidney with degenerate and necrotic cells in the interstitial tissue and tubules of

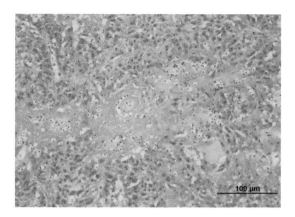

Fig. 4.2. Advanced focal necrosis in the liver of redfin perch. There is widespread dissociation of hepatocytes, many of which are in early stages of degeneration and contain amphophilic intracytoplasmic inclusion bodies. A liver with lesions of this type appears grossly to be darkly discoloured with multiple pale white spots. Stained with haematoxylin and eosin.

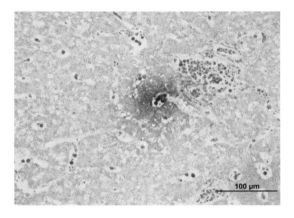

Fig. 4.3. Early focal necrosis in the liver of redfin perch. There is positive staining for EHNV (*Epizootic haematopoietic necrosis virus*) antigen in a focus of hepatocellular degeneration and necrosis adjacent to a large vessel. There is also staining of individual leucocytes within vessels and sinusoids, as well as of endothelial elements. Stained with an immunohistochemistry stain.

the kidney (Ogawa *et al.*, 1990; Pozet *et al.*, 1992). A generalized destruction of vascular endothelial cells in both types of fish was also evident, which resulted in diffuse congestion and haemorrhage in the internal organs. As for EHNV, viral antigen can be demonstrated associated with the lesions using indirect fluorescent antibody or immunohistochemical stains (Figs 4.4 and 4.5).

There are often scattered, individual necrotic cells within blood vessels and degenerate vascular endothelial cells. Other lesions include diffuse splenic necrosis, gastrointestinal epithelial necrosis, necrosis of atrial trabeculae, hyperplasia and necrosis of branchial epithelium, focal pancreatic necrosis, oedema and necrosis of the swim bladder and

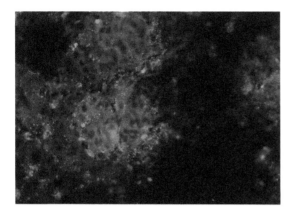

Fig. 4.4. Kidney of black bullhead infected with ECV (European catfish virus). There is bright apple green fluorescence associated with viral antigen in the cells. Stained with indirect fluorescent antibody. Image courtesy of Gobbo F. and Bovo G., Istituto Zooprofilattico Sperimentale delle Venezie, Italy.

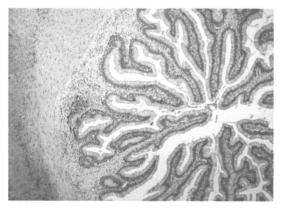

Fig. 4.5. Gut of black bullhead infected with ECV (European catfish virus). There is a large focus of cells containing ECV antigen in the submucosa. Stained with an immunohistochemistry stain. Image courtesy of Gobbo F. and Bovo G., Istituto Zooprofilattico Sperimentale delle Venezie, Italy.

ulcerative dermatitis (Langdon and Humphrey, 1987; Reddacliff and Whittington, 1996). Perivascular mononuclear cell infiltrates may be found in survivors (Becker et al., 2013).

Fishes susceptible to EHNV or ECV only by experimental infection include several native Australian freshwater species as well as pike (*Esox lucius*) and pike-perch (*Sander lucioperca*). The pathology of experimentally infected fishes was similar to that in naturally infected fishes (Jensen et al., 2009, 2011; Becker et al., 2013).

4.3.1 Pathophysiology

The pathogenesis of EHNV and ECV infection has not been studied in detail. Natural outbreaks of disease occur during warmer periods when large numbers of juveniles are present (Langdon et al., 1988; Whittington et al., 1994; Bigarre et al., 2008, Gobbo et al., 2010). Experimental studies have confirmed that transmission and pathogenicity are influenced by water temperature (Whittington and Reddacliff, 1995; Jensen et al., 2009, 2011; Gobbo et al., 2010; Leimbach et al., 2014). Mortality occurred in redfin perch at 12 to 21°C. Disease did not occur below 12°C and an increase in water temperature did not lead to disease in these fish. Rainbow trout were refractory to EHNV bath challenges at 10 and 15°C, with moderate mortalities (14–24%) occurring at 20°C (Langdon, 1989; Whittington and Reddacliff, 1995; Ariel and Jensen, 2009). The incubation period was negatively correlated with water temperature: in redfin perch, it was 10–28 days at 12–18°C compared with 10–11 days at 19–21°C; in rainbow trout, it was 14–32 days at 8–10°C compared with 3–10 days at 19–21°C following intraperitoneal (IP) challenge (Whittington and Reddacliff, 1995). Temperature-dependent replication of EHNV was demonstrated in pike with a peak viral load 7 days post bath challenge at 12°C and after only 3 days at 22°C (Holopainen et al., 2011).

Host differences in susceptibility may exist in redfin perch due to different genetics and culture conditions. Australian stocks were highly susceptible to challenge by bath and IP injection, with 100% mortality for both routes at 20°C, and 50% mortality induced by IP injection at 10°C (Langdon, 1989; Whittington and Reddacliff, 1995). Challenge trials of European redfin perch with EHNV via bath exposure did not induce mortality at 10, 15 or 20°C, and IP inoculation induced 47–80% mortality at 15°C and 28–42% mortality at 20°C (Ariel and Jensen, 2009). In another trial, mortality in juvenile redfin perch was 16% at 11–13°C and 24% at 20–22°C following bath exposure to EHNV (Borzym and Maj-Paluch, 2015). Experimental infections indicated different disease outcomes for EHNV and different isolates of ECV and ESV in the same host species (Jensen et al., 2009; Gobbo et al., 2010; Leimbach et al., 2014). ESV from sheatfish was highly pathogenic to sheatfish, with 100% mortality at 15°C, while ECV only induced 8% mortality in this species (Leimbach et al., 2014). Conversely, ECV from catfish was pathogenic to catfish with 82% mortality at 25°C and 30% mortality at 15°C, but this isolate only induced 5 or 1% mortality in sheatfish at these temperatures (Gobbo et al., 2010). These findings indicate that although the two ECV isolates were genetically similar, they acted differently in different host species. The dose of virus required to initiate an infection depends on the host species. For instance, Australian bred redfin perch are highly susceptible to EHNV with doses as low as 0.08 $TCID_{50}$ (50% tissue culture infectious dose)/ ml by bath exposure, whereas rainbow trout are resistant to bath exposure ($10^{2.2}$ $TCID_{50}$/ml) and IP infection is required for disease expression (Whittington and Reddacliff, 1995). These findings match the divergent clinical presentation of these species in nature.

The prominence of gastrointestinal lesions in fish naturally infected with EHNV and ECV (Ogawa et al., 1990; Pozet et al., 1992) (Fig. 4.5) is consistent with an oral route of infection (Reddacliff and Whittington, 1996). Viraemia is likely to result in dispersal of the virus, which is reflected in necrosis of circulating leucocytes and vascular endothelium. These processes lead to tissue necrosis at small and large scales, and to the signs of gross pathology (Langdon and Humphrey, 1987; Langdon et al., 1988; Whittington et al., 1994; Reddacliff and Whittington, 1996).

Currently, there is no evidence of natural infection of any species other than redfin perch or rainbow trout with EHNV, and of catfish or sheatfish with ECV and ESV. However, a number of fish species are susceptible to experimental infection. Ranaviruses have a global distribution and a few viruses that primarily infect amphibians are capable of inter-class transmission (natural or experimental), including to fish (Moody and Owens, 1994; Ariel and Owens, 1997; Duffus

et al., 2015). For example, seemingly identical ranaviruses were isolated from sympatric, free-living, clinically affected, three-spine stickleback (Gasterostelus aculeatus) and red-legged frog (Rana aurora) tadpoles (Mao et al., 1999). Nevertheless, reports of disease in fish caused by amphibian-associated ranaviruses are scarce (Waltzek et al., 2014), with only limited data on the non-piscine host range of EHNV and ECV. The European common frog (R. temporaria) is susceptible to the fish ranavirus, pike-perch iridovirus (PPIV), but not to EHNV and ECV (Bayley et al., 2013). Thus, experimental evidence and field observations provide support for and against a model in which reservoir species and accidental hosts exchange ranavirus infection.

4.4 Protective and Control Strategies

There is no effective therapy for diseases caused by EHNV and ECV. Therefore, biosecurity measures derived from general best practice recommendations are the focus of disease control (OIE, 2009). These measures include the importation of fish certified free from infection, disinfection of water and control of vectors in endemic regions. Quarantine of incoming fish should be for at least as long as the disease incubation period, which in the most susceptible species can be 32 days (Whittington and Reddacliff, 1995). Active surveillance can be conducted to identify zones that are free from infection to facilitate the trade of live fish (OIE, 2010). Methods for targeted surveillance of aquaculture facilities for pathogens including EHNV are prescribed for the EU (Commission of the European Communities, 2001). Passive surveillance of unusual mortality in wild fish with laboratory tests to identify outbreaks of EHN or ECV disease can be undertaken.

4.4.1 Vaccination

Vaccines are not available for EHNV and there has been limited investigation of the efficacy of alternative prevention measures. Vaccine development is feasible and is practised commercially for iridoviruses in the genus *Megalocytivirus* (Kurita and Nakajima, 2012). Methods to detect a specific EHNV antibody response are available, although further investigation of the immune response is required (Whittington and Reddacliff, 1995; Whittington et al., 1999).

4.4.2 Control or elimination of reservoir hosts

Exposure to apparently healthy carrier fish, including survivors of disease outbreaks and potential carrier species, should be prevented. With increasing evidence for a reservoir of many ranaviruses in different taxonomic classes, aquaculture biosecurity requires the rigorous exclusion of non-cultured species from farms.

4.4.3 Environmental modifications to interrupt transmission

The wide thermal range for the disease limits the applicability of temperature manipulation. Mortality might be reduced in cooler water through behavioural self-selection, as in adult redfin perch, which avoid the disease in thermally stratified environments, while juvenile fish succumb to the disease due to feeding in warmer shallow water (Whittington and Reddacliff, 1995).

The importance of the continuous disinfection of intake water is supported by EHNV infection models in which transmission between redfin perch and rainbow trout occurred without direct contact (Langdon, 1989; Ariel and Jensen, 2009). Carry-over contamination from an outbreak can be prevented by adequate cleaning and disinfection of aquaculture facilities. General guidelines for decontamination provided by the OIE (2009) can be interpreted using disinfection efficacy data specific for EHNV and related ranaviruses: sodium hypochlorite 200 mg/l; 70% ethanol; 150 mg/l chlorhexidine (0.75% Nolvasan®) for 1 min; 200 mg/l potassium peroxymonosulfate (1% Virkon®); or heating to 60°C for 15 min (Langdon, 1989; Bryan *et al.*, 2009, La Fauce *et al.*, 2012).

4.4.4 Animal husbandry practices

Optimal husbandry in aquaculture and the maintenance of healthy waterways for wild fish can reduce the impacts of disease in endemic regions. Lower stocking rates and improved water quality reduced mortality in farmed rainbow trout (Whittington *et al.*, 1994, 1999). Efforts to diagnose and control bacterial and external parasitic infections are likely to improve disease outcome due to EHNV and ECV (Ahne *et al.*, 1991; Whittington *et al.*, 2010).

4.5 Conclusions and Suggestions for Future Research

EHNV and ECV cause infrequent but spectacular disease in a few freshwater fishes that are important to aquaculture and wild fisheries. The disease is readily diagnosed using routine histopathological methods together with laboratory tests specific for the pathogens. Surveillance for the viruses and the application of effective biosecurity are the mainstays of disease control because effective treatments and preventive measures are not available. For this reason, the diseases are notifiable in many jurisdictions and EHNV is listed by the OIE to regulate the trade of fish between disease control zones. This is facilitated by an international reference laboratory for EHNV, by recommended and validated laboratory tests and by the international availability of reagents, controls and recommended laboratory protocols (OIE, 2016b). Tests are available for application in high-throughput formats suitable for the certification of freedom from infection at population level.

Experimental studies have demonstrated that EHNV and ECV have the potential to affect many fishes, including those important to food production and conservation. Additionally, the very broad host range of many ranaviruses indicates the possibility for as yet unidentified reservoir and disease hosts for EHNV and ECV. Consequently, these pathogens pose a threat to aquaculture and conservation. Further research is required to develop improved disease control strategies, including effective vaccines. The application of molecular techniques is required to trace the sources of disease outbreaks, routes of transmission and mechanisms of regional and global spread. The possibility of reservoir hosts among fishes, amphibians and reptiles needs to be resolved. Education of the public to the disease threats posed to lower vertebrates via unrestricted trade in ornamental species and baitfish is urgently needed (Jancovich *et al.*, 2005; Whittington and Chong, 2007).

References

Ahne, W., Schlotfeldt, H.J. and Thomsen, I. (1989) Fish viruses: isolation of an icosahedral cytoplasmic deoxyribovirus from sheatfish (*Silurus glanis*). *Zentralblatt für Veterinärmedizin, Reihe B/Journal of Veterinary Medicine, Series B* 36, 333–336.

Ahne, W., Schlotfeldt, H.J. and Ogawa, M. (1991) Iridovirus infection of adult sheatfish (*Silurus glanis*). *Bulletin of the European Association of Fish Pathologists* 11, 97–98.

Ahne, W., Bearzotti, M., Bremont, M. and Essbauer, S. (1998) Comparison of European systemic piscine and amphibian iridoviruses with *Epizootic haematopoietic necrosis virus* and *Frog virus 3*. *Journal of Veterinary Medicine, Series B: Infectious Diseases and Veterinary Public Health* [now *Zoonoses and Public Health*] 45, 373–383.

Ariel, E. and Jensen, B.B. (2009) Challenge studies of European stocks of redfin perch, *Perca fluviatilis* L., and rainbow trout, *Oncorhynchus mykiss* (Walbaum), with epizootic haematopoietic necrosis virus. *Journal of Fish Diseases* 32, 1017–1025.

Ariel, E. and Owens, L. (1997) Epizootic mortalities in tilapia *Oreochromis mossambicus*. *Diseases of Aquatic Organisms* 29, 1–6.

Ariel, E., Owens, L. and Moody, N.J.G. (1995) A barramundi bioassay for iridovirus refractory to cell culture. In: Shariff, R.P., Subasinghe, R.P. and Arthur, J.R. (eds) *Diseases in Asian Aquaculture II. Fish Health Section*. Asian Fisheries Society, Manila, pp. 355–367.

Ariel, E., Nicolajsen, N., Christophersen, M.B., Holopainen, R., Tapiovaara, H. and Jensen, B.B. (2009) Propagation and isolation of ranaviruses in cell culture. *Aquaculture* 294, 159–164.

Ariel, E., Holopainen, R., Olesen, N.J. and Tapiovaara, H. (2010) Comparative study of ranavirus isolates from cod (*Gadus morhua*) and turbot (*Psetta maxima*) with reference to other ranaviruses. *Archives of Virology* 155, 1261–1271.

Bayley, A.E., Hill, B.J. and Feist, S.W. (2013) Susceptibility of the European common frog *Rana temporaria* to a panel of ranavirus isolates from fish and amphibian hosts. *Diseases of Aquatic Organisms* 103, 171–183.

Becker, J.A., Tweedie, A., Gilligan, D., Asmus, M. and Whittington, R.J. (2013) Experimental infection of Australian freshwater fish with *Epizootic haematopoietic necrosis virus* (EHNV). *Journal of Aquatic Animal Health* 25, 66–76.

Bigarre, L., Cabon, J., Baud, M., Pozet, F. and Castric, J. (2008) Ranaviruses associated with high mortalities in catfish in France. *Bulletin of the European Association of Fish Pathologists* 28, 163–168.

Borzym, E. and Maj-Paluch, J. (2015) Experimental infection with epizootic haematopoietic necrosis virus (EHNV) of rainbow trout (*Oncorhynchus mykiss* Walbaum) and European perch (*Perca fluviatilis* L.). *Bulletin of the Veterinary Institute in Pulawy* 59, 473–477.

Borzym, E., Karpinska, T.A. and Reichert, M. (2015) Outbreak of ranavirus infection in sheatfish, *Silurus glanis* (L.), in Poland. *Polish Journal of Veterinary Sciences* 18, 607–611.

Bovo, G., Comuzzi, M., De Mas, S., Ceschia, G., Giorgetti, G. *et al.* (1993) Isolation of an irido-like viral agent from breeding cat fish (*Ictalurus melas*). *Bollettino Societa Italiana di Patologia Ittica* 11, 3–10.

Brenes, R., Miller, D.L., Waltzek, T.B., Wilkes, R.P., Tucker, J.L. *et al.* (2014) Susceptibility of fish and turtles to three ranaviruses isolated from different ectothermic vertebrate classes. *Journal of Aquatic Animal Health* 26, 118–126.

Bryan, L.K., Baldwin, C.A., Gray, M.J. and Miller, D.L. (2009) Efficacy of select disinfectants at inactivating ranavirus. *Diseases of Aquatic Organisms* 84, 89–94.

Chinchar, V.G. and Waltzek, T.B. (2014) Ranaviruses: Not Just for frogs. *PLoS Pathogens* 10(1): e1003850.

Chinchar, V.G., Yu, K.H. and Jancovich, J.K. (2011) The molecular biology of *Frog Virus 3* and other iridoviruses infecting cold-blooded vertebrates. *Viruses* 3, 1959–1985.

Commission of the European Communities (2001) Commission Decision of 22 February 2001. Laying down the sampling plans and diagnostic methods for the detection and confirmation of certain fish diseases and repealing Decision 92/532/EEC. 2001/183/EC. *Official Journal of the European Communities* L67, 44, 65–76.

Crane, M.S.J., Young, J. and Williams, L. (2005) Epizootic haematopoietic necrosis virus (EHNV): growth in fish cell lines at different temperatures. *Bulletin of the European Association of Fish Pathologists* 25, 228–231.

Duffus, A.L., Waltzek, T.B., Stöhr, A.C., Allender, M.C., Gotesman, M. *et al.* (2015) Distribution and host range of ranaviruses In: Gray, M.J. and Chincahr, V.G. (eds) *Ranaviruses: Lethal Pathogens of Ectothermic Vertebrates*. Springer, Cham, Switzerland, pp. 9–57.

Eaton, B.T., Hyatt, A.D. and Hengstberger, S. (1991) Epizootic haematopoietic necrosis virus – purification and classification. *Journal of Fish Diseases* 14, 157–169.

Epstein, B. and Storfer, A. (2015) Comparative genomics of an emerging amphibian virus. *G3: Genes, Genomes, Genetics* 6, 15–27.

Gobbo, F., Cappellozza, E., Pastore, M.R. and Bovo, G. (2010) Susceptibility of black bullhead *Ameiurus melas* to a panel of ranavirus isolates. *Diseases of Aquatic Organisms* 90, 167–174.

Gray, M.J., Miller, D.L. and Hoverman, J.T. (2009) Ecology and pathology of amphibian ranaviruses. *Diseases of Aquatic Organisms* 87, 243–266.

Grayfer, L., Andino, F.D., Chen, G.C., Chinchar, G.V. and Robert, J. (2012) Immune evasion strategies of ranaviruses and innate immune responses to these emerging pathogens. *Viruses* 4, 1075–1092.

He, J.G., Deng, M., Weng, S.P., Li, Z., Zhou, S.Y. *et al.* (2001) Complete genome analysis of the mandarin fish infectious spleen and kidney necrosis iridovirus. *Virology* 291, 126–139.

He, J.G., Lu, L., Deng, M., He, H.H., Weng, S.P. *et al.* (2002) Sequence analysis of the complete genome of an iridovirus isolated from the tiger frog. *Virology* 292, 185–197.

Holopainen, R., Ohlemeyer, S., Schuetze, H., Bergmann, S.M. and Tapiovaara, H. (2009) Ranavirus phylogeny and differentiation based on major capsid protein, DNA polymerase and neurofilament triplet H1-like protein genes. *Diseases of Aquatic Organisms* 85, 81–91.

Holopainen, R., Honkanen, J., Jensen, B.B., Ariel, E. and Tapiovaara, H. (2011) Quantitation of ranaviruses in cell culture and tissue samples. *Journal of Virological Methods* 171, 225–233.

Hossain, M., Song, J.Y., Kitamura, S.I., Jung, S.J. and Oh, M.J. (2008) Phylogenetic analysis of lymphocystis disease virus from tropical ornamental fish species based on a major capsid protein gene. *Journal of Fish Diseases* 31, 473–479.

Huang, Y., Huang, X., Liu, H., Gong, J., Ouyang, Z. *et al.* (2009) Complete sequence determination of a novel reptile iridovirus isolated from soft-shelled turtle and evolutionary analysis of Iridoviridae. *BMC Genomics* 10:224.

Hyatt, A.D., Gould, A.R., Zupanovic, Z., Cunningham, A.A., Hengstberger, S. *et al.* (2000) Comparative studies of piscine and amphibian iridoviruses. *Archives of Virology* 145, 301–331.

ISO (2005) ISO/IEC 17025:2005. General requirements for the competence of testing and calibration laboratories. International Organization for Standardization, Geneva, Switzerland. Available at: http://www.iso.org/iso/catalogue_detail?csnumber=39883 (accessed 26 October 2016).

Jancovich, J.K., Mao, J., Chinchar, V.G., Wyatt, C., Case, S.T. *et al.* (2003) Genomic sequence of a ranavirus (family *Iridoviridae*) associated with salamander mortalities in North America. *Virology* 316, 90–103.

Jancovich, J.K., Davidson, E.W., Parameswaran, N., Mao, J., Chinchar, V.G. *et al.* (2005) Evidence for emergence of an amphibian iridoviral disease because of human-enhanced spread. *Molecular Ecology* 14, 213–224.

Jancovich, J.K., Bremont, M., Touchman, J.W. and Jacobs, B.L. (2010) Evidence for multiple recent host species shifts among the ranaviruses (Family *Iridoviridae*). *Journal of Virology* 84, 2636–2647.

Jancovich, J.K., Chinchar, V.G., Hyatt, A., Miyazaki, T., Williams, T. and Zhang, Q.Y. (2012) Family *Iridoviridae*. In: King, A.M.Q., Adams, M.J., Carstens, E.B. and Lefkowitz, E.J. (eds) *Virus Taxonomy: Ninth Report of the International Committee on Taxonomy of Viruses*. Elsevier/Academic Press, San Diego, California, pp. 193–210.

Jancovich, J.K., Steckler, N.K. and Waltzek, T.B. (2015) Ranavirus taxonomy and phylogeny. In: Gray, M.J. and Chincahr, V.G (eds) *Ranaviruses: Lethal Pathogens of Ectothermic Vertebrates*. Springer, Cham, Switzerland, pp. 59–70.

Jaramillo, D., Tweedie, A., Becker, J.A., Hyatt, A., Crameri, S. and Whittington, R.J. (2012) A validated quantitative polymerase chain reaction assay for the detection of ranaviruses (Family *Iridoviridae*) in fish tissue and cell cultures, using EHNV as a model. *Aquaculture* 356, 186–192.

Jensen, B.B., Ersboll, A.K. and Ariel, E. (2009) Susceptibility of pike *Esox lucius* to a panel of ranavirus isolates. *Diseases of Aquatic Organisms* 83, 169–179.

Jensen, B.B., Holopainen, R., Tapiovaara, H. and Ariel, E. (2011) Susceptibility of pike-perch *Sander lucioperca* to a panel of ranavirus isolates. *Aquaculture* 313, 24–30.

Juhasz, T., Woynarovichne, L.M., Csaba, G., Farkas, L.S. and Dan, A. (2013) Isolation of ranavirus causing mass mortality in brown bullheads (*Ameiurus nebulosus*) in Hungary. *Magyar Allatorvosok Lapja* 135, 763–768.

Kurita, J. and Nakajima, K. (2012) Megalocytiviruses. *Viruses* 4, 521–538.

La Fauce, K., Ariel, E., Munns, S., Rush, C. and Owens, L. (2012) Influence of temperature and exposure time on the infectivity of *Bohle iridovirus*, a ranavirus. *Aquaculture* 354, 64–67.

Langdon, J.S. (1989) Experimental transmission and pathogenicity of epizootic haematopoietic necrosis virus (EHNV) in redfin perch, *Perca fluviatilis* L., and 11 other teleosts. *Journal of Fish Diseases* 12, 295–310.

Langdon, J.S. and Humphrey, J.D. (1987) Epizootic haematopoietic necrosis, a new viral disease in redfin perch *Perca fluviatilis* L. in Australia. *Journal of Fish Diseases* 10, 289–298.

Langdon, J.S., Humphrey, J.D., Williams, L.M., Hyatt, A.D. and Westbury, H.A. (1986) First virus isolation from Australian fish: an iridovirus-like pathogen from redfin perch, *Perca fluviatilis* L. *Journal of Fish Diseases* 9, 263–268.

Langdon, J.S., Humphrey, J.D. and Williams, L.M. (1988) Outbreaks of an EHNV-like iridovirus in cultured rainbow trout, *Salmo gairdneri* Richardson, in Australia. *Journal of Fish Diseases* 11, 93–96.

Lei, X.Y., Ou, T., Zhu, R.L. and Zhang, Q.Y. (2012) Sequencing and analysis of the complete genome of Rana grylio virus (RGV). *Archives of Virology* 157, 1559–1564.

Leimbach, S., Schutze, H. and Bergmann, S.M. (2014) Susceptibility of European sheatfish *Silurus glanis* to a panel of ranaviruses. *Journal of Applied Ichthyology* 30, 93–101.

Mao, J.H., Green, D.E., Fellers, G. and Chinchar, V.G. (1999) Molecular characterization of iridoviruses isolated from sympatric amphibians and fish. *Virus Research* 63, 45–52.

Marsh, I.B., Whittington, R.J., O'Rourke, B., Hyatt, A.D. and Chisholm, O. (2002) Rapid differentiation of Australian, European and American ranaviruses based on variation in major capsid protein gene sequence. *Molecular and Cellular Probes* 16, 137–151.

Mavian, C., López-Bueno, A., Fernández Somalo, M.P., Alcamí, A. and Alejo, A. (2012) Complete genome sequence of the European sheatfish virus. *Journal of Virology* 86, 6365–6366.

Moody, N.J.G. and Owens, L. (1994) Experimental demonstration of the pathogenicity of a frog virus, *Bohle iridovirus*, for a fish species, barramundi *Lates calcarifer*. *Diseases of Aquatic Organisms* 18, 95–102.

Ogawa, M., Ahne, W., Fischer-Scherl, T., Hoffmann, R.W. and Schlotfeldt, H.J., (1990) Pathomorphological alterations in sheatfish fry *Silurus glanis* experimentally infected with an iridovirus-like agent. *Diseases of Aquatic Organisms* 9, 187–191.

OIE (2009) Chapter 1.1.3. Methods for disinfection of aquaculture establishments. In: *Manual of Diagnostic Tests for Aquatic Animals*, 6th edn. World Organisation for Animal Health, Paris, pp. 31–42. Available at: http://web.oie.int/eng/normes/fmanual/1.1.3_DISINFECTION.pdf (accessed 26 October 2016).

OIE (2010) Chapter 1.4. Aquatic animal health surveillance. In: *Aquatic Animal Health Code, 2010*. World Organisation for Animal Health, Paris. Available at: http://web.oie.int/eng/normes/fcode/en_chapitre_1.1.4.pdf (accessed 15 December 2015).

OIE (2015) Chapter 1.2. Criteria for listing aquatic animal diseases. In: *Aquatic Animal Health Code, 2015*. World Organisation for Animal Health, Paris. Available at: http://www.oie.int/fileadmin/Home/eng/Health_standards/aahc/2010/chapitre_criteria_diseases.pdf (accessed 26 October 2016).

OIE (2016a) Chapter 2.1.2 Infection with ranavirus. In: *Manual of Diagnostic Tests for Aquatic Animals*. World Organisation for Animal Health, Paris. Available at: http://www.oie.int/fileadmin/Home/eng/Health_standards/aahm/current/chapitre_ranavirus.pdf (accessed 26 October 2016).

OIE (2016b) Chapter 2.3.1 Epizootic haematopoietic necrosis. *In Manual of Diagnostic Tests for Aquatic Animals* (World Organisation for Animal Health) Available at: http://www.oie.int/fileadmin/Home/eng/Health_standards/aahm/current/chapitre_ehn.pdf (accessed 26 October 2016).

OIE (2016c) Chapter 10.1. Epizootic haematopoietic necrosis. In: World Organisation for Animal Health, *Aquatic Animal Health Code, 2016*. Available at: http://www.oie.int/fileadmin/Home/eng/Health_standards/aahc/current/chapitre_vhs.pdf (accessed 10 December 2015).

Pallister, J., Gould, A., Harrison, D., Hyatt, A., Jancovich, J. and Heine, H. (2007) Development of real-time PCR assays for the detection and differentiation of Australian and European ranaviruses. *Journal of Fish Diseases* 30, 427–438.

Peeler, E.J., Afonso, A., Berthe, F.C.J., Brun, E. *et al.* (2009) Epizootic haematopoietic necrosis virus – an assessment of the likelihood of introduction and establishment in England and Wales. *Preventive Veterinary Medicine* 91, 241–253.

Pozet, F., Morand, M., Moussa, A., Torhy, C. and De, K.P. (1992) Isolation and preliminary characterization of a pathogenic icosahedral deoxyribovirus from the catfish *Ictalurus melas*. *Diseases of Aquatic Organisms* 14, 35–42.

Reddacliff, L.A. and Whittington, R.J. (1996) Pathology of epizootic haematopoietic necrosis virus (EHNV) infection in rainbow trout (*Oncorhynchus mykiss* Walbaum) and redfin perch (*Perca fluviatilis* L). *Journal of Comparative Pathology* 115, 103–115.

Rimmer, A.E., Becker, J.A., Tweedie, A. and Whittington, R.J. (2012) Validation of high throughput methods for tissue disruption and nucleic acid extraction for ranaviruses (family *Iridoviridae*). *Aquaculture* 338, 23–28.

Song, W.J., Oin, Q.W., Qiu, J., Huang, C.H., Wang, F. and Hew, C.L. (2004) Functional genomics analysis of Singapore grouper iridovirus: complete sequence determination and proteomic analysis. *Journal of Virology* 78, 12576–12590.

Tamura, K., Stecher, G., Peterson, D., Filipski, A. and Kumar, S. (2013) MEGA6: Molecular Evolutionary Genetics Analysis Version 6.0. *Molecular Biology and Evolution* 30, 2725–2729.

Tan, W.G., Barkman, T.J., Chinchar, V.G. and Essani, K. (2004) Comparative genomic analyses of *Frog virus 3*, type species of the genus *Ranavirus* (family *Iridoviridae*). *Virology* 323, 70–84.

Teng, Y., Hou, Z.W., Gong, J., Liu, H., Xie, X.Y. *et al.* (2008) Whole-genome transcriptional profiles of a novel marine fish iridovirus, Singapore grouper iridovirus (SGIV) in virus-infected grouper spleen cell cultures and in orange-spotted grouper, *Epinephulus coioides*. *Virology* 377, 39–48.

Tidona, C.A. and Darai, G. (1997) The complete DNA sequence of lymphocystis disease virus. *Virology* 230, 207–216.

Tsai, C.T., Ting, J.W., Wu, M.H., Wu, M.F., Guo, I.C. and Chang, C.Y. (2005) Complete genome sequence of the grouper iridovirus and comparison of genomic organization with those of other iridoviruses. *Journal of Virology* 79, 2010–2023.

Vesely, T., Cinkova, K., Reschova, S., Gobbo, F., Ariel, E. *et al.* (2011) Investigation of ornamental fish entering the EU for the presence of ranaviruses. *Journal of Fish Diseases* 34, 159–166.

Waltzek, T.B., Miller, D.L., Gray, M.J., Drecktrah, B., Briggler, J.T. *et al.* (2014) New disease records for hatchery-reared sturgeon. I. Expansion of *Frog virus 3* host range into *Scaphirhynchus albus*. *Diseases of Aquatic Organisms* 111, 219–227.

Whittington, R.J. and Chong, R. (2007) Global trade in ornamental fish from an Australian perspective: the case for revised import risk analysis and management strategies. *Preventive Veterinary Medicine* 81, 92–116.

Whittington, R.J. and Reddacliff, G.L. (1995) Influence of environmental temperature on experimental infection of redfin perch (*Percus fluviatilis*) and rainbow trout (*Oncorhynchus mykiss*) with *Epizootic hematopoietic necrosis virus*, an Australian iridovirus. *Australian Veterinary Journal* 72, 421–424.

Whittington, R.J. and Steiner, K.A. (1993) *Epizootic hae-matopoietic necrosis virus* (EHNV): improved ELISA for detection in fish tissues and cell cultures and an efficient method for release of antigen from tissues. *Journal of Virological Methods* 43, 205–220.

Whittington, R.J., Philbey, A., Reddacliff, G.L. and Macgown, A.R. (1994) Epidemiology of epizootic haematopoietic necrosis virus (EHNV) infection in farmed rainbow trout, *Oncorhynchus mykiss* (Walbaum): findings based on virus isolation, antigen capture ELISA and serology. *Journal of Fish Diseases* 17, 205–218.

Whittington, R.J., Kearns, C., Hyatt, A.D., Hengstberger, S. and Rutzou, T. (1996) Spread of *Epizootic haematopoietic necrosis virus* (EHNV) in redfin perch (*Perca fluviatilis*) in southern Australia. *Australian Veterinary Journal* 73, 112–114.

Whittington, R.J., Reddacliff, L.A., Marsh, I., Kearns, C., Zupanovic, Z. and Callinan, R.B. (1999) Further observations on the epidemiology and spread of *Epizootic haematopoietic necrosis virus* (EHNV) in farmed rainbow trout *Oncorhynchus mykiss* in south-eastern Australia and a recommended sampling strategy for surveillance. *Diseases of Aquatic Organisms* 35, 125–130.

Whittington, R.J., Becker, J.A. and Dennis, M.M. (2010) Iridovirus infections in finfish – critical review with emphasis on ranaviruses. *Journal of Fish Diseases* 33, 95–122.

Williams, T., Barbosa-Solomieu, V. and Chinchar, V.G. (2005) A decade of advances in iridovirus research. In: Maramorosch, K. and Shatkin, A.J. (eds) *Advances in Virus Research.* Elsevier/Academic Press, San Diego, California, pp. 173–248.

P. Hick *et al.*

5

Oncogenic Viruses: *Oncorhynchus masou* Virus and Cyprinid Herpesvirus

MAMORU YOSHIMIZU,[1]* HISAE KASAI,[1] YOSHIHIRO SAKODA,[2] NATSUMI SANO[3] AND MOTOHIKO SANO[4]

[1]*Faculty of Fisheries Sciences, Hokkaido University, Minato, Hakodate, Japan;* [2]*Graduate School of Veterinary Medicine, Hokkaido University, Sapporo, Japan;* [3]*Graduate School of Bioresources, Mie University, Tsu, Japan;* [4]*Faculty of Marine Science, Tokyo University of Marine Science and Technology, Tokyo, Japan*

5.1 Introduction

Due to their distinctive appearance and obvious pathological nature, tumours of fish have been recognized for centuries. Publications on fish tumours are widely scattered in the scientific literature (Walker, 1969; Anders and Yoshimizu, 1994). The largest registry of tumours in lower animals was established at the Smithsonian Institution in Washington, DC, in 1965. Pathologies range from benign epidermal papillomas to metastatic melanomas and hepatocellular carcinomas in more than 300 species of fish.

Depending upon the season and geographic location, certain types of skin tumours may be prevalent in wild European eel (*Anguilla anguilla*), dab (*Limanda limanda*), European smelt (*Osmerus eperlanus*) (Anders, 1989) and northern pike (*Esox lucius*) from north-east Atlantic coastal areas. However, the occurrence of 'carp-pox' lesion in cultured cyprinids has decreased in importance (Anders and Yoshimizu, 1994). Tumours with a suspected viral aetiology are well documented in mammals, birds, reptiles and fish (Anders and Yoshimizu, 1994). A viral aetiology of papillomas in fish was first suggested by Keysselitz (1908). Owing to the frequent epizootic occurrence of fish tumours, an infectious viral aetiology was suggested (Winqvist *et al.*, 1968; Walker, 1969; Mulcahy and O'Leary, 1970; Anders, 1989; McAllister and Herman, 1989; Lee and Whitfield, 1992; Anders and Yoshimizu, 1994), even though viral particles were not always evident. In these cases, evidence was usually based on the exclusion of other potential causative factors. In about 50% of all cases where electron microscopy and virological methods have been applied, viruses or virus-like particles were identified in tumour tissue. Benign tumours such as epidermal hyperplasia, papilloma and fibroma, were mostly caused by herpesviruses infection, and much less related to retroviruses, papovaviruses or adenoviruses. Moreover, retroviruses supposed to cause malignant forms such as sarcomas and lymphosarcomas. Typically, the tumours had just one viral type, in rare cases, different skin tumour types associated with different viruses occurred in the same specimen (Anders and Yoshimizu, 1994).

The significance of these viruses and virus-like particles for tumour induction is mostly speculative and will not be included in the present discussion. Oncogenicity has been clearly demonstrated only with herpesviruses from masu salmon (*Onchorhynchus masou*) (see Section 5.2) and the Japanese Asagi variety of koi carp (itself a domesticated variety of the common carp, *Cyprinus carpio*) (Section 5.3). Although in certain other cases tumour formation was induced experimentally via inoculation of

---

*Corresponding author e-mail: yosimizu@fish.hokudai.ac.jp

cell-free filtrates and/or live tumour cells, attempts to isolate viruses from tumours in cell culture were unsuccessful (Peters and Waterman, 1979). For the two herpesviruses, pathogenicity and oncogenicity have been clearly verified by successful isolation of the causative virus in cell culture and this fulfils River's postulates, but nothing is known about possible oncogenes in these viruses.

5.2 *Oncorhynchus masou* Virus

5.2.1 Introduction

Oncorhynchus masou virus disease (OMVD) is an oncogenic and skin ulcerative condition coupled with hepatitis that occurs among salmonid fishes in Japan. It is caused by *Salmonid herpesvirus 2* (SalHV-2) and was first described from *Oncorhynchus masou* (Kimura *et al.*, 1981a,b). The virus is more commonly known as *Oncorhynchus masou* virus (OMV), but other synonyms include the nerka virus Towada Lake, Akita and Amori Prefecture

(NeVTA; Sano 1976), yamame tumor virus (YTV; Sano *et al.*, 1983), *Oncorhynchus kisutch* virus (OKV; Horiuchi *et al.*, 1989), coho salmon tumor virus (COTV; Yoshimizu *et al.*, 1995), coho salmon herpesvirus (CHV; Kumagai *et al.*, 1994), rainbow trout (*O. mykiss*) kidney virus (RKV; Suzuki 1993) and rainbow trout herpesvirus (RHV; Yoshimizu *et al.*, 1995). SalHV-2 is a recognized species in the genus *Salmonivirus*, family *Alloherpesviridae*.

5.2.2 The disease agent

Biophysical and biochemical properties

At or near 15°C, cells infected with OMV show a distinctive cytopathic effect (CPE) within 5 to 7 days. This CPE is characterized by rounded cells followed by syncytium formation, and eventual lysis of RTG (rainbow trout gonad)-2 and other salmonid cell lines (Fig. 5.1A). Cells from non-salmonid species are refractory to infection (Yoshimizu *et al.*, 1988b). The maximum infectious

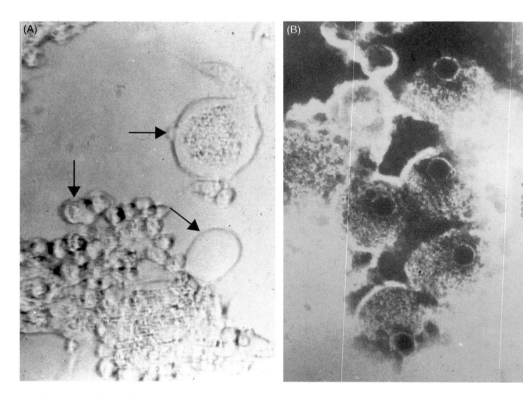

Fig. 5.1. Cytopathic effects of infection by *Oncorhyncus masou* virus (OMV), showing the presence of rounded cells followed by syncytium formation (arrows) produced by OMV in RTG-2 cells incubated at 15°C for 9 days; (B) electron micrograph of negatively stained enveloped virions provided by Dr T. Sano (Yoshimizu and Kasai, 2011).

M. Yoshimizu *et al.*

titre of culture-grown virus is about 10^6 TCID$_{50}$ (50% tissue culture infectious dose)/ml, with some variations depending on the cell line. OMV is heat-, ether- and acid (pH 3)-labile and does not haemagglutinate salmonid blood cells or human O-cells. It is inactivated by ultraviolet (UV) irradiation of 3.0×10^3 µW/s/cm^2 (mJ). Replication is inhibited by 50 µg/ml of the pyrimidine analogue, 5-iodo-2′-deoxyuridine (IUdR) and by anti-herpesvirus agents such as phosphonoacetate (PA), acyclovir (ACV; 9-(2-hydroxyethoxymethyl) guanine), (E)-5-(2-bromovinyl)-2′-deoxyuridine (BVdU), and 1-β-D-arabino-furanosylcytosine (Ara-C) (Kimura et al., 1981a, 1983b,c; Suzuki et al., 1992).

Electron microscopy of infected cells reveals that the intranuclear hexagonal capsids are 115 nm in diameter. An abundance of budding and enveloped virions that are 200 × 240 nm in diameter (Fig. 5.1B) are evident at the surface of and inside cytoplasmic vesicles. The calculated number of capsomeres of negatively stained virions is 162. The optimal growth temperature is 15°C with replication at 18°C and no growth at 20°C or higher. This psychrophilic nature of OMV differs from the temperature sensitivity of channel catfish herpesvirus and other amphibian herpesviruses (Kimura et al., 1981a).

Serological relationships

OMV is neutralized by homologous antiserum but not by antisera prepared against other salmonid viruses

(e.g. IPNV, infectious pancreatic necrosis virus; IHNV, infectious hematopoietic necrosis virus; CSV, chum salmon virus; and Herpesvirus salmonis). All 177 OMV strains isolated from ovarian fluid or tumour tissue of mature masu salmon in hatcheries located in northern Japan (collected from 1978 to 1986) were neutralized by anti-OMV OO-7812 serum, a rabbit antiserum against the reference OMV strain. The ND$_{50}$ (50% neutralization dose) ranged from 1:40 to 1:80 (Yoshimizu et al., 1988b). Eleven herpesvirus strains were compared serologically using serum cross-neutralization tests with polyclonal rabbit antisera: NeVTA from kokanee salmon (landlocked sockeye salmon, O. nerka); three strains of OMV and YTV from masu salmon; CSTV (coho salmon tumor virus), COTV and two strains of OKV from coho salmon (O. kisutch); RKV and RHV from rainbow trout; and H. salmonis. The herpesvirus strains in Japan were neutralized by antisera against these viruses and were closely related to Salmonid herpesvirus 2, reference strain OMV OO-7812 (Table 5.1). These strains, however, were clearly distinguished from H. salmonis (Salmonid herpesvirus 1) and OMV is designated as Salmonid herpesvirus 2 (Yoshimizu et al., 1995, after Hedrick et al., 1987).

Viral protein and genome

The general properties of OMV are similar to those of H. salmonis, Salmonid herpesvirus 1, although

Table 5.1. Serological relationship of herpesvirus strains from salmonid fishes using the 1/r (relatedness) value[a] based on serum cross-neutralization tests.

| Species | Virus[b] | Antiserum[b] | | | | | | | | |
| | | OMV | | | | | | | | |
		I	II	III	YTV	NeVTA	COTV	OKV (M)	RKV	HS
Masu salmon	OMV I	1.00	1.30	0.92	1.42	1.12	1.53	0.85	1.22	>3.16
	OMV II		1.00	0.80	1.00	1.22	1.47	0.85	0.67	>3.47
	OMV III			1.00	1.21	0.94	1.41	1.00	0.83	>3.47
	YTV				1.00	0.84	1.33	0.70	1.19	>5.69
Kokanee salmon	NeVTA					1.00	1.39	0.83	0.89	>4.89
Coho salmon	COTV						1.00	0.96	0.51	>3.16
	OKV (M)							1.00	0.83	>3.47
Rainbow trout	RKV								1.00	>3.47
	HS									1.00

[a]A value of 1 indicates serological identity and greater or lesser values indicate increasing differences (1/r: Archetti and Horsfall, 1950).
[b]Key: COTV, coho salmon tumor virus; HS, Herpesvirus salmonis; NeVTA, nerka virus Towada Lake, Akita and Amori Prefecture; OKV (M), Oncorhynchus kisutch virus (M); OMV, Oncorhynchus masou virus; RKV, rainbow trout kidney virus; YTV, yamame tumor virus.

OMV differs in virion size and in optimal growth temperature. Furthermore, OMV is different from other known fish herpesviruses with respect to the viral-induced polypeptide patterns; 34 polypeptides appear in OMV-infected cells that are virus specific. These polypeptides have molecular weights between 19,000 and 227,000. By contrast, *H. salmonis* induces 25 polypeptides with molecular weights between 19,500 and 250,000 (Kimura and Yoshimizu, 1989). CCV (channel catfish virus) induces 32 polypeptides (Dixon and Farber, 1980), which are distinct from those of OMV. Differences in the electrophoretic migration of two of 34 OMV-specific polypeptides led to the classification of 12 OMV strains into six groups (Kimura and Yoshimizu, 1989).

Restriction endonuclease cleavage patterns of OMV DNAs are different from those of *H. salmonis*. Seven representative OMV strains from ovarian fluids and tumour tissues of wild masu salmon in Hokkaido and Aomori prefectures were analysed with restriction endonuclease. The restriction patterns of OMV strain DNAs were divided into four groups. The restriction profiles of high-passage strains were different from those of low-passage strains when digested with *Bam*HI, *Hin*dIII and *Sma*I. However, no differences were observed between the high- and low-passage viral DNA with *Eco*RI (Hayashi *et al.*, 1987). By using ^{32}P-labelled DNA from standard OMV (strain OO-7812) as a probe, most fragments of other OMV DNAs were hybridized (Gou *et al.*, 1991). From the results of the DNA homologies, OMV and YTV were considered to be the same virus, while NeVTA was similar but distinct (Eaton *et al.*, 1991).

The genome sequence of OMV strain OO-7812 was determined (Yoshimizu *et al.*, 2012) and phylogenetic relationships among fish and amphibian herpesviruses were predicted using amino acid sequences from parts of the DNA polymerase and terminase genes. Strain OO-7812 showed 100% identity to strains YTV and NeVTA, and OMV is clearly distinguished from SalHV-1 and SalHV-3 (Fig. 5.2).

DNA polymerase activities were determined in both tumour and normal tissues of masu salmon. High DNA polymerase α-activity was detected in OMV-infected tumour tissues but not in normal tissues, which indicates that the OMV DNA in tumour cells was replicating well. DNA polymerase activity was the same in both tumour and normal tissues. This is the first evidence of the detection of herpesvirus DNA polymerase in tumour tissue in association with herpesvirus (Suzuki *et al.*, 1992).

Survivability and immunity

A significant reduction in the infectious titre of OMV occurred within 3 and 7 days in water at 15°C and 10°C, respectively. However, infectivity remained for 7–14 days when the water temperature was below 5°C (Yoshimizu *et al.*, 2005) because the activity of bacteria producing antiviral substance(s) in the water decreased at low temperatures (Yoshimizu *et al.*, 2014).

5.2.3 Geographical distribution

In the early 1960s, eggs of masu salmon were collected from the rivers around Sea of Japan coast of Hokkaido, and transported to Honshu. With unrestricted fish movements, the virus spread to Gifu, Yamanashi and Niigata Prefectures in Honshu where the first cancerous disease, basal cell carcinoma of masu salmon, was detected (Kimura, 1976). High mortality of kokanee salmon fry occurred in 1972 and 1974, and virus was isolated from moribund fish The virus was classified as a member of the *Herpesviridae* and was named the nerka virus in Towada Lake, Akita and Aomori Prefecture (NeVTA) (Sano, 1976). In 1978, a similar herpesvirus was isolated from the ovarian fluid of masu salmon cultured in the Otobe Salmon Hatchery on the Sea of Japan coast in Hokkaido and named *Oncorhynchus masou* virus (OMV) after the host fish (Kimura *et al.*, 1981a,b). OMV showed pathogenicity and oncogenicity towards salmonid fish, masu salmon, chum salmon (*O. keta*), coho salmon and rainbow trout.

From 1978 to 2015, six species of mature salmonid fish, masu salmon, chum salmon, pink salmon (*O. gorbuscha*), kokanee salmon, sockeye salmon (*O. nerka*) and rainbow trout (46,788 females) were collected and surveyed for viral infections in Hokkaido and in Aomori and Iwate prefectures in northern Honshu (Yoshimizu *et al.*, 1993; Kasai *et al.*, 2004). Herpesvirus was isolated from masu salmon at 13 hatcheries, excepting one hatchery where 60 specimens could not collected. All of the isolates were neutralized with anti-OMV rabbit serum (Yoshimizu *et al.*, 1993).

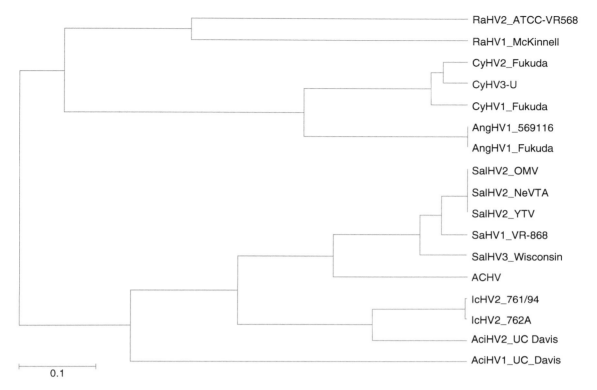

Fig. 5.2. Phylogram depicting the relationship among fish and amphibian herpesviruses based on the concatenated partial deduced amino acid (AA) sequences of the DNA polymerase and terminase genes (247 AA characters including gaps). Branch lengths are based on the number of inferred substitutions, as indicated by the scale bar. Key (in order downwards): RaHV, *Ranid herpesvirus* (frog); CyHV, *Cyprinid herpesvirus* (carp and gold fish); AngHV, *Anguillid herpesvirus* (eel); SalHV, *Salmonid herpesvirus*; ACHV, *Atlantic cod herpesvirus*; IcHV, *Ichalurus herpesvirus* (channel catfish); AcHV, *Acipenserid herpesvirus* (white sturgeon).

In 1981, a similar herpesvirus was isolated from the tissues of a basal cell carcinoma on the mouth of yamame (another name for masu salmon) cultured at Koide Branch, Niigata Prefectural Inland Fisheries Experimental Station in Honshu. This virus was named yamame tumour virus (YTV, Sano *et al.*, 1983). Since 1988, herpesvirus has been isolated from pond and pen-cultured coho salmon in Miyagi Prefecture, Tohoku district, Honshu (Kimura and Yoshimizu, 1989). Also, since 1992, a herpesvirus has been isolated from cultured rainbow trout in Hokkaido; this virus was tentatively named rainbow trout kidney herpesvirus (RKV) by Suzuki (1993). From 2000 to 2001, epizootics occurred in rainbow trout weighing 12–1.5 kg at 18 fish farms in Nagano, Shizuoka and Gifu prefectures in central Japan (Honshu). All of the viruses isolated from diseased fish were identified as OMV.

5.2.4 Economic importance of the disease

Many strains of herpesvirus identified as OMV have been isolated from ovarian fluid and tumours of wild and cultured masu salmon in northern Japan (Yoshimizu *et al.*, 1993). Mortality among kokanee salmon ranges from 80 to 100%, while there are no reports of mortality in masu salmon, but the presence of tumours reduces their commercial value. OMVD has been a major problem in the pen culture of coho salmon in the Tohoku district since 1988 and coho salmon culture has been economically damaged by this disease. OMV was found in 1992 pond cultures of rainbow trout in Hokkaido and it has occurred in central Japan since 2000. OMVD is currently regulated and controlled in kokanee salmon, masu salmon and coho salmon (see Section 5.2.7). OMVD outbreaks in rainbow trout remained a major problem in production farms until 2005, as demonstrated by the high proportion of low-weight dead fish found in an outbreak in 18 fish farms in Nagano Prefecture in 2000–2001 (see Fig. 5.3D; Furihata *et al.*, 2003).

5.2.5 Diagnosis

The infectivity of the virus remains unchanged for 2 weeks at 0–5°C, but 99.9% of the infectivity is lost within 17 days at −20°C. Viral isolation should be from fish transported on ice to the laboratory (Yoshimizu *et al.*, 2005). For the filtration of OMV, a 0.40 μm nucleopore filter (polycarbonate) is recommended because cellulose acetate membranes trap >99% of the viral particles. For virological surveys, ovarian fluid is collected as described by Yoshimizu *et al.* (1985), diluted with the same volume of antibiotic (Amos, 1985) and incubated at 5°C, overnight. In the case of tumours, a sample of tissue is excised, disinfected with iodophor (50 mg/l, 15 min) washed with Hank's BSS (balanced salt solution) and transported in an antibiotic solution to the laboratory. Tumour tissue is prepared for primary culture or co-culture with RTG-2 cells. After one subculture in primary cells, the culture medium should be inspected for virus (Yoshimizu, 2003b).

Rabbit antiserum or monoclonal antibody against OMV is used in a fluorescent antibody test (Hayashi *et al.*, 1993) and a DNA probe is used to detect viral genome (Gou *et al.*, 1991). PCR using an F10 primer, GTACCGAAACTCCGAGTC, and R05 primer, AACTTGAACTACTCCGGGG, amplified a 439 base pair segment of DNA from OMV strains from the liver, kidney, brain and nervous tissues of masu salmon, coho salmon and rainbow trout. The size of the amplified DNA of OMV and *H. salmonis* is different, so OMV and *H. salmonis* can be distinguished by their agarose gel profiles (Aso *et al.*, 2001).

Presumptive diagnosis

Certain key features such as life cycle, stage, species and stock of fish, as well as water temperature, clinical signs (see Section 5.2.6), and disease history of the facility are evaluated. To isolate OMV, tissue and reproductive fluids are used in standard cell culture techniques. Processed specimens must be inoculated on to RTG-2 or chinook salmon embryo cells (CHSE-214). Cytopathic effects include rounded cells and the formation of giant syncytium. Plaque assay procedures (Kamei *et al.*, 1987), which use a methylcellulose overlay, are also used to isolate and enumerate OMVD.

Confirmatory diagnosis

Confirmation of OMV is accomplished using neutralization tests with either a polyclonal rabbit antisera or monoclonal antibody An antigen detection ELISA (Yoshimizu, 2003a), and a fluorescent antibody technique (FAT) for OMV have been developed using either polyclonal or monoclonal antiserum (Hayashi *et al.*, 1993). The FAT is specific, reacts with all isolates of OMVD and requires less time for confirmatory diagnosis. PCR can also be used to confirm OMV grown in cell cultures (Aso *et al.*, 2001).

Procedures for detecting subclinical infection

The detection of OMV in carrier fish is difficult, but the virus replicates and appears in ovarian fluid at spawning. Antibody detection by a neutralization test or ELISA (Yoshimizu, 2003a) is available for epizootiological studies.

5.2.6 Pathology
Pathogenicity and host susceptibility

The susceptibility of salmonid fry to OMV was studied via immersion in water containing 100 TCID$_{50}$/ml OMV at 10°C for 1 h (Kimura *et al.*, 1983a). Kokanee salmon (1 month old) were most sensitive (100% mortality). Masu and chum salmon were also sensitive (87 and 83% mortality, respectively) while coho salmon and rainbow trout were the least sensitive (39 and 29% mortality, respectively). The cumulative mortality of just hatchling chum salmon was 35%, but it was more than 80% in 1–5 month old fry, and in 3-month-old fry it was 98%. At 6 and 7 months old, susceptibility was reduced to 7 and 2%, respectively. There was no mortality among 8-month-old fry that were immersed in virus and then injected intraperitoneally with 200 TCID$_{50}$/fish. By contrast, 1-month-old masu salmon fry were most sensitive and sustained a cumulative mortality of 87%, whereas 3- and 5-month-old masu salmon fry had cumulative mortalities of 65 and 24%, respectively.

Clinical signs and transmission

From June to September of every year since 1970, high mortality (about 80%) of kokanee salmon fry has occurred in Japan. Clinical signs in fish include a darkened body colour, sluggish behaviour and inappetance. Syncytium-forming virus was isolated from moribund fish in RTG-2 cells incubated at 10°C in 1972 and 1974. The virus was named the nerka virus in Towada Lake, Akita and Aomori Prefecture (NeVTA) (Sano, 1976).

In 1978, a herpesvirus was isolated from the ovarian fluid of an apparently healthy mature masu salmon, cultured in the Otobe salmon hatchery in Hokkaido. This virus was named *Oncorhynchus masou* virus (Kimura *et al.*, 1981a, 1981b). The general properties of OMV are similar to those of *H. salmonis* and NeVTA, but it differs in virion size and its optimal growth temperature. It is also distinct from *H. salmonis* with respect to its viral-induced polypeptide patterns, serological properties and in PCR (Kimura and Yoshimizu, 1989; Aso *et al.*, 2001). OMV is pathogenic and more significantly, it is oncogenic in masu salmon and several other salmonids (Kimura *et al.*, 1981a,b). One-month-old kokanee salmon are most sensitive to the virus. Masu and chum salmon are also highly susceptible to OMV infection, whereas coho salmon and rainbow trout are less susceptible (Tanaka *et al.*, 1984). The incidence of tumour-bearing fish approached more than 60%. There were epithelial tumours on 12–100 % of surviving chum, coho and masu salmon, and on rainbow trout beginning at 4 months and persisting for at least a year postinfection (Yoshimizu *et al.*, 1987).

Since 1988, herpesvirus had been isolated from the liver, kidney, and developing neoplasia in pond and pen-cultured coho salmon in Miyagi Prefecture (Kimura and Yoshimizu, 1989). Disease signs included white spots on livers, ulcers on the skin and/or neoplasia around the mouth or body surface. The herpesviruses isolated from coho salmon were tentatively named as coho salmon tumor virus (CSTV) by Igari *et al.* (1991), *O. kisutch* virus (OKV) by Horiuchi *et al.* (1989), coho salmon tumor virus (COTV) by Yoshimizu *et al.* (1995) and coho salmon herpesvirus (CHV) by Kumagai *et al.* (1994). These viruses were neutralized by anti-OMV or anti-NeVTA rabbit sera (Yoshimizu *et al.*, 1995), and the oncogenicity of CSTV, OKV and COTV was confirmed experimentally. In addition, the restriction endonuclease profiles of CSTV were similar to those of NeVTA and YTV (Igari *et al.*, 1991). CHV is highly pathogenic to coho salmon.

Massive mortalities, ranging from 13 to 78%, have occurred among cultured 1-year-old rainbow trout in Hokkaido since 1992. Diseased fish exhibited hardly any external clinical signs, although some fish had ulcerative lesions on their skin. Internally, intestinal haemorrhage and white spots on the liver were observed. No bacteria, fungi or parasites were found and a herpesvirus was isolated from the kidney, liver and skin ulcers. The herpesvirus was tentatively named rainbow trout kidney herpesvirus (RKV) by Suzuki (1993). RKV is highly pathogenic to marketable-size rainbow trout and masu salmon (Sung *et al.*, 1996a,b). Epizootics occurred in rainbow trout weighing 1.2–1.5 kg at 18 fish farms in Nagano Prefecture from February 2000 to January 2001. A virus was isolated from diseased fish in RTG-2 cells with CPE syncytia. High infectivity titres (about 10^8 $TCID_{50}$/g) were demonstrated in the internal organs and multiple necrotic foci were observed in the liver (Fig. 5.3A,B,C). The virus was identified as OMV using serological tests and PCR. Based on these results, the epizootic was diagnosed as OMVD. In more than 80% of these cases, outbreaks were linked to introductions of live fish. (Furihata *et al.*, 2003, 2004).

The horizontal transmission of OMV was accomplished via cohabitation in specific pathogen-free 5-month-old chum salmon fry. The resulting mortality was similar to the results from the immersion infection of 3 to 7-month-old fry (see above). Clinical signs in infected fish include inappetence, exophthalmia (Fig. 5.4A) and petechiae on the body surface (Fig. 5.4B), especially beneath the lower jaw. Internally, the liver has white lesions (Fig. 5.4C and D), and in advanced cases the whole liver becomes pearly white. In some fish, the spleen may be swollen, and the intestine is devoid of food (Kimura *et al.*, 1981a, 1983a).

Histopathology

The kidney of OMV-infected 1- and 3-month old masu salmon, 1-month-old coho salmon and 2-month-old chum salmon is the principal target organ for the virus. Necrosis of epithelial cells and kidney were observed in the early moribund fry, while partial necrosis of the liver, spleen and pancreas was recognized later in moribund 1-month-old masu salmon, with necrosis of the kidney haematopoietic tissue in 3-month-old masu salmon. It was suggested the principal target organ had

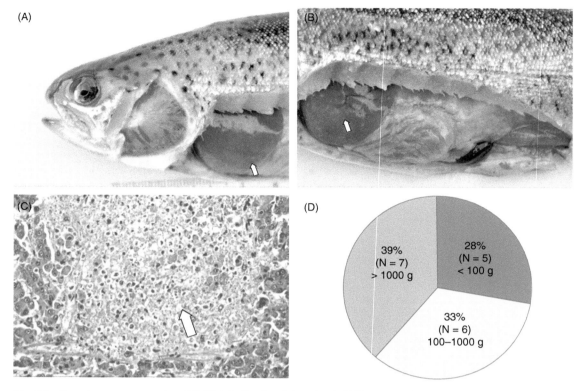

Fig. 5.3. *Oncorhyncus masou* virus (OMV) infection in rainbow trout: (A,B) arrows show necrotic liver symptoms in whole fish, in (B) with white spot lesions; (C) multiple foci of severe necrosis in the liver (arrow); (D) chart showing body weight of dead rainbow trout. Photos and figure were provided by Dr M. Furihata (see Furihata *et al.*, *Fish Pathology*, 2003).

moved from the kidney to the liver, and marked histopathological changes were observed in the later stages of the disease. The foci of necrosis in the liver became more severe with longer incubation periods. Hepatocytes showing margination of chromatin were present. Cellular degeneration in the spleen, pancreas, cardiac muscle and brain was also observed (Yoshimizu *et al.*, 1988a).

Histopathological changes in rainbow trout, coho salmon and chum salmon were similar to those in masu salmon (Kumagai *et al.*, 1994; Furihata *et al.*, 2004). In rainbow trout, high infectivity titres were measured in the internal organs and multiple necrotic foci were evident in the liver. The definitive changes were necrosis of OMV-infected cells in the spleen and haematopoietic tissues in the kidney, liver, intestine, heart, gill filaments, epidermis and lateral musculature. In particular, the intestine showed severe necrosis and haemorrhage in the epithelium and underlying

tissues of rainbow trout with OMVD (Furihata *et al.*, 2004).

Tumour induction

Tumour formation could be induced experimentally via waterborne infection of OMV. At about 4 months postinfection and persisting for at least 1 year, 12–100% of surviving masu, chum and coho salmon, as well as rainbow trout, developed oral epithelial tumours (Fig. 5.5A,B; Kimura *et al.*, 1981b). Histopathologically, tumours were composed of proliferative, well-differentiated epithelial cells supported by fine connective tissue stroma. OMV was recovered from the culture medium of one passage of the transplanted tumour cells in primary cultures (Yoshimizu *et al.*, 1987). The perioral site was the most frequent area for tumour development. As control fish held under the same conditions showed no tumours, OMV is presumed

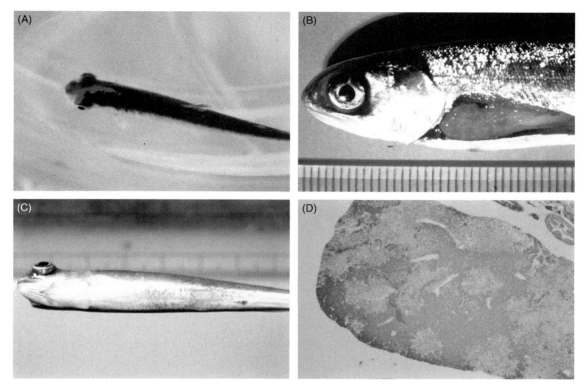

Fig. 5.4. Three-month-old chum salmon fry exposed to *Oncorhyncus masou* virus (OMV). (A) Exophthalmia; (B) white spot lesions on the liver; (C) petechiae in the body surface; and (D) multiple foci of severe necrosis in a liver section (haematoxylin and eosin (H&E) stain). The photographs and figure were provided by Dr T. Kimura (see Kimura *et al.*, Fish Pathology, 1981a).

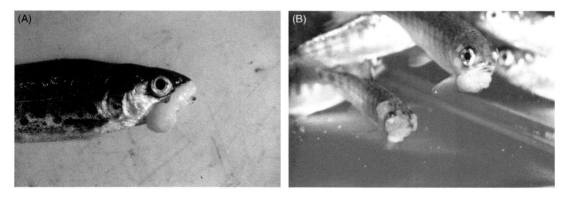

Fig. 5.5. Experimentally induced tumours caused by *Oncorhyncus masou* virus (OMV) in (A) masu salmon and (B) coho salmon. Photos were provided by M. Yoshimizu.

to have caused tumour development. This neoplasia may persist for one year postinfection.

The tumour cells appear to be of epithelial origin. There are several layers of epithelial cells in a papillomatous array and supported by fine connective tissue stroma. Abundant mitotic figures suggested a highly proliferative nature. Tumours appearing on the caudal fin, operculum, body surface, corneas of the eye and kidney had similar characteristics to those of the mouth (Kimura *et al.*, 1981b,c). Tumours in the kidney, as well as those from other sites, such as on the eye, under the operculum and the caudal fin, had similar histopathological lesions.

Electron micrographs revealed that tumour cells had the typical neoplastic feature of variable nuclear size and a loose intracellular connection. However, OMV particles were not found in the nuclei or in the cytoplasm of the tumour cells (Kimura *et al.*, 1981b,c; Yoshimizu *et al.*, 1987). The virus was isolated from eroded tumour tissue in one fish 9 months after infection, and from a primary culture of tumour cells from another fish 10 months after infection. The primary cultures exhibited continuous growth for 4 days, followed by CPE-like changes. At this time, OMV was isolated from the culture medium. Neutralizing antibody against OMV was detected in individual fish and in pooled sera from tumour-bearing fish.

5.2.7 Protective control strategies

Epizootiology

In the 1980s, OMV was distributed widely among masu salmon in northern Japan. In 1988, OMVD was diagnosed in net-pen cultured coho salmon in the marine environment of Tohoku district, central Japan. The virus in coho salmon was successfully controlled by selecting moribund fish just after transportation from fresh water to seawater, and testing their sera using FAT. If the fish were positive, farmers would disinfect the hatchery pond and equipment before eyed-eggs were transplanted in the following year (Kumagai *et al.*, 1994). OMVD was found in rainbow trout in 1991 in Hokkaido, and from 2000, it had become a major problem in pond culture on the mainland. Natural and experimental infections indicated that fish between 1 and 5 months old are the most susceptible. From 2000 to 2005, epizootics were reported in juvenile, yearling and mature rainbow trout (Furihata *et al.*, 2003), with morality exceeding 80%. Most OMVD occurs in fresh water at 15°C or lower.

Virucidal effects of disinfectants and antiviral chemotherapy

The virucidal effects of six disinfectants were examined against OMV (Hatori *et al.*, 2003). At 15°C for 20 min, the minimum concentrations showing 100% plaque reduction of OMV by iodophore, and by solutions of sodium hypochlorite, benzalkonium chloride, saponated cresol, formaldehyde and potassium permanganate were, 50, 100, 100, 3500 and 16 ppm, respectively.

ACV had high efficacy against the OMV, *Herpesvirus salmonis* and CCV. CPEs induced by 100 $TCID_{50}$/ml OMV in RTG-2 cells was inhibited by 2.5 μg/ml ACV. ACV was more effective than other compounds such as 9-β-D-arabinofuranosyladenine (Ara-A), IUdR and PA. The growth of RTG-2 cells was considerably inhibited by ACV at 25 μg/ml, but no morphological changes were observed in the cells. The replication of OMV in RTG-2 cells inoculated with 100 $TCID_{50}$/ml was completely suppressed by 2.5 μg/ml of ACV. The addition of ACV within 4 days postinfection reduced OMV replication. In order to be effective, ACV must be present continuously (Kimura *et al.*, 1983a).

The therapeutic efficacy of ACV was evaluated using OMV and chum salmon fry. Experimentally infected fish were treated with ACV either orally or by immersion. Daily immersion of fish into ACV solution (25 μg/ml, 30 min/day, 15 times) reduced the mortality of the infected fish. Oral administration of the drug (25 μg/fish a day, 60 times) did not affect the survival of chum salmon. In contrast, the group of chum salmon administered IUdR orally showed a higher survival than the ACV-administered group. This study suggests that an effective level of ACV was not maintained in fish that were medicated orally. The daily immersion of infected fish into ACV solution (25 μg/ml 30 min/day, 60 times) considerably suppressed the development of tumours induced by OMV (Kimura *et al.*, 1983b).

Vaccination

Mature rainbow trout vaccinated with formalin-inactivated OMV had anti-OMV IgM that showed neutralization activity in the brood and could reduce the ratio of OMV isolation from ovarian fluid (Yoshimizu and Kasai, 2011). Vaccination using formalin-inactivated OMV is also very effective in protecting from OMV infection at the fry stage (Furihata, 2008). Unfortunately, a commercial vaccine for OMVD is not available in Japan.

Control strategy

General sanitation is practised in hatcheries to control and prevent fish diseases (Yoshimizu, 2003b).

Special care should be taken to avoid the movement of equipment from one pond to another and all equipment should be routinely disinfected after use. Methods to sanitize a hatching unit should be carefully developed with respect to chemical toxicity for fish. Workers might be responsible for transferring pathogens; consequently, the proper disinfection of hands and boots is required to prevent the dissemination of the virus. Although it may be difficult to sanitize hatching and rearing units during use, raceways and ponds should routinely be disinfected with chlorine (Yoshimizu, 2009). Pathogen-free water supplies are often essential for success in aquaculture. The water that is commonly used in hatcheries, which comes from rivers or lakes, may contain fish pathogens. Such open water supplies should not be used without filtration and treatment to eliminate and kill fish pathogens. Fish viruses are divided into two groups based on sensitivity to UV irradiation. The sensitive viruses include OMV, IHNV, LCDV (Lymphocystis disease virus) and HIRRV (*Hirame rhabdovirus*), which are inactivated by UV at 10^4 μW s/cm (Yoshimizu et al., 1986).

Effective monitoring and management is very important during the collection of eggs at spawning. As some viruses are transmitted vertically from the adult to progeny via contaminated eggs or sperm, disinfection of the surface of fertilized eggs and eyed eggs can break the infection cycle for herpesvirus and rhabdovirus (Yoshimizu *et al.*, 1989; Yoshimizu, 2009). Health inspections of mature fish are conducted to ensure that fish are free from certifiable pathogens. Routine inspections and specialized diagnostic techniques are required to ensure specific pathogen-free brood stock. For salmonid fishes, ovarian fluid is collected by the method of Yoshimizu *et al.* (1985) and routinely inspected following cell culture. Fertilized eggs are disinfected with iodophore at 25 mg/l for 20 min or 50 mg/l for 15 min. The region inside the egg membrane of eyed eggs is considered pathogen free (Yoshimizu *et al.*, 1989, 2002). Since 1983, iodophore treatment to disinfect eggs in all hatcheries in Hokkaido has helped to eliminate outbreaks of OMV (Yoshimizu *et al.*, 1993; Kasai *et al.*, 2004; Yoshimizu, 2009).

Fry with abnormal swimming or disease signs should be removed immediately and brought to the laboratory for analysis. Moreover, health monitoring should be done regularly using cell culture isolation, FAT, the immunoperoxidase technique

(stain) (IPT), antigen-detecting ELISA and PCR tests (Yoshimizu *et al.*, 2005).

Breeding

Tetraploid female rainbow trout and sex-reversed diploid male brown trout (*Salmo trutta*) were crossed in Nagano Prefectural Fisheries Experimental Station to obtain triploid salmon (Kohara and Denda, 2008), which was named 'Shinsyu Salmon'. These triploid salmon display fast growth, good survival rates during the spawning season and are resistant to infection by both OMV and IHNV.

5.3 Cyprinid herpesvirus 1 (CyHV-1)

5.3.1 Introduction

Whitish or pinkish papillomatous lesions are sometimes seen on the skin and fin of carp, and these have been recognized since the Middle Ages in Europe as 'carp pox'. This condition has a worldwide distribution. The prevalence of these tumours may differ among strains or varieties of carp. A report by Calle *et al.* (1999) found a higher prevalence of tumours in carp with no scales, although this report constitutes the only evidence of this. The tumour tissue is organized with well differentiated cells and epidermal peg interdigitates with the papillae of the dermal connective tissue and capillary vessels (Sano *et al.*, 1985a).

5.3.2 The disease agent

The disease agent of carp pox was originally named *Herpesvirus cyprini* (CHV). It is a cyprinid herpesvirus that was first isolated in fathead minnow (FHM) cells from the papillomatous tissue of the Asagi variety of koi carp (*C. carpio*) reared in Japan (Sano *et al.*, 1985b), and it is currently classified as the species *Cyprinid herpesvirus 1* (CyHV-1) in the genus *Cyprinivirus*, family *Alloherpesviridae*, order *Herpesvirales*. The enveloped virion ranges from 153 to 234 nm in diameter with a capsid 93–126 nm in diameter (Sano *et al.*, 1985a). The virus grows in FHM cells at 10–25°C, but not at 30°C, and maximum infectious titres of 10^4–10^5 $TCID_{50}$/ml were recorded at 15 and 20°C (Sano *et al.*, 1985a, 1993a). Other cell lines derived from carp and cyprinid fishes such as koi fin (KF-1), common carp (*C. carpio*) brain (CCB) and epithelioma papulosum cyprini (EPC) cell lines are susceptible to the

---

Oncogenic Viruses

virus (Sano *et al.*, 1985a; Adkison *et al.*, 2005). CPE is characterized by the formation of cytoplasmic vacuolation and also of Cowdry type A intranuclear inclusion bodies. The sequence of the whole viral genome, which is a linear double-stranded DNA of 291 kbp has been deposited in GenBank as JQ815363 (Davison *et al.*, 2013). Restriction endonuclease cleavage profiles of isolates from Japan showed minor differences, suggesting that there is variation of the genome DNA sequence of the virus (Sano *et al.*, 1991a).

5.3.3 Diagnosis

Whitish or pinkish papillomatous lesions on the fin, skin or mandible of common carp or koi might be suspected as viral papilloma caused by CyHV-1. Isolation of the virus from the papillomatous tissue in cell culture using cell lines such as FHM or KF-1 is difficult because the appearance of the infectious virus depends on the developmental stage of the papilloma. The indirect immunofluorescent antibody test (IFAT) using specific rabbit antiserum can detect viral antigens in tissues (Sano *et al.*, 1991b), and *in situ* hybridization using a DNA probe has also been developed (Sano *et al.*, 1993b). The whole genome sequence of the virus is known, and some of the genes, including DNA polymerase, have been annotated (Davison *et al.*, 2013), so that specific primers for PCR detection can be designed after alignment of the sequences of related cyprinid herpesvirus, including CyHV-3 and CyHV-2, which cause mortality of the goldfish *Carassius auratus* and gibelio carp *C. auratus gibelio*.

5.3.4 Pathology

Pathogenicity and oncogenicity

Sano *et al.* (1985a, 1990) first reported the pathogenicity and oncogenicity of CyHV-1 to carp, although the virus did not cause mortality in the grass carp, *Ctenopharyngodon idella*, crucian carp, *Carassius auratus*, or willow shiner, *Gnathopogon elongatus*. Experimental infection with CyHV-1 showed high and weak lethality to 2-week-old and 4-week-old carp fry, respectively, but no mortality occurred in 8-week-old fry (Sano *et al.*, 1991b). Furthermore, infected carp fry (2 week old) showed cumulative mortality of 60, 16 and 0% at 15, 20 and 25°C, respectively, (Sano *et al.*, 1993a). During

acute disease, the virus was detected in the gills, liver, kidney and intestine, demonstrating the occurrence of systemic infection, but the detection rate and concentration of the virus in the infected fish at 25°C were lower than those in the fish at 15 and 20°C. Subsequently, papillomas (Fig. 5.6) on the skin, fin or mandible were induced in 35, 72.5 and 27.5% of the survivors at 15, 20 and 25°C, respectively (Sano *et al.*, 1993a). Histopathological lesions in experimentally induced papillomas (Fig. 5.7A) were similar to those in natural infections (Sano *et al.*, 1991b).

Viral antigens and infectious titres could not be detected in fish infected with CyHV-1 using polyclonal antibodies to the virus or FHM cells 8 weeks after the inoculation when the water temperature was around 25°C or higher. However, viral DNA was detected in the brain, spinal cord, liver and subcutaneous tissues of the survivors using *in situ* hybridization with a fragment of the virus genome (Sano *et al.*, 1992, 1993b). This shows that CyHV-1 can latently infect carp at high water temperatures. When the water temperature was lowered, papillomas appeared and the viral genome was detected both in the papilloma and in normal epidermis and subcutaneous tissue, spinal nerves and the liver.

Davison *et al.* (2013) reported that the JUNB family, which encodes a transcription factor involved in oncogenesis, is confirmed in the viral genome. Further study is needed to determine the gene(s) responsible for the induction of papillomas.

Fig. 5.6. Cyprinid herpesvirus 1 (CyHV-1)-induced papilloma on the mandible of carp that had survived from an infection with the cultured virus. Photograph provided by M. Sano.

M. Yoshimizu *et al.*

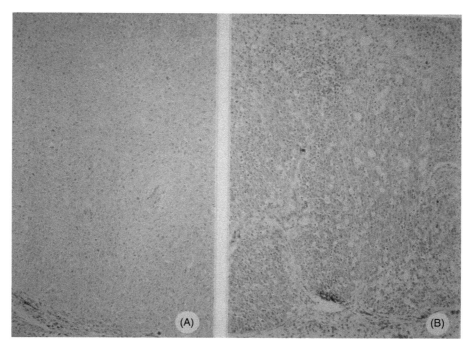

Fig. 5.7. Histopathology of (A) induced papilloma on the mandible of carp surviving infection by cyprinid herpesvirus 1 (CyHV-1) at 2 weeks old, and (B) virus-induced papilloma on the mandible of the fish 7 days after being treated with a water temperature shifting from 10 to 20°C for the induction of regression of the disease. Photographs were provided by N. Sano.

Regression and recurrence of tumours

Spontaneous regression of papillomas occurs in the spring as the water temperature rises. The regression also occurs in experimentally CyHV-1-induced papillomas when the water temperature was increased from 14 to 20, 25 and 30°C. At higher water temperatures (20–30°C), the regression is rapid, resulting in desquamation on the surface of the skin of all fish by 9 days.

After regression, papillomas frequently recur spontaneously (Sano *et al.*, 1991b); they recurred in 83% of survivors 7.5 months after desquamation because the virus is present latently. Subsequently, the virus may reactivate and induce papilloma development in the following year (Sano *et al.*, 1993a).

Leucocyte inflammation and oedema occur in the first stage of papilloma regression (see Fig. 5.7B). The subsequent stages include massive inflammation and oedema, cell necrosis that results in sponge-like tissues and, finally, desquamation from the epidermis. The injection of anti-carp peripheral blood lymphocyte (PBL) rabbit serum, which depressed the *in vitro* cytotoxic activity of normal carp PBLs, retarded the regression in experimentally CyHV-1 induced papilloma at 20°C (Morita and Sano, 1990). Moreover, the injection of anti-PBL serum induced recurrence of papillomas within 10 days of injection at 20°C on three out of eight fish whose papillomas had naturally desquamated 3 months earlier (N. Sano, unpublished data). This suggests that the cytotoxic activity of carp PBLs is important to the development and regression of papillomas (Morita and Sano, 1990).

5.3.5 Control strategies

In papillomas on sub-adult and adult fish, the mode of transmission of the virus seems to be horizontal from fish bearing the papillomas. Separation of papilloma-bearing fish can be a realistic way to control the occurrence of papillomas at the farm level. Also, in koi farms, treatment with elevated temperature is carried out routinely to induce desquamation of the papillomas (see above) before fish are shipped. However, little effort has been expended on developing and implementing hatchery-based control strategies as there is no mass mortality of fry caused by CyHV-1 in farms. The effects of disinfectants and physiological treatments (such as heat and UV) on the nature of the infection and survival of the virus are not known.

5.4 Conclusions and Suggestions for Future Research

Oncorhynchus masou virus disease (OMVD) is an oncogenic and skin ulcerative condition coupled with hepatitis among salmonid fishes in Japan. The causative agent is the *Salmonid herpesvirus 2* (SalHV-2), which was first described as an oncogenic virus isolated from *O. masou*. Other synonyms (acronyms) include OMV, NeVTA, YTV, OKV, COTV, CHV, RKV and RHV. SalHV-2 belongs to the genus *Salmonivirus* and family, *Alloherpesviridae*. The main susceptible fish species are masu salmon, coho salmon and rainbow trout. Economic losses caused by this virus have been recognized among coho salmon and rainbow trout. OMVD is successfully controlled by disinfection of all equipment and eggs with iodophor just after fertilization and again at the early-eyed stage, and culture in a clean facility using well water or disinfected river water. Pathogenicity and oncogenicity are different among the OMV strains isolated from masu salmon, coho salmon and rainbow trout. Further studies are needed to determine the similarities and differences of these strains; their pathogenicity and oncogenicity, and to elucidate the mechanisms of infection and tumour formation, and their gene sequences.

The oncogenicity of cyprinid herpesvirus 1 (CyHV-1) has been confirmed and viral dynamics, including systemic infection, latent infection and papilloma development and regression in fish have been demonstrated in experimental studies. The fish immune system can also play an important role in the development and regression of tumours induced by the virus. Unlike experimental infections using cultured virus, there is no mass mortality of fry caused by CyHV-1 in farms in Japan. Latent or persistent infection occurs in adult carp during the warmer season even when fish have no apparent tumours, and these carp can be a major source of infection. Further epizootiological studies are needed of CyHV-1, especially of how its life cycle relates to the development of papillomas. As the viral genome sequence predicts the presence of oncogenic genes, further studies on the identification

and function(s) of the proteins encoded in the genes should be carried out. The development and regression of the papillomas seem to rely on the state of the immune system, so it would be productive to more precisely determine the roles of that system in the host–pathogen relationship and ways to exploit it to the benefit of the host. It would also be helpful to better understand the disease process and to promote breeding programmes to develop less susceptible fish because on an empirical basis there may be differences in susceptibility to the virus among koi strains on farms.

References

Adkison, M.A., Gilad, O. and Hedrick, R.P. (2005) An enzyme linked immunosorbent assay (ELISA) for detection of antibodies to the koi herpesvirus (KHV) in the serum of koi *Cyprinus carpio*. *Fish Pathology* 40, 53–62.

Amos, K.H. (1985) *Procedures for the Detection and Identification of Certain Fish Pathogens*, 3rd edn. Fish Health Section, American Fisheries Society, Corvallis, Oregon.

Anders, K. (1989) A herpesvirus associated with an epizootic epidermal papillomatosis in European smelt (*Osmerus eperlanus*). In: Ahne, W. and Kurstak, E. (eds) *Viruses of Lower Vertebrates*. Springer, Berlin, pp. 184–197.

Anders, K. and Yoshimizu, M. (1994) Role of viruses in the induction of skin tumours and tumour-like proliferations of fish. *Diseases of Aquatic Organisms* 19, 215–232.

Archetti, I. and Horsfall, F.L. (1950) Persistent antigenic variation of influenza A viruses after incomplete neutralization *in vivo* with heterologous immune serum. *Journal of Experimental Medicine* 92, 441–462.

Aso, Y., Wani, J., Antonio, S-K.D. and Yoshimizu, M. (2001) Detection and identification of *Oncorhynchus masou* virus (OMV) disease by polymerase chain reaction (PCR). *Bulletin of Fisheries Sciences, Hokkaido University* 52, 111–116.

Calle, P.P., McNamara, T. and Kress, Y. (1999) Herpesvirus-associated papillomas in koi carp (*Cyprinus carpio*). *Journal of Zoo and Wildlife Medicine* 30, 165–169.

Davison, A.J., Kurobe, T., Gatherer, D., Cunningham, C., Korf, I. et al. (2013) Comparative genomics of carp herpesviruses. *Journal of Virology* 87, 2908–2922.

Dixon, R. and Farber, F. (1980) Channel catfish virus: physiochemical properties of viral genome and identification of viral polypeptides. *Virology* 103, 267–278.

Eaton, W.D., Wingfield, W.H. and Hedrick, R.P. (1991) Comparison of the DNA homologies of five salmonid herpesvirus. *Fish Pathology* 26, 183–187.

M. Yoshimizu *et al.*

Furihata, M. (2008) Study on the *Salmonid herpesvirus* infection of rainbow trout. *Bulletin of Nagano Prefectural Fisheries Experimental Station* 10, 1–41.

Furihata, M., Hosoe, A., Takei, K., Kohara, M., Nakamura, J. *et al.* (2003) Outbreak of salmonid herpesviral disease in cultured rainbow trout. *Fish Pathology* 38, 23–25.

Furihata, M., Suzuki, K., Hosoe, A. and Miyazaki, T. (2004) Histopathological study on *Oncorhynchus masou* virus disease (OMVD) of cultured rainbow trout in natural outbreaks, artificial infection. *Fish Pathology* 40, 161–167.

Gou, D.F., Kubota, H., Onuma, M. and Kodama, H. (1991) Detection of salmonid herpesvirus (*Oncorhynchus masou* virus) in fish by Southern-blot technique. *Journal of Veterinary Medical Science* 53, 43–48.

Hatori, S., Motonishi, A., Nishizawa, T. and Yoshimizu, M. (2003) Virucidal effect of disinfectants against *Oncorhynchus masou* virus (OMV). *Fish Pathology* 38, 185–187.

Hayashi, Y., Kodama, H., Mikami, T. and Izawa, H. (1987) Analysis of three salmonid herpesvirus DNAs by restriction endonuclease cleavage patterns. *Japanese Journal of Veterinary Science* 49, 251–260.

Hayashi, Y., Izawa, H., Mikami, T. and Kodama, H. (1993) A monoclonal antibody cross-reactive with three salmonid herpesviruses. *Journal of Fish Diseases* 16, 479–486.

Hedrick, R.P., McDowell, T., Eaton, W.D., Kimura, T. and Sano, T. (1987) Serological relationships of five herpesviruses isolated from salmonid fishes. *Journal of Applied Ichthyology* 3, 87–92.

Horiuchi, M., Miyazawa, M., Nakata, M., Iida, K. and Nishimura, S. (1989) A case of herpesvirus infection of fresh water-reared coho salmon *Oncorhynchus kisutch* in Japan. *Suisan-Zoushoku* 36, 297–305.

Igari, T., Fukuda, H. and Sano, T. (1991) The restriction endonuclease cleavage patterns of the salmonid herpesvirus strain's DNAs. *Fish Pathology* 26, 45–46.

Kamei, Y., Yoshimizu, M. and Kimura, T. (1987) Plaque assay of *Oncorhynchus masou* virus (OMV). *Fish Pathology* 22, 147–152.

Kasai, H., Nomura, T. and Yoshimizu, M. (2004) Surveillance and control of salmonid viruses of wild salmonid fish returning to the northern part of Japan, from 1976 to 2002. In: *Proceedings of the Japan–Korea Joint Seminar on Fisheries Sciences.* Hokkaido University, Hakodate, Japan, pp. 142–147.

Keysselitz, G. (1908) Über ein Epithelioma der Barden. *Archiv für Protistenkunde* 11, 326–333.

Kimura, I. (1976) Tumor of lower vertebrates. In: Sugiyama, T. and Yamamoto, Y. (eds) *Cancer.* Iwanami-Shoten, Tokyo, pp. 270–283.

Kimura, T. and Yoshimizu, M. (1989) Salmon herpesvirus: OMV, *Oncorhynchus masou* virus. In: Ahne, W. and Kurstak, E. (eds) *Viruses of Lower Vertebrates.* Springer, Berlin, pp. 171–183.

Kimura, T., Yoshimizu, M., Tanaka, M. and Sannohe, H. (1981a) Studies on a new virus (OMV) from *Oncorhynchus masou* virus (OMV) I. Characteristics and pathogenicity. *Fish Pathology* 15, 143–147.

Kimura, T., Yoshimizu, M. and Tanaka, M. (1981b) Studies on a new virus (OMV) from *Oncorhynchus masou* II. Oncogenic nature. *Fish Pathology* 15, 149–153.

Kimura, T., Yoshimizu, M. and Tanaka, M. (1981c) Fish viruses: tumor induction in *Oncorhynchus keta* by the herpesvirus. In: Dawe, C.J., Harshbarger, J.C., Kondo, S., Sugimura, T. and Takayama, S. (eds) *Phyletic Approaches to Cancer.* Japan Scientific Societies Press, Tokyo, pp. 59–68.

Kimura, T., Yoshimizu, M. and Tanaka, M. (1983a) Susceptibility of different fry stages of representative salmonid species to *Oncorhynchus masou* virus (OMV). *Fish Pathology* 17, 251–258.

Kimura, T., Suzuki, S. and Yoshimizu, M. (1983b) *In vitro* antiviral effect of 9-(2-hydroxyethoxymethyl) guanine on the fish herpesvirus *Oncorhynchus masou* virus (OMV). *Antiviral Research* 3, 93–101.

Kimura, T., Suzuki, S. and Yoshimizu, M. (1983c) *In vivo* antiviral effect of 9-(2-hydroxyethoxymethyl) guanine on experimental infection of chum salmon (*Oncorhynchus keta*) fry with *Oncorhynchus masou* virus (OMV). *Antiviral Research* 3, 103–108.

Kohara, M. and Denda, I. (2008) Production of allotriploid Shinsyu Salmon, by chromosome manipulation. *Fish Genetics and Breeding Science* 37, 61–65.

Kumagai, A., Takahashi, K. and Fukuda, H. (1994) Epizootics caused by salmonid herpesvirus type 2 infection in maricultured coho salmon. *Fish Pathology* 29, 127–134.

Lee, S. and Whitfield, P.J. (1992) Virus-associated spawning papillomatosis in smelt, *Osmerus eperlanus* L., in the River Thames. *Journal of Fish Biology* 40, 503–510.

McAllister, P.E. and Herman, R.L. (1989) Epizootic mortality in hatchery-reared lake trout *Salvelinus namaycush* caused by a putative virus possibly of the herpesvirus group. *Diseases of Aquatic Organisms* 6, 113–119.

Morita, N. and Sano, T. (1990) Regression effect of carp, *Cyprinus carpio* L., peripheral blood lymphocytes on CHV-induced carp papilloma. *Journal of Fish Diseases* 13, 505–511.

Mulcahy, M.F. and O'Leary, A. (1970) Cell-free transmission of lymphosarcomas in the northern pike (*Esox lucius*). *Cellular and Molecular Life Sciences* 26, 891.

Peters, N. and Waterman, B. (1979) Three types of skin papillomas of flatfishes and their causes. *Marine Ecology – Progress Series* 1, 269–276.

Sano, N., Hondo, R., Fukuda, H. and Sano, T. (1991a) *Herpesvirus cyprini*: restriction endonuclease cleavage profiles of the viral DNA. *Fish Pathology* 26, 207–208.

Sano, N., Sano, M., Sano, T. and Hondo, R. (1992) *Herpesvirus cyprini*: detection of the viral genome by *in situ* hybridization. *Journal of Fish Diseases* 15, 153–162.

Sano, N., Moriwake, M. and Sano, T. (1993a) *Herpesvirus cyprini*: thermal effects on pathogenicity and oncogenicity. *Fish Pathology* 28, 171–175.

Sano, N., Moriwake, M., Hondo, R. and Sano, T. (1993b) *Herpesvirus cyprini*: a search for viral genome in infected fish by *in situ* hybridization. *Journal of Fish Diseases* 16, 495–499.

Sano, T. (1976) Viral diseases of cultured fishes in Japan. *Fish Pathology* 10, 221–226.

Sano, T., Fukuda, H., Okamoto, N. and Kaneko, F. (1983) Yamame tumor virus: lethality and oncogenicity. *Bulletin of the Japanese Society of Scientific Fisheries* 49, 1159–1163.

Sano, T., Fukuda, H. and Furukawa, M. (1985a) *Herpesvirus cyprini*: biological and oncogenic properties. *Fish Pathology* 20, 381–388.

Sano, T., Fukuda, H., Furukawa, M., Hosoya, H. and Moriya, Y. (1985b) A herpesvirus isolated from carp papilloma in Japan. In: Ellis, A.E. (ed.) *Fish and Shellfish Pathology*. Academic Press, London, pp. 307–311.

Sano, T., Morita, N., Shima, N. and Akimoto, M. (1990) A preliminary report of pathogenicity and oncogenicity of cyprinid herpesvirus. *Bulletin of the European Association of Fish Pathologists* 10, 11–13.

Sano, T., Morita, N., Shima, N. and Akimoto, M. (1991b) *Herpesvirus cyprini*: lethality and oncogenicity. *Journal of Fish Diseases* 14, 533–543.

Sung, J.T., Yoshimizu, M., Nomura, T. and Ezura, Y. (1996a) *Oncorhynchus masou* virus: serological relationships among salmonid herpesvirus isolated from kokanee salmon, masu salmon, coho salmon and rainbow trout. *Scientific Report of the Hokkaido Salmon Hatchery* 50, 139–144.

Sung, J-T., Yoshimizu, M., Nomura, T. and Ezura, Y. (1996b) *Oncorhynchus masou* virus: pathogenicity of salmonid herpesvirus-2 strains against masu salmon. *Scientific Report of the Hokkaido Salmon Hatchery* 50, 145–148.

Suzuki, K. (1993) A new viral disease on rainbow trout. *Shikenkenkyuwa-Ima* 165, 1–2.

Suzuki, S., Yoshimizu, M. and Saneyoshi, M. (1992) Detection of viral DNA polymerase activity in salmon tumor tissue induced by herpesvirus, *Oncorhynchus masou* virus. *Acta Virologica* 36, 326–328.

Tanaka, M., Yoshimizu, M. and Kimura, T. (1984) *Oncorhynchus masou* virus: pathological changes in masu salmon (*Oncorhynchus masou*), chum salmon (*O. keta*) and coho salmon (*O. kisutch*) fry infected with OMV by immersion method. *Nippon Suisan Gakkaishi* 50, 431–437.

Walker, R. (1969) Virus associated with epidermal hyperplasia in fish. *National Cancer Institute Monograph* 31, 195–207.

Winqvist, G., Ljungberg, O. and Hellstroem, B. (1968) Skin tumours of northern pike (*Esox lucius* L.) II. Viral particles in epidermal proliferations. *Bulletin – Office International des Épizooties* 69, 1023–1031.

Yoshimizu, M. (2003a) Control strategy for viral diseases of salmonids and flounder. In: Lee, C.S. and Bryen, P.J.O (eds) *Biosecurity in Aquaculture Production Systems: Exclusion of Pathogens and Other Undesirables*. World Aquaculture Society, Baton Rouge, Louisiana, pp. 35–41.

Yoshimizu, M. (2003b) Chapter 2.1.3. *Oncorhynchus masou* virus disease. In: *Manual of Diagnostic Tests for Aquatic Animals*, 4th edn. World Organisation for Animal Health, Paris, pp. 100–107. Available at: http://www.oie.int/doc/ged/D6505.PDF (accessed 27 October 2016).

Yoshimizu, M. (2009) Control strategy for viral diseases of salmonid fish, flounders and shrimp at hatchery and seeds production facility in Japan. *Fish Pathology* 44, 9–13.

Yoshimizu, M. and Kasai, H. (2011) Oncogenic viruses and *Oncorhynchus masou* virus. In: Woo, P.T.K. and Bruno, D.D. (eds) *Fish Diseases and Disorders. Volume 3, Viral, Bacterial and Fungal Infections*, 2nd edn. CAB International, Wallingford, UK, pp. 276–301.

Yoshimizu, M., Kimura, T. and Winton, J.R. (1985) An improved technique for collecting reproductive fluid samples from salmonid fishes. *Progressive Fish-Culturist* 47, 199–200.

Yoshimizu, M., Takizawa, H. and Kimura, T. (1986) U.V. susceptibility of some fish pathogenic viruses. *Fish Pathology* 21, 47–52.

Yoshimizu, M., Kimura, T. and Tanaka, M. (1987) *Oncorhynchus masou* virus (OMV): incidence of tumor development among experimentally infected representative salmonid species. *Fish Pathology* 22, 7–10.

Yoshimizu, M., Tanaka, M. and Kimura, T. (1988a) Histopathological study of tumors induced by *Oncorhynchus masou* virus (OMV) infection. *Fish Pathology* 23, 133–138.

Yoshimizu, M., Kamei, M., Dirakubusarakom, S. and Kimura, T. (1988b) Fish cell lines: susceptibility to salmonid viruses. In: Kuroda, Y., Kurstak, K. and Maramorosch, K. (eds) *Invertebrate and Fish Tissue Culture*. Japan Scientific Societies Press, Tokyo/Springer, Berlin, pp. 207–210.

Yoshimizu, M., Sami, M. and Kimura, T. (1989) Survivability of infectious hematopoietic necrosis virus (IHNV) in fertilized eggs of masu (*Oncorhynchus masou*) and chum salmon (*O. keta*). *Journal of Aquatic Animal Health* 1, 13–20.

Yoshimizu, M., Nomura, T., Awakura, T., Ezura, Y. and Kimura, T. (1993) Surveillance and control of infectious hematopoietic necrosis virus (IHNV) and *Oncorhynchus masou* virus (OMV) of wild salmonid fish returning to the northern part of Japan 1976–1991. *Fisheries Research* 17, 163–173.

M. Yoshimizu *et al.*

Yoshimizu, M., Fukuda, H., Sano, T. and Kimura, T. (1995) Salmonid herpesvirus 2. Epidemiology and serological relationship. *Veterinary Research* 26, 486–492.

Yoshimizu, M., Furihata, M. and Motonishi, A. (2002) Control of *Oncorhynchus masou* virus disease (OMVD), salmonid disease notify [?notification] to OIE. In: *Report for the Disease Section, Japan Fisheries Resource Conservation.* Tokyo, pp. 101–108.

Yoshimizu, M., Yoshinaka, T., Hatori, S. and Kasai, H. (2005) Survivability of fish pathogenic viruses in environmental water, and inactivation of fish viruses.

Bulletin of Fisheries Research Agency, Suppl. No. 2, 47–54.

Yoshimizu, M., Kasai, H. and Sakoda, Y. (2012) Salmonid herpesvirus disease (*Oncorhynchus masou* virus disease; OMVD). In: *Proceedings of the 46th Annual Summer Symposium of Japanese Society for Virology Hokkaido Branch*, July 21–22, Ootaki, Hokkaido, Japan, p. 6.

Yoshimizu, M., Kasai, H. and Watanabe, K. (2014) Approaches to probiotics for aquaculture – Prevention of fish viral diseases using antiviral intestinal bacteria. *Journal of Intestinal Microbiology* 28, 7–14.

6 Infectious Salmon Anaemia

KNUT FALK* AND MARIA AAMELFOT

The Norwegian Veterinary Institute, Oslo, Norway

6.1 Introduction

Infectious salmon anaemia (ISA) is a significant infectious viral disease of farmed Atlantic salmon, *Salmo salar* L., that was first reported in Norway during 1984 (Thorud and Djupvik, 1988). Outbreaks of ISA have an impact on the economy of the Atlantic salmon aquaculture industry, and this has led to the implementation of large-scale biosecurity measures. Outbreaks have now been reported in most Atlantic salmon farming areas, including the east coast of Canada and the USA, Scotland, Norway, the Faroe Islands and Chile (Rimstad *et al.*, 2011). In Chile and the Faroe Islands, the disease caused major economic setbacks and left the entire industry with an uncertain future (Mardones *et al.*, 2009; Christiansen *et al.*, 2011) in a manner similar to that in Norway in and after 1989 (Håstein *et al.*, 1999; Rimstad *et al.*, 2011). In the early 1990s, ISA was listed as a notifiable disease by the World Organisation for Animal Health (OIE) (Håstein *et al.*, 1999; OIE, 2015a).

Following its initial detection, ISA spread rapidly throughout the Norwegian aquaculture industry, resulting in a peak with more than 90 outbreaks in 1990 (Håstein *et al.*, 1999). Although the causative virus was not identified until 1995 (Dannevig *et al.*, 1995), biosecurity measures were introduced to combat and control the disease. These measures included early detection of the disease and the slaughter of diseased populations, regulations on transport, the disinfection of offal and waste from slaughterhouses, year class separation at farming sites and improved health control and certification. The effect was remarkable as only two new outbreaks were reported in 1994, demonstrating that ISA can be controlled without the use of drugs and vaccines, and even without knowledge of the aetiological agent, the epidemiology or the pathogenesis of the disease. ISA still occurs at a low prevalence in Norway, while the other previously affected countries seem to have controlled the disease with no, or only a few sporadically occurring, disease outbreaks. Epidemiological information on ISA may be found in the OIE's World Animal Health Information System (WAHIS) Database – WAHID (http://www.oie.int/wahis_2/public/wahid.php/Wahidhome/Home).

ISA does not usually cause high mortalities at the start of an outbreak, and if proper control measures are implemented, an outbreak may be controlled with only minor mortalities. Even so, economic losses may be significant due to the various measures and restrictions implemented. However, if inadequate measures are implemented, the outbreak may develop into a serious disease problem both in the salmon farm concerned and in adjacent farms (Lyngstad *et al.*, 2008). Indeed, there are multiple examples from Norway where single ISA outbreaks have developed into small endemics. The only solution in such cases has been to fallow entire areas.

The causative agent of ISA is infectious salmon anaemia virus (ISAV), the only member of the genus *Isavirus*, family *Orthomyxoviridae* (Palese and Shaw, 2007). The two glycoproteins embedded in the ISAV envelope, the haemagglutinin esterase (HE) glycoprotein and the fusion (F) glycoprotein, are important for virus uptake and cell tropism (Falk *et al.*, 2004; Aspehaug *et al.*, 2005; Aamelfot *et al.*, 2012) . In addition, the virion is formed by two other major structural proteins, the nucleoprotein (NP) (Aspehaug *et al.*, 2004; Falk *et al.*, 2004) and the matrix (M) protein (Biering *et al.*, 2002; Falk *et al.*, 2004). The HE proteins bind the cellular

---

*Corresponding author e-mail: knut.falk@vetinst.no

receptors of ISAV, which are 4-O-acetylated sialic acids (Hellebo *et al.*, 2004), expressed on endothelial cells and red blood cells (RBCs) in the host (Aamelfot *et al.*, 2012).

The segmented ISAV genome is highly conserved. The two gene segments with the highest variability are those coding for the HE and F proteins. Phylogenetic analyses of ISAV isolates, based on the HE gene, revealed two major clades, one European and the other North American. In addition, ISAV has been characterized and typed based on the amino acid patterns of a highly polymorphic region (HPR) consisting of 11–35 amino acid residues in HE (Rimstad *et al.*, 2011), just upstream of the transmembrane domain. The virulent HPR variants may be explained as differential deletions (Mjaaland *et al.*, 2002) of a putative non-virulent full-length ancestral sequence (HPR0). The HPR0 variant was first identified in wild salmon in Scotland (Cunningham *et al.*, 2002). Whereas all ISAV isolates from ISA disease outbreaks have deletions in the HPR region, which are often denoted HPR-deleted, the HPR0 subtype has not been associated with clinical or pathological signs of ISA (Cunningham *et al.*, 2002; Mjaaland *et al.*, 2002; Cook-Versloot *et al.*, 2004; McBeath *et al.*, 2009; Christiansen *et al.*, 2011). Epidemiological studies show that ISAV HPR0 variants occur frequently in sea-reared Atlantic salmon. The HPR0 strain seems to be more seasonal and transient in nature and displays a cell and tissue tropism with prevalence on gill epithelial cells (Christiansen *et al.*, 2011; Lyngstad *et al.*, 2012; Aamelfot *et al.*, 2016) and possibly also on the skin (Aamelfot *et al.*, 2016). A peculiarity of the ISAV HPR0 type is that, unlike HPR-deleted ISAV, it cannot be replicated in currently available cell cultures (Christiansen *et al.*, 2011). The risk of emergence of pathogenic HPR-deleted ISAV variants from a reservoir of HPR0 ISAV is considered to be low, but not negligible (Christiansen *et al.*, 2011; EFSA Panel on Animal Health and Welfare (AHAW), 2012; Lyngstad *et al.*, 2012).

Other gene segments may also be important for the development of ISA. A putative virulence marker has been identified in the F protein. Here, a single amino acid substitution, or a sequence insertion, near the putative cleavage activation site of the protein, is a prerequisite for virulence (Kibenge *et al.*, 2007; Markussen *et al.*, 2008; Fourrier *et al.*, 2015). Indeed, Fourrier *et al.* (2015) recently demonstrated that the combination of deletions in the HPR and certain amino acid substitutions

in the F protein may influence the proteolytic activation and activity of the F protein.

Outbreaks of ISA have only been detected in farmed Atlantic salmon, and the majority of cases have occurred during the seawater stage of the salmon life cycle. However, experimental infection trials have demonstrated that Atlantic salmon of different developmental stages are equally susceptible to the infection in both fresh water and salt water. Viral replication without clinical disease has been demonstrated experimentally in other fish species, including brown trout (*S. trutta* L.), rainbow trout (*Oncorhynchus mykiis* (Walbaum)), Arctic charr (*Salvelinus alpinus* L.), chum salmon (*O. keta* (Walbaum)), coho salmon (*O. kisutch* (Walbaum)), herring (*Clupea harengus* L.) and Atlantic cod (*Gadus morhua* L.) (Rimstad *et al.*, 2011). ISAV has also been detected in healthy wild Atlantic salmon and sea trout (*S. trutta* L.) (Raynard *et al.*, 2001; Plarre *et al.*, 2005). These wild fish or fish of other species may act as carriers of, or reservoirs for, the virus.

The infectivity of ISAV is anticipated to be retained for a long time outside the host. In an experiment using both fresh water and seawater at different temperatures, Tapia *et al.* (2013) found the virus survives for 5 days in seawater at 20°C up to 70 days in fresh water at 10°C, but there is no knowledge of viral survival under natural conditions, where it may adhere to, and be protected by, organic matters. Waterborne transmission has been demonstrated in cohabitation experiments, indicating this is an important route for the spread of ISA within and between nearby farms (Thorud and Djupvik, 1988; Lyngstad *et al.*, 2008). The virus may be shed into the water via various routes, including the skin, mucus, faeces, urine and blood, and within waste from dead fish (Totland *et al.*, 1996). The main route of entry is thought to be the gills (Rimstad *et al.*, 2011), although recent studies have demonstrated that early replication occurs in both the gills and the skin (Aamelfot *et al.*, 2015).

Horizontal spread of the virus is well documented during outbreaks both within and between farms, but in most cases, investigators could not document how the infection was initially introduced. Infections/outbreaks after the transfer of infected smolt have been demonstrated in a few cases, while transfer by various farming equipment, including well-boats, is important (Vågsholm *et al.*, 1994; Jarp and Karlsen, 1997; Murray *et al.*, 2002).

The transition of virulent HPR-deleted virus from the frequently occurring HPR0 type has been suggested (Cunningham *et al.*, 2002; Mjaaland *et al.*, 2002; Lyngstad *et al.*, 2012), though the significance and frequency of this has not been documented. A reservoir in the marine environment is also a possibility, as several fishes can have subclinical infections.

Finally, a number of reports have suggested the possibility of vertical transmission (Melville and Griffiths, 1999; Nylund *et al.*, 2007; Vike *et al.*, 2009; Marshall *et al.*, 2014), but there is no confirmation of this (Rimstad *et al.*, 2011). Experience-based information from Norwegian salmon farming operations has suggested that vertical transmission is not a significant concern.

6.2 Diagnosis

Field outbreaks of ISA in Atlantic salmon vary considerably in the development of the disease, its clinical signs and histological changes. This reflects the complex interaction between the virus, host and environment. A peculiarity of ISA is that clinical disease often spreads in a non-systematic way, slowly from net pen to net pen within a farm, possibly reflecting the extended time from infection to the development of severe anaemia and clinical disease. The incubation period in natural outbreaks varies from a few weeks to several months (Vågsholm *et al.*, 1994; Jarp and Karlsen, 1997).

Diseased fish are usually lethargic, and often display abnormal swimming behaviour. Daily mortality is typically 0.05 to 0.1%. If nothing is done to limit disease development, the disease may spread, and the accumulated mortality in a farm can reach more than 80% over several months. In a disease outbreak with low mortality, the clinical signs and macroscopic pathological changes may be limited to anaemia and circulatory disturbances, including haemorrhages. The chronic disease phase with low mortality can easily be overlooked. Occasionally, episodes of acute, high mortality over a couple of weeks may ensue, especially if no measures are taken. The pathology is then more severe, with ascites and haemorrhages dominating. The disease appears throughout the year though outbreaks are more frequently detected in spring or early summer and in late autumn.

The most prominent clinical signs of ISA include pale gills (except in the case of blood stasis in the gills), exophthalmia, distended abdomen, blood in the anterior eye chamber and, sometimes, skin haemorrhages, especially of the abdomen (Fig. 6.1), as well as scale pocket oedema. Severe anaemia

Fig. 6.1. Atlantic salmon with infectious salmon anaemia (ISA) showing typical skin bleedings.

with haematocrit values below 10% is common (Thorud and Djupvik, 1988; Evensen *et al.*, 1991; Thorud, 1991; Rimstad *et al.*, 2011).

The diagnosis of ISA is based on the combined evaluation of clinical signs, macroscopic lesions and histological changes, supplemented with immunohistochemical (IHC) examinations for endothelial infection (see Fig. 6.4). IHC examination is of particular importance as this method may establish a direct link between the virus and the disease. Positive IHC findings are confirmed using quantitative (real-time) reverse-transcription PCR (qRT-PCR) and viral isolation. Usually, the HE and F genes are also sequenced for use in epidemiological evaluations and to determine the viral type. A description of the methods used is given in the OIE *Manual of Diagnostic Tests for Aquatic Animals* (OIE, 2015b).

Differential diagnoses include other anaemic and haemorrhagic conditions, including erythrocytic inclusion body syndrome, winter ulcer and septicaemias caused by infections with *Moritella viscosa*. Disease cases in Atlantic salmon with haematocrit values below 10% are not a unique finding for ISA; nevertheless, cases with low haematocrits without any obvious cause should always be tested for ISAV (OIE, 2015b).

6.3 Pathology

Fish infected with HPR-deleted ISAV may vary from showing no gross pathological changes to having severe lesions, depending on the size of the infective dose, viral strain, water temperature, and age and immune status of the fish. The most prominent sign is anaemia, often with a haematocrit below 10%, and circulatory disturbances. External signs include pale gills, localized haemorrhages of the eyes and skin (Fig. 6.1), exophthalmia, and scale oedema. However, the disease may appear in different manifestations, and development may appear as an acute form or as a slowly developing chronic disease (Rimstad *et al.*, 2011).

Ascites, swollen spleen, oedema and petechial bleeding on the serosa are common. More variable, but very obvious when present, are severe haemorrhagic lesions in the liver, kidney, gut or gills. The 'classical' liver manifestation of ISA is characterized by dark liver due to haemorrhagic necrosis (Evensen *et al.*, 1991).

Clinical signs and pathology may be more subtle in the slowly developing chronic form of the disease. The liver may appear pale or yellowish and the anaemia may not be as severe as in the acute disease. Less ascites fluid is found than in the acute form, but haemorrhages in the skin and swim bladder and oedema in the scale pockets and swim bladder can be more pronounced than in acutely diseased fish (Evensen *et al.*, 1991; Rimstad *et al.*, 2011).

Histological changes in the liver include zonal hepatocellular degeneration and necrosis (Fig. 6.2) (Evensen *et al.*, 1991; Speilberg *et al.*, 1995; Simko *et al.*, 2000). The lesions result in extensive congestion of the liver with dilated sinusoids, and in later stages, the appearance of blood-filled spaces. Speilberg *et al.* (1995) investigated the liver pathology of experimentally infected salmon, and detected sinusoidal endothelial degeneration and loss preceded by the degeneration of hepatocytes and multifocal haemorrhagic necrosis; the virus was not observed in affected hepatocytes. These observations were confirmed using IHC examinations, during which no virus was detected in the hepatocytes in the pathological lesions (Aamelfot *et al.*, 2012).

The haematopoietic tissue in the kidney is primarily affected by the haemorrhages but, in severe cases, the kidney tubules become necrotic (Fig. 6.3) (Byrne *et al.*, 1998). The kidney haemorrhages are most easily detected via histology as the kidney is diffusely dark in colour. Intestinal haemorrhages may resemble haemorrhagic enteritis. In fresh specimens, there is no blood in the gut lumen, but histological sections are characterized by extensive bleeding within the lamina propria with no inflammation. In some cases, the only pathological findings are anaemia and general circulatory disturbances. Although

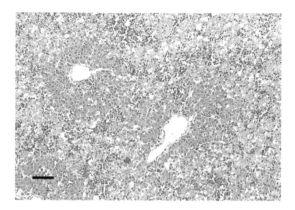

Fig. 6.2. Histological section of Atlantic salmon with infectious salmon anaemia (ISA) showing liver zonal haemorrhagic necrosis (haematoxylin–eosin stain). Scale bar = 50 μm. Courtesy of Dr Agnar Kvellestad.

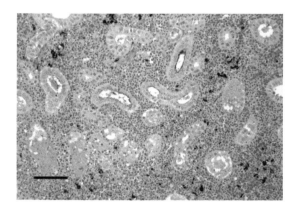

Fig. 6.3. Histological section of Atlantic salmon with infectious salmon anaemia (ISA) showing kidney haemorrhages and tubular necrosis (haematoxylin–eosin stain). Scale bar = 50 μm.

virus-infected endothelial cells are in all organs, a striking feature is the limited inflammatory cellular response (Aamelfot *et al.*, 2012).

Haemorrhages may be divided into different manifestations, i.e. liver, kidney, gut and gill manifestations, while the spleen is more constantly swollen and dark (Rimstad *et al.*, 2011; Aamelfot, 2012). The kidney manifestation is characterized by moderately swollen kidneys with interstitial haemorrhaging and some tubular necrosis (Byrne *et al.*, 1998; Simko *et al.*, 2000). The gut manifestation is characterized by a dark red gut due to haemorrhaging within the intestinal wall, but not in the lumen (in fresh specimens). The gill manifestation is an exception to the pale anaemic gills as blood has accumulated, especially in the central venous sinus of the gill filaments. The haemorrhagic organ lesions are visible on autopsy in organs such as the liver and gut, but are less obvious in the gills and kidney. In any particular ISA outbreak, one of the haemorrhagic organ manifestations may dominate, whereas in other outbreaks all manifestations can be found, even within the same fish. Outbreaks dominated by either the liver or kidney manifestation are most common. However, the haemorrhagic organ lesions can be absent or very rare in the initial stages of an outbreak, leaving only the anaemia and the more subtle circulatory disturbances as initial clues to the aetiology of the disease.

6.4 Pathogenesis

The disease is a generalized and lethal condition of farmed Atlantic salmon, with terminal ISA characterized by anaemia, bleeding and circulatory disturbances. The exact mechanisms behind this pathology are still somewhat obscure, though possible factors are starting to emerge (Aamelfot *et al.*, 2014). Significant differences in disease appearance and severity are common, even with the assumed same isolate. Thus, when evaluating the pathogenesis of ISA, interactions between the host, the infectious agent and the environment (i.e. the aetiological triad) must be considered.

Thorud (1991) who examined anaemic fish from ISA field outbreaks, found increased RBC fragility, increased numbers of immature RBCs and increased proportions of RBC ghost (or smudge) cells, all indicating a haemolytic anaemia. However, neither jaundice nor other signs of excess haemoglobin breakdown were observed, possibly because the anaemia occurs in the terminal stages of the disease. Leucopenia, including lymphocytopenia and thrombocytopenia, were observed, and these were attributed to general stress due to the disease. Other plasma parameters revealed an increase in aspartate aminotransferase, alanine aminotransferase and lactate dehydrogenase, which indicates organ damage. Finally, the increase in plasma osmolality indicated impaired osmoregulation.

Viral virulence is multifactorial as it depends on factors such as receptor binding, cellular uptake, replication rate and shedding of new virions, modulation of the host immune response and the ability to spread to new hosts (McBeath *et al.*, 2007;

Purcell *et al.*, 2009; Medina and Garcia-Sastre, 2011; Peñaranda *et al.*, 2011; Wargo and Kurath, 2012; Cauldwell *et al.*, 2014; McBeath *et al.*, 2015). Thus, evaluating pathogenesis may involve several features in addition to host and environmental factors. By comparison to the related influenza viruses, ISAV has equivalent functional features, including viral receptor binding, fusion activity, receptor destroying activity, replication efficiency promoted by the viral polymerases, and the ability to modulate the host immune response, all of which are important factors for disease development (Palese and Shaw, 2007; Rimstad *et al.*, 2011).

Major target cells are the endothelial cells lining the blood vessels of all organs, including the sinusoids, endocardium (Fig. 6.4), scavenger endothelial cells in the anterior kidney and, possibly, the endothelium of the secondary vessel system (Rummer *et al.*, 2014) and RBCs. Evensen *et al.* (1991) had earlier suggested endothelial cells as possible target cells. Using electron microscopy, it was demonstrated that endothelial damage followed by hepatocellular degeneration preceded anaemia in ISAV-infected salmon. In addition, Evensen *et al.* (1991) found increased splenic phagocytosis. This was corroborated by Aamelfot *et al.* (2012), who detected generalized haemophagocytosis, in addition to the *in situ* haemadsorption of RBCs to ISAV-infected endothelial cells. Aamelfot *et al.* (2012) also demonstrated intact endothelium with no cytopathic effects or significant perivascular infiltration of leucocytes, i.e. inflammation and

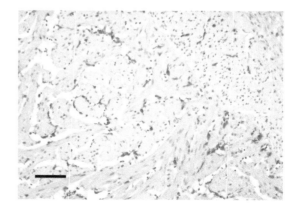

Fig. 6.4. Immunostained histological section of heart of Atlantic salmon with infectious salmon anaemia (ISA). The red coloured endothelial cells are infected with infectious salmon anaemia virus (ISAV). Scale bar = 50 µm.

K. Falk and M. Aamelfot

apoptosis were not detected, indicating that ISAV infection does not incite inflammation or cause cytopathic effects in infected cells. This is not consistent with findings reported from systemic influenza in poultry, where apoptosis and inflammation are common (Kobayashi *et al.*, 1996; Schultz-Cherry *et al.*, 1998). Similar findings were reported for other endotheliotropic viruses such as filoviruses (i.e. Ebola and Marburg), in which infection of the endothelial cells does not appear to disrupt the cell architecture (Geisbert *et al.*, 2003). Nevertheless, the permeability of the infected endothelial cells may still be compromised.

Aamelfot *et al.* (2012) also investigated the pathological lesions in the liver and kidney of severely anaemic moribund Atlantic salmon from confirmed ISA outbreaks. The necrotic lesions of the parenchyma appeared not to be infected by ISAV. A possible explanation is that the lesions are not directly caused by the infection. They may rather be indirect or secondary and caused by the host response, which destroys the RBCs and creates hypoxia in severely anaemic fish. Such hypoxic conditions may start a vicious circle leading to hypovolaemic shock and death, similar to that suggested for Ebola virus endothelial infection (Schnittler *et al.*, 1993).

While classical ISA is characterized by an endothelial infection of the circulatory system, the cell tropism during infection with non-pathogenic ISAV HPR0 is different. Although this virus is in both the gills and skin during infection, only epithelial cells become infected (Aamelfot *et al.*, 2016). The transition from ISAV HPR0 to HPR-deleted virus results in a modified fusion protein activity (Fourrier *et al.*, 2015), which highlights the fusion protein as a factor determining cell tropism and virulence during ISAV infections, in addition to the role of the receptor-binding HE protein. In addition, viral fusion activity also influences virus cell uptake and replication efficiency.

ISAV uptake has been studied via immersion experiments using viruses of high and low virulence (Aamelfot *et al.*, 2015; McBeath *et al.*, 2015). In both cases, a transient initial infection of epithelial cells was demonstrated; with the low virulence virus, early epithelial infection was more pronounced and lasted longer. It also disseminated more quickly to internal organs and induced a more rapid systemic host immune response. This response may have offered some protection to the fish, and at least partly explains the lower virulence and mortality caused by this virus isolate.

Viruses of the family *Orthomyxoviridae* can bind to or haemagglutinate RBCs. ISAV haemagglutinates RBCs from Atlantic salmon, rainbow trout, horse, donkey and rabbit, but not RBCs from brown trout. The expected elution of the *in vitro* haemagglutination reaction due to the viral esterase (i.e. the virus receptor-destroying enzyme on the HE protein), is observed on RBCs from all species examined, except Atlantic salmon (Falk *et al.*, 1997).

ISAV binds extensively to circulating RBCs, and RBCs attach to infected endothelial cells, which may be regarded as *in vivo* haemadsorption (Aamelfot *et al.*, 2012). These observations may have significant importance for disease pathogenesis and may explain several observations. First, the attachment of virus to RBCs may compromise cell membrane integrity, which increases RBC fragility as observed by Thorud (1991). This is supported by experimental data demonstrating that the occurrence of large amounts of virus on RBCs coincides with increased RBC fragility (K. Falk, 2016, unpublished results). Secondly, virus-coated RBCs may be regarded as foreign by the host scavenger system, which explains the increased haemophagocytosis observed. Indeed, canine RBCs with attached influenza virus survive only a fraction of their expected lifespan (Stewart *et al.*, 1955). Thirdly, the thrombi observed in early reports of ISA pathology (Evensen *et al.*, 1991) and the attachment of RBCs to ISAV-infected cells (Aamelfot *et al.*, 2012) can also be explained by *in situ* haemagglutination and haemadsorption. Fourthly, the viral receptors on the RBC surface may act as decoy receptors, and interact with virus dissemination, as noted by Baum *et al.* (2002). Whether this will promote spread of infection, as reported for HIV (Baum *et al.*, 2002), or delay the infection, as reported for parvovirus (Traving and Schauer, 1998), needs further examination. In addition, ISAV is not released from Atlantic salmon RBCs under in *vitro* conditions (Falk *et al.*, 1997). Although the significance of this latter observation is unknown, we have identified a couple of ISAV isolates that were eluted under *in vitro* conditions from Atlantic salmon RBCs. When one of these isolates was tested in an experimental infection, the isolate had low virulence, resulting in lower mortality and less pathology than a highly virulent virus (McBeath *et al.*, 2015).

Briefly, these observations indicate that the interaction between ISAV and RBCs is important in pathogenesis, and particularly explains the anaemia, but potentially also other clinical signs. Further examination of this interaction may contribute to understanding of the pathogenesis of other hae-magglutinating viruses.

Both innate and adaptive cellular and humoral immune responses against ISAV have been experimentally demonstrated in Atlantic salmon. This includes specific antibody responses (Falk and Dannevig, 1995; Lauscher et al., 2011) and interferon-related responses (McBeath et al., 2007; LeBlanc et al., 2010; McBeath et al., 2015). Furthermore, the ISAV non-structural proteins have interferon antagonistic activities (McBeath et al., 2006; Garcia-Rosado et al., 2008), though the significance of this in relation to pathogenesis and virulence is not known.

6.5 Protective and Control Strategies

ISA outbreaks often develop slowly and the disease spreads locally separated in time and space, which suggests horizontal transmission (Lyngstad et al., 2008, 2011). However, the origin of the infection in the farmed salmon population is unknown. Unidentified reservoirs, the maintenance in vectors such as salmon lice (Lepeophtheirus salmonis) (Krøyer, 1837), subclinical infections and/or vertical transmission are all possible. In addition, the emergence of virulent HPR-deleted ISAV through transitions from the prevalently occurring non-virulent HPR0 ISAV is considered likely.

The incidence and impact of ISA may be greatly reduced by the implementation of general biosecurity measures that reduce horizontal transfer and infection pressure. These measures include early detection, the isolation and slaughter of diseased fish, general restrictions on transport, the disinfection of offal and waste from slaughterhouses, year-class separation at farm sites, and improved health control and certification (Håstein et al., 1999). In Norway, the result of these actions was an improvement in the sanitary situation in the fish farming industry. Together with significant improvements in husbandry practices, better laboratory identification and subsequent restrictions imposed on farms with ISA, a remarkable and rapid reduction in the number of ISA outbreaks was obtained (Håstein et al., 1999).

Hence, ISA can be controlled if it is detected early, taken seriously and the correct measures are implemented. There are numerous examples showing that if initial outbreaks are not treated correctly, they will develop into devastating epidemics that are very costly. Both the Faroese and the Chilean Atlantic salmon farming industry have experienced significant production and economic setbacks due to ISA outbreaks (Rimstad et al., 2011).

Norway, Scotland, the east coast of Canada, the Faroe Islands and Chile have experienced major ISA epidemics. The strategies adopted to control or combat the disease are slightly different in these countries. In Norway, ISA is accepted as endemic in most areas, with 2–20 annual outbreaks. When detected, strict control measures, including slaughtering and zoning, are implemented to limit dissemination. Recently, vaccination has become more common. Following the ISA epidemic in Scotland in 1998, an eradication procedure was applied, and Scotland is still ISA free. Canada, the Faroe Islands and Chile have implemented a combination of biosecurity procedures and vaccination following their outbreaks, and are currently experiencing either no outbreaks or very few occasional outbreaks.

As ISA is slow to develop, the biosecurity is important for control, and an efficient vaccine will supplement other control strategies. The currently available vaccines are inactivated whole virus grown in cell culture added to mineral oil adjuvants. These vaccines are commonly used on the east coast of Canada, in the Faroe Islands and Chile and, to some extent, in Norway. Due to European Union (EU) regulations, vaccination in European aquaculture is not allowed in areas that have been declared free of ISA and Scotland is an example.

Reports on ISA vaccines are scanty in the scientific literature. Jones et al. (1999) reported a relative percentage survival (RPS) of 84–95% in experimental trials, using an adjuvanted, inactivated ISA virus from culture preparations. More recently, Lauscher et al. (2011) also tested a virus cell culture preparation against experimental infections, and found a RPS of 86%. Protection depended on the amount of vaccine antigen injected and correlated with the anti-ISAV antibodies. There are also several conference abstracts and Internet publications that support these findings.

Vaccination against ISAV was first attempted in Canada in 1999, where this was combined with control programmes and general biosecurity measures. However, ISA outbreaks were recorded in vaccinated populations, so the efficacy of the vaccine was questioned. There are no reports documenting the field performance of ISAV vaccines. Another example of the use of ISA vaccine includes the successful control programme following multiple ISA outbreaks in the Faroe Islands from 2000 to 2005. Almost all the Faroese farms were fallowed, and a strict control regime based on general hygienic principles was

K. Falk and M. Aamelfot

implemented. Most fish were vaccinated with an oil-adjuvanted inactivated cell culture grown ISA vaccine (Christiansen *et al.*, 2011). No ISA outbreaks have been recorded since 2005, though an extensive surveillance programme has frequently found the non-virulent ISAV HPR0 type. It was also speculated that vaccination might have prevented the transition of the ISAV HPR0 virus to the virulent HPR-deleted virus (Christiansen *et al.*, 2011).

Although commercial vaccine development has mainly concentrated on inactivated ISAV cell culture preparations, vaccine research using molecular principles has been reported. Mikalsen *et al.* (2005) injected plasmids coding for the HE protein intramuscularly (IM) three times at 3 week intervals; there was modest protection, with an RPS of 40–60%. After vaccination, the fish were challenged via intraperitoneal (IP) injection, which may explain the poor protection achieved. Wolf *et al.* (2013) vaccinated fish by IM injection of an ISAV haemagglutinin-esterase-expressing salmonid alphavirus replicon, followed by cohabitation challenge. The RPS ranged from 65 to 69%, compared with 80% in controls immunized with an inactivated ISAV cell culture antigen.

More interesting are two reports from Chilean research groups that describe oral vaccination. Both groups based their vaccines on chitosan-based encapsulation. Rivas-Aravena *et al.* (2015) used cell culture grown virus and also included a novel adjuvant – the DNA coding for the replicase of alphavirus – in the chitosan particles. They demonstrated a RPS of 77%. Caruffo *et al.* (2016) used recombinant expressed ISAV HE and F proteins for the chitosan encapsulation, and reported an RPS of 64%. Thus, in both reports, the vaccines were targeting the mucosal surface of the gut. However, in both studies, the fish were challenged via injection, which might have reduced the efficacy of the vaccines. The advantages of oral vaccination are: (i) it avoids serious side effects related to oil-adjuvanted vaccines; and (ii) it allows for booster vaccination.

There are no reports on the use of antiviral agents, chemotherapeutics or other pharmaceutical treatments to treat ISA.

6.6 Conclusions and Suggestions for Future Research

Infectious salmon anaemia is a slowly developing, generalized and lethal disease that may cause detrimental effects in farmed Atlantic salmon. It is an OIE reportable disease, and control programmes have been implemented in most Atlantic salmon farming regions. The diagnosis is usually made based on pathological examination, including IHC and RT-PCR.

The disease is mainly characterized by anaemia, with infection of endothelial cells throughout the circulatory system. The mechanism of pathogenesis is still obscure; however, the observed binding of virus to RBCs is suggested to be an important factor.

Outbreaks are relatively easy to control if they are detected early and appropriate and strict biosecurity measures are implemented immediately. Although experimental vaccines work well under laboratory conditions, their efficacy has not been documented in the field. The production of vaccines based on cultured virus has limitations because production of virus in cell cultures is generally low and this limits the amount of antigen in the vaccine. Vaccines based on ISAV virulence factors should be developed and their efficiencies in field situations should be tested and documented.

How ISA is introduced into, and maintained in, the salmon population need further study. Possible mechanisms include the existence of marine reservoir(s), vertical and/or horizontal transmission, the occurrence of low-virulence viruses causing undetected subclinical infections and the development of virulent ISA virus from the frequently occurring and non-virulent HPR0 type.

The current major challenges associated with ISA control are questions related to risk and factors affecting the transition of non-virulent HPR0 ISAV to virulent HPR-deleted ISAV. These questions have become important not only to the understanding of ISA epidemiology, but also in relation to the international Atlantic salmon trade, as both virus types are notifiable (OIE, 2015a). Thus, the detection of ISAV HPR0 can be, and has already been, used to limit salmon export. To address and solve the key question of the transition of non-virulent HPR0 ISAV to virulent HPR-deleted ISAV we need to elucidate how and why the transition occurs. As this transition is expected to be a stepwise process involving several mutations, there is also a need to know more about ISAV virulence disease mechanism(s).

References

Aamelfot, M. (2012) Tropism of Infectious salmon anaemia virus and distribution of the 4-*O*-acetylated sialic acid receptor. PhD thesis, Norwegian School of Veterinary Science, Oslo, Norway.

Aamelfot, M., Dale, O.B., Weli, S.C., Koppang, E.O. and Falk, K. (2012) Expression of the infectious salmon anemia virus receptor on Atlantic salmon endothelial cells correlates with the cell tropism of the virus. *Journal of Virology* 86, 10571–10578.

Aamelfot, M., Dale, O.B. and Falk, K. (2014) Infectious salmon anaemia – pathogenesis and tropism. *Journal of Fish Diseases* 37, 291–307.

Aamelfot, M., McBeath, A., Christiansen, D.H., Matejusova, I. and Falk, K. (2015) Infectious salmon anaemia virus (ISAV) mucosal infection in Atlantic salmon. *Veterinary Research* 46, 120.

Aamelfot, M., Christiansen, D.H., Dale, O.B., McBeath, A., Benestad, S.L. and Falk, K. (2016) Localised infection of Atlantic salmon epithelial cells by HPR0 infectious salmon anaemia virus. *PLoS ONE* 11(3): e0151723.

Aspehaug, V., Falk, K., Krossoy, B., Thevarajan, J., Sanders, L. *et al.* (2004) Infectious salmon anemia virus (ISAV) genomic segment 3 encodes the viral nucleoprotein (NP), an RNA-binding protein with two monopartite nuclear localization signals (NLS). *Virus Research* 106, 51–60.

Aspehaug, V., Mikalsen, A.B., Snow, M., Biering, E. and Villoing, S. (2005) Characterization of the infectious salmon anemia virus fusion protein. *Journal of Virology* 79, 12544–12553.

Baum, J., Ward, R.H. and Conway, D.J. (2002) Natural selection on the erythrocyte surface. *Molecular Biology and Evolution* 19, 223–229.

Biering, E., Falk, K., Hoel, E., Thevarajan, J., Joerink, M. *et al.* (2002) Segment 8 encodes a structural protein of infectious salmon anaemia virus (ISAV); the colinear transcript from Segment 7 probably encodes a non-structural or minor structural protein. *Diseases of Aquatic Organisms* 49, 117–122.

Byrne, P.J., MacPhee, D.D., Ostland, V.E., Johnson, G. and Ferguson, H.W. (1998) Haemorrhagic kidney syndrome of Atlantic salmon, *Salmo salar* L. *Journal of Fish Diseases* 21, 81–91.

Caruffo, M., Maturana, C., Kambalapally, S., Larenas, J. and Tobar, J.A. (2016) Protective oral vaccination against infectious salmon anaemia virus in *Salmo salar*. *Fish and Shellfish Immunology* 54, 54–59.

Cauldwell, A.V., Long, J.S., Moncorge, O. and Barclay, W.S. (2014) Viral determinants of influenza A host range. *Journal of General Virology* 95, 1193–1210.

Christiansen, D.H., Østergaard, P.S., Snow, M., Dale, O.B. and Falk, K. (2011) A low-pathogenic variant of infectious salmon anemia virus (ISAV-HPR0) is highly prevalent and causes a non-clinical transient infection in farmed Atlantic salmon (*Salmo salar* L.) in the Faroe Islands. *Journal of General Virology* 92, 909–918.

Cook-Versloot, M., Griffiths, S., Cusack, R., McGeachy, S. and Ritchie, R. (2004) Identification and characterisation of infectious salmon anaemia virus (ISAV) haemagglutinin gene highly polymorphic region (HPR) type 0 in North America. *Bulletin of the European Association of Fish Pathologists* 24, 203–208.

Cunningham, C.O., Gregory, A., Black, J., Simpson, I. and Raynard, R.S. (2002) A novel variant of the infectious salmon anaemia virus (ISAV) haemagglutinin gene suggests mechanisms for virus diversity. *Bulletin of the European Association of Fish Pathologists* 22, 366–374.

Dannevig, B.H., Falk, K. and Namork, E. (1995) Isolation of the causal virus of infectious salmon anaemia (ISA) in a long-term cell line from Atlantic salmon head kidney. *Journal of General Virology* 76, 1353–1359.

EFSA Panel on Animal Health and Welfare (AHAW) (2012) Scientific opinion on infectious salmon anaemia (ISA). *EFSA Journal* 10(11):2971.

Evensen, O., Thorud, K.E. and Olsen, Y.A. (1991) A morphological study of the gross and light microscopic lesions of infectious anaemia in Atlantic salmon (*Salmo salar*). *Research in Veterinary Science* 51, 215–222.

Falk, K. and Dannevig, B.H. (1995) Demonstration of a protective immune response in infectious salmon anaemia (ISA)-infected Atlantic salmon *Salmo salar*. *Diseases of Aquatic Organisms* 21, 1–5.

Falk, K., Namork, E., Rimstad, E., Mjaaland, S. and Dannevig, B.H. (1997) Characterization of infectious salmon anemia virus, an orthomyxo-like virus isolated from Atlantic salmon (*Salmo salar* L.). *Journal of Virology* 71, 9016–9023.

Falk, K., Aspehaug, V., Vlasak, R. and Endresen, C. (2004) Identification and characterization of viral structural proteins of infectious salmon anemia virus. *Journal of Virology* 78, 3063–3071.

Fourrier, M., Lester, K., Markussen, T., Falk, K., Secombes, C.J. *et al.* (2015) Dual mutation events in the haemagglutinin-esterase and fusion protein from an infectious salmon anaemia virus HPR0 genotype promote viral fusion and activation by an ubiquitous host protease. *PLoS ONE* 10(10): e0142020.

Garcia-Rosado, E., Markussen, T., Kileng, O., Baekkevold, E.S., Robertsen, B. *et al.* (2008) Molecular and functional characterization of two infectious salmon anaemia virus (ISAV) proteins with type I interferon antagonizing activity. *Virus Research* 133, 228–238.

Geisbert, T.W., Young, H.A., Jahrling, P.B., Davis, K.J., Larsen, T. *et al.* (2003) Pathogenesis of Ebola hemorrhagic fever in primate models – evidence that hemorrhage is not a direct effect of virus-induced cytolysis of endothelial cells. *American Journal of Pathology* 163, 2371–2382.

Håstein, T., Hill, B.J. and Winton, J.R. (1999) Successful aquatic animal disease emergency programmes. *Revue Scientifique et Technique* 18, 214–227.

Hellebo, A., Vilas, U., Falk, K. and Vlasak, R. (2004) Infectious salmon anemia virus specifically binds to and hydrolyzes 4-O-acetylated sialic acids. *Journal of Virology* 78, 3055–3062.

Jarp, J. and Karlsen, E. (1997) Infectious salmon anaemia (ISA) risk factors in sea-cultured Atlantic salmon *Salmo salar. Diseases of Aquatic Organisms* 28, 79–86.

Jones, S.R.M., Mackinnon, A.M. and Salonius, K. (1999) Vaccination of freshwater-reared Atlantic salmon reduces mortality associated with infectious salmon anaemia virus. *Bulletin of the European Association of Fish Pathologists* 19, 98–101.

Kibenge, F.S., Kibenge, M.J., Wang, Y., Qian, B., Hariharan, S. and McGeachy, S. (2007) Mapping of putative virulence motifs on infectious salmon anemia virus surface glycoprotein genes. *Journal of General Virology* 88, 3100–3111.

Kobayashi, Y., Horimoto, T., Kawaoka, Y., Alexander, D.J. and Itakura, C. (1996) Pathological studies of chickens experimentally infected with two highly pathogenic avian influenza viruses. *Avian Pathology* 25, 285–304.

Lauscher, A., Krossoy, B., Frost, P., Grove, S., Konig, M. *et al.* (2011) Immune responses in Atlantic salmon (*Salmo salar*) following protective vaccination against Infectious salmon anemia (ISA) and subsequent ISA virus infection. *Vaccine* 29, 6392–6401.

LeBlanc, F., Laflamme, M. and Gagne, N. (2010) Genetic markers of the immune response of Atlantic salmon (*Salmo salar*) to infectious salmon anemia virus (ISAV). *Fish and Shellfish Immunology* 29, 217–232.

Lyngstad, T.M., Jansen, P.A., Sindre, H., Jonassen, C.M., Hjortaas, M.J. *et al.* (2008) Epidemiological investigation of infectious salmon anaemia (ISA) outbreaks in Norway 2003–2005. *Preventive Veterinary Medicine* 84, 213–227.

Lyngstad, T.M., Hjortaas, M.J., Kristoffersen, A.B., Markussen, T., Karlsen, E.T. *et al.* (2011) Use of molecular epidemiology to trace transmission pathways for infectious salmon anaemia virus (ISAV) in Norwegian salmon farming. *Epidemics* 3, 1–11.

Lyngstad, T.M., Kristoffersen, A.B., Hjortaas, M.J., Devold, M., Aspehaug, V. *et al.* (2012) Low virulent infectious salmon anaemia virus (ISAV-HPR0) is prevalent and geographically structured in Norwegian salmon farming. *Diseases of Aquatic Organisms* 101, 197–206.

Mardones, F.O., Perez, A.M. and Carpenter, T.E. (2009) Epidemiologic investigation of the re-emergence of infectious salmon anemia virus in Chile. *Diseases of Aquatic Organisms* 84, 105–114.

Markussen, T., Jonassen, C.M., Numanovic, S., Braaen, S., Hjortaas, M. *et al.* (2008) Evolutionary mechanisms involved in the virulence of infectious salmon anaemia virus (ISAV), a piscine orthomyxovirus. *Virology* 374, 515–527.

Marshall, S.H., Ramírez, R., Labra, A., Carmona, M. and Muñoz, C. (2014) Bona fide evidence for natural vertical transmission of infectious salmon anemia virus (ISAV) in freshwater brood stocks of farmed Atlantic salmon (*Salmo salar*) in southern Chile. *Journal of Virology* 88, 6012–6018. 10.1128/JVI.03670-13.

McBeath, A.J.[A.], Collet, B., Paley, R., Duraffour, S., Aspehaug, V. *et al.* (2006) Identification of an interferon antagonist protein encoded by segment 7 of infectious salmon anaemia virus. *Virus Research* 115, 176–184.

McBeath, A.J.[A.], Snow, M., Secombes, C.J., Ellis, A.E. and Collet, B. (2007) Expression kinetics of interferon and interferon-induced genes in Atlantic salmon (*Salmo salar*) following infection with infectious pancreatic necrosis virus and infectious salmon anaemia virus. *Fish and Shellfish Immunology* 22, 230–241.

McBeath, A.J.A., Bain, N. and Snow, M. (2009) Surveillance for infectious salmon anaemia virus HPR0 in marine Atlantic salmon farms across Scotland. *Diseases of Aquatic Organisms* 87, 161–169.

McBeath, A.[J.A.], Aamelfot, M., Christiansen, D.H., Matejusova, I., Markussen, T. *et al.* (2015) Immersion challenge with low and highly virulent infectious salmon anaemia virus reveals different pathogenesis in Atlantic salmon, *Salmo salar* L. *Journal of Fish Diseases* 38, 3–15.

Medina, R.A. and Garcia-Sastre, A. (2011) Influenza A viruses: new research developments. *Nature Reviews: Microbiology* 9, 590–603.

Melville, K.J. and Griffiths, S.G. (1999) Absence of vertical transmission of infectious salmon anemia virus (ISAV) from individually infected Atlantic salmon *Salmo salar. Diseases of Aquatic Organisms* 38, 231–234.

Mikalsen, A.B., Sindre, H., Torgersen, J. and Rimstad, E. (2005) Protective effects of a DNA vaccine expressing the infectious salmon anemia virus hemagglutinin-esterase in Atlantic salmon. *Vaccine* 23, 4895–4905.

Mjaaland, S., Hungnes, O., Teig, A., Dannevig, B.H., Thorud, K. and Rimstad, E. (2002) Polymorphism in the infectious salmon anemia virus hemagglutinin gene: importance and possible implications for evolution and ecology of infectious salmon anemia disease. *Virology* 304, 379–391.

Murray, A.G., Smith, R.J. and Stagg, R.M. (2002) Shipping and the spread of infectious salmon anemia in Scottish aquaculture. *Emerging Infectious Diseases* 8, 1–5.

Nylund, A., Plarre, H., Karlsen, M., Fridell, F., Ottem, K.F. *et al.* (2007) Transmission of infectious salmon anaemia virus (ISAV) in farmed populations of Atlantic salmon (*Salmo salar*). *Archives of Virology* 152, 151–179.

OIE (2015a) Chapter 10.4. Infection with infectious salmon anaemia virus. In: *Aquatic Animal Health Code.* World Organisation for Animal Health, Paris. Updated 2016 version available at: http://www.oie.int/

index.php?id=171&L=0&htmfile=chapitre_isav.htm (accessed 28 October 2016).

OIE (2015b) Chapter 2.3.5. Infection with infectious salmon anaemia virus. In: *Manual of Diagnostic Tests for Aquatic Animals*. World Organisation for Animal Health, Paris. Updated 2016 version available at: http://www.oie.int/fileadmin/Home/eng/Health_ standards/aahm/current/chapitre_isav.pdf (accessed 28 October 2016).

Palese, P. and Shaw, M.L. (2007) *Orthomyxoviridae*: the viruses and their replication. In: Knipe, D.M. and Howley, P.M. (eds) *Fields Virology*, 5th edn. Lippincott Williams & Wilkins, Philadelphia, Pennsylvania, pp. 1647–1689.

Peñaranda, M.M., Wargo, A.R. and Kurath, G. (2011) *In vivo* fitness correlates with host-specific virulence of *Infectious hematopoietic necrosis virus* (IHNV) in sockeye salmon and rainbow trout. *Virology* 417, 312–319.

Plarre, H., Devold, M., Snow, M. and Nylund, A. (2005) Prevalence of infectious salmon anaemia virus (ISAV) in wild salmonids in western Norway. *Diseases of Aquatic Organisms* 66, 71–79.

Purcell, M.K., Garver, K.A., Conway, C., Elliott, D.G. and Kurath, G. (2009) Infectious haematopoietic necrosis virus genogroup-specific virulence mechanisms in sockeye salmon, *Oncorhynchus nerka* (Walbaum), from Redfish Lake, Idaho. *Journal of Fish Diseases* 32, 619–631.

Raynard, R.S., Murray, A.G. and Gregory, A. (2001) Infectious salmon anaemia virus in wild fish from Scotland. *Diseases of Aquatic Organisms* 46, 93–100.

Rimstad, E., Dale, O.B., Dannevig, B.H. and Falk, K. (2011) Infectious salmon anaemia. In: Woo, P.T.K. and Bruno, D.W. (eds) *Fish Diseases and Disorders, Volume 3: Viral, Bacterial and Fungal Infections*, 2nd edn. CAB International, Wallingford, UK, pp. 143–165.

Rivas-Aravena, A., Fuentes, Y., Cartagena, J., Brito, T., Poggio, V. *et al.* (2015) Development of a nanoparticle-based oral vaccine for Atlantic salmon against ISAV using an alphavirus replicon as adjuvant. *Fish and Shellfish Immunology* 45, 157–166.

Rummer, J.L., Wang, S., Steffensen, J.F. and Randall, D.J. (2014) Function and control of the fish secondary vascular system, a contrast to mammalian lymphatic systems. *Journal of Experimental Biology* 217, 751–757.

Schnittler, H.J., Mahner, F., Drenckhahn, D., Klenk, H.D. and Feldmann, H. (1993) Replication of Marburg virus in human endothelial cells. A possible mechanism for the development of viral hemorrhagic disease. *Journal of Clinical Investigation* 91, 1301–1309.

Schultz-Cherry, S., Krug, R.M. and Hinshaw, V.S. (1998) Induction of apoptosis by influenza virus. *Seminars in Virology* 8, 491–495.

Simko, E., Brown, L.L., MacKinnon, A.M., Byrne, P.J., Ostland, V.E. and Ferguson, H.W. (2000) Experimental infection of Atlantic salmon, *Salmo salar* L., with infectious salmon anaemia virus: a histopathological study. *Journal of Fish Diseases* 23, 27–32.

Speilberg, L., Evensen, O. and Dannevig, B.H. (1995) A sequential study of the light and electron microscopic liver lesions of infectious anaemia in Atlantic salmon (*Salmo salar* L.). *Veterinary Pathology* 32, 466–478.

Stewart, W.B., Petenyi, C.W. and Rose, H.M. (1955) The survival time of canine erythrocytes modified by influenza virus. *Blood* 10, 228–234.

Tapia, E., Monti, G., Rozas, M., Sandoval, A., Gaete, A. *et al.* (2013) Assessment of the *in vitro* survival of the infectious salmon anaemia virus (ISAV) under different water types and temperature. *Bulletin of the European Association of Fish Pathologists* 33, 3–12.

Thorud, K.E. (1991) Infectious salmon anaemia – transmission trials. Haematological, clinical and morphological investigations. PhD thesis, Norwegian College of Veterinary Medicine, Oslo, Norway.

Thorud, K. and Djupvik, H.O. (1988) Infectious anaemia in Atlantic salmon (*Salmo salar* L.). *Bulletin of the European Association of Fish Pathologists* 8, 109–111.

Totland, G.K., Hjeltnes, B.K. and Flood, P.R. (1996) Transmission of infectious salmon anaemia (ISA) through natural secretions and excretions from infected smelts of Atlantic salmon Salmo salar during their presymptomatic phase. *Diseases of Aquatic Organisms* 26, 25–31.

Traving, C. and Schauer, R. (1998) Structure, function and metabolism of sialic acids. *Cellular and Molecular Life Sciences* 54, 1330–1349.

Vågsholm, I., Djupvik, H.O., Willumsen, F.V., Tveit, A.M. and Tangen, K. (1994) Infectious salmon anaemia (ISA) epidemiology in Norway. *Preventive Veterinary Medicine* 19, 277–290.

Vike, S., Nylund, S. and Nylund, A. (2009) ISA virus in Chile: evidence of vertical transmission. *Archives of Virology* 154, 1–8.

Wargo, A.R. and Kurath, G. (2012) Viral fitness: definitions, measurement, and current insights. *Current Opinion in Virology* 2, 538–545.

Wolf, A., Hodneland, K., Frost, P., Braaen, S. and Rimstad, E. (2013) A hemagglutinin-esterase-expressing salmonid alphavirus replicon protects Atlantic salmon (*Salmo salar*) against infectious salmon anemia (ISA). *Vaccine* 31, 661–669.

K. Falk and M. Aamelfot

7 Spring Viraemia of Carp

PETER DIXON AND DAVID STONE*

Centre for Environment, Fisheries and Aquaculture Science (Cefas) Weymouth Laboratory, Weymouth, UK

7.1 Introduction

Spring viraemia of carp (SVC) is often a fatal haemorrhagic disease of common carp, *Cyprinus carpio*, and other fishes. It is caused by the spring viraemia of carp virus (SVCV), a rhabdovirus (Fig. 7.1) of the type species, *Carp sprivivirus*, in the genus *Sprivivirus* (Stone *et al.*, 2013; Adams *et al.*, 2014). SVC is reportable to the World Organisation for Animal Health (OIE), but the virus is serologically related to other non-reportable viruses.

7.2 Hosts

SVC has been described in numerous hosts, although some early reports of the disease may actually have been carp erythrodermatitis, a bacterial disease. In addition to the common and koi carp (*C. carpio*), hosts include the following cyprinids (where no citation is referenced, check Dixon, 2008): bighead carp, *Aristichthys nobilis*; crucian carp, *Carassius carassius*; grass carp, *Ctenopharyngodon idella*; silver carp, *Hypophthalmichthys molitrix*; goldfish, *Carassius auratus*; orfe, *Leuciscus idus*; tench, *Tinca tinca*; roach, *Rutilus rutilus*; bream, *Abramis brama* (Basic *et al.*, 2009); and emerald shiner, *Notropis atherinoides* (Cipriano *et al.*, 2011). Non-cyprinid hosts include: sheatfish (other names, European catfish, wels), *Silurus glanis*; pike, *Esox lucius*; Siberian sturgeon, *Acipenser baerii* (Vicenova *et al.*, 2011); largemouth bass, *Micropterus salmoides*; and bluegill sunfish, *Lepomis macrochirus* (Cipriano *et al.*, 2011; Phelps *et al.*, 2012).

SVCV has been isolated from rainbow trout, *Oncorhynchus mykiss* (Stone *et al.*, 2003; Jeremic *et al.*, 2006; Haghighi Khiabanian Asl *et al.*, 2008a; I. Shchelkunov, personal communication) and SVCV nucleotides have been found in trout (Shchelkunov *et al.*, 2005). The virus was not pathogenic to rainbow trout following experimental bath infection and the virus isolated by Jeremic *et al.* (2006) was also not pathogenic to rainbow trout following intraperitoneal (IP) injection (P.F. Dixon, J. Munro and D.M. Stone, unpublished data), though both strains of the virus were pathogenic to common carp. In contrast to the aforementioned results, a recent study (Emmenegger *et al.*, 2016) has shown that rainbow trout were moderately susceptible to a North American isolate of SVCV following IP injection.

Viruses isolated from the black bullhead catfish, *Ictalurus melas* (preferred name *Ameiurus melas*) (Selli *et al.*, 2002) and Nile tilapia, *Sarotherodon niloticus* (preferred name *Oreochromis niloticus*), were identified as SVCV based on tissue immunohistochemistry; Soliman *et al.*, 2008). As antibodies against SVCV will cross-react with other viruses, the aforementioned species should not be considered hosts without more robust confirmation. A reverse transcription polymerase chain reaction (RT-PCR) apparently identified nucleotide sequences of SVCV in tissues from three diseased Indian carp species (rohu, *Labeo rohita*; merigal, *Cirrhinus mrigala*; catla, *Catla catla*), but the virus was not isolated (Haghighi Khiabanian Asl *et al.*, 2008b). However, the nucleotide sequence data deposited at Genbank shares <80% nucleotides with known SVCV nucleotide sequence (D.M. Stone, unpublished data).

A rhabdovirus of penaeid shrimps was isolated from *Litopenaeus stylirostris* and *L. vannamei* in

*Corresponding author e-mail: david.stone@cefas.co.uk

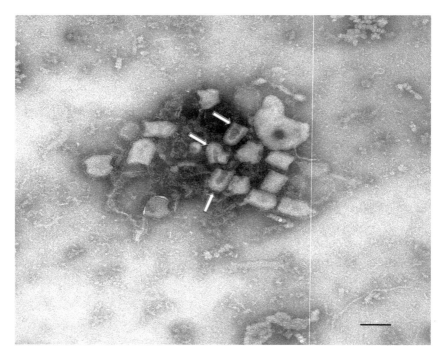

Fig. 7.1. Negatively stained spring viraemia of carp virus (SVCV) seen by transmission electron microscopy showing numerous bullet-shaped particles with a nucleocapsid surrounded by an envelope with prominent spikes on the surface. Bar = 100 nm. Image courtesy Dr J V Warg.

Hawaii, and replicated in both species (see Dixon, 2008). The nucleotide sequence of its G (glycoprotein) gene is identical to that of SVCV, and the two viruses should be considered genuine SVCV. Hence, SVCV has been in Hawaii even though the disease has not been reported. It is also not known whether the virus is still present there.

Other fish species are susceptible to SVCV under experimental conditions, but because there is insufficient supporting data, these will not be discussed. The zebrafish, *Danio rerio*, is important in that it is increasingly used to model SVCV–host interactions (Sanders *et al.*, 2003).

7.3 Geographic Range

SVC was for a long time only recorded in the carp-producing countries of Europe (see Dixon, 2008; Rexhepi *et al.*, 2011), but since 2000, the geographic range of SVCV isolations has expanded, though the number of isolations has not mirrored the occurrence of the disease. For example, the virus has been isolated from cultured common and koi carp in China and found in wild carp in Canada, but the

disease has not been observed in either country (Garver *et al.*, 2007; Zhang *et al.*, 2009). However, SVCV has been associated with mortality in common and koi carp in the USA (see Warg *et al.*, 2007; Phelps *et al.*, 2012). SVCV antigens were detected in goldfish tissues using an enzyme-linked immunosorbent assay (ELISA) in Brazil, but viral isolation was not attempted (Alexandrino *et al.*, 1998), but further confirmation is needed before Brazil is included within the geographic range of SVC.

7.4 Mode of Transmission

SVC has a wide temperature range but field mortalities in carp usually occur between 5 and 18°C (Fijan, 1988; Ahne *et al.*, 2002). Progress of the disease is often rapid above 10°C, and it occurs in spring, particularly following a cold winter. Experimentally, a rise in temperature is not needed to cause disease, but higher mortalities or a more rapid disease course occurred in fish that had overwintered compared with those infected in autumn; the poor condition of the overwintered fish was a possible risk factor (Baudouy *et al.*, 1980a,b,c).

P. Dixon and D. Stone

In aquaculture, 9–12 and 21–24 month old carp are routinely affected, although all ages are susceptible (Fijan, 1988). In experimental studies, grass carp and common carp became more resistant with age (Shchelkunov and Shchelkunova, 1989), but adult wild common carp was affected (Marcotegui et al., 1992; Dikkeboom et al., 2004; Phelps et al., 2012). Also, under experimental conditions, wild common carp were more susceptible to SCV than farmed carp (Hill, 1977).

The transmission of SVCV is horizontal (Fijan, 1988). The virus has been isolated from ovarian fluid (Békési and Csontos, 1985), but vertical transmission has not been demonstrated. SVCV appears to enter via the gills (see Dixon, 2008) and then spreads to the kidney, liver, heart, spleen and alimentary tract. During disease outbreaks, high titres of virus are detected in the liver and kidneys of infected fish, whereas lower titres are found in the gills, spleen and brain (Fijan et al., 1971).

The virus is released into the environment from carcasses and by excretion from infected fish (see Dixon, 2008). There is an inverse relationship between temperature and the duration of viral survival outside the host, e.g. 35 days in river water at 10°C and 42 days in pond sludge at 4°C (Ahne, 1982a,b).

SVCV shed by survivors is probably the main way the virus is transmitted, but not much is known about persistence of the virus in infected fish, the duration of viral shedding or amounts of virus shed. Virus may be shed following a stressful event, particularly from fish in poor condition in the spring following a harsh winter (Fijan, 1988).

The mechanical transmission of SVCV by herons (Peters and Neukirch, 1986), carp louse, *Argulus foliaceus*, and leech, *Piscicola geometra* (Ahne, 1978, 1985a; Pfeil-Putzien, 1978) has been demonstrated experimentally, but is not known whether such transmission occurs in nature. Pfeil-Putzien and Baath (1978) isolated SVCV from carp lice taken from naturally diseased fish, which indicates uptake of the virus from an infected host. Pike preying upon SVCV-infected pike fry became infected (Ahne, 1985b). SVCV may be translocated through the movement of bait fish (Goodwin et al., 2004; Misk et al., 2016), and invertebrates may be reservoirs of the virus, but there is no evidence to support either mode of transmission.

7.5   Impacts on Fish Production

SVC has caused major losses in carp production, particularly in European countries and in states of the former USSR (Ahne et al., 2002), although precise figures are hard to obtain because epizootics are sporadic and losses fluctuate annually (Fijan, 1999). A historic figure from 1980 estimated that SVC caused an annual loss of 4000 t of 1-year-old carp (10–15% of that age group) in Europe (see Sano et al., 2011). In some countries, carp are bred for sport fisheries, which is a lucrative leisure industry. Estimated figures from the UK collated in 1997 showed that the cost of the disease was £20,000–230,000 for fish farms, upwards of £30,000 for fisheries and £20,000–30,000 for retail outlets (see Taylor et al., 2013). Further losses of production occur in control programmes that require the slaughter of infected fish (Taylor et al., 2013), and any control programme will have a detrimental financial impact (see Dixon, 2008).

7.6   Clinical Signs of Disease

The clinical signs of SVC are non-specific, and not all fish will exhibit all of the signs. Initially there may be increasing mortality, often with a rapid onset. Infected fish may swim slowly and erratically, lose balance and swim on their sides. Two of the most obvious and consistent features are abdominal distension and haemorrhages. The latter may occur on the skin, fin bases, eyes and gills, which may be pale (Fig. 7.2). The skin may darken and exophthalmia is often observed. The vent may be swollen, inflamed and trail mucoid casts. The abdomen is usually filled with a clear fluid, sometimes bloodstained. The spleen is often enlarged, most internal organs are usually oedematous, and organs adhere to each other and to the peritoneum. There are petechial haemorrhages in the musculature and the swim bladder (see Dixon, 2008), though these are uncommon in SVC caused by Asian strains of SVCV (Goodwin, 2003; Dikkeboom et al., 2004).

Gaafar et al. (2011) conducted a detailed histopathological study in common carp and compared their findings with those of other authors, and Misk et al. (2016) recently reported the histopathology of SVC in three experimental hosts. Histopathological studies commonly reveal oedema, inflammation, haemorrhage and necrosis in the liver, pancreas, intestine, spleen, kidney, heart, muscles and air bladder, but again, not all fish will exhibit each feature.

Early in disease development, there is hyperplasia in the spleen, especially of the reticuloendothelium,

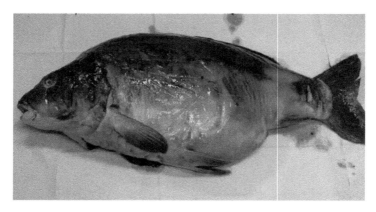

Fig. 7.2. Gross appearance of a common carp (*Cyprinus carpio*) showing the abdominal distension, extensive haemorrhage and exophthalmia consistent with an infection of spring viraemia of carp virus (SVCV).

and the pulp can fill with erythrocytes, while in the later stages of the disease there is dystrophy and cellular degeneration. There can be increased pigmentation, because of larger numbers of iron-containing macrophages (Fig. 7.3A), and degenerative and proliferative changes in the kidney. In severe disease, there is marked oedema, with dissociation of the cells and liquid in the renal glomeruli and clogging in some tubuli. There is focal exfoliation of the epithelium of the urinary tract and nephrosis and peritubular oedema. Haemopoietic tissue is oedematous and its cells show dystrophy and degeneration with haemorrhaging (Fig. 7.3B). In less severe cases, there is an increase in blood filling the kidney and hyperplasia of the haemopoietic tissue, with increased numbers of melanomacrophages.

Periglandular oedema occurs in the islets of the pancreas with infiltration by leucocytes. Hyperplasia of the hepatocytes, and abnormal hepatocytes may be present. There can also be proliferative changes. In liver oedema of the parenchyma, focal or diffuse erythrodiapedesis (extravasation of erythrocytes) and fatty degeneration of hepatocytes are often seen. The hepatic cells are enlarged and exhibit granular or vacuolated dystrophy. The intestine may have degenerative and proliferative changes (Fig. 7.3C). There may be perivascular inflammation of intramural vessels, mucoid degeneration, necrosis and desquamation of the epithelium. Late in infection, the villi may be atrophied. Focal cellular infiltration may occur in the heart in the epicardium and myocardium, followed by focal degeneration and necrosis (Fig. 7.3D). Muscle bundles are friable with the elimination of striations.

The blood vessels of the brain may enlarge with pericellular oedema of the neurons. Eosinic inclusions have been observed in a small number of glial cells. Inflammation of the peritoneum can occur. There can be dilated lymph vessels filled with detritus, macrophages and lymphocytes. Internal mucus may thicken and contain foci of tissue degeneration where the epithelium is exfoliated and desquamated. The gills may show degenerative and proliferative changes, necrosis and infiltrations. There may be diffuse lamellar fusion and hyperplasia of the epithelial lining at the base of cellular lamellae and diffuse branchial necrosis. The numbers of melanomacrophages may increase.

7.7 Diagnosis

Definitive diagnosis of SVC is made by isolation of the virus in cell culture followed by its identification using serological and/or molecular methods. The OIE, in its *Manual of Diagnostic Tests for Aquatic Animals*, includes comprehensive methodology to diagnose and identify SVC (OIE, 2015a). It recommends either the epithelioma papulosum cyprini (EPC) (Fijan *et al.*, 1983) or fathead minnow (FHM) (ATCC CCL-42) cell lines for isolation. When there are high levels of virus in clinical fish, their tissues can be directly tested using serological or molecular methods to provide a rapid presumptive diagnosis while awaiting the results of viral isolation. The isolation of SVCV from survivors, or those with a sublethal infection, is not readily achieved, which can make tracking the distribution of virus-exposed fish difficult. The viral

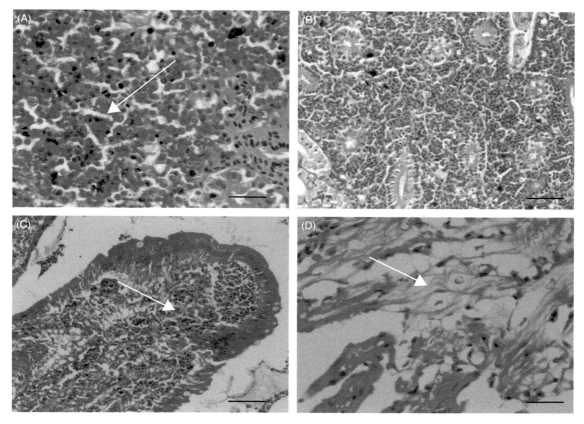

Fig. 7.3. Haematoxylin and eosin stained tissue sections from carp infected with spring viraemia of carp virus (SVCV) showing: (A) a region of cardiomyocytic necrosis (arrow) with loss of cytoplasmic staining and nuclear degeneration; (B) diffuse haemorrhaging within the renal interstitium; (C) haemorrhaging within the lamina propria of an intestinal villus (arrow); and (D) increased numbers of melanomacrophages in the renal interstitial tissue (arrow), Bar = 50 μm in panels A–C and 25 μm in panel D. Images courtesy Dr S.W. Feist.

screening of apparently healthy carp in China and Canada has identified SVCV in both countries (Liu *et al.*, 2004; Garver *et al.*, 2007). Viral screening in cell culture after each of two consecutive 7 day periods could be required for concluding that a population is negative. Although there is no evidence that viral propagation takes more than 7 days, the cultures can also be checked using molecular methods to enhance the robustness of a negative result.

Virus isolated in cell culture must be identified, but some antisera cross-react with other viruses, such as pike fry rhabdovirus (PFRV). Antisera may also react with some isolates and not others (see Dixon, 2008). Hence definitive identification of SVCV in culture is best achieved by using RT-PCR-based methods. The assay adopted by the OIE targets a partial nucleotide sequence of the G gene of a wide range of SVCV isolates and other closely related viral isolates that could be misidentified as SVCV serologically (Stone *et al.*, 2003). According to Stone *et al.* (2003), there are four genogroups: genogroup I comprises SVCV isolates, genogroup II is a single isolate from grass carp, genogroup III comprises the reference PFRV isolate, and genogroup IV, also called the tench rhabdovirus group, comprises unassigned isolates and isolates previously identified as PFRV. However, the International Committee for Taxonomy of Viruses (ICTV) recognizes only two species, SVCV and PFRV, with the grass carp rhabdovirus and tench rhabdovirus considered to belong within the PFRV species (Adams *et al.*, 2014).

According to Stone *et al.* (2003), isolates of geno-group I have significant nucleotide sequence divergence, and can be divided into four subgroups by geographic origin. Subgroup Ia isolates are from Asia, the USA and Canada (Garver *et al.*, 2007; Phelps *et al.*, 2012). Subgroup Ib is from Moldova and Ukraine. Subgroup Ic is from Ukraine and Russia. Subgroup Id is from Europe, plus one isolate from Moldova and one from Ukraine. A potential fifth subgroup, Ie, comprises isolates from Austria (D.M. Stone unpublished data). Phylogenetic comparisons of isolates based on the G gene may become a valuable tool for tracking the source of SVCV isolates (Stone *et al.*, 2003).

Additional conventional RT-PCR, quantitative real-time PCR and loop-mediated isothermal amplification assays are available for detecting and confirming SVCV infections (Koutná *et al.*, 2003; Liu *et al.*, 2008a,b; Yue *et al.*, 2008; Zhang *et al.*, 2009; Shimahara *et al.*, 2016). With the exception of the conventional PCR assay developed by Shimahara *et al.* (2016), the rest of these assays were not validated against representatives from each of the recognized SVCV genogroups, and in the case of the real-time assays (Liu *et al.*, 2008a; Yue *et al.*, 2008), they failed to detect isolates from at least one subgroup (D.M. Stone, unpublished data). A generic primer set based on the polymerase gene also identifies viruses from both the *Sprivivirus* and *Perhabdovirus* genera, and this can be used to screen a virus culture (Ruane *et al.*, 2014).

7.8 Pathophysiology

Infected carp produce a temperature-dependent humoral response against SVCV. Carp infected at 13–14°C responded more slowly and had lower neutralizing antibody titres than those at 25°C (Fijan *et al.*, 1977). At 10–12°C, mortalities due to SVCV in carp reached 90% with no detectable neutralizing antibody, while there were no mortalities at 20–22°C and neutralizing antibody was detected 30 days postinfection (Ahne, 1980). Interferon was produced within 24 h of infection (Baudouy, 1978). In separate experiments, IP injection of carp with SVCV resulted in: (i) the upregulation of several antiviral molecules (Feng *et al.*, 2011); (ii) significant increases in the expression of the natural killer cell enhancing factor beta gene in the blood cells, gills, intestine and spleen (Huang *et al.*, 2009); and (iii) significant upregulation of prothymosin alpha transcripts in the kidney,

peripheral blood, spleen and intestine, and upregulation of thymosin beta transcripts in the intestine, peripheral blood, liver and spleen (Xiao *et al.*, 2015) – both of these peptides stimulate the immune response.

The pathophysiology of SVC is not well studied. What is known was derived from studies of natural infections and from one experimental infection. Řehulka (1996) studied the development of SVC in carp that had been taken from a State Fishery overwintering pond in the Czech Republic and found to be susceptible to SVC at the end of their hibernation. They were transferred to an experimental pond early in April, fed on pellets, and the course of disease development studied until early June. In diseased carp (as confirmed by virus isolation and neutralization), there was a decrease in lymphocytes, an increase in neutrophils, particularly myelocytes and metamyelocytes, and an increase in monocytes, These changes were associated with the worsening health of the fish and were accompanied by a change in morphology of the cells, particularly vacuolization of the nucleus and cytoplasm. There were also indications of anaemia. Overall, the disease caused almost 60% mortality. During the course of the disease, Osadchaya and Rudenko (1981) reported a marked decrease in glycogen synthesis in the liver as indicated by a practically complete absence of glycogen granules in hepatocytes and a very small quantity of iron-containing pigment in the splenic pulp. According to Fijan (1999), the multiplication of SVCV in capillary epithelium, and in haematopoietic and excretory kidney tissues affected osmoregulation, which could be fatal.

Jeney *et al.* (1990) showed that following experimental infection of fingerling sheatfish, haematocrit values and haemoglobin concentrations significantly decreased, suggesting that haemopoesis was affected. Serum glutamic oxaloacetic transaminase and glutamic pyruvic transaminase levels increased significantly following infection, indicative of tissue necrosis. In black carp, *Mylopharyngodon piceus*, there was an increase in mitochondria antiviral signalling protein (part of the innate immune response) mRNA in the intestine, muscle and liver at 33 h postinjection, but a decrease occurred in the spleen (Zhou *et al.*, 2015).

7.9 Prevention and Control

Both the prevention and control of SVC are currently enacted through legislation and good management.

Prevention has also been attempted by vaccination and the use of resistant strains of carp. For example, Kirpichnikov *et al.* (1993) reported that the hybrid line of Ukrainian-Ropsha carp had increased resistance to SVC compared with other stocks or strains of carp.

Commercial vaccines underwent field trials in the former Czechoslovakia in the 1970s and 1980s (see Fijan, 1984, 1988; Dixon, 1997; Sano *et al.*, 2011). However, the vaccine is no longer available, and problems with interpretation of the trials have been discussed by Dixon (1997). Oral vaccination using low virulence SVCV, or virus attenuated by passage through cell cultures, has been attempted. Laboratory experiments showed promise, but a large-scale field trial in pond-cultured fish was unsuccessful (Fijan, 1988). Oral vaccination utilizing different attenuated viruses was also investigated (Kölbl, 1980, 1990). Survival in field trials was encouraging (15% mortality in vaccinated fish versus 49% mortality in control fish). The vaccine was modified by an unspecified method and vaccination in the field was successful as judged by no SVC being observed along with an increase in yield; for example, there was an increase from a low yield of 15 t at one site which was attributed to SVC, up to 50 t in the first year following vaccination, and to 70 t the year after; the vaccine did not go on to be commercialized.

Different approaches to vaccine production have subsequently been undertaken, but none have produced a commercial vaccine. Recombinant SVCV glycoprotein injected into carp produced neutralizing antibodies but did not protect fish against SVCV (see Dixon, 1997). A genetically engineered *Lactobacillus plantarum* with a surface-expressing SVCV glycoprotein and cyprinid herpesvirus-3 (CyHV-3) ORF81 protein was incorporated into pelleted feed as an oral vaccine (Cui *et al.*, 2015). Under experimental conditions, there was 29% mortality in vaccinated carp compared with 78 and 89% mortality in control fish challenged with SVCV. The vaccine also protected koi carp against CyHV-3 (see Chapter 10). There have been laboratory trials with DNA vaccines based on the G gene of SVCV. Different combinations of SVCV DNA plasmids containing partial or complete G gene fragments were compared (Kanellos *et al.*, 2006). Two vaccinated groups of common carp had relative percentage survival (RPS) values of 33 and 48% in challenges that produced >60% mortality in control fish. Emmenegger and Kurath (2008) used a single DNA vaccine to immunize koi carp against SVCV and the RPS ranged from 50 to 70%.

Interferon-inducer double-stranded RNA injected into carp prior to challenge with SVCV reduced mortality from 100% to 22–40% (Masycheva *et al.*, 1995; Alikin *et al.*, 1996). The use of liposomes containing poly(I:C) (polyinosinic–polycytidylic acid, a synthetic interferon inducer) and bacterial lipopolysaccharide as an immunostimulant prior to SVCV infection has also been investigated (Ruyra *et al.*, 2014). The administration of the liposomes to zebrafish by immersion or IP injection reduced mortality (RPS of 33.3 and 42.3%, respectively) following challenge. Methisoprinol (an antiviral agent) inhibited the replication of SVCV *in vitro*, but its efficacy was not tested in fish (Siwicki *et al.*, 2003).

Other approaches to preventing or ameliorating SVC include the identification of multipath genes in zebrafish surviving experimental SVCV infection in order to develop drugs that might prevent the disease (Encinas *et al.*, 2013), investigating autophagy or RNA inhibition as a method to suppress SVCV replication (Garcia-Valtanen *et al.*, 2014a; Gotesman *et al.*, 2015; Liu *et al.*, 2015), investigating the use of plasmid constructs expressing glycoproteins of viral haemorrhagic septicaemia virus as a molecular adjuvant for an SVCV G gene DNA vaccine (Martinez-Lopez *et al.*, 2014), and investigating the immunomodulatory properties of zebrafish β-defensin and its role as a DNA vaccine molecular adjuvant (Garcia-Valtanen *et al.*, 2014b).

Without commercial vaccines or therapeutic substances, biosecurity strategies remain the most effective ways of preventing SVC. These range from national or international legislation or standards to good management practices at the farm level. The latter include on-site quarantine of new fish stocks, the disinfection of equipment, use of footbaths, and reducing stress and other diseases, etc.

Several studies have identified disinfectants and other chemical or physical means for inactivating the virus (see Kiryu *et al.*, 2007; Dixon, 2008; Dixon *et al.*, 2012), although the results from these studies must be interpreted with care. For example, 99.9% of SVCV was inactivated in hydrochloric acid after 2 h at pH 3.0 (Ahne, 1976), but in formic acid at pH 4.0 (to mimic the procedure used to dispose of dead fish), the virus survived for at least 28 days (Dixon *et al.*, 2012).

Legislation to control the introduction and spread of disease within a country is the most important way of preventing and controlling SVC. Such legislation is often strengthened by international standards or agreements. The OIE's *Aquatic Animal Health Code* (OIE, 2015b) outlines approaches to biosecurity that can be implemented from the national level down to a production site. Håstein *et al.* (2008) have outlined an overview of the implementation of biosecurity strategies, which uses SVC as a disease-specific case study. Oidtmann *et al.* (2011) reviewed the international standards relating to the movement of fish and fish products, particularly with emphasis on national measures to combat fish disease. The UK has implemented a fish health strategy to establish freedom from SVC, which has been outlined by Taylor *et al.* (2013). Legislative controls may have led to freedom from SVC in Croatia, and reduced the incidence of SVC in Hungary (Molnár and Csaba, 2005; Oraic and Zrncic, 2005). However, the epizootiology of the virus in individual countries will determine the feasibility of establishing an SVC-free country. For example, SVC occurs predominantly in wild fish in the USA (Phelps *et al.*, 2012), where establishing freedom from disease is difficult to achieve. Jeremic *et al.* (2004) were also pessimistic about an SVC-free Serbia because of repeated transfer of the disease from cultivated to wild carp and vice versa. None the less, even though SVC is widespread within a country, it may still be possible to establish disease-free zones, provided that fish movements are strictly controlled. Legislation to prevent the introduction of SVC via contaminated fish or fish products is an important approach. All movements, whether they involve live fish or processed contaminated products, would be strictly controlled. Non-approved movements of such products may have introduced or reintroduced SVC into the USA and the UK (Håstein *et al.*, 2008; Taylor *et al.*, 2013). Therefore, continual surveillance of imported fish and fish products is essential.

7.10 Conclusions and Future Research

SVC has caused major losses in carp production, but obtaining accurate figures is complicated by the sporadic nature of the disease and annual fluctuations in losses. Clinical signs are non-specific and diagnosis is via isolation of the virus and subsequent identification using serological and/or molecular biology methods. The OIE provides comprehensive methodology for the diagnosis and identification of SVC using the EPC or FHM cell lines cultured at 20°C for isolation of the virus. The confirmation of SVCV is best achieved using RT-PCR-based methods followed by sequence analysis. Analysis of the partial G gene sequence allows the assignment of SVCV into one of four distinct genogroups.

Effective prevention and control of the disease are currently best achieved by a combination of legislation and good management practices. Vaccination and the use of resistant strains of carp are possible, but no commercial vaccine is available. Non-approved movement of fish products may have been responsible for the introduction or reintroduction of SVC into the USA and the UK, and emphasizes the need for the continual surveillance of imported fish and fish products.

The disease has not been reported in China, although the virus has been detected on several occasions. Apparently healthy SVCV-infected common carp were also found in Canada, which is in stark contrast to high mortality attributed to the Asian SVCV in common and koi carp in the USA. This suggests a significant degree of variation in the pathogenicity of Asian SVCV strains, which needs further investigation.

The transmission of SVCV and the permissive temperatures for this are generally well understood, but the survival and persistence of SVCV is quite variable and needs further study. Host age and genetics, the water temperature and general welfare may determine the progression of the disease.

The use of *in situ* hybridization and real-time qPCR techniques on archived material (e.g. formalin fixed organs) will aid future investigations into the persistence of SVC in carrier fish and the potential re-emergence of the disease. These tools are also invaluable for detecting potential alternative hosts.

References

Adams, M.J., Lefkowitz, E.J., King, A.M.Q. and Carstens, E.B. (2014) Ratification vote on taxonomic proposals to the International Committee on Taxonomy of Viruses (2014). *Archives of Virology* 159, 2831–2841.

Ahne, W. (1976) Untersuchungen über die Stabilität des karpfenpathogenen Virusstammes 10/3. *Fisch und Umwelt* 2, 121–127.

Ahne, W. (1978) Untersuchungsergebnisse über die akute Form der Bauchwassersucht (Frühlingsvirämie) der Karpfen. *Fischwirt* 28, 46–47.

Ahne, W. (1980) Rhabdovirus carpio-Infektion beim Karpfen (*Cyprinus carpio*): Untersuchungen über Reaktionen des Wirtsorganismus. *Fortschritte in der Veterinärmedizin* 30, 180–183.

Ahne, W. (1982a) Untersuchungen zur Tenazität der Fischviren. *Fortschritte in der Veterinärmedizin* 35, 305–309.

Ahne, W. (1982b) Vergleichende Untersuchungen über die Stabilität von vier fischpathogenen Viren (VHSV, PFR, SVCV, IPNV). *Zentralblatt für Veterinärmedizin, Reihe B* 29, 457–476.

Ahne, W. (1985a) *Argulus foliaceus* L. and *Piscicola geometra* L. as mechanical vectors of spring viraemia of carp virus (SVCV). *Journal of Fish Diseases* 8, 241–242.

Ahne, W. (1985b) Viral infection cycles in pike (*Esox lucius* L.). *Journal of Applied Ichthyology* 1, 90–91.

Ahne, W., Bjorklund, H.V., Essbauer, S., Fijan, N., Kurath, G. et al. (2002) Spring viremia of carp (SVC). *Diseases of Aquatic Organisms* 52, 261–272.

Alexandrino, A.C., Ranzani-Paiva, M.J.T. and Romano, L.A. (1998) Identificación de viremia primaveral de la carpa (VPC) *Carrassius auratus* en San Pablo, Brasil. *Revista Ceres* 45, 125–137.

Alikin, Y.S., Shchelkunov, I.S., Shchelkunova, T.I., Kupinskaya, O.A., Masycheva, V.I. et al. (1996) Prophylactic treatment of viral diseases in fish using native RNA linked to soluble and corpuscular carriers. *Journal of Fish Biology* 49, 195–205.

Basic, A., Schachner, O., Bilic, I. and Hess, M. (2009) Phylogenetic analysis of spring viremia of carp virus isolates from Austria indicates the existence of at least two subgroups within genogroup Id. *Diseases of Aquatic Organisms* 85, 31–40.

Baudouy, A.-M. (1978) Relation hôte-virus au cours de la virémie printanière de la carpe. *Comptes Rendus Hebdomadaires des Séances de l'Académie des Sciences. Serie D: Sciences Naturelles* 286, 1225–1228.

Baudouy, A.-M., Danton, M. and Merle, G. (1980a) Experimental infection of susceptible carp fingerlings with spring viremia of carp virus, under wintering environmental conditions. In: Ahne, W. (ed.) *Fish Diseases. Third COPRAQ-Session*. Springer, Berlin, pp. 23–27.

Baudouy, A.-M., Danton, M. and Merle, G. (1980b) Virémie printanière de la carpe. Résultats de contaminations expérimentales effectuées au printemps. *Annales de Recherches Vétérinaires* 11, 245–249.

Baudouy, A.-M., Danton, M. and Merle, G. (1980c) Virémie printanière de la carpe: étude expérimentale de l'infection évoluant à différentes températures. *Annales de l'Institut Pasteur/Virologie* 131E, 479–488.

Békési, L. and Csontos, L. (1985) Isolation of spring viraemia of carp virus from asymptomatic broodstock carp, *Cyprinus carpio* L. *Journal of Fish Diseases* 8, 471–472.

Cipriano, R.C., Bowser, P.R., Dove, A., Goodwin, A. and Puzach, C. (2011) Prominent emerging diseases within the United States. In: *Bridging America and Russia with Shared Perspectives on Aquatic Animal Health. Proceedings of the Third Bilateral Conference between Russia and the United States, 12–20 July,* *2009 Shepherdstown, W V.* Khaled bin Sultan Living Oceans Foundation, Landover, Maryland, pp. 6–17.

Cui, L.C., Guan, X.T., Liu, Z.M., Tian, C.Y. and Xu, Y.G. (2015) Recombinant lactobacillus expressing G protein of spring viremia of carp virus (SVCV) combined with ORF81 protein of koi herpesvirus (KHV): a promising way to induce protective immunity against SVCV and KHV infection in cyprinid fish via oral vaccination. *Vaccine* 33, 3092–3099.

Dikkeboom, A.L., Radi, C., Toohey-Kurth, K., Marcquenski, S.V., Engel, M. et al. (2004) First report of spring viremia of carp virus (SVCV) in wild common carp in North America. *Journal of Aquatic Animal Health* 16, 169–178.

Dixon, P.F. (1997) Immunization with viral antigens: viral diseases of carp and catfish. In: Gudding, R., Lillehaug, A., Midtlyng, P.J. and Brown, F. (eds) *Fish Vaccinology*. Karger, Basel, pp. 221–232.

Dixon, P.F. (2008) Virus diseases of cyprinids. In: Eiras, J.C., Segner, H., Wahli, T. and Kapoor, B.G. (eds) *Fish Diseases*. Science Publishers, Enfield, New Hampshire, pp. 87–184.

Dixon, P.F., Smail, D.A., Algoet, M., Hastings, T.S., Bayley, A. et al. (2012) Studies on the effect of temperature and pH on the inactivation of fish viral and bacterial pathogens. *Journal of Fish Diseases* 35, 51–64.

Emmenegger, E.J. and Kurath, G. (2008) DNA vaccine protects ornamental koi (*Cyprinus carpio koi*) against North American spring viremia of carp virus. *Vaccine* 26, 6415–6421.

Emmenegger, E.J., Sanders, G.E., Conway, C.M., Binkowski, F.P., Winton, J.R. et al. (2016) Experimental infection of six North American fish species with the North Carolina strain of spring viremia of carp virus. *Aquaculture* 450, 273–282.

Encinas, P., Garcia-Valtanen, P., Chinchilla, B., Gomez-Casado, E., Estepa, A. et al. (2013) Identification of multipath genes differentially expressed in pathway-targeted microarrays in zebrafish infected and surviving spring viremia carp virus (SVCV) suggest preventive drug candidates. *PLoS ONE* 8(9): e73553.

Feng, H., Liu, H., Kong, R.Q., Wang, L., Wang, Y.P. et al. (2011) Expression profiles of carp IRF-3/-7 correlate with the up-regulation of RIG-I/MAVS/TRAF3/TBK1, four pivotal molecules in RIG-I signaling pathway. *Fish and Shellfish Immunology* 30, 1159–1169.

Fijan, N. (1984) Vaccination of fish in European pond culture: prospects and constraints. In: Oláh, J. (ed.) *Fish, Pathogens and Environment in European Polyculture*. Akadémiai Kiadó, Budapest, pp. 233–241.

Fijan, N. (1988) Vaccination against spring viraemia of carp. In: Ellis, A.E. (ed.) *Fish Vaccination*. Academic Press, London, pp. 204–215.

Fijan, N. (1999) Spring viraemia of carp and other viral diseases and agents of warm-water fish. In: Woo, P.T.K. and Bruno, D.W. (eds) *Fish Diseases and Disorders,*

Volume 3: Viral, Bacterial and Fungal Infections, 1st edn. CAB International, Wallingford, UK, pp. 177–244.

Fijan, N., Petrinec, Z., Sulimanovic, D. and Zwillenberg, L.O. (1971) Isolation of the viral causative agent from the acute form of infectious dropsy of carp. *Veterinarsky Arhiv* 41, 125–138.

Fijan, N., Petrinec, Z., Stancl, Z., Kezic, N. and Teskeredzic, E. (1977) Vaccination of carp against spring viraemia: comparison of intraperitoneal and peroral application of live virus to fish kept in ponds. *Bulletin de l'Office International des Epizooties* 87, 441–442.

Fijan, N., Sulimanovic, D., Bearzotti, M., Muzinic, D., Zwillenberg, L.O. *et al.* (1983) Some properties of the *Epithelioma papulosum cyprini* (EPC) cell line from carp *Cyprinus carpio*. *Annales de l'Institut Pasteur/Virologie* 134, 207–220.

Gaafar, A.Y., Veselý, T., Nakai, T., El-Manakhly, E.M., Soliman, M.K. *et al.* (2011) Histopathological and ultrastructural study of experimental spring viraemia of carp (SVC) infection of common carp with comparison between different immunohistodignostic techniques efficacy. *Life Science Journal* 8, 523–533.

Garcia-Valtanen, P., Ortega-Villaizan, M.D., Martinez-Lopez, A., Medina-Gali, R., Perez, L. *et al.* (2014a) Autophagy-inducing peptides from mammalian VSV and fish VHSV rhabdoviral G glycoproteins (G) as models for the development of new therapeutic molecules. *Autophagy* 10, 1666–1680.

Garcia-Valtanen, P., Martinez-Lopez, A., Ortega-Villaizan, M., Perez, L., Coll, J.M. *et al.* (2014b) In addition to its antiviral and immunomodulatory properties, the zebrafish beta-defensin 2 (zfBD2) is a potent viral DNA vaccine molecular adjuvant. *Antiviral Research* 101, 136–147.

Garver, K.A., Dwilow, A.G., Richard, J., Booth, T.F., Beniac, D.R. *et al.* (2007) First detection and confirmation of spring viraemia of carp virus in common carp, *Cyprinus carpio* L., from Hamilton Harbour, Lake Ontario, Canada. *Journal of Fish Diseases* 30, 665–671.

Goodwin, A.E. (2003) Differential diagnosis: SVCV vs KHV in koi. *Fish Health Newsletter, American Fisheries Society, Fish Health Section* 31, 9–13.

Goodwin, A.E., Peterson, J.E., Meyers, T.R. and Money, D.J. (2004) Transmission of exotic fish viruses: the relative risks of wild and cultured bait. *Fisheries* 29, 19–23.

Gotesman, M., Soliman, H., Besch, R. and El-Matbouli, M. (2015) Inhibition of spring viraemia of carp virus replication in an *Epithelioma papulosum cyprini* cell line by RNAi. *Journal of Fish Diseases* 38, 197–207.

Haghighi Khiabanian Asl, A., Bandehpour, M., Sharifnia, Z. and Kazemi, B. (2008a) The first report of spring viraemia of carp in some rainbow trout propagation and breeding by pathology and molecular techniques in Iran. *Asian Journal of Animal and Veterinary Advances* 3, 263–268.

Haghighi Khiabanian Asl, A., Azizzadeh, M., Bandehpour, M., Sharifnia, Z. and Kazemi, B. (2008b) The first report of SVC from Indian carp species by PCR and histopathologic methods in Iran. *Pakistan Journal of Biological Sciences* 11, 2675–2678.

Håstein, T., Binde, M., Hine, M., Johnsen, S., Lillehaug, A. *et al.* (2008) National biosecurity approaches, plans and programmes in response to diseases in farmed aquatic animals: evolution, effectiveness and the way forward. *Revue Scientifique et Technique* 27, 125–145.

Hill, B.J. (1977) Studies on SVC virulence and immunization. *Bulletin de l'Office International des Epizooties* 87, 455–456.

Huang, R., Gao, L.-Y., Wang, Y.-P., Hu, W. and Guo, Q.-L. (2009) Structure, organization and expression of common carp (*Cyprinus carpio* L.) NKEF-B gene. *Fish and Shellfish Immunology* 26, 220–229.

Jeney, G., Jeney, Z., Oláh, J. and Fijan, N. (1990) Effect of rhabdovirus infections on selected blood parameters of wels (*Silurus glanis* L.). *Aquacultura Hungarica* 6, 153–160.

Jeremic, S., Dobrila, J.-D. and Radosavljevic, V. (2004) Dissemination of spring viraemia of carp (SVC) in Serbia during the period 1992–2002. *Acta Veterinaria (Beograd)* 54, 289–299.

Jeremic, S., Ivetic, V. and Radosavljevic, V. (2006) *Rhabdovirus carpio* as a causative agent of disease in rainbow trout (*Oncorhynchus mykiss*-Walbaum). *Acta Veterinaria (Beograd)* 56, 553–558.

Kanellos, T., Sylvester, I.D., D'Mello, F., Howard, C.R., Mackie, A. *et al.* (2006) DNA vaccination can protect *Cyprinus carpio* against spring viraemia of carp virus. *Vaccine* 24, 4927–4933.

Kirpichnikov, V.S., Ilyasov, J.I., Shart, L.A., Vikhman, A.A., Ganchenko, M.V. *et al.* (1993) Selection of Krasnodar common carp (*Cyprinus carpio* L.) for resistance to dropsy: principal results and prospects. *Aquaculture* 111, 7–20.

Kiryu, I., Sakai, T., Kurita, J. and Iida, T. (2007) Virucidal effect of disinfectants on spring viremia of carp virus. *Fish Pathology* 42, 111–113.

Kölbl, O. (1980) Diagnostic de la virémie printanière de la carpe et essais d'immunisation contre cette maladie. *Bulletin de l'Office International des Epizooties* 92, 1055–1068.

Kölbl, O. (1990) Entwicklung eines Impfstoffes gegen die Fruhjahrsviramie der Karpfen (spring viraemia of carp, SVC); Feldversuchserfahrungen. *Tierärztliche Umschau* 45, 624–649.

Koutná, M., Vesely, T., Psikal, I. and Hulová, J. (2003) Identification of spring viraemia of carp virus (SVCV) by combined RT-PCR and nested PCR. *Diseases of Aquatic Organisms* 55, 229–235.

Liu, H., Gao, L., Shi, X., Gu, T., Jiang, Y. *et al.* (2004) Isolation of spring viraemia of carp virus (SVCV) from cultured koi (*Cyprinus carpio koi*) and common carp

(*C. carpio carpio*) in PR China. *Bulletin of the European Association of Fish Pathologists* 24, 194–202.

Liu, L.Y., Zhu, B.B., Wu, S.S., Lin, L., Liu, G.X. *et al.* (2015) Spring viraemia of carp virus induces autophagy for necessary viral replication. *Cellular Microbiology* 17, 595–605.

Liu, Z., Teng, Y., Liu, H., Jiang, Y., Xie, X. *et al.* (2008a) Simultaneous detection of three fish rhabdoviruses using multiplex real-time quantitative RT-PCR assay. *Journal of Virological Methods* 149, 103–109.

Liu, Z., Teng, Y., Xie, X., Li, H., Lv, J. *et al.* (2008b) Development and evaluation of a one-step loop-mediated isothermal amplification for detection of spring viraemia of carp virus. *Journal of Applied Microbiology* 105, 1220–1226.

Marcotegui, M.A., Estepa, A., Frías, D. and Coll, J.M. (1992) First report of a rhabdovirus affecting carp in Spain. *Bulletin of the European Association of Fish Pathologists* 12, 50–52.

Martinez-Lopez, A., Garcia-Valtanen, P., Ortega-Villaizan, M., Chico, V., Gomez-Casado, E. *et al.* (2014) VHSV G glycoprotein major determinants implicated in triggering the host type I IFN antiviral response as DNA vaccine molecular adjuvants. *Vaccine* 32, 6012–6019.

Masycheva, V.I., Alikin, Y.S., Klimenko, V.P., Fadina, V.A., Shchelkunov, I.S. *et al.* (1995) Comparative antiviral effects of dsRNA on lower and higher vertebrates. *Veterinary Research* 26, 536–538.

Misk, E., Garver, K.A., Nagy, E., Isaacs, S., Tubbs, L. *et al.* (2016) Pathogenesis of spring viraemia of carp virus in emerald shiner *Notropis atherinoides* Rafinesque, fathead minnow *Pimephales promelas* Rafinesque and white sucker *Catostomus commersonii* (Lacepede). *Journal of Fish Diseases* 39, 729–739.

Molnár, K. and Csaba, G. (2005) Sanitary management in Hungarian aquaculture. *Veterinary Research Communications* 29, 143–146.

Oidtmann, B.C., Thrush, M.A., Denham, K.L. and Peeler, E.J. (2011) International and national biosecurity strategies in aquatic animal health. *Aquaculture* 320, 22–33.

OIE (2015a) Spring viraemia of carp. In: *Manual of Diagnostic Tests for Aquatic Animals*. World Organisation for Animal Health, Paris. Updated 2016 version (Chapter 2.3.9.) available at: http://www.oie.int/index.php?id=2439&L=0&htmfile=chapitre_svc.htm (accessed 31 October 2016).

OIE (2015b) Spring viraemia of carp. In: *Aquatic Animal Health Code*. World Organisation for Animal Health, Paris. Updated 2016 version (Chapter 10.9.) available at: http://www.oie.int/index.php?id=171&L=0&htmfile=chapitre_svc.htm (accessed 31 October 2016).

Oraic, D. and Zrncic, S. (2005) An overview of health control in Croatian aquaculture. *Veterinary Research Communications* 29, 139–142.

Osadchaya, E.F. and Rudenko, A.P. (1981) Patogennost virusov, vydelennyh pri krasnuhe (vesennej viremii) karpov i kliniko-morfologicheskaja harakteristika estestvennogo techenija bolezni i v eksperimente [Pathogenicity of viruses isolated from carp with infectious dropsy (red-spot disease, spring viraemia) and clinical and morphological characteristics of the course of the natural and experimental disease]. *Rybnoe Khozyaistvo (Kiev)* 32, 66–71.

Peters, F. and Neukirch, M. (1986) Transmission of some fish pathogenic viruses by the heron, *Ardea cinerea*. *Journal of Fish Diseases* 9, 539–544.

Pfeil-Putzien, C. (1978) Experimentelle Übertragung der Frühjahrsvirämie (spring viraemia) der Karpfen durch Karpfenläuse (*Argulus foliaceus*). *Zentralblatt für Veterinarmedizin, Reihe B* 25, 319–323.

Pfeil-Putzien, C. and Baath, C. (1978) Nachweis einer Rhabdovirus-carpio-Infektion bei Karpfen im Herbst. *Berliner und Münchener Tierärztliche Wochenschrift* 91, 445–447.

Phelps, N.B.D., Armien, A.G., Mor, S.K., Goyal, S.M., Warg, J.V. *et al.* (2012) Spring viremia of carp virus in Minnehaha Creek, Minnesota. *Journal of Aquatic Animal Health* 24, 232–237.

Řehulka, J. (1996) Blood parameters in common carp with spontaneous spring viremia (SVC). *Aquaculture International* 4, 175–182.

Rexhepi, A., Bërxholi, K., Scheinert, P., Hamidi, A. and Sherifi, K. (2011) Study of viral diseases in some freshwater fish in the Republic of Kosovo. *Veterinarsky Arhiv* 81, 405–413.

Ruane, N.M., Rodger, H.D., McCarthy, L.J., Swords, D., Dodge, M. *et al.* (2014) Genetic diversity and associated pathology of rhabdovirus infections in farmed and wild perch *Perca fluviatilis* in Ireland. *Diseases of Aquatic Organisms* 112, 121–130.

Ruyra, A., Cano-Sarabia, M., Garcia-Valtanen, P., Yero, D., Gibert, I. *et al.* (2014) Targeting and stimulation of the zebrafish (*Danio rerio*) innate immune system with LPS/dsRNA-loaded nanoliposomes. *Vaccine* 32, 3955–3962.

Sanders, G.E., Batts, W.N. and Winton, J.R. (2003) Susceptibility of zebrafish (*Danio rerio*) to a model pathogen, spring viremia of carp virus. *Comparative Biochemistry and Physiology – Part C: Toxicology and Pharmacology* 53, 514–521.

Sano, M., Nakai, K. and Fijan, N. (2011) Viral diseases and agents of warmwater fish. In: Woo, P.T.K. and Bruno, D.W. (eds) *Fish Diseases and Disorders, Volume 3: Viral, Bacterial and Fungal Infections*, 2nd edn. CAB International, Wallingford, UK, pp. 166–244.

Selli, L., Manfrin, A., Mutinelli, F., Giacometti, P., Cappellozza, E. *et al.* (2002) Isolamento, da pesce gatto (*Ictalurus melas*), di un agente virale sierologicamente correlato al virus della viremia primaverile della carpa (SVCV). *Bollettino Societa Italiana di Patologia Ittica* 14, 3–13.

Shchelkunov, I.S. and Shchelkunova, T.I. (1989) Rhabdovirus carpio in herbivorous fishes: isolation, pathology and comparative susceptibility of fishes. In: Ahne, W. and Kurstak, E. (eds) *Viruses of Lower Vertebrates*. Springer, Berlin, pp. 333–348.

Shchelkunov, I.S., Oreshkova, S.F., Popova, A.G., Nikolenko, G.N., Shchelkunova, T.I. *et al.* (2005) Development of PCR-based techniques for routine detection and grouping of spring viraemia of carp virus. In: Cipriano, R.C., Shchelkunov, I.S. and Faisal, M. (eds) *Health and Diseases of Aquatic Organisms: Bilateral Perspectives*. Michigan State University Press, East Lansing, Michigan, pp. 260–284.

Shimahara, Y., Kurita, J., Nishioka, T., Kiryu, I., Yuasa, K. *et al.* (2016) Development of an improved RT-PCR for specific detection of spring viraemia of carp virus. *Journal of Fish Diseases* 39, 269–275.

Siwicki, A.K., Pozet, F., Morand, M., Kazun, B., Trapkowska, S. *et al.* (2003) Influence of methisoprinol on the replication of rhabdoviruses isolated from carp (*Cyprinus carpio*) and catfish (*Ictalurus melas*): *in vitro* study. *Polish Journal of Veterinary Sciences* 6, 47–50.

Soliman, M.K., Aboeisa, M.M., Mohamed, S.G. and Saleh, W.D. (2008) First record of isolation and identification of spring viraemia of carp virus from *Oreochromis niloticus* in Egypt. In: *Eighth International Symposium on Tilapia in Aquaculture. Proceedings of the ISTA 8, October 12–14, 2008, Cairo, Egypt. Final Papers Submitted to ISTA 8*, pp. 1287–1306. Available at: http://ag.arizona.edu/azaqua/ista/ISTA8/FinalPapers/12/4%20Viral%20Infection/Magdy%20Khalil%20paper%20final.doc (accessed 31 October 2016).

Stone, D.M., Ahne, W., Denham, K.D., Dixon, P.F., Liu, C.T.Y. *et al.* (2003) Nucleotide sequence analysis of the glycoprotein gene of putative spring viraemia of carp virus and pike fry rhabdovirus isolates reveals four genogroups. *Diseases of Aquatic Organisms* 53, 203–210.

Stone, D.M., Kerr, R.C., Hughes, M., Radford, A.D. and Darby, A.C. (2013) Characterisation of the genomes of four putative vesiculoviruses: tench rhabdovirus, grass carp rhabdovirus, perch rhabdovirus and eel rhabdovirus European X. *Archives of Virology* 158, 2371–2377.

Taylor, N.G.H., Peeler, E.J., Denham, K.L., Crane, C.N., Thrush, M.A. *et al.* (2013) Spring viraemia of carp (SVC) in the UK: the road to freedom. *Preventive Veterinary Medicine* 111, 156–164.

Vicenova, M., Reschova, S., Pokorova, D., Hulova, J. and Vesely, T. (2011) First detection of pike fry-like rhabdovirus in barbel and spring viraemia of carp virus in sturgeon and pike in aquaculture in the Czech Republic. *Diseases of Aquatic Organisms* 95, 87–95.

Warg, J.V., Dikkeboom, A.L., Goodwin, A.E., Snekvik, K. and Whitney, J. (2007) Comparison of multiple genes of spring viremia of carp viruses isolated in the United States. *Virus Genes* 35, 87–95.

Xiao, Z.G., Shen, J., Feng, H., Liu, H., Wang, Y.P. *et al.* (2015) Characterization of two thymosins as immune-related genes in common carp (*Cyprinus carpio* L.). *Developmental and Comparative Immunology* 50, 29–37.

Yue, Z., Teng, Y., Liang, C., Xie, X., Xu, B. *et al.* (2008) Development of a sensitive and quantitative assay for spring viremia of carp virus based on real-time RT-PCR. *Journal of Virological Methods* 152, 43–48.

Zhang, N.Z., Zhang, L.F., Jiang, Y.N., Zhang, T. and Xia, C. (2009) Molecular analysis of spring viraemia of carp virus in China: a fatal aquatic viral disease that might spread in East Asian. *PLoS ONE* 4(7): e6337.

Zhou, W., Zhou, J.J., Lv, Y., Qu, Y.X., Chi, M.D. *et al.* (2015) Identification and characterization of MAVS from black carp *Mylopharyngodon piceus*. *Fish and Shellfish Immunology* 43, 460–468.

8

Channel Catfish Viral Disease

LARRY A. HANSON[1]* AND LESTER H. KHOO[2]

[1]*Department of Basic Sciences, College of Veterinary Medicine, Mississippi State University, Mississippi, USA; [2]Thad Cochran Warmwater Aquaculture Center, Stoneville, Mississippi, USA*

8.1 Introduction

Channel catfish viral disease (CCVD) is an acute viraemia that occurs primarily among young (0–4 month old) channel catfish (*Ictalurus punctatus*) in aquaculture. CCVD outbreaks occur almost exclusively in the summer when water temperatures exceed 25°C and may exceed 90% mortality in less than 2 weeks. Older fish may experience a more chronic outbreak, often with secondary *Flavobacterium columnare* or *Aeromonas* infections that can mask the underlying CCVD (Plumb, 1978). Pond-to-pond spread is often reported within fingerling production facilities. The disease was first described by Fijan *et al.* (1970) and the most notable clinical signs were exophthalmia, abdominal distension, disoriented swimming and rapidly increasing mortality.

The causative agent, *Ictalurid herpesvirus 1* (IcHV1, commonly known as channel catfish virus, CCV), was characterized by Wolf and Darlington (1971). This virus is the type species of the *Ictalurivirus* genus and it is one of the best characterized members of the family *Alloherpesviridae*, which includes most herpesviruses of fishes and amphibians. IcHV1 is similar in virion structure and replication to other members of the order *Herpesvirales* (Booy *et al.*, 1996). However, at the molecular level (genome sequence), IcHV1 is substantially different from members of other alloherpesvirus genera and shows almost no homology to members of the *Herpesviridae* (Davison, 1992; Waltzek *et al.*, 2009; Doszpoly *et al.*, 2011a).

A closely related virus in the genus *Ictalurivius* is *Ictalurid herpesvirus 2* (IcHV2), which causes a disease similar to CCVD that has devastated the aquaculture of the black bullhead catfish (*Ameiurus melas*) in Italy (Alborali *et al.*, 1996; Doszpoly *et al.*, 2008, 2011b; Roncarati *et al.*, 2014). Channel catfish fingerlings and juveniles are extremely sensitive to IcHV2, and in experimental infections they develop a CCVD-like disease at 24–25°C, which is cooler than the temperatures commonly seen with IcHV1 infections (Hedrick *et al.*, 2003).

IcHV1 is host specific and only produces infections in channel catfish, blue catfish (*Ictalurus furcatus*) and channel × blue hybrids (Plumb, 1989). Early fears that CCVD would limit the catfish industry have not been realized. Producers have adjusted management by reductions of stocking densities and by minimizing stress during the critical first summer of production. Consequently, the industry-wide impact of CCVD mortality has become manageable. Nevertheless, when sporadic outbreaks do occur, the individual operational losses can be devastating. Such outbreaks often occur in multiple ponds and this can result in catastrophic loss of production. Furthermore, CCVD survivors have reduced growth (McGlamery and Gratzek, 1974).

8.1.1 Prevalence and transmission

IcHV1 is present in most locations where channel catfish are produced commercially. The virus establishes latency, is vertically transmitted and can remain undetected (with no sign of disease) for many generations. Using polymerase chain reaction (PCR), Thompson *et al.* (2005) evaluated the prevalence of IcHV1 in 3–5 day old fry in commercial hatcheries in Mississippi. They found that

---

*Corresponding author e-mail: hanson@cvm.msstate.edu

breeding populations at all five hatcheries tested had the virus. These hatcheries produced 20% of the fingerlings for the industry. The prevalence of latent IcHV1 in the fry ranged from 11.7 to 26.7%, but replicating virus was not detected by cell culture and CCVD did not occur within the production facilities during the year of the study. When fish populations were tracked during their first production season, latent IcHV1 infection increased from 13 to >35%. Still, no disease occurred and viral replication remained undetected by cell culture. These results suggested that vertical transmission commonly occurs between parents and their progeny, and that subclinical horizontal transmission is common within fingerling populations.

8.1.2 Factors that promote CCVD outbreaks

Factors that influence CCVD outbreaks include high temperatures, stress, population density and age (Plumb, 1978, 1989). The effect of temperature was experimentally demonstrated in infection studies; increasing the water temperature from 19 to 28°C had a profound effect on the progression of CCVD, whereas reducing the temperature from 28 to 19°C substantially curbed losses (Plumb, 1973a).

Circumstantial evidence indicates susceptibility is inversely related to age and almost all outbreaks in commercial production systems occur in young of the year populations. The apparent effect of age may be related to the development of immunity due to sublethal natural exposure or to changes in behaviour as the fish age. Within the most susceptible period, experimental evidence suggests that very young fry are more resistant to CCVD than are fish >1 month of age, which might be related to the presence of maternal antibodies (Hanson et al., 2004). Also, differences in the susceptibility of channel catfish strains have been demonstrated (Plumb et al., 1975).

The influence of stress on CCVD outbreaks is less clear. Circumstantial evidence suggests that poor water quality and handling stress correlate with CCVD outbreaks in natural systems, but low water stress and the oral administration of cortisol did not influence experimentally induced CCVD losses (Davis et al., 2002, 2003). The effect of stress hormones may be more related to the amount of virus recrudescence than it is to the susceptibility of fish to immersion infections. In cold temperature-exposed adult channel catfish, the administration of the synthetic corticosteroid dexamethazone

resulted in a higher level of viral recovery from isolated leucocytes (Bowser et al., 1985). Arnizaut and Hanson (2011) found that the administration of dexamethazone to carrier juvenile channel catfish caused an increase on viral DNA loads and increased viral gene expression. An increase in circulating IcHV1-specific antibodies suggested that some viral recrudescence had occurred, although infectious virus was not detected. Also, many outbreaks have occurred without a clear predisposing stress event.

The importance of IcHV1 infection to large juvenile and adult channel catfish is less well understood. Hedrick et al. (1987) demonstrated that naive adult channel catfish are susceptible to CCVD in experimental immersion infections. Additionally, seasonal increases of antiviral antibody suggested that at least some antigen expression occurs in adults during the summer (Bowser and Munson, 1986). IcHV1 has been isolated from diseased adult channel catfish in the winter when the water temperature was 8°C; the virus was also isolated from non-diseased fish in the population (Bowser et al., 1985). The detection of replicating virus in fish during the winter is very rare and the associated disease is economically insignificant.

8.2 Diagnosis

Presumptive diagnosis of CCVD is usually based on rapidly progressing mortality of young channel catfish with the fish displaying the typical clinical signs (abdominal distension, exophthalmia and disoriented swimming), gross pathology, histopathology and the production of the typical cytopathic effect (CPE) of IcHV1 in cell culture.

IcHV1 is easily cultured in catfish cell lines but does not produce CPE on most non-catfish cell lines. Briefly, the culture is started by homogenizing whole fry, viscera from small fingerlings or posterior kidney from larger fingerlings in serum-free cell culture medium. Then microbial components of the homogenates are removed using a 0.22 μm filter and this filtrate is diluted in cell culture medium at a 1:100 final dilution of tissue to medium on channel catfish ovary (CCO) and/or brown bullhead (BB) cell lines at 28–30° C. Both cell lines are available from the American Type Culture Collection (Manassas, Virginia). Cultures from acutely CCVD-affected fish produce high levels of infectious virus and develop a rapidly progressive CPE within 24–48 h. The characteristic

CPE in positive cultures is syncytial formation due to fusion of cells. These syncytia subsequently dislodge themselves from the culture flask leaving radiating cytoplasmic projections to attachment points (see Fig. 8.1). Cultures without CPE should be observed daily for 7 days and then blind passaged for an additional 7 days before they are considered negative for IcHV1.

The channel catfish reovirus (Amend *et al.*, 1984) and IcHV2 are capable of infecting North American catfish species and will also cause CPE on CCO and BB cells at 28–30° C. The catfish reovirus causes limited syncytia but the CPE spreads more slowly. This virus is not associated with high losses in channel catfish. However, IcHV2 is of more concern because it can cause CCVD-like disease in experimental infections of North American catfish. It also produces a CPE that is similar to that of IcHV1 (Hedrick *et al.*, 2003). In the original description of IcHV2 disease outbreaks, the CPE was seen in the bluegill cell line, BF2, and the EPC (epithelioma papulosum cyprini) cell line of fathead minnow origin (Alborali *et al.*, 1996). This demonstrates that ICHV2 has a broader cell host range that may be of diagnostic value.

The confirmation of virus isolated using cell culture requires serology or molecular analysis. The use of neutralizing monoclonal antibody (Arkush *et al.*, 1992) distinguishes IcHV1 from IcHV2 (Hedrick *et al.*, 2003). Polyclonal antibodies are also useful, but IcHV1 induces neutralizing antibodies poorly in rabbits. Additional neutralizing monoclonal and monospecific polyclonal antibodies have been developed (Wu *et al.*, 2011; Liu *et al.*, 2012), but their ability to distinguish IcHV1 from IcHV2 has not been determined. These diagnostic antibodies are not commercially available.

The most common molecular method for confirming a presumptive diagnosis is the PCR assay. Several sensitive traditional PCR assays have been described (Boyle and Blackwell, 1991; Baek and Boyle, 1996; Gray *et al.*, 1999; Thompson *et al.*, 2005), but their ability to distinguish IcHV1 from IcHV2 has not been reported. A hydrolysis probe (TaqMan)-based quantitative PCR has been developed specifically for IcHV2 (Goodwin and Marecaux, 2010). We use two different TaqMan-based quantitative PCRs (qPCRs) that distinguish between two genomovars of IcHV1 and do not detect IcHV2 (see Table 8.1) (Hanson, 2016). As with any PCR assay, these methods are very sensitive to genomic contamination from previous PCR products, so assays should be set up in separate locations and use separate equipment. Furthermore, negative control cultures must be processed to ensure that contamination did not occur during sample preparation.

The detection of latent infections is needed to identify carrier populations of fish. Direct cell culture from carrier fish has no diagnostic value because no infectious virus is produced when the virus is latent. Although the use of dexamethazone injection to induce recrudescence and the subsequent co-culture of leucocytes with CCO cells has been successful in demonstrating the presence of IcHV1 in adult catfish (Bowser *et al.*, 1985), this was only successful in fish from cold water during the winter. Other workers have been unsuccessful in reactivating culturable latent IcHV1 in known carrier fish using the same protocol; consequently, this is not a reliable diagnostic method (Arnizaut and Hanson, 2011). The detection of antibodies to IcHV1 in brood fish has been used to successfully identify carrier populations (Plumb, 1973b; Amend and McDowell, 1984; Bowser and Munson, 1986; Crawford *et al.*, 1999). The enzyme-linked immunosorbent assay (ELISA) is more sensitive than serum neutralization (Crawford *et al.*, 1999). Serological assays are an indirect detection method; their reliability depends not only on the presence of the pathogen but also on the immune status of the fish. IcHV-1-specific antibody concentrations in carrier fish have been shown to vary substantially between fish within a population (Arnizaut and Hanson, 2011). What is more, the mean concentrations vary

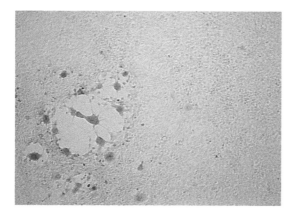

Fig. 8.1. A plaque caused by *Ictalurid herpesvirus 1* (IcHV1) on a monolayer of the Channel catfish ovary cell line. Note the contracting syncytium leaving cytoplasmic projections (100× magnification).

Channel Catfish Viral Disease

Table 8.1. Primers, probes and amplification parameters for qPCR assays of *Ictalurid herpesvirus 1* (IcHV) and *Ictalurid herpesvirus 2* (IcHV2).[a]

Assay, primers and probes	Sequence#	Cycle parameters
IcHV1 TaqMan		
Upper primer	CTCCGAGCGATGACACCAC	60 s at 64°C
Lower primer	TGTGTTCAGAGGAGCGTCG	15 s at 72°C
IcHV1a probe[b]	FAM-CCCATCCCTTCCCTCCTCCCTG-BHQ-1[c]	15 s at 95°C
IcHV1b probe[b]	FAM-CCCATCCTTCCCCTCCTCCCTG-BHQ-1	
IcHV2 TaqMan		
Upper primer	ATACATCGGTCTCACTCAAGAGG	45 s at 59°C
Lower primer	TAATGGGTATTGGTACAAATCTTCATC	45 s at 72°C
Probe	FAM-CGC+CTG+AGA+ACC+GAGCA –BHQ-1[d]	30 s at 95°C

[a]The IcHV2 assay is from Goodwin and Marecaux (2010).
[b]Probe IcHV1a detects genome type A, Probe IcHV1b detects genome type B. Both assays use the same primers and reaction conditions.
[c]FAM designates 6-FAM (6-carboxyfluorescein) fluorescent dye and BHQ-1 indicates black hole quencher.
[d]+ indicates the use of locked bases to increase the melting point.

substantially over the year, with the highest levels of antibody seen primarily in the autumn (Bowser and Munson, 1986). In addition, catfish that carry IcHV1 latently as a result of vertical transmission, but have not experienced virus antigen expression, have no detectable antibodies. Because of the variation in antibody levels, direct detection of the latent genome in carrier fish is preferred.

The conventional PCR and qPCR methods described above are very sensitive and can detect latent IcHV1 in the tissues of carriers. However, even though the use of PCR to detect carrier populations is very sensitive, the concentration of IcHV1 genome in latently infected tissue can be close to the minimal limit of detection. The tissues sampled for PCR evaluation depend on the life stage and the health status of the fish. Sac fry should be processed whole, but without their yolk (Thompson *et al.*, 2005). To evaluate clinically diseased fish for direct CCVD diagnosis, posterior kidney tissue is preferred. For latency evaluation of larger fish, caudal fin biopsies are reliable (Arnizaut, 2002). Samples can be prepared using commercially available DNA extraction kits such as the Gentra Puregene Tissue Kit (Qiagen). When using qPCR on 1 µg DNA from tissues, the number of threshold cycles for a CCVD clinically affected fish or an infected cell culture is below 27, whereas for latently infected fish it is 33–38. In studies among brood stock, we found that PCR-negative parents could still produce IcHV1-positive offspring (Y. Habte and L. Hanson, unpublished). Although PCR assays are useful in detecting carrier populations,

they are not sensitive enough to definitively cull positive individuals from a population.

8.3 Pathology

8.3.1 Clinical signs and gross lesions

The original descriptions and reviews of CCVD (Fijan *et al.*, 1970; Plumb, 1978, 1986; Hanson *et al.*, 2011; Plumb and Hanson, 2011) have provided detailed descriptions of the clinical signs as well as the gross pathology of the disease. Clinical signs include abnormal swimming behaviour (i.e. rotating along the longitudinal axis) and convulsions followed by quiescence when the affected fish breathe rapidly at the bottom of the water column. Fish may then posture themselves vertically in the water column with their heads at the surface shortly before they expire. This vertical posture in the water was once thought to be diagnostic for the disease. Grossly, the affected fish may show external haemorrhages at the base of the fins and ventral abdomen, exophthalmia, abdominal distention, and pale or haemorrhagic gills. Recent clinical submissions of naturally affected CCVD fish to our laboratory usually have the bilateral exophthalmia and distended abdomens, but often lack the external haemorrhagic lesions (Fig. 8.2A).

Unfortunately, with clinical submissions, there is often no accurate means of determining the chronicity of the disease, and the lack of haemorrhagic external lesions may reflect a different point/phase of the disease continuum. Historically described

L.A. Hanson and L.H. Khoo

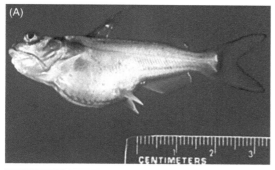

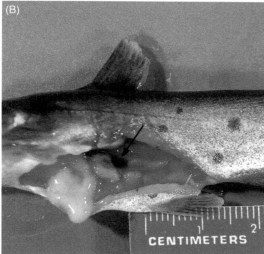

Fig. 8.2. Images of channel catfish fingerlings with channel catfish viral disease (CCVD). (A) External gross pathology; note the markedly distended abdomen. (B) Internal gross lesions; the arrow indicates a congested spleen that is adjacent to a slightly swollen posterior kidney, the gastrointestinal tract is slightly hyperaemic and the organs are glistening from the ascites that flowed out of the incised abdomen.

gross internal lesions include haemorrhagic areas in the musculature, liver and kidneys. Pale livers, pale enlarged posterior kidneys, an enlarged congested spleen and gastrointestinal tracts devoid of feed but filled instead with a yellow mucoid fluid are lesions that have been ascribed to this disease. Plumb (1986) described a general hyperaemia in the visceral or coelomic cavity as well as a straw/yellow-coloured ascites which caused the abdominal distension. The ascites, exophthalmia and congested spleen are almost always seen in clinical submissions. Haemorrhage in the musculature,

liver and posterior kidney, and renomegaly, are less apparent and may be underappreciated owing to the diminutive size of the affected fish (Fig. 8.2B).

8.3.2 Histopathology

Early descriptions of the histopathological lesions were based mainly on experimentally infected fish that were injected with the virus (Wolf et al., 1972; Plumb et al., 1974). Kidneys (anterior and posterior) were often the most severely affected tissue, followed by the gastrointestinal tract, liver and skeletal muscle. Plumb et al. (1974) detected increased numbers of lymphoid cells in the posterior kidney in addition to haemorrhage and necrosis of epithelial cells of the proximal tubules. There was also extensive necrosis of the hematopoietic tissue of the kidney. Necrosis of the epithelial cells surrounding the pancreatic acinar cells, with limited or minimal pancreatic acinar cell necrosis, was evident. The marked haemorrhage/congestion in the spleen obscured the white pulp. Haemorrhaging was also visible in the submucosa and villi of the gastrointestinal tract.

Major et al. (1975) studied both experimentally and naturally infected fish. Lesions were more severe in the younger naturally infected fish. They also found lesions in the brain and pancreas in addition to the lesions that were reported by Wolf et al. (1972) and Plumb et al. (1974). In addition, Major et al. (1975) described vacuolation of neurons with oedema of the surrounding nerve fibres and necrosis of the pancreatic acinar cells. Oedematous changes were not limited to the brain but were also present in the heart, spinal cord, gills, kidneys and gastrointestinal tract. Hepatocellular necrosis was present in younger naturally infected fish, while oedema and multifocal necrosis were more common in older fish. They also detected eosinophilic intracytoplasmic inclusion bodies in hepatocytes of experimentally infected fish, which were less frequent in naturally infected fish.

The naturally infected CCVD fish that have been submitted to our laboratory have histopathological lesions consistent with those described by Major et al. (1975). In addition, these fish also have multifocal to locally extensive necrosis of the hematopoietic elements within the anterior kidney (Fig. 8.3). It is difficult to appreciate the increased lymphoid elements in these samples as compared with those in experimentally infected fish (Plumb et al., 1974). Similarly, there is moderate-to-severe multifocal necrosis of the interstitial portions of the posterior kidney (Fig. 8.4), which only occasionally extends to the renal tubular epithelium. Rare multinucleated syncytia are sometimes present in these necrotic foci (Fig. 8.4A). Eosinophilic intranuclear inclusions are sometimes present in cells presumed to be lymphocytes (the nucleus is markedly distended with only a thin rim of basophilic, and these cells appear to lack cytoplasm) within the renal vessels of the posterior kidney, and

rarely in other organs such as the spleen (Fig. 8.4B). Intranuclear inclusion bodies are rarely seen.

The hepatocellular necrosis is multifocal and mild, and the pancreatic involvement is minimal when present (Fig. 8.5). The severe congestion/haemorrhage of the spleen masks the normal splenic architecture as well as any necrosis that may be present (Fig. 8.6). There is necrosis present in the glandular portions of the stomach (Fig. 8.7A) which can extend into the proximal intestine. In the intestinal mucosa, the necrosis is multifocal; it is regionally extensive in the submucosa. Small aggregates of mononuclear inflammatory cells are sometimes present at the base of the villi (Fig. 8.7B). There may also be local extensive congestion in portions of the gastrointestinal tract. No significant microscopic lesions are seen in the other tissues including the brain, muscle and heart. Lesions in clinical cases correlate well with what has already been reported in the literature (Plumb *et al.*, 1974; Wolf *et al.*, 1972; Major *et al.*, 1975) Differences may reflect the chronicity of the disease, varying environmental conditions, and perhaps the strain of virus.

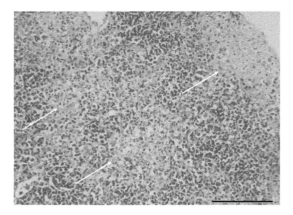

Fig. 8.3. Histopathology of the anterior kidney of a channel catfish viral disease (CCVD)-affected channel catfish fingerling. The arrows point to the more eosinophilic necrotic foci with pyknotic nuclei and karyorrhectic debris (haematoxylin and eosin (H and E) stain; Bar ~ 100 μm)

8.4 Pathophysiology

The kidney appears to be one of the major organs affected by CCVD, and renal damage results in

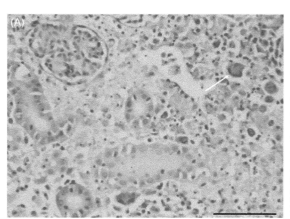
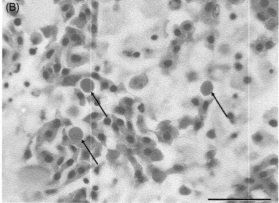

Fig. 8.4. Histopathology of the posterior kidney of a channel catfish viral disease (CCVD)-affected fingerling revealing: (A) interstitial necrosis with a multinucleated syncytium as indicated by the arrow (haematoxylin and eosin (H and E) stain; Bar ~ 50 μm); and (B) cells, presumably lymphocytes, with eosinophilic inclusion bodies and only a thin rim of chromatin as indicated by the arrows (H and E stain; Bar ~ 20 μm).

L.A. Hanson and L.H. Khoo

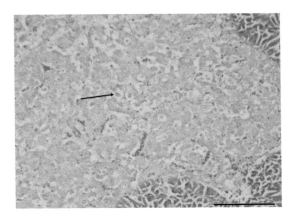

Fig. 8.5. Focus of necrosis within the liver of a channel catfish viral disease (CCVD)-affected channel catfish fingerling. The arrow points to one of several slightly more eosinophilic shrunken hepatocytes (haematoxylin and eosin (H and E) stain; Bar ~ 100 μm).

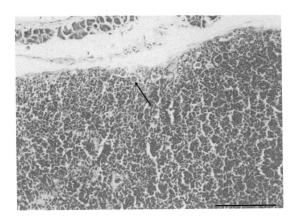

Fig. 8.6. Histopathology of the severely congested spleen of a channel catfish viral disease (CCVD)-affected channel catfish fingerling. The arrow points to a cell with an intranuclear inclusion, as seen in Fig. 8.4A. The white pulp is obscured by the red blood cells. (haematoxylin and eosin (H and E) stain; Bar ~ 100 μm).

fluid imbalances, which would explain the gross lesions (e.g. ascites, which produces abdominal distension, and exophthalmia). However, the colour of the ascites is more indicative that this may be a modified transudate with leakage from the vasculature. Lesions in the gastrointestinal tract and liver may also contribute to the ascites and oedema due to decreased absorption and protein synthesis and thus reduced oncotic pressure of the

plasma. The pale gills are most likely the result of both haemorrhaging and congestion in multiple organs, but especially in the spleen. Additionally, there is necrosis within the anterior kidney and in the interstitium of the posterior kidney, which would affect haematopoiesis and cause anaemia. While the cellular tropism of IcHV1 has not been established, most alloherpesviruses are epitheliotropic (reviewed in Hanson et al., 2011). The necrosis of the mucosal epithelium of the gastrointestinal tract and the epithelium surrounding the pancreas indicates that IcHV1 has some epitheliotropism. Virus quantification and tracking experiments demonstrate peak and earliest virus production in renal tissue, followed by the liver, intestine and skin (Plumb, 1971; Nusbaum and Grizzle, 1987; Kancharla and Hanson, 1996). Cell culture assays demonstrate that IcHV1 replicates well in fibroblast-derived cell lines as well as B lymphocytes and causes CPE in other leucocytes (Bowser and Plumb, 1980; Chinchar et al., 1993). Peak virus production in the kidney occurs at 3–4 days postexposure to the virus at 28°C in both sublethal and lethal doses, and this is also the period of peak mortality (Kancharla and Hanson, 1996).

The specific pathogenic mechanisms of IcHV1 have not been well defined. The genome encodes over 76 genes (Davison, 1992), and most are non-essential for replication. The non-essential genes that have roles in virulence include the thymidine kinase (TK) gene and gene 50. Virally encoded TK probably facilitates virus replication in non-replicating cells (Zhang and Hanson, 1995). The TK mutant of CCV replicates to similar levels as the wild type virus and in experimentally infected fish also had early kinetics that were similar to those of the wild type virus, but the mutant virus is cleared more quickly (Zhang and Hanson, 1995; Kancharla and Hanson, 1996). The product of gene 50 is a secreted mucin of unknown function, and viruses that are mutated in this gene are attenuated in virulence (Vanderheijden et al., 1996, 1999, 2001).

Temperature and dose effects due to crowding are the most important factors that influence CCVD outbreaks. The influence of temperature appears to be a common factor that regulates the expression of alloherpesviruses and promotes or limits the associated viral diseases. Temperature has also been shown to regulate disease outbreaks caused by Cyprinid herpesvirus 1, Cyprinid herpesvirus 3, Salmonid herpesvirus 3 (epizootic epitheliotropic disease of lake trout), Percid herpesvirus 1 (diffuse epidermal

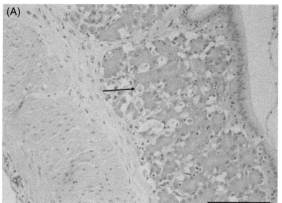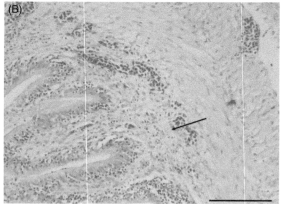

Fig. 8.7. Pathology associated with channel catfish viral disease (CCVD) in the gastrointestinal tract of an affected channel catfish fingerling. (A) Section of the stomach; the arrow points to one of the many necrotic cells within the glandular portion of the stomach (haematoxylin and eosin (H and E) stain; Bar ~ 100 µm). (B) Section of a portion of the proximal intestine; the arrow points to the focus of inflammation necrosis in the submucosa. The necrotic regions in the mucosa are within the more vacuolated areas and there are small aggregates of mononuclear inflammatory cells at the base of the villi. Several of the blood vessels are also congested in this figure (H and E stain; Bar ~ 100 µm).

hyperplasia of walleye) and Esocid herpesvirus 1 (blue spot disease of northern pike) (reviewed in Hanson *et al.*, 2011; Plumb and Hanson, 2011). The effect of temperature is best characterized in diseases caused by *Cyprinid herpesvirus 3* infections: koi herpesviral disease is limited to temperatures of 18–28°C and temperatures outside this range appear to cause the virus to establish a latent infection; this has been demonstrated in cell cultures as well as in the host (Gilad *et al.*, 2003; Dishon *et al.*, 2007; Ilouze *et al.*, 2012).

8.5 Protective and Control Strategies

The relatively low incidence of CCVD outbreaks in catfish fingerling production facilities is due to active and intentional management. Production ponds are typically stocked at less than 250,000 fry/ha. Fingerling producers avoid crowding or harvesting fingerlings when water temperatures are above 25°C. Ponds are carefully monitored and aerated to keep oxygen levels above 3 mg/l. Chloride levels are generally maintained above 100 mg/l to avoid nitrite toxicity and only quality fresh feed is provided. When producers vaccinate fry against other diseases, temperature and oxygen levels are carefully monitored. Fry are generally stocked only in the cool morning hours and any batches of fry that show unusual elevated mortality

are discarded and their tanks are disinfected. Many producers avoid collecting eggs and hatching fry late in the season to avoid the elevated water temperatures that would occur when the fingerlings are ready to be stocked into ponds.

Vaccination, the selection of resistant strains and establishing IcHV1-free breeding stock are prevention and control methods that show promise, but are not currently implemented. Attenuated virus vaccine candidates include a natural mutant developed by culturing IcHV1 in a *Clarias bactrachus* cell line (Noga and Hartmann, 1981), a recombinant TK-deleted IcHV1 (Zhang and Hanson, 1995) and a gene 50-deleted recombinant IcHV1 (Vanderheijden *et al.*, 2001). Both of the recombinant attenuated viruses showed the potential to express foreign genes and induce immunity to the gene products, which indicates that they could be used as vaccine vectors (Zhang and Hanson, 1996; Vanderheijden *et al.*, 2001). The development of bacterial artificial chromosomes with IcHV1 inserts allows researchers to use efficient bacteria-based recombination and selection methods to produce recombinant IcHV1 (Kunec *et al.*, 2008, 2009). A limiting factor in the implementation of CCVD vaccines is the need to vaccinate fish at an age when they are immunologically competent and can develop protective immunity (Petrie-Hanson and Ainsworth, 2001), but still young enough to be most susceptible to CCVD. Moreover, the current management methods preclude applying immersion vaccines after 8 days of age because this is when fry are stocked into the fingerling ponds. Other immune-potentiating treatments that increase channel catfish resistance are the application of interferon inducers such as poly(I:C) (polyinosinic–polycytidylic acid) (Plant *et al.*, 2005), or pre-exposure to catfish reovirus as has been demonstrated in cell culture (Chinchar *et al.*, 1998). Practical application methods of these immune stimulants and their utility in populations of channel catfish have not been studied.

As some strains of channel catfish are more resistant to CCVD, selective breeding could be used to control outbreaks (Plumb *et al.*, 1975). Early studies suggested that channel × blue hybrids may be more resistant to CCVD (Plumb and Chappell, 1978), but later studies did not support this finding; it was suggested that the strain of channel catfish used influences the susceptibility of the hybrid (Silverstein *et al.*, 2008). The selection of IcHV1-free brood stock to produce virus-free fingerlings is possible using aggressive screening and culling methods. However, the fish must be raised in more biosecure environments than are used in current aquaculture facilities. The virus is most likely endemic because brood fish have not been routinely screened for the virus.

If a CCVD outbreak occurs, there are a few practical measures that reduce losses. If practicable, reducing water temperatures will reduce mortality. Otherwise, management after an outbreak is limited to preventing transmission by avoidance. For example, fish, water or equipment from an infected area should not be used in another pond. Interactions with other wildlife, especially predators and scavengers, should also be limited. Dead fish should be quickly removed from culture ponds and their carcasses should be buried or incinerated to prevent scavengers from reintroducing the disease. If the disease occurs in the hatchery before the fish are transferred to ponds, farmers typically euthanize all fish in the affected trough and disinfect the equipment. There is relatively little investment at this point and farmers tend to take this drastic action rather than risking spread within the hatchery. The survivors of CCVD outbreaks are often reared to market without obvious negative effects, but these fish should be avoided in brood stock selection because they are likely to carry the virus and facilitate vertical transmission. After an epizootic

has occurred, ponds should be depopulated, sterilized by drying and hydrated lime applied to wet areas before any restocking occurs (Camus, 2004).

8.6 Conclusions and Suggestions for Future Research

Channel catfish virus disease has caused sporadic high losses within channel catfish fingerling operations for >45 years and this virus is present subclinically in fish throughout the industry. Because CCVD outbreaks are associated with elevated water temperatures and high loading densities, the industry has optimized aquaculture practices to minimize the impact of the disease during the critical first summer of growth. Catfish production has evolved to use high-density systems. When combined with increasing temperatures and variable weather patterns, these densities are likely to make CCVD a greater threat. To minimize the impact of CCVD, it is critical to identify triggers for viral activation so that effective control measures can be developed. Research into the molecular basis of immune system regulation and effective viral recombination systems (Kunec *et al.*, 2008) can elucidate the intricate host/pathogen relationships that are typical of herpesviruses. Furthermore, the establishment of virus-free fish stocks and use of effective biosecurity measures can help to achieve more successful prevention of CCVD. Methods to prevent vertical transmission should be evaluated, and these could be integrated into *in vitro* fertilization protocols for the production of hybrid catfish. The identification of the mechanisms responsible for the resistance of blue catfish to CCVD and the application of this information to optimize the genetics of hybrid stocks could further alleviate the deleterious impacts of CCVD within the catfish industry.

References

Alborali, L., Bovo, G., Lavazza, A., Cappellaro, H. and Guadagnini, P.F. (1996) Isolation of an herpesvirus in breeding catfish (*Ictalurus melas*). *Bulletin of the European Association of Fish Pathologists* 16, 134–137.

Amend, D.F. and Mcdowell, T. (1984) Comparison of various procedures to detect neutralizing antibody to the channel catfish virus in California brood channel catfish. *Progressive Fish Culturist* 46, 6–12.

Amend, D.F., McDowell, T. and Hedrick, R.P. (1984) Characteristics of a previously unidentified virus from channel catfish (*Ictalurus punctatus*). *Canadian Journal of Fisheries and Aquatic Sciences* 41, 807–811.

Arkush, K.D., Mcneill, C. and Hedrick, R.P. (1992) Production and characterization of monoclonal antibodies against channel catfish virus. *Journal of Aquatic Animal Health* 4, 81–89.

Arnizaut, A. (2002) *Comparison of Channel Catfish Virus and Thymidine Kinase Negative Recombinant Channel Catfish Virus Latency and Recrudescence*. PhD, Mississippi State University, Starkville, Mississippi.

Arnizaut, A.B. and Hanson, L.A. (2011) Antibody response of channel catfish after channel catfish virus infection and following dexamethasone treatment. *Diseases of Aquatic Organisms* 95, 189–201.

Baek, Y.-S. and Boyle, J.A. (1996) Detection of channel catfish virus in adult channel catfish by use of nested polymerase chain reaction. *Journal of Aquatic Animal Health* 8, 97–103.

Booy, F.P., Trus, B.L., Davison, A.J. and Steven, A.C. (1996) The capsid architecture of channel catfish virus, an evolutionarily distant herpesvirus, is largely conserved in the absence of discernible sequence homology with herpes simplex virus. *Virology* 215, 134–141.

Bowser, P.R. and Munson, A.D. (1986) Seasonal variation in channel catfish virus antibody titers in adult channel catfish. *The Progressive Fish Culturist* 48, 198–199.

Bowser, P.R. and Plumb, J.A. (1980) Channel catfish virus: comparative replication and sensitivity of cell lines from channel catfish ovary and the brown bullhead. *Journal of Wildlife Diseases* 16, 451–454.

Bowser, P.R., Munson, A.D., Jarboe, H.H., Francis-Floyd, R. and Waterstrat, P.R. (1985) Isolation of channel catfish virus from channel catfish, *Ictalurus punctatus* (Rafinesque), broodstock. *Journal of Fish Diseases* 8, 557–561.

Boyle, J. and Blackwell, J. (1991) Use of polymerase chain reaction to detect latent channel catfish virus. *American Journal of Veterinary Research* 52, 1965–1968.

Camus, A.C. (2004) *Channel Catfish Virus Disease*. SRAC Publication No. 4702, Southern Regional Aquaculture Center, Stoneville, Mississippi.

Chinchar, V.G., Rycyzyn, M., Clem, L.W. and Miller, N.W. (1993) Productive infection of continuous lines of channel catfish leukocytes by channel catfish virus. *Virology* 193, 989–992.

Chinchar, V.G., Logue, O., Antao, A. and Chinchar, G.D. (1998) Channel catfish reovirus (CRV) inhibits replication of channel catfish herpesvirus (CCV) by two distinct mechanisms: viral interference and induction of an antiviral factor. *Diseases of Aquatic Organisms* 33, 77–85.

Crawford, S.A., Gardner, I.A. and Hedrick, R.P. (1999) An enzyme linked immunosorbent assay (ELISA) for detection of antibodies to channel catfish virus (CCV) in channel catfish. *Journal of Aquatic Animal Health* 11, 148–153.

Davis, K.B., Griffin, B.R. and Gray, W.L. (2002) Effect of handling stress on susceptibility of channel catfish *Ictalurus punctatus* to *Ichthyophthirius multifiliis* and channel catfish virus infection. *Aquaculture* 214, 55–66.

Davis, K.B., Griffin, B.R. and Gray, W.L. (2003) Effect of dietary cortisol on resistance of channel catfish to infection by *Ichthyophthirius multifiliis* and channel catfish virus disease. *Aquaculture* 218, 121–130.

Davison, A.J. (1992) Channel catfish virus: a new type of herpesvirus. *Virology* 186, 9–14.

Dishon, A., Davidovich, M., Ilouze, M. and Kotler, M. (2007) Persistence of cyprinid herpesvirus 3 in infected cultured carp cells. *Journal of Virology* 81, 4828–4836.

Doszpoly, A., Kovacs, E.R., Bovo, G., Lapatra, S.E., Harrach, B. *et al.* (2008) Molecular confirmation of a new herpesvirus from catfish (*Ameiurus melas*) by testing the performance of a novel PCR method, designed to target the DNA polymerase gene of alloherpesviruses. *Archives of Virology* 153, 2123–2127.

Doszpoly, A., Somogyi, V., Lapatra, S.E. and Benkő, M. (2011a) Partial genome characterization of *Acipenserid herpesvirus 2*: taxonomical proposal for the demarcation of three subfamilies in *Alloherpesviridae*. *Archives of Virology* 156, 2291–2296.

Doszpoly, A., Benko, M., Bovo, G., Lapatra, S.E. and Harrach, B. (2011b) Comparative analysis of a conserved gene block from the genome of the members of the genus *Ictalurivirus*. *Intervirology* 54, 282–289.

Fijan, N.N., Welborn, T.L.J. and Naftel, J.P. (1970) An acute viral disease of channel catfish. Technical Paper 43, US Fish and Wildlife Service, Washington, DC.

Gilad, O., Yun, S., Adkison, M.A., Way, K., Willits, N.H. *et al.* (2003) Molecular comparison of isolates of an emerging fish pathogen, koi herpesvirus, and the effect of water temperature on mortality of experimentally infected koi. *Journal of General Virology* 84, 2661–2667.

Goodwin, A.E. and Marecaux, E. (2010) Validation of a qPCR assay for the detection of *Ictalurid herpesvirus-2* (IcHV-2) in fish tissues and cell culture supernatants. *Journal of Fish Diseases* 33, 341–346.

Gray, W.L., Williams, R.J., Jordan, R.L. and Griffin, B.R. (1999) Detection of channel catfish virus DNA in latently infected catfish. *Journal of General Virology* 80, 1817–1822.

Hanson, L. (2016) Ictalurid herpesvirus 1. In: Liu, D. (ed.) *Molecular Detection of Animal Viral Pathogens*. CRC Press/Taylor and Francis Group, Boca Raton, Florida, pp. 797–806.

Hanson, L.A., Rudis, M.R. and Petrie-Hanson, L. (2004) Susceptibility of channel catfish fry to channel catfish virus (CCV) challenge increases with age. *Diseases of Aquatic Organisms* 62, 27–34.

Hanson, L., Dishon, A. and Kotler, M. (2011) Herpesviruses that infect fish. *Viruses* 3, 2160–2191.

Hedrick, R.P., Groff, J.M. and Mcdowell, T. (1987) Response of adult channel catfish to waterborne exposures of channel catfish virus. *The Progressive Fish Culturist* [now *North American Journal of Aquaculture*] 49, 181–187.

Hedrick, R.P., Mcdowell, T.S., Gilad, O., Adkison, M. and Bovo, G. (2003) Systemic herpes-like virus in catfish *Ictalurus melas* (Italy) differs from *Ictalurid herpesvirus 1* (North America). *Diseases of Aquatic Organisms* 55, 85–92.

Ilouze, M., Dishon, A. and Kotler, M. (2012) Down-regulation of the cyprinid herpesvirus-3 annotated genes in cultured cells maintained at restrictive high temperature. *Virus Research* 169, 289–295.

Kancharla, S.R. and Hanson, L.A. (1996) Production and shedding of channel catfish virus (CCV) and thymidine kinase negative CCV in immersion exposed channel catfish fingerlings. *Diseases of Aquatic Organisms* 27, 25–34.

Kunec, D., Hanson, L.A., van Haren, S., Nieuwenhuizen, I.F. and Burgess, S.C. (2008) An over-lapping bacterial artificial chromosome system that generates vector-less progeny for channel catfish herpesvirus. *Journal of Virology* 82, 3872–3881.

Kunec, D., van Haren, S., Burgess, S.C. and Hanson, L.A. (2009) A Gateway® recombination herpesvirus cloning system with negative selection that produces vectorless progeny. *Journal of Virological Methods* 155, 82–86.

Liu, Y., Yuan, J., Wang, W., Chen, X., Tang, R. *et al.* (2012) Identification of envelope protein ORF10 of channel catfish herpesvirus. *Canadian Journal of Microbiology* 58, 271–277.

Major, R.D., McCraren, J.P. and Smith, C.E. (1975) Histopathological changes in channel catfish (*Ictalurus punctatus*) experimentally and naturally infected with channel catfish virus disease. *Journal of the Fisheries Board of Canada* 32, 563–567.

McGlamery, M.H. Jr and Gratzek, J. (1974) Stunting syndrome associated with young channel catfish that survived exposure to channel catfish virus. *The Progressive Fish-Culturist* [now North American Journal of Aquaculture] 36, 38–41.

Noga, E.J. and Hartmann, J.X. (1981) Establishment of walking catfish (*Clarias batrachus*) cell lines and development of a channel catfish (*Ictalurus punctatus*) virus vaccine. *Canadian Journal of Fisheries and Aquatic Sciences* 38, 925–929.

Nusbaum, K.E. and Grizzle, J.M. (1987) Uptake of channel catfish virus from water by channel catfish and bluegills. *American Journal of Veterinary Research* 48, 375–377.

Petrie-Hanson, L. and Ainsworth, A.J. (2001) Ontogeny of channel catfish lymphoid organs. *Veterinary Immunology and Immunopathology* 81, 113–127.

Plant, K.P., Harbottle, H. and Thune, R.L. (2005) Poly I:C induces an antiviral state against *Ictalurid herpesvirus 1* and Mx1 transcription in the channel catfish (*Ictalurus punctatus*). *Developmental and Comparative Immunology* 29, 627–635.

Plumb, J.A. (1971) Tissue distribution of channel catfish virus. *Journal of Wildlife Diseases* 7, 213–216.

Plumb, J.A. (1973a) Effects of temperature on mortality of fingerling channel catfish (*Ictalurus punctatus*) experimentally infected with channel catfish virus. *Journal of the Fisheries Research Board of Canada* 30, 568–570.

Plumb, J.A. (1973b) Neutralization of channel catfish virus by serum of channel catfish. *Journal of Wildlife Diseases* 9, 324–330.

Plumb, J.A. (1978) Epizootiology of channel catfish virus disease. *Marine Fisheries Review* 3, 26–29.

Plumb, J.A. (1986) *Channel Catfish Virus Disease*. Fish Disease Leaflet 73, US Department of the Interior, Washington, DC.

Plumb, J. (1989) Channel catfish herpesvirus. In: Ahne, W. and Kurstak, E. (eds) *Viruses of Lower Vertebrates*. Springer, Berlin, pp. 198–216.

Plumb, J.A. and Chappell, J. (1978) Susceptibility of blue catfish to channel catfish virus. *Proceedings of the Annual Conference of the Southeastern Association of Fish and Wildlife Agencies* 32, 680–685.

Plumb, J.A. and Hanson, L.A. (2011) *Health Maintenance and Principal Microbial Diseases of Cultured Fishes*. Wiley, Ames, Iowa.

Plumb, J.A., Gaines, J.L., Mora, E.C. and Bradley, G.G. (1974) Histopathology and electron microscopy of channel catfish virus in infected channel catfish, *Ictalurus punctatus* (Rafinesque). *Journal of Fish Biology* 6, 661–664.

Plumb, J.A., Green, O.L., Smitherman, R.O. and Pardue, G.B. (1975) Channel catfish virus experiments with different strains of channel catfish. *Transactions of the American Fisheries Society* 104, 140–143.

Roncarati, A., Mordenti, O., Stocchi, L. and Melotti, P. (2014) Comparison of growth performance of 'Common Catfish Ameiurus melas, Rafinesque 1820', reared in pond and in recirculating aquaculture system. *Journal of Aquaculture Research and Development* 5:218.

Silverstein, P.S., Bosworth, B.G. and Gaunt, P.S. (2008) Differential susceptibility of blue catfish, *Ictalurus furcatus* (Valenciennes), channel catfish, *I. punctatus* (Rafinesque), and blue × channel catfish hybrids to channel catfish virus. *Journal of Fish Diseases* 31, 77–79.

Thompson, D.J., Khoo, L.H., Wise, D.J. and Hanson, L.A. (2005) Evaluation of channel catfish virus latency on fingerling production farms in Mississippi. *Journal of Aquatic Animal Health* 17, 211–215.

Vanderheijden, N., Alard, P., Lecomte, C. and Martial, J.A. (1996) The attenuated V60 strain of channel catfish virus possesses a deletion in ORF50 coding for a potentially secreted glycoprotein. *Virology* 218, 422–426.

Vanderheijden, N., Hanson, L.A., Thiry, E. and Martial, J.A. (1999) Channel catfish virus gene 50 encodes a secreted, mucin-like glycoprotein. *Virology* 257, 220–227.

Vanderheijden, N., Martial, J.A. and Hanson, L.A. (2001) *Channel Catfish Virus Vaccine*. US Patent 6, 322, 793. Available at: https://www.google.com.au/patents/US6322793 (accessed 1 November 2016).

Waltzek, T.B., Kelley, G.O., Alfaro, M.E., Kurobe, T., Davison, A.J. *et al.* (2009) Phylogenetic relationships in the family *Alloherpesviridae*. *Diseases of Aquatic Organisms* 84, 179–194.

Wolf, K. and Darlington, R.W. (1971) Channel catfish virus: a new herpesvirus of Ictalurid fish. *Journal of Virology* 8, 525–533.

Wolf, K., Herman, R.L. and Carlson, C.P. (1972) Fish viruses: histopathologic changes associated with experimental channel catfish virus disease. *Journal of the Fisheries Research Board of Canada* 29, 149–150.

Wu, B., Zhu, B., Luo, Y., Wang, M., Lu, Y. *et al.* (2011) Generation and characterization of monoclonal antibodies against channel catfish virus. *Hybridoma* 30, 555–558.

Zhang, H.G. and Hanson, L.A. (1995) Deletion of thymidine kinase gene attenuates channel catfish herpesvirus while maintaining infectivity. *Virology* 209, 658–663.

Zhang, H.G. and Hanson, L.A. (1996) Recombinant channel catfish virus (*Ictalurid herpesvirus 1*) can express foreign genes and induce antibody production against the gene product. *Journal of Fish Diseases* 19, 121–128.

9 Largemouth Bass Viral Disease

RODMAN G. GETCHELL[1]* AND GEOFFREY H. GROOCOCK[2]

[1]*Veterinary Medical Center, Department of Microbiology and Immunology, Cornell University, Ithaca, New York, USA; [2]Transit Animal Hospital, Depew, New York, USA*

9.1 Introduction

The first *Ranavirus* was isolated from leopard frogs, *Rana pipiens*, in the eastern USA (Granoff *et al.*, 1965). Thirty years later, the first outbreak of largemouth bass viral disease was reported at Santee-Cooper Reservoir, South Carolina by Plumb *et al.* (1996), who isolated the virus concerned on fathead minnow, *Pimephales promelas* (FHM) cells inoculated with filtered homogenates from two infected adult largemouth bass, *Micropterus salmoides*, collected during the mortality event. Icosahedral virus particles (enveloped virions about 174 nm in diameter) were transmitted to uninfected largemouth bass by experimental transmission by injection. The isolate was tentatively classified as belonging to the family *Iridoviridae* and largemouth bass virus (LMBV) proposed as its name (Plumb *et al.*, 1996).

Previously, iridovirus infections in fish were limited to *Lymphocystis disease virus 1* (LCDV-1), the type species of the genus *Lymphocystivirus* (Weissenberg, 1965; Wolf, 1988). Mao *et al.* (1997) showed that six fish iridoviruses were more closely related to *Ranavirus* than to LCDV; ranaviruses had originally been thought to only infect amphibians (Hedrick *et al.*, 1992). These new fish iridoviruses caused systemic disease with high morbidity and mortality in fishes in Australia, Japan and Europe (Langdon *et al.*, 1986, 1988; Ahne *et al.*, 1989; Inouye *et al.*, 1992; Hedrick and McDowell, 1995; Go *et al.*, 2006).

Between August 1997 and November 1998, the Southeastern Cooperative Fish Disease Laboratory (Auburn University, Alabama) investigated LMBV outbreaks and surveyed 78 locations in eight US states. Virus was isolated from largemouth bass collected from six reservoirs on four different river systems (Plumb *et al.*, 1999). The gross pathological signs noted in fish with LMBV were enlarged swim bladders (Fig. 9.1) and erythematous gas glands (Plumb *et al.*, 1996, 1999; Hanson *et al.*, 2001a,b). Sequence analysis showed that the virus from these surveys was identical to the LMBV found earlier at Santee-Cooper Reservoir in South Carolina.

The growth of LMBV in five fish cell lines had an optimum replication temperature of 30°C (Piaskoski *et al.*, 1999). Bluegill fry-2 (BF-2) and FHM cells performed best, demonstrating an early onset cytopathic effect (CPE), rapid viral replication, and high titres of LMBV. In the initial study, clinical signs or mortality were not seen in adult largemouth bass experimentally infected with LMBV (Plumb *et al.*, 1996), but juvenile fish were susceptible and sustained up to 60% mortality (Plumb and Zilberg, 1999b). Similarly, treated juvenile striped bass, *Morone saxatilis*, suffered 63% mortality.

Mao *et al.* (1999) confirmed that LMBV belonged to the genus *Ranavirus* and showed that it was similar to viruses from the doctor fish, *Labroides dimidatus* (doctor fish virus, DFV) and guppies, *Poecilia reticulata* (guppy virus 6, GV6) (Hedrick and McDowell, 1995). A retrospective investigation by Grizzle *et al.* (2002) genetically compared a 1991 iridovirus from a clinically normal largemouth bass from Lake Weir, Florida, with the 1995 Santee-Cooper isolate. The Lake Weir isolate was associated with sporadic fish kills in that lake. Restriction fragment length polymorphism (RFLP) analysis and determination of the DNA sequence of a portion of the major capsid protein (MCP) gene of the two isolates showed that they were identical. The authors noted that it was unknown whether

---

*Corresponding author e-mail: rgg4@cornell.edu

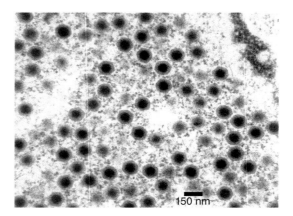

Fig. 9.2. Transmission electron microscope image of LMBV in cell culture. Photo courtesy of Andrew Goodwin; original from the Collection of John Plumb, Auburn University.

Fig. 9.1. Largemouth bass infected with largemouth bass virus (LMBV) showing an over-inflated swim bladder (arrow). Photo original courtesy of John Grizzle, Auburn University.

LMBV was a newly introduced pathogen spreading throughout the Southeastern USA, or whether it was a resident virus that was found because of increased testing. LMBV continued to be found throughout the eastern USA (Grizzle and Brunner, 2003). Many infected fish were clinically healthy, so infection was not always fatal (Plumb *et al.*, 1999; Hanson *et al.*, 2001b).

9.1.1 Description of LMBV

The family *Iridoviridae* consists of five genera of double-stranded DNA-viruses – *Iridovirus* and *Chloriridovirus* – infecting invertebrates, and *Ranavirus*, *Megalocytivirus* and *Lymphocystivirus*, which infect poikilotherms. Both *Ranavirus* and *Megalocytivirus* are important pathogens of marine and freshwater fishes (MacLachlan and Dubovi, 2011). Ranaviruses are genetically diverse, infecting bony fishes, amphibians and reptiles (Chinchar, 2002). LMBV was only one of many new ranaviruses identified during mortalities of fish during the mid-1980s and 1990s (Williams *et al.*, 2005). Its present accepted name, as designated by the International Committee on Taxonomy of Viruses (ICTV) and after its original site of isolation, is *Santee-Cooper ranavirus*.

Electron microscopy of LMBV-infected FHM cells (Fig. 9.2) revealed cytoplasmic, icosahedral virions that averaged 132 nm from facet to facet

and 145 nm from corner to corner (Plumb *et al.*, 1996), within the *Iridoviridae* range of 120–200 nm diameter (MacLachlan and Dubovi, 2011). Infectious virions are either non-enveloped or enveloped, depending upon whether they are released from the cell by lysis or budded from the plasma membrane. *Ranavirus* infections are accompanied by marked CPE and cultured cells infected with ranaviruses undergo apoptosis (Chinchar *et al.*, 2005).

9.1.2 Mode of transmission

The transmission of LMBV can occur when susceptible fish are immersed in water with the viral agent (Plumb and Zilberg, 1999b) or fed with LMBV-infected prey (Woodland *et al.*, 2002a). Five of 24 young largemouth bass gavaged (force-fed) with LMBV-infected guppies tested positive, though none showed clinical signs of disease. Vertical transmission has not been demonstrated and there is no evidence of it occurring in hatchery fish (Woodland *et al.*, 2002b). The virus is sometimes present in the cutaneous mucus of infected experimental fish (Woodland *et al.*, 2002a), which could allow direct transmission. Several warm-water fishes such as smallmouth bass, *Micropterus dolomieu*, chain pickerel, *Esox niger*, bluegill, *Lepomis macrochirus*, and redear sunfish, *Lepomis microlophus*, can be infected but without clinical signs (Grizzle *et al.*, 2003). Humans may spread LMBV unknowingly because the virus survives in water, such as in the live wells of fishing boats or tanks on stocking trucks (Grizzle and Brunner, 2003). Johnson and Brunner (2014) suggested that microbial and zooplankton communities may rapidly inactivate ranaviruses in pond water and thereby minimize environmental transmission. Host behaviour, density and contact rates also play critical roles in shaping transmission dynamics.

9.1.3 Geographical distribution

Since the first reports of the virus (Plumb *et al.*, 1996, 1999), LMBV has been documented throughout the eastern USA. In 1998, LMBV-related mortality was found in the Sardis Reservoir, Mississippi (Hanson *et al.*, 2001b) and, in 1999, in wild fish populations in Tunica Cutoff and Lake Ferguson, also in Mississippi, as well as in waters in Texas and Louisiana. These new outbreaks suggested that LMBV was spreading or that conditions had become more optimal for LMBV-induced disease. Over 1400 largemouth bass were evaluated for LMBV from 17 different fisheries in Mississippi over 2 years; there was an increase in the prevalence of LMBV in seven fisheries in the second year at the same time that other fisheries saw a decrease (Hanson *et al.*, 2001a). The most obvious result of the study was that LMBV was prevalent in Mississippi, occurring at detectable levels in 12 of 17 of largemouth bass populations (Hanson *et al.*, 2001a).

The exotic origin of LMBV was suggested by the pattern of fish kills and viral isolations in North America, where it was first found in Lake Weir, Florida, in 1991 (Grizzle *et al.*, 2002), with subsequent isolations in states to the north and west. In 2000, a survey of Texas waters in which 2876 adult largemouth bass were tested, detected LMBV in 15 of 50 reservoirs and eight of 13 major river basins (Southard *et al.*, 2009). The prevalence of LMBV detected ranged from 1.7 to 13.3% in infected reservoirs. The virus also was found during a long-term mortality study of smallmouth bass brood stock (Southard *et al.*, 2009). In 2002, LMBV-infected largemouth bass were found in Lake Champlain, Vermont. The virus was also found in largemouth bass from Lake St. Clair, which lies between Ontario (Canada) and Michigan (the USA), in 2003, and in the Bay of Quinte, Lake Ontario, Canada, in 2000, although no large fish kills were reported in the Great Lakes basin (US EPA and Environment Canada, 2009). Surveys of

37 water bodies in New York (2004 and 2005) verified the presence of LMBV in 23 of 283 largemouth bass and five of eight smallmouth bass when assayed using quantitative PCR (Groocock *et al.*, 2008). Age, sex and season were not significantly associated with LMBV prevalence. In the summer of 2006, a 3-month-long LMBV-induced die-off at a 22 ha Arkansas private impoundment demonstrated that 7% of the estimated population of largemouth bass collected were LMBV positive (Neal *et al.*, 2009).

To date, LMBV has been detected in at least 17 fish species from 30 states. Attempts have been made to detect LMBV in states west of the current known range (Fig. 9.3), but infected fish were found only in Arizona in 2010 (Silverwood and McMahon, 2012). The first report of LMBV in the invasive northern snakehead, *Channa argus*, involved collections from tidal tributaries within Chesapeake Bay, which is bordered by Maryland and Virginia (Iwanowicz *et al.*, 2013). Centrarchids are part of the northern snakehead diet, so exposure to LMBV-infected prey was a likely source of infection. LMBV was detected with bacterial pathogens in large-scale fish kills of smallmouth bass in the Susquehanna and Potomac River watersheds. The role of LMBV in these fish kills is unclear (Blazer *et al.*, 2010).

Ranaviruses were isolated from largemouth bass cultured in Guangdong Province, China, that died with skin and muscle ulcerations, as well as a swollen spleen and kidneys (Deng *et al.*, 2011). Investigators were able to transmit the infection by intramuscular injection of the isolate into healthy bass and mandarinfish, *Siniperca chuatsi*. Amplification, sequencing and phylogenetic analysis of the MCP gene revealed that the virus was identical to DFV and closely related to LMBV (Deng *et al.*, 2011). Also, using virus isolation, electron microscopy, PCR, MCP sequencing (99.9% identity) and transmission studies, George *et al.* (2015) confirmed that LMBV was the cause of mortalities of koi, *Cyprinus carpio*, in fish farms on the south-east coast of India.

Scientists have reported disease outbreaks attributed to other iridoviruses, which may suggest an increase in prevalence of LMBV. However, it was not clear whether this was due to an actual increase of iridovirus disease, or a result of the collective ability to detect iridovirus infections in more vertebrates (Williams *et al.*, 2005). The first detections of an iridovirus in largemouth bass outside the USA were reported from Taiwan, where mortality rarely

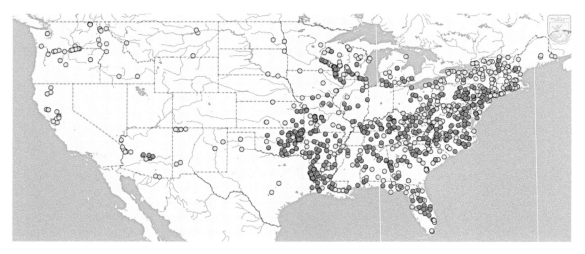

Fig. 9.3. National Wild Fish Health Survey (NWFHS) of the distribution of largemouth bass virus (LMBV) in the USA as on 2 November 2016. The red circles indicate LMBV detected and the yellow circles are largemouth bass (LMB) samples that tested negative. From US Fish and Wildlife Service (2015).

exceeded 30–50%. Sequence comparisons of the MCP initially showed they were closely related to grouper iridovirus in Taiwan (TGIV), but different from LMBV (Chao *et al.*, 2002). Subsequent research designated this virus a *Megalocytivirus* (Chao *et al.*, 2004; Huang *et al.*, 2011).

9.1.4 Impacts of LMBV on fish populations

In 1999, 15 state hatcheries in ten states in the south-eastern USA were surveyed for LMBV. The virus was found in largemouth bass brood stock at five hatcheries in Florida, Louisiana, Tennessee and Texas. There were differences in the management practices at these hatcheries, so it was impossible to discriminate the factors that affected transmission. Juvenile largemouth bass were infected at one Texas location (Woodland *et al.*, 2002b), but the vertical transmission of LMBV from infected brood fish was not established. The discovery of LMBV in juvenile fish in a hatchery does indicate that fish stocking contributes to virus spread. Signs of LMBV disease were not seen at any hatchery during the survey.

Another impact study occurred through repeated sampling at Sardis Reservoir, Mississippi, after a die-off of about 3000 adult largemouth bass when the surface waters were at temperatures of 29–32°C. Sampling on five separate occasions (at 1, 2, 4, 7 and 13 months) during the following year revealed that 53, 57, 42, 57 and 32%, respectively, of the

171 bass were positive for LMBV. The authors did not view a reduction of 3000 adult largemouth bass in the Sardis Reservoir population to have a major impact, even with evidence of LMBV persistence (Hanson *et al.*, 2001b). Annual fishing losses in the lake exceed 30,000 largemouth bass per year. The authors suggested that further evaluations were needed to determine the impact that LMBV has on largemouth bass populations. No correlation was found with gender, size, handling injuries or the presence of skin lesions, although 97% of the 36 fish with swim bladder lesions were positive for LMBV by viral culture. LMBV was isolated from juvenile largemouth bass, confirming that the infection was not limited to adults. Hanson *et al.* (2001b) did not find LMBV infection in sympatric fish species, including bluegill, white crappie, *Pomoxis annularis*, white bass, *Morone chrysops*, and gizzard shad, *Dorosoma cepedianum*, 4–7 months after the outbreak.

Grizzle and Brunner (2003) emphasized that LMBV was associated with 24 largemouth bass die-offs, but sometimes the virus was found with no clinical signs. As decreases in bass populations after die-offs had not been documented, additional research was needed to determine whether LMBV had reduced the number of trophy-sized largemouth bass. Different geographical isolates of LMBV may exhibit major differences in pathogenicity and virulence (Goldberg *et al.*, 2003). The mortality rate from LMBV infection and viral titre

R.G. Getchell and G.H. Groocock

in experimental survivors differed according to the origin of the isolate. The median survival times of bass injected with the three different isolates were 11.0, 7.5 and 4.5 days. The most virulent viral strain also replicated to the highest level in fish (Goldberg *et al.*, 2003). The authors suggested that strain variation could explain clinical differences in responses of bass populations to LMBV.

9.2 Diagnosis

The gross lesions induced by *Ranavirus* are not unique to this virus. The visualization of 120–300 nm non-enveloped icosahedral particles within the cytoplasm of infected cells establishes probable infection by an iridovirus. Procedures for the isolation of LMBV are outlined in the American Fisheries Society (AFS) Fish Health Section (FHS) *FHS Blue Book: Suggested Procedures for the Detection and Identification of Certain Finfish and Shellfish Pathogens* (AFS-FHS, 2014). The organs used for LMBV testing are the swim bladder, trunk kidney and spleen. Beck *et al.* (2006) showed that in the first few days after an immersion exposure, the gills and swim bladder would probably contain LMBV, and that the swim bladder and trunk kidney would have higher titres of virus. Pooled organ samples are recommended because a single organ may not always have detectable levels of LMBV (Beck *et al.*, 2006). Currently, LMBV is isolated in cell culture and is then confirmed using PCR. Several fish cell lines promote the replication of LMBV (Piaskoski *et al.*, 1999; McClenahan *et al.*, 2005b). The FHM and BF-2 cell lines were optimal for

LMBV isolation from both cell-culture fluids and homogenized organ samples. Frozen tissues from dead or moribund fish that were held at –10°C were stable if the fish were fresh when frozen or assayed within 5-months (Plumb and Zilberg, 1999a). Largemouth bass virus CPE in BF-2 cells began with pyknosis, rounding of the cells across the monolayer, followed by cell lysis and detachment. LMBV-inoculated FHM cells with CPE had foci of infection that became plaques devoid of cells (Fig. 9.4). High viral titres were reached in 24 h or less with BF-2 or FHM cells. A new cell line derived from the ovary of largemouth bass was recently developed and characterized (Getchell *et al.*, 2014), though LMBV titres were tenfold lower in this than in BF-2 cells.

Cytoplasmic inclusion bodies are not visible without staining (AFS-FHS, 2014). Hanson *et al.* (2001b) evaluated the gills, spleen and swim bladder for the virus using cell culture. Gill tissue was rarely infected, whereas the highest frequency of viral isolation was from the swim bladder. The use of both the swim bladder and the spleen increased a positive diagnosis in 10% more fish than culturing of the swim bladder alone. Orally exposed largemouth bass that became infected with LMBV had varying viral titres in cutaneous mucus, swim bladder, head kidney, trunk kidney, spleen, gonads and intestine (Woodland *et al.*, 2002a). The swim bladder produced the highest titres at $10^{5.5-9.5}$ $TCID_{50}$ (50% tissue culture infectious dose)/g, but lesions therein were not observed.

Mao *et al.* (1997, 1999) found that both PCR and reverse transcription PCR (RT-PCR) successfully

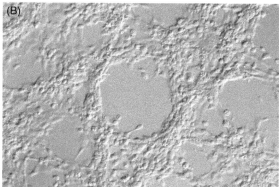

Fig. 9.4. (A) Control fathead minnow cells (FHM) and (B) largemouth bass virus (LMBV)-induced plaque formation on FHM cells. Photo original courtesy of John Grizzle, Auburn University.

amplified virus-specific nucleic acid and demonstrated that those piscine iridoviruses examined belonged within the genus *Ranavirus*. Electrophoretic analysis of proteins from virus-infected cells and RFLP analysis of viral DNA clarified the relatedness of these isolates. Later studies used PCR to confirm LMBV infection (Grizzle *et al.*, 2002; Grizzle and Brunner, 2003). When there was not enough LMBV for cell culture, PCR could be used because of its greater sensitivity (Grizzle *et al.*, 2003; McClenahan *et al.*, 2005b). The PCR method can be simplified by adding diluted culture fluid from a presumptively positive cell culture to the PCR mixture, rather than using extracted DNA for the PCR template (McClenahan *et al.*, 2005a). The specific LMBV288F-535R primer pair developed by Grizzle *et al.* (2003) amplified a 248 bp DNA fragment of LMBV from Santee-Cooper Reservoir, but did not amplify DNA from four other ranaviruses (*Frog virus 3*, FV3; DFV; GV6; *Epizootic haematopoietic necrosis virus*, EHNV) or from red sea bream iridovirus. This primer pair amplified the correct size fragment from 30 viral isolates that were presumptively identified as LMBV based on viral isolation and other methods. Later, Ohlemeyer *et al.* (2011) cloned and sequenced the complete MCP gene of DFV, GV6 and LMBV. The DFV and GV6 sequences were identical, and the LMBV sequence was 99.21% related. These authors then developed a Santee-Cooper-specific PCR that would not amplify related grouper iridoviruses. They concluded that these three viruses might not belong to the *Ranavirus* genus (Hyatt *et al.*, 2000; Whittington *et al.*, 2010).

Real-time PCR (qPCR) techniques to detect a fragment of the LMBV MCP were developed by Goldberg *et al.* (2003) and Getchell *et al.* (2007). The qPCR assay offered significant advantages for monitoring pathogen prevalence in fish populations, including high throughput capability and reduced contamination issues. Comparison of the qPCR assay with the plaque assay confirmed the extended linear range of the qPCR when dilutions of LMBV were tested and showed that the real-time qPCR assay was approximately 100 times more sensitive than the plaque assay when infected cell culture fluids were tested. Another qPCR assay developed to detect many fish ranaviruses was not tested against LMBV, but worked well with DFV and GV6 (Holopainen *et al.*, 2011).

Serological tests are not sufficient to identify closely related ranaviruses due to extensive cross-reactivity (Hedrick *et al.*, 1992; Marsh *et al.*, 2002).

In situ hybridization targeted at MCP gene sequences was successfully employed to identify an iridovirus, although the tissues of the Malabar grouper, *Epinephelus malabaricus*, were fixed in formalin (Huang *et al.*, 2004). PCR and RFLP were used to differentiate between species of *Ranavirus* (Marsh *et al.*, 2002). Most recently, amplified fragment length polymorphism (AFLP) and qPCR were applied to detect and differentiate between geographical isolates of LMBV (Goldberg *et al.*, 2003). Miller *et al.* (2015) do an excellent job in suggesting which diagnostic tools should be used to achieve specific investigational results. Ideally, multiple samples should be submitted and multiple tests performed.

9.3 Pathology

An enlarged swim bladder and an erythematous gas gland are gross pathological signs (Plumb *et al.*, 1996) of LMBV infection. Zilberg *et al.* (2000) described the lesions in several juvenile largemouth bass injected intraperitoneally (IP) with LMBV. Clinical signs and external lesions included inflammation at the site of injection, distended abdomen, corkscrew swimming, lateral recumbency and lethargy. Internal lesions included focally pale livers, bright red spleens and reddened intestinal caecae. Microscopic lesions included acute fibrinous peritonitis and exudate present in the ventral aspect of the swim bladder. In another study (Hanson *et al.*, 2001b), LMBV infections appeared to involve only the swim bladder and resulted in the accumulation of a yellow wax-like material in the lumen of the bladder (Fig. 9.5). The material consisted of erythrocytes, cellular debris and eosinophils in a fibrin clot. Grant *et al.* (2003) reported internal gross pathology such as exudative polyserositis, pneumocystitis and colour changes to various visceral organs, particularly the liver. Clinical signs observed in moribund infected juvenile koi included skin darkening, loss of scales, vertical hanging, uncoordinated swimming, turning upside down, lateral rotation, intermittent surfacing, settling at the bottom on their sides and death (George *et al.*, 2015).

9.4 Pathophysiology

Fishing tournaments targeting largemouth bass occur in 48 out of 50 US states (Schramm *et al.*, 1991). The specific effects of recreational fishing and environmental stressors, and their impact on

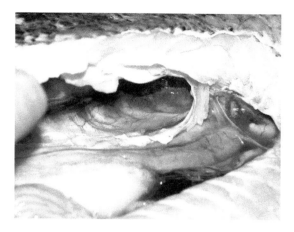

Fig. 9.5. Yellow exudate in the swim bladder of a largemouth bass infected with largemouth bass virus (LMBV). Photo original courtesy of Andrew Goodwin, US Fish and Wildlife Service.

the physiology of largemouth bass, may be associated with LMBV outbreaks (Schramm *et al.*, 2006). The interaction of human-induced and environmental stressors with viral infection may precipitate fish kills. These physiological changes may contribute to increased pathogenicity of LMBV. LMBV-associated fish kills have been shown to typically occur during the hot summer months (Plumb *et al.*, 1996; Hanson *et al.*, 2001b). Environmental stressors affecting physiology during the summer months include elevated water temperature, low dissolved oxygen content and increased angling pressure (Grant *et al.*, 2003). These may impede the immune response and permit increased viral replication. The identification of specific factors that initiate LMBV fish kills is important for managing the disease. Temperatures outside the normal thermal ranges have immunosuppressive effects in fish (Bly and Clem, 1992). Pickering and Pottinger (1989) also demonstrated that low chronic stress could lower disease resistance.

9.4.1 Effects on the endocrine system and osmoregulation

Catch-and-release angling is associated with physiological changes that affect homeostasis (Gustaveson *et al.*, 1991; Suski *et al.*, 2003; Cooke *et al.*, 2004) and could result in immunosuppression (Anderson, 1990). Gustaveson *et al.* (1991) determined that the length of time a largemouth

bass was played prior to landing the fish correlated with increases in blood cortisol and plasma lactate. Stress hormone production did not occur in largemouth bass hooked and played for 1–5 min in cold water (11–13°C), was moderate at 16–20°C and was severe in fish angled for 5 min at 28–30°C. Suski *et al.* (2003) showed the plasma cortisol and glucose concentrations and plasma osmolarity in tournament-caught bass were significantly greater than those in control groups. They suggested that factors such as live well confinement, handling and air exposure during weighing in might play an important role in the metabolic disturbances.

9.4.2 Effects on growth

Large changes in the metabolic status of largemouth bass sampled following fishing tournament weighing in included major reductions in the muscle energy stores and large increases in lactate concentrations (Suski *et al.*, 2003). These may make largemouth bass more susceptible to LMBV. The Texas Parks and Wildlife Department staff measured bass growth parameters at two Texas impoundments where LMBV-attributed die-offs had occurred (Bister *et al.*, 2006). No drop in mean length at age 3, relative weight or angling success at either reservoir was seen, whereas at five Alabama reservoirs, Maceina and Grizzle (2006) noted slower largemouth bass growth and lower relative weights associated with LMBV infections. This occurred at the same time as the decline in memorable length bass recorded in tournament and electrofishing catches. Beck *et al.* (2006) suggested that LMBV-induced hyperbuoyancy might cause exhaustion as the fish tried to maintain submergence.

9.4.3 Disease mechanism and bioenergetic cost(s) of LMBV

Grant *et al.* (2003) showed that largemouth bass infected with LMBV by IP injection at 30°C had higher mortality rates than bass injected and held at 25°C. They measured a higher viral load using qPCR in moribund and dead fish versus bass that survived the trial, and suggested that the higher temperature was probably more optimum for viral replication and more stressful for the host. There was a significant association between viral load and the presence of internal lesions. However, a similar challenge with the same source of largemouth bass but with high LMBV titre did not find the same association (Goldberg *et al.*, 2003). Juvenile largemouth bass kept in high-density captivity and infected with LMBV experienced higher mortality, increased viral loads and reduced body conditions compared with LMBV-inoculated bass raised at a lower density (Inendino *et al.*, 2005). The conclusion was that 'Strategies that maximize the quality of the physical environment in which fish live while simultaneously minimizing sociobehavioural stress should most effectively increase health and productivity'.

Insights into the molecular mechanisms of pathogenesis were recently determined by studying the cell signalling events involved in LMBV-caused cell death (Huang *et al.*, 2014). LMBV infections of epithelioma papulosum cyprini (EPC) cells produced apoptosis mediated by both intrinsic and extrinsic pathways involving caspase-8 and caspase-9. Phosphatidylinositol 3-kinase (PI3K) and extracellular-signal-regulated kinase (ERK) signalling pathways were involved in LMBV replication as well as infection-induced apoptosis. Identifying these mechanisms will contribute to the development of potential therapeutic targets for all iridovirus infections (Reshi *et al.*, 2016).

9.5 Protective and Control Strategies

Oh *et al.* (2015) showed that fish vaccinated with a live red sea bream iridovirus following poly(I:C) (polyinosinic–polycytidylic acid) administration were protected, although DNA vaccines were less protective. No commercial vaccines are available to prevent LMBV.

Catch-and-release angling practices done by placing uninfected and infected largemouth bass in the same live well or holding tank facilitated LMBV transmission (Grant *et al.*, 2005). The rapid uptake (1 h) of LMBV, combined with tank mates shedding virus, could mean that even a short exposure time could infect other fish in a live well (Beck *et al.*, 2006). Compartmentalization of these tanks or decreasing their usage could reduce transmission, and lowering the temperature in live wells would be beneficial as higher temperatures would facilitate viral replication (Grant *et al.*, 2003). Schramm *et al.* (2006) recommended that tournament-caught bass should not be retained or be confined with fish from infected populations. *Pseudomonas fluorescens* biofilms protected LMBV against hypochlorite and iodophor but not alcohol-based disinfectants (Nath *et al.*, 2010). The virus was not detected in biofilms

or water from ponds that had contained LMBV-positive fish 2 months earlier.

Concerns about the impacts of LMBV by fisheries managers, anglers and the sportfishing industry prompted the Bass Anglers Sportsman Society to arrange a multi-agency, multi-sector collaboration that facilitated research and guided action to control LMBV (Terre *et al.*, 2008). Recommendations from workshops remind anglers and boaters to take the following steps to help prevent the spread of the virus:

1. Clean boats, trailers and other fishing equipment thoroughly between fishing trips.
2. Do not move fish or fish parts from one body of water to another, and do not release live bait into any water body.
3. Handle bass as gently as possible if you intend to release them and release them as quickly as possible.
4. Refrain from hauling the fish for long periods in live wells if you intend to release them.
5. Minimize the targeting of largemouth bass during the period from mid-July to mid-August, especially during exceptionally hot weather conditions.
6. Report dead or dying adult largemouth bass fish to natural resource agency offices.
7. Volunteer to help agencies collect bass for LMBV monitoring.
8. Educate other anglers about LMBV.

9.6 Conclusions and Suggestions for Future Research

9.6.1 Identifying gaps in our knowledge

The lack of a complete genome sequence for LMBV limits advances that will have an impact on control efforts of LMBV. Understanding how phenotypic and genotypic variation in LMBV affects virulence is important for understanding the biology, evolution and control of this disease. However, Goldberg *et al.* (2003) concluded that factors other than the inherent virulence of the pathogen, such as environmental and host-related factors, must contribute significantly to the clinical manifestations of LMBV infection in the field.

9.6.2 Suggestions for future studies

Reviewing the past research efforts on FV3 and the present work on other ranaviruses, as well as that on iridoviruses from other genera such as *Megalocytivirus*,

R.G. Getchell and G.H. Groocock

suggests that researchers will identify viral genes that play important roles in virulence and ultimately facilitate the creation of effective anti-ranavirus vaccines. Understanding how these viruses evade the antiviral immune responses of lower vertebrates will, it is hoped, be fruitful (Jancovich *et al.*, 2015).

References

AFS-FHS (2014) Section 2. USFWS standard procedures for aquatic animal health inspections. In: *FHS Blue Book: Suggested Procedures for the Detection and Identification of Certain Finfish and Shellfish Pathogens*, 2014 edn. American Fisheries Society-Fish Health Section, Bethesda, Maryland. Updated 2016 edition available at: http://afs-fhs.org/bluebook/inspection-index.php (accessed 1 November 2016).

Ahne, W., Schlotfeldt, H.J. and Thomsen, I. (1989) Fish viruses: isolation of an icosahedral cytoplasmic deoxyribovirus from sheatfish (*Silurus glanis*). *Journal of Veterinary Medicine Series B* [now *Zoonoses and Public Health*] 36, 333–336.

Anderson, D.P. (1990) Immunological indicators: effects of environmental stress on immune protection and disease outbreaks. In: Adams, S.M. (ed.) *Biological Indicators of Stress in Fish*. American Fisheries Society Symposium 8. American Fisheries Society, Bethesda, Maryland, pp. 38–50.

Beck, B.H., Bakal, R.S., Brunner, C.J. and Grizzle, J.M. (2006) Virus distribution and signs of disease after immersion exposure to largemouth bass virus. *Journal of Aquatic Animal Health* 18, 176–183.

Bister, T.J., Myers, R.A. Driscoll, M.T. and Terre, D.R. (2006) Largemouth bass population trends in two Texas reservoirs with LMBV-attributed die-offs. *Proceedings of the Annual Conference of Southeastern Association of Fish and Wildlife Agencies* 60, 101–105.

Blazer, V.S., Iwanowicz, L.R., Starliper, C.E., Iwanowicz, D.D., Barbash, P. *et al.* (2010) Mortality of centrarchid fishes in the Potomac drainage: survey results and overview of potential contributing factors. *Journal of Aquatic Animal Health* 22, 190–218.

Bly, J.E. and Clem, L.W. (1992) Temperature and teleost immune functions. *Fish and Shellfish Immunology* 2, 159–171.

Chao, C.B., Yang, S.C., Tsai, H.Y., Chen, C.Y., Lin, C.S. *et al.* (2002) A nested PCR for the detection of grouper iridovirus in Taiwan (TGIV) in cultured hybrid grouper, giant sea perch and largemouth bass. *Journal of Aquatic Animal Health* 14, 104–113.

Chao, C.B., Chen, C.Y., Lai, Y.Y., Lin, C.S. and Huang, H.T. (2004) Histological, ultrastructural, and *in situ* hybridization study on enlarged cells in grouper *Epinephelus* hybrids infected by grouper iridovirus in Taiwan (TGIV). *Diseases of Aquatic Organisms* 58, 127–142.

Chinchar, V.G. (2002) Ranaviruses (family Iridoviridae): emerging cold-blooded killers. *Archives of Virology* 147, 447–470.

Chinchar, V.G., Essbauer, S., He, J.G., Hyatt, A, Miyazaki, T. *et al.* (2005) Iridoviridae. In: Fauquet, C.M., Mayo, M.A., Maniloff, J., Desselberger, U. and Ball, L.A. (eds) *Virus Taxonomy: Eighth Report of the International Committee on the Taxonomy of Viruses*. Elsevier, San Diego, California and London, pp. 163–175.

Cooke, S.J., Bunt, C.M., Ostrand, K.G., Philipp, D.P. and Wahl, D.H. (2004) Angling-induced cardiac disturbance of free-swimming largemouth bass (*Micropterus salmoides*) monitored with heart rate telemetry. *Journal of Applied Ichthyology* 20, 28–36.

Deng, G., Li, S., Xie, J., Bai, J., Chen, K. *et al.* (2011) Characterization of a ranavirus isolated from cultured largemouth bass (*Micropterus salmoides*) in China. *Aquaculture* 312, 198–204.

George, M.R., John, K.R., Mansoor, M.M., Saravanakumar, R., Sundar, P. *et al.* (2015) Isolation and characterization of a ranavirus from koi, *Cyprinus carpio* L., experiencing mass mortalities in India. *Journal of Fish Diseases* 38, 389–403.

Getchell, R.G., Groocock, G.H., Schumacher, V.L., Grimmett, S.G., Wooster, G.A. *et al.* (2007) Quantitative polymerase chain reaction assay for largemouth bass virus. *Journal of Aquatic Animal Health* 19, 226–233.

Getchell, R.G., Groocock, G.H., Cornwell, E.R., Schumacher, V.L., Glasner, L.I. *et al.* (2014) Development and characterization of a largemouth bass cell line. *Journal of Aquatic Animal Health* 26, 194–201.

Go, J., Lancaster, M., Deece, K., Dhungyel, O. and Whittington, R. (2006) The molecular epidemiology of iridovirus in Murray cod (*Maccullochella peelii peelii*) and dwarf gourami (*Colisa lalia*) from distant biogeographical regions suggests a link between trade in ornamental fish and emerging iridoviral diseases. *Molecular and Cellular Probes* 20, 212–222.

Goldberg, T.L., Coleman, D.A., Inendino, K.R., Grant, E.C. and Philipp, D.P. (2003) Strain variation in an emerging iridovirus of warm water fishes. *Journal of Virology* 77, 8812–8818.

Granoff, A., Came, P.E., Keen, A. and Rafferty, J. (1965) The isolation and properties of viruses from *Rana pipiens*: their possible relationship to the renal adenocarcinoma of the leopard frog. *Annals of the New York Academy of Sciences* 126, 237–255.

Grant, E.C., Philipp, D.P., Inendino, K.R. and Goldberg, T.L. (2003) Effects of temperature on the susceptibility of largemouth bass to largemouth bass virus. *Journal of Aquatic Animal Health* 15, 215–220.

Grant, E.C., Philipp, D.P., Inendino, K.R. and Goldberg, T.L. (2005) Effects of practices related to catch-and-release angling on mortality and viral transmission in juvenile largemouth bass infected with largemouth bass virus. *Journal of Aquatic Animal Health* 17, 315–322.

Grizzle, J.M. and Brunner, C.J. (2003) Review of large-mouth bass virus. *Fisheries* 28, 10–14.

Grizzle, J.M., Altinok, I., Fraser, W.A. and Francis-Floyd, R. (2002) First isolation of largemouth bass virus. *Diseases of Aquatic Organisms* 50, 233–235.

Grizzle, J.M., Altinok, I. and Noyes, A.D. (2003) PCR method for detection of largemouth bass virus. *Diseases of Aquatic Organisms* 54, 29–33.

Groocock, G.H., Grimmett, S.G., Getchell, R.G., Wooster, G.A. and Bowser, P.R. (2008) A survey to determine the presence and distribution of largemouth bass virus in wild freshwater bass in New York State. *Journal of Aquatic Animal Health* 20, 158–164.

Gustaveson, A.W., Wydoski, R.S. and Wedemeyer, G.A. (1991) Physiological response of largemouth bass to angling stress. *Transactions of the American Fisheries Society* 120, 629–636.

Hanson, L.A., Hubbard, W.D. and Petrie-Hanson, L. (2001a) Distribution of largemouth bass virus may be expanding in Mississippi. *Fish Health Newsletter, Fish Health Section/American Fisheries Society* (Bethesda, Maryland) 29(4), 10–12. Available at: http://www.afs-fhs.org/communications/newsletter/V29-4_2001.PDF (accessed 29 November 2015).

Hanson, L.A., Petrie-Hanson, L., Meals, K.O., Chinchar, V.G. and Rudis, M. (2001b) Persistence of largemouth bass virus infection in a northern Mississippi reservoir after a die-off. *Journal of Aquatic Animal Health* 13, 27–34.

Hedrick, R.P. and McDowell, T.S. (1995) Properties of iridoviruses from ornamental fish. *American Journal of Veterinary Research* 26, 423–437.

Hedrick, R.P., McDowell, T.S., Ahne, W., Torhy, C. and de Kinkelin, P. (1992) Properties of three iridovirus-like agents associated with systemic infections of fish. *Diseases of Aquatic Organisms* 13, 203–209.

Holopainen, R., Honkanen, J., Bang Jensen, B., Ariel, E. and Tapiovaara, H. (2011) Quantitation of ranaviruses in cell culture and tissue samples. *Journal of Virological Methods* 171, 225–233.

Huang, C., Zhang, X., Gin, K.Y.H. and Qin, Q.W. (2004) *In situ* hybridization of a marine fish virus, Singapore grouper iridovirus with a nucleic acid probe of major capsid protein. *Journal of Virological Methods* 117, 123–128.

Huang, S.-M., Tu, C., Tseng, C.-H., Huang, C.-C., Chou, C.-C. *et al.* (2011) Genetic analysis of fish iridoviruses isolated in Taiwan during 2001–2009. *Archives of Virology* 156, 1505–1515.

Huang, X., Wang, W., Huang, Y., Xu, L. and Qin, Q. (2014) Involvement of the PI3K and ERK signaling pathways in largemouth bass virus-induced apoptosis and viral replication. *Fish and Shellfish Immunology* 41, 371–379.

Hyatt, A.D., Gould, A.R., Zupanovic, Z., Cunningham, A.A., Hengstberger, S. *et al.* (2000) Comparative studies of piscine and amphibian iridoviruses. *Archives of Virology* 145, 301–331.

Inendino, K.R., Grant, E.C., Philipp, D.P. and Goldberg, T.L. (2005) Effects of factors related to water quality and population density on the sensitivity of juvenile largemouth bass to mortality induced by viral infection. *Journal of Aquatic Animal Health* 17, 304–314.

Inouye, K., Yamano, K., Maeno, Y., Nakajima, K., Matsuoka, M. *et al.* (1992) Iridovirus infection of cultured red sea bream, *Pagrus major*. *Fish Pathology* 27, 19–27 (in Japanese with English abstract).

Iwanowicz, L., Densmore, C., Hahn, C., McAllister, P. and Odenkirk, J. (2013) Identification of largemouth bass virus in the introduced northern snakehead inhabiting the Chesapeake Bay watershed. *Journal Aquatic Animal Health* 25, 191–196.

Jancovich, J.K., Qin, Q., Zhang, C.-Y. and Chinchar, V.G. (2015) Ranavirus replication: molecular, cellular, and immunological events. In: Gray, M.J. and Chinchar, V.G. (eds) *Ranaviruses: Lethal Pathogens of Ectothermic Vertebrates*. Springer, New York, pp. 105–139.

Johnson, A.F. and Brunner, J.L. (2014) Persistence of an amphibian ranavirus in aquatic communities. *Diseases of Aquatic Organisms* 111, 129–138.

Langdon, J.S., Humphrey, J.D., Williams, L.M., Hyatt, A.D. and Westbury, H.A. (1986) First virus isolation from Australian fish: an iridovirus-like pathogen from redfin perch, *Perca fluviatilis* L. *Journal of Fish Diseases* 9, 263–268.

Langdon, J.S., Humphrey, J.D. and Williams, L.M. (1988) Outbreaks of an EHNV-like iridovirus in cultured rainbow trout, *Salmo gairdneri* Richardson, in Australia. *Journal of Fish Diseases* 11, 93–96.

Maceina, M.J. and Grizzle, J.M. (2006) The relation of largemouth bass virus to largemouth bass population metrics in five Alabama reservoirs. *Transactions of the American Fisheries Society* 135, 545–555.

MacLachlan, N.J. and Dubovi, E.J. (2011) Asfarviridae and Iridoviridae. In: MacLachlan, N.J. and Dubovi, E.J. (eds) *Fenner's Veterinary Virology*, 4th edn. Elsevier/Academic Press, London, pp. 272–277.

Mao, J. Hedrick, R.P. and Chinchar, V.G. (1997) Molecular characterization, sequence analysis and taxonomic position of newly isolated fish iridoviruses. *Virology* 229, 212–220.

Mao, J., Wang, J., Chinchar, G.D. and Chinchar, V.G. (1999) Molecular characterization of a ranavirus isolated from largemouth bass *Micropterus salmoides*. *Diseases of Aquatic Organisms* 37, 107–114.

Marsh, I.B., Whittington, R.J., O'Rouke, B., Hyatt, A.D. and Chisholm, O. (2002) Rapid identification of Australian, European, and American ranaviruses based on variation in major capsid protein gene sequences. *Molecular and Cellular Probes* 16, 137–151.

McClenahan, S.D., Grizzle, J.M. and Schneider, J.E. (2005a) Evaluation of unpurified cell culture supernatant as template for the polymerase chain reaction (PCR) with largemouth bass virus. *Journal of Aquatic Animal Health* 17, 191–196.

McClenahan, S.D., Beck, B.H. and Grizzle, J.M. (2005b) Evaluation of cell culture methods for detection of largemouth bass virus. *Journal of Aquatic Animal Health* 17, 365–372.

Miller, D.L., Pessier, A.P., Hick, P. and Whittington, R.J. (2015) Comparative pathology of ranaviruses and diagnostic techniques. In: Gray, M.J. and Chinchar, V.G. (eds) *Ranaviruses: Lethal Pathogens of Ectothermic Vertebrates.* Springer, New York, pp. 171–208.

Nath, S., Aron, G.M., Southard, G.M. and McLean, R.J.C. (2010) Potential for largemouth bass virus to associate with and gain protection from bacterial biofilms. *Journal of Aquatic Animal Health* 22, 95–101.

Neal, J.W., Eggleton, M.A. and Goodwin, A.E. (2009) The effects of largemouth bass virus on a quality largemouth bass population in Arkansas. *Journal of Wildlife Diseases* 45, 766–771.

Oh, S.-Y., Kim, W.-S., Oh, M.-J. and Nishizawa, T. (2015) Quantitative change of red seabream iridovirus (RSIV) in rock bream *Oplegnathus fasciatus*, following Poly(I:C) administration. *Aquaculture International* 23, 93–98.

Ohlemeyer, S., Holopainen, R., Tapiovaara, H., Bergmann, S.M. and Schütze, H. (2011) Major capsid protein gene sequence analysis of the Santee-Cooper ranaviruses DFV, GV6, and LMBV. *Diseases of Aquatic Organisms* 96, 195–207.

Piaskoski, T.O., Plumb, J.A. and Roberts, S.R. (1999) Characterization of the largemouth bass virus in cell culture. *Journal of Aquatic Animal Health* 11, 45–51.

Pickering, A.D. and Pottinger, T.G. (1989) Stress response and disease resistance in salmonid fish: effects of chronic elevation of plasma cortisol. *Fish Physiology and Biochemistry* 7, 253–258.

Plumb, J.A. and Zilberg, D. (1999a) Survival of largemouth bass iridovirus in frozen fish. *Journal of Aquatic Animal Health* 11, 94–96.

Plumb, J.A. and Zilberg, D. (1999b) The lethal dose of largemouth bass virus in juvenile largemouth bass and the comparative susceptibility of striped bass. *Journal of Aquatic Animal Health* 11, 246–252.

Plumb, J.A., Grizzle, J.M., Young, H.E. and Noyes, A.D. (1996) An iridovirus isolated from wild largemouth bass. *Journal of Aquatic Animal Health* 8, 265–270.

Plumb, J.A., Noyes, A.D., Graziano, S., Wang, J., Mao, J. and Chinchar, V.G. (1999) Isolation and identification of viruses from adult largemouth bass during a 1997–1998 survey in the Southeastern United States. *Journal of Aquatic Animal Health* 11, 391–399.

Reshi, L., Wu, J.-L., Wang, H.-V. and Hong, J.-R. (2016) Aquatic viruses induce host cell death pathways and its application. *Virus Research* 211, 133–144.

Schramm, H.L., Armstrong, M.L., Funicelli, N.A., Green, D.M., Lee, D.P. *et al.* (1991) The status of competitive sport fishing in North America. *Fisheries* 16, 4–12.

Schramm, H.L., Walters, A.R., Grizzle, J.M., Beck, B.H., Hanson, L.A. *et al.* (2006) Effects of live-well conditions on mortality and largemouth bass virus prevalence in largemouth bass caught during summer tournaments. *North American Journal of Fisheries Management* 26, 812–825.

Silverwood, K. and McMahon, T. (2012) *2012 – Arizona Risk Analysis: Largemouth Bass Virus (LMBV).* Arizona Game and Fish Department, Phoenix, Arizona. Available at: https://portal.azgfd.stagingaz. gov/PortalImages/files/fishing/InvasiveSpecies/RA/ Largemouth%20Bass%20Virus%202012RAforAZ. pdf (accessed 2 November 2016).

Southard, G.M., Fries, L.T. and Terre, D.R. (2009) Largemouth bass virus in Texas: distribution and management issues. *Journal of Aquatic Animal Health* 21, 36–42.

Suski, C.D., Killen, S.S., Morrissey, M.B., Lund, S.G. and Tufts, B.L. (2003) Physiological changes in the largemouth bass caused by live-release angling tournaments in southeastern Ontario. *North American Journal of Fisheries Management* 23, 760–769.

Terre, D.R., Schramm, H.L. and Grizzle, J.M. (2008) Dealing with largemouth bass virus: benefits of multisector collaboration. *Proceedings of the Annual Conference of Southeastern Association of Fish and Wildlife Agencies* 62, 115–119.

US EPA and Environment Canada (2009) *Nearshore Areas of the Great Lakes, 2009. State of the Lakes Ecosystem Conference 2008: Background Paper.* [Prepared for the SOLEC Conference at Niagara Falls, Ontario in 2008] by the Governments of Canada and the United States of America. Prepared by Environment Canada and the US Environmental Protection Agency. Available at: https://binational.net// wp-content/uploads/2014/05/SOGL_2009_nearshore_ en.pdf (accessed 1 November 2016).

US Fish and Wildlife Service (2015) National Wild Fish Health Survey, US Fish and Wildlife Service, Washington, DC. Available at: http://ecos.fws.gov/wild-fishsurvey/database/nwfhs/ (accessed 2 November 2016).

Weissenberg, R. (1965) Fifty years of research on the lymphocystis virus disease of fishes (1914–1964). *Annals of the New York Academy of Sciences* 126, 362–374.

Whittington, R.J., Becker, J.A. and Dennis, M.M. (2010) Iridovirus infections in finfish: critical review with emphasis on ranaviruses. *Journal of Fish Diseases* 33, 95–122.

Williams, T., Barbarosa-Solomieu, V. and Chinchar, V.G. (2005) A decade of advances in iridovirus research. *Advanced Virus Research* 65, 173–248.

Wolf, K. (1988) *Fish Viruses and Fish Viral Diseases.* Cornell University Press, Ithaca, New York.

Woodland, J.E., Brunner, C.J., Noyes, A.D. and Grizzle, J.M. (2002a) Experimental oral transmission of largemouth bass virus. *Journal of Fish Diseases* 25, 669–672.

Woodland, J.E., Noyes, A.D. and Grizzle, J.M. (2002b) A survey to detect largemouth bass virus among fish from hatcheries in the southeastern USA. *Transactions of the American Fisheries Society* 131, 308–311.

Zilberg, D., Grizzle, J.M. and Plumb, J.A. (2000) Preliminary description of lesions in juvenile largemouth bass injected with largemouth bass virus. *Diseases of Aquatic Organisms* 39, 143–146.

10 Koi Herpesvirus Disease

KEITH WAY* AND PETER DIXON

Centre for Environment, Fisheries and Aquaculture Science (Cefas) Weymouth Laboratory, Weymouth, UK

10.1 Introduction

Koi herpesvirus disease (KHVD) is a herpesvirus infection (Hedrick *et al.*, 2000) that induces a lethal acute viraemia that is highly contagious in common carp (*Cyprinus carpio*) and varieties of *C. carpio* such as koi carp and ghost carp (koi × common carp) (Haenen *et al.*, 2004). The causative agent is classified as *Cyprinid herpesvirus 3* (CyHV-3), a member of the family *Alloherpesviridae* and one of ten alloherpesviruses that infect fishes (Boutier *et al.*, 2015a).

The transmission of CyHV-3 is horizontal and can occur directly or indirectly. Uchii *et al.* (2014) suggested that CyHV-3 in recovered fish reactivates periodically when the water temperature increases and transmits to naive fish when they are in close contact, such as at spawning. Direct transmission occurs by skin-to-skin contact between infected and naive carp, and through the cannibalistic and necrophagous behaviour of carp. Several vectors may facilitate the indirect transmission of CyHV-3. These include faeces, aquatic sediments, plankton and aquatic invertebrates (Boutier *et al.*, 2015a). Contaminated water represents the major abiotic transmission medium as it can contain virulent virus excreted in urine and shed via faeces, gills and skin mucus. Infectious water is a highly efficient mode of transmission (Boutier *et al.*, 2015b), although in the absence of hosts, CyHV-3 rapidly inactivates in water (Boutier *et al.*, 2015a). There is currently no published evidence for vertical transmission.

Following the first reports of KHVD in Germany in 1997, and in Israel and the USA in 1998, the geographical range of the disease became extensive. Worldwide trade in koi carp is generally responsible for the spread of CyHV-3 and the disease now occurs or has been reported in fish imported into at least 30 countries (OIE, 2012; Boutier *et al.*, 2015a). In Asia, the first outbreak of KHVD with mass mortalities of cultured koi carp occurred in Indonesia in 2002. In 2003, it was reported in Japan following mass mortalities of cage-cultured common carp (Haenen *et al.*, 2004). CyHV-3 has since been detected in Taiwan, China, South Korea, Singapore, Malaysia and Thailand (Boutier *et al.*, 2015a). In Africa, KHVD has only been reported in South Africa (Haenen *et al.*, 2004).

In North America, the first KHVD outbreaks involved koi dealers in 1998 and 1999 (Haenen *et al.*, 2004), but subsequently CyHV-3 caused mass mortalities of wild common carp in the USA and Canada (Boutier *et al.*, 2015a). In Europe, the disease has been recorded in 18 countries, with widespread mass mortalities reported in carp in Germany, Poland and the UK (OIE, 2012; Boutier *et al.*, 2015a).

CyHV-3 causes a highly virulent and contagious disease that induces massive mortalities and major economic losses in common and koi carp. The disease has been listed as notifiable by the World Organisation for Animal Health (OIE) since 2007 and is seasonal, occurring mostly at water temperatures between 17 and 28°C. Mortality occurs within 6–22 days postinfection (dpi), peaking between 8 and 12 dpi (Ilouze *et al.*, 2006). Naturally occurring CyHV-3 infections have only been recorded from common carp and varieties of this species such as koi (OIE, 2012). However, *Carassius* spp. (crucian carp) hybridized with common carp may be susceptible to KHVD. Mortality reported in experimental infections has varied from 91 to 100% in crucian carp × koi and 35 to 42% in goldfish (*Carassius*

© P.T.K. Woo and R.C. Cipriano 2017. *Fish Viruses and Bacteria: Pathobiology and Protection* (P.T.K. Woo and R.C. Cipriano)

115

auratus) × koi hybrids (Bergmann *et al.*, 2010a), to only 5% in goldfish × common carp hybrids (Hedrick *et al.*, 2006).

Goldfish are susceptible to CyHV-3 infection (OIE, 2012), but they do not develop disease. There is also increasing evidence that other cyprinid and non-cyprinid species (e.g. grass carp, *Ctenopharyngodon idella*; catfish, Siluriformes) are potential carriers that transmit the infection to carp (Boutier *et al.*, 2015a).

From 1998 to 2000, KHVD spread to 90% of Israeli carp farms at an estimated cost of US$3 million a year. In Indonesia, between April and November 2002 the disease spread from East Java to four other major islands and the losses exceeded US$15 million by 2003 (Haenen *et al.*, 2004). Undoubtedly, CyHV-3 had been spread globally before regulators became aware of the disease and when detection methods were unavailable. CyHV-3 DNA was detected in archived histological specimens collected during unexplained mass mortalities of carp in the UK in 1996 (Haenen *et al.*, 2004) and in South Korea in 1998 (Lee *et al.*, 2012). In Israel, piscivorous birds are suspected to spread CyHV-3 among farms (Ilouze *et al.*, 2011). Other routes include the mixing of fish in the same tanks at koi shows, selling infected fish below the market price and the release of infected fish into public waters (Boutier *et al.*, 2015a).

10.2 Diagnostics

The diagnosis of KHVD in clinically affected fish uses a range of tests, but few are fully validated. The chapter on KHVD in the *Manual of Diagnostic Tests for Aquatic Animals* (OIE, 2012) advises that the diagnosis of KHVD should rely on a combination of tests, including clinical examination as well as viral detection. The final diagnosis of KHVD must rely on the direct detection of viral DNA or the isolation and identification of CyHV-3 using immunological and molecular techniques.

10.2.1 Behavioural changes

The most evident behavioural sign of KHVD is lethargy. Fish will separate from the shoal and lie at the bottom of the tank or pond or hang in a head-down position. They may also gasp at the surface near the water inlet and other aerated areas or at the sides of the pond (OIE, 2012). Some fish may experience loss of equilibrium and disorientation

but, at the same time, may become hyperactive (Hedrick *et al.*, 2000).

10.2.2 External gross pathology

There are no pathognomonic gross lesions of KHVD. The most consistent gross pathology is pale, irregular patches on the skin, associated with excess mucus secretion, and also the underproduction of mucus, resulting in patches of skin with a sandpaper-like texture (Haenen *et al.*, 2004; Fig. 10.1A). The skin may also show hyperaemia, haemorrhages and ulceration (Boutier *et al.*, 2015a; Fig. 10.1B). As the disease progresses, fish may experience focal or extensive loss of skin epithelium. In less acutely affected carp, commonly reported clinical signs include anorexia and enophthalmia (eyes sunken into their sockets) (Boutier *et al.*, 2015a; Fig. 10.1C). The gross pathology in the gills is the most consistent feature in clinical KHVD. The pathology varies from pale necrotic patches to extensive discoloration (bleaching) with severe necrosis and inflammation (OIE, 2012, Fig. 10.2).

10.2.3 Internal gross pathology

There may be accumulation of abdominal fluid and abdominal adhesions, and organs may be enlarged, darker and/or mottled, but these lesions are not pathognomonic of KHVD (Boutier *et al.*, 2015a).

The appearance of gross lesions may also be complicated by heavy ectoparasite infections in diseased fish, particularly in the common carp. Eight parasite genera (e.g. *Gyrodactylus* sp.) are commonly associated with KHVD-affected carp; secondary infections with gill monogean parasites and infections with a range of bacteria (e.g. *Flavobacterium* sp.) have also been reported (OIE, 2012).

10.2.4 Sampling

Young carp (<1 year) are generally more susceptible to clinical KHVD and should be selected for sampling. Common carp, or varieties such as koi or ghost carp, are most susceptible to the disease, followed by any common carp × *Carassius* spp. hybrids (OIE, 2012). Moribund or freshly dead fish with clinical signs are suitable for most of the immunological and/or molecular tests that can be used. Tissues from decomposed carp may only be suitable for testing using polymerase chain reaction (PCR)-based methods. Samples from apparently

Fig. 10.1. External gross pathology of fish with koi herpesvirus disease (KHVD): (A) irregular patches on the skin mainly associated with the underproduction of mucus where patches of skin have a sandpaper-like texture; (B) hyperaemia and haemorrhages on the skin of a common carp; (C) koi displaying enophthalmia.

healthy carp, in a suspect population, are most reliably tested using PCR-based assays (OIE, 2012).

Organ samples should be collected immediately after the carp has been selected (OIE, 2012). Whole fish may be submitted for testing packed in ice and tissues preserved in viral transport medium or in 80–100% ethanol. Samples preserved in alcohol or those received frozen are only suitable for PCR-based tests.

Pooled samples should be avoided or restricted to a maximum of two fish per pool. The *Report from Meeting on Sampling and Diagnostic Procedures for the Surveillance and Confirmation of KHV Disease* held in Copenhagen in 2014 (Olesen *et al.*, 2014) advises that, in acute cases, the tissues of five fish can be pooled. Recommended tissues include gill, kidney and spleen (OIE, 2012), as these contain the greatest DNA concentrations (Gilad *et al.*, 2004).

10.2.5 Direct immunodiagnostic methods

A method for the direct detection of CyHV-3 from kidney imprints using an indirect fluorescent antibody test (IFAT) is available (OIE, 2012). Immunoperoxidase staining has also been used to detect the CyHV-3 antigen, but it is prone to producing false-positive staining (Pikarsky *et al.*, 2004).

10.2.6 Detection using PCR-based assays

PCR-based assays are generally the most sensitive and reliable methods to detect CyHV-3 in tissues. The OIE (2012) recommends one assay that incorporates a primer set targeting the thymidine kinase (TK) gene (Bercovier *et al.*, 2005) and another assay that is an improvement of the Gray SpH primer set protocol (Yuasa *et al.*, 2005).

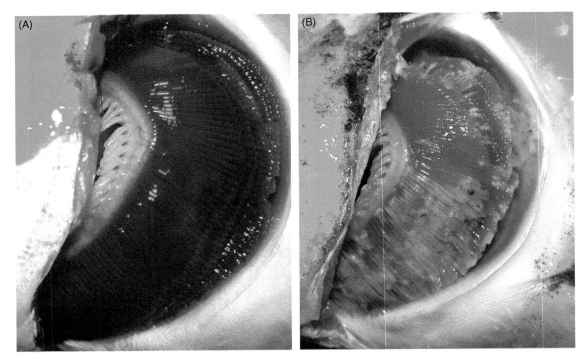

Fig. 10.2. External gross pathology of carp gill with the operculum removed: (A) normal healthy gill; (B) gill of fish with koi herpesvirus disease (KHVD) displaying inflammation and necrosis.

Alternatively, many diagnosticians favour quantitative (real-time) PCR (qPCR) assays over conventional PCR. The most common qPCR for the detection of CyHV-3 is the Gilad Taqman real-time PCR assay (Gilad *et al.*, 2004) that detects and can quantitatively assess very low copy numbers of target nucleic acid sequences (Boutier *et al.*, 2015a). Real-time PCR avoids much of the contamination risk inherent in conventional PCR (OIE, 2012); it also targets shorter DNA sequences and is more likely to detect degraded viral DNA in decomposing tissues.

Loop-mediated isothermal amplification (LAMP) is a rapid single-step PCR assay that is widely adapted for pond-side diagnosis because it does not require a thermal cycler. LAMP of the TK gene has been developed to detect CyHV-3 and is equal to or more sensitive than conventional PCR (Yoshino *et al.*, 2009). An assay incorporating DNA hybridization technology and antigen–antibody reactions in combination with LAMP has also been developed and had improved sensitivity and specificity (Soliman and El-Matbouli, 2010). LAMP has

potential for the non-destructive testing of clinically affected fish, but the OIE does not elaborate upon LAMP assays for KHVD because none have been submitted for assessment and registration.

The most sensitive PCR method is the Gilad Taqman assay mentioned above (OIE, 2012). In a comparison of PCR-based assays Bergmann *et al.* (2010b) showed that conventional PCR assays which include a second round with nested primers were as sensitive as real-time assays. However, nested PCR is more susceptible to cross-contamination by PCR products from previous tests, so producing false positives.

10.2.7 Histopathology

The histopathology of the disease is non-specific and variable. Alterations are found in the gills, skin, kidneys, heart, spleen, liver, gut and brain in KHVD-affected fish (OIE, 2012; Boutier *et al.*, 2015a). Evidence of herpesvirus infection may be most readily observed in the skin, gills and kidney, as described in Section 10.3 Pathology.

10.2.8 Electron microscopy

The examination of tissues from clinically infected carp for viral particles using transmission electron microscopy (TEM) is not reliable unless the fish has a heavy infection (OIE, 2012).

10.2.9 Virus isolation in cell culture

The common carp brain (CCB) and koi fin (KF)-1 cell lines are recommended for the isolation of CyHV-3, but the former is more susceptible (OIE, 2012; J. Savage, UK, personal communication). Boutier *et al.* (2015a) list other cell lines that are susceptible to CyHV-3. Before clinical signs appear, viral levels are higher in gill than in kidney tissues (Gilad *et al.*, 2004; Yuasa *et al.*, 2012), and the virus is most reliably isolated from gill tissue. However, virus isolation in cell culture is not as reliable or sensitive as PCR-based methods for detecting CyHV-3 DNA (OIE, 2012).

10.2.10 Other diagnostic assays, including non-lethal methods

In situ hybridization (ISH) and IFAT have been used to detect and identify CyHV-3 in fish leucocytes (Boutier *et al.*, 2015a). Although these methods have not been fully compared with other techniques, they are non-lethal and may facilitate diagnosis (OIE, 2012). Non-lethal samples such as blood, gill swabs, gill biopsy and mucus scrapes are also suitable substrates for diagnosis (Olesen *et al.*, 2014). Other non-lethal immunodiagnostic methods include an ELISA developed to detect CyHV-3 antigen in fish faeces (Dishon *et al.*, 2005). In addition, a lateral flow device that detects CyHV-3 glycoprotein (the FASTest Koi HV kit) in a 15 min pond-side test and works best with gill swabs (Vrancken *et al.*, 2013).

10.3 Pathology

The gills and skin of CyHV-3-infected fish exhibit prominent clinical signs of the disease, which is reflected in the histopathology. Hyperplasia and hypertrophy are common in the gills (Boutier *et al.*, 2015a; Fig. 10.3A). Pikarsky *et al.* (2004) observed pathological changes in the gills of experimentally infected fish at 2 dpi, including the loss of lamellae and a mixed inflammatory infiltrate in some filaments. Hyperplasia, severe inflammation and congestion

of the central venous sinus became more pronounced at 6 dpi. There was subepithelial inflammation and congestion of blood vessels of the gill rakers, accompanied by reduction in their height, and sloughing of the surface epithelium; haemorrhages were sometimes observed in the lamellar capillaries. Enlarged branchial epithelial cells had nuclear swelling accompanied by margination of the chromatin and pale diffuse eosinophilic inclusions – the 'signet ring' appearance, also termed intranuclear inclusion bodies (Fig. 10.3B). Such cells usually contain CyHV-3 when observed using TEM. Inclusion bodies have also been observed in the heart, kidney (Fig. 10.3C), spleen, liver, intestine, stomach, brain and fin epidermis. The secondary lamellae are often fused with the hyperplastic branchial epithelium, and necrosis is often seen, particularly at the tips (Boutier *et al.*, 2015a; Miwa *et al.*, 2015).

Infiltration of cells under the epidermis of the fin is observed from 1 dpi onwards and epidermal integrity is disrupted from 2 dpi onwards. Loss of epidermis is also noted from 3 dpi onwards and the epidermis of the head is often sloughed. The number of goblet cells is reduced by 50% in infected fish, and they appear thin, which suggests that mucus is released and not replenished (Adamek *et al.*, 2013). At later stages, erosion of the skin epidermis is frequently observed and is often the severest pathological change observed (Adamek *et al.*, 2013; Miwa *et al.*, 2015).

The kidneys exhibit notable pathology, starting with a peritubular inflammatory infiltrate from 2 dpi, followed at 6 dpi by a heavy interstitial inflammatory infiltrate, with congestion of the blood vessels. At or beyond 8 dpi there is hyperplasia or degeneration of the tubular epithelium and intraepithelial lymphocytes are present. Necrotic cells are observed; in severe disease haematopoietic cells are necrotic. Chronic glomerulitis, periglomerular fibrosis and interstitial nephritis with the loss of haemopoietic cells are observed (Boutier *et al.*, 2015a; Miwa *et al.*, 2015).

The liver exhibits mild inflammatory infiltrates, mainly in the parenchyma, with focal necrosis in some fish. Proliferation in the columnar epithelial cells of the bile duct may occur (Pikarsky *et al.*, 2004; Cheng *et al.*, 2011).

Hyperplasia in the epithelium lining the gastric gland occludes the lumen of the stomach and there is hyperplasia of the intestinal villi. Intestinal epithelial cells are sloughed into the lumen (El Din, 2011).

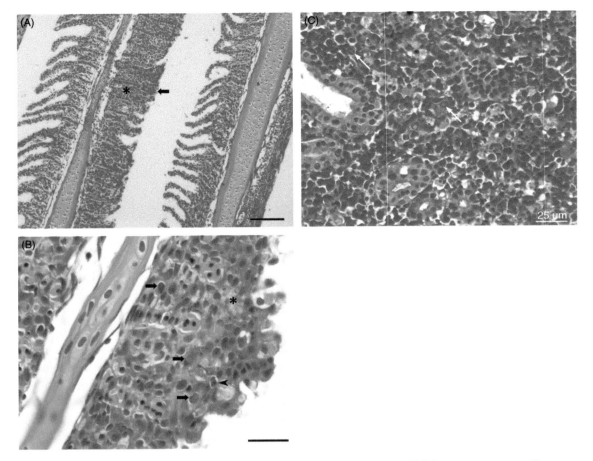

Fig. 10.3. Microscopic lesions in tissue sections (stained with haemotoxylin and eosin) from common carp with koi herpesvirus disease (KHVD); (A) Hypertrophy and hyperplasia of the branchial epithelial cells (*) and fusion of secondary lamellae (arrow) of the gill; (B) necrosis and inflammation (*), apoptosis (arrowhead) and intranuclear inclusions (arrows) in the branchial epithelium; (C) interstitial haematopoietic cell necrosis and nuclei with inclusions (arrows) in the kidney. Scale bars: (A) 100 μm; (B) 20 μm; (C) 25 μm.

The parenchyma of the spleen contains foci of necrosis, and there is necrosis of pancreatic acinar cells in the spleen and kidney (Hedrick *et al.*, 2000; Cheng *et al.*, 2011).

In the brain, there is focal meningeal and parameningeal inflammation and the congestion of capillaries and small veins (Boutier *et al.*, 2015a).

Many myocardial cells exhibit nuclear degeneration and myofibrils are dilated or coagulated with the disappearance of cross-striations. Later in the course of the disease, macrophages and lymphocytes infiltrate the myocardium and necrosis of the heart muscle is observed (Miyazaki *et al.*, 2008; Cheng *et al.*, 2011).

10.4 Pathophysiology

The major portal of entry for CyHV-3 in carp has been found to be the skin, not the gills and intestines as initially proposed (Costes *et al.*, 2009). However, gill lesions are observed early in infection (Hedrick *et al.*, 2000; Pikarsky *et al.*, 2004), and viral DNA is detected in the gills and intestines as early as 1 dpi (Gilad *et al.*, 2004). Costes *et al.* (2009) used a luciferase (LUC)-expressing recombinant CyHV-3 to induce KHVD in common carp that was indistinguishable from disease induced by the parental wild type strain. Furthermore, imaging of naturally infected carp using this LUC-expressing

K. Way and P. Dixon

recombinant revealed that CyHV-3 entered the fish via the skin and not the gills. Early viral replication occurred in the skin epithelium, principally in the fins (Fig. 10.4). CyHV-3 RNA expression was shown in the skin 12 h postinfection (hpi) (Adamek et al., 2013), and viral DNA was detected in infected cells of fin epithelium by 2 dpi (Miwa et al., 2015). Immersion infection of carp with an attenuated CyHV-3 also demonstrated dispersion from the skin to other organs. (Boutier et al., 2015a). Using similar techniques, Fournier et al. (2012) showed that the pharyngeal periodontal mucosa was a major portal of entry after oral infection (Fig. 10.4). What is more, the spread of the virus to other organs and progress of clinical disease was comparable to that of infection via immersion.

Following infection, CyHV-3 spreads rapidly to other organs that represent secondary sites of replication. The tropism of CyHV-3 for white blood cells may explain the rapid spread of the virus via the blood (Boutier et al., 2015a). Viral DNA can be detected in the blood, gill, liver, spleen, kidney, intestine and brain tissue at 1 dpi (Gilad et al., 2004; Pikarsky et al., 2004). In acutely infected freshly dead carp, CyHV-3 DNA copy numbers range from 10^9 to $10^{11}/10^6$ host cells in gill, intestine and kidney tissue (Gilad et al., 2004). These osmoregulatory organs undergo marked pathological changes during the course of the disease, and Gilad et al. (2004) suggested that the loss of osmoregulatory function contributes to death. Negenborn et al. (2015) showed that electrolyte levels, mainly sodium ions in the urine, and a concomitant decrease in serum electrolyte levels occurred during experimental infection. Changes in electrolyte levels were paralleled by severe pathology in the kidney and gills, providing further evidence that severe osmoregulatory dysfunction could cause death. In experimental CyHV-3 infections, Miwa et al. (2015) reported extensive damage to the skin epidermis at 5–8 dpi. Serum osmolality was also very low just before death and suggested that hypo-osmotic shock from skin damage was a likely cause of death.

10.4.1 Innate immune response

Investigators have suggested a strong and rapid innate immune response to CyHV-3 infection, as evidenced by the early upregulation of complement-associated and C-reactive proteins (Pionnier et al., 2014). Interferon (IFN) plays an important mediation role and studies of the response in the skin and intestine revealed activation of IFN Class I pathways (Adamek et al., 2013; Syakuri et al., 2013). However, unlike spring viraemia of carp virus (SVCV), CyHV-3 can inhibit the IFN type-I pathway and inhibit the activity of stimulated macrophages and the proliferative response of lymphocytes (Boutier et al., 2015a). Furthermore, CyHV-3 does not induce apoptosis and stimulation of the apoptosis intrinsic pathway is delayed (Miest et al., 2015). Genes encoding claudins (the tight-junction proteins), mucin and beta defensin antimicrobial peptides are downregulated during CyHV-3 infection. The disruption of these important components of the skin mucosal barrier contribute to the disintegration of the skin (Adamek et al., 2013).

10.4.2 Adaptive immune response and immune evasion

Carp produce a strong, temperature-dependent, protective antibody against CyHV-3. Antibody response to the virus is slow at 12–14°C compared with a rapid response at 31°C. At temperatures permissive for viral infection, antibodies can be detected between 7 and 14 dpi, with peaks between 20 and 40 dpi, and remain detectable for at least 65 weeks (Perelberg et al., 2008; St-Hilaire et al., 2009).

In carp infected via immersion, an attenuated CyHV-3 recombinant virus was detectable in the skin mucosa and induced a strong protective mucosal immune response against wild type CyHV-3 (Boutier et al., 2015b). This may be related to the stimulation of B cells secreting IgT, an immunoglobulin isotype that is involved in mucosal immunity in teleosts (Boutier et al., 2015a).

In silico, in vitro and in vivo studies have suggested that CyHV-3 may express proteins involved in immune evasion (see Boutier et al., 2015a) and that this might explain the acute and dramatic clinical signs associated with KHVD.

10.4.3 Latent infection

The survivors of KHVD outbreaks in wild and cultured common carp are persistently infected with CyHV-3 (Baumer et al., 2013; Uchii et al., 2014), although latency has not been demonstrated conclusively for the virus (Boutier et al., 2015a). Seasonal reactivation is proposed to explain how CyHV-3 persists in convalescent populations of

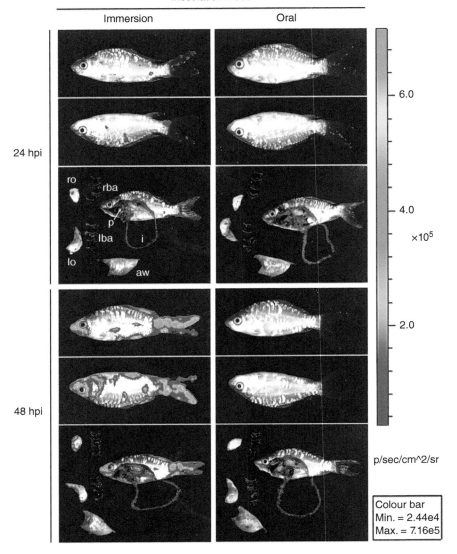

Fig. 10.4. The portals of entry of CyHV-3 in carp analysed by *in vivo* bioluminescent imaging. Two groups of fish (mean weight 10 g) were infected with a recombinant CyHV-3 strain expressing luciferase as a reporter gene, either by bathing them in water containing the virus (immersion, left column) or by feeding them with food pellets contaminated with the virus (oral, right column). At the indicated times (h) postinfection (hpi), six fish per group were analysed by a bioluminescence *in vivo* imaging system (IVIS). Each fish was analysed lying on its right and left sides. The internal signal was analysed after euthanasia and dissection. Dissected fish and isolated organs were analysed for *ex vivo* bioluminescence. One representative fish is shown for each time point and inoculation mode. Images collected over the course of the experiment were normalized using an identical pseudocolour scale ranging from violet (least intense) to red (most intense) using Living Image 3.2 software. Key: aw, abdominal wall; i, intestine; lba, left branchial arches; lo, left operculum; p, pharynx; rba, right branchial arches; ro, right operculum. Units: Min. and Max., standardized minimum and maximum threshold value for photon flux/; p/sec/cm^2/sr = photons/s/cm²/steradian. Reproduced with permission from Fournier *et al*. (2012); original publisher BioMed Central.

K. Way and P. Dixon

carp. RNA expression of virus replication-related genes was detected in the brain of seropositive fish, while other fish expressed only latency-related genes (Uchii *et al.*, 2014); this suggests that reactivation is a transient phenomenon within a population. The reactivation of CyHV-3 may be triggered by temperature stress (Eide *et al.*, 2011) and netting stress (Bergmann and Kempter, 2011). Viral DNA was detected in white blood cells (WBCs), in the absence of clinical signs and detectable infectious virus particles, among koi carp that have been previously exposed to CyHV-3 and in wild common carp with no previous history of KHVD (Eide *et al.*, 2011; Xu *et al.*, 2013). The main WBC type involved in persistence was the IgM+ B cell, in which roughly 20× more DNA copies are found than in the remaining WBC fraction (Reed *et al.*, 2014). The virus has been found in various tissues of long-term infected fish, and especially in the brain. The nervous system may represent an additional site of latency, as in other herpesviruses (Boutier *et al.*, 2015a).

10.5 Prevention and Control

10.5.1 Resistant strains of carp

Strains of carp differ in their susceptibility to CyHV-3. In cross-breeding experiments, some wild strains of carp, such as the Sassan and Amur strains, confer resistance to KHVD when crossed with domestic strains (Shapira *et al.*, 2005; Dixon *et al.*, 2009; Piackova *et al.*, 2013). The domestic Israeli strain Dor-70 has also shown high resistance to infection (Zak *et al.*, 2007). In contrast, in Japan, indigenous common carp were more severely affected by KHVD than domesticated Eurasian common carp and koi (Ito *et al.*, 2014). The analysis of disease resistance at the molecular level has identified links with the polymorphism of genes involved in the immune response, including the MHC (major histocompatibility complex) class II B genes (Rakus *et al.*, 2012) and the carp IL-10 (interleukin 10) gene (Kongchum *et al.*, 2011). The results from these studies have not been conclusive but do indicate that there are genetic markers for resistance that could enhance selective breeding.

10.5.2 Vaccination

Early attempts to vaccinate carp involved cohabiting them with CyHV-3-infected fish for 3–5 days at 22–23°C and then transferring them to ponds at 30°C for 30 days (Ronen *et al.*, 2003). The fish were more resistant to the virus following challenge with CyHV-3. This method reduced the mortality in CyHV-3 endemic areas within Israel (H. Bercovier personal communication to Dixon, 2008) but the disadvantages included high costs and disease recurrence, the treated fish were reservoirs of infection and mortality remained significant.

CyHV-3 attenuated by serial passage in cell culture and injected intraperitoneally (IP) into carp conferred complete protection when they were cohabited with CyHV-3-infected fish (Ronen *et al.*, 2003). The administration of attenuated virus by 10 min bath immersion induced protection, but the efficacy declined with time and was ineffective after 4 h in the water (Perelberg *et al.*, 2005). The attenuated virus was irradiated with ultraviolet (UV) light to induce mutations that might prevent reversion to virulence. UV-treated virus protected carp against CyHV-3 infection, but no data were presented to confirm that the virus would remain avirulent. Research has continued into developing an attenuated vaccine (Perelberg *et al.*, 2008; O'Connor *et al.*, 2014; Weber *et al.*, 2014). Recently, the efficacy and safety of this vaccine were examined in the USA (Weber *et al.*, 2014) and it was found to be safe and efficacious, particularly in fish weighing >87g. Vaccinated koi carp (weight not stated) had 36% mortality compared with 100% mortality in non-vaccinated control fish when challenged at 13 months postvaccination (O'Connor *et al.*, 2014).

An attenuated vaccine (Cavoy, manufactured by KoVax Ltd), is available in Israel and was authorized in 2012 for use in the USA as an immersion vaccine for carp weighing >100 g. However, the vaccine was withdrawn from sale after a year (Boutier *et al.*, 2015b). The latter authors produced a recombinant CyHV-3 lacking ORF56 and ORF57, which replicated at lower levels and spread less efficiently within the host compared to the parental virus. Common carp were vaccinated with the recombinant virus via immersion and there were no mortalities 20 days later. After cohabitation challenge with carp infected with the parental virus, 80% survival was reported in the vaccinated groups compared with 0% survival among mock vaccinates. There was low-level transmission of recombinant virus when vaccinated fish were cohabited with sentinel fish, but none by water alone.

Yasumoto et al. (2006) inactivated CyHV-3 with formalin and entrapped it within liposomes. The CyHV-3-liposomes were sprayed on to dry-pelleted feed which was fed to common carp for 3 days, with normal pelleted feed fed thereafter. CyHV-3 virus was then administered to the gills of experimental fish, resulting in 23% mortality in two groups of vaccinated fish and 66 and >80% mortality in two groups of control fish.

Research groups are trying to develop DNA vaccines or recombinant vaccines that have no residual virulence or even the potential to revert to virulence. Prototype DNA vaccines based on the CyHV-3 glycoprotein gene (Nuryati et al., 2010) and the ORF81 and ORF25 genes (Boutier et al., 2015a) have provided varying degrees of protection to carp following injection, but none have been field tested. Vaccines based on recombinant viruses containing deletions in the TK, ribonucleotide reductase or deoxyuridine triphosphatase genes (Fuchs et al., 2011) have given inconsistent results in immersion vaccination trials. Recently, genetically engineered *Lactobacillus plantarum* expressing both CyHV-3 ORF81 protein and SVCV glycoprotein was incorporated into a pelleted feed to produce an oral vaccine (Cui et al., 2015). Under experimental conditions, there was 47% mortality in vaccinated common carp and 85 and 93% mortality in control fish following challenge with CyHV-3. The vaccine also protected koi carp against SVCV (see Chapter 7).

10.5.3 Management and biosecurity strategies

In the absence of proven commercial vaccines or therapeutic agents, management practices and biosecurity strategies are the main ways to prevent the spread of CyHV-3. These strategies range from national or international legislation and/or standards to good management practices at the farm level. The latter include on-site quarantine of new stocks of fish, the disinfection of equipment and use of footbaths, and reducing stress and other diseases, etc. New fish brought on to a site should be quarantined for a minimum of 4 weeks to 2 months before mixing with susceptible species, but the procedures undertaken and their impact will depend on the scale of the aquaculture undertaking. The rates of contagion and mortality can be rapid and limit the use of management practices, though increasing the water temperature to above

26–28°C and reducing stocking densities may reduce mortalities (Gilad et al., 2003; Ronen et al., 2003; Sunarto et al., 2005).

There has only been one study on the inactivation of CyHV-3 by chemical and physical treatments (Kasai et al., 2005), and the results were not conclusive. The minimum concentrations of the chemicals to effect 100% plaque reduction after 30 s and 20 min, respectively, at 15°C were iodophor 200 mg/l (both time periods), sodium hypochlorite >400 and 200 mg/l, benzalkonium chloride 60 mg/l (both time periods) and ethyl alcohol 40 and 30%. An ultraviolet dose of 4×10^3 µWs/cm^2 gave 100% plaque reduction in a virus solution containing 1×10^5 pfu/ml (time of exposure not stated) and 1.6×10^4 pfu/ml. CyHV-3 was inactivated at temperatures above 50°C for 1 min.

Yoshida et al. (2013) produced a laboratory-scale effluent treatment system in which bacterial substances inhibitory to CyHV-3 were adsorbed on to a porous carrier contained within a column. Samples of effluent water seeded with CyHV-3 taken before and after passage through the column were injected into carp and resulted in >80 and 0% mortality, respectively. The authors are working to engineer an operational system that could be used to treat aquaculture effluent water.

A majority of countries have enacted legislation to control the introduction and spread of diseases, including fish diseases. This is currently the most important way to prevent and control KHVD. Such legislation is often strengthened by international agreements. The OIE has produced the *International Aquatic Animal Health Code* (OIE, 2015), which outlines approaches to biosecurity that can be applied nationally and locally on production sites. Håstein et al. (2008) provided an overview of approaches to implementing biosecurity strategies, and Oidtmann et al. (2011) have reviewed international standards relating to movements of fish and fish products.

10.6 Conclusions

KHVD occurs worldwide and is an OIE notifiable disease. The virus, CyHV-3, induces a highly contagious acute viraemia that is responsible for severe financial losses in the common carp and koi culture industries.

Although there are no pathognomonic gross lesions, the rapid onset and severity of external clinical signs and numbers of affected carp provide good indications of KHVD disease. The likely cause of death is osmoregulatory dysfunction or osmotic shock. Rapid identification of the virus is possible using PCR-based methods, and pond-side, non-lethal tests have been developed. The virus induces a strong innate immune and protective antibody response in carp, but the availability of commercial vaccines is limited.

Survivors of KHVD are persistently infected with CyHV-3 and there is evidence for seasonal and stress-induced reactivation. More studies are needed to confirm and identify both the site of latency and the latency-associated viral gene transcripts that might be useful targets for virus surveillance. Also, research is needed to assess how temperature affects and possibly regulates the switch between lytic and latent infection.

Some strains of carp are resistant to KHVD. Continuation of research into the genetic basis for resistance/susceptibility to CyHV-3 may enhance selective breeding programmes for resistance to KHVD.

Good management and biosecurity practices are the main ways to prevent the spread of CyHV-3. The last 15 years has seen improvements in these practices in cold-water aquaculture industries worldwide, many as a result of the threat from KHVD.

References

Adamek, M., Syakuri, H., Harris, S., Rakus, K.L., Brogden, G. et al. (2013) Cyprinid herpesvirus 3 infection disrupts the skin barrier of common carp (Cyprinus carpio L.). Veterinary Microbiology 162, 456–470.

Baumer, A., Fabian, M., Wilkens, M.R., Steinhagen, D. and Runge, M. (2013) Epidemiology of cyprinid herpesvirus-3 infection in latently infected carp from aquaculture. Diseases of Aquatic Organisms 105, 101–108.

Bercovier, H., Fishman, Y., Nahary, R., Sinai, S., Zlotkin, A. et al. (2005) Cloning of the koi herpesvirus (KHV) gene encoding thymidine kinase and its use for a highly sensitive PCR based diagnosis. BMC Microbiology 5:13.

Bergmann, S.M. and Kempter, J. (2011) Detection of koi herpesvirus (KHV) after re-activation in persistently infected common carp (Cyprinus carpio L.) using non-lethal sampling methods. Bulletin of the European Association of Fish Pathologists 31, 92–100.

Bergmann, S.M., Sadowski, J., Kielpinski, M., Bartlomiejczyk, M., Fichtner, D. et al. (2010a) Susceptibility of koi × crucian carp and koi × goldfish hybrids to koi herpesvirus (KHV) and the development of KHV disease (KHVD). Journal of Fish Diseases 33, 267–272.

Bergmann, S.M., Riechardt, M., Fichtner, D., Lee, P. and Kempter, J. (2010b) Investigation on the diagnostic sensitivity of molecular tools used for detection of koi herpesvirus. Journal of Virological Methods 163, 229–233.

Boutier, M., Ronsmans, M., Rakus, K., Jazowiecka-Rakus, J., Vancsok, C. et al. (2015a) Cyprinid herpesvirus 3, an archetype of fish alloherpesviruses. In: Kielian, M., Maramorosch, K. and Mettenleiter, T.C. (eds) Advances in Virus Research, Volume 93. Elsevier/Academic Press, Waltham, Massachussetts/San Diego, California/London/Kidlington, Oxford, UK, pp. 161–256.

Boutier, M., Ronsmans, M., Ouyang, P., Fournier, G., Reschner, A. et al. (2015b) Rational development of an attenuated recombinant cyprinid herpesvirus 3 vaccine using prokaryotic mutagenesis and in vivo bioluminescent imaging. PLoS Pathogens 11(2): e1004690.

Cheng, L., Chen, C.Y., Tsai, M.A., Wang, P.C., Hsu, J.P. et al. (2011) Koi herpesvirus epizootic in cultured carp and koi, Cyprinus carpio L., in Taiwan. Journal of Fish Diseases 34, 547–554.

Costes, B., Stalin Raj, V., Michel, B., Fournier, G., Thirion, M. et al. (2009) The major portal of entry of koi herpesvirus in Cyprinus carpio is the skin. Journal of Virology 83, 2819–2830.

Cui, L.C., Guan, X.T., Liu, Z.M., Tian, C.Y. and Xu, Y.G. (2015) Recombinant lactobacillus expressing G protein of spring viremia of carp virus (SVCV) combined with ORF81 protein of koi herpesvirus (KHV): a promising way to induce protective immunity against SVCV and KHV infection in cyprinid fish via oral vaccination. Vaccine 33, 3092–3099.

Dishon, A., Perelberg, A., Bishara-Shieban, J., Ilouze, M., Davidovich, M. et al. (2005) Detection of carp interstitial nephritis and gill necrosis virus in fish droppings. Applied and Environmental Microbiology 71, 7285–7291.

Dixon, P.F., Joiner, C.L., Way, K., Reese, R.A., Jeney, G. et al. (2009) Comparison of the resistance of selected families of common carp, Cyprinus carpio (L.), to koi herpesvirus: preliminary study. Journal of Fish Diseases 32, 1035–1039S.

Eide, K.E., Miller-Morgan, T., Heidel, J.R., Kent, M.L., Bildfell, R.J. et al. (2011) Investigation of koi herpesvirus latency in koi. Journal of Virology 85, 4954–4962.

El Din, M.M.M. (2011) Histopathological studies in experimentally infected koi carp (Cyprinus carpio koi) with koi herpesvirus in Japan. World Journal of Fish and Marine Science 3, 252–259.

Fournier, G., Boutier, M., Raj, V.S., Mast, J., Parmentier, E. et al. (2012) Feeding Cyprinus carpio with infectious materials mediates cyprinid herpesvirus 3 entry through infection of pharyngeal periodontal mucosa. Veterinary Research 43:6.

Fuchs, W., Fichtner, D., Bergmann, S.M. and Mettenleiter, T.C. (2011) Generation and characterization of koi herpesvirus recombinants lacking viral enzymes of nucleotide metabolism. *Archives of Virology* 156, 1059–1063.

Gilad, O., Yun, S., Adkison, M.A., Way, K., Willits, N.H. et al. (2003) Molecular comparison of isolates of an emerging fish pathogen, koi herpesvirus, and the effect of water temperature on mortality of experimentally infected koi. *Journal of General Virology* 84, 2661–2668.

Gilad, O., Yun, S., Zagmutt-Vergara, F.J., Leutenegger, C.M., Bercovier, H. et al. (2004) Concentrations of a koi herpesvirus (KHV) in tissues of experimentally infected *Cyprinus carpio koi* as assessed by real-time TaqMan PCR. *Diseases of Aquatic Organisms* 60, 179–187.

Haenen, O.L.M., Way, K., Bergmann, S.M. and Ariel, E. (2004) The emergence of koi herpesvirus and its significance to European aquaculture. *Bulletin of the European Association of Fish Pathologists* 24, 293–307.

Håstein, T., Binde, M., Hine, M., Johnsen, S., Lillehaug, A. et al. (2008) National biosecurity approaches, plans and programmes in response to diseases in farmed aquatic animals: evolution, effectiveness and the way forward. *Revue Scientifique et Technique* 27, 125–145.

Hedrick, R.P., Gilad, O., Yun, S., Spangenberg, J.V., Marty, G.D. et al. (2000) A herpesvirus associated with mass mortality of juvenile and adult koi, a strain of common carp. *Journal of Aquatic Animal Health* 12, 44–57.

Hedrick, R.P., Waltzek, T.B. and McDowell, T.S. (2006) Susceptibility of koi carp, common carp, goldfish, and goldfish × common carp hybrids to cyprinid herpesvirus-2 and herpesvirus-3. *Journal of Aquatic Animal Health* 18, 26–34.

Ilouze, M., Dishon, A. and Kotler, M. (2006) Characterization of a novel virus causing a lethal disease in carp and koi. *Microbiology and Molecular Biology Reviews* 70, 147–156.

Ilouze, M., Davidovich, M., Diamant, A., Kotler, M. and Dishon, A. (2011) The outbreak of carp disease caused by CyHV-3 as a model for new emerging viral diseases in aquaculture: a review. *Ecological Research* 28, 885–892.

Ito, T., Kurita, J. and Yuasa, K. (2014) Differences in the susceptibility of Japanese indigenous and domesticated Eurasian common carp (*Cyprinus carpio*), identified by mitochondrial DNA typing, to cyprinid herpesvirus 3 (CyHV-3). *Veterinary Microbiology* 171, 31–40.

Kasai, H., Muto, Y. and Yoshimizu, M. (2005) Virucidal effects of ultraviolet, heat treatment and disinfectants against koi herpesvirus (KHV). *Fish Pathology* 40, 137–138.

Kongchum, P., Sandel, E., Lutzky, S., Hallerman, E.M., Hulata, G. et al. (2011) Association between IL-10a single nucleotide polymorphisms and resistance to cyprinid herpesvirus-3 infection in common carp (*Cyprinus carpio*). *Aquaculture* 315, 417–421.

Lee, N.S., Jung, S.H., Park, J.W. and Do, J.W. (2012) *In situ* hybridization detection of koi herpesvirus in paraffin-embedded tissues of common carp *Cyprinus carpio* collected in 1998 in Korea. *Fish Pathology* 47, 100–103.

Miest, J.J., Adamek, M., Pionnier, N., Harris, S., Matras, M. et al. (2015) Differential effects of alloherpesvirus CyHV-3 and rhabdovirus SVCV on apoptosis in fish cells. *Veterinary Microbiology* 176, 19–31.

Miwa, S., Kiryu, I., Yuasa, K., Ito, T. and Kaneko, T. (2015) Pathogenesis of acute and chronic diseases caused by cyprinid herpesvirus-3. *Journal of Fish Diseases* 38, 695–712.

Miyazaki, T., Kuzuya, Y., Yasumoto, S., Yasuda, M. and Kobayashi, T. (2008) Histopathological and ultrastructural features of koi herpesvirus (KHV)-infected carp *Cyprinus carpio*, and the morphology and morphogenesis of KHV. *Diseases of Aquatic Organisms* 80, 1–11.

Negenborn, J., van der Marel, M.C., Ganter, M. and Steinhagen, D. (2015) Cyprinid herpesvirus-3 (CyHV-3) disturbs osmotic balance in carp (*Cyprinus carpio* L.) – a potential cause of mortality. *Veterinary Microbiology* 177, 280–288.

Nuryati, S., Alimuddin, Sukenda, Soejoedono, R.D., Santika, A. et al. (2010) Construction of a DNA vaccine using glycoprotein gene and its expression towards increasing survival rate of KHV-infected common carp (*Cyprinus carpio*). *Jurnal Natur Indonesia* 13, 47–52.

O'Connor, M.R., Farver, T.B., Malm, K.V., Yun, S.C., Marty, G.D. et al. (2014) Protective immunity of a modified-live cyprinid herpesvirus 3 vaccine in koi (*Cyprinus carpio koi*) 13 months after vaccination. *American Journal of Veterinary Research* 75, 905–911.

Oidtmann, B.C., Thrush, M.A., Denham, K.L. and Peeler, E.J. (2011) International and national biosecurity strategies in aquatic animal health. *Aquaculture* 320, 22–33.

OIE (2012) Chapter 2.3.6. Koi herpesvirus disease. In: *Manual of Diagnostic Tests for Aquatic Animals.* World Organisation for Animal Health, Paris. Updated 2016 version available as Chapter 2.3.7. at: http://www.oie.int/index.php?id=2439&L=0&htmfile=chapitre_koi_herpesvirus.htm (accessed 2 November 2016).

OIE (2015) *International Aquatic Animal Health Code.* World Organisation for Animal Health, Paris. Updated 2016 version available at: http://www.oie.int/en/international-standard-setting/aquatic-code/access-online/ (accessed 2 November 2016).

Olesen, N.J., Mikkelsen, S.S., Vendramin, N., Bergmann, S., Way, K. and Engelsma, M. (2014) *Report from*

Meeting on Sampling and Diagnostic Procedures for the Surveillance and Confirmation of KHV Disease, Copenhagen, February 25–26th 2014. Available at: http://www.eurl-fish.eu/-/media/Sites/EURL-FISH/english/diagnostic%20manuals/khv_disease/MEETING-REPORT-11-03-14-Final.ashx?la=da (accessed 2 November 2016).

Perelberg, A., Ronen, A., Hutoran, M., Smith, Y. and Kotler, M. (2005) Protection of cultured *Cyprinus carpio* against a lethal viral disease by an attenuated virus vaccine. *Vaccine* 23, 3396–3403.

Perelberg, A., Ilouze, M., Kotler, M. and Steinitz, M. (2008) Antibody response and resistance of *Cyprinus carpio* immunized with cyprinid herpes virus 3 (CyHV-3). *Vaccine* 26, 3750–3756.

Piackova, V., Flajshans, M., Pokorova, D., Reschova, S., Gela, D. *et al.* (2013) Sensitivity of common carp, *Cyprinus carpio* L., strains and crossbreeds reared in the Czech Republic to infection by cyprinid herpesvirus 3 (CyHV-3; KHV). *Journal of Fish Diseases* 36, 75–80.

Pikarsky, E., Ronen, A., Abramowitz, J., Levavi-Sivan, B., Hutoran, M. *et al.* (2004) Pathogenesis of acute viral disease induced in fish by carp interstitial nephritis and gill necrosis virus. *Journal of Virology* 78, 9544–9551.

Pionnier, N., Adamek, M., Miest, J.J., Harris, S.J., Matras, M. *et al.* (2014) C-reactive protein and complement as acute phase reactants in common carp *Cyprinus carpio* during CyHV-3 infection. *Diseases of Aquatic Organisms* 109, 187–199.

Rakus, K.L., Irnazarow, I., Adamek, M., Palmeira, L., Kawana, Y. *et al.* (2012) Gene expression analysis of common carp (*Cyprinus carpio* L.) lines during cyprinid herpesvirus 3 infection yields insights into differential immune responses. *Developmental and Comparative Immunology* 37, 65–76.

Reed, A.N., Izume, S., Dolan, B.P., LaPatra, S., Kent, M. *et al.* (2014) Identification of B cells as a major site for cyprinid herpesvirus 3 latency. *Journal of Virology* 88, 9297–9309.

Ronen, A., Perelberg, A., Abramowitz, J., Hutoran, M., Tinman, S. *et al.* (2003) Efficient vaccine against the virus causing a lethal disease in cultured *Cyprinus carpio*. *Vaccine* 21, 4677–4684.

Shapira, Y., Magen, Y., Zak, T., Kotler, M., Hulata, G. *et al.* (2005) Differential resistance to koi herpes virus (KHV)/carp interstitial nephritis and gill necrosis virus (CNGV) among common carp (*Cyprinus carpio* L.) strains and crossbreds. *Aquaculture* 245, 1–11.

Soliman, H. and El-Matbouli, M. (2010) Loop mediated isothermal amplification combined with nucleic acid lateral flow strip for diagnosis of cyprinid herpes virus-3. *Molecular and Cellular Probes* 24, 38–43.

St-Hilaire, S., Beevers, N., Joiner, C., Hedrick, R.P. and Way, K. (2009) Antibody response of two populations of common carp, *Cyprinus carpio* L., exposed to koi herpesvirus. *Journal of Fish Diseases* 32, 311–320.

Sunarto, A., Rukyani, A. and Itami, T. (2005) Indonesian experience on the outbreak of koi herpesvirus in koi and carp (*Cyprinus carpio*). *Bulletin of Fisheries Research Agency* [Japan], Supplement No. 2, 15–21.

Syakuri, H., Adamek, M., Brogden, G., Rakus, K.L., Matras, M. *et al.* (2013) Intestinal barrier of carp (*Cyprinus carpio* L.) during a cyprinid herpesvirus 3-infection: molecular identification and regulation of the mRNA expression of claudin encoding genes. *Fish and Shellfish Immunology* 34, 305–314.

Uchii, K., Minamoto, T., Honjo, M.N. and Kawabata, Z. (2014) Seasonal reactivation enables cyprinid herpesvirus 3 to persist in a wild host population. *Fems Microbiology Ecology* 87, 536–542.

Vrancken, R., Boutier, M., Ronsmans, M., Reschner, A., Leclipteux, T. *et al.* (2013) Laboratory validation of a lateral flow device for the detection of CyHV-3 antigens in gill swabs. *Journal of Virological Methods* 193, 679–682.

Weber, E.P.S., Malm, K.V., Yun, S.C., Campbell, L.A., Kass, P.H. *et al.* (2014) Efficacy and safety of a modified-live cyprinid herpesvirus 3 vaccine in koi (*Cyprinus carpio koi*) for prevention of koi herpesvirus disease. *American Journal of Veterinary Research* 75, 899–904.

Xu, J.R., Bently, J., Beck, L., Reed, A., Miller-Morgan, T. *et al.* (2013) Analysis of koi herpesvirus latency in wild common carp and ornamental koi in Oregon, USA. *Journal of Virological Methods* 187, 372–379.

Yasumoto, S., Kuzuya, Y., Yasuda, M., Yoshimura, T. and Miyazaki, T. (2006) Oral immunization of common carp with a liposome vaccine fusing koi herpesvirus antigen. *Fish Pathology* 41, 141–145.

Yoshida, N., Sasaki, R.K., Kasai, H. and Yoshimizu, M. (2013) Inactivation of koi-herpesvirus in water using bacteria isolated from carp intestines and carp habitats. *Journal of Fish Diseases* 36, 997–1005.

Yoshino, M., Watari, H., Kojima, T., Ikedo, M. and Kurita, J. (2009) Rapid, sensitive and simple detection method for koi herpesvirus using loop-mediated isothermal amplification. *Microbiology and Immunology* 53, 375–383.

Yuasa, K., Sano, M., Kurita, J., Ito, T. and Iida, T. (2005) Improvement of a PCR method with the Sph 1-5 primer set for the detection of koi herpesvirus (KHV). *Fish Pathology* 40, 37–39.

Yuasa, K., Sano, M. and Oseko, N. (2012) Effective procedures for culture isolation of koi herpesvirus (KHV). *Fish Pathology* 47, 97–99.

Zak, T., Perelberg, A., Magen, I., Milstein, A. and Joseph, D. (2007) Heterosis in the growth rate of Hungarian–Israeli common carp crossbreeds and evaluation of their sensitivity to koi herpes virus (KHV) disease. *Israeli Journal of Aquaculture* 59, 63–72.

11 Viral Encephalopathy and Retinopathy

ANNA TOFFAN*

OIE Reference Centre for Viral Encephalopathy and Retinopathy, Istituto Zooprofilattico Sperimentale delle Venezie, Legnaro (Padova), Italy

11.1 Introduction

Viral encephalopathy and retinopathy (VER), also known as viral nervous necrosis (VNN) is a severe neuropathological disease caused by RNA viruses of the genus *Betanodavirus* (Family: *Nodaviridae*). This infectious agent, detected in the late 1980s, spread worldwide, became endemic and came to represent a major limiting factor for mariculture in several countries. The disease has recently been included among the most significant viral pathogens of finfish, given the expanding host range and the lack of properly effective prophylactic measures (Rigos and Katharios, 2009; Walker and Winton, 2010; Shetty *et al.*, 2012).

11.2 The Infectious Agents

The causative agent of the disease is a small (25–30 nm diameter), spherical, non-enveloped virion, with a bi-segmented genome made of two single-stranded positive-sense RNA molecules. The name *Nodaviridae* originates from the Japanese village of Nodamura, where the prototype virus was first isolated form mosquitos (*Culex tritaeniorhynchus*). In 1992, a different nodavirus was isolated from larvae of the striped jack (*Pseudocaranx dentex*), which accounts for the name of striped jack nervous necrosis virus (Mori *et al.*, 1992). Subsequent molecular studies separated these viruses into two different genera: *Alphanodavirus* and *Betanodavirus*, which infect insects and fish, respectively (King *et al.*, 2011). A third genus, Gammanodavirus, which has not yet been accepted by the International Committee on Taxonomy of Viruses (ICTV), was recently detected in prawns (*Macrobranchium rosenbergii*) in India (NaveenKumar *et al.*, 2013).

The *Betanodavirus* genome is formed by two open reading frames (ORFs); the RNA1 segment (3.1 kb) encodes for the RNA-dependent RNA-polymerase (RdRp) and the RNA2 segment (1.4 kb) encodes for the viral capsid protein (King *et al.*, 2011). The transcription of the RNA1 segment apparently occurs at the beginning of the viral cycle, whereas the expression and production of the capsid, as well as the increase of infective viral particles, occurs at a later stage (Lopez-Jimena *et al.*, 2011). A further subgenomic transcript of 0.4 kb, called RNA3, is cleaved from the RNA1 molecule during active viral replication and encodes for the B1 and B2 proteins, which antagonize host cell RNA interference mechanisms (Iwamoto *et al.*, 2005; Fenner *et al.*, 2006a; Chen *et al.*, 2009). The synthesis of RNA3 in cell cultures is much more abundant than the transcription of RNA1 at an earlier point in time after infection (Sommerset and Nerland, 2004). The B2 non-structural protein is detected only at an early stage of infection, both in infected cell cultures and, recently, in infected Atlantic halibut *(Hippoglossus hippoglossus)*, while the *Betanodavirus* capsid protein is also present in chronically infected fish (Mézeth *et al.*, 2009).

According to Nishizawa *et al.* (1997), there are four species of betanodaviruses based on the phylogenetic analysis of the T4 variable region within the RNA2 segment: the striped jack nervous necrosis virus (SJNNV), the tiger puffer nervous necrosis virus

*E-mail: atoffan@izsvenezie.it

(TPNNV), the barfin flounder nervous necrosis virus (BFNNV) and the redspotted grouper nervous necrosis virus (RGNNV). Intra- and intergenotype reassortment among betanodaviruses have both been detected (Panzarin *et al.*, 2012; He and Teng, 2015), and the two reassortant viruses are RGNNV/SJNNV and SJNNV/RGNNV (Toffolo *et al.*, 2007; Olveira *et al.*, 2009; Panzarin *et al.*, 2012). These reassortant strains may have occurred due to the coexistence of two different viral species in the same host, presumably in wild fish, and may have resulted from a single reassortment event in the 1980s (Sakamoto *et al.*, 2008; Lopez-Jimena *et al.*, 2010; He and Teng, 2015). Other *Betanodavirus*, i.e. the Atlantic cod nervous necrosis virus (ACNNV), Atlantic halibut nervous necrosis virus (AHNV), turbot nodavirus (TNV) and many others, have been described, but they still have to be recognized by the ICTV (King *et al.*, 2011).

When polyclonal antibodies are used, betanodaviruses are grouped into three distinct serotypes: serotype A (the SJNNV genotype), serotype B (the TPNNV genotype) and serotype C (the RGNNV and BFNNV genotypes) (Mori *et al.*, 2003). The antigenic diversity between the RGNNV and the SJNNV genotypes had already been assumed (Skliris *et al.*, 2001; Chi *et al.*, 2003; Costa *et al.*, 2007) and has recently been confirmed using reverse genetic viruses (Panzarin *et al.*, 2016). The C-terminal protruding domain of the capsid protein seems to be responsible for the different immunoreactivity and may contain host-specific determinants (Iwamoto *et al.*, 2004; Ito *et al.*, 2008; Bandín and Dopazo, 2011; Souto *et al.*, 2015a). The reassortant viruses mentioned above still remain in the same antigenic group as the RNA2 donor strain. The BFNNV genotype was recently allocated to the B serotype (Panzarin *et al.*, 2016) in contrast to the previous report by Mori *et al.* (2003). Further studies are needed to confirm the immunoreactivity and the molecular determinant(s) of *Betanodavirus*.

Genetically different betanodaviruses behave differently in response to environmental temperatures: BFNNV and the TPNNV are considered to be 'cold water VNN' viruses, because they are the most psychrophilic, with optimal culture temperatures from 15 to 20°C. However, SJNNV replicates best at 25°C, while RGNNV has tremendous temperature tolerance (15–35°C) with an optimum temperature between 25 and 30°C (Iwamoto *et al.*, 2000; Hata *et al.*, 2010). The reassortants behave like the RNA1 donor strain, which proves that by

codifying for the polymerase, the RNA1 gene regulates the temperature dependency of fish betanodaviruses (Panzarin *et al.*, 2014).

11.2.1 Geographical distribution, host range and transmission routes

Since its first description in 1985, VNN appeared in the 1990s almost simultaneously in Asia, Australia and southern Europe. By the beginning of 2000, the disease had spread across North America and northern Europe, and VNN is now present almost worldwide, affecting practically all farmed species of marine fishes (Shetty *et al.*, 2012).

The geographic distribution of different betanodaviruses reflects their temperature dependency. RGNNV is the most common VNN virus and has caused clinical disease worldwide (Ucko *et al.*, 2004; Sakamoto *et al.*, 2008; Chérif *et al.*, 2009; Gomez *et al.*, 2009; Panzarin *et al.*, 2012; Ransangan and Manin, 2012; Shetty *et al.*, 2012; Binesh and Jithendran, 2013). The SJNNV strain ranks second (Maeno *et al.*, 2004; García-Rosado *et al.*, 2007; Sakamoto *et al.*, 2008), followed by the reassortant RGNNV/SJNNV, which is common in the Iberian Peninsula and the Mediterranean Sea (Olveira *et al.*, 2009; Panzarin *et al.*, 2012; Souto *et al.*, 2015a). It is noteworthy that VNN outbreaks from reassortant strains have been described only in the Mediterranean Sea (He and Teng, 2015). BFNNV is found only in the cold waters of the northern Atlantic Ocean, the North Sea and the Sea of Japan (Nguyen *et al.*, 1994; Grotmol *et al.*, 2000; Nylund *et al.*, 2008), whereas TPNNV is only found in Japan (Nishizawa *et al.*, 1995; Furusawa *et al.*, 2007).

VNN has been detected in 160 fish species belonging to 79 families and 24 orders (see Table 11.1). Among the susceptible species listed in Table 11.1, the most frequently reported belong to the families: Carangidae, Percichthydae, Serranidae, Sciaenidae, Pleuronectidae, Mugilidae, Sebastidae and Gadidae. Indeed, the most commonly and severely affected species are sea bass (*Dicentrarchus labrax* and *Lates calcarifer*), grouper (*Ephinephelus* spp.), flatfish (*Solea* spp., *Scophtalmus maximus*, *Paralichthys olivaceus*), striped jack (*Pseudocaranx dentex*, *Trachinotus* spp.) and drum (*Umbrina cirrosa*, *Argyrososmus regius*, *Scienops ocellatus*, *Atractoscion nobilis*).

An increasing number of outbreaks of VNN have recently been reported in freshwater fishes (Vendramin *et al.*, 2012; Binesh, 2013; Pascoli

Table 11.1. Fish species that are susceptible to viral nervous necrosis (VNN).[a]

Order	Family	Species	Reference
Acipenseriformes	Acipenseridae	Russian sturgeon (*Acipenser gueldenstaedti*)[b]	Maltese and Bovo, 2007
Anguilliformes	Anguillidae	European eel (*Anguilla anguilla*)[b]	
	Muraenesocidae	Daggertooth pike conger (*Muraenesox cinereus*)[c]	Baeck *et al.*, 2007
	Murenidae	Ribbon moray (*Rhinomuraena quaesita*)[c]	Gomez *et al.*, 2006
Atheriniformes	Melanotaeniidae	Dwarf rainbowfish (*Melanotaenia praecox*)[b,d]	Furusawa *et al.*, 2007
		Threadfin rainbowfish (*Iriatherina werneri*)[b,d]	
	Telmatherinidae	Celebes rainbowfish (*Marosatherina ladigesi*)[b,d]	
Batrachoidiformes	Batrachoididae	Lusitanian toadfish (*Halobatrachus didactylus*)[c]	Moreno *et al.*, 2014
Beloniformes	Adrianichthyidae	Medaka (*Oryzias latipes*)[b,d]	Furusawa *et al.*, 2007
	Belonidae	Garfish (*Belone belone*)[c]	Ciulli *et al.*, 2006a
Beryciformes	Monocentridae	Pineconefish (*Monocentris japonica*)[c]	Gomez *et al.*, 2006
	Trachichthyidae	Mediterranean slimehead (*Hoplostethus mediterraneus*)[c]	Giacopello *et al.*, 2013
Characiformes	Serrasalmidae	Red piranha (*Pygocentrus nattereri*)[b]	Gomez *et al.*, 2006
Clupeiformes	Clupeidae	European pilchard (*Sardina pilchardus*)[c]	Ciulli *et al.*, 2006a
	Engraulidae	Japanese anchovy (*Engraulis japonicus*)[c]	Gomez *et al.*, 2006
Cypriniformes	Cyprinidae	Goldfish (*Carassius auratus*)[b]	Binesh, 2013
		Zebrafish (*Danieo rerio*)[b]	Lu *et al.*, 2008
Cyprinodontiformes	Poeciliidae	Guppy (*Poecilia reticulata*)[b]	Maltese and Bovo, 2007
Gadiformes	Gadidae	Atlantic cod (*Gadus morhua*)	Munday *et al.*, 2002
		Pacific cod (*Gadus macrocephalus*)	
		Haddock (*Melanogrammus aeglefinus*)[c]	Maltese and Bovo, 2007
		European hake (*Merluccius merluccius*)[c]	Ciulli *et al.*, 2006a
		Whiting (*Merlangius merlangus*)[c]	
		Poor cod (*Trisopterus minutus*)[c]	
	Macrouridae	Glasshead grenadier (*Hymenocephalus italicus*)[c]	Giacopello *et al.*, 2013
		Spearnose grenadier (*Caelorinchus multispinulosus*)[c]	Baeck *et al.*, 2007
Gonorynchiformes	Chanidae	Milkfish (*Chanos chanos*)	OIE, 2013
Heterodontiformes	Heterodontidae	Japanese bullhead shark (*Heterodontus japonicus*)[c]	Gomez *et al.*, 2004
Lophiiformes	Lophiidae	Yellow goosefish (*Lophius litulon*)[c]	Baeck *et al.*, 2007
Notacanthiformes	Notacanthidae	Shortfin spiny eel (*Notacanthus bonaparte*)[c]	Giacopello *et al.*, 2013
Perciformes	Acanthuridae	Convict surgeonfish (*Acanthurus triostegus*)	OIE, 2013
		Yellow tang (*Zebrasoma flavescens*)[c]	Gomez *et al.*, 2006
	Acropomatidae	Black throat seaperch (*Doederleinia berycoides*)[c]	Baeck *et al.*, 2007
	Anabantidae	Climbing perch (*Anabas testudineus*)[b,d]	Furusawa *et al.*, 2007
	Anarhichadidae	Wolf fish (*Anarchichas minor*)	OIE, 2013
	Apogonidae	Narrowstrip cardinalfish (*Apogon exostigma*)	OIE, 2013
		Indian perch (*Apogon lineatus*)[c]	Baeck *et al.*, 2007
	Blenniidae	Freshwater blenny (*Salaria fluviatitis*)[b]	Vendramin *et al.*, 2012
	Carangidae	Striped jack (*Pseudocaranx dentex*)	Munday *et al.*, 2002
		Greater amberjack (*Seriola dumerili*)	
		Yellow-wax pompano (*Trachinotus falcatus*)	
		Snub nose pompano (*Trachinotus blochii*)	Maltese and Bovo, 2007
		Mackerel (*Trachurus* spp.)[c]	Ciulli *et al.*, 2006a
		Japanese jack mackerel (*Trachurus japonicus*)[c]	Baeck *et al.*, 2007
		Black scraper (*Trachurus modestus*)[c]	Gomez *et al.*, 2004
		Japanese scad (*Decapterus maruadsi*)[c]	
		Lookdown (*Selene vomer*)[c]	Gomez *et al.*, 2006
	Callionymidae	Moon dragonet (*Callionymus lunatus*)[c]	Baeck *et al.*, 2007
	Centrarchidae	Largemouth bass (*Micropterus salmoides*)[b]	Bovo *et al.*, 2011

Continued

A. Toffan

Table 11.1. Continued.

Order	Family	Species	Reference
	Cichlidae	Tilapia (*Oreochromis niloticus*)[b]	OIE, 2013
		Angelfish (*Pterophyllum scalare*)[b,d]	Furusawa *et al.*, 2007
		Blue streak hap (*Labidochromis caeruleus*)[b,d]	
		Golden mbuna (*Melanochromis auratus*)[b,d]	
		Kenyi cichlid (*Maylandia lombardoi*)[b,d]	
	Eleotridae	Sleepy cod (*Oxyeleotris lineolata*)	Munday *et al.*, 2002
	Ephippidae	Orbicular batfish (*Platax orbicularis*)	OIE, 2013
	Epigonidae	Cardinal fish (*Epigonus telescopus*)[c]	Giacopello *et al.*, 2013
	Gobidae	Black goby (*Gobius niger*)[c]	Ciulli *et al.*, 2006a
	Kyphosidae	Stripey (*Microcanthus strigatus*)[c]	Gomez *et al.*, 2004
	Lateolabracidae	Japanese sea bass (*Lateolabrax japonicus*)	Maltese and Bovo, 2007
	Latidae	Asian sea bass/barramundi (*Lates calcarifer*)	Munday *et al.*, 2002
	Latridae	Striped trumpeter (*Latris lineata*)	
	Leiognathidae	Silver ponyfish (*Leiognathus nuchalis*)[c]	Baeck *et al.*, 2007
	Lutjanidae	Crimson snapper (*Lutjanus erythropterus*)	Maltese and Bovo, 2007
		Mangrove red snapper (*Lutjanus argentimaculatus*)[d]	Maeno *et al.*, 2007
	Malacanthidae	Horsehead tilefish (*Branchiostegus japonicus*)	OIE, 2013
	Moronidae	Striped bass × white bass (*Morone saxatilis* × *M. chrysops*)[b]	Bovo *et al.*, 2011
		European sea bass (*Dicentrarchus labrax*)	Munday *et al.*, 2002
	Mugilidae	Grey mullet (*Mugil cephalus*)	Maltese and Bovo, 2007
		Golden grey mullet (*Liza aurata*)	
		Red mullet (*Mullus barbatus*)	OIE, 2013
		Surmullet (*Mullus surmuletus*)[c]	Panzarin *et al.*, 2012
		Thicklip grey mullet (*Chelon labrosus*)[c]	
		Sharpnose grey mullet (*Liza saliens*)	Zorriehzahra *et al.*, 2014
		Thinlip grey mullet (*Liza ramada*)[c]	Ciulli *et al.*, 2006a
	Oplegnathidae	Barred knifejaw (*Oplegnathus fasciatus*)	Munday *et al.*, 2002
		Spotted knifejaw (*Oplegnathus punctatus*)	
	Osphronemidae	Honey gourami (*Trichogaster chuna*)[b,d]	Furusawa *et al.*, 2007
		Three spot gourami (*Trichopodus trichopterus*)[b,d]	
		Pygmy gourami (*Trichopsis pumila*)[b,d]	
		Siamese fighting fish (*Betta splendens*),[b,d]	
	Percichthydae	Australian bass (*Macquaria novemaculeata*)	Moody *et al.*, 2009
		Macquarie perch (*Macquaria australasica*)[d]	Munday *et al.*, 2002
		Murray cod (*Maccullochella pelii*)[d]	
	Percidae	Pike-perch (*Sander lucioperca*)[b]	Bovo *et al.*, 2011
	Polycentridae	Amazon leaffish (*Monocirrhus polyacanthus*)[b]	Gomez *et al.*, 2006
	Pomacentridae	Sebae clownfish (*Amphiprion sebae*)	Binesh *et al.*, 2013
		Neon damselfish (*Pomacentrus coelestis*)[c]	Gomez *et al.*, 2004
		Threespot dascyllus (*Dascyllus trimaculatus*)[c]	Gomez *et al.*, 2008b
	Rachycentridae	Cobia (*Rachycentron canadum*)	Maltese and Bovo, 2007
	Sciaenidae	Red drum (*Sciaenops ocellatus*)	
		Shi drum (*Umbrina cirrosa*)	
		White sea bass (*Atractoscion nobilis*)	Munday *et al.*, 2002
		Meagre (*Argyrososmus regius*)	Thiéry *et al.*, 2004
		White croaker (*Pennahia argentata*)[c]	Baeck *et al.*, 2007
	Scombridae	Pacific bluefin tuna (*Thunnus orientalis*)	OIE, 2013
		Chub mackerel (*Scomber japonicus*)[c]	Baeck *et al.*, 2007
	Sebastidae	White weakfish (*Sebastes oblongus*)	Maltese and Bovo, 2007
		Korean rockfish (*Sebastes schlegeli*)	Gomez *et al.*, 2004
		Spotbelly rockfish (*Sebastes pachycephalus*)	
		Black rockfish (*Sebastes inermis*)[c]	
		False kelpfish (*Sebastiscus marmoratus*)[c]	

Continued

Table 11.1. Continued.

Order	Family	Species	Reference
	Serranidae	Redspotted grouper (*Epinephelus akaara*)	Munday *et al.*, 2002
		Yellow grouper (*Epinephelus awoara*)	
		Sevenband grouper (*Epinephelus septemfasciatus*)	
		Blackspotted grouper (*Epinephelus fuscoguttatus*)	
		Brownspotted grouper (*Epinephelus malabaricus*)	
		Dusky grouper (*Epinephelus marginatus*)	
		Kelp grouper (*Epinephelus moara*)	
		Greasy grouper (*Epinephelus tauvina*)	
		Dragon grouper (*Epinephelus lanceolatus*)	
		Humpback grouper (*Chromileptes altivelis*)	
		White grouper (*Epinephelus aeneus*)	OIE, 2013
		Orangespotted grouper (*Epinephelus coioides*)	
		Golden grouper (*Epinephelus costae*)	Vendramin *et al.*, 2013
		Spotted coral grouper (*Pletropomus maculatus*)	Pirarat *et al.*, 2009a
	Sillaginidae	Japanese sillago (*Sillago japonica*)[c]	Baeck *et al.*, 2007
	Sparidae	Gilthead sea bream (*Sparus aurata*)	Munday *et al.*, 2002
		Common pandora (*Pagellus erythrinus*)[c]	Ciulli *et al.*, 2006a
		Axillary sea bream (*Pagellus acarne*)[c]	
		Common sea bream (*Pagrus pagrus*)	García-Rosado *et al.*, 2007
		Redbanded sea bream (*Pagrus auriga*)	
		Red sea bream (*Pagrus major*)[c]	Baeck *et al.*, 2007
		White sea bream (*Diplodus sargus*)	Dalla Valle *et al.*, 2005
		Black sea bream (*Spondyliosoma cantharus*)[c]	Moreno *et al.*, 2014
		Bogue (*Boops boops*)[c]	Ciulli *et al.*, 2006a
	Terapontidae	Silver perch (*Bidyanus bidyanus*)[d]	Munday *et al.*, 2002
	Zanclidae	Moorish idol (*Zanclus cornutus*)[c]	Gomez *et al.*, 2004
	Zoarcidae	Blotched eel pout (*Zoarces gilli*)[c]	Baeck *et al.*, 2007
Pleuronectiformes	Cynoglossidae	(Half-smooth) Tongue sole (*Cynoglossus semilaevis*)	Li *et al.*, 2014
		Robust tonguefish (*Cynoglossus robustus*)[c]	Baeck *et al.*, 2007
	Paralichthyidae	Japanese flounder (*Paralichthys olivaceus*)	Munday *et al.*, 2002
		Fivespot flounder (*Pseudorhombus pentophthalmus*)[c]	Baeck *et al.*, 2007
	Pleuronectidae	Barfin flounder (*Verasper moseri*)	Munday *et al.*, 2002
		Winter flounder (*Pleuronectes americanus*)	
		Atlantic halibut (*Hippoglossus hippoglossus*)	
		Marbled sole (*Pleuronectes yokohamae*)[c]	Baeck *et al.*, 2007
	Scophthalmidae	Turbot (*Scophtalmus maximus*)	Munday *et al.*, 2002
	Soleidae	Common (Dover) sole (*Solea solea*)	Maltese and Bovo, 2007
		Senegalese sole (*Solea senegalensis*)	Ito *et al.*, 2008
Rajiformes	Rajidae	Ocellate spot skate (*Okamejei kenojei*)[c]	Baeck *et al.*, 2007
Salmoniformes	Salmonidae	Sea trout (*Salmo trutta trutta*)[c]	Panzarin *et al.*, 2012
		Atlantic salmon (*Salmo salar*)[d]	Korsnes *et al.*, 2005
Scorpaeniformes	Platycephalidae	Bartail flathead (*Platycephalus indicus*)	Munday *et al.*, 2002
	Scorpaenidae	Luna lion fish (*Pterois lunulata*)[c]	Gomez *et al.*, 2004
	Synanceiidae	Devil stinger (*Inimicus japonicus*)	
	Triglidae	Spiny red gurnard (*Chelidonichthys spinosus*)[c]	Baeck *et al.*, 2007
		Tub gurnard (*Chelidonichthys lucerna*)[c]	Ciulli *et al.*, 2006a
Siluriformes	Siluridae	Chinese catfish (*Parasilurus asotus*)[b]	Maltese and Bovo, 2007
		Australian catfish (*Tandanus tandanus*)[b]	Munday *et al.*, 2002
Syngnathiformes	Centriscidae	Razorfish (*Aeoliscus strigatus*)[c]	Gomez *et al.*, 2006
Tetraodontiformes	Balistidae	Triggerfish (*Balistapus* spp.)[c]	Panzarin *et al.*, 2010

Continued

A. Toffan

Table 11.1. Continued.

Order	Family	Species	Reference
	Diodontide	Longspined porcupinefish (*Diodon holocanthus*)[c]	Gomez *et al.*, 2004
	Monacanthidae	Korean black scraper (*Thamnaconus modestus*)[c]	
		Threadsail filefish (*Stephanolepis cirrhifer*)	Pirarat *et al.*, 2009b
	Tetraodontidae	Tiger puffer (*Takifugu rubripes*)	Munday *et al.*, 2002
		Grass puffer (*Takifugu niphobles*)[c]	Baeck *et al.*, 2007
		Panther puffer (*Takifugu pardalis*)[c]	Gomez *et al.*, 2004
		Moontail puffer (*Lagocephalus lunaris*)[c]	Baeck *et al.*, 2007
Zeiformes	Zeidae	John dory (*Zeus faber*)[c]	Baeck *et al.*, 2007

[a]Classification of fish based on www.fishbase.org
[b]Freshwater species
[c]*Betanodavirus* detected in wild asymptomatic fish.
[d]Experimentally infected.

et al., 2016). Betanodaviruses can cause disease and infect a great number of fish species, and this is independent of water salinity (Furusawa *et al.*, 2007; Maeno *et al.*, 2007; Bovo *et al.*, 2011). The detection of *Betanodavirus* in wild fish with no clinical signs is also extensive (as seen by the large number of species included in this category listed in Table 11.1), which may account for its widespread nature. Finally, the increased death rate in wild fish (especially groupers, *Epinephelus* spp.) with nervous disorders accompanied by the detection of the virus in these fish, is a concern (Gomez *et al.*, 2009; Vendramin *et al.*, 2013; Haddad-Boubaker *et al.*, 2014; Kara *et al.*, 2014).

Transmission of the disease occurs horizontally through direct contact with infected fish, contaminated water and/or contaminated equipment. The disease can easily be experimentally reproduced by bath exposure as well as by intramuscular or intraperitoneal injection (Munday *et al.*, 2002; Maltese and Bovo, 2007). Evidence of oral transmission through the ingestion of infected fish or contaminated feed has been suggested (Shetty *et al.*, 2012). The virus has been detected in several marine invertebrates (Gomez *et al.*, 2006, 2008a; Panzarin *et al.*, 2012; Fichi *et al.*, 2015). Indeed, the virus has a very high resistance to chemical and physical agents (i.e. heat, pH and disinfectant) and, therefore, it cannot only contaminate marine water, invertebrates and microorganisms, but also nets, pens, tanks and other equipment (Maltese and Bovo, 2007).

Wild and farmed fish that have survived the disease are the most likely source of infections in 'VNN-free' farms. Disease and viral shedding in infected fish with no clinical signs may reactivate several times after the onset of the disease due to stress or water temperature variations (Johansen *et al.*, 2004; Rigos and Katharios, 2009; Lopez-Jimena *et al.*, 2010; Souto *et al.*, 2015b).

Vertical transmission had been described in the most susceptible fish species (Shetty *et al.*, 2012) and the virus has been found in gonads and seminal fluids.

11.3 Diagnosis of the Infection

11.3.1 Clinical signs

Clinical signs of VNN infection include variations in skin colour, anorexia, lethargy, abnormal swimming behaviour and nervous signs caused by lesions in the brain and retina. Generally, younger fish are most susceptible to infection. In larvae/juveniles, the onset of disease can be hyperacute, and the only apparent sign is a sharp increase in mortality. In older animals, the onset of disease can be slower and the cumulative mortality lower. Infected fish have reduced growth, resulting in uneven weight/size (Vendramin *et al.*, 2014); this represents an indirect but significant economic loss, which is often underestimated. Hyperinflation of the swim bladder is another common clinical sign (Maltese and Bovo, 2007; Pirarat *et al.*, 2009a; Hellberg *et al.*, 2010; Vendramin *et al.*, 2013).

In European sea bass (*D. labrax*) clinical signs include anorexia, darkening of the skin and abnormal swimming behaviour, which is characterized by swirling, circular movements that alternate with periods of lethargy, abnormal bathymetry and anomalous vertical positions in a water column. Fish appear blind and can display hyperexcitability when disturbed (Péducasse *et al.*, 1999;

Athanassopoulou *et al.*, 2003). Traumatic lesions to the jaw, head, eyes and nose (Fig. 11.1) are a natural consequence of impaired swimming capacity (Shetty *et al.*, 2012). Hyperinflation of the swim bladder also contributes to the abnormal swimming behaviour, causing the fish to sink or float (Lopez-Jimena *et al.*, 2011). In sea bass larvae and juveniles, congestion of the brain (sometimes perceptible through the skull) or of the whole head can be observed, as well as a typical 'sickle position' due to muscular hypercontraction associated with high mortality (Bovo, 2010, personal communication). Mortality varies according to water temperature and age, and outbreaks in hatcheries can be devastating, with extremely high mortalities (80–100%). Older fish are generally less affected, even though severe losses have been described in mature animals (Munday *et al.*, 2002; Chérif *et al.*, 2009). In sea bass, the disease is caused almost exclusively by the RGNNV genotype, which occurs at temperatures >23–25°C, and mortality decreases when temperatures fall below 18–22°C (Bovo *et al.*, 1999; Breuil *et al.*, 2001; Chérif *et al.*, 2009). European sea bass may also be infected with other VNN strains, and when this happens, the clinical signs are milder (Vendramin *et al.*, 2014; Souto *et al.*, 2015a).

Darkening of the skin, clustering of the fish at the surface of the water, and a cumulative mortality of 60–100% was observed in naturally infected Asian sea bass larvae (*Lates calcarifer*). Larvae also displayed anorexia, pale grey pigmentation of the body, loss of equilibrium and corkscrew-like swimming prior to death. The mortality rate at all ages is generally >50% (Maeno *et al.*, 2004; Azad *et al.*, 2005; Parameswaran *et al.*, 2008).

Groupers are among the most VNN-susceptible fishes; the disease occurs at any age, in both farmed and wild fish, and is mainly caused by RGNNV. The typical signs of disease include loss of equilibrium, swimming in a corkscrew fashion and lethargy associated with abnormal response to stimulation. Hyperinflation of the swim bladder and corneal opacity are the most relevant clinical signs (Fig. 11.2). Spinal deformities and exophthalmos have also been reported (Sohn *et al.*, 1998; Gomez *et al.*, 2009; Pirarat *et al.*, 2009a; Vendramin *et al.*, 2013; Kara *et al.*, 2014).

Clinical signs in flatfish (order Pleuronectiformes) are less obvious; they remain at the bottom of the tank and bend their bodies with the head and tail raised, sometimes upside down on the bottom. They may tremble or drop to the bottom of the tank like a 'falling leaf' (Maltese and Bovo, 2007). Skin discoloration has been reported in the Atlantic halibut (*H. hippoglossus*) (Grotmol *et al.*, 1997). During a natural outbreak, the common Dover sole (*Solea solea*) appeared either darker or paler than usual and at times anorexic, with cumulative mortality that may reach 100% (Starkey *et al.*, 2001). The Senegalese soil (*S. senegalensis*), may also have

Fig. 11.1. A juvenile farmed European sea bass (*Dicentrachus labrax*) with viral nervous necrosis (VNN) showing skin erosion and congestion of the head.

Fig. 11.2. A wild dusky grouper (*Epinephelus marginatus*) from the Mediterranean Sea showing clinical signs of viral nervous necrosis (VNN): head skin erosion, corneal opacity and panophthalmitis (inflammation of all coats of the eye, including intraocular structures).

A. Toffan

skin ulcers (Olveira *et al.*, 2008) and is extremely susceptible to RGNNV/SJNNV, with a mortality rate of 100% at 22°C, although at 16°C, mortality is reduced to 8%. VNN can also cause a persistent infection in the sole, and the disease can be reactivated by an increase of the water temperature to 22°C (Souto *et al.*, 2015a,b).

Sciaenidae, such as the shi drum (*Umbrina cirrosa*) and red drum (*Sciaenops ocellatus*) are highly susceptible to the disease, especially juvenile fish (Bovo *et al.*, 1999; Oh *et al.*, 2002; Katharios and Tsigenopoulos, 2010). In addition to the classical clinical signs, white sea bass (*Atractoscion nobilis*) have marked hyperinflation of the swim bladder at 12–15°C (Curtis *et al.*, 2003); in contrast, infected wild meagre (*Argyrososmus regius*) did not show any particular clinical signs (Lopez-Jimena *et al.*, 2010).

Some cold-water fish, such as cod (*Gadus morua*) and the Atlantic halibut (*H. hippoglossus*), can be infected at 6–15°C and display lethargy, anorexia, darkening of the skin, nervous clinical signs, swim bladder inflammation and corneal opacity (Grotmol *et al.*, 1997; Patel *et al.*, 2007; Hellberg *et al.*, 2010).

Sparidae are often resistant to clinical disease. For example, gilthead sea bream (*Sparus aurata*) reared in cohabitation with *Betanodavirus*-infected European sea bass never showed mortality or clinical signs (Castric *et al.*, 2001; Aranguren *et al.*, 2002; Ucko *et al.*, 2004). However, sea bream are susceptible to experimental infection (intramuscular injection with RGNNV viruses), which can provoke mortality in juveniles or enable them to act as non-clinical contagious hosts for sea bass (Castric *et al.*, 2001; Aranguren *et al.*, 2002). Increased outbreaks of VER with high mortality in sea bream larvae have recently been observed (Beraldo *et al.*, 2011; Toffan, 2015, unpublished results). The viruses isolated from sea bream are always RGNNV/SJNNV, which indicates a special adaptation of this reassortant to sea bream, as reported for the Senegalese sole (Souto *et al.*, 2015a).

Clinical signs in naturally or experimentally infected freshwater fish species are similar to those in marine fishes.

11.3.2 Laboratory diagnosis

VNN isolation is generally in continuous cell cultures, and the most commonly used cell lines are SSN-1 cells from snakeheads (*Ophicephalus striatus*) (Frerichs *et al.*, 1996) and their derived clone E-11 (Iwamoto *et al.*, 2000). The high susceptibility of these two cell lines has been attributed to their persistent infection with the snakehead SnRV retrovirus (Lee *et al.*, 2002; Nishizawa *et al.*, 2008). Other continuous fish cell lines (for examples the GF-1 and other cell lines from the orange-spotted grouper *Epinephelus coioides*; the SISS and ASBB cell lines from *L. calcarifer*; and the SAF-1 cell line from *S. aurata*) have been developed and used successfully, but most of these are not available commercially; also, their diagnostic performances and applications are unknown (Hasoon *et al.*, 2011; Sano *et al.*, 2011).

Suitable incubation temperatures depend upon the VNN genotype (Iwamoto *et al.*, 2000; Ciulli *et al.*, 2006b; Hata *et al.*, 2010; Panzarin *et al.*, 2014). A typical cytopathic effect (CPE) is the darkening and contraction of the infected cells. The appearance of clusters of vacuoles in the cytoplasm of infected cells, cell rounding up and detachment from the monolayer, and evolution into extended necrotic foci are typical signs. The cultivated virus can be identified by serum neutralization with monoclonal or polyclonal antibodies (Mori *et al.*, 2003; V. Panzarin, Italy, 2015, personal communication), fluorescein-conjugated antibodies (Péducasse *et al.*, 1999; Thiéry *et al.*, 1999; Castric *et al.*, 2001; Mori *et al.*, 2003), enzyme-linked immunosorbent assay (ELISA) (Fenner *et al.*, 2006b) and by molecular biology techniques. ELISA can be used directly with central nervous system (CNS) tissue of the infected fish (Breuil *et al.*, 2001; Nuñez-Ortiz *et al.*, 2016), as can indirect fluorescent antibody testing (IFAT) (Nguyen *et al.*, 1997; Totland *et al.*, 1999; Curtis *et al.*, 2001; Johansen *et al.*, 2003).

The rapid diagnosis of VNN is now possible using reverse transcriptase polymerase chain reaction (RT-PCR) and real-time RT-PCR (qRT-PCR). The F2-R3 primer set, targeting the T4 variable region of the RNA2 segment, has been widely used for diagnostic purposes (Nishizawa *et al.*, 1994). Several other PCR-based protocols have been developed to increase the sensitivity and specificity of *Betanodavirus* diagnostic assays (Grotmol *et al.*, 2000; Gomez *et al.*, 2004; Dalla Valle *et al.*, 2005; Cutrín *et al.*, 2007). Validated qRT-PCR methods are also available for detecting all known VNN genotypes (Grove *et al.*, 2006a; Panzarin *et al.*, 2010; Hick *et al.*, 2011; Hodneland *et al.*, 2011; Baud *et al.*, 2015).

Viral isolation, histopathology or immunostaining can be used for the diagnosis and confirmation of VNN, whereas a combination of at least two different molecular assays (i.e. RT-PCR protocols targeting different regions of the viral genome, or RT-PCR followed by sequencing) is advisable when the disease is identified for the first time (OIE, 2013).

Antibodies in experimentally immunized/infected fish can be detected using an ELISA test or a serum neutralization assay. ELISA is the most common test used (Breuil *et al.*, 2000; Watanabe *et al.*, 2000; Grove *et al.*, 2006b; Scapigliati *et al.*, 2010), while serum neutralization is often used to evaluate the humoral response of fish (Skliris and Richards, 1999; Tanaka *et al.*, 2001; Kai *et al.*, 2010).

Antibodies in vaccinated as well as in naturally exposed fish can be detected for over a year, although important intra-fish species variations exist (Breuil and Romestand, 1999; Breuil *et al.*, 2000; Grove *et al.*, 2006b; Kai *et al.*, 2010). However, there was a significant drop in serum-neutralizing antibody titres in groupers 75 days postvaccination (Pakingking *et al.*, 2010). In temperate fish, higher antibody titres are more frequent in the summer than in the winter, which indicates that serum samples should be collected in the summer (Breuil *et al.*, 2000). Due to the lack of robust research, the detection of specific antibodies has not been considered as a routine screening method for assessing the viral status of fish (OIE, 2013).

11.4 Pathology and Pathophysiology

11.4.1 Gross and microscopic lesions

Congestion, abrasion and sometimes necrosis of the nose, jaws and head are clinical signs of VNN in *D. labrax*. Corneal opacity has been reported in several species, as well as skin erosions on the body and fins caused by an impaired swimming ability (Maltese and Bovo, 2007; Shetty *et al.*, 2012). At necropsy, the hyperinflation of the swim bladder is common in almost all susceptible species, with the congestion of the CNS and meninges being the most relevant internal lesions (Fig 11.3).

The encephalitis in the CNS is characterized by multiple intracytoplasmic vacuolations; numerous empty areas of 5–10 µm diameter, clearly separated from the surrounding areas, are present in the grey matter of the olfactory bulb, telencephalon, diencephalon, mesencephalon, cerebellum, medulla oblongata, spinal cord and retina (Munday *et al.*,

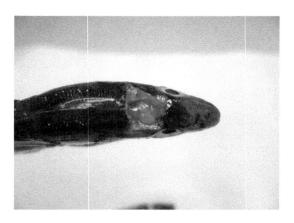

Fig. 11.3. European sea bass (*Dicentrachus labrax*) with viral nervous necrosis (VNN) showing congestion of the brain and meninges.

2002; Grove *et al.*, 2003; Mladineo, 2003; Maltese and Bovo, 2007; Lopez-Jimena *et al.*, 2011). Severity of the vacuolization depends on the fish species, age and stage of infection but, notably, it is a consistent (almost pathognomonic) finding in the case of VNN. Pyknosis, karyorrhexis, neuronal degeneration and inflammatory infiltration have been described in all the nervous tissues of infected fish. The olfactory bulb, the optic tectum of the mesencephalon, the granular and Purkinje cell layers of cerebellum and the motor neurons of the spinal cord are generally the most affected areas (Totland *et al.*, 1999; Pirarat *et al.*, 2009a,b). In the preclinical phase of the infection of Asian sea bass, darkly stained and actively dividing neuronal zones are present in the CNS and spinal cord without vacuolization. Vacuoles appear later, together with the nervous signs of the disease (Azad *et al.*, 2006). Congestion of the blood vessels in the encephalic parenchyma and meninges is frequent and can evolve into mild or extensive haemorrhages in the brain (Mladineo, 2003; Korsnes *et al.*, 2005; Pirarat *et al.*, 2009b).

Immunohistochemistry (IHC) or *in situ* hybridization (ISH) can be used to detect viral antigens surrounding vacuoles in the CNS, optic tectum and cerebellum (Fig. 11.4), but also when vacuoles are not present (Grove *et al.*, 2003; Mladineo, 2003; Pirarat *et al.*, 2009a,b; Katharios and Tsigenopoulos, 2010; Lopez-Jimena *et al.*, 2011). Transmission electron microscopy (TEM) generally shows crystalline arrays of membrane-bound nodaviruses in the cytoplasm of infected brain and retina cells

A. Toffan

(Grotmol *et al.*, 1997; Tanaka *et al.*, 2004; Ucko *et al.*, 2004).

Vacuolar lesions (confirmed using IHC) have been observed in the dendritic cell of the spinal cord, particularly in the cranial part (Grotmol *et al.*, 1997). Spinal cord lesions can appear before or after the occurrence of encephalic lesions (Nguyen *et al.*, 1996; Pirarat *et al.*, 2009b).

The retina of infected fish may also have major lesions, which have marked histopathological changes, such as massive necrosis of small round cells and spongiotic vacuolization in the outer and inner nuclear layers, as well as in the ganglion cell layer (Fig. 11.5). In some cases, vacuoles are also present in the internal plexiform layer (Pirarat *et al.*, 2009a; Katharios and Tsigenopoulos, 2010; Lopez-Jimena *et al.*, 2011). The optic nerve may show various degrees of alteration from extensive vacuolization to no apparent lesions (Mladineo, 2003; Lopez-Jimena *et al.*, 2011).

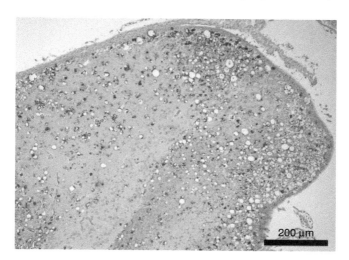

Fig. 11.4. Immunohistochemical detection of *Betanodavirus* in sea bass (*Dicentrachus labrax*) brain (10×) showing abundant immunoprecipitates and severe vacuolization of the olfactory lobe. Courtesy of Dr Pretto Tobia.

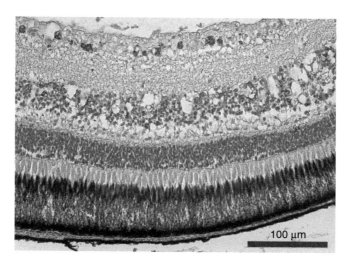

Fig. 11.5. Immunohistochemical detection of *Betanodavirus* in sea bass (*Dicentrachus labrax*) retina (25×). Immunoprecipitates and vacuolizations are evident in the internal nuclear layer and in the ganglion cell layer. Courtesy of Dr Pretto Tobia.

Viral particles were detected using IHC/ISH or PCR in non-nervous tissues such as the gills, fins, heart, anterior and posterior intestine, stomach, spleen, liver, kidney and gonads of several fish species (Grotmol *et al.*, 1997; Nguyen *et al.*, 1997; Johansen *et al.*, 2003; Mladineo, 2003; Grove *et al.*, 2006a; Korsnes *et al.*, 2009; Lopez-Jimena *et al.*, 2011; Mazelet *et al.*, 2011). These organs are not generally believed to play an important role in the disease pathogenesis, unlike nervous tissues, which are the primary sites for viral replication. The digestive tract can also play a role in disease transmission (Nguyen *et al.*, 1996; Totland *et al.*, 1999). The strong presence of *Betanodavirus* antigen in the gastrointestinal cells of infected threadsail filefish (*Stephanolepis cirrhifer*) suggests that this may be a particularly active site for viral replication and spread (Pirarat *et al.*, 2009b). Also, immunostaining in the reproductive tissues, such as the ovary, supports the vertical transmission of the disease (Nguyen *et al.*, 1997; Azad *et al.*, 2005, 2006).

Despite the recurrent and irreversible hyperinflation of the swim bladder in fish with clinical signs of VNN, this is not considered to be a target organ for nodaviruses because histopathological lesions have never been observed, except for a slight congestion or the presence of eosinophilic intranuclear inclusions (Pirarat *et al.*, 2009a).

11.4.2 Pathogenesis

Despite several recent studies on the pathogenesis of VNN, the important viral mechanisms for the development of the disease remain unclear. The heat shock protein GHSC70 commonly expressed by the GF-1 cell line is a good VNN receptor candidate (Chang and Chi, 2015). Nevertheless, there may be other membrane proteins that act as viral receptors. The neurotropism of *Betanodavirus* is irrefutable, but the route to the nervous tissues is still a matter of discussion. Several tissues, such as the nasal cavity, the intestinal epithelium and intact skin, may be entry portals. Tanaka *et al.* (2004) infected sevenband groupers (*Epinephelus septemfasciatus*) intranasally and assumed that the virus first penetrated the nasal epithelium, reached the olfactory nerve and bulb, and finally invaded the olfactory lobe, where they replicated. This route was also suggested in naturally infected sevenband groupers, spotted coral groupers (*Pletropomus maculatus*) and European sea bass larvae in which pathological changes first developed in the olfactory

lobes (Mladineo, 2003; Banu *et al.*, 2004; Pirarat *et al.*, 2009a).

The epithelial cells in the skin and the intestinal epithelium may also be possible entry portals (Totland *et al.*, 1999; Grove *et al.*, 2003). Positive IHC reactions in the skin and lateral line had led some authors to suggest that VNN can enter through the epithelial cells of the skin (Nguyen *et al.*, 1997; Péducasse *et al.*, 1999; Totland *et al.*, 1999; Azad *et al.*, 2006; Kuo *et al.*, 2011), but this assumption needs further investigation.

Once in its host, the *Betanodavirus* multiplies and spreads to the target organs through a viraemic phase, as has been extensively demonstrated by detection of the viruses in the blood (Lu *et al.*, 2008; Olveira *et al.*, 2008; Korsnes *et al.*, 2009; Lopez-Jimena *et al.*, 2011). Recurrent endocardial lesions, probably due to viraemia, have also been reported in the Atlantic halibut (Grotmol *et al.*, 1997). Another hypothesis is that the virus reaches the CNS directly by axonal transport through the cranial nerves (Grotmol *et al.*, 1999; Húsgaro *et al.*, 2001; Tanaka *et al.*, 2004), where it then actively replicates into the target nervous organs; here, the manifestation of vacuolization is generally associated with the appearance of clinical signs (Azad *et al.*, 2006).

Betanodavirus infection strongly induces an innate immunity (Lu *et al.*, 2008; Scapigliati *et al.*, 2010; Chen *et al.*, 2014), and following this first, short-term, non-specific immune response, the development of specific circulating anti-VNN antibodies has been detected in different fishes. These induced antibodies neutralize the viral activity, but it is still debated whether they can overcome the haemato–encephalic (blood–brain) barrier to protect the CNS from damage and clear the infection. In fact, *Betanodavirus* tries to avoid the protective systems of the host by hiding in the nervous tissue, where the viruses generate latent infections (Chen *et al.*, 2014). Clearly, the onset of the carrier status needs further investigation.

The transfer of maternal immunity has been described in the eggs and fry of groupers from vaccinated broodstock and has been hypothesized in the European sea bass (Breuil *et al.*, 2001; Kai *et al.*, 2010), but, again, further investigations are needed to understand its duration and efficacy.

11.5 Protective and Control Strategies

Some promising experimental vaccines against VNN range from inactivated whole viruses to recombinant capsid proteins and virus-like particles (produced in different vector microorganisms), which are administered by intramuscular or intraperitoneal injection (Gomez-Casado et al., 2011; Shetty et al., 2012; Chen et al., 2014).

Good results have been obtained with a live vaccine in grouper cultured at low temperatures (17°C); however, due to the ability of Betanodavirus to cause a persistent infection, the vaccination cannot guarantee the clearance of latent infections (Nishizawa et al., 2012; Oh et al., 2013).

As indicated in Section 11.3.1, larval and juvenile fish are more susceptible to the disease, and these are the life stages when it is most impractical to inject a vaccine. To overcome this problem, a vaccine formulation for bath and oral administration has recently been tested, but the data that are available do not confirm the efficacy of either approach (Gomez-Casado et al., 2011; Nishizawa et al., 2012; Kai et al., 2014; Wi et al., 2015).

As no commercially available vaccines against Betanodavirus are available, the spread of the disease will be exponential in dense fish farming areas (e.g. the Mediterranean Basin, the Japanese and Taiwanese coasts and South-east Asia) and availability of non-carrier fish could be reduced, as already observed for grouper in Taiwan (Chen et al., 2014).

In the absence of an efficient vaccine, the best way to prevent the introduction and/or the spread of the disease in fish farms is to comply with strict biosecurity rules. Biosecurity must play a primary role in hatcheries, where the accidental introduction of VNN can devastate stocks. The most relevant safety measures include the disinfection of inlet water, the division of the hatchery into sectors with dedicated personnel and equipment, the periodic disinfection of tanks and equipment, periodic fallowing, quarantine and testing of newly introduced broodstock and control of frozen or live feeds (Munday et al., 2002; Maltese and Bovo, 2007; Shetty et al., 2012). Obviously, such measures are not applicable in open environments such as lagoons and sea cages. None the less, in open environments, biosecurity precautions (e.g. the fast and proper disposal of carcasses, reduction of stock densities, reduction of feeding rates, application of regular fallowing) and attentive management may help to reduce stress and minimize losses.

It is important to emphasize that betanodaviruses are extremely resistant to various environmental conditions (Arimoto et al., 1996; Munday

et al., 2002; Liltved et al., 2006). They can tolerate extreme variations of pH and survive a 6 week storage at pH 3–7 and over 24 h at pH 2–11, as well as endure extreme temperature conditions for more than a month at 25°C and more than a year at 15°C (Frerichs, 2000). VNN inactivation is possible through a heat treatment at 60°C or higher for at least 30 min (Arimoto et al., 1996), which makes thermal inactivation inapplicable as a first choice cleaning method. Iodophors and hypochlorite solutions have the best virucidal effects and are fully effective at concentrations of 50 ppm for 10 min at 20°C (Munday et al., 2002), while formalin is less efficient because it requires a combination of concentration, temperature and exposure time that is difficult to apply in practice (Arimoto et al., 1996; Frerichs, 2000; Maltese and Bovo, 2007; Kai and Chi, 2008).

UV irradiation and ozone treatment are feasible disinfection measures (Arimoto et al., 1996; Frerichs, 2000). Particular attention should be paid to the disinfection of embryonated eggs. The application of ozone (4 mg O_3/l for 0.5 min) to eggs from experimentally infected halibut neutralized the virus with practically no impact on hatchability (Grotmol and Totland, 2000). Ozone was also successful in disinfecting haddock (Melanogrammus aeglefinus) eggs that were experimentally infected (Buchan et al., 2006), but the safe dosage of ozone varies among species (Grotmol et al., 2003). Nevertheless, ozonation may not always be efficacious for the complete inactivation of VNN viruses and discrepancies in results between studies are due to the lack of standard procedures to conduct inactivation experiments and to determine oxidant residues in seawater (Liltved et al., 2006). More trials under different environmental conditions are necessary to obtain a safe and efficient protocol that can be routinely applied in the field.

As there is vertical transmission of the virus in several fish species, the introduction of new broodstock into a disease-free farm always needs serious consideration. Some combination of serology and molecular biology techniques on ovary biopsies or on eggs themselves may provide criteria by which hatchery personnel can select virus-free broodstock (Mushiake et al., 1992; Breuil and Romestand, 1999; Watanabe et al., 2000; Mazelet et al., 2011).

The genetic selection of resistance genes is largely applied to fight viral infection in freshwater species (i.e. salmon), and not much research has been conducted on this approach in *Betanodavirus*. An extremely high estimated within-strain heritability of survival in VNN-challenged Atlantic cod was observed, although the data structure was not optimal for obtaining conclusive results (Ødegård et al., 2010).

11.6 Conclusions and Suggestions for Future Research

At present, the major constraint to controlling VNN is the lack of immunizing strategies. There is an urgent need for safe and inexpensive commercial vaccines, especially against the most prevalent *Betanodavirus* genotypes (RGNNV and SJNNV).

The availability of reliable serology tests will be critical to shedding more light on the fish humoral response and evaluating the degree of protection elicited by any vaccine. It will also serve as a non-invasive diagnostic test to complement biomolecular protocols for confirming virus-free brood stocks. Standardized protocols for sampling, the certification of specific pathogen-free stocks and farm disinfections need to be developed, shared and applied.

More attention should be addressed to the interactions and exchanges of pathogens between farmed and wild fish populations in order to assess the risk of transmission of the infection from one environment to another, and to better understand the role of wild fish in ecology of *Betanodavirus*.

Finally, genetic selection programmes are needed to increase the natural resistance of different fish to the disease. Such a selection programme should be considered a synergistic tool to the use of vaccines to efficiently arrest the spread and impact of this disease.

References

Aranguren, R., Tafalla, C., Novoa, B. and Figueras, A. (2002) Experimental transmission of encephalopathy and retinopathy induced by nodavirus to sea bream, *Sparus aurata* L., using different infection models. *Journal of Fish Diseases* 25, 317–324.

Arimoto, M., Sato, J., Maruyama, K., Mimura, G. and Furusawa, I. (1996) Effect of chemical and physical treatments on the inactivation of striped jack nervous necrosis virus (SJNNV). *Aquaculture* 143, 15–22.

Athanassopoulou, F., Billinis, C., Psychas, V. and Karipoglou, K. (2003) Viral encephalopathy and retinopathy of *Dicentrarchus labrax* (L.) farmed in fresh water in Greece. *Journal of Fish Diseases* 26, 361–365.

Azad, I.S., Shekhar, M.S., Thirunavukkarasu, A.R., Poornima, M., Kailasam, M. *et al.* (2005) Nodavirus infection causes mortalities in hatchery produced larvae of *Lates calcarifer*: first report from India. *Diseases of Aquatic Organisms* 63, 113–118.

Azad, I.S., Jithendran, K.P. and Shekhar, M.S. (2006) Immunolocalisation of nervous necrosis virus indicates vertical transmission in hatchery produced Asian sea bass (*Lates calcarifer* Bloch) – a case study. *Aquaculture* 255, 39–47.

Baeck, G.W., Gornez, D.K., Oh, K.S., Kim, J.H., Choresca, C.H. *et al.* (2007) Detection of piscine nodaviruses from apparently healthy wild marine fish in Korea. *Bulletin of the European Association of Fish Pathologists* 27, 116–122.

Bandín, I. and Dopazo, C.P. (2011) Host range, host specificity and hypothesized host shift events among viruses of lower vertebrates. *Veterinary Research* 42:67.

Banu, G., Mori, K., Arimoto, M., Chowdhury, M. and Nakai, T. (2004) Portal entry and progression of betanodaviruses causing viral nervous necrosis in sevenband grouper *Ephinephelus septemfasciatus*. *Bangladesh Journal of Veterinary Medicine* 2, 83–87.

Baud, M., Cabon, J., Salomoni, A., Toffan, A., Panzarin, V. and Bigarré, L. (2015) First generic one step real-time Taqman RT-PCR targeting the RNA1 of betanodaviruses. *Journal of Virological Methods* 211, 1–7.

Beraldo, P., Panzarin, V., Galeotti, M. and Bovo, G. (2011) Isolation and molecular characterization of viral encephalopathy and retinopathy virus from gilthead sea bream larvae (*Sparua aurata*) showing mass mortalities. In *15th EAFP International Conference on Diseases of Fish and Shellfish*, Split, Croatia. *Conference Abstract Book*, p. 351. European Association of Fish Pathologists. Conference website available at: https://eafp.org/15th-eafp-spilt-2011/ (accessed 3 November 2016).

Binesh, C.P. (2013) Mortality due to viral nervous necrosis in zebrafish *Danio rerio* and goldfish *Carassius auratus*. *Diseases of Aquatic Organisms* 104, 257–260.

Binesh, C.P. and Jithendran, K.P. (2013) Genetic characterization of betanodavirus isolates from Asian seabass *Lates calcarifer* (Bloch) in India. *Archives of Virology* 158, 1543–1546.

Binesh, C.P., Renuka, K., Malaichami, N. and Greeshma, C. (2013) First report of viral nervous necrosis-induced mass mortality in hatchery-reared larvae of clownfish, *Amphiprion sebae* Bleeker. *Journal of Fish Diseases* 36, 1017–1020.

Bovo, G., Nishizawa, T., Maltese, C., Borghesan, F., Mutinelli, F. *et al.* (1999) Viral encephalopathy and retinopathy of farmed marine fish species in Italy. *Virus Research* 63, 143–146.

Bovo, G., Gustinelli, A., Quaglio, F., Gobbo, F., Panzarin, V. *et al.* (2011) Viral encephalopathy and retinopathy outbreak in freshwater fish farmed in Italy. *Diseases of Aquatic Organisms* 96, 45–54.

Breuil, G. and Romestand, B. (1999) A rapid ELISA method for detecting specific antibody level against nodavirus in the serum of the sea bass, *Dicentrarchus labrax* (L.): application to the screening of spawners in a sea bass hatchery. *Journal of Fish Diseases* 22, 45–52.

Breuil, G., Pepin, J.F., Castric, J., Fauvel, C. and Thiéry, R. (2000) Detection of serum antibodies against nodavirus in wild and farmed adult sea bass: application to the screening of broodstock in sea bass hatcheries. *Bulletin of the European Association of Fish Pathologists* 20, 95–100.

Breuil, G., Mouchel, O., Fauvel, C. and Pepin, J.F. (2001) Sea bass *Dicentrarchus labrax* nervous necrosis virus isolates with distinct pathogenicity to sea bass larvae. *Diseases of Aquatic Organisms* 45, 25–31.

Buchan, K.A.H., Martin-Robichaud, D.J., Benfey, T.J., MacKinnon, A.M. and Boston, L. (2006) The efficacy of ozonated seawater for surface disinfection of haddock (*Melanogrammus aeglefinus*) eggs against piscine nodavirus. *Aquacultural Engineering* 35, 102–107.

Castric, J., Thiéry, R., Jeffroy, J., de Kinkelin, P. and Raymond, J.C. (2001) Sea bream *Sparus aurata*, an asymptomatic contagious fish host for nodavirus. *Diseases of Aquatic Organisms* 47, 33–38.

Chang, J.S. and Chi, S.C. (2015) GHSC70 is involved in the cellular entry of nervous necrosis virus. *Journal of Virology* 89, 61–70.

Chen, L.J., Su, Y.C. and Hong, J.R. (2009) Betanodavirus non-structural protein B1: a novel anti-necrotic death factor that modulates cell death in early replication cycle in fish cells. *Virology* 385, 444–454.

Chen, Y.M., Wang, T.Y. and Chen, T.Y. (2014) Immunity to betanodavirus infections of marine fish. *Developmental and Comparative Immunology* 43, 174–183.

Chérif, N., Thiéry, R., Castric, J., Biacchesi, S., Brémont, M. et al. (2009) Viral encephalopathy and retinopathy of *Dicentrarchus labrax* and *Sparus aurata* farmed in Tunisia. *Veterinary Research Communications* 33, 345–353.

Chi, S.C., Shieh, J.R. and Lin, S.J. (2003) Genetic and antigenic analysis of betanodaviruses isolated from aquatic organisms in Taiwan. *Diseases of Aquatic Organisms* 55, 221–228.

Ciulli, S., Di Marco, P., Natale, A., Galletti, E., Battilani, M. et al. (2006a) Detection and characterization of *Betanodavirus* in wild fish from Sicily, Italy. *Ittiopatologia* 3, 101–112.

Ciulli, S., Gallardi, D., Scagliarini, A., Battilani, M., Hedrick, R.P. et al. (2006b) Temperature-dependency of *Betanodavirus* infection in SSN-1 cell line. *Diseases of Aquatic Organisms* 68, 261–265.

Costa, J.Z., Adams, A., Bron, J.E., Thompson, K.D., Starkey, W.G. et al. (2007) Identification of B-cell epitopes on the betanodavirus capsid protein. *Journal of Fish Diseases* 30, 419–426.

Curtis, P.A., Drawbridge, M., Iwamoto, T., Nakai, T., Hedrick, R.P. et al. (2001) Nodavirus infection of juvenile white seabass, *Atractoscion nobilis*, cultured in southern California: first record of viral nervous necrosis (VNN) in North America. *Journal of Fish Diseases* 24, 263–271.

Curtis, P.A., Drawbridge, M., Okihiro, M.S., Nakai, T., Hedrick, R.P. et al. (2003) Viral nervous necrosis (VNN) in white seabass, *Atractoscion nobilis*, cultured in Southern California, and implications for marine fish aquaculture. *Journal of Fish Diseases* 24, 263–271.

Cutrín, J.M., Dopazo, C.P., Thiéry, R., Leao, P., Olveira, J.G. et al. (2007) Emergence of pathogenic betanodaviruses belonging to the SJNNV genogroup in farmed fish species from the Iberian Peninsula. *Journal of Fish Diseases* 30, 225–232.

Dalla Valle, L., Toffolo, V., Lamprecht, M., Maltese, C., Bovo, G. et al. (2005) Development of a sensitive and quantitative diagnostic assay for fish nervous necrosis virus based on two-target real-time PCR. *Veterinary Microbiology* 110, 167–179.

Fenner, B., Thiagarajan, R., Chua, H., and Kwang, J. (2006a) Betanodavirus B2 is an RNA interference antagonist that facilitates intracellular viral RNA accumulation. *Journal of Virology* 80, 85–94.

Fenner, B.J., Du, Q., Goh, W., Thiagarajan, R., Chua, H.K. et al. (2006b) Detection of betanodavirus in juvenile barramundi, *Lates calcarifer* (Bloch), by antigen capture ELISA. *Journal of Fish Diseases* 29, 423–432.

Fichi, G., Cardeti, G., Perrucci, S., Vanni, A., Cersini, A. et al. (2015) Skin lesion-associated pathogens from *Octopus vulgaris*: first detection of *Photobacterium swingsii*, *Lactococcus garvieae* and betanodavirus. *Diseases of Aquatic Organisms* 115, 147–156.

Frerichs, G. (2000) Temperature, pH and electrolyte sensitivity, and heat, UV and disinfectant inactivation of sea bass (*Dicentrarchus labrax*) neuropathy nodavirus. *Aquaculture* 185, 13–24.

Frerichs, G.N., Rodger, H.D. and Peric, Z. (1996) Cell culture isolation of piscine neuropathy nodavirus from juvenile sea bass, *Dicentrarchus labrax*. *Journal of General Virology* 77, 2067–2071.

Furusawa, R., Okinaka, Y., Uematsu, K. and Nakai, T. (2007) Screening of freshwater fish species for their susceptibility to a betanodavirus. *Diseases of Aquatic Organisms* 77, 119–125.

García-Rosado, E., Cano, I., Martín-Antonio, B., Labella, A., Manchado, M. et al. (2007) Co-occurrence of viral and bacterial pathogens in disease outbreaks affecting newly cultured sparid fish. *International Microbiology* 10, 193–199.

Giacopello, C., Foti, M., Bottari, T., Fisichella, V. and Barbera, G. (2013) Detection of viral encephalopathy and retinopathy virus (VERV) in wild marine fish species of the south Tyrrhenian Sea (central Mediterranean). *Journal of Fish Diseases* 36, 819–821.

Gomez, D.K., Sato, J., Mushiake, K., Isshiki, T., Okinaka, Y. et al. (2004) PCR-based detection of betanodaviruses from cultured and wild marine fish with no clinical signs. *Journal of Fish Diseases* 27, 603–608.

Gomez, D.K., Lim, D.J., Baeck, G.W., Youn, H.J., Shin, N.S. et al. (2006) Detection of betanodaviruses in apparently healthy aquarium fishes and invertebrates. *Journal of Veterinary Science* 7, 369–374.

Gomez, D.K., Baeck, G.W., Kim, J.H., Choresca, C.H. and Park, S.C. (2008a) Molecular detection of betanodaviruses from apparently healthy wild marine invertebrates. *Journal of Invertebrate Pathology* 97, 197–202.

Gomez, D.K., Baeck, G.W., Kim, J.H., Choresca, C.H. and Park, S.C. (2008b) Molecular detection of betanodavirus in wild marine fish populations in Korea. *Journal of Veterinary Diagnostic Investigation* 20, 38–44.

Gomez, D.K., Matsuoka, S., Mori, K., Okinaka, Y., Park, S.C. et al. (2009) Genetic analysis and pathogenicity of betanodavirus isolated from wild redspotted grouper *Epinephelus akaara* with clinical signs. *Archives of Virology* 154, 343–346.

Gomez-Casado, E., Estepa, A. and Coll, J.M. (2011) A comparative review on European-farmed finfish RNA viruses and their vaccines. *Vaccine* 29, 2657–2671.

Grotmol, S. and Totland, G.K. (2000) Surface disinfection of Atlantic halibut *Hippoglossus hippoglossus* eggs with ozonated sea-water inactivates nodavirus and increases survival of the larvae. *Diseases of Aquatic Organisms* 39, 89–96. Available at: http://www.ncbi.nlm.nih.gov/pubmed/10715814.

Grotmol, S., Totland, G.K., Thorud, K. and Hjeltnes, B.K. (1997) Vacuolating encephalopathy and retinopathy associated with a nodavirus-like agent: a probable cause of mass mortality of cultured larval and juvenile Atlantic halibut *Hippoglossus hippoglossus*. *Disease of Aquatic Organisms* 29, 85–97.

Grotmol, S., Bergh, O. and Totland, G.K. (1999) Transmission of viral encephalopathy and retinopathy (VER) to yolk-sac larvae of the Atlantic halibut *Hippoglossus hippoglossus*: occurrence of nodavirus in various organs and a possible route of infection. *Diseases of Aquatic Organisms* 36, 95–106.

Grotmol, S., Nerland, A.H., Biering, E., Totland, G.K. and Nishizawa, T. (2000) Characterization of the capsid protein gene from a nodavirus strain affecting the Atlantic halibut *Hippoglossus hippoglossus* and design of an optimal reverse-transcriptase polymerase chain reaction (RT-PCR) detection assay. *Diseases of Aquatic Organisms* 39, 79–88.

Grotmol, S., Dahl-Paulsen, E. and Totland, G.K. (2003) Hatchability of eggs from Atlantic cod, turbot and Atlantic halibut after disinfection with ozonated sea-water. *Aquaculture* 221, 245–254.

Grove, S., Johansen, R., Dannevig, B.H., Reitan, L.J. and Ranheim, T. (2003) Experimental infection of Atlantic halibut *Hippoglossus hippoglossus* with nodavirus: tissue distribution and immune response. *Diseases of Aquatic Organisms* 53, 211–221.

Grove, S., Faller, R., Soleim, K.B. and Dannevig, B.H. (2006a) Absolute quantitation of RNA by a competitive real-time RT-PCR method using piscine nodavirus as a model. *Journal of Virological Methods* 132, 104–112.

Grove, S., Johansen, R., Reitan, L.J., Press, C.M. and Dannevig, B.H. (2006b) Quantitative investigation of antigen and immune response in nervous and lymphoid tissues of Atlantic halibut (*Hippoglossus hippoglossus*) challenged with nodavirus. *Fish Shellfish Immunology* 21, 525–539.

Haddad-Boubaker, W., Sondès Boughdir, W., Sghaier, S., Souissi, J., Aida Megdich, B. et al. (2014) Outbreak of viral nervous necrosis in endangered fish species *Epinephelus costae* and *E. marginatus* in Northern Tunisian coasts. *Fish Pathology* 49, 53–56.

Hasoon, M.F., Daud, H.M., Abdullah, A.A., Arshad, S.S. and Bejo, H.M. (2011) Development and partial characterization of new marine cell line from brain of Asian sea bass *Lates calcarifer* for virus isolation. *In Vitro Cellular and Developmental Biology – Animal* 47, 16–25.

Hata, N., Okinaka, Y., Iwamoto, T., Kawato, Y., Mori, K.I. et al. (2010) Identification of RNA regions that determine temperature sensitivities in betanodaviruses. *Archives of Virology* 155, 1597–1606.

He, M. and Teng, C.B. (2015) Divergence and codon usage bias of *Betanodavirus*, a neurotropic pathogen in fish. *Molecular Phylogenetics and Evolution* 83, 137–142.

Hellberg, H., Kvellestad, A., Dannevig, B., Bornø, G., Modahl, I. et al. (2010) Outbreaks of viral nervous necrosis in juvenile and adult farmed Atlantic cod, *Gadus morhua* L., in Norway. *Journal of Fish Diseases* 33, 75–81.

Hick, P., Tweedie, A. and Whittington, R.J. (2011) Optimization of *Betanodavirus* culture and enumeration in striped snakehead fish cells. *Journal of Veterinary Diagnostic Investigation* 23, 465–475.

Hodneland, K., García, R., Balbuena, J.A., Zarza, C. and Fouz, B. (2011) Real-time RT-PCR detection of betanodavirus in naturally and experimentally infected fish from Spain. *Journal of Fish Diseases* 34, 189–202.

Húsgaro, S., Grotmol, S., Hjeltnes, B.K., Rødseth, O.M. and Biering, E. (2001) Immune response to a recombinant capsid protein of *Striped jack nervous necrosis virus* (SJNNV) in turbot *Scophthalmus maximus* and Atlantic halibut *Hippoglossus hippoglossus*, and evaluation of a vaccine against SJNNV. *Diseases of Aquatic Organisms* 45, 33–44.

Ito, Y., Okinaka, Y., Mori, K.-I., Sugaya, T., Nishioka, T. et al. (2008) Variable region of betanodavirus RNA2 is sufficient to determine host specificity. *Diseases of Aquatic Organisms* 79, 199–205.

Iwamoto, T., Nakai, T., Mori, K., Arimoto, M. and Furusawa, I. (2000) Cloning of the fish cell line SSN-1 for piscine nodaviruses. *Diseases of Aquatic Organisms* 43, 81–89.

Iwamoto, T., Okinaka, Y., Mise, K., Mori, K.-I., Arimoto, M. et al. (2004) Identification of host-specificity determinants in betanodaviruses by using reassortants between striped jack nervous necrosis virus and sevenband grouper nervous necrosis virus. *Journal of Virology* 78, 1256–1262.

Iwamoto, T., Mise, K., Takeda, A., Okinaka, Y., Mori, K.I. et al. (2005) Characterization of *Striped jack nervous necrosis virus* subgenomic RNA3 and biological activities of its encoded protein B2. *Journal of General Virology* 86, 2807–2816.

Johansen, R., Amundsen, M., Dannevig, B.H. and Sommer, A.I. (2003) Acute and persistent experimental nodavirus infection in spotted wolffish *Anarhichas minor*. *Diseases of Aquatic Organisms* 57, 35–41.

Johansen, R., Sommerset, I., Tørud, B., Korsnes, K., Hjortaas, M.J. et al. (2004) Characterization of nodavirus and viral encephalopathy and retinopathy in farmed turbot, *Scophthalmus maximus* (L.). *Journal of Fish Diseases* 27, 591–601.

Kai, Y.-H. and Chi, S.C. (2008) Efficacies of inactivated vaccines against betanodavirus in grouper larvae (*Epinephelus coioides*) by bath immunization. *Vaccine* 26, 1450–1457.

Kai, Y.H., Su, H.M., Tai, K.T. and Chi, S.C. (2010) Vaccination of grouper broodfish (*Epinephelus tukula*) reduces the risk of vertical transmission by nervous necrosis virus. *Vaccine* 28, 996–1001.

Kai, Y.H., Wu, Y.C. and Chi, S.C. (2014) Immune gene expressions in grouper larvae (*Epinephelus coioides*) induced by bath and oral vaccinations with inactivated betanodavirus. *Fish and Shellfish Immunology* 40, 563–569.

Kara, H.M., Chaoui, L., Derbal, F., Zaidi, R., de Boisséson, C. et al. (2014) Betanodavirus-associated mortalities of adult wild groupers *Epinephelus marginatus* (Lowe) and *Epinephelus costae* (Steindachner) in Algeria. *Journal of Fish Diseases* 37, 237–238.

Katharios, P. and Tsigenopoulos, C.S. (2010) First report of nodavirus outbreak in cultured juvenile shi drum, *Umbrina cirrosa* L., in Greece. *Aquaculture Research* 42, 147–152.

King, A., Adams, M., Carstens, E. and Lefkowitz, E. (2011) *Virus Taxonomy. Ninth Report of the International Committee on Taxonomy of Viruses*. Elsevier/Academic Press, London/Waltham, Massachussetts, San Diego, California, pp. 1061–1066.

Korsnes, K., Devold, M., Nerland, A.H. and Nylund, A. (2005) Viral encephalopathy and retinopathy (VER) in Atlantic salmon *Salmo salar* after intraperitoneal challenge with a nodavirus from Atlantic halibut *Hippoglossus hippoglossus*. *Diseases of Aquatic Organisms* 68, 7–15.

Korsnes, K., Karlsbakk, E., Devold, M., Nerland, A.H. and Nylund, A. (2009) Tissue tropism of nervous necrosis virus (NNV) in Atlantic cod, *Gadus morhua* L., after intraperitoneal challenge with a virus isolate

from diseased Atlantic halibut, *Hippoglossus hippoglossus* (L.). *Journal of Fish Diseases* 32, 655–665.

Kuo, H.C., Wang, T.Y., Chen, P.P., Chen, Y.M., Chuang, H.C. et al. (2011) Real-time quantitative PCR assay for monitoring of nervous necrosis virus infection in grouper aquaculture. *Journal of Clinical Microbiology* 49, 1090–1096.

Lee, K.W., Chi, S.C. and Cheng, T.M. (2002) Interference of the life cycle of fish nodavirus with fish retrovirus. *Journal of General Virology* 83, 2469–2474.

Li, J., Shi, C.Y., Huang, J., Geng, W.G., Wang, S.Q. and Su, Z.D. (2014) Complete genome sequence of a betanodavirus isolated from half-smooth tongue sole (*Cynoglossus semilaevis*). *Genome Announcements* 2, 3–4.

Liltved, H., Vogelsang, C., Modahl, I. and Dannevig, B.H. (2006) High resistance of fish pathogenic viruses to UV irradiation and ozonated seawater. *Aquacultural Engineering* 34, 72–82.

Lopez-Jimena, B., Cherif, N., Garcia-Rosado, E., Infante, C., Cano, I. et al. (2010) A combined RT-PCR and dot-blot hybridization method reveals the coexistence of SJNNV and RGNNV betanodavirus genotypes in wild meagre (*Argyrosomus regius*). *Journal of Applied Microbiology* 109, 1361–1369.

Lopez-Jimena, B., Alonso, M.D.C., Thompson, K.D., Adams, A., Infante, C. et al. (2011) Tissue distribution of red spotted grouper nervous necrosis virus (RGNNV) genome in experimentally infected juvenile European seabass (*Dicentrarchus labrax*). *Veterinary Microbiology* 154, 86–95.

Lu, M.W., Chao, Y.M., Guo, T.C., Santi, N., Evensen, O. et al. (2008) The interferon response is involved in nervous necrosis virus acute and persistent infection in zebrafish infection model. *Molecular Immunology* 45, 1146–1152.

Maeno, Y., de La Peña, L.D. and Cruz-Lacierda, E.R. (2004) Mass mortalities associated with viral nervous necrosis in hatchery-reared sea bass *Lates calcarifer* in the Philippines. *Japan Agricultural Research Quarterly* 38, 69–73.

Maeno, Y., de La Peña, L.D. and Cruz-Lacierda, E.R. (2007) Susceptibility of fish species cultured in mangrove brackish area to piscine nodavirus. *Japan Agricultural Research Quarterly* 41, 95–99.

Maltese, C. and Bovo, G. (2007) Monografie. Viral encephalopathy and retinopathy/Encefalopatia e retinopatia virale. *Ittiopatologia* 4, 93–146.

Mazelet, L., Dietrich, J. and Rolland, J.L. (2011) New RT-qPCR assay for viral nervous necrosis virus detection in sea bass, *Dicentrarchus labrax* (L.): application and limits for hatcheries sanitary control. *Fish and Shellfish Immunology* 30, 27–32.

Mézeth, K.B., Patel, S., Henriksen, H., Szilvay, A.M. and Nerland, A.H. (2009) B2 protein from betanodavirus is expressed in recently infected but not in chronically infected fish. *Diseases of Aquatic Organisms* 83, 97–103.

Mladineo, I. (2003) The immunohistochemical study of nodavirus changes in larval, juvenile and adult sea bass tissue. *Journal of Applied Ichthyology* 19, 366–370.

Moody, N.J.G., Horwood, P.F., Reynolds, A., Mahony, T.J., Anderson, I.G. *et al.* (2009) Phylogenetic analysis of betanodavirus isolates from Australian finfish. *Diseases of Aquatic Organisms* 87, 151–160.

Moreno, P., Olveira, J.G., Labella, A., Cutrín, J.M., Baro, J.C. *et al.* (2014) Surveillance of viruses in wild fish populations in areas around the Gulf of Cadiz (South Atlantic Iberian Peninsula). *Applied and Environmental Microbiology* 80, 6560–6571.

Mori, K., Nakai, T., Muroga, K., Arimoto, M., Mushiake, K. *et al.* (1992) Properties of a new virus belonging to nodaviridae found in larval striped jack (*Pseudocaranx dentex*) with nervous necrosis. *Virology* 187, 368–371.

Mori, K., Mangyoku, T., Iwamoto, T., Arimoto, M., Tanaka, S. *et al.* (2003) Serological relationships among genotypic variants of betanodavirus. *Diseases of Aquatic Organisms* 57, 19–26.

Munday, B.L., Kwang, J. and Moody, N. (2002) Betanodavirus infections of teleost fish: a review. *Journal of Fish Diseases* 25, 127–142.

Mushiake, K., Arimoto, M., Furusawa, T., Furusawa, I., Nakai, T. *et al.* (1992) Detection of antibodies against striped jack nervous necrosis virus (SJNNV) from broodstocks of striped jack. *Nippon Suisan Gakkaishi* 58, 2351–2356.

NaveenKumar, S., Shekar, M., Karunasagar, I. and Karunasagar, I. (2013) Genetic analysis of RNA1 and RNA2 of *Macrobrachium rosenbergii* nodavirus (MrNV) isolated from India. *Virus Research* 173, 377–385.

Nguyen, H.D., Mekuchi, T., Imura, K., Nakai, T., Nishizawa, T. *et al.* (1994) Occurrence of viral nervous necrosis (VNN) in hatchery-reared juvenile Japanese flounder *Paralichthys olivaceus. Fisheries Science* 60, 551–554.

Nguyen, H.D., Nakai, T. and Muroga, K. (1996) Progression of striped jack nervous necrosis virus (SJNNV) infection in naturally and experimentally infected striped jack *Pseudocaranx dentex* larvae. *Diseases of Aquatic Organisms* 24, 99–105.

Nguyen, H.D., Mushiake, K., Nakai, T. and Muroga, K. (1997) Tissue distribution of striped jack nervous necrosis virus (SJNNV) in adult striped jack. *Diseases of Aquatic Organisms* 28, 87–91.

Nishizawa, T., Toshihiro, K.M., Iwao, N. and Muroga, K. (1994) Polymerase chain reaction (PCR) amplification of RNA of striped jack nervous necrosis virus (SJNNV). *Diseases of Aquatic Organisms* 18, 103–107.

Nishizawa, T., Mori, K., Furuhashi, M., Nakai, T., Furusawa, I. *et al.* (1995) Comparison of the coat protein genes of five fish nodaviruses, the causative agents of viral nervous necrosis in marine fish. *Journal of General Virology* 76, 1563–1569.

Nishizawa, T., Furuhashi, M., Nagai, T., Nakai, T. and Muroga, K. (1997) Genomic classification of fish nodaviruses by molecular phylogenetic analysis of the coat protein gene. *Applied and Environmental Microbiology* 63, 1633–1636.

Nishizawa, T., Kokawa, Y., Wakayama, T., Kinoshita, S. and Yoshimizu, M. (2008) Enhanced propagation of fish nodaviruses in BF-2 cells persistently infected with snakehead retrovirus (SnRV). *Diseases of Aquatic Organisms* 79, 19–25.

Nishizawa, T., Gye, H.J., Takami, I. and Oh, M.J. (2012) Potentiality of a live vaccine with nervous necrosis virus (NNV) for sevenband grouper *Epinephelus septemfasciatus* at a low rearing temperature. *Vaccine* 30, 1056–1063.

Nuñez-Ortiz, N., Stocchi, V., Toffan, A, Pascoli, F., Sood, N. *et al.* (2016) Quantitative immunoenzymatic detection of viral encephalopathy and retinopathy virus (betanodavirus) in sea bass *Dicentrarchus labrax. Journal of Fish Diseases* 39, 821–831.

Nylund, A., Karlsbakk, E., Nylund, S., Isaksen, T.E., Karlsen, M. *et al.* (2008) New clade of betanodaviruses detected in wild and farmed cod (*Gadus morhua*) in Norway. *Archives of Virology* 153, 541–547.

Ødegård, J., Sommer, A.-I. and Præbel, A.K. (2010) Heritability of resistance to viral nervous necrosis in Atlantic cod (*Gadus morhua* L.). *Aquaculture* 300, 59–64.

Oh, M.-J., Jung, S.-J., Kim, S.-R., Rajendran, K.V, Kim, Y.-J. *et al.* (2002) A fish nodavirus associated with mass mortality in hatchery-reared red drum, *Sciaenops ocellatus. Aquaculture* 211, 1–7.

Oh, M.-J., Gye, H.J. and Nishizawa, T. (2013) Assessment of the sevenband grouper *Epinephelus septemfasciatus* with a live nervous necrosis virus (NNV) vaccine at natural seawater temperature. *Vaccine* 31, 2025–2027.

OIE (2013) Chapter 2.3.11.Viral encephalopathy and retinopathy. In: *Manual of Diagnostic Tests for Aquatic Animals*. World Organization for Animal Health, Paris. Updated 2016 version available as Chapter 2.3.12. at: http://www.oie.int/fileadmin/Home/eng/Health_standards/ aahm/current/chapitre_viral_encephalopathy_ retinopathy.pdf (accessed 4 November 2016).

Olveira, J.G., Soares, F., Engrola, S., Dopazo, C.P. and Bandín, I. (2008) Antemortem versus postmortem methods for detection of betanodavirus in Senegalese sole (*Solea senegalensis*). *Journal of Veterinary Diagnostic Investigation* 20, 215–219.

Olveira, J.G., Souto, S., Dopazo, C.P., Thiéry, R., Barja, J.L. *et al.* (2009) Comparative analysis of both genomic segments of betanodaviruses isolated from epizootic outbreaks in farmed fish species provides evidence for genetic reassortment. *Journal of General Virology* 90, 2940–2951.

Pakingking, R., Bautista, N.B., de Jesus-Ayson, E.G. and Reyes, O. (2010) Protective immunity against viral nervous necrosis (VNN) in brown-marbled grouper (*Epinephelus fuscogutattus*) following vaccination with inactivated betanodavirus. *Fish and Shellfish Immunology* 28, 525–533.

Panzarin, V., Patarnello, P., Mori, A., Rampazzo, E., Cappellozza *et al.* (2010) Development and validation of a real-time TaqMan PCR assay for the detection of betanodavirus in clinical specimens. *Archives of Virology* 155, 1193–1203.

Panzarin, V., Fusaro, A., Monne, I., Cappellozza, E., Patarnello, P. *et al.* (2012) Molecular epidemiology and evolutionary dynamics of betanodavirus in southern Europe. *Infection, Genetics and Evolution* 12, 63–70.

Panzarin, V., Cappellozza, E., Mancin, M., Milani, A., Toffan, A. *et al.* (2014) *In vitro* study of the replication capacity of the RGNNV and the SJNNV betanodavirus genotypes and their natural reassortants in response to temperature. *Veterinary Research* 45, 56.

Panzarin, V., Toffan, A., Abbadi, M., Buratin, A., Mancin, M. *et al.* (2016) Molecular basis for antigenic diversity of genus *Betanodavirus*. *PLoS One* 11(7): e0158814.

Parameswaran, V., Kumar, S.R., Ahmed, V.P.I. and Hameed, A.S.S. (2008) A fish nodavirus associated with mass mortality in hatchery-reared Asian Sea bass, *Lates calcarifer*. *Aquaculture* 275, 366–369.

Pascoli, F., Serra, M., Toson, M., Pretto, T. and Toffan, A. (2016) Betanodavirus ability to infect juvenile European sea bass, *Dicentrarchus labrax*, at different water salinity. *Journal of Fish Diseases* 39, 1061–1068.

Patel, S., Korsnes, K., Bergh, Ø., Vik-Mo, F., Pedersen, J. *et al.* (2007) Nodavirus in farmed Atlantic cod *Gadus morhua* in Norway. *Diseases of Aquatic Organisms* 77, 169–173.

Péducasse, S., Castric, J., Thiéry, R., Jeffroy, J., Le Ven, A. *et al.* (1999) Comparative study of viral encephalopathy and retinopathy in juvenile sea bass *Dicentrarchus labrax* infected in different ways. *Diseases of Aquatic Organisms* 36, 11–20.

Pirarat, N., Ponpornpisit, A., Traithong, T., Nakai, T., Katagiri, T. *et al.* (2009a) Nodavirus associated with pathological changes in adult spotted coralgroupers (*Plectropomus maculatus*) in Thailand with viral nervous necrosis. *Research in Veterinary Science* 87, 97–101.

Pirarat, N., Katagiri, T., Maita, M., Nakai, T. and Endo, M. (2009b) Viral encephalopathy and retinopathy in hatchery-reared juvenile thread-sail filefish (*Stephanolepis cirrhifer*). *Aquaculture* 288, 349–352.

Ransangan, J. and Manin, B.O. (2012) Genome analysis of *Betanodavirus* from cultured marine fish species in Malaysia. *Veterinary Microbiology* 156, 16–44.

Rigos, G. and Katharios, P. (2009) Pathological obstacles of newly-introduced fish species in Mediterranean mariculture: a review. *Reviews in Fish Biology and Fisheries* 20, 47–70.

Sakamoto, T., Okinaka, Y., Mori, K.I., Sugaya, T., Nishioka, T. *et al.* (2008) Phylogenetic analysis of Betanodavirus RNA2 identified from wild marine fish in oceanic regions. *Fish Pathology* 43, 19–27.

Sano, M., Nakai, T. and Fijan, N. (2011) Viral diseases and agents of warmwater fish. In: Woo, P. and Bruno, D.W. (eds) *Fish Disease and Disorders. Volume 3: Viral, Bacterial and Fungal Infections*. CAB International, Wallingford, UK, pp. 198–207.

Scapigliati, G., Buonocore, F., Randelli, E., Casani, D., Meloni, S. *et al.* (2010) Cellular and molecular immune responses of the sea bass (*Dicentrarchus labrax*) experimentally infected with betanodavirus. *Fish and Shellfish Immunology* 28, 303–311.

Shetty, M., Maiti, B., Shivakumar Santhosh, K., Venugopal, M.N. and Karunasagar, I. (2012) Betanodavirus of marine and freshwater fish: distribution, genomic organization, diagnosis and control measures. *Indian Journal of Virology* 23, 114–123.

Skliris, G.P. and Richards, R.H. (1999) Induction of nodavirus disease in seabass, *Dicentrarchus labrax*, using different infection models. *Virus Research* 63, 85–93.

Skliris, G.P., Krondiris, J.V, Sideris, D.C., Shinn, A.P., Starkey, W.G. *et al.* (2001) Phylogenetic and antigenic characterization of new fish nodavirus isolates from Europe and Asia. *Virus Research* 75, 59–67.

Sohn, S.-G., Park, M.-A., Oh, M.-J. and Cho, S.-K. (1998) A fish nodavirus isolated from cultured sevenband grouper, *E. septemfasciatus*. *Journal of Fish Pathology* 11, 97–104.

Sommerset, I. and Nerland, A.H. (2004) Complete sequence of RNA1 and subgenomic RNA3 of Atlantic halibut nodavirus (AHNV). *Diseases of Aquatic Organisms* 58, 117–125.

Souto, S., Lopez-Jimena, B., Alonso, M.C., García-Rosado, E. and Bandín, I. (2015a) Experimental susceptibility of European sea bass and Senegalese sole to different betanodavirus isolates. *Veterinary Microbiology* 177, 53–61.

Souto, S., Olveira, J.G. and Bandín, I. (2015b) Influence of temperature on Betanodavirus infection in Senegalese sole (*Solea senegalensis*). *Veterinary Microbiology* 179, 162–167.

Starkey, W.G., Ireland, J.H., Muir, K.F., Jenknis, M.E., Roy, W.J. *et al.* (2001) Nodavirus infection in Atlantic cod and Dover sole in the UK. *Veterinary Record* 149, 179–181.

Tanaka, S., Mori, K., Arimoto, M., Iwamoto, T. and Nakai, T. (2001) Protective immunity of sevenband grouper, *Epinephelus septemfasciatus* Thunberg, against experimental viral nervous necrosis. *Journal of Fish Diseases* 24, 15–22. doi:10.1046/j.1365-2761.2001.00259.x.

Tanaka, S., Takagi, M. and Miyazaki, T. (2004) Histopathological studies on viral nervous necrosis of sevenband grouper, *Epinephelus septemfasciatus* Thunberg, at the grow-out stage. *Journal of Fish Diseases* 27, 385–399. doi:10.1111/j.1365-2761.2004.00559.x.

Thiéry, R., Raymond, J.C. and Castric, J. (1999) Natural outbreak of viral encephalopathy and retinopathy in juvenile sea bass, *Dicentrarchus labrax*: study by nested reverse transcriptase-polymerase chain reaction. *Virus Research* 63, 11–17.

Thiéry, R., Cozien, J., de Boisséson, C., Kerbart-Boscher, S. and Névarez, L. (2004) Genomic classification of new betanodavirus isolates by phylogenetic analysis of the coat protein gene suggests a low host-fish species specificity. *Journal of General Virology* 85, 3079–3087.

Toffolo, V., Negrisolo, E., Maltese, C., Bovo, G., Belvedere, P. *et al.* (2007) Phylogeny of betanodaviruses and molecular evolution of their RNA polymerase and coat proteins. *Molecular Phylogenetics and Evolution* 43, 298–308.

Totland, G.K., Grotmol, S., Morita, Y., Nishioka, T. and Nakai, T. (1999) Pathogenicity of nodavirus strains from striped jack *Pseudocaranx dentex* and Atlantic halibut *Hippoglossus hippoglossus*, studied by waterborne challenge of yolk-sac larvae of both teleost species. *Diseases of Aquatic Organisms* 38, 169–175.

Ucko, M., Colorni, A. and Diamant, A. (2004) Nodavirus infections in Israeli mariculture. *Journal of Fish Diseases* 27, 459–469.

Vendramin, N., Padrós, F., Pretto, T., Cappellozza, E., Panzarin, V. *et al.* (2012) Viral encephalopathy and retinopathy outbreak in restocking facilities of the endangered freshwater species, *Salaria fluviatilis* (Asso). *Journal of Fish Diseases* 35, 867–871.

Vendramin, N., Patarnello, P., Toffan, A., Panzarin, V., Cappellozza, E. *et al.* (2013) Viral encephalopathy and retinopathy in groupers (*Epinephelus* spp.) in southern Italy: a threat for wild endangered species? *BMC Veterinary Research* 9:20.

Vendramin, N., Toffan, A., Mancin, M., Cappellozza, E., Panzarin, V. *et al.* (2014) Comparative pathogenicity study of ten different betanodavirus strains in experimentally infected European sea bass, *Dicentrarchus labrax* (L.). *Journal of Fish Diseases* 37, 371–383.

Walker, P.J. and Winton, J.R. (2010) Emerging viral diseases of fish and shrimp. *Veterinary Research* 41, 51.

Watanabe, K.I., Nishizawa, T. and Yoshimizu, M. (2000) Selection of brood stock candidates of barfin flounder using an ELISA system with recombinant protein of barfin flounder nervous necrosis virus. *Diseases of Aquatic Organisms* 41, 219–223.

Wi, G.R., Hwang, J.Y., Kwon, M.G., Kim, H.J., Kang, H.A. *et al.* (2015) Protective immunity against nervous necrosis virus in convict grouper *Epinephelus septemfasciatus* following vaccination with virus-like particles produced in yeast *Saccharomyces cerevisiae*. *Veterinary Microbiology* 177, 214–218.

Zorriehzahra, M.J., Nazari, A., Ghasemi, M., Ghiasi, M., Karsidani, S.H. *et al.* (2014) Vacuolating encephalopathy and retinopathy associated with a nodavirus-like agent: a probable cause of mass mortality of wild Golden grey mullet (*Liza aurata*) and Sharpnose grey mullet (*Liza saliens*) in Iranina waters of the Caspian sea. *Virus Diseases* 25, 430–436.

A. Toffan

12 Iridoviral Diseases: Red Sea Bream Iridovirus and White Sturgeon Iridovirus

Yasuhiko Kawato,[1] Kuttichantran Subramaniam,[2] Kazuhiro Nakajima,[1] Thomas Waltzek[2] and Richard Whittington[3]*

[1]National Research Institute of Aquaculture, Japan Fisheries Research and Education Agency, Nakatsuhamaura, Minami-Ise, Mie, Japan; [2]Department of Infectious Diseases and Pathology, University of Florida, Gainesville, Florida, USA; [3]Faculty of Veterinary Science, University of Sydney, Camden, New South Wales, Australia

12.1 Red Sea Bream Iridovirus

12.1.1 Introduction

The red sea bream iridovirus (RSIV) has a double-stranded DNA genome in an icosahedral virion capsid that is 200 nm in diameter. According to the International Committee on the Taxonomy of Viruses (ICTV), it is in the genus *Megalocytivirus* within the family *Iridoviridae*, although it has not yet been approved as a species within that genus (Jancovich *et al.*, 2012).

Red sea bream iridoviral disease (RSIVD) was first detected causing mass mortality among cultured red sea bream (*Pagrus major*) in the summer of 1990 in Japan, (Inouye *et al.*, 1992). The outbreak ceased in November when the water temperature fell below 20°C. The disease recurred in the next summer, not only in red sea bream, but also in other marine cultured species such as yellowtail or Japanese amberjack (*Seriola quinqueradiata*), greater amberjack (*S. dumerili*), sea bass (*Lateolabrax* sp.), striped jack (*Pseudocaranx dentex*) and barred knifejaw (*Oplegnathus fasciatus*) (Matsuoka *et al.*, 1996). Since then, RSIVD has spread to more than 30 species of cultured marine fishes and has occurred every summer, causing substantial economic losses in mariculture within Japan (Kawakami and Nakajima, 2002). Similar viral diseases have been reported in many cultured fishes in East and South-east Asia since the 1990s, and the causative viruses are also megalocytiviruses (Kurita and Nakajima, 2012). These viruses include the Taiwan grouper iridovirus (TGIV) from a species of grouper (*Epinephelus* sp.) (Chou *et al.*, 1998; Chao *et al.*, 2004), ISKNV from the freshwater mandarin fish (*Siniperca chuatsi*) (He *et al.*, 2000), rock bream iridovirus (RBIV) from the barred knifejaw (Jung and Oh, 2000) and turbot reddish body iridovirus (TRBIV) from the turbot (*Scophthalmus maximus*) (Shi *et al.*, 2004). *Infectious spleen and kidney necrosis virus* (ISKNV) is the type species of *Megalocytivirus* (Jancovich *et al.*, 2012).

These viruses are genetically classified into three groups according to the nucleotide sequence of the major capsid protein gene (Kurita and Nakajima, 2012). These are: the RSIV type, which are mainly reported in marine fishes (Inouye *et al.*, 1992; Jung and Oh, 2000; Chen *et al.*, 2003; Gibson-Kueh *et al.*, 2004; Lü *et al.*, 2005; Dong *et al.*, 2010; Huang *et al.*, 2011); the ISKNV type, which are reported in both marine and freshwater fish

---

*Corresponding author e-mail: richard.whittington@sydney.edu.au

(He *et al.*, 2000, 2002; Weng *et al.*, 2002; Chao *et al.*, 2004; Wang *et al.*, 2007; Fu *et al.*, 2011; Huang *et al.*, 2011); and the TRBIV type, which are reported in flatfish (Shi *et al.*, 2004; Kim *et al.*, 2005; Do *et al.*, 2005a; Oh *et al.*, 2006). The RSIV-type viruses are widely distributed in East and South-east Asia, and all viruses in Japan are of the RSIV type. Viruses of the ISKNV type are from countries in South-east Asia, China and Taiwan, and reports of the TRBIV type are limited to China and South Korea (Kurita and Nakajima, 2012). Currently, RSIVD caused by the RSIV and ISKNV type viruses is listed by the World Organisation for Animal Health (OIE), but TRBIV is not included because the pathogenicity of the virus is still debatable (OIE, 2012).

An outbreak of RSIVD often spreads from one net pen to adjacent pens. Transmission is by cohabitation, which has been confirmed experimentally (He *et al.*, 2002). However, RSIVD has not been reported in hatchery fish; hence, the risk of vertical transmission through eggs or sperm is low (Nakajima and Kurita, 2005). Once the disease has occurred, outbreaks recur every summer in the same area. This is in accord with the observation that the optimal *in vitro* growth temperature of RSIV is 25°C (Nakajima and Sorimachi, 1994). RSIV-infected fish can recover when they are reared at low water temperatures (<18°C) for more than 100 days, and survivors become resistant to RSIVD (Oh *et al.*, 2014; Jung *et al.*, 2015). In addition, RSIVD is most often reported among the yearlings of many species (Matsuoka *et al.*, 1996; Kawakami and Nakajima, 2002). Annual outbreaks of RSIVD in fish newly introduced into net pens are probably caused by horizontal transmission from older recovered fish that still harbour the virus. Ito *et al.*, (2013) found the viral genome in the spleen, kidneys, heart, gills, intestine and caudal fin of surviving Japanese amberjacks that were experimentally infected with RSIV. Around aquaculture areas, the viral genome was detected using PCR from many wild fishes (six orders, 18 families and 39 species) that had no clinical signs (Wang *et al.*, 2007); this implies that such fish can also be virus carriers. Recently, Jin *et al.* (2014) suggested that some bivalves living around fish culture sites were potential vectors.

Megalocytiviruses are not host specific (Table 12.1); for example, they infect ornamental fishes such as the African lampeye (*Aplocheilichthys normani*) and dwarf gourami (*Colisa lalia*) in South-east Asia

(Sudthongkong *et al.*, 2002a, Jeong *et al.*, 2008a, Weber *et al.*, 2009; Kim *et al.*, 2010; Sriwanayos *et al.*, 2013; Nolan *et al.*, 2015). The histopathology of these diseases is similar to those of RSIVD and the causative agents belong to the ISKNV type (Sudthongkong *et al.*, 2002b; Weber *et al.*, 2009; Kurita and Nakajima, 2012; Sriwanayos *et al.*, 2013). In addition, a new *Megalocytivirus* to be formally known as the threespine stickleback iridovirus (TSIV) has been reported from the three-spine stickleback (*Gasterosteus aculeatus*) (Waltzek *et al.*, 2012).

12.1.2 Diagnosis

Infected fish are dark in colour, lethargic and often have conspicuous respiratory movements owing to severe anaemia. Slight exophthalmos (protrusion of the eyeballs) and haemorrhage from the skin are occasionally observed. Common signs include severe anaemia, petechia in the gills, an enlarged spleen, the presence of a haemorrhagic exudate in the pericardial cavity and pale internal organs (Inouye *et al.*, 1992). In natural infections, the disease is usually chronic and cumulative mortality often reaches 20–60% in 1–2 months (Nakajima *et al.*, 1999).

Abnormally enlarged cells, which are typical of RSIVD, can be confirmed by microscopic examinations of Giemsa-stained impression smears of the spleen. This is the simplest and most rapid method for presumptive diagnosis (Fig. 12.1), with the caution that enlarged cells are rarely found in some infected fishes.

Immunofluorescence staining using specific monoclonal antibody (M10) was developed for the diagnosis of RSIVD (Nakajima and Sorimachi, 1995; Nakajima *et al.*, 1995). Although the TRBIV type has not been examined, M10 antibody reacts with many megalocytiviruses, including RSIV and ISKNV, but does not react with other iridoviruses, such as *Frog virus 3* (FV3), *Epizootic haematopoietic necrosis virus* (EHNV), European sheatfish virus (ESV) and Singapore grouper iridovirus (SGIV) (Nakajima and Sorimachi, 1995; Nakajima *et al.*, 1998). The indirect fluorescent antibody test (IFAT) using M10 antibody is listed in the OIE *Manual of Diagnostic Tests for Aquatic Animals* (OIE, 2012) as a confirmatory diagnostic method for RSIVD and is widely used in Japan because it is rapid and reliable (Matsuoka *et al.*, 1996; Kawakami and Nakajima, 2002).

Table 12.1. Fish host range of red sea bream iridovirus (RSIV) and other megalocytiviruses.[a]

Order	Family	No. species	Representative fish species	Virus	OIE[b]	References
Characiformes	Serrasalmidae	1	*Metynnis argenteus*	Unknown		Nolan et al., 2015
Cyprinodontiformes	Nothobranchiidae	1	*Aphyosemion gardneri*	Unknown		Nolan et al., 2015
	Poeciliidae	4	*Aplocheilichthys normani*	ISKNV, unknown		Paperna et al., 2001; Sudthongkong et al., 2002a; Nolan et al., 2015
Mugiliformes	Mugilidae	1	*Mugil cephalus*	ISKNV	+	Gibson-Kueh et al., 2004
Perciformes	Apogonidae	1	*Pterapogon kauderni*	ISKNV		Weber et al., 2009
	Belontiidae	2	*Trichogaster leeri*	ISKNV		Jeong et al., 2008a
	Carangidae	7	*Seriola quinqueradiata*	RSIV	+	Kawakami and Nakajima, 2002
	Centrarchidae	1	*Micropterus salmoides*	ISKNV	+	He et al., 2002
	Cichlidae	8	*Pterophyllum scalare*	Unknown		Armstrong and Ferguson, 1989; Rodger et al., 1997; Nolan et al., 2015
	Eleotridae	1	*Oxyeleotris marmoratus*	ISKNV		Wang et al., 2011
	Ephippidae	1	*Platax orbicularis*	ISKNV		Sriwanayos et al., 2013
	Haemulidae	2	*Parapristipoma trilineatum*	RSIV	+	Kawakami and Nakajima, 2002
	Helostomatidae	1	*Helostoma temminckii*	Unknown		Nolan et al., 2015
	Kyphosidae	1	*Girella punctata*	RSIV	+	Kawakami and Nakajima, 2002
	Lateolabracidae	2	*Lateolabrax japonicus*	RSIV	+	Kawakami and Nakajima, 2002; Do et al., 2005b
	Latidae	1	*Lates calcarifer*	RSIV, ISKNV	+	Huang et al., 2011
	Lethrinidae	2	*Lethrinus haematopterus*	RSIV	+	Kawakami and Nakajima, 2002
	Moronidae	1	*Morone saxatilis × M. chrysops*	RSIV	+	Kurita and Nakajima, 2012
	Oplegnathidae	2	*Oplegnathus fasciatus*	RSIV	+	Kawakami and Nakajima, 2002
	Osphronemidae	3	*Colisa lalia*	ISKNV, unknown		Paperna et al., 2001; Sudthongkong et al., 2002a; Kim et al., 2010,
	Percichthyidae	2	*Siniperca chuatsi*	ISKNV	+	He et al., 2000; Go et al., 2006
	Rachycentridae	1	*Rachycentron canadum*	RSIV	+	Kawakami and Nakajima, 2002
	Sciaenidae	2	*Larimichthys crocea*	RSIV, ISKNV	+	Weng et al., 2002; Chen et al., 2003
	Scombridae	3	*Thunnus thynnus*	RSIV	+	Kawakami and Nakajima, 2002
	Serranidae	10	*Epinephelus septemfasciatus*	RSIV, ISKNV	+	Kawakami and Nakajima, 2002; Sudthongkong et al., 2002b; Chao et al., 2004; Gibson-Kueh et al., 2004; Huang et al., 2011
	Sparidae	4	*Pagrus major*	RSIV	+	Inouye et al., 1992; Kawakami and Nakajima, 2002; Kurita and Nakajima, 2012
Pleuronectiformes	Paralichthyidae	1	*Paralichthys olivaceus*	RSIV, TRBIV	+	Kawakami and Nakajima, 2002; Do et al., 2005a
	Pleuronectidae	2	*Verasper variegatus*	RSIV, TRBIV		Kawakami and Nakajima, 2002; Won et al., 2013
	Scophthalmidae	1	*Scophthalmus maximus*	TRBIV		Shi et al., 2004
Scorpaeniformes	Sebastidae	2	*Sebastes schlegeli*	RSIV	+	Kim et al., 2002
Tetraodontiformes	Monacanthidae	1	*Stephanolepis cirrhifer*	RSIV		This chapter
	Tetraodontidae	1	*Takifugu rubripes*	RSIV	+	Kawakami and Nakajima, 2002

[a]ISKNV, *Infectious spleen and kidney necrosis virus*; RSIV, red sea bream iridovirus; TRBIV, turbot reddish body iridovirus.
[b]+ indicates listed in the OIE Manual (OIE, 2012).

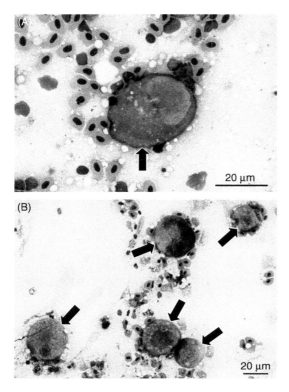

Fig. 12.1. Large, virus-infected cells in an impression smear of the spleen of red sea bream iridovirus (RSIV)-infected fish: (A) from the red sea bream (*Pagrus major*); (B) from the greater amberjack (*Seriola dumerili*). Arrowheads indicate presumptive virus-infected cells. Giemsa stain.

Several PCR primers have been developed (Kurita *et al.*, 1998; Oshima *et al.*, 1998; Jeong *et al.*, 2004; Go *et al.*, 2006; Rimmer *et al.*, 2012), and the polymerase chain reaction (PCR) assay is used for diagnosis and surveys in wild fishes (Wang *et al.*, 2007). Real-time PCR for the quantitative detection of RSIV is rapid, specific and sensitive (Caipang *et al.*, 2003; Wang *et al.*, 2006; Gias *et al.*, 2011; Rimmer *et al.*, 2012; Oh and Nishizawa, 2013).

RSIV was first isolated from splenic homogenates of diseased fish using RTG-2 (rainbow trout gonad), CHSE-214 (Chinook salmon embryo), FHM (fathead minnow), BF-2 (blue gill fry) and KRE-3 (kelp and red-spotted grouper embryo) cells (Inouye *et al.*, 1992). The cytopathic effect (CPE) of RSIV is characterized by rounding and enlargement of the infected cells. However, the sensitivity of

these cell lines is low and viral titres gradually decrease through serial passages (Nakajima and Sorimachi, 1994). The Grunt fin (GF) cell line is more sensitive to RSIV and it is often used for the isolation and propagation of the virus (Jung *et al.*, 1997; Nakajima and Maeno, 1998; Ito *et al.*, 2013). Nevertheless, complete CPE may not occur in RSIV-inoculated cells, even in GF cells, and it is difficult to recognize any CPE when the viral concentration is low (Ito *et al.*, 2013). Several other fish cell lines are sensitive to RSIV and ISKNV, (Imajoh *et al.*, 2007; Dong *et al.*, 2008; Wen *et al.*, 2008), but they are not available from the American Type Culture Collection (ATCC) or the European Collection of Authenticated Cell Cultures (ECACC), and are not otherwise publicly available.

12.1.3 Pathology

The most typical histopathological change is the presence of enlarged cells in the spleen, heart, kidneys, liver or gills. The cells are basophilically stained with haematoxylin in paraffin sections (Fig. 12.2), whereas they are often stained heterogeneously with May–Grunwald–Giemsa stain in impression smears (Fig. 12.1). A large nucleus-like structure is often seen in moderately enlarged cells. These abnormally enlarged cells are infected. They have been reported in RSIVD (Inouye *et al.*, 1992; Jung *et al.*, 1997; He *et al.*, 2000; Jung and Oh, 2000) and other megalocytiviral infections (Sudthongkong *et al.*, 2002a, Shi *et al.*, 2004; Weber *et al.*, 2009; Sriwanayos *et al.*, 2013). The most severe lesions are usually in the spleen, where there are often a large number of the virus-infected enlarged cells, diffuse tissue necrosis and a decrease in melanomacrophage centres. Many abnormally enlarged cells also appear in the lymphoid tissues of the kidney interstitium and glomeruli, endocardium and central venous sinus, and beneath the epithelium of the gill filaments (Inouye *et al.*, 1992). These enlarged cells stain positively using the IFAT (Nakajima *et al.*, 1995) or by *in situ* hybridization (ISH) (Chao *et al.*, 2004). In ISH, the hybridization signal appears first in the nucleus in an RSIV-infected cell, and the cytoplasm becomes positive in the later stages of infection (Chao *et al.*, 2004).

In electron microscopy, icosahedral virions (approximately 200 nm in diameter) are found in the cytoplasm of enlarged cells. Each virion consists

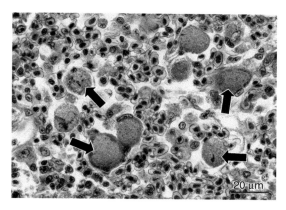

Fig. 12.2. The spleen of red sea bream (*Pagrus major*) with naturally occurring red sea bream iridovirus disease (RSIVD). Arrowheads indicate presumptive virus-infected cells. Haematoxylin and eosin stain.

of a central electron-dense core (120 nm) and an electron-translucent zone. In the virus-infected cells, enlargement of the cell and nuclear degeneration are observed before virion formation (Inouye *et al.*, 1992).

12.1.4 Pathophysiology

RSIV has apparent tissue tropism to the spleen, which is where enlarged cells first appear (Chao *et al.*, 2004) and the number of viral genomes first increases (Zhang *et al.*, 2012; Ito *et al.*, 2013), but enlarged cells are rarely observed in the brain, spinal cord and ganglia (Jung *et al.*, 1997; Chao *et al.*, 2004). Multiple organ failure may occur in later stages of the disease when large numbers of viral genomes are detected in the spleen, kidneys, heart, gills, fins and intestine (Ito *et al.*, 2013). Nevertheless, the parenchyma of these organs, except for the spleen, appears intact. The virus-infected, enlarged cells are often in haematopoietic tissues such as the spleen and kidney, and throughout the blood vessels of various organs. These observations, together with the fact that virus-infected cells are not found among the cells composing tissue parenchyma, suggest that the targets of RSIV are haematopoietic cells. By damaging these cells, RSIV probably causes aplastic anaemia, which contributes to death. Further investigation is needed to identify the target cells of RSIV, and to clarify the cause of death.

12.1.5 Protective and control strategies

Vaccination against RSIVD is available but there is no chemotherapeutic agent. The vaccine for red sea bream has been sold in Japan since 1999 and is the first commercially available vaccine against a viral disease in marine fish. It consists of a formalin-inactivated culture supernatant of GF cells inoculated with RSIV. The effectiveness of the vaccine has been confirmed experimentally and in field trials (Nakajima *et al.*, 1997, 1999). Currently, the vaccine is used in Japan for red sea bream, yellowtail, greater amberjack, striped jack, Malabar grouper (*Epinephelus malabaricus*), orange-spotted grouper (*E. coioides*), longtooth grouper (*E. bruneus*) and sevenband grouper (*E. septemfasciatus*). The frequency of outbreaks of RSIVD in Japan has dramatically decreased since the vaccine was approved. The cost of the vaccine increases the expense of fish production and, consequently, more economic recombinant subunit vaccines and DNA vaccines for RSIV have been studied, but they are not commercially available (Caipang *et al.*, 2006; Kim *et al.*, 2008; Shimmoto *et al.*, 2010; Fu *et al.*, 2012, 2014) and their efficacies are not satisfactory. Recently, another inactivated vaccine has been studied that was produced in a fish cell line (from mandarin fish) which yields more virus (Dong *et al.*, 2013a,b). There is another problem in that for the vaccine currently sold in Japan, the efficacy in barred knifejaw and the spotted knifejaw (*O. punctatus*) is still inadequate. To overcome this problem, live vaccine has been suggested because virus-inoculated fish develop immunity when they are reared at low water temperatures (<18°C) (Oh *et al.*, 2014; Jung *et al.*, 2015).

Fasting for more than 10 days reduces RSIVD mortality (Tanaka *et al.*, 2003). Even though the mechanism of this reduction is still unclear, fasting is often practised in fish farms after the onset of RSIVD, but it leads to considerable weight loss.

RSIVD most often occurs in yearling fish and fish seem to acquire some resistance as they grow larger (Matsuoka *et al.*, 1996; Kawakami and Nakajima, 2002). Therefore, rapid growth of juveniles before the summer helps to avoid severe RSIVD.

12.1.6 Conclusions

RSIVD suddenly occurred in cultured red sea bream in a small area in Japan in 1990. It spread rapidly to other areas and other fish species in subsequent years (Matsuoka *et al.*, 1996). All of the viral isolates in Japan are of the RSIV type, suggesting that the virus is likely to have been introduced into Japan via imported fish. It currently occurs in coastal south-western Japan, and in East and South-east Asia, and its lack of host specificity leads to disease in many species whenever the water temperatures are conducive for the virus.

Megalocytiviruses have been reported in numerous fishes (Table 12.1), including ornamental fishes from South-east Asia (Sudthongkong *et al.*, 2002a, Jeong *et al.*, 2008a; Weber *et al.*, 2009; Kim *et al.*, 2010; Sriwanayos *et al.*, 2013; Subramaniam *et al.*, 2014; Nolan *et al.*, 2015). A similar viral infection (characterized by the formation of basophilic enlarged cells) in ornamental fish was confirmed before RSIVD was found in Japan, though the virus was not identified (Armstrong and Ferguson, 1989). Therefore, the possibility of the transmission of megalocytivirus by ornamental fishes through international trade is worrisome (Jeong *et al.*, 2008a). For example, an outbreak of megalocytivirus infection occurred in cultured Murray cod (*Maccullochella peelii peelii*) via the importation of dwarf gourami into Australia (Go *et al.*, 2006; Go and Whittington, 2006). Also, the megalocytivirus from pearl gourami (*Trichogaster leeri*) was highly pathogenic to barred knifejaw (Jeong *et al.*, 2008b).

12.2 White Sturgeon Iridovirus

12.2.1 Introduction

Over the last century, increased demand for caviar (traditionally from wild sturgeon in the Black and Caspian seas) and related habitat degradation (e.g. pollution and river damming) have resulted in the International Union for Conservation of Nature (IUCN) listing of several species, notably sturgeons and paddlefishes, as critically endangered (Birstein 1993; Pikitch *et al.*, 2005). The development of global sturgeon aquaculture for enhancing the food and stock availability has sought to alleviate pressures on these diminished wild stock.

The emergence of a lethal viral disease of the integument, gills, and upper alimentary tract in cultured juvenile white sturgeon (*Acipenser transmontanus*), known as white sturgeon iridovirus (WSIV), has had a severe impact on commercial farms in the Pacific Northwest of North America

since 1988 (Hedrick *et al.*, 1990; LaPatra *et al.*, 1994, 1999; Raverty *et al.*, 2003; Drennan *et al.*, 2006; Kwak *et al.*, 2006). WSIV was first reported in juvenile white sturgeon reared on farms in northern and central California where it produced annual losses of up to 95% (Hedrick *et al.*, 1990, 1992). Subsequently, WSIV was reported in cultivated white sturgeon from the lower Columbia River in Oregon and Washington, the Snake River in southern Idaho and the Kootenai River in northern Idaho (LaPatra *et al.*, 1994, 1999). In 2001, white sturgeon fingerlings from an aquaculture facility in British Columbia, Canada, were infected, which are likely to have originated from brood stock obtained from the Fraser River drainage basin in that province (Raverty *et al.*, 2003).

Experimental studies suggested that the horizontal transmission of WSIV can occur via contaminated water (Hedrick *et al.*, 1990, 1992). Mortality in juvenile white sturgeon begins at 10 days and peaks between 15 to 30 days after viral exposure. Although adult white sturgeon appear to be resistant to clinical disease, persistently infected adults are believed to vertically transmit WSIV to their offspring. LaPatra *et al.* (1994) suggested that the sources of WSIV in epizootics among white sturgeon hatcheries in Idaho and Oregon were wild brood stock. The epizootiology of WSIV on a Californian sturgeon farm support vertical transmission to be more important than tank-to-tank transmission (Georgiadis *et al.*, 2001). Finally, lake sturgeon (*A. fluvescens*) can be experimentally infected, but do not express clinical disease (Hedrick *et al.*, 1992).

A similar lethal virus of captive juvenile pallid sturgeon (*Scaphirhynchus albus*) and the shovelnose sturgeon (*S. platorhynchus*), known as the Missouri River sturgeon iridovirus (MRSIV), has negatively affected stock enhancement programmes in hatcheries within the Missouri River Basin (Kurobe *et al.*, 2010, 2011). Additional epizootics characterized by strikingly similar pathology have been reported in other cultured sturgeon species, including the Russian sturgeon (*A. gueldenstaedtii*) in northern Europe (Adkison *et al.*, 1998), lake sturgeon in Manitoba, Canada (Clouthier *et al.*, 2013) and shortnose sturgeon (*A. brevirostrum*) in New Brunswick, Canada (LaPatra *et al.*, 2014).

WSIV has been tentatively classified as a member of the family *Iridoviridae* based on its histopathological and ultrastructural resemblance to iridoviruses (Hedrick *et al.*, 1990; Jancovich *et al.*, 2012).

Y. Kawato *et al.*

However, the WSIV major capsid protein (MCP) lacks significant sequence homology to that of iridoviruses (Kwak *et al.*, 2006). Preliminary phylogenetic analyses relying solely on the MCP support WSIV and the other sturgeon irido-like viruses as a new branch within the family *Mimiviridae* (Kwak *et al.*, 2006; Kurobe *et al.*, 2010; Clouthier *et al.*, 2013, 2015; Waltzek *et al.*, unpublished).

12.2.2 Diagnosis

Cultivated white sturgeon fry and fingerlings are highly susceptible to WSIV at water temperatures between 12 and 20°C (Hedrick *et al.*, 1992; LaPatra *et al.*, 1994; Watson *et al.*, 1998b). Typical WSIV epizootics manifest as a chronic wasting syndrome in which anorexic juveniles linger at the bottom of the tank and gradually become emaciated, leading to impaired growth with up to 95% mortality (Hedrick *et al.*, 1990; Raverty *et al.*, 2003). WSIV is epitheliotropic, resulting in pathognomonic microscopic lesions in external tissues, including the integument, gills and upper alimentary tract (Hedrick *et al.*, 1990; LaPatra *et al.*, 1994; Watson *et al.*, 1998a). Histological examination reveals characteristic hypertrophic epithelial cells that stain amphophilic to basophilic (Hedrick *et al.*, 1990). An immunohistochemical assay employing a monoclonal antibody against WSIV can confirm the virus in tissue sections (OIE, 2003).

Viral isolation is challenging as WSIV grows slowly in white sturgeon spleen (WSS-2) and skin cells (WSSK-1) (Hedrick *et al.*, 1992). Pooled gill and skin tissue homogenates are inoculated on to cell cultures, maintained at 20°C and examined for 30 days for CPE; a blind passage is performed with 15 additional days of observation if no CPE is observed (OIE, 2003). Although CPE may be detected as early as 1 week post inoculation, changes may not be observed for 2–3 weeks. Infected cells often occur singly or in small groups and are recognizable by their rounded, enlarged and refractile appearance (Hedrick *et al.*, 1992; Watson *et al.*, 1998a; Fig. 12.3). Infected cells eventually die and detach, but complete destruction of the monolayer only occurs in the most heavily infected cultures. Cultures with a suspicious CPE can be confirmed using a serum neutralization assay, indirect fluorescent antibody test or transmission electron microscopy (TEM) (Hedrick *et al.*, 1990; OIE, 2003).

The development of a conventional PCR assay targeting the WSIV major capsid protein has

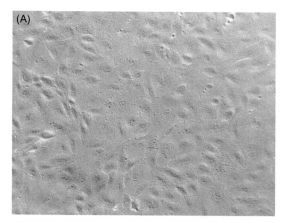

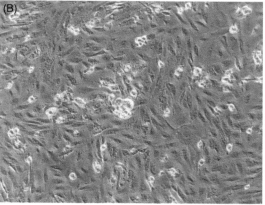

Fig. 12.3. White sturgeon (*Acipenser transmontanus*) spleen cells (WSS-2) infected with white sturgeon iridovirus (WSIV): (A) uninfected control; (B) infected cells scattered throughout the monolayer singly or in small groups and recognizable by their rounded, enlarged and refractile appearance. Images courtesy of Ronald P. Hedrick.

provided a much needed specific, sensitive and rapid diagnostic tool (Kwak *et al.*, 2006). The assay does not amplify other iridoviruses and detects down to 1 fg of target plasmid DNA. The sampling of tissue fragments of the pectoral fins is non-lethal and eliminates the need to sacrifice large numbers of valuable cultured or threatened populations of wild white sturgeon (Drennan *et al.*, 2007a). Recently, a conventional PCR assay and two separate quantitative PCR assays targeting conserved regions of the MCP were designed to amplify all of the North American sturgeon nucleocytoplasmic large DNA viruses, including WSIV (Clouthier *et al.*, 2015).

12.2.3 Pathology

The epitheliotropic nature of WSIV results in gross lesions on the skin and gills (Hedrick *et al.*, 1990; LaPatra *et al.*, 1994). The ventral scutes of infected fish are often red or haemorrhagic (Fig. 12.4A) and the gills may appear swollen and pale. Infected juveniles typically appear emaciated and stunted. Internally, a pale liver, an empty gastrointestinal tract and reduced fat stores may be observed.

Histopathological examination of the skin and gills typically reveals focal to diffuse epithelial hyperplasia with an associated enlargement of some epithelial cells that stain amphophilic to basophilic (Fig. 12.4.B). These hypertrophic epithelial cells are pathognomonic for WSIV infection and have been observed in the integument (skin and fins), olfactory organs (barbels and nares), oropharyngeal cavity (hard palate, tongue, lips,

opercular flap), upper alimentary tract (oesophagus) and gills (Hedrick *et al.*, 1990; Watson *et al.*, 1998a). In advanced disease, both hyperplasia of the branchial epithelium and necrosis of the pillar cells lining the lamellar vascular channels leading to small haemorrhages may be observed (Hedrick *et al.*, 1990). The histological or cytological evaluation of WSIV-positive fish may reveal secondary pathogens, including bacteria (*Flavobacterium* sp.), protists (*Trichodina* sp.) and oomycetes (*Saprolegnia* sp.) (Hedrick *et al.*, 1990; Watson *et al.*, 1998a).

TEM of infected tissues (e.g. skin or gills) or cultures can confirm the diagnosis. Aggregations of distinctive large and complex icosahedral-shaped virions can be observed around the viral assembly site within the cytoplasm of hypertrophic epithelial cells (Watson *et al.*, 1998a). The average diameter of these virions is 262–273 nm between opposite sides

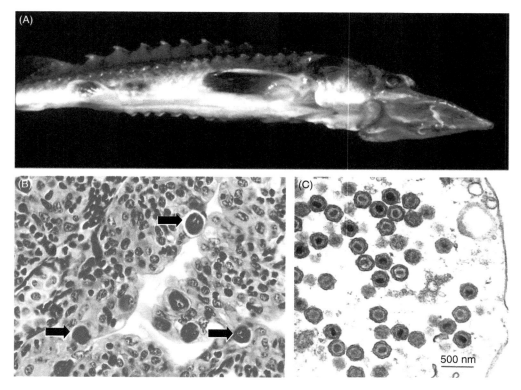

Fig. 12.4. Juvenile white sturgeon (*Acipenser transmontanus*) infected with white sturgeon iridovirus (WSIV): (A) small haemorrhages present along the ventral surfaces, including the oral region and lateral scutes; (B) WSIV-infected gill tissue section revealing numerous enlarged amphophilic cells (arrows), haematoxylin and eosin stain; (C) transmission electron photomicrograph illustrating the complex virions of WSIV as found directly from infected cells in the gill of white sturgeon. Images courtesy of Joseph M. Groff and Ronald P. Hedrick.

and 299–302 nm between opposite vertices (Hedrick *et al.*, 1990; Watson *et al.*, 1998a; Fig. 12.4C).

12.2.4 Pathophysiology

The pathophysiology of white sturgeon WSIV infection has been investigated (Watson *et al.*, 1998a,b). Lesions in the integument and gills are believed to alter the respiratory and osmoregulatory capacities of infected sturgeons. Experimentally, infected juvenile white sturgeon had a decreased haematocrit and decreased plasma protein concentrations at higher water temperatures (Watson *et al.*, 1998b). WSIV infection of the sensory epithelia (nares, barbels, lips, tongue) in juvenile white sturgeon is believed to result in anorexia, leading to a lethal wasting syndrome (Watson *et al.*, 1998a). Watson *et al.* (1998b) found that juvenile white sturgeon experimentally infected with WSIV were unable to maintain normal hepatosomatic indices at lower water temperatures, which is indicative of food deprivation (Hung *et al.*, 1997) and reduced dietary glucose (Fynn-Aikens *et al.*, 1992). Finally, the WSIV induced disruption of mucosal barriers is hypothesized to predispose juvenile white sturgeon to secondary invaders (Watson *et al.*, 1998b).

12.2.5 Protective and control strategies

Factors contributing to the occurrence or severity of WSIV epizootics under hatchery conditions include high stocking density, physical and procedural stressors, fluctuating or low water temperatures, and lapses in biosecurity (LaPatra *et al.*, 1994, 1996; Watson *et al.*, 1998b; Georgiadis *et al.*, 2001; Drennan *et al.*, 2005). Reducing stock densities and increasing the water flow reduced mortality (LaPatra *et al.*, 1994). Experimental studies have demonstrated significantly higher cumulative mortality in groups held at higher versus lower stocking densities (LaPatra *et al.*, 1996; Drennan *et al.*, 2005). The physical and procedural stressors that can increase the frequency of WSIV outbreaks include handling, transportation and fluctuations in temperature (LaPatra *et al.*, 1994, 1996). Stress can trigger disease onset in subclinical carrier fish, which subsequently spread the virus to other individuals (Georgiadis *et al.*, 2001). An experimental infection study (Watson *et al.*, 1998b) demonstrated that juvenile white sturgeon held at a lower water temperature (10°C) suffered significantly higher cumulative mortality and had more severe

microscopic lesions than those held at a higher water temperature (23°C). Lower water temperatures promoted *in vitro* WSIV replication (Hedrick *et al.*, 1992) and the virus may depress the immune response of sturgeon (Watson *et al.*, 1998b). Although rearing juvenile white sturgeon at higher water temperatures reduces cumulative mortality, secondary bacterial (*Flavobacterium* sp.) infections are more problematic at these elevated temperatures (Watson *et al.*, 1998b).

White sturgeon juveniles that survive a WSIV outbreak appear to be protected even following stressful events. The survivors develop individual or herd immunity, which protects them when they are re-exposed to WSIV (Georgiadis *et al.*, 2001). Anti-WSIV serum and mucosal antibodies have been confirmed in survivors (Drennan *et al.*, 2007b). Thus, production strategies involving the development of resistant brood stock or stocking with fish that have survived an outbreak have been proposed (Georgiadis *et al.*, 2000, 2001). Additionally, farmers should practise stringent biosecurity, including quarantining new brood stock in a separate building and screening them for WSIV prior to spawning. The development of an economic WSIV vaccine and delivery system capable of stimulating mucosal immunity is needed to mitigate the effect of WSIV epizootics in sturgeon aquaculture.

12.2.6 Conclusions

In the last couple of decades, progress has been made: (i) in our understanding of WSIV pathology, pathophysiology and epizootiology; and (ii) in the availability of diagnostic tools for detecting WSIV. Physical and procedural stressors influence the occurrence and severity of epizootics. Through proper biosecurity and management practices, the effects of WSIV can be minimized; however, an economic vaccine would benefit the aquaculture of white sturgeon.

References

Adkison, M.A., Cambre, M. and Hedrick, R.P. (1998) Identification of an iridovirus in Russian sturgeon (*Acipenser guldenstadi*) from northern Europe. *Bulletin of the European Association of Fish Pathologists* 18, 29–32.

Armstrong, R.D. and Ferguson, H.W. (1989) Systemic viral disease of the chromide cichlid *Etropus maculatus*. *Diseases of Aquatic Organisms* 7, 155–157.

Birstein, V.J. (1993) Sturgeons and paddlefishes: threatened fishes in need of conservation. *Conservation Biology* 7, 773–787.

Caipang, C.M., Hirono, I. and Aoki, T. (2003) Development of a real-time PCR assay for the detection and quantification of red seabream iridovirus (RSIV). *Fish Pathology* 38, 1–7.

Caipang, C.M., Takano, T., Hirono, I. and Aoki, T. (2006) Genetic vaccines protect red seabream, *Pagrus major*, upon challenge with red seabream iridovirus (RSIV). *Fish and Shellfish Immunology* 21, 130–138.

Chao, C.B., Chen, C.Y., Lai, Y.Y., Lin, C.S. and Huang, H.T. (2004) Histological, ultrastructural, and *in situ* hybridization study on enlarged cells in grouper *Epinephelus* hybrids infected by grouper iridovirus in Taiwan (TGIV). *Diseases of Aquatic Organisms* 58, 127–142.

Chen, X.H., Lin, K.B. and Wang, X.W. (2003) Outbreaks of an iridovirus disease in maricultured large yellow croaker, *Larimichthys crocea* (Richardson), in China. *Journal of Fish Diseases* 26, 615–619.

Chou, H.Y., Hsu, C.C. and Peng, T.Y. (1998) Isolation and characterization of a pathogenic *Iridovirus* from cultured grouper (*Epinephelus* sp.) in Taiwan. *Fish Pathology* 33, 201–206.

Clouthier, S.C., VanWalleghem, E., Copeland, S., Klassen, C., Hobbs, G. *et al.* (2013) A new species of nucleo-cytoplasmic large DNA virus (NCLDV) associated with mortalities in Manitoba lake sturgeon *Acipenser fulvescens*. *Disease of Aquatic Organisms* 102, 195–209.

Clouthier, S.C., VanWalleghem, E. and Anderson, E.D. (2015) Sturgeon nucleo-cytoplasmic large DNA virus phylogeny and PCR tests. *Diseases of Aquatic Organisms* 117, 93–106.

Do, J.W., Cha, S.J., Kim, J.S., An, E.J., Lee, N.S. *et al.* (2005a) Phylogenetic analysis of the major capsid protein gene of iridovirus isolates from cultured flounders *Paralichthys olivaceus* in Korea. *Diseases of Aquatic Organisms* 64, 193–200.

Do, J.W., Cha, S.J., Kim, J.S., An, E.J., Park, M.S. *et al.* (2005b) Sequence variation in the gene encoding the major capsid protein of Korean fish iridoviruses. *Archives of Virology* 150, 351–359.

Dong, C., Weng, S., Shi, X., Xu, X., Shi, N. and He, J. (2008) Development of a mandarin fish *Siniperca chuatsi* fry cell line suitable for the study of infectious spleen and kidney necrosis virus (ISKNV). *Virus Research* 135, 273–281.

Dong, C., Weng, S., Luo, Y., Huang, M., Ai, H. *et al.* (2010) A new marine megalocytivirus from spotted knifejaw, *Oplegnathus punctatus*, and its pathogenicity to freshwater mandarinfish, *Siniperca chuatsi*. *Virus Research*, 147, 98–106.

Dong, C.F., Xiong, X.P., Luo, Y., Weng, S., Wang, Q. and He, J. (2013a) Efficacy of a formalin-killed cell vaccine against infectious spleen and kidney necrosis virus (ISKNV) and immunoproteomic analysis of its major immunogenic proteins. *Veterinary Microbiology* 162, 419–428.

Dong, Y., Weng, S., He, J. and Dong, C. (2013b) Field trial tests of FKC vaccines against RSIV genotype *Megalocytivirus* in cage-cultured mandarin fish (*Siniperca chuatsi*) in an inland reservoir. *Fish and Shellfish Immunology* 35, 1598–1603.

Drennan, J.D., Ireland, S., LaPatra, S.E., Grabowski, L., Carrothers, T.K. and Cain, K.D. (2005) High-density rearing of white sturgeon *Acipenser transmontanus* (Richardson) induces white sturgeon iridovirus disease among asymptomatic carriers. *Aquaculture Research* 36, 824–827.

Drennan, J.D., LaPatra, S.E., Siple, J.T., Ireland, S. and Cain, K.D. (2006) Transmission of white sturgeon iridovirus in Kootenai River white sturgeon *Acipenser transmontanus*. *Diseases of Aquatic Organisms* 70, 37–45.

Drennan, J.D., LaPatra, S.E., Samson, C.A., Ireland, S., Eversman, K.F. and Cain, K.D. (2007a) Evaluation of lethal and non-lethal sampling methods for the detection of white sturgeon iridovirus infection in white sturgeon, *Acipenser transmontanus* (Richardson). *Journal of Fish Diseases* 30, 367–379.

Drennan, J.D., LaPatra, S.E., Swan, C.M., Ireland, S. and Cain, K.D. (2007b) Characterization of serum and mucosal antibody responses in white sturgeon (*Acipenser transmontanus* Richardson) following immunization with WSIV and a protein hapten antigen. *Fish and Shellfish Immunology* 23, 657–669.

Fu, X., Li, N., Liu, L., Lin, Q., Wang, F. *et al.* (2011) Genotype and host range analysis of infectious spleen and kidney necrosis virus (ISKNV). *Virus Genes* 42, 97–109.

Fu, X., Li, N., Lai, Y., Liu, L., Lin, Q., Shi, C., Huang, Z. and Wu, S. (2012) Protective immunity against iridovirus disease in mandarin fish, induced by recombinant major capsid protein of infectious spleen and kidney necrosis virus. *Fish & Shellfish Immunology* 33, 880–885.

Fu, X., Li, N., Lin, Q., Guo, H., Zhang, D. *et al.* (2014) Protective immunity against infectious spleen and kidney necrosis virus induced by immunization with DNA plasmid containing mcp gene in Chinese perch *Siniperca chuatsi*. *Fish & Shellfish Immunology* 40, 259–266.

Fynn-Aikens, K., Hung, S.S.O., Liu, W. and Hongbin, L. (1992) Growth, lipogenesis and liver composition of juvenile white sturgeon fed different levels of D-glucose. *Aquaculture* 105, 61–72.

Georgiadis, M.P., Hedrick, R.P., Johnson, W.O., Yun, S. and Gardner, I.A. (2000) Risk factors for outbreaks of disease attributable to white sturgeon iridovirus and white sturgeon herpesvirus-2 at a commercial sturgeon farm. *American Journal of Veterinary Research* 61, 1232–1240.

Georgiadis, M.P., Hedrick, R.P., Carpenter, T.E. and Gardner, I.A. (2001) Factors influencing transmission, onset and severity of outbreaks due to white sturgeon iridovirus in a commercial hatchery. *Aquaculture* 194, 21–35.

Gias, E., Johnston, C., Keeling, S., Spence, R.P. and McDonald, W.L. (2011) Development of real-time PCR assays for detection of megalocytiviruses in imported ornamental fish. *Journal of Fish Diseases* 34, 609–618.

Gibson-Kueh, S., Ngoh-Lim, G.H., Netto, P., Kurita, J., Nakajima, K. and Ng, M.L. (2004) A systematic iridoviral disease in mullet, *Mugil cephalus* L., and tiger grouper, *Epinephelus fuscoguttatus* Forsskal: a first report and study. *Journal of Fish Diseases* 27, 693–699.

Go, J. and Whittington, R. (2006) Experimental transmission and virulence of a megalocytivirus (Family *Iridoviridae*) of dwarf gourami (*Colisa lalia*) from Asia in Murray cod (*Maccullochella peelii peelii*) in Australia. *Aquaculture* 258, 140–149.

Go, J., Lancaster, M., Deece, K., Dhungyel, O. and Whittington, R. (2006) The molecular epidemiology of iridovirus in Murray cod (*Maccullochella peelii peelii*) and dwarf gourami (*Colisa lalia*) from distant biogeographical regions suggests a link between trade in ornamental fish and emerging iridoviral diseases. *Molecular and Cellular Probes* 20, 212–222.

He, J.G., Wang, S.P., Zeng, K., Huang, Z.J. and Chan, S.-M. (2000) Systemic disease caused by an iridovirus-like agent in cultured mandarin fish, *Siniperca chuatsi* (Basilewsky), in China. *Journal of Fish Diseases* 23, 219–222.

He, J.G., Zeng, K., Weng, S.P. and Chan, S.-M. (2002) Experimental transmission, pathogenicity and physical-chemical properties of infectious spleen and kidney necrosis virus (ISKNV). *Aquaculture* 204, 11–24.

Hedrick, R.P., Groff, J.M., McDowell, T. and Wingfield, W.H. (1990) An iridovirus infection of the integument of the white sturgeon *Acipenser transmontanus*. *Diseases of Aquatic Organisms* 8, 39–44.

Hedrick, R.P., McDowell, T.S., Groff, J.M., Yun, S. and Wingfield, W.H. (1992) Isolation and some properties of an iridovirus-like agent from white sturgeon *Acipenser transmontanus*. *Diseases of Aquatic Organisms* 12, 75–81.

Huang, S.M., Tu, C., Tseng, C.H., Huang, C.C., Chou, C.C. et al. (2011) Genetic analysis of fish iridoviruses isolated in Taiwan during 2001–2009. *Archives of Virology* 156, 1505–1515.

Hung, S.O., Liu, W., Hongbin, L., Storebakken, S. and Cui, Y. (1997) Effect of starvation on some morphological and biochemical parameters in white sturgeon *Acipenser transmontanus*. *Aquaculture* 150, 357–363.

Imajoh, M., Ikawa, T. and Oshima, S. (2007) Characterization of a new fibroblast cell line from a tail fin of red sea bream, *Pagrus major*, and phylogenetic relationships of a recent RSIV isolate in Japan. *Virus Research* 126, 45–52.

Inouye, K., Yamano, K., Maeno, Y., Nakajima, K., Matsuoka, M. et al. (1992) Iridovirus infection of cultured red sea bream, *Pagrus major*. *Fish Pathology* 27, 19–27.

Ito, T., Yoshiura, Y., Kamaishi, T., Yoshida, K. and Nakajima, K. (2013) Prevalence of red sea bream iridovirus (RSIV) among organs of Japanese amberjack (*Seriola quinqueradiata*) exposed to cultured RSIV. *Journal of General Virology* 94, 2094–2101.

Jancovich, J.K., Chinchar, V.G., Hyatt, A., Miyazaki, T., Williams, T. and Zhang, Q.Y. (2012) Family *Iridoviridae*. In: King, A.M.Q., Adams, M.J., Carstens, E.B. and Lefkowitz, E.J. (eds) *Virus Taxonomy: Ninth Report of the International Committee on Taxonomy of Viruses*. Elsevier/Academic Press, London/Waltham, Massachussetts/San Diego, California, pp. 193–210. Available at: http://www.trevorwilliams.info/ictv_iridoviridae_2012.pdf (accessed 7 November 2016).

Jeong, J.B., Park, K.H., Kim, H.Y., Hong, S.H., Chung, J.-K. et al. (2004) Multiplex PCR for the diagnosis of red sea bream iridoviruses isolated in Korea. *Aquaculture* 235, 139–152.

Jeong, J.B., Kim, H.Y., Jun, L.J., Lyu, J.H., Park, N.G. et al. (2008a) Outbreaks and risks of infectious spleen and kidney necrosis virus disease in freshwater ornamental fishes. *Diseases of Aquatic Organisms* 78, 209–215.

Jeong, J.B., Cho, H.J., Jun, L.J., Hong, S.H., Chung, J.K. and Jeong, H.D. (2008b) Transmission of iridovirus from freshwater ornamental fish (pearl gourami) to marine fish (rock bream). *Diseases of Aquatic Organisms* 82, 27–36.

Jin, J.W., Kim, K.I., Kim, J.K., Park, N.G. and Jeong, H.D. (2014) Dynamics of megalocytivirus transmission between bivalve molluscs and rock bream *Oplegnathus fasciatus*. *Aquaculture* 428/429, 29–34.

Jung, M.H., Jung, S.J., Vinay, T.N., Nikapitiya, C., Kim, J.O. et al. (2015) Effects of water temperature on mortality in *Megalocytivirus*-infected rock bream *Oplegnathus fasciatus* (Temminck et Schlegel) and development of protective immunity. *Journal of Fish Diseases* 38, 729–737.

Jung, S., Miyazaki, T., Miyata, M., Danayadol, Y. and Tanaka, S. (1997) Pathogenicity of iridovirus from Japan and Thailand for the red sea bream *Pagrus major* in Japan, and histopathology of experimentally infected fish. *Fisheries Science* 63, 735–740.

Jung, S.J. and Oh, M.J. (2000) Iridovirus-like infection associated with high mortalities of striped beakperch, *Oplegnathus fasciatus* (Temminck et Schlegel), in southern coastal areas of the Korean Peninsula. *Journal of Fish Diseases* 23, 223–226.

Kawakami, H. and Nakajima, K. (2002) Cultured fish species affected by red sea bream iridoviral disease from 1996 to 2000. *Fish Pathology* 37, 45–47.

Kim, T.J., Jang, E.J. and Lee, J.I. (2008) Vaccination of rock bream, *Oplegnathus fasciatus* (Temminck and Schlegel), using a recombinant major capsid protein of fish iridovirus. *Journal of Fish Diseases* 31, 547–551.

Kim, W.S., Oh, M.J., Jung, S.J., Kim, Y.J. and Kitamura, S.I. (2005) Characterization of an iridovirus detected from cultured turbot *Scophthalmus maximus* in Korea. *Diseases of Aquatic Organisms* 64, 175–180.

Kim, W.S., Oh, M.J., Kim, J.O., Kim, D., Jeon, C.H. and Kim, J.H. (2010) Detection of megalocytivirus from imported tropical ornamental fish, paradise fish *Macropodus opercularis*. *Diseases of Aquatic Organisms* 90, 235–239.

Kim, Y.J., Jung, S.J., Choi, T.J., Kim, H.R., Rajendran, K.V. and Oh, M.J. (2002) PCR amplification and sequence analysis of irido-like virus infecting fish in Korea. *Journal of Fish Diseases* 25, 121–124.

Kurita, J. and Nakajima, K. (2012) Megalocytiviruses: a review. *Viruses* 4, 521–538.

Kurita, J., Nakajima, K., Hirono, I. and Aoki, T. (1998) Polymerase chain reaction (PCR) amplification of DNA of red sea bream iridovirus (RSIV). *Fish Pathology* 33, 17–23.

Kurobe, T., Kwak, K.T., MacConnell, E., McDowell, T.S., Mardones, F.O. and Hedrick, R.P. (2010) Development of PCR assays to detect iridovirus infections among captive and wild populations of Missouri River sturgeon. *Diseases of Aquatic Organisms* 93, 31–42.

Kurobe, T., MacConnell, E., Hudson, C., Mardones, F.O. and Hedrick, R.P. (2011) Iridovirus infections among Missouri River sturgeon: initial characterization, transmission and evidence for establishment of a carrier state. *Journal of Aquatic Animal Health* 23, 9–18.

Kwak, K.T., Gardner, I.A., Farver, T.B. and Hedrick, R.P. (2006) Rapid detection of white sturgeon iridovirus (WSIV) using a polymerase chain reaction (PCR) assay. *Aquaculture* 254, 92–101.

LaPatra, S.E., Groff, J.M., Jones, G.R., Munn, B., Patterson, T.L. et al. (1994) Occurrence of white sturgeon iridovirus infections among cultured white sturgeon in the Pacific Northwest. *Aquaculture* 126, 201–210.

LaPatra, S.E., Groff, J.M., Patterson, T.L., Shewmaker, W.K., Casten, M. et al. (1996) Preliminary evidence of sturgeon density and other stressors on manifestation of white sturgeon iridovirus disease. *Journal of Applied Aquaculture* 6, 51–58.

LaPatra, S.E., Ireland, S.C., Groff, J.M., Clemens, K.M. and Siple, J.T. (1999) Adaptive disease management strategies for the endangered population of Kootenai River white sturgeon. *Fisheries* 24, 6–13.

LaPatra, S.E., Groff, J.M., Keith, I., Hogans, W.E. and Groman, D. (2014) Case report: concurrent herpesviral and presumptive iridoviral infection associated with disease in cultured shortnose sturgeon, *Acipenser brevirostrum* (L.), from the Atlantic coast of Canada. *Journal of Fish Diseases* 37, 141–147.

Lü, L., Zhou, S.Y., Chen, C., Weng, S.P., Chan, S.-M. and He, J.G. (2005) Complete genome sequence analysis of an iridovirus isolated from the orange-spotted grouper, *Epinephelus coioides*. *Virology* 339, 81–100.

Matsuoka, S., Inouye, K. and Nakajima, K. (1996) Cultured fish species affected by red sea bream iridoviral disease from 1991 to 1995. *Fish Pathology* 31, 233–234.

Nakajima, K. and Kurita, J. (2005) Red sea bream iridoviral disease. *Uirusu* 55, 115–126.

Nakajima, K. and Maeno, Y. (1998) Pathogenicity of red sea bream iridovirus and other fish iridoviruses to red sea bream. *Fish Pathology* 33, 143–144.

Nakajima, K. and Sorimachi, M. (1994) Biological and physico-chemical properties of the iridovirus isolated from cultured red sea bream, *Pagrus major*. *Fish Pathology* 29, 29–33.

Nakajima, K. and Sorimachi, M. (1995) Production of monoclonal antibodies against red sea bream iridovirus. *Fish Pathology* 30, 47–52.

Nakajima, K., Maeno, Y., Fukudome, M., Fukuda, Y., Tanaka, S. et al. (1995) Immunofluorescence test for the rapid diagnosis of red sea bream iridovirus infection using monoclonal antibody. *Fish Pathology* 30, 115–119.

Nakajima, K., Maeno, Y., Kurita, J. and Inui, Y. (1997) Vaccination against red sea bream iridoviral disease in red sea bream. *Fish Pathology* 32, 205–209.

Nakajima, K., Maeno, Y., Yokoyama, K., Kaji, C. and Manabe, S. (1998) Antigen analysis of red sea bream iridovirus and comparison with other fish iridoviruses. *Fish Pathology* 33, 73–78.

Nakajima, K., Maeno, Y., Honda, A., Yokoyama, K., Tooriyama, T. and Manabe, S. (1999) Effectiveness of a vaccine against red sea bream iridoviral disease in a field trial test. *Diseases of Aquatic Organisms* 36, 73–75.

Nolan, D., Stephens, F., Crockford, M., Jones, J.B. and Snow, M. (2015) Detection and characterization of viruses of the genus *Megalocytivirus* in ornamental fish imported into an Australian border quarantine premises: an emerging risk to national biosecurity. *Journal of Fish Diseases* 38, 187–195.

Oh, M.J., Kitamura, S.I., Kim, W.S., Park, M.K., Jung, S.J. et al. (2006) Susceptibility of marine fish species to a megalocytivirus, turbot iridovirus, isolated from turbot, *Psetta maximus* (L.). *Journal of Fish Diseases* 29, 415–421.

Oh, S.Y. and Nishizawa, T. (2013) Optimizing the quantitative detection of the red seabream iridovirus (RSIV) genome from splenic tissues of rock bream *Oplegnathus fasciatus*. *Fish Pathology* 48, 21–24.

Oh, S.Y., Oh, M.J. and Nishizawa, T. (2014) Potential for a live red seabream iridovirus (RSIV) vaccine in rock bream *Oplegnathus fasciatus* at a low rearing temperature. *Vaccine*, 32, 363–368.

OIE (2003) *Manual of Diagnostic Tests for Aquatic Animals*. World Organisation for Animal Health, Paris. Updated version 2016 available at: http://www.oie.int/international-standard-setting/aquatic-manual/access-online/ (accessed 7 November 2016).

OIE (2012) Red sea bream iridoviral disease. In: *Manual of Diagnostic Tests for Aquatic Animals 2012*. World Organisation for Animal Health, Paris, pp. 345–356. Updated version 2016 available at: http://www.oie.int/index.php?id=2439&L=0&htmfile=chapitre_rsbid.htm (accessed 7 November 2016).

Oshima, S., Hata, J., Hirasawa, N., Ohtaka, T., Hirono, I. et al. (1998) Rapid diagnosis of red sea bream iridovirus infection using the polymerase chain reaction. *Diseases of Aquatic Organisms* 32, 87–90.

Paperna, I., Vilenkin, M. and de Matos, A.P. (2001) Iridovirus infections in farm-reared tropical ornamental fish. *Diseases of Aquatic Organisms* 48, 17–25.

Pikitch, E.K., Doukakis, P., Lauck, L., Chakrabarty, P. and Erickson, D.L. (2005) Status, trends and management of sturgeon and paddlefish fisheries. *Fish and Fisheries* 6, 233–265.

Raverty, S., Hedrick, R., Henry, J. and Saksida, S. (2003) Diagnosis of sturgeon iridovirus infection in farmed white sturgeon in British Columbia. *Canadian Veterinary Journal* 44, 327–328.

Rimmer, A.E., Becker, J.A., Tweedie, A. and Whittington, R.J. (2012) Development of a quantitative polymerase chain reaction (qPCR) assay for the detection of dwarf gourami iridovirus (DGIV) and other megalocytiviruses and comparison with the Office International des Epizooties (OIE) reference PCR protocol. *Aquaculture* 358/359, 155–163.

Rodger, H.D., Kobs, M., Macartney, A. and Frerichs, G.N. (1997) Systemic iridovirus infection in freshwater angelfish, *Pterophyllum scalare* (Lichtenstein). *Journal of Fish Diseases* 20, 69–72.

Shi, C.Y., Wang, Y.G., Yang, S.L., Huang, J. and Wang, Q.Y. (2004) The first report of an iridovirus-like agent infection in farmed turbot, *Scophthalmus maximus*, in China. *Aquaculture* 236, 11–25.

Shimmoto, H., Kawai, K., Ikawa, T. and Oshima, S. (2010) Protection of red sea bream *Pagrus major* against red sea bream iridovirus infection by vaccination with a recombinant viral protein. *Microbiology and Immunology* 54, 135–142.

Sriwanayos, P., Francis-Floyd, R., Stidworthy, M.F., Petty, B.D., Kelley, K. and Waltzek, T.B. (2013) Megalocytivirus infection in orbiculate batfish *Platax orbicularis*. *Diseases of Aquatic Organisms* 105, 1–8.

Subramaniam, K., Shariff, M., Omar, A.R., Hair-Bejo, M. and Ong, B.L. (2014) Detection and molecular characterization of infectious spleen and kidney necrosis virus from major ornamental fish breeding states in Peninsular Malaysia. *Journal of Fish Diseases* 37, 609–618.

Sudthongkong, C., Miyata, M. and Miyazaki, T. (2002a) Iridovirus disease in two ornamental tropical freshwater fishes: African lampeye and dwarf gourami. *Diseases of Aquatic Organisms* 48, 163–173.

Sudthongkong, C., Miyata, M. and Miyazaki, T. (2002b) Viral DNA sequences of genes encoding the ATPase and the major capsid protein of tropical iridovirus isolates which are pathogenic to fishes in Japan, South China Sea and Southeast Asian countries. *Archives of Virology* 147, 2089–2109.

Tanaka, S., Aoki, H., Inoue, M. and Kuriyama, I. (2003) Effectiveness of fasting against red sea bream iridoviral disease in red sea bream. *Fish Pathology* 38, 67–69.

Waltzek, T.B., Marty, G.D., Alfaro, M.E., Bennett, W.R. Garver, K.A. *et al.* (2012) Systemic iridovirus from three-spine stickleback *Gasterosteus aculeatus* represents a new megalocytivirus species (family *Iridoviridae*). *Diseases of Aquatic Organisms* 98, 41–56.

Wang, Q., Zeng, W.W., Li, K.B., Chang, O.Q., Liu, C. *et al.* (2011) Outbreaks of an iridovirus in marbled sleepy goby, *Oxyeleotris marmoratus* (Bleeker), cultured in southern China. *Journal of Fish Diseases* 34, 399–402.

Wang, X.W., Ao, J.Q., Li, Q.G. and Chen, X.H. (2006) Quantitative detection of a marine fish iridovirus isolated from large yellow croaker, *Pseudosciaena crocea*, using a molecular beacon. *Journal of Virological Methods* 133, 76–81.

Wang, Y.Q., Lu, L., Weng, S.P., Huang, J.N., Chan, S.M. and He, J.G. (2007) Molecular epidemiology and phylogenetic analysis of a marine fish infectious spleen and kidney necrosis virus-like (ISKNV-like) virus. *Archives of Virology* 152, 763–773.

Watson, L.R., Groff, J.M. and Hedrick, R.P. (1998a) Replication and pathogenesis of white sturgeon iridovirus (WSIV) in experimentally infected white sturgeon *Acipenser transmontanus* juveniles and sturgeon cell lines. *Diseases of Aquatic Organisms* 32, 173–184.

Watson, L.R., Milani, A. and Hedrick, R.P. (1998b) Effects of water temperature on experimentally-induced infections of juvenile white sturgeon (*Acipenser transmontanus*) with the white sturgeon iridovirus (WSIV). *Aquaculture* 166, 213–228.

Weber, E.S., Waltzek, T.B., Young, D.A., Twitchell, E.L., Gates, A.E. *et al.* (2009) Systemic iridovirus infection in the Banggai cardinalfish (*Pterapogon kauderni* Koumans 1933). *Journal of Veterinary Diagnostic Investigation* 21, 306–320.

Wen, C.M., Lee, C.W., Wang, C.S., Cheng, Y.H. and Huang, H.Y. (2008) Development of two cell lines from *Epinephelus coioides* brain tissue for characterization of betanodavirus and megalocytivirus infectivity and propagation. *Aquaculture* 278, 14–21.

Weng, S.P., Wang, Y.Q., He, J.G., Deng, M., Lu, L. *et al.* (2002) Outbreaks of an iridovirus in red drum, *Sciaenops ocellata* (L.), cultured in southern China. *Journal of Fish Diseases* 25, 681–685.

Won, K.M, Cho, M.Y., Park, M.A., Jee, B.Y., Myeong, J.I. and Kim, J.W. (2013) The first report of a megalocytivirus infection in farmed starry flounder, *Platichthys stellatus*, in Korea. *Fisheries and Aquatic Sciences* 16, 93–99.

Zhang, M., Xiao, Z.Z., Hu, Y.H. and Sun, L. (2012) Characterization of a megalocytivirus from cultured rock bream, *Oplegnathus fasciatus* (Temminck & Schlege), in China. *Aquaculture Research* 43, 556–564.

13 Alphaviruses in Salmonids

MARIUS KARLSEN[1]* AND RENATE JOHANSEN[2]

[1]PHARMAQ AS, Oslo, Norway; [2]PHARMAQ Analytiq, Bergen, Norway

13.1 Introduction

Alphavirus is a genus of RNA viruses belonging to the family *Togaviridae*. Most known alphaviruses are mosquito borne and cause diseases in terrestrial hosts such as birds, rodents and larger mammals, including humans (Strauss and Strauss, 1994). Infections may lead to diverse symptoms, such as rashes, gastrointestinal problems, arthritis/muscular inflammation and encephalitis (Kuhn, 2007; Steele and Twenhafel, 2010). Salmon pancreas disease virus, which is commonly named Salmonid alphavirus and abbreviated SAV (Weston *et al.*, 2002), is the only known alphavirus that has fish as a natural host (Powers *et al.*, 2001). SAV is distantly related to other members of the genus, but it still causes pathology that may resemble some of that seen in mammals (McLoughlin and Graham, 2007; Biacchesi *et al.*, 2016). The first isolation of SAV in cell culture was reported in 1995 from marine farmed Atlantic salmon (*Salmo salar*) in Ireland suffering from pancreas disease (PD) (Nelson *et al.*, 1995). This occurred at the same time as the isolation of a similar virus from freshwater rainbow trout (*Oncorhynchus mykiss*) suffering from sleeping disease (SD) in France (Castric *et al.*, 1997). Even though PD and SD are caused by the same virus species and are characterized by the same histopathology (Boucher and Baudin Laurencin, 1996; Weston *et al.*, 2002), the two different names are still commonly used. This is probably for historical reasons, and because SD is mainly associated with the production of smaller sized (*c.* 0.3–2 kg at slaughter) rainbow trout in fresh water, whereas PD is associated with infections in the marine Atlantic salmon and rainbow trout industry.

The genome of SAV is an 11.9 kb large single-stranded RNA molecule in messenger sense, with two large open reading frames (Weston *et al.*, 2002). These encode the four non-structural proteins (nsP1–4) that replicate the genome, and the structural proteins (capsid, E1, E2, E3 and 6K/TF). These latter proteins build an icosahedral-shaped virus particle that contains an inner capsid flanked by a membrane derived from the host (Villoing *et al.*, 2000a). Embedded into this membrane are numerous glycoprotein spikes. The glycoprotein E2 constitutes the majority of the surface of the particle (Voss *et al.*, 2010; Karlsen *et al.*, 2015), and is also a main antigenic target for the immune system (Moriette *et al.*, 2005).

SAVs from different areas of Europe show considerable genetic diversity, and six genetic subtypes, SAV1–6, have been described (Fringuelli *et al.*, 2008; see Fig. 13.1). Subtypes 1, 2, 4, 5 and 6 have been found in Atlantic salmon and rainbow trout around the British Isles (Graham *et al.*, 2012; Karlsen *et al.*, 2014); SAV3 has only been described from Norway (Hodneland *et al.*, 2005); and SAV2 appears to be more diverse than the other subtypes, and contains two distinct lineages, one that causes PD in marine farms in Scotland and Norway, and another that is the cause of SD epizootics in Continental Europe, with occasional outbreaks in Scotland and England (Graham *et al.*, 2012; Hjortaas *et al.*, 2013; Karlsen *et al.*, 2014).

Although it remains unknown how SAV was initially introduced into farmed fish, a diverse wild reservoir of SAV may exist in or around the North Sea. Phylogenetic dating studies have shown that each of the subtypes must represent separate introduction events from this reservoir (Karlsen *et al.*, 2014). Evidence for the presence of SAV in wild common dab (*Limanda limanda*), long rough dab (*Hippoglossoides platessoides*) and plaice (*Platessa*

*Corresponding author e-mail: marius.karlsen@pharmaq.no

© P.T.K. Woo and R.C. Cipriano 2017. *Fish Viruses and Bacteria: Pathobiology and Protection*
(P.T.K. Woo and R.C. Cipriano)

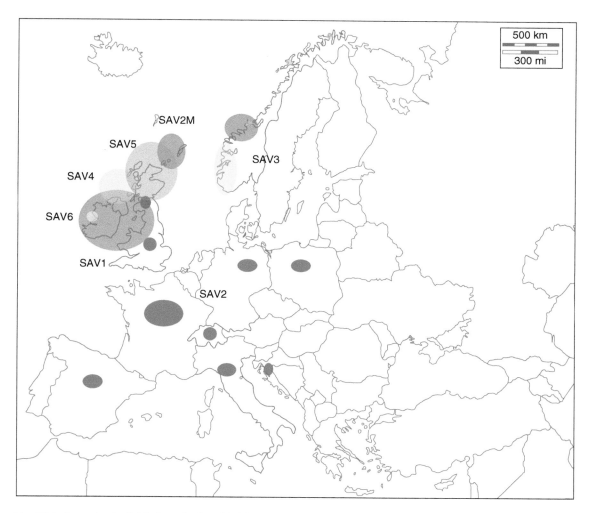

Fig. 13.1. Approximate distribution of salmonid alphavirus (SAV) subtypes in European salmonid farms. The geographical distribution is roughly indicated by the coloured spheres for each subtype. For subtype 2, the distributions of a freshwater form (SAV2) and a marine form (SAV2M) are included. Details of the distribution of freshwater SAV2 are poorly known, and the figure merely illustrates occurrences for this particular subtype at a country level. The map used as a basis for the figure was obtained from www.d-maps.com.

platessa) has been presented (Snow *et al.*, 2010; Bruno *et al.*, 2014; McCleary *et al.*, 2014; Simons *et al.*, 2016), and the prevalence of SAV in the common dab appears to be relatively high around the British Isles. It is still not clear whether these species of flatfish constitute the original wild reservoir of SAV, or if they have been infected as a result of spillover from salmon infections (Karlsen, 2015; Simons *et al.*, 2016). Experimental transmission of the virus in brown trout (*Salmo trutta*) has shown that this species is also susceptible to SAV (Boucher *et al.*, 1995).

Infection with SAV may develop into a high-titre viraemia (Desvignes *et al.*, 2002) and fish shed considerable amounts of infectious virus particles into the water during the viraemic phase (Andersen *et al.*, 2010; Andersen, 2012). Shedding to water is significantly reduced, or stops, when the host develops a strong antibody response and the viraemia is cleared. The shed virus particles infect nearby fish (McLoughlin *et al.*, 1996). It is likely that marine SAVs persist in farmed populations of fish via horizontal transmission between farms in the same

---

Alphaviruses in Salmonids

current network (Viljugrein *et al.*, 2009). The transport of fish and/or equipment between such networks can enhance transmission, and is the most probable explanation for occasional long-distance transmissions (Karlsen *et al.*, 2014). Viral RNA has also been detected in salmon lice (*Lepeophtheirus salmonis*) that feed on infected fish, but active replication in the lice has not been demonstrated (Karlsen *et al.*, 2006; Petterson *et al.*, 2009), nor has it been possible to infect *L. salmonis* in the laboratory (Karlsen *et al.*, 2015). The spread of SAV2/SD in the smaller sized rainbow trout industry is probably connected with trade practices that involve the transport of live fish or ova, and vertical transmission cannot be excluded (Castric *et al.*, 2005; Borzym *et al.*, 2014). The epizootiology of SD has been only poorly studied.

In Europe, the economic impact of losses due to SAV is significant (Aunsmo *et al.*, 2010); when comparing pathogen-related losses in the Norwegian salmon industry, those from SAV are probably only second to those from sea lice infestations. Biomass lost through mortalities contributes to one part of the loss, but poor growth and a reduction in fillet quality are also major consequences of infection. In Norway, during 2007 it was estimated that SAV infections increased the production costs of Atlantic salmon by 6.0 Norwegian kroner (NOK) per kilo (from 25–30 NOK/kg). An outbreak at a 500,000 smolt production site was estimated to cost about 14.4 million NOK (Aunsmo *et al.*, 2010). These numbers should not be taken as representative of SAV2 epizootics in the rainbow trout industry, where outbreaks affect smaller fish. Because outbreaks of SD are not reported on a regular basis, the cost due to SAV in rainbow trout is difficult to estimate.

13.2 Diagnosis

SAV can be detected either indirectly by observing effects on the host, or more directly by identifying virus-specific molecules such as the viral genome, viral proteins or the infectious virus particle within the host. A preliminary and indirect diagnosis can be based on gross clinical signs and histopathology (McLoughlin *et al.*, 2002; Taksdal *et al.*, 2007), but because other pathogens may produce similar signs, it is always appropriate to identify viral molecules to confirm SAV infection.

Clinical disease occurs during the whole year, but it is more common in marine farms when the water temperature is higher in the summer, probably reflecting that the virus has an optimum temperature of 10–15°C (Villoing *et al.*, 2000a; Graham *et al.*, 2008). The first clinical sign is usually a drop in appetite. Fish later aggregate towards the surface of the pen and swim sluggishly, sometimes with their dorsal fins above the water surface. This behaviour often begins in one net pen and spreads to others. Abundant yellowish faecal casts are frequently observed in the water during a clinical outbreak of the disease (McVicar, 1987; McLoughlin *et al.*, 2002; Taksdal *et al.*, 2007). Most SAV infections result in an increased rate of mortality.

A study of 23 Norwegian sites experiencing PD due to SAV3 in 2006–2007 suggested that the mean mortality was 6.9% during the outbreak, and that elevated mortality lasted an average of 2.8 months (Jansen *et al.*, 2010). Mortalities due to SAV2 in Norway have been suggested to be lower (Jansen *et al.*, 2015). Mortality rates from outbreaks around the British Isles have been reported to be very variable (Crockford *et al.*, 1999), but it is difficult to exclude the contributions of other pathogens in extreme cases.

The gross pathology due to SAV includes a miscoloured liver, petechiae on the pyloric caeca and visceral fat, and ascites in the intraperitoneal cavity (Fig. 13.2). The intestine can be either empty or filled with yellowish faeces (McLoughlin and Graham, 2007). Damage in the fillet, melanization and miscoloured areas are other signs that are associated with ongoing or previous SAV infection (Taksdal *et al.*, 2012).

Fig. 13.2. Atlantic salmon parr suffering from pancreas disease (PD) after experimental infection by salmonid alphavirus (SAV). The typical gross pathology of clinical PD includes: (a) ascites; (b) a discoloured liver; and (c) petechiae on the pyloric caeca. Photo by Rolf Hetlelid Olsen/PHARMAQ AS.

M. Karlsen and R. Johansen

In the late stage of outbreaks, fish with very low K-factor (low body weight relative to length) can be observed (McLoughlin and Graham, 2007).

SAV does not always manifest as an acute clinical disease. The infection may persist following an acute outbreak (Graham et al., 2010), and it can also occur without an acute phase and with few observable signs (Graham et al., 2006b; Jansen et al., 2010). Such subclinical infections probably transmit virus, and this underscores the importance of not relying on clinical signs, but rather detecting SAV in the farm with a more specific method (Graham et al., 2006b, 2010; Hodneland and Endresen, 2006).

A conclusive diagnosis of PD is usually based on histopathology and the detection of viral RNA. Both the gross pathology and the histopathology due to SAV bear similarities to those of other viral diseases of salmon, e.g. heart and skeletal muscle inflammation (HSMI), cardiomyopathy syndrome (CMS) and infectious pancreatic necrosis (IPN) (McLoughlin and Graham, 2007). The combination of pancreatic lesions and pathology in the heart and skeletal muscle does separate PD from these conditions. SAV is easily detected in fish during an ongoing infection, and real-time RT-PCR is commonly used due to its superior sensitivity, specificity and ability to rapidly test many samples (Graham et al., 2006a; Hodneland and Endresen, 2006). Several commercial laboratories in Norway and the UK (e.g. Patogen Analyse, PHARMAQ Analytiq, Fish Vet Group) provide this service. Viral RNA can be detected in most, if not all, organs during viraemia, but the amount of viral RNA is usually higher in the heart (Andersen et al., 2007). The temporal duration of the PCR signal is also longer in this organ. PCR-based methods can be used to separate the different SAV subtypes (Hodneland and Endresen, 2006) and consequently these methods do provide more detailed information on epizootiology than most other detection methods.

Other methods of diagnosis include isolation of the virus in cell culture from serum or tissue homogenates during the viraemic phase, before an antibody response has been activated (Nelson et al., 1995; Castric et al., 1997; Desvignes et al., 2002). Most fish cell lines appear to be susceptible to the virus (Graham et al., 2008), but a cytopathic effect (CPE) can be weak or absent in some cases (Karlsen et al., 2006). Subsequent confirmation of in vitro SAV replication by specific molecular methods, e.g. the immunofluorescence antibody test (IFAT) (Todd et al., 2001; Moriette et al., 2005) is,

therefore recommended. Viral infection can also be detected in situ using antibody-mediated assays such as immunohistochemistry (Villoing et al., 2000b) or hybridization to RNA probes (Cano et al., 2015). These methods are useful for studying tissue tropism and pathogenesis, but are less sensitive and more laborious than PCR. As a result, they are not routinely used for diagnosis. Late-stage or completely cleared infections can be detected using a neutralizing antibody assay. A virus neutralization test has been developed, and is used particularly in Ireland and the UK to screen fish that have had previous exposure to SAV (Graham et al., 2003).

13.3 Pathology

13.3.1 Mortality in laboratory challenges

Early experimental infection studies with SAV did not lead to increased mortality, and resembled the subclinical infections sometimes reported from the field (Boucher et al., 1995; Boucher and Baudin Laurencin, 1996; McLoughlin et al., 1996; Andersen et al., 2007; Christie et al., 2007). Later studies have, though, demonstrated significant mortalities accompanied by clinical signs after experimental infection (Boscher et al., 2006; Moriette et al., 2006; Karlsen et al., 2012; Xu et al., 2012; Hikke et al., 2014a). Experimental infections with SAV do therefore adequately span the range of clinical signs and observations that have been reported from the field. Clinical disease can be reproduced through all the infection routes that have been tested: intraperitoneal and intramuscular injections, bathing and cohabitation. If the average SAV3 field mortality of 6.9% calculated by Jansen et al. (2010) is representative, then one would expect only two dead fish in a group of 30. Hence, many of the experimental infections reported in the literature lacked the statistical power to detect a mortality rate of this small size.

Factors such as host status, challenge titre and virus strain affect the severity of the disease in laboratory experiments. It is, however, possible to induce mortality rates higher than even extreme field cases, especially in injection models using high doses of virulent strains (Karlsen et al., 2012).

13.3.2 Histopathological changes

The name pancreas disease comes from the initial description of the disease, which reported pancreatic necrosis (Munro *et al.*, 1984). It later became evident that myopathy in the heart, skeletal muscle and oesophagus were also significant features of the disease (Ferguson *et al.*, 1986). These lesions develop in a sequential manner, with pancreatic necrosis followed by myopathy and inflammation of the heart (Boucher and Baudin Laurencin, 1994; McLoughlin *et al.*, 1996; Weston *et al.*, 2002). The lesions in skeletal muscle develop later.

The first histological observations in experimental infections are the necrosis of single cells in the heart (Fig. 13.3b) and pancreas (Fig. 13.4b) (see Mcloughlin and Graham, 2007). The necrotic myocardial cells develop a characteristic hyaline eosinophilic look with haematoxylin–eosin (HE) staining, and are often referred to as PD cells, and the process designated single cell necrosis (SCN). At this stage, there is little or no clinical sign of the disease. During clinical outbreaks in the field, the most common findings are severe inflammation in the ventricle of the heart (Fig. 13.3c) and necrosis and or atrophy of the pancreas (Figs 13.4b and 13.4c). Inflammation in the skeletal muscle may also be observed, especially in red muscle along the lateral side of the fish (Fig. 13.3d). Liver necrosis is often observed in fish with reduced capability to oxygenate the blood, and may be observed in fish with PD if the heart is severely affected (Fig. 13.4d).

The pancreas can fully regenerate in fish that survive the disease. Damage to the heart and skeletal muscle takes longer to heal and can reduce fillet quality (Taksdal *et al.*, 2012).

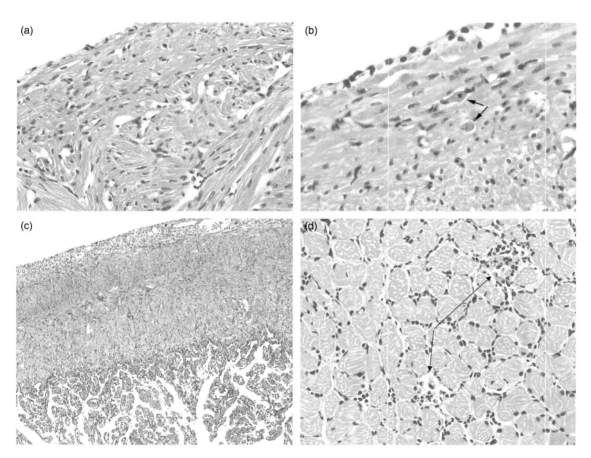

Fig. 13.3. Histological images from salmon using haematoxylin–eosin staining: (a) normal heart tissue; (b) ventricle in the early stage of a salmonid alphavirus (SAV) infection with single cell necrosis (arrows); (c) ventricle with severe inflammation from the clinical stage of a pancreas disease (PD)-outbreak; and (d) focal inflammation (arrows) in the red skeletal muscle of salmon with PD.

M. Karlsen and R. Johansen

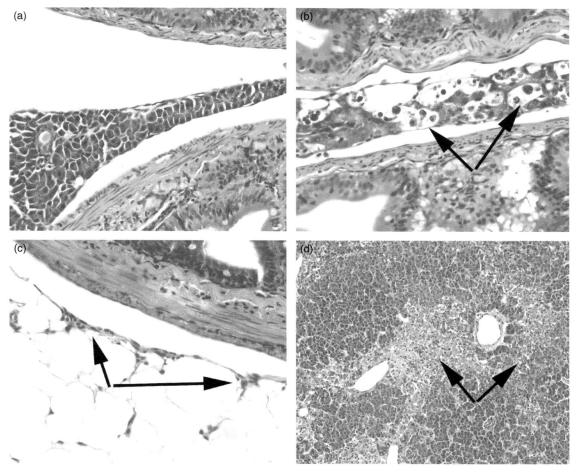

Fig. 13.4. Histological images from salmon using haematoxylin–eosin staining: (a) normal pancreatic tissue in between the small intestines; (b) necrosis of pancreatic cells in the early stages of pancreas disease (PD) (arrows); (c) pancreatic atrophy has led to the absence of pancreatic tissue in the clinical stage of PD (arrows); and (d) liver necrosis in a fish with heart failure due to salmonid alphavirus (SAV) infection (arrows).

13.3.3   Tissue tropism and replication *in vivo*

Alphaviruses in general are not very host cell specific. The replication apparatus of SAV is functional under diverse conditions, and in cells from fish, mammals, insects and crustaceans (Graham *et al.*, 2008; Olsen *et al.*, 2013; Hikke *et al.*, 2014b). Although the choice of host cell can be influenced by interactions between the replication apparatus and the host proteins, tissue tropism is probably decided more by the structural proteins, in particular E2, which covers most of the surface of the viral particle (Kuhn, 2007; Voss *et al.*, 2010; Karlsen *et al.*, 2015). It is assumed, based on homology

with other alphaviruses, that E2 is the receptor binding protein of SAV (Villoing *et al.*, 2000a). While there is considerable genetic diversity in E2 in the different SAV subtypes (Fringuelli *et al.*, 2008), they all seem to have qualitatively the same tissue tropism (Graham *et al.*, 2011). It is possible, however, that they may differ somewhat in quantitative terms in their choice of target organ.

The route of entry of SAV into the host is unknown, but the virus transmits via water contact (McLoughlin *et al.*, 1996). Cells in the intestine, gills and the skin are, therefore, plausible entry sites. Following the initial infection, SAV rapidly amplifies in the host and spreads systemically to several organs through a high-titre viraemia (Desvignes *et al.*, 2002; Andersen *et al.*, 2007). Viral RNA can be detected in all blood-rich organs in this phase. Andersen *et al.* (2007) demonstrated that in the later stages of infection (up to 190 days post intra-peritoneal injection), viral RNA can still be detected in the pseudobranch, gills, heart and kidney. It is not clear whether this persistence results in limited shedding of infectious particles, but laboratory experiments have linked the main shedding period to the viraemic phase of the infection (Andersen, 2012). Attempts to provoke the recurrence of a clinical outbreak in such carriers have not succeeded (Andersen *et al.*, 2007). Active replication has been shown to occur in pancreatic tissue, the heart, skeletal muscle and leucocytes (Houghton, 1995; Villoing *et al.*, 2000b; Moriette *et al.*, 2005; Andersen, 2012; Cano *et al.*, 2014). Recently, it was also demonstrated that muscle satellite cells are target cells for the virus in rainbow trout (Biacchesi *et al.*, 2016). Several alphaviruses are associated with neurotrophic infections and neuropathology (Strauss and Strauss, 1994), but this seems not to be a significant feature of SAV infections in salmonids (McLoughlin and Graham, 2007).

13.3.4 Virulence differences among strains

The strains of SAV do not appear to differ qualitatively in the pathology they cause, but they do show significant differences in the quantity of tissue damage that they cause (Weston *et al.*, 2002; Christie *et al.*, 2007; Graham *et al.*, 2011; Taksdal *et al.*, 2015). Although it is not clear whether differences in virulence are subtype specific, or if larger variation exists within each subtype, an initial comparison between SAV2 and SAV3 suggested that SAV3 is generally more virulent than SAV2 (Taksdal *et al.*, 2015).

SAV adapts to cell culture through mutations that lead to lower virulence in fish (Moriette *et al.*, 2006; Merour *et al.*, 2013; Petterson *et al.*, 2015), which has been observed for the strains SAV2 and SAV3. The changes in virulence are caused by mutations in E2, and a single substitution is enough to attenuate a pathogenic SAV2 in rainbow trout (Merour *et al.*, 2013). This demonstrates that even though SAV acts as a rather stable and predictable virus in its pathology, it has the capacity to rapidly change its characteristics. Substitutions in E2 have also been observed in field isolates of SAV3. One

substitution, from proline to serine in E2 position 206 now dominates in contemporary strains of the SAV3 subtype, but it has been suggested that it could affect the rate at which the virus transmits rather than its virulence (Karlsen *et al.*, 2015).

13.4 Pathophysiology

The main target organs of SAV (the exocrine pancreas, heart muscle and skeletal muscle) may be severely affected by the infection. Necrosis of pancreatic tissue probably reduces the ability to digest food, and this may be a reason behind the reduced growth and the production of abundant faecal casts that are often observed during outbreaks (McLoughlin and Graham, 2007). Another reason for reduced growth is the loss of appetite, which is a common general reaction of fish to severe infections.

Heart inflammation leads to reduced circulatory capacity and could explain some of the autopsy findings, such as discoloured liver, bleeding and ascites. Liver necrosis could be a consequence of the reduced capacity to oxygenate blood and, thus, possibly a secondary finding due to heart failure. Fish suffering from PD also commonly carry infections with heterologous pathogens in the gills, and this could reduce gas exchange (Nylund *et al.*, 2011). Circulatory failure could be particularly dangerous for fish with such co-infections.

Discoloration of the fillet is probably a consequence of the inflammatory response to SAV. The red muscle is also sensitive to low oxygen levels. Accordingly, circulatory failure may explain why the red muscle often is more affected than the white anaerobic muscle.

An interesting aspect of SAV infection at the cellular level is the ability of the virus to modulate transcription and/or translation (Xu *et al.*, 2010). Alphaviruses have several different strategies for achieving this, including the interactions of nsP2 and the capsid with cellular proteins. It has been shown that fish cells that express the SAV capsid cannot continue with cell division (Karlsen *et al.*, 2010b). It is likely that many of these transcriptomic changes are transient, and it is not clear how this influences the organism.

13.5 Control Strategies

In Norway, the driving reservoir of both SAV3 and SAV2 appears to be farmed salmonids (Viljugrein

M. Karlsen and R. Johansen

et al., 2009; Karlsen et al., 2014; Hjortaas et al., 2016). Mitigation strategies have, therefore, focused on this reservoir. It is likely that water contact between marine farms is one major route of transmission (Viljugrein et al., 2009). A second route of unknown importance is mechanical transmission via a vector, for example well-boats or other equipment that is shared between farms (Karlsen et al., 2014). The mitigation of SAV3 was intensified in Norway in 2007 when PD became a notifiable disease, and an endemic zone was defined on the west coast. Fish positive for SAV, or showing other signs of PD, are not allowed out of this zone. Sites within the zone are allowed to keep SAV-positive fish until slaughter, but the site has to be disinfected and fallowed before restocking. Infected fish outside the zone are immediately slaughtered and the farms fallowed to prevent further establishment of the disease. With the introduction of SAV2 north of the SAV3-endemic zone (Hjortaas et al., 2013), an SAV2-specific regulation was made to define an SAV2-endemic zone north of the SAV3-endemic zone. Marine farms that receive smolts from this zone must show freedom from SAV by PCR testing 2 and 4 months after transfer out of the zone. This regulation also specifies other actions, such as restrictions on the transport of SAV2-positive fish, and details for the disinfection of well-boats and net pens.

One strategy used by the industry to reduce economic losses due to PD is prescheduled harvesting. In this strategy, farmers take advantage of information generated using PCR-based monitoring of the infection status in the farm. As disease prevalence increases, it can be expected that a clinical outbreak is near and farmers may choose to slaughter fish to avoid the costs associated with mortality, reduced growth and poor fillet quality. It has been estimated that this strategy becomes cost-effective when the fish weigh >3.2 kg (Pettersen et al., 2015).

13.5.1 Vaccines

A monovalent inactivated water-in-oil formulated whole-virus vaccine based on the SAV1 type-strain F93-125 has been commercially available from MSD Animal Health in Ireland and in Norway since 2003 and in the UK since 2005. This vaccine has been used extensively in geographical regions where PD commonly appears. Although it has been reported to have a positive effect on some production parameters, its efficacy has not been sufficient to eliminate SAV from farmed fish populations, and vaccinated fish have continued to experience outbreaks of PD (Bang et al., 2012). Two multivalent vaccines based on F93-125 became available for sale in 2015 (also from MSD Animal Health).

Laboratory experiments, as well as commercial-scale field trials, have identified promising new vaccine candidates based on inactivated whole-virus technology (Karlsen et al., 2012), DNA plasmids (Hikke et al., 2014a; Simard and Horne, 2014), live attenuated strains (Moriette et al., 2006) and recombinantly expressed subunit antigens (Xu et al., 2012). Inactivated whole-virus vaccines have the benefit of following a relatively predictable regulatory pathway for approval. Their efficacies can be improved, for example through the addition of CpG oligonucleotides or poly(I:C) (polyinosinic–polycytidylic acid) to the formulation in order to activate a cellular immune response (Strandskog et al., 2011; Thim et al., 2012, 2014). Inactivated antigens also have an advantage over other vaccine technologies in that they are easier to include in multivalent vaccines, together with bacterial and viral antigens protecting against other diseases.

DNA vaccines expressing the structural proteins of SAV appear to give significant protection against infection provided that all of the glycoproteins are expressed together to ensure proper folding (Xu et al., 2012; Hikke et al., 2014a; Simard and Horne, 2014). This vaccine technology is already used to protect salmonids against infectious hematopoietic necrosis virus (IHNV) in Canada, but the registration process of DNA vaccines for food animals in Europe is less clear than that for inactivated whole-virus vaccines. Recently, however, a DNA vaccine against SAV was recommended to be granted marketing authorization in Europe.

An infection with SAV provides strong protection against subsequent infection (Lopez-Doriga et al., 2001). Live vaccines are, therefore, likely to provide very good efficacy, and an attenuated strain of SAV2 has been shown to protect rainbow trout fry against subsequent infections by the wild type virus (Moriette et al., 2006). The challenge with such technology is to ensure safe attenuation without loss of efficacy. The attenuated strain of SAV2 mentioned is attenuated through only one or two mutations (Merour et al., 2013). A safer strategy could be to delete parts of the genome, or to replace them with homologous parts from other alphaviruses. This has been done for alphaviruses that are pathogenic to humans, and is likely to provide a stronger hurdle for reversion to virulence. Reverse genetics systems that can be used to construct such attenuated strains have been made for the SAV2 and SAV3 strains (Moriette *et al.*, 2006; Karlsen *et al.*, 2010a; Guo *et al.*, 2014). In addition to regulatory requirements, several patents cover aspects of vaccine development against SAV, which may reduce the freedom to operate for pharmaceutical companies.

13.5.2   Other tools to mitigate SAV

Initiatives have been undertaken to breed salmon with improved resistance to SAV infection. This strategy has been successfully applied to improve the control of infectious pancreatic necrosis virus (IPNV), and it is likely that this has contributed significantly to the reduction of IPNV outbreaks in recent years. It appears that resistance towards PD also has a moderate-to-high heritability in Atlantic salmon, and a quantitative trait locus (QTL) affecting the resistance of the species to SAV infection was recently identified (Gonen *et al.*, 2015). It is still rather too preliminary to conclude that introduction of this QTL into populations of farmed salmon will reduce PD to the same degree as breeding for improved resistance to IPN, but major producers of ova offer eggs that have been selected on the basis of PD resistance.

'Functional feeds' designed to have beneficial effects during viral infection are marketed by feed companies such as EWOS, Skretting and Biomar. These are meant to either reduce inflammatory responses or to be more easily digested (e.g. Alne *et al.*, 2009; Martinez-Rubio *et al.*, 2012). Although considerable amounts of these feeds are used by the industry, few published studies have addressed their effect against PD, so it remains unclear how effective they are.

Disinfectants commonly used in aquaculture (Virkon S, Virex, Halamid, FAM30 and Buffodine) have been reported to be generally effective in inactivating SAV, but Virkon S and Virex were found to perform better under conditions with increased organic load or varying temperature (Graham *et al.*, 2007).

13.6   Conclusions and Future Perspectives

SAV is a major viral problem in European Atlantic salmon farming, and it also affects the rainbow trout industry. Infection causes pathology in the pancreas, heart and skeletal muscle, and may lead to reduced growth and fillet quality, as well as mortality. Several lines of work could help to improve mitigation of the disease. Improved preventive methods such as better vaccines or breeds of fish with improved resistance could potentially reduce transmission and improve control by challenge pressure within local transmission networks. In relation to this, it would be useful to document these networks better, and to monitor virus spread more closely using molecular techniques such as PCR and sequencing. This could greatly aid decision making on when and where to transfer new cohorts of fish.

One likely development in the future diagnostics of SAV and other RNA viruses is the increased focus on sequences, which contain a much richer amount of information than PCR signals. SAV evolves, like most RNA viruses, with a rather high rate of substitution (Karlsen *et al.*, 2014). A parental strain of SAV in farm A can give rise to strains of SAV in farms B and C that will be separated by several substitutions in their genome sequences just a year after the initial transmission event (Karlsen, 2015). This phylogenetic signal can be very useful for tracking transmission chains and identifying more accurately the mechanisms behind transmission.

Conflict of Interest Statement

The authors are employees of PHARMAQ and PHARMAQ Analytiq, subsidiaries of Zoetis, companies with financial interests in fish vaccines and diagnostics. MK is a shareholder in Zoetis.

References

Alne, H., Thomassen, M.S., Takle, H., Terjesen, B., Grammes, B.F. *et al.* (2009) Increased survival by dietary tetradecylthioacetic acid (TTA) during a natural outbreak of heart and skeletal muscle inflammation (HSMI) in S0 Atlantic salmon. *Journal of Fish Diseases* 32, 953–961.

Andersen, L. (2012) *Alphavirus Infection in Atlantic Salmon, Salmo salar L. – Viral Pathogenesis.* University of Bergen, Bergen, Norway.

Andersen, L., Bratland, A., Hodneland, K. and Nylund, A. (2007) Tissue tropism of salmonid alphaviruses (subtypes SAV1 and SAV3) in experimentally challenged Atlantic salmon (*Salmo salar* L.). *Archives of Virology* 152, 1871–1883.

Andersen, L., Hodneland, K. and Nylund, A. (2010) No influence of oxygen levels on pathogenesis and virus shedding in salmonid alphavirus (SAV)-challenged Atlantic salmon (*Salmo salar* L.). *Virology Journal* 7:198.

Aunsmo, A., Valle, P.S., Sandberg, M., Midtlyng, P.J. and Bruheim, T. (2010) Stochastic modelling of direct costs of pancreas disease (PD) in Norwegian farmed Atlantic salmon (*Salmo salar* L.). *Preventive Veterinary Medicine* 93, 233–241.

Bang, J.B., Kristoffersen, A.B., Myr, C. and Brun, E. (2012) Cohort study of effect of vaccination on pancreas disease in Norwegian salmon aquaculture. *Diseases of Aquatic Organisms* 102, 23–31.

Biacchesi, S., Jouvion, G., Mérour, E., Boukadiri, A., Desdouits, M. *et al.* (2016) Rainbow trout (*Oncorhynchus mykiss*) muscle satellite cells are targets of salmonid alphavirus infection. *Veterinary Research* 47:9.

Borzym, E., Maj-Paluch, J., Stachnik, M., Matras, M. and Reichert, M. (2014) First laboratory confirmation of salmonid alphavirus type 2 (SAV2) infection in Poland. *Bulletin of the Veterinary Institute in Pulawy* 58, 341–345.

Boscher, S.K., McLoughlin, M., Le Ven, A., Cabon, J., Baud, M. *et al.* (2006) Experimental transmission of sleeping disease in one-year-old rainbow trout, *Oncorhynchus mykiss* (Walbaum), induced by sleeping disease virus. *Journal of Fish Diseases* 29, 263–273.

Boucher, P. and Baudin Laurencin, F. (1994) Sleeping disease (SD) of salmonids. *Bulletin of the European Association Fish Pathologists* 14, 179–180.

Boucher, P. and Baudin Laurencin, F. (1996) Sleeping disease and pancreas disease: comparative histopathology and acquired cross protection. *Journal of Fish Diseases* 19, 303–310.

Boucher, P., Raynard, R.S., Hughton, G. and Baudin Laurencin, F. (1995) Comparative experimental transmission of pancreas disease in Atlantic salmon, rainbow trout and brown trout. *Diseases of Aquatic Organisms* 22, 19–24.

Bruno, D.W., Noguera, P.A., Black, J., Murray, W., Macqueen, D.J. *et al.* (2014) Identification of a wild reservoir of salmonid alphavirus in common dab *Limanda limanda*, with emphasis on virus culture and sequencing. *Aquaculture Environment Interactions* 5, 89–98.

Cano, I., Joiner, C., Bayley, A., Rimmer, G., Bateman, K. *et al.* (2015) An experimental means of transmitting pancreas disease in Atlantic salmon *Salmo salar* L. fry in freshwater. *Journal of Fish Diseases* 38, 271–281.

Castric, J., Baudin Laurencin, F., Brémont, M., Jeffroy, J., Le Ven, A. *et al.* (1997) Isolation of the virus responsible for sleeping disease in experimentally infected rainbow trout (*Oncorhynchus mykiss*). *Bulletin of the European Association Fish Pathologists* 17, 27–30.

Castric, J., Cabon, J. and Le Ven, A. (2005) Experimental study of vertical transmission of sleeping disease virus (SDV) in rainbow trout (*Oncorhynchus mykiss*). Conference poster presented at: *EAFP European Association of Fish Pathologists, 12th International Conference "Diseases of Fish and Shellfish", 11–16 September 2005, Copenhagen, Denmark, Volumes 1–2.* DIS Congress Service Copenhagen A/S, p. 95 [vol. not known].

Christie, K.E., Graham, D.A., McLoughlin, M.F., Villoing, S., Todd, D. *et al.* (2007) Experimental infection of Atlantic salmon *Salmo salar* pre-smolts by i.p. injection with new Irish and Norwegian salmonid alphavirus (SAV) isolates: a comparative study. *Diseases of Aquatic Organisms* 75, 13–22.

Crockford, T., Menzies, F.D., McLoughlin, M.F., Wheatley, S.B. and Goodall, E.A. (1999) Aspects of the epizootiology of pancreas disease in farmed Atlantic salmon *Salmo salar* in Ireland. *Diseases of Aquatic Organisms* 36, 113–119.

Desvignes, L., Quentel, C., Lamour, F. and Le, V.A. (2002) Pathogenesis and immune response in Atlantic salmon (*Salmo salar* L.) parr experimentally infected with salmon pancreas disease virus (SPDV). *Fish and Shellfish Immunology* 12, 77–95.

Ferguson, H.W., Roberts, R.J., Richards, R.H., Collins, R.O. and Rice, D.A. (1986) Severe degenerative cardiomyopathy associated with pancreas disease in Atlantic salmon, *Salmo salar* L. *Journal of Fish Diseases* 20, 95–98.

Fringuelli, E., Rowley, H.M., Wilson, J.C., Hunter, R., Rodger, H. *et al.* (2008) Phylogenetic analyses and molecular epidemiology of European salmonid alphaviruses (SAV) based on partial E2 and nsP3 gene nucleotide sequences. *Journal of Fish Diseases* 31, 811–823.

Gonen, S., Baranski, M., Thorland, I., Norris, A., Grove, H. et al. (2015) Mapping and validation of a major QTL affecting resistance to pancreas disease (salmonid alphavirus) in Atlantic salmon (*Salmo salar*). *Heredity* 115, 405–414.

Graham, D.A., Jewhurst, V.A., Rowley, H.M., McLoughlin, M.F. and Todd, D. (2003) A rapid immunoperoxidase-based virus neutralization assay for salmonid alphavirus used for a serological survey in Northern Ireland. *Journal of Fish Diseases* 26, 407–413.

Graham, D.A., Taylor, C., Rodgers, D., Weston, J., Khalili, M. *et al.* (2006a) Development and evaluation of a one-step real-time reverse transcription polymerase chain reaction assay for the detection of salmonid alphaviruses in serum and tissues. *Diseases of Aquatic Organisms* 70, 47–54.

Graham, D.A., Jewhurst, H., McLoughlin, M.F., Sourd, P., Rowley, H.M. *et al.* (2006b) Sub-clinical infection of farmed Atlantic salmon *Salmo salar* with salmonid alphavirus – a prospective longitudinal study. *Diseases of Aquatic Organisms* 72, 193–199.

Graham, D.A., Cherry, K., Wilson, C.J. and Rowley, H.M. (2007) Susceptibility of salmonid alphavirus to a range of chemical disinfectants. *Journal of Fish Diseases* 30, 269–277.

Graham, D.A., Wilson, C., Jewhurst, H. and Rowley, H. (2008) Cultural characteristics of salmonid alphaviruses – influence of cell line and temperature. *Journal of Fish Diseases* 31, 859–868.

Graham, D.A., Fringuelli, E., Wilson, C., Rowley, H.M., Brown, A. et al. (2010) Prospective longitudinal studies of salmonid alphavirus infections on two Atlantic salmon farms in Ireland; evidence for viral persistence. *Journal of Fish Diseases* 33, 123–135.

Graham, D.A., Frost, P., McLaughlin, K., Rowley, H.M., Gabestad, I. et al. (2011) A comparative study of marine salmonid alphavirus subtypes 1–6 using an experimental cohabitation challenge model. *Journal of Fish Diseases* 34, 273–286.

Graham, D.A., Fringuelli, E., Rowley, H.M., Cockerill, D., Cox, D.I. et al. (2012) Geographical distribution of salmonid alphavirus subtypes in marine farmed Atlantic salmon, *Salmo salar* L., in Scotland and Ireland. *Journal of Fish Diseases* 10, 755–765.

Guo, T.C., Johansson, D.X., Haugland, O., Liljestrom, P. and Evensen, Ø. (2014) A 6K-deletion variant of salmonid alphavirus is non-viable but can be rescued through RNA recombination. *PLoS ONE* 9(7): e100184.

Hikke, M.C., Braaen, S., Villoing, S., Hodneland, K., Geertsema, C. et al. (2014a) Salmonid alphavirus glycoprotein E2 requires low temperature and E1 for virion formation and induction of protective immunity. *Vaccine* 32, 6206–6212.

Hikke, M.C., Verest, M., Vlak, J.M. and Pijlman, G.P. (2014b) Salmonid alphavirus replication in mosquito cells: towards a novel vaccine production system. *Microbial Biotechnology* 7, 480–484.

Hjortaas, M.J., Skjelstad, H.R., Taksdal, T., Olsen, A.B., Johansen, R. et al. (2013) The first detections of subtype 2-related salmonid alphavirus (SAV2) in Atlantic salmon, *Salmo salar* L., in Norway. *Journal of Fish Diseases* 36, 71–74.

Hjortaas, M.J., Bang, J.B., Taksdal, T., Olsen, A.B., Lillehaug, A. et al. (2016) Genetic characterization of salmonid alphavirus in Norway. *Journal of Fish Diseases* 39, 249–257.

Hodneland, K. and Endresen, C. (2006) Sensitive and specific detection of salmonid alphavirus using real-time PCR (TaqMan). *Journal of Virological Methods* 131, 184–192.

Hodneland, K., Bratland, A., Christie, K.E., Endresen, C. and Nylund, A. (2005) New subtype of salmonid alphavirus (SAV), *Togaviridae*, from Atlantic salmon *Salmo salar* and rainbow trout *Oncorhynchus mykiss* in Norway. *Diseases of Aquatic Organisms* 66, 113–120.

Houghton, G. (1995) Kinetics of infection of plasma, blood leucocytes and lymphoid tissue from Atlantic salmon *Salmo salar* experimentally infected with pancreas disease. *Diseases of Aquatic Organisms* 22, 193–198.

Jansen, M.D., Taksdal, T., Wasmuth, M.A., Gjerset, B., Brun, E. et al. (2010) Salmonid alphavirus (SAV) and pancreas disease (PD) in Atlantic salmon, *Salmo salar* L., in freshwater and seawater sites in Norway from 2006 to 2008. *Journal of Fish Diseases* 33, 391–402.

Jansen, M.D., Jensen, B.B. and Brun, E. (2015) Clinical manifestations of pancreas disease outbreaks in Norwegian marine salmon farming – variations due to salmonid alphavirus subtype. *Journal of Fish Diseases* 38, 343–353.

Karlsen, M. (2015) *Salmonid Alphavirus Subtype 3 – Characterization by Phylogenetics and Reverse Genetics*. University of Bergen. Bergen, Norway.

Karlsen, M., Hodneland, K., Endresen, C. and Nylund, A. (2006) Genetic stability within the Norwegian subtype of salmonid alphavirus (family *Togaviridae*). *Archives of Virology* 151, 861–874.

Karlsen, M., Villoing, S., Ottem, K.F., Rimstad, E. and Nylund, A. (2010a) Development of infectious cDNA clones of salmonid alphavirus subtype 3. *BMC Research Notes* 3:241.

Karlsen, M., Yousaf, M.N., Villoing, S., Nylund, A. and Rimstad, E. (2010b) The amino terminus of the salmonid alphavirus capsid protein determines subcellular localization and inhibits cellular proliferation. *Archives of Virology* 155, 1281–1293.

Karlsen, M., Tingbo, T., Solbakk, I.T., Evensen, O., Furevik, A. et al. (2012) Efficacy and safety of an inactivated vaccine against salmonid alphavirus (family *Togaviridae*). *Vaccine* 30, 5688–5694.

Karlsen, M., Gjerset, B., Hansen, T. and Rambaut, A. (2014) Multiple introductions of salmonid alphavirus from a wild reservoir have caused independent and self-sustainable epizootics in aquaculture. *Journal of General Virology* 95, 52–59.

Karlsen, M., Andersen, L., Blindheim, S.H., Rimstad, E. and Nylund, A. (2015) A naturally occurring substitution in the E2 protein of salmonid alphavirus subtype 3 changes viral fitness. *Virus Research* 196, 79–86.

Kuhn, R.J. (2007) *Togaviridae*: the viruses and their replication. In: Knipe, D.M. and Howley, P.M. (eds) *Field's Virology*, 5th edn. Wolters Kluwer Health/Lippincott Williams & Wilkins, Philadelphia, Pennsylvania, pp. 1001–1022.

Lopez-Doriga, M.V., Smail, D.A., Smith, R.J., Domenech, A., Castric, J. et al. (2001) Isolation of salmon pancreas disease virus (SPDV) and its ability to protect against infection by the "wild-type" agent. *Fish and Shellfish Immunology* 11, 505–522.

Martinez-Rubio, L., Morais, S., Evensen, O., Wadsworth, S., Ruohonen, K. et al. (2012) Functional feeds reduce heart inflammation and pathology in Atlantic salmon (*Salmo salar* L.) following experimental challenge with Atlantic salmon reovirus (ASRV) *PLoS ONE* 7(11): e40266.

McCleary, S., Giltrap, M., Henshilwood, K. and Ruane, N.M. (2014) Detection of salmonid alphavirus RNA in Celtic and Irish Sea flatfish. *Diseases of Aquatic Organisms* 109, 1–7.

McLoughlin, M.F. and Graham, D.A. (2007) Alphavirus infections in salmonids – a review. *Journal of Fish Diseases* 30, 511–531.

McLoughlin, M.F., Nelson, R.T., Rowley, H.M., Cox, D.I. and Grant, A.N. (1996) Experimental pancreas disease in Atlantic salmon *Salmo salar* post-smolts induced by salmon pancreas disease virus (SPDV). *Diseases of Aquatic Organisms* 26, 117–124.

McLoughlin, M.F., Nelson, R.T., McCormack, J.E., Rowley, H.M. and Bryson, D.B. (2002) Clinical and histopathological features of naturally occurring pancreas disease in farmed Atlantic salmon, *Salmo salar* L. *Journal of Fish Diseases* 25, 33–43.

McVicar, A.H. (1987) Pancreas disease of farmed Atlantic Salmon, *Salmo salar*, in Scotland: epidemiology and early pathology. *Aquaculture* 67, 71–78.

Merour, E., Lamoureux, A., Bernard, J., Biacchesi, S. and Brémont, M. (2013) A fully attenuated recombinant salmonid alphavirus becomes pathogenic through a single amino acid change in the E2 glycoprotein. *Journal of Virology* 87, 6027–6030.

Moriette, C., LeBerre, M., Boscher, S.K., Castric, J. and Brémont, M. (2005) Characterization and mapping of monoclonal antibodies against the sleeping disease virus, an aquatic alphavirus. *Journal of General Virology* 86, 3119–3127.

Moriette, C., LeBerre, M., Lamoureux, A., Lai, T.L. and Brémont, M. (2006) Recovery of a recombinant salmonid alphavirus fully attenuated and protective for rainbow trout. *Journal of Virology* 80, 4088–4098.

Munro, A.L.S., Ellis, A.E., McVicar, A.H., McLay, H.A. and Needham, E.A. (1984) An exocrine pancreas disease of farmed Atlantic salmon in Scotland. *Helgoländer Meeresuntersuchungen* 37, 571–586.

Nelson, R.T., McLoughlin, M.F., Rowley, H.M., Platten, M.A. and McCormick, J.I. (1995) Isolation of a toga-like virus from farmed Atlantic salmon *Salmo salar* with pancreas disease. *Diseases of Aquatic Organisms* 22, 25–32.

Nylund, S., Andersen, L., Saevareid, I., Plarre, H., Watanabe, K. *et al.* (2011) Diseases of farmed Atlantic salmon *Salmo salar* associated with infections by the microsporidian *Paranucleospora theridion*. *Diseases of Aquatic Organisms* 16, 41–57. doi: 10.3354/dao02313.

Olsen, C.M., Pemula, A.K., Braaen, S., Sankaran, K. and Rimstad, E. (2013) Salmonid alphavirus replicon is functional in fish, mammalian and insect cells and *in vivo* in shrimps (*Litopenaeus vannamei*). *Vaccine* 31, 5672–5679.

Pettersen, J.M., Rich, K.M., Jensen, B.B. and Aunsmo, A. (2015) The economic benefits of disease triggered early harvest: a case study of pancreas disease in farmed Atlantic salmon from Norway. *Preventive Veterinary Medicine* 121, 314–324.

Petterson, E., Sandberg, M. and Santi, N. (2009) Salmonid alphavirus associated with *Lepeophtheirus salmonis* (Copepoda: Caligidae) from Atlantic salmon, *Salmo salar* L. *Journal of Fish Diseases* 32, 477–479.

Petterson, E., Guo, T.C., Evensen, O., Haugland, O. and Mikalsen, A.B. (2015) *In vitro* adaptation of SAV3 in cell culture correlates with reduced *in vivo* replication capacity and virulence to Atlantic salmon (*Salmo salar* L.) parr. *Journal of General Virology* 96, 3023–3034.

Powers, A.M., Brault, A.C., Shirako, Y., Strauss, E.G., Kang, W. *et al.* (2001) Evolutionary relationships and systematics of the alphaviruses. *Journal of Virology* 75, 10118–10131.

Simard, N. and Horne, M. (2014) Salmonid alphavirus and uses thereof. Patent Publication No. WO2014041189 A1, Application No. PCT/EP2013/069241. Available at: https://www.google.com/patents/WO2014041189A1?cl=zh-TW (accessed 8 November 2016).

Simons, J., Bruno, D.W., Ho, Y.M., Murray, W. and Matejusova, I. (2016) Common dab, *Limanda limanda* (L.), as a natural carrier of salmonid alphavirus (SAV) from waters off north–west Ireland. *Journal of Fish Diseases* 39, 507–510.

Snow, M., Black, J., Matejusova, I., McIntosh, R., Baretto, E. *et al.* (2010) Detection of salmonid alphavirus RNA in wild marine fish: implications for the origins of salmon pancreas disease in aquaculture. *Diseases of Aquatic Organisms* 91, 177–188.

Steele, K.E. and Twenhafel, N.A. (2010) Review paper: pathology of animal models of alphavirus encephalitis. *Veterinary Pathology* 47, 790–805.

Strandskog, G., Villoing, S., Iliev, D.B., Thim, H.L., Christie, K.E. *et al.* (2011) Formulations combining CpG containing oliogonucleotides and poly I:C enhance the magnitude of immune responses and protection against pancreas disease in Atlantic salmon. *Developmental and Comparative Immunology* 35, 1116–1127.

Strauss, J.H. and Strauss, E.G. (1994) The alphaviruses: gene expression, replication, and evolution. *Microbiological Reviews* 58, 491–562.

Taksdal, T., Olsen, A.B., Bjerkas, I., Hjortaas, M.J., Dannevig, B.H. *et al.* (2007) Pancreas disease in farmed Atlantic salmon, *Salmo salar* L., and rainbow trout, *Oncorhynchus mykiss* (Walbaum), in Norway. *Journal of Fish Diseases* 30, 545–558.

Taksdal, T., Wiik-Nielsen, J., Birkeland, S., Dalgaard, P. and Morkore, T. (2012) Quality of raw and smoked fillets from clinically healthy Atlantic salmon, *Salmo salar* L., following an outbreak of pancreas disease (PD). *Journal of Fish Diseases* 35, 897–906.

Taksdal, T., Bang, J.B., Bockerman, I., McLoughlin, M.F., Hjortaas, M.J. *et al.* (2015) Mortality and weight loss of Atlantic salmon, *Salmo salar* L., experimentally infected with salmonid alphavirus subtype 2 and subtype 3 isolates from Norway. *Journal of Fish Diseases* 38, 1047–1061.

Thim, H.L., Iliev, D.B., Christie, K.E., Villoing, S., McLoughlin, M.F. *et al.* (2012) Immunoprotective activity of a salmonid alphavirus vaccine: comparison of the immune responses induced by inactivated whole virus antigen formulations based on CpG class B oligonucleotides and poly I:C alone or combined with an oil adjuvant. *Vaccine* 30, 4828–4834.

Thim, H.L., Villoing, S., McLoughlin, M., Christie, K.E., Grove, S. *et al.* (2014) Vaccine adjuvants in fish vaccines make a difference: comparing three adjuvants (Montanide ISA763A Oil, CpG/Poly I:C Combo and VHSV Glycoprotein) alone or in combination formulated with an inactivated whole salmonid alphavirus antigen. *Vaccines (Basel)* 2, 228–251.

Todd, D., Jewhurst, V.A., Welsh, M.D., Borghmans, B.J., Weston, J.H. *et al.* (2001) Production and characterisation of monoclonal antibodies to salmon pancreas disease virus. *Diseases of Aquatic Organisms* 46, 101–108.

Viljugrein, H., Staalstrom, A., Molvaelr, J., Urke, H.A. and Jansen, P.A. (2009) Integration of hydrodynamics into a statistical model on the spread of pancreas disease (PD) in salmon farming. *Diseases of Aquatic Organisms* 88, 35–44.

Villoing, S., Béarzotti, M., Chilmonczyk, S., Castric, J. and Brémont, M. (2000a) Rainbow trout sleeping disease virus is an atypical alphavirus. *Journal of Virology* 74, 173–183.

Villoing, S., Castric, J., Jeffroy, J., Le Ven, A., Thiery, R. *et al.* (2000b) An RT-PCR-based method for the diagnosis of the sleeping disease virus in experimentally and naturally infected salmonids. *Diseases of Aquatic Organisms* 40, 19–27.

Voss, J.E., Vaney, M.C., Duquerroy, S., Vonrhein, C., Girard-Blanc, C. *et al.* (2010) Glycoprotein organization of Chikungunya virus particles revealed by X-ray crystallography. *Nature* 468, 709–712.

Weston, J., Villoing, S., Brémont, M., Castric, J., Pfeffer, M. *et al.* (2002) Comparison of two aquatic alphaviruses, salmon pancreas disease virus and sleeping disease virus, by using genome sequence analysis, monoclonal reactivity, and cross-infection. *Journal of Virology* 76, 6155–6163.

Xu, C., Guo, T.C., Mutoloki, S., Haugland, O., Marjara, I.S. *et al.* (2010) Alpha interferon and not gamma interferon inhibits salmonid alphavirus subtype 3 replication *in vitro*. *Journal of Virology* 84, 8903–8912.

Xu, C., Mutoloki, S. and Evensen, O. (2012) Superior protection conferred by inactivated whole virus vaccine over subunit and DNA vaccines against salmonid alphavirus infection in Atlantic salmon (*Salmo salar* L.). *Vaccine* 30, 3918–3928.

14 *Aeromonas salmonicida* and *A. hydrophila*

BJARNHEIDUR K. GUDMUNDSDOTTIR[1]* AND BRYNDIS BJORNSDOTTIR[2]

[1]*Faculty of Medicine, University of Iceland, Reykjavik, Iceland;*
[2]*Matís, Reykjavik, Iceland*

14.1 Introduction

Aeromonas belongs to the family *Aeromonadales* within the class *Gammaproteobacteria* (Colwell *et al.*, 1986). Aeromonads occur in freshwater, estuarine and marine environments, invertebrates, vertebrates and soils (Janda and Abbott, 2010). The type species is the motile *A. hydrophila*, an animal pathogen; in contrast, the species *A. salmonicida*, a fish pathogen, is non-motile. Aeromonads induce furunculosis, atypical furunculosis, ulcerative diseases, motile *Aeromonas* septicaemia (MAS) and tail and fin rot in fishes (Cipriano and Austin, 2011). *A. hydrophila* and other motile species (e.g. *A. veronii* biovar. *sobria*, *A. bestiarum*, *A. dhakensis*) cause diseases in aquaculture and are potentially zoonotic pathogens (Rahman *et al.*, 2002; Janda and Abbott, 2010; Austin and Austin, 2012a; Colston *et al.*, 2014). The intraspecies classification of *A. salmonicida* and *A. hydrophila* needs reassessment (Martin-Carnahan and Joseph, 2005). Genomic sequencing of some aeromonads has revealed the potential for horizontal gene transfer, which enhances bacterial adjustment to different environments and hosts (Piotrowska and Popowska, 2015).

A. salmonicida is among the most important fish pathogens worldwide. It consists of four psychrophilic subspecies, *A. s. salmonicida*, *A. s. achromogenes*, *A. s. masoucida* and *A. s. smithia*, and the mesophilic subspecies *A. s. pectinolytica* (Pavan *et al.*, 2000; Martin-Carnahan and Joseph, 2005). *A. s.* subsp. *salmonicida* is typical for the species and produces systemic furunculosis that affects many cold-water fishes. The disease had an economic impact on the farming of salmonids internationally until oil-adjuvanted furunculosis vaccines were marketed (Gudding and van Muiswinkel, 2013). Other atypical strains of *A. salmonicida* are heterogeneous both genetically and phenotypically (Austin *et al.*, 1998).

Atypical *A. salmonicida* strains also cause significant problems in aquaculture (Gudmundsdottir, 1998; Wiklund and Dalsgaard, 1998). *A. salmonicida* subsp. *pectinolytica* is the only non-pathogenic subspecies (Austin and Austin, 2012b). Vaccination against atypical furunculosis is problematic and the disease currently threatens the farming of Atlantic cod (*Gadus morhua*); there is also a problem among farmed Arctic charr (*Salvelinus alpinus*), Atlantic halibut (*Hippoglossus hippoglossus*), Atlantic wolffish (*Anarhichas lupus*), spotted wolffish (*Anarhichas minor*) and various 'cleaner fish' – wrasses (*Labridae*) and the lumpsucker (*Cyclopterus lumpus*) (Gulla *et al.*, 2015b). Ulcerative diseases caused by atypical *A. salmonicida* occur in goldfish (*Carassius auratus*), carp (*Cyprinus carpio*), flounder (*Platichthys flesus*) and many more species (Trust *et al.*, 1980; Wiklund and Dalsgaard, 1998).

Infection by *A. salmonicida* is transmitted horizontally through contaminated water, surface-infected eggs, vertebrate and non-vertebrate carriers, and equipment and clothing; vertical transmission has not been confirmed (Nomura, 1993; Wiklund, 1995). *A. salmonicida* attaches to the gills and skin/mucus regions of fish (Ferguson *et al.*, 1998). Except for New Zealand, atypical *A. salmonicida* infections occur worldwide. Typical *A. salmonicida*

*Corresponding author e-mail: bjarngud@hi.is

infections have not been reported within Australia, Chile and New Zealand.

The second species covered in this chapter, *A. hydrophila*, infects warm- and cool-blooded animals, including humans. It causes serious mortalities in farmed warm-water fishes (Janda and Abbott, 2010; Cipriano and Austin, 2011; Colston *et al.*, 2014). Highly virulent *A. hydrophila* of Asian origin has caused epizootics among channel catfish (*Ictalurus punctatus*) in the south-eastern USA since 2009 (Hossain *et al.*, 2014). *A. hydrophila* has two subspecies, *A. h. hydrophila* (Seshadri *et al.*, 2006) and *A. h. ranae* (Huys *et al.*, 2003); a third subspecies, *A. h. decolorationis*, has been proposed (Ren *et al.*, 2006). Furthermore, *A. hydrophila* belongs to three DNA hybridization groups (HGs) (Martin-Carnahan and Joseph, 2005). Most reports of *A. hydrophila* infections do not define the subspecies classification. The species is opportunistic and disease occurrence is commonly related to stress (e.g. high fish density, elevated temperature). Mortalities can also peak at low temperatures (Cipriano and Austin, 2011). Vaccination is problematic owing to strain variation and the nature of the different hosts (Austin and Austin, 2012b). *A. hydrophila* spreads horizontally through contaminated water, carrier fish, external parasites, equipment and clothing (Rusin *et al.*, 1997; Udeh, 2004; Austin and Austin, 2012a).

14.2 Diagnosis

Diagnosis is based on clinical signs, the history of the facility involved, and the isolation and identification of the bacterium concerned. Infections often result in septicaemias, in which the bacteria or bacterial products are found in numerous organs and with skin ulcerations, but mortalities also occur without any detectable pathology. Disease signs may include a darkened colour, lethargy, abnormal swimming behaviour, inappetance, pale gills, dermal ulcerations, fin and tail rot, erythrodermatitis, haemorrhages, septicaemia, bloating, and the protrusion or loss of scales. Bacterial identification is based on phenotypic characterization and molecular techniques (Martin-Carnahan and Joseph, 2005; Tenover, 2007; Austin and Austin, 2012a,b).

Aeromonads are Gram-negative diplobacilli 0.3–1.0 μm in diameter and 1.0–3.5 μm long. Coccoid forms occur and staining may be bipolar. They are cytochrome oxidase positive, facultative anaerobes that ferment glucose with or without the production of gas and are resistant to the vibriostatic agent O/129. The initial isolation is usually from the head kidney, skin lesions or gills. *A. salmonicida* is non-motile and does not grow at 37°C, whereas *A. hydrophila* is motile and does grow at this temperature (Martin-Carnahan and Joseph, 2005). The bacterium is cultured on tryptone soya agar (TSA) or brain heart infusion agar (BHIA), but many atypical *A. salmonicida* require blood agar (Cipriano and Bertolini, 1988; Wiklund, 1990; Gulla *et al.*, 2015b). The latter strains are very fastidious and are easily overgrown by other organisms. Therefore, other techniques have been developed to identify these, including enzyme-linked immunosorbent assay (ELISA), agglutination with specific antibodies (Adams and Thompson, 1990; Gilroy and Smith, 2003; Saleh *et al.*, 2011) and DNA sequence-based methods using 16S rRNA gene sequencing, PCR amplification and real-time quantitative PCR (qRT-PCR) techniques (Gustafson *et al.*, 1992; Byers *et al.*, 2002; Clarridge, 2004; Balcázar *et al.*, 2007; Beaz-Hidalgo *et al.*, 2013). The intraspecies grouping of *A. salmonicida* and *A. hydrophila* is complex and requires bioinformatic comparisons of genes or genomes (Colston *et al.*, 2014).

A. salmonicida has a protein surface A-layer protein (VapA) that gives the bacterium increased hydrophobicity and its autoagglutination characteristic (Chart *et al.*, 1984; Belland and Trust, 1985). The addition of Coomassie Brilliant Blue R-250 dye in media helps to identify A-layer positive isolates (Cipriano and Bertolini, 1988). Incubation at 15–20°C is recommended, because temperatures above 20°C may enhance loss of the A-layer (Moki *et al.*, 1995). Recent data indicate that *vapA* typing, based on sequence variation, separates *A. salmonicida* subspecies and recognizes undescribed subtypes (Gulla *et al.*, 2015a). The optimal growth temperature of *A. salmonicida* is 22–25°C and most strains do not grow at 37°C (Martin-Carnahan and Joseph, 2005). However, there is a motile biogroup that grows at 37°C and motility is induced by cultivation at 30–37°C (Mcintosh and Austin, 1991). Furthermore, some strains may be oxidase negative (Wiklund *et al.*, 1994). Typical *A. salmonicida* generally produce a water-soluble brown pigment when grown in tryptone media, but achromogenic strains do exist (Wiklund *et al.*, 1993). If atypical strains do produce brown pigment, the production of the pigment is slower. Schwenteit *et al.* (2011) reported that pigment production is regulated by quorum sensing (Fig. 14.1), and this may correlate with virulence.

A. hydrophila is motile and has a polar flagellum. It is a heterogeneous group whose members differ serologically and genotypically. Consequently, 16S rRNA gene sequences, in addition to phenotypic characterizations, are necessary for identification (MacInnes *et al.*, 1979; Valera and Esteve, 2002; Martin-Carnahan and Joseph, 2005).

14.3 Pathology

Furunculosis is a systemic disease caused by *A. salmonicida* subsp. *salmonicida* (McCarthy and Roberts, 1980), whereas atypical furunculosis is caused by other subspecies (Gudmundsdottir, 1998). Another form of disease caused by *Aeromonas* is skin ulcer disease, which is induced by some atypical *A. salmonicida* strains. Mortality is generally less severe in fish with the ulcerative disease than in systemic furunculosis (Austin and Austin, 2012b).

The systemic disease caused by *A. salmonicida* can be peracute, acute, subacute or chronic (McCarthy, 1975). Peracute disease is common in young fish that die without signs other than darkening and exophthalmos (protruding eyes). Fish that survive often develop lesions with hyperaemia and punctuate haemorrhages. Focal haemorrhages

and gill congestion may be evident. Bacteria are found in the anterior kidney, gills, spleen and myocardium, where necrosis of the cardiac tissue may occur. The disease progresses rapidly, with minimal host response. Acute furunculosis is manifested by general septicaemia; mortality is high and fish may die without detectable pathology. Hyperaemia occurs in all serosal surfaces and the spleen is often enlarged and red. The kidneys become soft and friable or liquefied. The liver is pale with subcapsular haemorrhages, or the surface may appear mottled owing to focal necrosis with ascites in the coelomic cavity. Skin lesions may develop, either as haemorrhagic patches along the body or as typical furuncles.

Toxic septicaemia changes are more severe in the acute than in the subacute and chronic forms of the systemic disease. Toxic haemopoieitic necrosis, and myocardial and renal tubule degeneration are seen irrespective of where the bacteria localize. The subacute and chronic forms of furunculosis are more common in older fish, which often survive and recover. Infected fish have a darkened skin and loss of appetite. Lethargy and furuncles are common. As shown in Fig. 14.2, the ulcers induced by typical strains of *A. salmonicida* extend deeper into

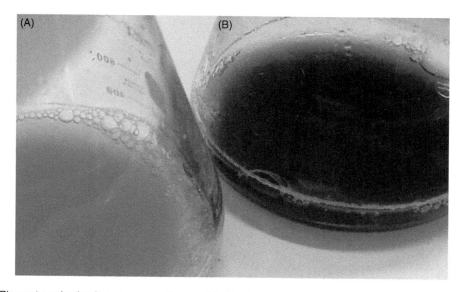

Fig. 14.1. Pigment production is quorum sensing regulated in *Aeromonas salmonicida*. The figures show cultures of *A. salmonicida* subsp. *achromogenes* strain Keldur 265-87 grown in brain heart infusion agar (BHIA) at 16°C for 190 h: (A) a Δ*asal* mutant (quorum-sensing negative); and (B) the wild type quorum-sensing positive strain. Photo: Bjarnheidur K Gudmundsdottir.

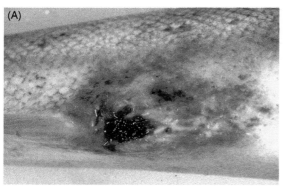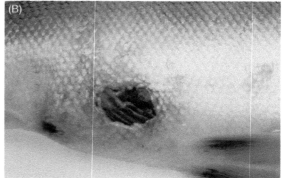

Fig. 14.2. Wild Atlantic salmon (*Salmo salar*; 2.5 kg average weight) from the river Ellidaár in Iceland suffering from chronic furunculosis induced by (A) *Aeromonas salmonicida* subsp. *salmonicida* and (B) *A. salmonicida* subsp. *achromogenes*, respectively. Photo: Ragnar Th. Sigurdsson.

the musculature, liquefactive necrosis is more prominent, and skin haemorrhages are more intensive than in fish infected with atypical strains. The gross pathology of subacute furunculosis is similar to the acute form but general visceral congestion and peritonitis are pronounced in the chronic form. The heart and spleen are frequently affected. Fish that survive epizootics become carriers that may spread the disease (McCarthy, 1975; Hellberg *et al.*, 1996; Gudmundsdottir, 1998; Bjornsdottir *et al.*, 2005).

The pathological features of Atlantic cod with systemic *A. salmonicida* subsp. *achromogenes* infection can be explained by distinct features of the cod immune system. The major histocompatibility complex class II (MHC class II) gene, which is fundamental for humoral response, is absent in cod (Star *et al.*, 2011). Therefore, cod rely on cellular and innate responses, in which granuloma formation (Fig. 14.3) is typical (Magnadottir *et al.*, 2002, 2013).

Ulcerative diseases induced by atypical *A. salmonicida* involve superficial pathology. The initial signs are small haemorrhages in skin that progress to multiple lesions commonly found on the flanks. Secondary infections by other bacteria are common. Fish can die without detectable bacteraemia, but bacteria are found systemically and within internal organs as the infection progresses (Mawdesley-Thomas, 1969; Bootsma *et al.*, 1977). Wiklund and Bylund (1993) divided ulcer development into three stages in the flounder. First, there is weak skin haemorrhage; secondly, this develops into white lesions rimmed by haemorrhagic inflammatory tissue;

finally, the ulcer manifests itself as eroded skin with exposed muscles. Other *A. salmonicida*-induced ulcer diseases include carp erythrodermatitis (Bootsma *et al.*, 1977), ulcer disease of salmonids (Paterson *et al.*, 1980; Bullock *et al.*, 1983), goldfish ulcer disease (Elliott and Shotts, 1980) and miscellaneous expressions in minnows (*Phoxinus phoxinus*) (Hastein *et al.*, 1978), Japanese eel (*Anguilla japonica*) (Kitao *et al.*, 1984) and rockfish (*Sebastes schlegeli*) (Han *et al.*, 2011).

The pathology of fish infected with *A. hydrophila* varies. Lesions may be restricted to the skin, but systemic infections also occur. Infections can result in acute, chronic and covert conditions (Cipriano and Austin, 2011). In acute MAS, fish may die before signs of the disease become visible. Signs may include exophthalmia, haemorrhage and necrosis of the skin, fins and oral cavity. Skin ulcers may develop in which necrosis extends into the muscle. Hepatic petechiation, enlargement of the spleen and swollen kidneys are common. The gills may haemorrhage and the scales protrude. Histopathologically, the skin and musculature may exhibit acute-to-chronic dermatitis and myositis with neutrophilic infiltrates. Alterations often involve necrosis in the kidney, liver, spleen and heart (Cipriano *et al.*, 1984; Cipriano and Austin, 2011). Pathological signs in carp are most severe in the liver and kidneys. Degeneration of the epithelium of the intestine and heart, and haemorrhages in the interstitial tissues of organs are common (Stratev *et al.*, 2015). Infections in the channel catfish can induce systemic and cutaneous disease (Ventura and Grizzle, 1988). In cutaneous infections,

B.K. Gudmundsdottir and B. Bjornsdottir

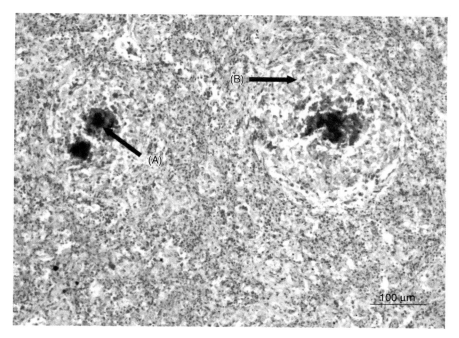

Fig. 14.3. Granulomas in the spleen of Atlantic cod (*Gadus morhua*) with atypical furunculosis immunostained with anti-*Aeromonas salmonicida* subsp. *achromogenes* antibodies. Granuloma centres consist of (A) bacterial colonies and cellular debris that are (B) surrounded by epitheloid cells. Haematoxylin staining with alkaline phosphatase staining of bound antibodies. Photo: Bergljot Magnadottir.

fish may have several types of concealed lesions that may be induced by bacterial toxins (Ventura and Grizzle, 1988).

14.4 Pathophysiology

14.4.1 Pathogenomics

The genomes of a few *Aeromonas* spp. are sequenced and available in the NCBI (US National Center for Biotechnology Information) database. Ten *A. salmonicida* genomes are available. The genome of the non-virulent A449 subspecies *salmonicida* is complete (Reith *et al.*, 2008) and consists of a circular chromosome and five plasmids. Uncompleted genomes of five typical strains and four atypical strains are also available, including the *A. salmonicida* subspecies *pectinolytica* (Pavan *et al.*, 2013), *masoucida* and *achromogenes* (Han *et al.*, 2013), although the strain identified as *achromogenes* may be subsp. *salmonicida* (Colston *et al.*, 2014). A microarray-based comparison of *A. salmonicida* virulence genes indicated subspecies variability (Nash *et al.*, 2006).

The genome (a single chromosome) of the *A. hydrophila* type strain ATCC 7966[T] was described by Seshadri *et al.* (2006). Analysis revealed that the strain has broad metabolic capabilities and numerous putative virulence genes and systems, which allows it to survive in diverse ecosystems. Reith *et al.* (2008) found and published a large number of dysfunctional genes in *A. salmonicida* subsp. *salmonicida* strain A449, which are restricted to this subspecies and include several putative virulence genes. The genome of A449 shows a distinctively high number of insertion sequence (IS) transposases. The majority of these transposases are within *A. salmonicida*-specific genes, suggesting that they may be used to exchange DNA through horizontal transfer or non-reproductive gene transfer (Studer *et al.*, 2013). The accumulation of pseudogenes has decreased the ability of *A. salmonicida* subsp. *salmonicida* to produce several organelles and synthesize some metabolic enzymes. These events are key to speciation and reflect its host specificity for salmonids (Reith *et al.*, 2008).

14.4.2   Virulence factors

The following is a summary of the virulence factors of *A. salmonicida* and *A. hydrophila*. For greater detail, see Tomás (2012), Beaz-Hidalgo and Figueras (2013) and Dallaire-Dufresne *et al.* (2014).

Surface adhesins

The crystalline surface layer of *Aeromonas* is the outermost cell envelope that is constructed from an auto-assembling single protein. In *A. salmonicida*, it is the A-layer, encoded by *vapA* (Chu *et al.*, 1991), and in *A. hydrophila* it is the S-layer, encoded by *ahsA* (Dooley and Trust, 1988). The A-layer is linked to lipopolysaccharide (LPS) and provides resistance to phagocytosis and protection against complement-mediated lysis; it also facilitates adhesion to host cells and immunoglobulins (Munn *et al.*, 1982; Phipps and Kay, 1988; Garduno *et al.*, 2000). Several other outer membrane proteins (OMPs) have been identified (Ebanks *et al.*, 2005), including complement-activating porin (Yu *et al.*, 2005).

The structure, biosynthesis and genomics of *A. salmonicida* and *A. hydrophila* LPS have been reviewed by Tomás (2012). The O-antigens of *A. salmonicida* are immunologically heterogeneous and the LPS core of each subspecies is unique (Wang *et al.*, 2007; Jimenez *et al.*, 2009). Merino *et al.* (2015) demonstrated that all strains of *A. salmonicida,* shared the same O-antigen biosynthesis gene cluster and that all subspecies, except for *pectinolytica*, produced an A-layer. Motile aeromonads, such as *A. hydrophila*, have a single polar flagellum (Canals *et al.*, 2007), but other strains also express lateral flagella in viscous environments. Both flagellum types are involved in adherence, invasion and biofilm formation (Kirov *et al.*, 2004). Non-motile *A. salmonicida* has disrupted gene sets for both flagella, which explains why they are not functional (Reith *et al.*, 2008). Genes encoding type I and three type IV pili systems are found in typical *A. salmonicida*, though no pili have been detected (Boyd *et al.*, 2008; Reith *et al.*, 2008). Type I pili in *A. salmonicida* aid adherence (Dacanay *et al.*, 2010) and type IV pili contribute to virulence (Masada *et al.*, 2002; Boyd *et al.*, 2008). Type IV pilus systems have also been identified in *A. hydrophila* (Seshadri *et al.*, 2006).

Secreted virulence components

Aeromonas produces and secretes toxins, lipases and peptidases (Pemberton *et al.*, 1997). Although typical and atypical *A. salmonicida* share most of their cell-associated factors, their secreted extracellular virulence factors vary (Gudmundsdottir *et al.*, 2003a; Nash *et al.*, 2006).

A. hydrophila produces both cytotoxic and cytotonic enterotoxins, such as aerolysin, Act, Ast and Alt (Seshadri *et al.*, 2006; Pang *et al.*, 2015). Aerolysin (AerA) in *A. salmonicida* and *A. hydrophila* acts as a pore-forming toxin and imparts the haemolytic and enterotoxic activities of *A. hydrophila* (Chakraborty *et al.*, 1987; Bucker *et al.*, 2011). Knockouts of three enterotoxins, Axt, Alt and Ast, were constructed in a clinical *A. hydrophila* diarrhoeal isolate (SSU) and each contributed to gastroenteritis by evoking fluid secretions in mice (Sha *et al.*, 2002). Several atypical *A. salmonicida* produce an aspzincin metalloendopeptidase (Gudmundsdottir *et al.*, 2003a), AsaP1, which is the major secreted virulence factor of subsp. *achromogenes*. AsaP1 is lethal, but a knockout of the peptidase gene impairs virulence (Arnadottir *et al.*, 2009). The typical *A. salmonicida* strain A449 does not produce a functional AsaP1 (Boyd *et al.*, 2008).

Several types of haemolysins and peptidases are associated with virulence, including a serine peptidase (AspA or P1) which causes muscle liquefaction and contributes to furuncle formation (Fyfe *et al.*, 1987). The peptidase also activates the lipase glycerophospholipid:cholesterol acyltransferase (GCAT). GCAT attacks membrane phospholipids, lyses fish erythrocytes (Ellis, 1991) and is produced by both *A. salmonicida* and *A. hydrophila* (Chacón *et al.*, 2002). Furthermore, GCAT complexes with LPS (GCAT-LPS), which is toxic for salmonids (Lee and Ellis, 1990). However, GCAT-deficient mutants are virulent (Vipond *et al.*, 1998).

Iron acquisition

The genome of *A. salmonicida* subsp. *salmonicida* has two complete gene clusters that encode siderophore synthesis and uptake. One encodes a putative catechol siderophore, anguibactin (or acinetobactin); the other encodes amonabactin, which is found in *A. hydrophila* (Stintzi and Raymond, 2000; Reith *et al.*, 2008). Recently, Balado *et al.* (2015) showed that several strains of subsp. *salmonicida* produce acinetobactin and amonabactin, but that some pathogenic strains only produced acinetobactin. They further concluded that acinetobactin type siderophores were limited to *A. salmonicida*, whereas amonabactin occurred throughout the genus. So far, only typical *A. salmonicida* produce siderophores but some atypical strains have siderophore-independent systems that sequester iron (Hirst *et al.*, 1991; Hirst and Ellis, 1996). Such independent mechanisms involve OMPs or receptors that specifically bind host proteins, such as haem, which then are internalized through a process dependent on TonB, a transport protein that interacts with OMPs. *A. salmonicida* and *A. hydrophila* have haem uptake genes (Ebanks *et al.*, 2004; Najimi *et al.*, 2008; Reith *et al.*, 2008). The expression of many genes involved in iron acquisition, both siderophore dependent and independent, is regulated by the ferric uptake regulator (Fur) iron-binding repressor protein (Braun *et al.*, 1998). When iron is limited, *A. salmonicida* exhibits iron-regulated outer membrane proteins, or IROMPs, including siderophores and haem receptors (Ebanks *et al.*, 2004; Menanteau-Ledouble *et al.*, 2014). Najimi *et al.* (2009) found significant genetic diversity of the iron-regulated genes in *A. salmonicida* subsp. *salmonicida*.

Secretion systems

Gram-negative bacteria have developed at least six secretion systems (T1SS–T6SS) which translocate proteins/virulence factors to host tissues (Costa *et al.*, 2015). *A. salmonicida* has T2SS and T3SS. The genes encoding a non-functional T6SS are present in strain A449 (Reith *et al.*, 2008; Vanden Bergh and Frey, 2014). The gene sets identified as the secretion system in subsp. *achromogenes* (ATCC 33659) are similar to the gene sets in typical strains (Nash *et al.*, 2006). *A. hydrophila* possesses T2SS, T3SS and TSS6, although T3SS is not present in all strains (Seshadri *et al.*, 2006; Pang *et al.*, 2015).

The T2SS is a two-step system, which secretes toxins and peptidases (Nivaskumar and Francetic, 2014). *Aeromonas* strains utilize the T2SS to secrete a range of factors, including aerolysin and GCAT (Brumlik *et al.*, 1997; Schoenhofen *et al.*, 1998).

T3SS secretes and translocates effector proteins or toxins in a single step (Portaliou *et al.*, 2015). Effector proteins disrupt the actin cytoskeleton, induce apoptosis, and prevent signal transduction and phagocytosis (Sierra *et al.*, 2010; Tosi *et al.*, 2013). Several *A. salmonicida* T3SS effector proteins have been described (Vanden Bergh and Frey, 2014) which are critical for pathogenicity (Burr *et al.*, 2005; Dacanay *et al.*, 2006). The genes

encoding the T3SS structural and regulatory proteins in *A. salmonicida* are located on a large plasmid (pAsa5), which when incubated above 22°C leads to loss of functional T3SS and avirulence (Daher *et al.*, 2011).

T6SS, like T3SS, delivers effector proteins directly into the host cytoplasm or competing bacteria (Cianfanelli *et al.*, 2016). *A. hydrophila* possesses T6SS, which induces cytotoxicity and contributes to innate immune evasion (Suarez *et al.*, 2010a,b; Sha *et al.*, 2013).

Quorum sensing

The quorum sensing systems in *Aeromonas* are homologous to LuxI/R (Jangid *et al.*, 2007), which is named AsaI/R and AhyI/R in *A. salmonicida* and *A. hydrophila*, respectively. The main autoinducer produced by both species is butanoyl-l-homoserine lactone, CH_4-AHL (Swift *et al.*, 1997; Schwenteit *et al.*, 2011; Tan *et al.*, 2015). Biofilm development (Lynch *et al.*, 2002) and the production of both serine protease and metalloprotease activities (Swift *et al.*, 1999) are regulated by quorum sensing in *A. hydrophila*. The production of the AsaP1 toxic metalloprotease, cytotoxicity and brown pigment is regulated by quorum sensing in *A. salmonicida* subsp. *achromogenes* (Schwenteit *et al.*, 2011). The virulence of both *A. salmonicida* and *A. hydrophila* to burbot (*Lota lota*) is quorum sensing regulated (Natrah *et al.*, 2012).

14.5 Protective and Control Strategies

14.5.1 Sound animal husbandry practices and disinfection

Quality rearing water, the education of personnel, good hygiene and sound animal husbandry limits disease on fish farms. Fish should be inspected and screened regularly for known pathogens. Disinfection is advised within hatcheries using recirculated water. Ozonation, filtration and ultraviolet irradiation are effective against aeromonads (Bullock and Stuckey, 1977; Colberg and Lingg, 1978). Fallowing is effective when tanks and gear can be desiccated. Fish eggs should be disinfected to kill bacteria on egg surfaces (Bergh *et al.*, 2001). When an aeromonad outbreak occurs, disinfection procedures may be necessary and isolation may be required before restocking (Torgersen and Hastein, 1995). Mainous *et al.* (2011) found that 1 min of contact with a number of agents (2% glutaraldehyde; 50 or 70% ethanol; 1% benzyl-4-chlorophenol/phenylphenol; 1% potassium peroxymonosulfate + sodium chloride; 50, 100, 200 or 50,000 mg/l sodium hypochlorite; 1:256 N-alkyl dimethyl benzyl ammonium chloride; 50 or 100 mg/l povidone–iodine) were active against *A. salmonicida* and *A. hydrophila* at 5–22°C. Formalin (250 mg/l) was ineffective at these temperatures. Chlorination (0.1 mg Cl_2/l soft water) killed 99.9% of *A. salmonicida* within 30 s, but the Cl_2 concentration needs to be doubled in hard water (Wedermeyer and Nelson, 1977).

14.5.2 Antibiotics and inhibition of bacterial pathogenicity

The control of aeromonad infections relies on the prudent use of effective antibiotics (Cabello, 2006). Requirements vary between countries, and culturists must consult the regulating authority in their country when a treatment is needed. Tetracyclines (oxytetracycline), quinolones (oxolinic acid, flumequine and enrofloxacin) and phenicols (florfenicol) are commonly used. Amoxicillin and sulfonamides may also be administered (Rodgers and Furones, 2009; Cipriano and Austin, 2011).

Resistance to antibiotics drives the need for new drugs. The seaweeds *Gracilaria folifera* and *Sargassum longifolium* have antibacterial activities against *A. salmonicida* (Thanigaivel *et al.*, 2015). Aniseed (*Pimpinella anisum*) is antagonistic to *A. salmonicida* and *A. hydrophila* (Parasa *et al.*, 2012). The plant *Dorycnium pentaphyllum* produces antibacterial activity against *A. hydrophila,* but does not inhibit *A. salmonicida* (Turker and Yıldırım, 2015).

The inhibition or inactivation of quorum sensing signal molecules, termed quorum quenching, attenuates processes regulated by quorum sensing and can potentially control diseases (Defoirdt *et al.*, 2011). *In vitro* studies have shown that quorum quenching can reduce the pathogenicity of *A. salmonicida* and *A. hydrophila* (Swift *et al.*, 1999; Rasch *et al.*, 2007; Schwenteit *et al.*, 2011), but this requires further study.

14.5.3 Non-specific immunostimulation

Probiotics

Probiotics not only improve fish growth and health, but combat aeromonad infections. A fluorescent pseudomonad strain limits the growth of *A. salmo-*

nicida by competing for free iron (Smith and Davey, 1993) and Gram-positive lactobacilli (LAB) reduce the mortality of rainbow trout (*Oncorhynchus mykiss*) challenged with *A. salmonicida* subsp. *salmonicida* (Nikoskelainen *et al.*, 2001). *A. hydrophila, Carnobacterium* sp., *Vibrio fluvialis* and an unidentified bacterium in the intestines of salmonids and turbot are antagonistic to *A. salmonicida* infections of rainbow trout (Irianto and Austin, 2002), and feeding rainbow trout with dead probiotic cells restricted the development of furunculosis (Irianto and Austin, 2003). Probiotic *V. alginolyticus* may induce resistance to furunculosis in Atlantic salmon (*Salmo salar*) (Austin *et al.*, 1995).

Vine *et al.* (2004) used the intestinal microbiota of clownfish (*Amphiprion percula*) to combat *A. hydrophila*. Feeding grass carp (*Ctenopharyngodon idellus*) for 28 days with probiotics based on *Shewanella xiamenensis* and *Aeromonas veronii*, reduced mortality following infection with *A. hydrophila* (Wu *et al.*, 2015). Cellular products from *Pseudomonas aeruginosa, Bacillus subtilis* and *Lactobacillus plantarum* also enhanced protection against *A. hydrophila* in rohu (*Labeo rohita*) (Giri *et al.*, 2015b).

Prebiotics and other immunostimulants

Feeding rainbow trout with mannanoligosaccharide (MOS)-enriched feed (2.5 g MOS/kg) improved growth and enhanced protection against *A. salmonicida* (Rodriguez-Estrada *et al.*, 2013), and feed enriched with beta-glucans can significantly enhance acute phase responses during *A. salmonicida* infection (Pionnier *et al.*, 2013).

Yersinia ruckeri flagellin has been used as a non-adjuvanted subunit vaccine that evoked non-specific immunoprotection in rainbow trout against *A. salmonicida* (Scott *et al.*, 2013). Immunostimulants of herbal origin have also gained increased attention. Ashagwanda (Indian ginseng, *Withania somnifera*) possesses several medicinal properties (Mirjalili *et al.*, 2009) and rohu resisted *A. hydrophila* infection when fed with root powder of ashagwanda (Sharma *et al.*, 2010). Feed containing garlic extract (0.5 or 1.0%) increased the resistance of rainbow trout to *A. salmonicida* infection, but higher doses were not effective and became detrimental to fish health (Breyer *et al.*, 2015). Feeding rohu with guava (*Psidium guajava*) leaves promoted growth and increased protection against *A. hydrophila* (Giri *et al.*, 2015a).

B.K. Gudmundsdottir and B. Bjornsdottir

Heat shock protein (Hsp)-enriched bacteria may control diseases in aquaculture (Sung, 2014). Enhancing the Hsp70 response by repeated handling during winter increased the resistance of gibel carp (*Carassius auratus gibelio*) and channel catfish to *A. hydrophila* (Yang *et al.*, 2015). Expression analysis of the genes encoding seven Hsp proteins during *A. hydrophila* infection in rohu has been reported (Das *et al.*, 2015); Hsp70 was downregulated, but HspA5 was upregulated in the liver, spleen and anterior kidney, indicating its potential immune regulatory role. HspA8 and Hsp70 may also mediate immune responses against *A. hydrophila* in Prenant's schizothoracin (*Schizothorax prenanti*) (Li *et al.*, 2015).

14.5.4 Vaccination

Vaccination has controlled furunculosis since the 1990s when bacterins of *A. salmonicida* subsp. *salmonicida* emulsified in oil adjuvant became commercially available. Currently, commercial vaccines contain inactivated bacterins and are most frequently licensed for salmonids. Parenteral furunculosis vaccines are used to vaccinate salmonids, turbot, koi carp (*Cyprinus carpio*) and several other species against typical and atypical furunculosis; they have varying success owing to the antigenic diversity among atypical *A. salmonicida* and differences in the immune responses of fishes. In some regions (e.g. Norway, the Faroe Islands and Iceland) vaccines against atypical furunculosis of salmonids and marine fishes, including the lumpsucker, are commercially available (Hastein *et al.*, 2005; Gudmundsdottir and Bjornsdottir, 2007; Coscelli *et al.*, 2015; see http://www.pharmaq.no/). Atlantic cod cannot be effectively vaccinated against pathogens with protective antigens that are T-cell dependent, and the lack of effective vaccines to protect cod against atypical furunculosis limits cod farming (FAO, 2014; Magnadottir, 2014).

There is significant correlation between circulating *A. salmonicida* antibodies and protection in Atlantic salmon (Romstad *et al.*, 2013), rainbow trout (Villumsen *et al.*, 2012), Arctic charr (Schwenteit *et al.*, 2014), Atlantic halibut (Gudmundsdottir *et al.*, 2003b), spotted wolffish (Ingilae *et al.*, 2000) and turbot (Bjornsdottir *et al.*, 2005; Coscelli *et al.*, 2015). Long-lasting protection against waterborne *A. salmonicida* in rainbow trout was induced by bath vaccination in a formalin-inactivated bacterin (Villumsen and Raida, 2013) and by intraperitoneal (IP) injection with bacterins emulsified in an oil adjuvant (Villumsen *et al.*, 2015). No elevation of circulating antibodies was detected among the immersion-vaccinated trout, but antibody levels correlated with protection in the IP-vaccinated fish.

The heterogeneity of *A. hydrophila* complicates the development of effective vaccines. There is no commercial vaccine, and current vaccines involve autogeneous preparations that are licensed within a restricted region (Yanong, 2008/9, rev. 2014). Vaccines based on native and recombinantly produced *A. hydrophila* proteins have induced immunoprotection. These include: an adhesin (Fang *et al.*, 2004); LPS (Sun *et al.*, 2012); OMPs (Khushiramani *et al.*, 2012; Maiti *et al.*, 2012; Sharma and Dixit, 2015); and haemolysin co-regulated protein (Hcp) (Wang *et al.*, 2015).

Vaccines based on live attenuated *A. hydrophila* have been constructed and evoke good protection in homologous challenges (Liu and Bi, 2007; Pridgeon *et al.*, 2013). Furthermore, the delivery of *A. hydrophila* antigens by live bacterial carriers (attenuated *Vibrio anguillarum*) has produced favourable vaccination results (Zhao *et al.*, 2011). Tu *et al.* (2010) used empty cell envelopes (ghosts) of *A. hydrophila* as well as a vaccine based on killed bacteria to orally vaccinate carp. The protective mucosal and systemic immune responses evoked by the ghosts were superior.

14.6 Conclusions and Suggestions for Future Research

Currently, around 350 species of finfish are farmed worldwide (FAO, 2014) and many are susceptible to aeromonad infections. The diversity of aeromonads in habitat, host susceptibility and pathogenic properties is considerable. Environmentally acceptable measures to control diseases are required for the continuous development of sustainable aquaculture, to prevent the transmission of diseases between wild and farmed fish, and to prevent the introduction of disease to new areas. In addition, the zoonotic spread of motile aeromonads and distribution of antibiotic resistance are major concerns.

The diagnosis of infection relies on accurate bacterial identification, but the taxonomy and identification of atypical *A. salmonicida* and motile aeromonads are uncertain and need clarification. Therefore, rapid and reliable methods to correctly identify and classify these bacteria are important for disease diagnosis and for epidemiological and prophylactic purposes. More whole genome sequencing is needed to establish multinational methods for the standardization of taxonomic and diagnostic tools. Furthermore, information on natural habitats of *A. salmonicida* is needed to reveal whether it has important reservoirs other than fishes.

Multidisciplinary measures are necessary to increase the prevention and control of disease caused by aeromonads. There still remain considerable gaps in understanding their pathogenicity, but it is hoped that the fast-growing knowledge of the genome sequences of pathogens and hosts, gene functions and their interactions will help to bridge these gaps.

References

Adams, A. and Thompson, K. (1990) Development of an enzyme-linked immunosorbent assay (ELISA) for the detection of *Aeromonas salmonicida* in fish tissue. *Journal of Aquatic Animal Health* 2, 281–288.

Arnadottir, H., Hvanndal, I., Andresdottir, V., Burr, S.E., Frey, J. and Gudmundsdottir, B.K. (2009) The AsaP1 peptidase of *Aeromonas salmonicida* subsp. *achromogenes* is a highly conserved deuterolysin metalloprotease (family M35) and a major virulence factor. *Journal of Bacteriology* 191, 403–410.

Austin, B. and Austin, D.A. (2012a) 4 *Aeromonadaceae* representatives (motile aeromonads). In: Austin, B. and Austin, D.A. *Bacterial Fish Pathogens: Disease of Farmed and Wild Fish*, 5th edn. Springer, Dordrecht, The Netherlands/Heidelberg, Germany/New York/London, pp. 119–146.

Austin, B. and Austin, D.A. (2012b) 5 *Aeromonadaceae* representative (*Aeromonas salmonicida*). In: Austin, B. and Austin, D.A. *Bacterial Fish Pathogens: Disease of Farmed and Wild Fish*, 5th edn. Springer, Dordrecht, The Netherlands/Heidelberg, Germany/New York/London, pp. 145–212.

Austin, B., Stuckey, L.F., Robertson, P.A.W., Effendi, I. and Griffith, D.R.W. (1995) A probiotic strain of *Vibrio alginolyticus* effective in reducing diseases caused by *Aeromonas salmonicida*, *Vibrio anguillarum* and *Vibrio ordalii*. *Journal of Fish Diseases* 18, 93–96.

Austin, B., Austin, D.A., Dalsgaard, I., Gudmundsdottir, B.K., Hoie, S. *et al.* (1998) Characterization of atypical *Aeromonas salmonicida* by different methods. *Systematic and Applied Microbiology* 21, 50–64.

Balado, M., Souto, A., Vences, A., Careaga, V.P., Valderrama, K. *et al.* (2015) Two catechol siderophores, acinetobactin and amonabactin, are simultaneously produced by *Aeromonas salmonicida* subsp. *salmonicida* sharing part of the biosynthetic pathway. *ACS Chemical Biology* 10, 2850–2860.

Balcázar, J.L., Vendrell, D., de Blas, I., Ruiz-Zarzuela, I., Gironés, O. and Múzquiz, J.L. (2007) Quantitative detection of *Aeromonas salmonicida* in fish tissue by real-time PCR using self-quenched, fluorogenic primers. *Journal of Medical Microbiology* 56, 323–328.

Beaz-Hidalgo, R. and Figueras, M.J. (2013) *Aeromonas* spp. whole genomes and virulence factors implicated in fish disease. *Journal of Fish Diseases* 36, 371–388.

Beaz-Hidalgo, R., Latif-Eugenin, F. and Figueras, M.J. (2013) The improved PCR of the *fstA* (ferric siderophore receptor) gene differentiates the fish pathogen *Aeromonas salmonicida* from other *Aeromonas* species. *Veterinary Microbiology* 166, 659–663.

Belland, R.J. and Trust, T.J. (1985) Synthesis, export, and assembly of *Aeromonas salmonicida* A-layer analyzed by transposon mutagenesis. *Journal of Bacteriology* 163, 877–881.

Bergh, O., Nilsen, F. and Samuelsen, O.B. (2001) Diseases, prophylaxis and treatment of the Atlantic halibut *Hippoglossus hippoglossus*: a review. *Diseases of Aquatic Organisms* 48, 57–74.

Bjornsdottir, B., Gudmundsdottir, S., Bambir, S.H. and Gudmundsdottir, B.K. (2005) Experimental infection of turbot, *Scophthalmus maximus* (L.), by *Aeromonas salmonicida* subsp *achromogenes* and evaluation of cross protection induced by a furunculosis vaccine. *Journal of Fish Diseases* 28, 181–188.

Bootsma, R., Fijan, N. and Blommaert, J. (1977) Isolation and identification of the causative agent of carp erythrodermatitis. *Veterinarski Arhiv* 47, 291–302.

Boyd, J.M., Dacanay, A., Knickle, L.C., Touhami, A., Brown, L.L. *et al.* (2008) Contribution of type IV pili to the virulence of *Aeromonas salmonicida* subsp *salmonicida* in Atlantic salmon (*Salmo salar* L.). *Infection and Immunity* 76, 1445–1455.

Braun, V., Hantke, K. and Koster, W. (1998) Bacterial iron transport: mechanisms, genetics, and regulation. *Metal Ions in Biological Systems* 35, 67–145.

Breyer, K.E., Getchell, R.G., Cornwell, E.R. and Wooster, A.A. (2015) Efficacy of an extract from garlic, *Allium sativum*, against infection with the furunculosis bacterium, *Aeromonas salmonicida*, in rainbow trout, *Oncorhynchus mykiss*. *Journal of the World Aquaculture Society* 46, 569–569.

Brumlik, M.J., van der Goot, F.G., Wong, K.R. and Buckley, J.T. (1997) The disulfide bond in the *Aeromonas hydrophila* lipase/acyltransferase stabilizes the structure but is not required for secretion or activity. *Journal of Bacteriology* 179, 3116–3121.

Bucker, R., Krug, S.M., Rosenthal, R., Gunzel, D., Fromm, A. *et al.* (2011) Aerolysin from *Aeromonas hydrophila* perturbs tight junction integrity and cell lesion repair in intestinal epithelial HT-29/B6 cells. *Journal of Infectious Diseases* 204, 1283–1292.

Bullock, G.L. and Stuckey, H.M. (1977) Ultraviolet treatment of water for destruction of five Gram-negative bacteria pathogenic to fishes. *Journal of the Fisheries Research Board of Canada* 34, 1244–1249.

Bullock, G.L., Cipriano, R.C. and Snieszko, S.F. (1983) Furunculosis and other diseases caused by *Aeromonas salmonicida*. Paper 133, US Fish and Wildlife Publications, US Fish and Wildlife Service, Washington, DC. Available at: http://digitalcommons.unl.edu/cgi/viewcontent.cgi?article=1132&context=usfwspubs (accessed 9 November 2016). Revised (2001) version available at: https://articles.extension.org/sites/default/files/w/b/b7/Furunculosis.pdf (accessed 9 November 2016).

Burr, S.E., Pugovkin, D., Wahli, T., Segner, H. and Frey, J. (2005) Attenuated virulence of an *Aeromonas salmonicida* subsp. *salmonicida* type III secretion mutant in a rainbow trout model. *Microbiology* 151, 2111–2118.

Byers, H.K., Cipriano, R.C., Gudkovs, N. and Crane, M.S. (2002) PCR-based assays for the fish pathogen *Aeromonas salmonicida*. II. Further evaluation and validation of three PCR primer sets with infected fish. *Diseases of Aquatic Organisms* 49, 139–144.

Cabello, F.C. (2006) Heavy use of prophylactic antibiotics in aquaculture: a growing problem for human and animal health and for the environment. *Environmental Microbiology* 8, 1137–1144.

Canals, R., Vilches, S., Wilhelms, M., Shaw, J.G., Merino, S. *et al.* (2007) Non-structural flagella genes affecting both polar and lateral flagella-mediated motility in *Aeromonas hydrophila*. *Microbiology* 153, 1165–1175.

Chacón, M.R., Castro-Escarpulli, G., Soler, L., Guarro, J. and Figueras, M.J. (2002) A DNA probe specific for *Aeromonas* colonies. *Diagnostic Microbiology and Infectious Disease* 44, 221–225.

Chakraborty, T., Huhle, B., Hof, H., Bergbauer, H. and Goebel, W. (1987) Marker exchange mutagenesis of the aerolysin determinant in *Aeromonas hydrophila* demonstrates the role of aerolysin in *A. hydrophila*-associated systemic infections. *Infection and Immunity* 55, 2274–2280.

Chart, H., Shaw, D.H., Ishiguro, E.E. and Trust, T.J. (1984) Structural and immunochemical homogeneity of *Aeromonas salmonicida* lipopolysaccharide. *Journal of Bacteriology* 158, 16–22.

Chu, S., Cavaignac, S., Feutrier, J., Phipps, B.M., Kostrzynska, M. *et al.* (1991) Structure of the tetragonal surface virulence array protein and gene of *Aeromonas salmonicida*. *The Journal of Biological Chemistry* 266, 15258–15265.

Cianfanelli, F.R., Monlezun, L. and Coulthurst, S.J. (2016) Aim, load, fire: the type VI secretion system, a bacterial nanoweapon. *Trends in Microbiology* 24, 51–62.

Cipriano, R.C. and Austin, B. (2011) Furunculosis and other aeromoniosis diseases. In: Woo, P.T.K. and Bruno, D.W. (eds) *Fish Diseases and Disorders, Volume 3: Viral,* *Bacterial and Fungal Infections*, 2nd edn. CAB International, Wallingford, UK.

Cipriano, R.C. and Bertolini, J. (1988) Selection for virulence in the fish pathogen *Aeromonas salmonicida*, using Coomassie Brilliant Blue agar. *Journal of Wildlife Diseases* 24, 672–678.

Cipriano, R.C., Bullock, G.L. and Pyle, S.W. (1984) *Aeromonas hydrophila* and motile aeromonad septicemias of fish. Paper 134, *US Fish and Wildlife Publications*, US Fish and Wildlife Service, Washington, DC. Available at: http://digitalcommons.unl.edu/usfwspubs/134/ (accessed 9 November 2016). Revised (2001) version available at: https://articles.extension.org/sites/default/files/w/1/1e/Aeromonas_hydrophila.pdf (accessed 9 November 2016).

Clarridge, J.E. 3rd (2004) Impact of 16S rRNA gene sequence analysis for identification of bacteria on clinical microbiology and infectious diseases. *Clinical Microbiology Reviews* 17, 840–862.

Colberg, P.J. and Lingg, A.J. (1978) Effect of ozoniation on microbial fish pathogens, ammonia, nitrate, nitrite and biological oxygen demand in simulated reuse hatchery water. *Journal of the Fisheries Research Board of Canada* 35, 1290–1296.

Colston, S.M., Fullmer, M.S., Beka, L., Lamy, B., Gogarten, J.P. *et al.* (2014) Bioinformatic genome comparisons for taxonomic and phylogenetic assignments using *Aeromonas* as a test case. *Mbio* 5(6):e02136.

Colwell, R.R., Macdonell, M.T. and Deley, J. (1986) Proposal to recognize the family *Aeromonadaceae* fam-nov. *International Journal of Systematic Bacteriology* 36, 473–477.

Coscelli, G.A., Bermúdez, R., Losada, A.P., Santos, A. and Quiroga, M.I. (2015) Vaccination against *Aeromonas salmonicida* in turbot (*Scophthalmus maximus* L.): study of the efficacy, morphological changes and antigen distribution. *Aquaculture* 445, 22–32.

Costa, T.R., Felisberto-Rodrigues, C., Meir, A., Prevost, M.S., Redzej, A. *et al.* (2015) Secretion systems in Gram-negative bacteria: structural and mechanistic insights. *Nature Reviews Microbiology* 13, 343–359.

Dacanay, A., Knickle, L., Solanky, K.S., Boyd, J.M., Walter, J.A. *et al.* (2006) Contribution of the type III secretion system (TTSS) to virulence of *Aeromonas salmonicida* subsp. *salmonicida*. *Microbiology* 152, 1847–1856.

Dacanay, A., Boyd, J.M., Fast, M.D., Knickle, L.C. and Reith, M.E. (2010) *Aeromonas salmonicida* type I pilus system contributes to host colonization but not invasion. *Diseases of Aquatic Organisms* 88, 199–206.

Daher, R.K., Filion, G., Tan, S.G., Dallaire-Dufresne, S., Paquet, V.E. *et al.* (2011) Alteration of virulence factors and rearrangement of pAsa5 plasmid caused by the growth of *Aeromonas salmonicida* in stressful conditions. *Veterinary Microbiology* 152, 353–360.

Dallaire-Dufresne, S., Tanaka, K.H., Trudel, M.V., Lafaille, A. and Charette, S.J. (2014) Virulence, genomic features, and plasticity of *Aeromonas salmonicida* subsp. *salmonicida*, the causative agent of fish furunculosis. *Veterinary Microbiology* 169, 1–7.

Das, S., Mohapatra, A. and Sahoo, P.K. (2015) Expression analysis of heat shock protein genes during *Aeromonas hydrophila* infection in rohu, *Labeo rohita*, with special reference to molecular characterization of Grp78. *Cell Stress Chaperones* 20, 73–84.

Defoirdt, T., Sorgeloos, P. and Bossier, P. (2011) Alternatives to antibiotics for the control of bacterial disease in aquaculture. *Current Opinion in Microbiology* 14, 251–258.

Dooley, J.S. and Trust, T.J. (1988) Surface protein composition of *Aeromonas hydrophila* strains virulent for fish: identification of a surface array protein. *Journal of Bacteriology* 170, 499–506.

Ebanks, R.O., Dacanay, A., Goguen, M., Pinto, D.M. and Ross, N.W. (2004) Differential proteomic analysis of *Aeromonas salmonicida* outer membrane proteins in response to low iron and *in vivo* growth conditions. *Proteomics* 4, 1074–1085.

Ebanks, R.O., Goguen, M., Mckinnon, S., Pinto, D.M. and Ross, N.W. (2005) Identification of the major outer membrane proteins of *Aeromonas salmonicida*. *Diseases of Aquatic Organisms* 68, 29–38.

Elliott, D.G. and Shotts, E.B. (1980) Aetiology of an ulcerative disease in goldfish *Carassius auratus* (L): microbiological examination of diseased fish from seven locations. *Journal of Fish Diseases* 3, 133–143.

Ellis, A.E. (1991) An appraisal of the extracellular toxins of *Aeromonas salmonicida* ssp. *salmonicida*. *Journal of Fish Diseases* 14, 265–277.

Fang, H.M., Ge, R.W. and Sin, Y.M. (2004) Cloning, characterisation and expression of *Aeromonas hydrophila* major adhesin. *Fish and Shellfish Immunology* 16, 645–658.

FAO (2014) *The State of World Fisheries and Aquaculture: Opportunities and Challenges*. Food and Agriculture Organization of the United Nations, Rome.

Ferguson, Y., Bricknell, I.R., Glover, L.A., Macgregor, D.M. and Prosser, J.I. (1998) Colonisation and transmission of lux-marked and wild-type *Aeromonas salmonicida* strains in Atlantic salmon (*Salmo salar* L.). *FEMS Microbiology Ecology* 27, 251–260.

Fyfe, L., Coleman, G. and Munro, A.L.S. (1987) Identification of major common extracellular proteins secreted by *Aeromonas salmonicida* strains isolated from diseased fish. *Applied and Environmental Microbiology* 53, 722–726.

Garduno, R.A., Moore, A.R., Olivier, G., Lizama, A.L., Garduno, E. *et al.* (2000) Host cell invasion and intracellular residence by *Aeromonas salmonicida*: role of the S-layer. *Canadian Journal of Microbiology* 46, 660–668.

Gilroy, D. and Smith, P. (2003) Application-dependent, laboratory-based validation of an enzyme-linked immunosorbent assay for *Aeromonas salmonicida*. *Aquaculture* 217, 23–38.

Giri, S.S., Sen, S.S., Chi, C., Kim, H.J., Yun, S. *et al.* (2015a) Effect of guava leaves on the growth performance and cytokine gene expression of *Labeo rohita* and its susceptibility to *Aeromonas hydrophila* infection. *Fish and Shellfish Immunology* 46, 217–224.

Giri, S.S., Sen, S.S., Chi, C., Kim, H.J., Yun, S. *et al.* (2015b) Effect of cellular products of potential probiotic bacteria on the immune response of *Labeo rohita* and susceptibility to *Aeromonas hydrophila* infection. *Fish and Shellfish Immunology* 46, 716–722.

Gudding, R. and van Muiswinkel, W.B. (2013) A history of fish vaccination: science-based disease prevention in aquaculture. *Fish and Shellfish Immunology* 35, 1683–1688.

Gudmundsdottir, B.K. (1998) Infections by atypical strains of the bacterium *Aeromonas salmonicida*: a review. *Icelandic Agricultural Sciences* 12, 61–72.

Gudmundsdottir, B.K. and Bjornsdottir, B. (2007) Vaccination against atypical furunculosis and winter ulcer disease of fish. *Vaccine* 25, 5512–5523.

Gudmundsdottir, B.K., Hvanndal, I., Bjornsdottir, B. and Wagner, U. (2003a) Analysis of exotoxins produced by atypical isolates of *Aeromonas salmonicida*, by enzymatic and serological methods. *Journal of Fish Diseases* 26, 15–29.

Gudmundsdottir, S., Lange, S., Magnadottir, B. and Gudmundsdottir, B.K. (2003b) Protection against atypical furunculosis in Atlantic halibut, *Hippoglossus hippoglossus*, comparison of fish vaccinated with commercial furunculosis vaccine and an autogenous vaccine based on the challenge strain. *Journal of Fish Diseases* 26, 331–338.

Gulla, S., Lund, V., Kristoffersen, A.B., Sørum, H. and Colquhoun, D.J. (2015a) *vapA* (A-layer) typing differentiates *Aeromonas salmonicida* subspecies and identifies a number of previously undescribed subtypes. *Journal of Fish Diseases* 39, 329–342.

Gulla, S., Duodu, S., Nilsen, A., Fossen, I. and Colquhoun, D.J. (2015b) *Aeromonas salmonicida* infection levels in pre- and post-stocked cleaner fish assessed by culture and an amended qPCR assay. *Journal of Fish Diseases* 39, 867–877.

Gustafson, C.E., Thomas, C.J. and Trust, T.J. (1992) Detection of *Aeromonas salmonicida* from fish by using polymerase chain reaction amplification of the virulence surface array protein gene. *Applied and Environmental Microbiology* 58, 3816–3825.

Han, H.J., Kim, D.Y., Kim, W.S., Kim, C.S., Jung, S.J. *et al.* (2011) Atypical *Aeromonas salmonicida* infection in the black rockfish, *Sebastes schlegeli* Hilgendorf, in Korea. *Journal of Fish Diseases* 34, 47–55.

Han, J.E., Kim, J.H., Shin, S.P., Jun, J.W., Chai, J.Y. *et al.* (2013) Draft genome sequence of *Aeromonas salmonicida* subsp. *achromogenes* AS03, an atypical strain isolated from crucian carp (*Carassius carassius*) in the Republic of Korea. *Genome Announcements* 1(5): e00791-13.

Hastein, T., Saltveit, S.J. and Roberts, R.J. (1978) Mass mortality among minnows *Phoxinus phoxinus* (L) in Lake Tveitevatn, Norway, due to an aberrant strain of *Aeromonas salmonicida*. *Journal of Fish Diseases* 1, 241–249.

Hastein, T., Gudding, R. and Evensen, O. (2005) Bacterial vaccines for fish: an update of the current situation worldwide. *Developmental Biology* 121, 55–74.

Hellberg, H., Moksness, E. and Hoie, S. (1996) Infection with atypical *Aeromonas salmonicida* in farmed common wolffish, *Anarhichas lupus* L. *Journal of Fish Diseases* 19, 329–332.

Hirst, I.D. and Ellis, A.E. (1996) Utilization of transferrin and salmon serum as sources of iron by typical and atypical strains of *Aeromonas salmonicida*. *Microbiology* 142, 1543–1550.

Hirst, I.D., Hastings, T.S. and Ellis, A.E. (1991) Siderophore production by *Aeromonas salmonicida*. *Journal of Genetic Microbiology* 137, 1185–1192.

Hossain, M.J., Sun, D.W., McGarey, D.J., Wrenn, S., Alexander, L.M. *et al.* (2014) An Asian origin of virulent *Aeromonas hydrophila* responsible for disease epidemics in United States-farmed catfish. *Mbio* 5, e00848–00814.

Huys, G., Pearson, M., Kampfer, P., Denys, R., Cnockaert, M. *et al.* (2003) *Aeromonas hydrophila* subsp. *ranae* subsp. nov., isolated from septicaemic farmed frogs in Thailand. *International Journal of Systematic and Evolutionary Microbiology* 53, 885–891.

Ingilae, M., Arnesen, J.A., Lund, V. and Eggset, G. (2000) Vaccination of Atlantic halibut *Hippoglossus hippoglossus* L., and spotted wolffish *Anarhichas minor* L., against atypical *Aeromonas salmonicida*. *Aquaculture* 183, 31–44.

Irianto, A. and Austin, B. (2002) Use of probiotics to control furunculosis in rainbow trout, *Oncorhynchus mykiss* (Walbaum). *Journal of Fish Diseases* 25, 333–342.

Irianto, A. and Austin, B. (2003) Use of dead probiotic cells to control furunculosis in rainbow trout, *Oncorhynchus mykiss* (Walbaum). *Journal of Fish Diseases* 26, 59–62.

Janda, J.M. and Abbott, S.L. (2010) The genus *Aeromonas*: taxonomy, pathogenicity, and infection. *Clinical Microbiology Reviews* 23, 35–73.

Jangid, K., Kong, R., Patole, M.S. and Shouche, Y.S. (2007) *LuxRI* homologs are universally present in the genus *Aeromonas*. *BMC Microbiology* 7:93.

Jimenez, N., Lacasta, A., Vilches, S., Reyes, M., Vazquez, J. *et al.* (2009) Genetics and proteomics of *Aeromonas salmonicida* lipopolysaccharide core biosynthesis. *Journal of Bacteriology* 191, 2228–2236.

Khushiramani, R.M., Maiti, B., Shekar, M., Girisha, S.K., Akash, N. *et al.* (2012) Recombinant *Aeromonas hydrophila* outer membrane protein 48 (Omp48) induces a protective immune response against *Aeromonas hydrophila* and *Edwardsiella tarda*. *Research in Microbiology* 163, 286–291.

Kirov, S.M., Castrisios, M. and Shaw, J.G. (2004) *Aeromonas* flagella (polar and lateral) are enterocyte adhesins that contribute to biofilm formation on surfaces. *Infection and Immunity* 72, 1939–1945.

Kitao, T., Yoshida, T., Aoki, T. and Fukudome, M. (1984) Atypical *Aeromonas salmonicida*, the causative agent of an ulcer disease of eel occurred in Kagoshima Prefecture. *Fish Pathology* 19, 113–117.

Lee, K.K. and Ellis, A.E. (1990) Glycerophospholipid:cholesterol acyltransferase complexed with lipopolysaccharide (LPS) is a major lethal exotoxin and cytolysin of *Aeromonas salmonicida*: LPS stabilizes and enhances toxicity of the enzyme. *Journal of Bacteriology* 172, 5382–5393.

Li, J.X., Zhang, H.B., Zhang, X.Y., Yang, S.Y., Yan, T.M. *et al.* (2015) Molecular cloning and expression of two heat-shock protein genes (HSC70/HSP70) from Prenant's schizothoracin (*Schizothorax prenanti*). *Fish Physiology and Biochemistry* 41, 573–585.

Liu, Y.J. and Bi, Z.X. (2007) Potential use of a transposon Tn916-generated mutant of *Aeromonas hydrophila* J-1 defective in some exoproducts as a live attenuated vaccine. *Preventive Veterinary Medicine* 78, 79–84.

Lynch, M.J., Swift, S., Kirke, D.F., Keevil, C.W., Dodd, C.E.R. *et al.* (2002) The regulation of biofilm development by quorum sensing in *Aeromonas hydrophila*. *Environmental Microbiology* 4, 18–28.

MacInnes, J.I., Trust, T.J. and Crosa, J.H. (1979) Deoxyribonucleic-acid relationships among members of the genus *Aeromonas*. *Canadian Journal of Microbiology* 25, 579–586.

Magnadottir, B. (2014) The immune response of Atlantic cod, *Gadus morhua* L. *Icelandic Agricultural Sciences* 27, 41–61.

Magnadottir, B., Bambir, S.H., Gudmundsdottir, B.K., Pilstrom, L. and Helgason, S. (2002) Atypical *Aeromonas salmonicida* infection in naturally and experimentally infected cod, *Gadus morhua* L. *Journal of Fish Diseases* 25, 583–597.

Magnadottir, B., Gudmundsdottir, B.K. and Groman, D. (2013) Immuno-histochemical determination of humoral immune markers within bacterial induced granuloma formation in Atlantic cod (*Gadus morhua* L.). *Fish and Shellfish Immunology* 34, 1372–1375.

Mainous, M.E., Kuhn, D.D. and Smith, S.A. (2011) Efficacy of common aquaculture compounds for disinfection of *Aeromonas hydrophila*, *A. salmonicida* subsp. *salmonicida*, and *A. salmonicida* subsp. *achromogenes* at various temperatures. *North American Journal of Aquaculture* 73, 456–461.

Maiti, B., Shetty, M., Shekar, M., Karunasagar, I. and Karunasagar, I. (2012) Evaluation of two outer membrane proteins, Aha1 and OmpW of *Aeromonas hydrophila* as vaccine candidate for common carp. *Veterinary Immunology and Immunopathology* 149, 298–301.

Martin-Carnahan, A. and Joseph, S.W. (2005) Family I. *Aeromononadaceae*. Order XII. *Aeromonadales* ord.

nov. In: Garrity, G.M., Brenner, D.J., Krieg, N.R. and Staley, J.R. (eds) *Bergey's Manual of Systematic Bacteriology. Volume 2: The Proteobacteria, Part B: The Gammaproteobacteria*, 2nd edn. Springer, New York, pp. 556–587.

Masada, C.L., Lapatra, S.E., Morton, A.W. and Strom, M.S. (2002) An *Aeromonas salmonicida* type IV pilin is required for virulence in rainbow trout *Oncorhynchus mykiss*. *Diseases of Aquatic Organisms* 51, 13–25.

Mawdesley-Thomas, L.E. (1969) Furunculosis in the goldfish *Carassius auratus* (L). *Journal of Fish Biology* 1, 19–23.

McCarthy, D.H. (1975) Fish furunculosis caused by *Aeromonas salmonicida* var. *achromogenes*. *Journal of Wildlife Diseases* 11, 489–493.

McCarthy, D.H. and Roberts, R.J. (1980) Furunculosis of fish – the present state of our knowledge. In: Droop, M.A. and Jannasch, H.W. (eds) *Advances in Aquatic Microbiology*. Academic Press, London.

McIntosh, D. and Austin, B. (1991) Atypical characteristics of the salmonid pathogen *Aeromonas salmonicida*. *Journal of General Microbiology* 137, 1341–1343.

Menanteau-Ledouble, S., Kattlun, J., Nobauer, K. and El-Matbouli, M. (2014) Protein expression and transcription profiles of three strains of *Aeromonas salmonicida* ssp. *salmonicida* under normal and iron-limited culture conditions. *Proteome Sciences* 12: 29.

Merino, S., Canals, R., Knirel, Y.A. and Tomás, J.M. (2015) Molecular and chemical analysis of the lipopolysaccharide from *Aeromonas hydrophila* strain AH-1 (Serotype O11). *Marine Drugs* 13, 2233–2249.

Mirjalili, M.H., Moyano, E., Bonfill, M., Cusido, R.M. and Palazon, J. (2009) Steroidal lactones from *Withania somnifera*, an ancient plant for novel medicine. *Molecules* 14, 2373–2393.

Moki, S.T., Nomura, T. and Yoshimizu, M. (1995) Effect of incubation temperature for isolation on autoagglutination of *Aeromonas salmonicida*. *Fish Pathology* 30, 67–68.

Munn, C.B., Ishiguro, E.E., Kay, W.W. and Trust, T.J. (1982) Role of surface components in serum resistance of virulent *Aeromonas salmonicida*. *Infection and Immunity* 36, 1069–1075.

Najimi, M., Lemos, M.L. and Osorio, C.R. (2008) Identification of siderophore biosynthesis genes essential for growth of *Aeromonas salmonicida* under iron limitation conditions. *Applied and Environmental Microbiology* 74, 2341–2348.

Najimi, M., Lemos, M.L. and Osorio, C.R. (2009) Identification of iron regulated genes in the fish pathogen *Aeromonas salmonicida* subsp. *salmonicida*: genetic diversity and evidence of conserved iron uptake systems. *Veterinary Microbiology* 133, 377–382.

Nash, J.H., Findlay, W.A., Luebbert, C.C., Mykytczuk, O.L., Foote, S.J. *et al.* (2006) Comparative genomics profiling of clinical isolates of *Aeromonas salmonicida* using DNA microarrays. *BMC Genomics* 7:43.

Natrah, F.M.I., Alam, M.I., Pawar, S., Harzevili, A.S., Nevejan, N. *et al.* (2012) The impact of quorum sensing on the virulence of *Aeromonas hydrophila* and *Aeromonas salmonicida* towards burbot (*Lota lota* L.) larvae. *Veterinary Microbiology* 159, 77–82.

Nikoskelainen, S., Ouwehand, A., Salminen, S. and Bylund, G. (2001) Protection of rainbow trout (*Oncorhynchus mykiss*) from furunculosis by *Lactobacillus rhamnosus*. *Aquaculture* 198, 229–236.

Nivaskumar, M. and Francetic, O. (2014) Type II secretion system: a magic beanstalk or a protein escalator. *Biochimica et Biophysica Acta* 1843, 1568–1577.

Nomura, T. (1993) The epidemiological study of furunculosis in salmon propagation. *Scientific Reports of the Hokaido Salmon Hatchery [Japan]* 47, 1–99.

Pang, M., Jiang, J., Xie, X., Wu, Y., Dong, Y. *et al.* (2015) Novel insights into the pathogenicity of epidemic *Aeromonas hydrophila* ST251 clones from comparative genomics. *[Nature] Scientific Reports* 5: 9833.

Parasa, L.S., Tumati, S.R., Prasad, C. and Kumar, C.A. (2012) *In vitro* antibacterial activity of culinary spices aniseed, star anise and cinnamon against bacterial pathogens of fish. *International Journal of Pharmacy and Pharmaceutical Sciences* 4, 667–670.

Paterson, W.D., Douey, D. and Desautels, D. (1980) Isolation and identification of an atypical *Aeromonas salmonicida* strain causing epizootic losses among Atlantic salmon (*Salmo salar*) reared in a Nova Scotian hatchery. *Canadian Journal of Fisheries and Aquatic Sciences* 37, 2236–2241.

Pavan, M.E., Abbott, S.L., Zorzopulos, J. and Janda, J.M. (2000) *Aeromonas salmonicida* subsp. *pectinolytica* subsp. nov., a new pectinase-positive subspecies isolated from a heavily polluted river. *International Journal of Systematic and Evolutionary Microbiology* 50, 1119–1124.

Pavan, M.E., Pavan, E.E., Lopez, N.I., Levin, L. and Pettinari, M.J. (2013) Genome sequence of the melanin-producing extremophile *Aeromonas salmonicida* subsp. *pectinolytica* strain 34melT. *Genome Announcements* 1(5): e00675-13.

Pemberton, J.M., Kidd, S.P. and Schmidt, R. (1997) Secreted enzymes of *Aeromonas*. *FEMS Microbiology Letters* 152, 1–10.

Phipps, B.M. and Kay, W.W. (1988) Immunoglobulin binding by the regular surface array of *Aeromonas salmonicida*. *The Journal of Biological Chemistry* 263, 9298–9303.

Pionnier, N., Falco, A., Miest, J., Frost, P., Irnazarow, I. *et al.* (2013) Dietary beta-glucan stimulate complement and C-reactive protein acute phase responses in common carp (*Cyprinus carpio*) during an *Aeromonas salmonicida* infection. *Fish and Shellfish Immunology* 34, 819–831.

Piotrowska, M. and Popowska, M. (2015) Insight into the mobilome of *Aeromonas* strains. *Frontiers in Microbiology* 6: 494.

B.K. Gudmundsdottir and B. Bjornsdottir

Portaliou, A.G., Tsolis, K.C., Loos, M.S., Zorzini, V. and Economou, A. (2015) Type III secretion: building and operating a remarkable nanomachine. *Trends in Biochemical Sciences* 41, 175–189.

Pridgeon, J.W., Klesius, P.H. and Yildirim-Aksoy, M. (2013) Attempt to develop live attenuated bacterial vaccines by selecting resistance to gossypol, proflavine hemisulfate, novobiocin, or ciprofloxacin. *Vaccine* 31, 2222–2230.

Rahman, M., Colque-Navarro, P., Kuhn, I., Huys, G., Swings, J. *et al.* (2002) Identification and characterization of pathogenic *Aeromonas veronii* biovar *sobria* associated with epizootic ulcerative syndrome in fish in Bangladesh. *Applied and Environmental Microbiology* 68, 650–655.

Rasch, M., Kastbjerg, V.G., Bruhn, J.B., Dalsgaard, I., Givskov, M. and Gram, L. (2007) Quorum sensing signals are produced by *Aeromonas salmonicida* and quorum sensing inhibitors can reduce production of a potential virulence factor. *Diseases of Aquatic Organisms* 78, 105–113.

Reith, M.E., Singh, R.K., Curtis, B., Boyd, J.M., Bouevitch, A. *et al.* (2008) The genome of *Aeromonas salmonicida* subsp. *salmonicida* A449: insights into the evolution of a fish pathogen. *BMC Genomics* 9:427.

Ren, S., Guo, J., Zeng, G. and Sun, G. (2006) Decolorization of triphenylmethane, azo, and anthraquinone dyes by a newly isolated *Aeromonas hydrophila* strain. *Applied Microbiology and Biotechnology* 72, 1316–1321.

Rodgers, C.J. and Furones, M.D. (2009) Antimicrobial agents in aquaculture: practice, needs and issues. In: Rogers, C. and Basurco, B. (eds) *The Use of Veterinary Drugs and Vaccines in Mediterranean Aquaculture. Options Méditerranéennes: Série A. Séminaires Méditerranéens* No. 86, CIHEAM (Centre International de Hautes Études Agronomiques Méditerranéennes), Montpeller, France/Zaragoza, Spain, pp. 41–59.

Rodriguez-Estrada, U., Satoh, S., Haga, Y., Fushimi, H. and Sweetman, J. (2013) Effects of inactivated *Enterococcus faecalis* and mannan oligosaccharide and their combination on growth, immunity, and disease protection in rainbow trout. *North American Journal of Aquaculture* 75, 416–428.

Romstad, A.B., Reitan, L.J., Midtlyng, P., Gravningen, K. and Evensen, O. (2013) Antibody responses correlate with antigen dose and *in vivo* protection for oil-adjuvanted, experimental furunculosis (*Aeromonas salmonicida* subsp *salmonicida*) vaccines in Atlantic salmon (*Salmo salar* L.) and can be used for batch potency testing of vaccines. *Vaccine* 31, 791–796.

Rusin, P.A., Rose, J.B., Haas, C.N. and Gerba, C.P. (1997) Risk assessment of opportunistic bacterial pathogens in drinking water. *Reviews of Environmental Contamination and Toxicology* 152, 57–83.

Saleh, M., Soliman, H., Haenen, O. and El-Matbouli, M. (2011) Antibody-coated gold nanoparticles immunoassay for direct detection of *Aeromonas salmonicida* in fish tissues. *Journal of Fish Diseases* 34, 845–852.

Schoenhofen, I.C., Stratilo, C. and Howard, S.P. (1998) An ExeAB complex in the type II secretion pathway of *Aeromonas hydrophila*: effect of ATP-binding cassette mutations on complex formation and function. *Molecular Microbiology* 29, 1237–1247.

Schwenteit, J., Gram, L., Nielsen, K.F., Fridjonsson, O.H., Bornscheuer, U.T. *et al.* (2011) Quorum sensing in *Aeromonas salmonicida* subsp *achromogenes* and the effect of the autoinducer synthase AsaI on bacterial virulence. *Veterinary Microbiology* 147, 389–397.

Schwenteit, J.M., Weber, B., Milton, D.L., Bornscheuer, U.T. and Gudmundsdottir, B.K. (2014) Construction of *Aeromonas salmonicida* subsp. *achromogenes* AsaP1-toxoid strains and study of their ability to induce immunity in Arctic char, *Salvelinus alpinus* L. *Journal of Fish Diseases* 38, 891–900.

Scott, C.J.W., Austin, B., Austin, D.A. and Morris, P.C. (2013) Non-adjuvanted flagellin elicits a non-specific protective immune response in rainbow trout (*Oncorhynchus mykiss*, Walbaum) towards bacterial infections. *Vaccine* 31, 3262–3267.

Seshadri, R., Joseph, S.W., Chopra, A.K., Sha, J., Shaw, J. *et al.* (2006) Genome sequence of *Aeromonas hydrophila* ATCC 7966[T]: jack of all trades. *Journal of Bacteriology* 188, 8272–8282.

Sha, J., Kozlova, E.V. and Chopra, A.K. (2002) Role of various enterotoxins in *Aeromonas hydrophila*-induced gastroenteritis: generation of enterotoxin gene-deficient mutants and evaluation of their enterotoxic activity. *Infection and Immunity* 70, 1924–1935.

Sha, J., Rosenzweig, J.A., Kozlova, E.V., Wang, S., Erova, T.E. *et al.* (2013) Evaluation of the roles played by Hcp and VgrG type 6 secretion system effectors in *Aeromonas hydrophila* SSU pathogenesis. *Microbiology* 159, 1120–1135.

Sharma, A., Deo, A.D., Riteshkumar, S.T., Chanu, T.I. and Das, A. (2010) Effect of *Withania somnifera* (L. Dunal) root as a feed additive on immunological parameters and disease resistance to *Aeromonas hydrophila* in *Labeo rohita* (Hamilton) fingerlings. *Fish and Shellfish Immunology* 29, 508–512.

Sharma, M. and Dixit, A. (2015) Identification and immunogenic potential of B cell epitopes of outer membrane protein OmpF of *Aeromonas hydrophila* in translational fusion with a carrier protein. *Applied Microbiology and Biotechnology* 99, 6277–6291.

Sierra, J.C., Suarez, G., Sha, J., Baze, W.B., Foltz, S.M. *et al.* (2010) Unraveling the mechanism of action of a new type III secretion system effector AexU from *Aeromonas hydrophila*. *Microbial Pathoglogy* 49, 122–134.

Smith, P. and Davey, S. (1993) Evidence for the competitive exclusion of *Aeromonas salmonicida* from fish with stress-inducible furunculosis by a fluorescent pseudomonad. *Journal of Fish Diseases* 16, 521–524.

Star, B., Nederbragt, A.J., Jentoft, S., Grimholt, U., Malmstrom, M. *et al.* (2011) The genome sequence of Atlantic cod reveals a unique immune system. *Nature* 477, 207–210.

Stintzi, A. and Raymond, K.N. (2000) Amonabactin-mediated iron acquisition from transferrin and lactoferrin by *Aeromonas hydrophila*: direct measurement of individual microscopic rate constants. *Journal of Biological Inorganic Chemistry* 5, 57–66.

Stratev, D., Stoev, S., Vashin, I. and Daskalov, H. (2015) Some varieties of pathological changes in experimental infection of carps (*Cyprinus carpio*) with *Aeromonas hydrophila*. *Journal of Aquaculture Engineering and Fisheries Research* 1, 191–202.

Studer, N., Frey, J. and Vanden Bergh, P. (2013) Clustering subspecies of *Aeromonas salmonicida* using IS630 typing. *BMC Microbiology* 13:36.

Suarez, G., Sierra, J.C., Erova, T.E., Sha, J., Horneman, A.J. *et al.* (2010a) A type VI secretion system effector protein, VgrG1, from *Aeromonas hydrophila* that induces host cell toxicity by ADP ribosylation of actin. *Journal of Bacteriology* 192, 155–168.

Suarez, G., Sierra, J.C., Kirtley, M.L. and Chopra, A.K. (2010b) Role of Hcp, a type 6 secretion system effector, of *Aeromonas hydrophila* in modulating activation of host immune cells. *Microbiology-SGM* 156, 3678–3688.

Sun, J.H., Wang, Q.K., Qiao, Z.Y., Bai, D.Q., Sun, J.F. *et al.* (2012) Effect of lipopolysaccharide (LPS) and outer membrane protein (OMP) vaccines on protection of grass carp (*Ctenopharyngodon idella*) against *Aeromonas hydrophila*. *The Israeli Journal of Aquaculture – Bamidgeh* 64, 1–8.

Sung, Y.Y. (2014) Heat shock proteins: an alternative to control disease in aquatic organism. *Journal of Marine Science Research and Development* 4, e126.

Swift, S., Karlyshev, A.V., Fish, L., Durant, E.L., Winson, M.K. *et al.* (1997) Quorum sensing in *Aeromonas hydrophila* and *Aeromonas salmonicida*: identification of the LuxRI homologs AhyRI and AsaRI and their cognate *N*-acylhomoserine lactone signal molecules. *Journal of Bacteriology* 179, 5271–5281.

Swift, S., Lynch, M.J., Fish, L., Kirke, D.F., Tomás, J.M. *et al.* (1999) Quorum sensing-dependent regulation and blockade of exoprotease production in *Aeromonas hydrophila*. *Infection and Immunity* 67, 5192–5199.

Tan, W.S., Yin, W.F. and Chan, K.G. (2015) Insights into the quorum sensing activity in *Aeromonas hydrophila* strain M013 as revealed by whole genome sequencing. *Genome Announcements* 3(1): e01372–14.

Tenover, F.C. (2007) Rapid detection and identification of bacterial pathogens using novel molecular technologies: infection control and beyond. *Clinical Infectious Diseases* 44, 418–423.

Thanigaivel, S., Hindu, S.V., Vijayakumar, S., Mukherjee, A., Chandrasekaran, N. *et al.* (2015) Differential solvent extraction of two seaweeds and their efficacy in controlling *Aeromonas salmonicida* infection in *Oreochromis mossambicus*: a novel therapeutic approach. *Aquaculture* 443, 56–64.

Tomás, J.M. (2012) The main *Aeromonas* pathogenic factors. *ISRN Microbiology* 2012: 256261.

Torgersen, Y. and Hastein, T. (1995) Disinfection in Aquaculture. *Revue Scientifique et Technique* 14, 419–434.

Tosi, T., Pflug, A., Discola, K.F., Neves, D. and Dessen, A. (2013) Structural basis of eukaryotic cell targeting by type III secretion system (T3SS) effectors. *Research in Microbiology* 164, 605–619.

Trust, T.J., Khouri, A.G., Austen, R.A. and Ashburner, L.D. (1980) First isolation in Australia of atypical *Aeromonas salmonicida*. *FEMS Microbiology Letters* 9, 39–42.

Tu, F.P., Chu, W.H., Zhuang, X.Y. and Lu, C.P. (2010) Effect of oral immunization with *Aeromonas hydrophila* ghosts on protection against experimental fish infection. *Letters in Applied Microbiology* 50, 13–17.

Turker, H. and Yıldırım, A.B. (2015) Screening for antibacterial activity of some Turkish plants against fish pathogens: a possible alternative in the treatment of bacterial infections. *Biotechnology and Biotechnological Equipment* 29, 281–288.

Udeh, P.J. (2004) *A Guide to Healthy Drinking Water*. iUniverse Inc., Lincoln, Nebraska.

Valera, L. and Esteve, C. (2002) Phenotypic study by numerical taxonomy of strains belonging to the genus *Aeromonas*. *Journal of Applied Microbiology* 93, 77–95.

Vanden Bergh, P. and Frey, J. (2014) *Aeromonas salmonicida* subsp. *salmonicida* in the light of its type-three secretion system. *Microbial Biotechnology* 7, 381–400.

Ventura, M.T. and Grizzle, J.M. (1988) Lesions associated with natural and experimental infections of *Aeromonas hydrophila* in channel catfish, *Ictalurus punctatus* (Rafinesque). *Journal of Fish Diseases* 11, 397–407.

Villumsen, K.R. and Raida, M.K. (2013) Long-lasting protection induced by bath vaccination against *Aeromonas salmonicida* subsp. *salmonicida* in rainbow trout. *Fish and Shellfish Immunology* 35, 1649–1653.

Villumsen, K.R., Dalsgaard, I., Holten-Andersen, L. and Raida, M.K. (2012) Potential role of specific antibodies as important vaccine induced protective mechanism against *Aeromonas salmonicida* in rainbow trout. *PLoS ONE* 7(10): e46733.

Villumsen, K.R., Koppang, E.O. and Raida, M.K. (2015) Adverse and long-term protective effects following oil-adjuvanted vaccination against *Aeromonas salmonicida* in rainbow trout. *Fish and Shellfish Immunology* 42, 193–203.

Vine, N.G., Leukes, W.D. and Kaiser, H. (2004) *In vitro* growth characteristics of five candidate aquaculture probiotics and two fish pathogens grown in fish intestinal mucus. *FEMS Microbiology Letters* 231, 145–152.

Vipond, R., Bricknell, I.R., Durant, E., Bowden, T.J., Ellis, A.E. *et al.* (1998) Defined deletion mutants demonstrate

B.K. Gudmundsdottir and B. Bjornsdottir

that the major secreted toxins are not essential for the virulence of *Aeromonas salmonicida*. *Infection and Immunity* 66, 1990–1998.

Wang, N.N., Wu, Y.F., Pang, M.D., Liu, J., Lu, C.P. *et al.* (2015) Protective efficacy of recombinant hemolysin co-regulated protein (Hcp) of *Aeromonas hydrophila* in common carp (*Cyprinus carpio*). *Fish and Shellfish Immunology* 46, 297–304.

Wang, Z., Liu, X., Dacanay, A., Harrison, B.A., Fast, M. *et al.* (2007) Carbohydrate analysis and serological classification of typical and atypical isolates of *Aeromonas salmonicida*: a rationale for the lipopolysaccharide-based classification of *A. salmonicida*. *Fish and Shellfish Immunology* 23, 1095–1106.

Wedermeyer, G.A. and Nelson, N.C. (1977) Survival of two bacterial fish pathogens (*Aeromonas salmonicida* and the enteric redmouth bacterium) in ozonated, chlorinated and untreated water. *Journal of the Fisheries Research Board of Canada* 34, 429–432.

Wiklund, T. (1990) Atypical *Aeromonas salmonicida* isolated from ulcers of pike, *Esox lucius* L. *Journal of Fish Diseases* 13, 541–544.

Wiklund, T. (1995) Survival of atypical *Aeromonas salmonicida* in water and sediment microcosms of different salinities and temperatures. *Diseases of Aquatic Organisms* 21, 137–143.

Wiklund, T. and Bylund, G. (1993) Skin ulcer disease of flounder *Platichthys flesus* in the northern Baltic Sea. *Diseases of Aquatic Organisms* 17, 165–174.

Wiklund, T. and Dalsgaard, I. (1998) Occurrence and significance of atypical *Aeromonas salmonicida* in non-salmonid and salmonid fish species: a review. *Diseases of Aquatic Organisms* 32, 49–69.

Wiklund, T., Lonnstrom, L. and Niiranen, H. (1993) *Aeromonas salmonicida* ssp. *salmonicida* lacking pigment production, isolated from farmed salmonids in Finland. *Diseases of Aquatic Organisms* 15, 219–223.

Wiklund, T., Dalsgaard, I., Eerola, E. and Olivier, G. (1994) Characteristics of atypical, cytochrome oxidase-negative *Aeromonas salmonicida* isolated from ulcerated flounders (*Platichthys flesus* (L.). *Journal of Applied Bacteriology* 76, 511–520.

Wu, Z.Q., Jiang, C., Ling, F. and Wang, G.X. (2015) Effects of dietary supplementation of intestinal autochthonous bacteria on the innate immunity and disease resistance of grass carp (*Ctenopharyngodon idellus*). *Aquaculture* 438, 105–114.

Yang, B., Wang, C., Tu, Y., Hu, H., Han, D. *et al.* (2015) Effects of repeated handling and air exposure on the immune response and the disease resistance of gibel carp (*Carassius auratus gibelio*) over winter. *Fish and Shellfish Immunology* 47, 933–941.

Yanong, R.P.E. (2008/9, rev. 2014) *Use of Vaccines in Finfish Aquaculture*. Publication No. FA156, revised August 2014, Program in Fisheries and Aquatic Sciences, School of Forest Resources and Conservation, Florida Cooperative Extension Service, IFAS (Institute of Food and Agricultural Sciences), University of Florida, Gainesville, Florida. Available at: http://edis.ifas.ufl.edu/pdffiles/FA/FA15600.pdf (accessed 8 November 2016).

Yu, H.B., Zhang, Y.L., Lau, Y.L., Yao, F., Vilches, S. *et al.* (2005) Identification and characterization of putative virulence genes and gene clusters in *Aeromonas hydrophila* PPD134/91. *Applied and Environmental Microbiology* 71, 4469–4477.

Zhao, Y., Liu, Q., Wang, X.H., Zhou, L.Y., Wang, Q.Y. *et al.* (2011) Surface display of *Aeromonas hydrophila* GAPDH in attenuated *Vibrio anguillarum* to develop a novel multivalent vector vaccine. *Marine Biotechnology* 13, 963–970.

15 *Edwardsiella* spp.

MATT J. GRIFFIN,* TERRENCE E. GREENWAY
AND DAVID J. WISE

*Thad Cochran National Warmwater Aquaculture Center,
Mississippi State University, Stoneville, Mississippi, USA*

15.1 Introduction

The *Edwardsiella* (family *Enterobacteriacae*) was originally described as a new genus of the *Enterobacteriaceae* in the mid-1960s; it represented 37 isolates recovered from open wounds, blood, urine and faeces of humans and animals in the USA, Brazil, Ecuador, Israel and Japan (Ewing *et al.*, 1965). In spite of this, the species of *Edwardsiella* are mostly considered to be pathogens of fish (Mohanty and Sahoo, 2007; Table 15.1). *E. tarda* was first reported from outbreaks in farmed channel catfish (*Ictalurus punctatus*) in Arkansas in the USA (Meyer and Bullock, 1973) and has become one of the most globally recognized fish pathogens, affecting both wild and cultured fish worldwide (Park *et al.*, 2012).

Similarly, *E. ictaluri* was described from farm-raised catfish in the south-eastern USA in the early 1980s. Disease signs differed from the emphesematous putrefactive disease originally attributed to *E. tarda* (Meyer and Bullock, 1973), and the new disease was deemed enteric septicaemia of catfish (ESC) (Hawke *et al.*, 1981). A third *Edwardsiella* sp. has been reported from birds and reptiles. This species, *E. hoshinae*, is phenotypically and genotypically distinct from other *Edwardsiella* spp. (Grimont *et al.*, 1980). It is most often recovered from avian and reptilian hosts and is not known to cause disease in humans, birds, reptiles or fish (Janda *et al.*, 1991; Yang *et al.*, 2012).

This classification remained constant until molecular methods with more resolution emerged that had greater discriminatory power. Intraspecific genetic differences placed *E. tarda* isolates into two distinct polyphyletic groups, giving rise to the designation of fish pathogenic and fish non-pathogenic *E. tarda* (Yamada and Wakabayashi, 1998; Yamada and Wakabayashi, 1999). This distinction was pushed further when researchers discovered fish pathogenic *E. tarda* could be differentiated based on motility (Matsuyama *et al.*, 2005), which also corresponded with genetic groupings based on fimbrial gene sequences (Sakai *et al.*, 2007, 2009c).

Later research further supported these findings, as studies in Europe and the USA determined that *E. tarda* represented two or more genetically distinct and phenotypically ambiguous species (Abayneh *et al.*, 2012; Griffin *et al.*, 2013). This led to the adoption of *E. piscicida* as a new species (Abayneh *et al.*, 2013) and the identification of a putative fifth and genetically distinct *E. piscicida*-like species, later identified as *E. anguillarum* (Shao *et al.*, 2015). In light of recent findings, comparisons of archived nucleotide data with recently released *Edwardsiella* genomes (Table 15.2) have identified *E. piscicida* and *E. anguillarum* as conspecific with isolates that were historically designated as 'typical, motile' and 'atypical, non-motile' fish pathogenic *E. tarda*, respectively.

One factor driving taxonomic ambiguity is the use of 16S rRNA for bacterial classification and a misplaced reliance on publicly accessible nucleotide databases as taxonomic authorities. The limitations of using 16S rRNA sequences to differentiate *Edwardsiella* spp. are well documented (Griffin *et al.*, 2013, 2014, 2015). In general, 16S rRNA does not adequately differentiate between *Edwardsiella* spp. because the five species previously mentioned share >99% similarity at this

---

*Corresponding author e-mail: griffin@cvm.msstate.edu

© P.T.K. Woo and R.C. Cipriano 2017. *Fish Viruses and Bacteria: Pathobiology and Protection*
(P.T.K. Woo and R.C. Cipriano)

Table 15.1. *Edwardsiella* spp. in fishes.

Host	Reference
***Edwardsiella anguillarum* (syn. *E. piscicida*-like sp.)**	
European eel (*Anguilla anguilla*)	Shao *et al.*, 2015
Japanese eel (*A. japonica*)	Shao *et al.*, 2015
Marbled eel (*A. marmorata*)	Shao *et al.*, 2015
Red sea bream (*Pagrus major*)	Oguro *et al.*, 2014
Sea bream (*Evynnis japonica*)	Abayneh *et al.*, 2012
Tilapia (*Oreochromis nilotica*)	Griffin *et al.*, 2014
White grouper (*Epinephelus aeneus*)	Ucko *et al.*, 2016
Edwardsiella ictaluri	
Ayu (*Plecoglossus altivelis*)	Nagai *et al.*, 2008
Bagrid catfish (*Pelteobagrus nudiceps*)	Sakai *et al.*, 2009b
Brown bullhead (*Amieurus nebulosus*)	Iwanowicz *et al.*, 2006
Channel catfish (*Ictalurus punctatus*)	Hawke *et al.*, 1981
Danio (*Danio devario*)	Waltman *et al.*, 1985
Green knife fish (*Eigemannia virescens*)	Kent and Lyons, 1982
Rainbow trout (*Oncorhynchus mykiss*)	Keskin *et al.*, 2004
Rosy barb (*Pethia conchonius*)	Humphrey *et al.*, 1986
Striped catfish (*Pangasius hypophthalmus*)	Crumlish *et al.*, 2002
Tadpole madtom (*Noturus gyrinus*)	Klesius *et al.*, 2003
Tilapia (*Oreochromis nilotica*)	Soto *et al.*, 2012
Walking catfish (*Clarias batrachus*)	Kasornchandra *et al.*, 1987
Yellow catfish (*Pelteobagrus fulvidraco*)	Ye *et al.*, 2009
Zebrafish (*Danio rerio*)	Hawke *et al.*, 2013
Edwardsiella piscicida	
Blotched fantail stingray (*Taeniura meyeni*)	Camus *et al.*, 2016
Blue catfish (*Ictalurus furcatus*)	Griffin *et al.*, 2014
Channel catfish (*I. punctatus*)	Griffin *et al.*, 2014
European eel (*A. anguilla*)	Abayneh *et al.*, 2013
Hybrid catfish (*I. punctatus* × *I. furcatus*)	Griffin *et al.*, 2014
Korean catfish (*Silurus asotus*)	Abayneh *et al.*, 2013
Largemouth bass (*Micropterus salmoides*)	Fogleson *et al.*, 2016
Olive flounder (*Paralichthys olivaceus*)	Oguro *et al.*, 2014
Serpae tetra (*Hyphessobrycon eques*)	Shao *et al.*, 2015
Striped catfish (*Pangasianodon hypophthalmus*)	Shetty *et al.*, 2014
Tilapia (*O. nilotica*)	Griffin *et al.*, 2014
Turbot (*Scophthalmus maximus*)	Abayneh *et al.*, 2013
Whitefish (*Coregonus lavaretus*)	Shafiei *et al.*, 2016
Edwardsiella tarda	
Asian catfish (*Clarias gariepinus*)	Abraham *et al.*, 2015
Asian swamp eel (*Monopterus albus*)	Shao *et al.*, 2016
Ayu (*Plecoglossus altivelis*)	Yamada and Wakabayashi, 1999
Barcoo grunter (*Scortum barcoo*)	Ye *et al.*, 2010
Barramundi (*Lates calcarifer*)	Humphrey and Langdon, 1986
Brook trout (*Salvelinus fontinalis*)	Uhland *et al.*, 2000
Channel catfish (*Ictalurus punctatus*)	Meyer and Bullock, 1973
Chinook salmon (*Oncorhynchus tshawytscha*)	Amandi *et al.*, 1982
Coloured carp (*Cyprinus carpio*)	Sae-oui *et al.*, 1984
Crimson sea bream (*E. japonica*)	Kusuda *et al.*, 1977
European eel (*A. anguilla*)	Alcaide *et al.*, 2006
Golden tiger barb (*Puntius tetrazona*)	Akinbowale *et al.*, 2006
Hybrid striped bass (*Morone chrysops* × *M. saxatilis*)	Griffin *et al.*, 2014
Japanese eel (*A. japonica*)	Yamada and Wakabayashi, 1999

Continued

Table 15.1. Continued.

Host	Reference
Japanese flounder (*Paralichthys olivaceus*)	Yamada and Wakabayashi, 1999
Korean catfish (*Silurus asotus*)	Yu *et al.*, 2009
Largemouth bass (*Micropterus salmoides*)	White *et al.*, 1973
Mullet (*Mugil cephalus*)	Kusuda *et al.*, 1976
Oscar fish (*Astronotus ocellatus*)	Wang *et al.*, 2011
Pangas catfish (*Pangasius pangasius*)	Nakhro *et al.*, 2013
Red sea bream (*Chrysophrus major*)	Yamada and Wakabayashi, 1999
Sea bass (*Dicentrarchus labrax*)	Blanch *et al.*, 1990
Siamese fighting fish (*Betta splendens*)	Humphrey *et al.*, 1986
Silver carp (*Hypophthalmichthys molitrix*)	Xu and Zhang, 2014
Striped bass (*Morone saxatilis*)	Herman and Bullock, 1986
Tilapia (*O. niloticus*)	Yamada and Wakabayashi, 1999

Table 15.2. Available *Edwardsiella* spp. genomes showing phylogroups and GenBank (US National Institutes of Health) accession numbers.

Strain	Name as received	Phylogroup (Shao *et al.*, 2015)	GenBank accession no.	Reference
Complete genomes				
080813	*E. anguillarum*	*E. anguillarum*	CP006664	Shao *et al.*, 2015
93-146	*E. ictaluri*	*E. ictalarui*	CP001600	Williams *et al.*, 2012
C07-187	*E. tarda*	*E. piscicida*	CP004141	Tekedar *et al.*, 2013
EA181011	*Edwardsiella* sp.	*E. anguillarum*	CP011364	Reichley *et al.*, 2015b
EIB202	*E. tarda*	*E. piscicida*	CP001135	Wang *et al.*, 2009
FL6-60	*E. tarda*	*E. piscicida*	CP002154	van Soest *et al.*, 2011
FL95-01	*E. tarda*	*E. tarda*	CP011359	Reichley *et al.*, 2015a
LADL 05-105	*Edwardsiella* sp.	*E. anguillarum*	CP011516	Reichley *et al.*, 2015c
Draft genomes (whole-genome shotgun contigs)				
ATCC #15947	*E. tarda*	*E. tarda*	PRJNA39897	Yang *et al.*, 2012
ATCC #23685	*E. tarda*	*E. tarda*	PRJNA28661	Turnbaugh *et al.*, 2007
ATCC #33202	*E. ictaluri*	*E. ictaluri*	PRJNA66365	Yang *et al.*, 2012
ATCC #33379	*E. hoshinae*	*E. hoshinae*	PRJDB228	Nite.go.jp
DT	*E. tarda*	*E. tarda*	PRJNA66369	Yang *et al.*, 2012
ET070829	*E. tarda*	*E. anguillarum*	PRJNA231705	Shao *et al.*, 2015
ET081126R	*E. tarda*	*E. anguillarum*	PRJNA231706	Shao *et al.*, 2015
ET883	*E. tarda*	*E. piscicida*	PRJNA259400	Shao *et al.*, 2015
JF1305	*E. piscicida*	*E. piscicida*	PRJDB1727	Oguro *et al.*, 2014
LADL 11-100	*E. ictaluri*	*E. ictaluri*	PRJ285663	Wang *et al.*, 2015
LADL 11-194	*E. ictaluri*	*E. ictaluri*	PRJNA285852	Wang *et al.*, 2015
RSB1309	*E. piscicida*	*E. anguillarum*	PRJDB1727	Oguro *et al.*, 2014

locus (Griffin *et al.*, 2015), and incomplete or fragmented 16S rRNA sequences can lead to erroneous classifications. Alternative markers, such as the B subunit of the DNA gyrase gene (*gyrB*), or an internal fragment of the iron-cofactored superoxide dismutase gene (*sodB*), give a more resolved identification. Molecular phylogenies based on *gyrB* (Griffin *et al.*, 2013, 2014, 2015) and *sodB* (Yamada and Wakabayashi, 1999) are congruent

with more robust molecular phylogenetic inferences (Fig. 15.1; Yang *et al.*, 2012; Abayneh *et al.*, 2013) and phylogenetic topographies based on *gyrB* and *sodB* largely agree with evolutionary histories determined by multilocus sequencing and comparative phylogenomics (Shao *et al.*, 2015; Fig. 15.2).

These phylogenomic studies have restructured the genus into five species (Fig. 15.2). As such, many historical reports of *E. tarda* are likely to

M.J. Griffin *et al.*

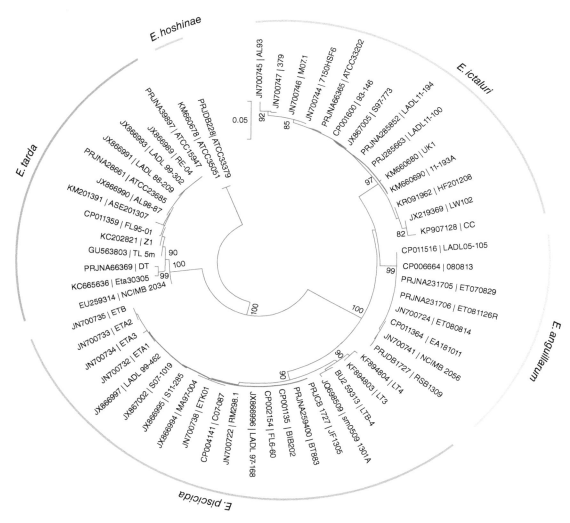

Fig. 15.1. Neighbour-joining evolutionary history of *Edwardsiella* spp. based on sequencing of the B subunit of the DNA gyrase gene (*gyrB*).

have been misclassified. Molecular surveys and assessments of archived nucleotide sequences and historical accounts suggest *E. ictaluri*, *E. piscicida* and *E. anguillarum* are more commonly associated with fish (Griffin *et al.*, 2014; Shao *et al.*, 2015). Hence, any discussion of pathobiology and disease associated with *E. tarda* infections in fish is likely to represent a mix of *E. tarda*, *E. piscicida* and *E. anguillarum*.

Numerous reports have demonstrated phenotypic characters alone are inadequate for the differentiation of *E. tarda*, *E. piscicida* and *E. anguillarum* (Abayneh *et al.*, 2012; Yang *et al.*, 2012; Abayneh *et al.*, 2013;

Griffin *et al.*, 2013, 2014; Shao *et al.*, 2015). At present, molecular analysis is required to accurately segregate the *E. tarda* taxa. Therefore, it is proposed that future *Edwardsiella* research be supplemented with appropriate molecular analysis to facilitate consistent reporting between different laboratories.

Lastly, *E. tarda* is considered a zoonotic agent (Mohanty and Sahoo, 2007). With recent escalations in the production and consumption of aquaculture products, the potential for contracting zoonotic infections from handling or ingesting these products increases. One of the principal pathogens acquired

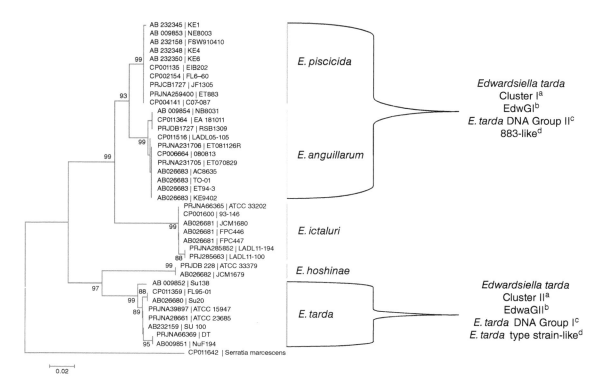

Fig. 15.2. Neighbour-joining evolutionary history of *Edwardsiella* spp. based on sequencing of an internal fragment of the iron-cofactored superoxide dismutase gene (*sodB*). Taxonomic assignments (coloured) are according to classifications reported by Shao *et al.* (2015). References to previously assigned phylogroups are designated by superscripts as follows: [a]Yamada and Wakabayashi, 1999; [b]Yang *et al.*, 2012; [c]Griffin *et al.*, 2013; [d]Abayneh *et al.*, 2013.

from fish or shellfish is *E. tarda* (Haenen *et al.*, 2013) which, though rare, can be fatal for humans (Hirai *et al.*, 2015). In addition, there are reports of human edwardsiellosis contracted from ornamental fish (Vandepitte *et al.*, 1983) and sport fishing (Clarridge *et al.*, 1980). Humans infected with *E. tarda* via puncture, open wounds or ingestion may develop necrotic skin lesions and gastroenteritis or, in severe cases, septicaemia leading to cholecystitis, osteomyelitis, meningitis and necrotizing fasciitis (Gilman *et al.*, 1971; Janda and Abbott, 1993; Matshushima *et al.*, 1996). It is, therefore, critical the scientific and medical community come to a consensus on the identification and reporting of edwardsielliosis.

15.2 *Edwardsiella tarda* Ewing *et al.* (1965)

Globally, *E. tarda* is one of the most economically and environmentally important pathogens of fish, with reports from all seven continents (Meyer and Bullock, 1973; Van Damme and Vandepitte, 1980; Clavijo *et al.*, 2002; Akinbowale *et al.*, 2006; Alcaide *et al.*, 2006; Leotta *et al.*, 2009; Joh *et al.*, 2011; Xu and Zhang, 2014). Small whitish grey colonies are visible on common bacterial media (brain heart infusion or tryptic soy blood agar) supplemented with 5% sheep blood after 24–36 h at 28–37°C, although the bacteria grow better at 37°C. The bacterium is a short, variably motile, Gram-negative bacillus (0.6 × 2.0 μm), is cytochrome oxidase negative and indole positive, ferments glucose with acid and gas production, and has a triple sugar iron (TSI) reaction of alkaline over acid (K/A) with hydrogen sulfide produced (Hawke and Khoo, 2004). Existing databases for phenotypic diagnostic test kits lag behind current taxonomic classifications. Given this phenotypic ambiguity, both *E. piscicida* and *E. anguillarum* key as *E. tarda*. However, confirmatory diagnosis can be made using species-specific PCR (Griffin *et al.*, 2014; Reichley *et al.*, 2015d) or the sequencing of relevant genetic markers.

15.2.1 Pathobiology

E. tarda is predominantly considered a warm-water, opportunistic fish pathogen. Environmental variables such as high temperature, poor water quality and high organic content contribute to the severity of infections (Meyer and Bullock, 1973; Hawke and Khoo, 2004; Mohanty and Sahoo, 2007; Park *et al.*, 2012). Wyatt *et al.* (1979) found *E. tarda* in 75% of domestic catfish [probably *I. punctatus*] pond water samples, 64% of catfish pond mud samples and 100% of the frogs, turtles and crayfish collected from catfish aquaculture ponds. *E. tarda* has also been confirmed from Chinese soft-shelled turtles (*Trionyx sinensis*) (Pan *et al.*, 2010).

In channel catfish, the disease begins as small cutaneous ulcers, progressing into large gas-filled abscesses in the musculature. In chronic infections, abscesses fill with malodorous gas and necrotic tissue (Meyer and Bullock, 1973), although a generalized septicaemia is more common. In some cases, disease manifestations mimic the symptoms of *E. ictaluri*, with the presence of a haemorrhagic ulcer in the top of the skull (Hawke and Khoo, 2004), though this ulcer differs from that produced by *E. ictaluri* in that the edges are more refined. Clinical signs include haemorrhaging in the musculature and head, exophthalmia (protruding eyes), opacity of the eyes, distended abdomen, bloody ascites, a protruded haemorrhagic anus, an enlarged spleen and kidneys, and small white nodules in the gills, kidneys, liver, spleen and, occasionally, intestines (Austin and Austin, 1993; Park *et al.*, 2012).

E. tarda was first described from disease outbreaks on catfish farms in Arkansas, Mississippi, Louisiana and Texas (Meyer and Bullock, 1973). These isolates have since been molecularly confirmed as *E. tarda* (Griffin, unpublished data). Outbreaks primarily occurred in fish of >450 g from July through to October, when the water temperature exceeded 30°C and the water had high levels of organic material. Catastrophic losses were not reported, and disease prevalence rarely exceeded 5%. Nevertheless, when infected fish were harvested and transferred to holding tanks, mortality could approach 50%. Experimental infections of juvenile catfish (5–10 cm) via intraperitoneal (IP) injections of *E. tarda* at doses from 8×10^5 to 8×10^7 colony forming units (cfu) killed four of five fish within 10 days at 27°C. Similar dosages at 13°C in brown trout (*Salmo trutta*) did not cause mortality.

Darwish *et al.* (2000) abraded the skin of channel catfish prior to an immersion exposure to mimic natural infection. The petechiae and cutaneous ulcers observed were consistent with lesions in naturally infected channel catfish (Meyer and Bullock, 1973) and largemouth bass (*Micropterus salmoides*) (Francis-Floyd *et al.*, 1993). Focal necrotic myositis was observed, but gas-filled abscesses in the muscle were absent. In a more recent comparative study, Reichley *et al.* (2015d) injected (IP) catfish with isolates identified molecularly as *E. tarda*, *E. piscicida* and *E. anguillarum*. The median lethal dose for *E. tarda* isolate FL95-01 was nearly 100× higher than that observed for *E. piscicida*. In addition, all mortality occurred within the first 24–36 h postinfection, which differed from the mortality curves described by Darwish *et al.* (2000). Moreover, diagnostic reports from the Aquatic Research and Diagnostic Laboratory in Stoneville, Massachusetts, suggested an increased occurrence of *E. tarda* in US aquaculture since the mid-2000s. However, a survey of archived isolates from this period identified a majority to be *E. piscicida* (Griffin *et al.*, 2014), highlighting a plausible discrepancy between historical records and current classifications.

E. tarda was also implicated in a largemouth bass kill in a lake in Florida (Francis-Floyd *et al.*, 1993). The 6 week mortality event occurred in early autumn, with losses of approximately 1500 fish. Nearly 50% of the fish sampled had deep cutaneous malodorous necrotic ulcers, but without putrefaction and gas release. Multifocal granulomas and scattered melanomacrophage centres were present in interstitial tissues of the posterior kidney, spleen and liver. Isolates phenotypically consistent with *E. tarda* were isolated from the internal organs of several fish, in line with previous reports of *E. tarda* recovered from reptiles, birds, fish and the surface waters of Florida lakes (White *et al.*, 1973).

Padrós *et al.* (2006) reported *E. tarda* from a disease outbreak in turbot (*Scophthalmus maximus*) on a fish farm in Spain, when water temperatures were 15–18°C. Losses ranged from 3 to 10%. Fish presented with tumefaction around the eyes attributed to excess purulent fluid. The musculature and head were haemorrhaged and the abdomen distended with ascites. The liver was haemorrhagic, the spleen and kidney were enlarged and there was congestion in the intestine, spleen and kidney. Histologically, distinct abscesses were observed in haematopoietic tissues of the head and trunk kidney, with significant inflammatory cell infiltrates surrounding these abscesses. The spleen harboured large numbers of macrophages, but very few bacteria were observed in extracellular spaces, blood vessels or macrophage centres. The liver was not extensively damaged and only small groups of macrophages containing small numbers of bacterial cells were embedded in the fibrin matrix.

Clinical isolates identified biochemically as *E. tarda* were also recovered from a mortality event in sea bass (*Dicentriarchus labrax*) at an aquaculture operation in Spain (Blanch *et al.*, 1990). Water temperature during the outbreak was ~24°C and the salinity ranged from 37 to 38%. Mortality was observed in fish injected with recovered isolates at a dose of ~1.6×10^8 cfu/fish. Affected fish presented with eroded caudal fins and haemorrhagic spotting of the head, opercular region, fins and body surface, with some fish presenting ulcerative lesions on the body surface and ascites.

E. tarda has also been reported from cold-water fishes, with isolates from moribund and dead pre-spawn wild adult autumn chinook salmon (*Oncorhynchus tshawytscha*) in Oregon (Amandi *et al.*, 1982). Recovered isolates had relatively low infectivity in chinook salmon and rainbow trout (*O. mykiss*), with LD_{50} values nearly ten times higher than channel catfish exposed under similar conditions. These isolates were phenotypically consistent with the *E. tarda* type strain from humans (ATCC #15947) and an *E. tarda* from largemouth bass (FL-77-4).

Similarly, Amandi *et al.* (1982) and Uhland *et al.* (2000) reported summer outbreaks of *E. tarda* infections in brook trout (*Salvelinus fontinalis*) on two farms in Quebec, Canada. Fish exhibited exophthalmia, petechial haemorrhages on the gills, blood-tinged fluid in the abdomen and haemorrhagic intestines. Isolates were identified biochemically as *E. tarda*, with an API20E profile of 4544000, consistent with the codes generated for both *E. piscicida* (LTB4, EIB202, TX1) and *E. tarda* (ATCC #15947, DT) (Wang *et al.*, 2011; Shao *et al.*, 2015).

E. tarda has also been reported from intensively reared rainbow trout in the Czech Republic (Řehulka *et al.*, 2012). Diseased fish were anorexic and lethargic, with internal hyperaemia and petechial haemorrhages on the liver, with chronic lymphocytic portal hepatitis, focal necrosis, steatosis, dilation of the blood sinuses and activation of sinusoidal cells. Bacteria were phenotypically consistent with *E. tarda* type strain CCM 2238 (= ATCC #15947). Experimentally, the isolate was not pathogenic to common carp (*Cyprinus carpio*), Prussian carp (*Carassius gibelio*) or tench (*Tinca tinca*), leading the authors to speculate that: (i) these fishes were unsuitable for pathogenicity determinations; (ii) the isolate lost virulence in cryogenic storage; or (iii) this was a non-pathogenic strain and disease was initiated by stress associated with heavy stocking density.

Lastly, *E. tarda* was recovered from disease outbreaks in cultured Asian swamp eels (*Monopterus albus*) in China (Shao *et al.*, 2016). Diseased eels displayed cutaneous and visceral haemorrhage as well as oedematous anal tissue. The LD_{50} of these isolates was 1×10^6 cfu by IP injection and infected fish presented clinical signs similar to natural disease. Genetically, the recovered isolate (ASE201307) shared 100% sequence similarity at *gyrB* to *E. tarda* isolates RE-04 and ATCC #15947 (Griffin *et al.*, 2014).

15.2.2 Control and prevention strategies

Treatment

A survey of isolates from the USA and Taiwan found *E. tarda* susceptible to the aminoglycosides, cephalosporins, penicillins, nitrofurantoin, sulfamethoxazole/trimethoprim and the quinolones. Meanwhile, resistance was observed to penicillin G, sulfadiazine, colistin, novobiocin, spectinomycin, ampicillin, tetracycline and chloramphenicol (Waltman and Shotts, 1986a). Moreover, the effect of common aquaculture chemicals topically applied to surfaces against *E. tarda* (isolate USFWS 9.36) found that ethyl alcohol (30, 50 or 70%), benzyl-4-chlorophenol/phenylphenol (1%), sodium hypochlorite (50, 100, 200 or 50,000 mg/l), N-alkyl dimethyl benzyl ammonium chloride (1:256), povidone iodine (50 or 100 mg/l), glutaraldehyde (2%) and potassium peroxymonosulfate/sodium chloride (1%) were effective disinfectants, while chloramine-T (15 mg/l) and formalin (250 mg/l) were not (Mainous *et al.*, 2010).

Oxytetracycline, florfenicol and oxolinic acid have been used against *E. tarda* in fish (Kusuda and Kawai, 1998), with tetracycline frequently used to treat *E. tarda* in Asia (Lo *et al.*, 2014). Meyer and Bullock (1973) reported that oxytetracycline given in the feed (55 mg/kg for 10 days) reduced mortality in catfish within 3 days. An outbreak in brook trout in Quebec, Canada, was treated using oxytetracycline mixed with the feed and vegetable oil (dosage 100 mg/kg live weight) (Uhland *et al.*, 2000). However, terramycin (oxytetracycline) resistance has been reported in *E. tarda* from channel catfish (Hilton and Wilson, 1980). Oxytetracycline and doxycycline resistance was detected in *E. tarda* from eels in Taiwan (Lo *et al.*, 2014), and *E. tarda* from golden tiger barb (*Puntius tetrazona*) in Australia was resistant to ampicillin, amoxycillin, erythromycin, chloramphenicol, tetracycline and oxytetracycline (Akinbowale *et al.*, 2006). A survey of *E. tarda* from fish, water and sediments in a freshwater culture system revealed that 78% of isolates had multidrug resistance (Acharya *et al.*, 2007), and plasmid mediated antibiotic resistance has been reported (Aoki *et al.*, 1987). In Japan, the emergence of antibiotic resistance has spurred the development of new therapeutics. Eels infected with drug-resistant strains of *E. tarda* were successfully treated with a 1:3 combination of ormetoprim/sulfamonomethoxine (25 mg/kg daily), oxolinic acid (12.5 mg/kg daily) and miloxacin at a dosage of 6.2 mg/kg daily (Aoki *et al.*, 1989).

Prevention

The economic importance of *E. tarda* has speared tremendous research efforts in vaccine development. In spite of these efforts, a broad-spectrum vaccine has been elusive owing to variation in *E. tarda* serotypes (Kawai *et al.*, 2004; Mohanty and Sahoo, 2007). Consequently, Costa *et al.* (1998) demonstrated atypical, non-motile *E. tarda* from sea bream (*Pagrus major* and *Evynnis japonica*) and typical *E. tarda* from Japanese eel (*Anguilla japonica*) and flounder (*Paralichthys olivaceus*) had the same O-serotype and similar surface antigens. In rohu (*Labeo rohita*), strong protection against *E. tarda* infection was obtained via bath immersions with an *E. tarda* bacterin, although the efficacy depended on the age of fish and duration of the bath. Fish less than 3 weeks old did not respond to immunization (Swain *et al.*, 2002). In eels, intramuscular immunization seemingly enhanced survival when challenged via intramuscular (IM) injections with virulent field isolates (Salati and Kusuda, 1985). Japanese flounder fed formalin-killed *E. tarda* cells, the immunostimulant Curdlan (bacterial β-1,3-glucan) and a quillaja saponin suspension daily for 3 weeks survived better than control fish (Ashida *et al.*, 1999). Also in flounder, vaccination with an *E. tarda* outer membrane protein via IP injection evoked strong protection against multiple *E. tarda* serotypes (Kawai *et al.*, 2004). Likewise, flounder vaccinated with a genetically modified double knockout vaccine *E. tarda* strain (NH1) conferred protection against challenge when exposed to wild-type NH1 (Choi and Kim, 2011). Additionally, successful immunization was achieved in turbot injected with live and formalin-killed *E. tarda* (Castro *et al.*, 2008).

In an attempt to develop a multivalent vaccine against edwardsiellosis in tilapia (*Oreochromis niloticus*), a recombinant GAPDH (glyceraldehyde-3-phosphate dehydrogenase) outer membrane protein of *E. ictaluri*, expressed in *E. coli* was injected with formalin-inactivated *E. ictaluri* cells. The GAPDH and killed bacterin combination resulted in 71.4% relative percentage survival (RPS) following challenge with *E. tarda* strain OT9805-27 (Trung Cao *et al.*, 2014). Interestingly, OT9805-27 was identified using a loop-mediated isothermal amplification PCR (LAMP) method (Savan *et al.*, 2004) based on an *Edwardsiella* sp. iron-regulated haemolysin gene sequence (GenBank D89876; Hirono *et al.*, 1997). This gene carries 99% identity with the *E. piscicida* genomes EIB202, FL6-60 and C07-087. Comparatively, a Blastn (nucleotide basic local alignment search tool) search of D89876 results in only fragmented hits with <80% similarity to *E. tarda* FL95-01.

15.3 *Edwardsiella piscicida* Abayneh *et al.* (2013)

The incubation of *E. piscicida* on blood agar base #2 (Oxoid, Adelaide, Australia) containing 5% bovine blood for 24 h at 30°C revealed colony morphology and biochemical characteristics similar to those of *E. tarda* (Abayneh *et al.*, 2013). Genetically, *E. piscicida* is synonymous with what has been described as typical, motile, fish pathogenic *E. tarda* (Griffin *et al.*, 2014). As such, *E. piscicida* is identified by most commercial phenotypic diagnostic kits as *E. tarda*. Confirmatory diagnosis can be made by using species-specific PCR (Griffin *et al.*, 2014; Reichley *et al.*, 2015d) or sequence comparisons of appropriate genetic markers.

15.3.1 Pathobiology

Comparisons of archived nucleotide sequences to recently released *E. piscicida* genomes (Tekedar *et al.*, 2013; Oguro *et al.*, 2014; Shao *et al.*, 2015) have facilitated inferences into the pathobiology of *E. piscicida*. For example, 16S rDNA sequence comparisons suggest outbreaks among turbot in China (Xiao *et al.*, 2008) were caused by *E. piscicida* and not *E. tarda* as originally reported. Subsequent infectivity challenges showed LD_{50} values via IM injections ranging from 4×10^3 to 4×10^5 cfu/g in swordtails (*Xiphophorus helleri*). Experimentally infected fish were haemorrhagic with necrotic lesions in the skin and muscle, and suppurative abscesses. Similarly, *Edwardsiella* sp. strain KE6 was recovered from diseased Korean catfish (*Silurus asotus*) showing abdominal distention and reddish foci on the skin. Internally, fish showed abdominal dropsy, enteritis, liver congestion, splenomegaly, renomegaly and abscesses with purulent fluid in the kidney (Yu *et al.*, 2009). *sodB* sequencing places KE6 within the *E. piscicida* phylogroup (Fig. 15.2).

Shafiei *et al.* (2016) reported *E. piscicida* from whitefish (*Coregonus lavaretus*) in Finland. Infected whitefish had generalized septicaemia, exophthalmia, haemorrhagic congestion on the skin and base of the fins, swelling and hyperaemia of the anal region, haemorrhages in the internal organs, gills and muscle tissue, and mottled livers, splenomegaly and renomegaly. There was generalized peritonitis, with inflammatory cells in most visceral organs. Gram-negative rod-shaped bacteria were identified in the peritoneal cavity of affected areas. Necrotic foci were also seen in the kidney and single cell necrosis with small necrotic foci was detected in the liver.

Similarly, *E. piscicida* was recovered during chronic mortality events in captive populations of largemouth bass. There were multifocal areas of necrosis scattered throughout the heart, liver, anterior kidney, posterior kidney and spleen. Many of the necrotic foci were encapsulated or replaced by discrete granulomas and associated with colonies of Gram-negative bacteria (Fogelson *et al.*, 2016).

E. piscicida was also recovered from a septic blotched fantail stingray (*Taeniura meyeni*) from a large display aquarium (Camus *et al.*, 2016). Upon necropsy, the peritoneal wall and serosal surfaces of the visceral organs and heart were white and thickened. Similar material was multifocally present on the endocardial surface of the atrium. The retroperitoneal musculature and perirenal connective tissues were oedematous and the caudal aspect of the left kidney contained a malodorous area of liquefaction. Microscopic findings were consistent with bacterial sepsis, including necrotizing myocarditis and cellulitis in the presence of bacterial rods, splenic vascular necrosis, branchial and renal vascular thrombi, and coelomitis.

In the original description of *E. piscicida*, pathogenicity studies were performed using five isolates from fish and the *E. tarda* type strain from humans (ATCC #15947). Three isolates were from the *E. piscicida* phylogroup, including the *E. piscicida* type strain ET883 (Abayneh *et al.*, 2013). Experimental infections in zebrafish (*Danio rerio*) consisted of intramuscular injections of ~3×10^3 cfu of each isolate followed by a 7 day observational period. Isolates ET883 and LTB4 resulted in 100% and 95% mortality, respectively, while *E. piscicida* isolate ETK01 demonstrated low pathogenicity (11.1% cumulative mortality). Clinical signs included erratic swimming, bottom dwelling and inappetence, with ulceration on dorsal surfaces and injection sites.

The pathogenesis of isolate LTB4 was also characterized in turbot (Lan *et al.*, 2008). Diseased fish presented with inappetence, lethargic swimming and failure to maintain neutral buoyancy, distended abdomens, exophthalmia and haemorrhages on and at the base of the fins, head, opercula, mouth, lower jaw and abdomen. Fish had enlarged livers and kidneys, with haemorrhagic and necrotic areas. Spleens were enlarged and the intestine was sometimes prolapsed from the anus. There were small white nodules on the liver. Experimentally infected turbot maintained in seawater at 19–20°C demonstrated the median lethal dose of LTB4 by IP injection to be 3×10^3 cfu/g fish.

Matsuyama *et al.* (2005) investigated the pathogenicity of typical, motile, fish pathogenic *E. tarda* (syn. *E. piscicida*) in yellowtail (*Seriola quinqueradiata*), flounder and red sea bream (*Pagrus major*). IP injection of isolate NUF806 gave LD_{50} values of 1.6×10^6 cfu in yellowtail, 1.4×10^3 cfu in flounder and 2.6×10^9 cfu in red sea bream at 25°C, suggesting variable levels of virulence in these different hosts.

In farm-raised catfish in the south-eastern USA, *E. piscicida* presents with a similar generalized septicaemia to that reported for *E. tarda*. Likewise, some cases present with a haemorrhagic ulcer in the top of the skull (Griffin, unpublished data). A recent survey of archived bacterial specimens collected from farm-raised catfish in Mississippi revealed that all isolates recovered from 2007 to 2012, which had been identified phenotypically as *E. tarda*, were actually *E. piscicida* (Griffin *et al.*, 2014). Furthermore, Reichley *et al.* (2015d) determined the LD_{50} of *E. piscicida*, *E. tarda* and *E. anguillarum* in channel catfish fingerlings.

The median lethal dose of *E. piscicida* isolate S11-285 was 3.9×10^5 via intracoelomic injection. This was 100–1000 times lower than the LD_{50} values of *E. tarda* (FL95-01) or *E. anguillarum* (syn. *E. piscicida*-like sp.) under similar conditions, supporting the assertion that *E. piscicida* is presently a greater threat to catfish aquaculture than *E. tarda*.

15.3.2 Prevention strategies

Takano *et al.* (2010) investigated the efficacy of five avirulent *Edwardsiella* spp. strains as a live vaccine against edwardsiellosis in Japanese flounder. Archived nucleotide sequences identified four isolates that belonged to the *E. tarda* phylogroup, and one (E22) clustered with *E. piscicida* (Fig. 15.1). Efficacy was evaluated after challenge with *E. piscicida* strain NUF806. *E. piscicida* E22 was the only isolate that significantly reduced mortality caused by NUF806, suggesting that the *E. tarda* phylogroup does not protect against *E. piscicida*. Conversely, exposure of Japanese flounder to the *E. tarda* type strain from humans (ATCC #15947) resulted in resistance against subsequent challenge with isolate TX1 (Cheng *et al.*, 2010), a member of the *E. piscicida* cluster (Yang *et al.*, 2012; Shao *et al.*, 2015). This would suggest that there is at least some degree of epitopic conservation between *E. piscicida* and *E. tarda*.

Further work in turbot investigated the application of a putative surface antigen Esa1 from isolate TX1 as a purified recombinant subunit vaccine. The recombinant protein was expressed in a *Pseudomonas* sp. delivered orally or injected, resulting in 52 and 79% RPS, respectively (Sun *et al.*, 2010). Esa1 was also as efficacious as a DNA vaccine, resulting in 57% RPS in fish immunized by injections of plasmid constructs carrying the *Esa1* inserts (Sun *et al.*, 2011a). Similarly to the Esa1 work, purified recombinant subunit and DNA vaccines exploited the antigenicity of the *Eta2* gene product (another surface antigen, Eta2) from TX1. The vaccination of flounder using injections of the purified recombinant subunit or the DNA vaccine produced RPS values of 83 and 67%, respectively. Moreover, passive immunization studies demonstrated sera from fish vaccinated with the recombinant vaccine resulted in stronger protection in naive fish (RPS = 57%) than sera from fish immunized with the DNA vaccine (RPS = 29%; Sun *et al.*, 2011b).

Likewise, immersion immunizations of turbot with a live *E. piscicida esrB* mutant (EsrB positively regulates the type III and VI secretion systems T3SS and T6SS) protected against subsequent *E. piscicida* challenge (Yang *et al.*, 2015). Although originally classified as *E. tarda*, the strain used in these studies, EIB202, has been identified as a member of the *E. piscicida* phylogroup (Griffin *et al.*, 2014; Shao *et al.*, 2015). Vaccination via IP injection using an EIB202 mutant lacking UDP-glucose dehydrogenase elicited significant protection (43.3-76.7% RPS) against the parental wild type isolate (Lv *et al.*, 2012). In addition, turbot vaccinated via IP injection with a recombinant EIB202 glyceraldehyde-3-phosphate dehydrogenase had lower cumulative mortality when compared with identically challenged control fish (Liang *et al.*, 2012).

Lastly, bacterial 'ghosts' have also been evaluated as vaccine candidates against *Edwardsiella* infections in fish. The isolate used in these studies (FSW910410) was recovered from a moribund olive flounder on a fish farm in Korea. Its *sodB* sequence identifies this isolate as a member of the *E. piscicida* tribe. In tilapia, two IP immunizations with $\sim 1 \times 10^6$ ghost cells, administered 2 weeks apart, resulted in significantly greater survival in comparison either with cohorts that received either formalin-killed cells or with non-vaccinated controls. Similarly, oral delivery of FSW910410 bacterial ghosts to flounder also resulted in enhanced survival compared with treatment groups receiving formalin-killed cells per os and non-vaccinated controls (Kwon *et al.*, 2006, 2007).

15.4 *Edwardsiella anguillarum* Shao *et al.* (2015)

E. anguillarum represents a group of genetically distinct isolates from eels and other brackish water fishes that do not fit within the other *Edwardsiella* lineages. It is phenotypically indistinguishable from *E. tarda*, therefore confirmation of the species requires species-specific PCR (Griffin *et al.*, 2014; Reichley *et al.*, 2015d) or sequencing of the relevant genetic markers. *E. anguillarum* is synonymous with the atypical, non-motile *E. tarda* described by Matsuyama *et al.* (2005) and Sakai *et al.* (2007, 2009c). However, in their description of *E. anguillarum*, Shao *et al.* (2015) cited the bacteria as motile by peritrichous flagella. Similarly, Griffin *et al.* (2013) observed motility at both 25 and 37°C for *E. anguillarum* isolate LADL 05-105 (Reichley *et al.*, 2015d), suggesting that motility is variable for this group.

15.4.1 Pathobiology

Matsuyama *et al.* (2005) also investigated the pathogenicity of atypical, non-motile, fish pathogenic *E. tarda* in yellowtail, flounder and red sea bream. The isolate FPC503 possesses genetic elements that share 98–100% sequence identity with three *E. anguillarum* genomes (LADL 05-105; 080813; EA181011) (Sakai *et al.*, 2009a; Reichley *et al.*, 2015b,2015c; Shao *et al.*, 2015). IP injection of FPC503 gave LD$_{50}$ values of 1.5×10^5 cfu in yellowtail, 1.4×10^6 cfu in flounder and 8.9×10^3 cfu in red sea bream at 25°C, which differed from the mortality trends observed for typical, motile, fish pathogenic *E. tarda* (syn. *E. piscicida*). *E. anguillarum* is pathogenic in turbot, with significant mortality at 15°C from IP injections of 1×10^2 cfu (Shao *et al.*, 2015). Similarly, IM injections of 5×10^2 cfu in 0.25 g zebrafish caused 100% mortality within 3 days at ~22°C (Yang *et al.*, 2012).

Abayneh *et al.* (2013) evaluated the pathogenicity of two isolates now recognized as members of the *E. anguillarum* phylogroup (Shao *et al.*, 2015). Isolate NCIMB 2034 from an unknown fish species was non-pathogenic to zebrafish via IM injection (3×10^3 cfu). Meanwhile, similar injections of strain ET080813 caused 87.5% mortality. Conversely, *E. anguillarum* (syn. *E. piscicida*-like sp.) strain LADL 05-105 caused only minimal mortality at ~30°C in channel catfish fingerlings receiving IP injections of 1×10^8 cfu (Reichley *et al.*, 2015c), thus supporting previous work that suggested variable levels of pathogenicity of these organisms in different hosts.

15.4.2 Prevention strategies

Given the recent adoption of *E. anguillarum* (Shao *et al.*, 2016), there is a paucity of research on the prevention and treatment of this newly recognized fish pathogen. That said, isolate FPC503 demonstrated protective efficacy as a formalin-killed vaccine against *E. anguillarum* infection in red sea bream (Takano *et al.*, 2011). IP injection of formalin-killed FPC503, with and without adjuvant, resulted in 85–100% RPS.

15.5 *Edwardsiella ictaluri* Hawke *et al.* (1981)

E. ictaluri, the causative agent of enteric septicaemia of catfish (ESC), as mentioned in Section 15.1, is a significant pathogen of commercially cultured catfish and other temperate freshwater fishes (Table 15.2) (Hawke *et al.*, 1981). The disease is endemic on most commercial catfish operations in the south-eastern USA (Plumb and Vinitnantharat, 1993) and can cause catastrophic losses, particularly in first-year fingerlings (Wise *et al.*, 2004).

Initial studies suggested isolates from different geographic locations were biochemically, biophysically and serologically homogeneous and the outer membrane proteins, isoenzymes and plasmids were largely conserved (Newton *et al.*, 1988; Plumb and Vinitnantharat, 1989; Hawke and Khoo, 2004). Genetically, *E. ictaluri* is mostly clonal in the catfish farming region of the south-eastern USA, although genotypic subtypes have been identified worldwide (Bader *et al.*, 1998; Griffin *et al.*, 2011; Bartie *et al.*, 2012). Vietnamese isolates from striped catfish (*Pangasianodon hypothalamus*) possessed unique plasmid profiles compared with those in channel catfish (Fernandez *et al.*, 2001) and did not react with a monoclonal antibody (Ed9) raised against the *E. ictaluri* type strain, ATCC #33202 (Rogge *et al.*, 2013). Similarly, *E. ictaluri* from recent outbreaks in laboratory populations of zebrafish had unique plasmid profiles, were not recognized by Ed9 and autoaggregated in broth (Hawke *et al.*, 2013). In Japan, *E. ictaluri* from wild ayu (*Plecoglossus altivelis*) were clonal with significantly different amplified fragment length polymorphism (AFLP) profiles than strains from the USA or Indonesia (Sakai *et al.*, 2009b). More recently, differences were detected among isolates from catfish, zebrafish and tilapia in genomic arrangement, plasmid confirmation and content, virulence factor genes, antimicrobial susceptibilities and serological profiles (Griffin *et al.*, 2015).

Confirmatory diagnosis of *E. ictaluri* is typically made by isolation and culture of the bacterium from kidney and brain. Culture on commonly available media (tryptic soy agar, Mueller–Hinton, or brain heart infusion agar) supplemented with 5% sheep blood for 36–48 h at 25–30°C reveals small, punctuate, whitish grey colonies. *E. ictaluri* does not grow well at 37°C and presumptive identification can be made by incubating duplicate cultures at 27 and 37°C. The bacteria are Gram-negative weakly motile baccili (0.75 × 1.25 µm), cytochrome oxidase and indole negative, ferment glucose and have a TSI reaction of K/A without H$_2$S production (Hawke *et al.*, 1981; Hawke and Khoo, 2004). Molecular confirmation can be made by PCR (Bilodeau *et al.*, 2003; Panangala *et al.*, 2007) or LAMP assays (Yeh *et al.*, 2005).

15.5.1 Pathobiology

Outbreaks of ESC in the catfish farming region of the south-eastern USA occur in late spring and early autumn, when water temperatures fall between 22 and 28°C (Francis-Floyd *et al.*, 1987; Wise *et al.*, 2004), but mortality can occur outside this range (Plumb and Shoemaker, 1995). Epizootics in catfish aquaculture are characterized by disease-induced inappetence with complete cessation of feeding at the height of severe epizootics. Fish become listless near the surface, accumulate along the leeward pond bank and often swim in circles. Presumptive diagnosis can be based on clinical signs, which range from multifocal, pinpoint red and white ulcers and petechial haemorrhages on the skin, to swollen abdomens, exophthalmia and meningoencephalitis. Meningoencephalitis can lead to erosion of the skin covering the fontanelle of the frontal bones, evidenced as a 'hole-in-the-head', the common colloquialism used for the disease. Cranial lesions have rough eroded margins, in contrast to the smoother and more defined margins of lesions associated with *E. tarda* and *E. piscicida*. Internally, the abdomen can fill with clear, yellow or bloody ascites accompanied by renomegaly and pale, necrotic, mottled livers (Thune *et al.*, 1993).

Initial studies suggested *E. ictaluri* is an obligate pathogen, with short-lived survival in ponds (Plumb and Quinlan, 1986). The bacterium survived in pond water for 10–15 days and in sediments for up to 3 months, although survival was temperature dependent. At 5°C, *E. ictaluri* survived <15 days, which would suggest that it may not overwinter in catfish ponds. The development of quantitative PCR (qPCR) protocols for the detection and quantification of *E. ictaluri* (Bilodeau *et al.*, 2003) affords the opportunity to evaluate pathogen levels in the environment (Griffin *et al.*, 2011). Recent work has found *E. ictaluri* DNA is present in nursery ponds after fish are removed in late winter and prior to restocking the following summer (Griffin, unpublished data). However, it is unclear from these data whether the organisms are viable. Contrary to previous findings, these results indicate *E. ictaluri* persists and survives in pond environments and/or within the presence of disease vectors apart from the catfish host.

Stress, poor nutrition, co-infection and suboptimal water quality increase ESC-related mortality.

As *E. ictaluri* is highly pathogenic to channel catfish, it can cause severe epizootics without predisposing conditions (Mqolomba and Plumb, 1992; Wise *et al.*, 1993; Ciembor *et al.*, 1995; Plumb and Shoemaker, 1995). There is no evidence of vertical disease transmission and water-borne exposure is considered the primary mode of transmission, either by ingestion of the bacteria with feed or direct entry through mucosal surfaces (Thune *et al.*, 1993; Wise *et al.*, 2004). Following experimental intubation, *E. ictaluri* crossed the intestinal epithelium and established acute disease (Baldwin and Newton, 1993).

Additional studies have demonstrated mortality in juvenile channel catfish increased with feeding compared to treatments where feed was withheld (Wise *et al.*, 2008). This relationship has also been demonstrated in experimental pond trials (Wise and Johnson, 1998). Moreover, *E. ictaluri* can enter a host through the nares and initiate pathogenicity within the olfactory sac. Once established, the bacteria migrate to the brain via the olfactory nerve causing meningoencephalitis (Miyazaki and Plumb, 1985; Shotts *et al.*, 1986). While other portals of entry have been reported (Menanteau-Ledouble *et al.*, 2011), oral ingestion of bacteria disseminating from the faeces of infected fish, or from the carcasses of dead and decomposing fish, is thought to drive the rapid spread of infection during an outbreak (Klesius, 1994; Hawke and Khoo, 2004; Wise *et al.*, 2008).

Enteric septicaemia of catfish can manifest as acute, subacute and chronic infections (Hawke and Khoo, 2004). Acute infections are characterized by rapid fulminating septicaemias with few clinical signs before mortality. Subacute disease is characterized by pronounced clinical signs involving petechial and ecchymotic haemorrhages on the skin, necrosis of the liver and kidneys, and intestinal haemorrhage. Clear-to-bloody fluid in the abdominal cavity and splenomegaly are often present. Surviving fish either recover and become refractive to subsequent infection, or develop chronic disease. In most chronic cases, bacteria can only be isolated from the brain or, occasionally, from necrotic livers.

There are conflicting reports concerning the role of piscivorous birds in the transmission of *E. ictaluri*. Surveys of intestinal and rectal smears from snowy egrets (*Egretta thula*), great egrets (*Casmerodius albus*), great blue herons (*Ardea herodias*) and double-crested cormorants (*Phalacrocorax auritus*) detected *E. ictaluri* in 53% of the sampled birds, including two samples in which the bacteria were viable (Taylor, 1992). Conversely, viable bacteria were not recovered from great blue herons fed *E. ictaluri*-infected fish, which suggested that transmission via the heron's gastrointestinal tract is unlikely (Waterstrat *et al.*, 1999). Even so, avian transport of infected fish between ponds could disseminate the disease.

E. ictaluri has also emerged in non-ictalurid fishes (e.g. zebrafish, tilapia, ayu). Outbreaks have been reported in laboratory populations of zebrafish, seemingly triggered by handling (Hawke *et al.*, 2013). Clinical signs were similar to those in catfish, with systemic disease characterized by necrotic spleen, kidneys, liver, gut and brain with large numbers of bacteria observed, often within macrophages. Inflammatory cell infiltrates were often observed in the kidney and spleen, with diffuse and severe inflammation and necrosis of the nasal pits and significant damage to the olfactory rosettes. In some cases, there is marked expansion of the mesencephalic ventricles by necrotic leucocytes and bacteria (Petrie-Hanson *et al.*, 2007; Hawke *et al.*, 2013). IM injections of *E. ictaluri* in zebrafish at doses exceeding 6×10^4 cfu resulted in >75% mortality within 6 days at 28°C. Comparable immersion infections resulted in <15% mortality, even with doses exceeding 2.0×10^7 cfu/ml (Petrie-Hanson *et al.*, 2007).

E. ictaluri has also been linked to mortality in cultured tilapia. Fingerlings had consistent multifocal nodulation of the spleen and head kidney, with hepatomegaly and reduced fat in the liver and peritoneum. Histologically, severe inflammatory infiltrates were seen in the spleen, head kidney and liver, as were multifocal areas of necrosis and granuloma formation. Injection and immersion challenges of 10^6 cfu/fish or 10^6 cfu/ml, respectively, caused mortalities of 100% within 3 days of injection and 8 days after immersion. Lower doses (10^3 cfu) caused <40% mortality and fish presented with bloody ascites, splenomegaly, renomegaly and multifocal white nodules on the anterior kidney and spleen (Soto *et al.*, 2012).

Late summer–early autumn mortality associated with *E. ictaluri* has been observed among wild ayu in Japanese rivers. Diseased fish had haemorrhagic ascites and exophthalmia with reddening of the body surface, anus or bases of fins. Experimental IP infections with recovered isolates demonstrated an LD_{50} of approximately 1.3×10^4 cfu at 17–20°C with clinical signs similar to those in natural outbreaks (Sakai *et al.*, 2008). Nagai and Nakai (2014)

demonstrated *E. ictaluri* was comparatively more virulent to ayu at 28 than at 20°C, which is consistent with earlier studies in channel catfish (Plumb and Shoemaker, 1995).

15.5.2 Control and prevention strategies

Control treatments

The survival of catfish during an ESC outbreak can be improved by restricting feeding to every other day or every third day, limiting the ingestion of *E. ictaluri* and preventing the faecal/oral route of contagion (Wise and Johnson, 1998; Wise *et al.*, 2004). Restricting feed when water temperatures are permissive for ESC is a common management practice in catfish aquaculture. In 2009, nearly 30% of fingerling operations reported that they withdrew feed to control ESC (USDA/APHIS/NAHMS, 2010); however, prophylactic feed restriction results in reduced production due to lost feed days (Wise and Johnson, 1998; Wise *et al.*, 2008).

Similarly to *E. tarda*, the surface application of ethyl alcohol (30, 50 or 70%), benzyl-4-chlorophenol/phenylphenol (1%), sodium hypochlorite (50, 100, 200 or 50,000 mg/l), N-alkyl dimethyl benzyl ammonium chloride (1:256), povidone iodine (50 or 100 mg/l), glutaraldehyde (2%), and potassium peroxymonosulfate/sodium chloride (1%) reduced or eliminated the number of detectable organisms within 1 min of contact time. Comparatively, chloramine-T (15 mg/l) and formalin (250 mg/l) did not (Mainous *et al.*, 2010). Moreover, *E. ictaluri* is susceptible to aminoglycosides, cephalosporins, penicillins, quinolones, tetracyclines, chloramphenicol, nitrofurantoin and potentiated sulfonamides (Waltman and Shotts, 1986b). In the USA, Romet® and Aquaflor® are approved for the control of *E. ictaluri* in catfish.

Romet® (Hoffman LaRoche) is a potentiated sulfonamide formulated as a 5:1 combination of sulfadimethoxine and ormetoprim, which effectively reduces ESC mortality in channel catfish fingerlings at a recommended dose of 50 mg/kg for 5 consecutive days (Plumb *et al.*, 1987). Aquaflor® (Merck Animal Health) is a broad-spectrum antibiotic of the phenicol class that is approved for use in a variety of fish species worldwide. It is delivered as a feed premix (50% w/w florfenicol (Gaunt *et al.*, 2004, 2006) at 10 mg/kg body weight/day for 10 days (Gaikowski *et al.*, 2003). Outside the USA, producers must consult their respective regulatory authorities before any treatment is considered.

M.J. Griffin *et al.*

E. ictaluri can develop plasmid-mediated antimicrobial resistance to tetracycline, sulfadimethoxine/ormetoprim and florfenicol (Waltman *et al.*, 1989; Welch *et al.*, 2008). Moreover, field isolates of *E. ictaluri* have been reported that are resistant to tetracycline, sulfadimethoxine/ormetoprim and florfenicol (Starliper *et al.*, 1993; Welch *et al.*, 2008).

Prevention

Significant research has been devoted to developing an effective vaccine against *E. ictaluri*. Initial efforts to vaccinate channel catfish focused on the administration of cellular components or killed vaccines, and were largely unsuccessful or impractical. The immune response of channel catfish to lipopolysaccharide (LPS) and whole-cell *E. ictaluri* vaccines was evaluated by Saeed and Plumb (1986), who showed catfish immunized by multiple injections of *E. ictaluri* LPS in Freund's complete adjuvant (FCA) were protected against subsequent challenge with *E. ictaluri*. None the less, given the logistics of catfish production in the south-eastern USA, injectable vaccines have not received strong industry support.

Considerable research has been invested in developing an efficacious immersible *E. ictaluri* vaccine that would permit vaccination before fish are transferred from the hatchery. In light of the success of bacterin vaccines in salmonid culture, early attempts at vaccinating channel catfish in a commercial setting targeted the oral and immersion delivery of *E. ictaluri* bacterins (Thune *et al.*, 1994; Shoemaker and Klesius, 1997), though these strategies did not gain industry acceptance.

Later attempts at vaccinating fish were conducted with live, attenuated immersible vaccines that were highly effective in laboratory trials (Lawrence *et al.*, 1997; Thune *et al.*, 1999; Abdelhamed *et al.*, 2013). However, these strategies have demonstrated limited effectiveness when delivered in accordance with industry production timelines, which target fish prior to or during transfer from the hatchery to nursery ponds at 7–10 days post hatch. The poor efficacy is attributed to lack of immunocompetency at this age, as catfish are not fully immunocompetent until 18–21 days post hatch (Petrie-Hanson and Ainsworth, 1999), well beyond the hatchery/nursery transfer window. While there are reports of successful vaccination using a commercially available immersible vaccine (Aquavac-ESC, Merck Animal Health) in 7-day-old

catfish fry (RPS 58.4–77.5%) and eyed channel catfish eggs (27.3–87.9%) (Shoemaker *et al.*, 1999, 2002, 2007), immersible vaccines have not received widespread industry adoption due to marginal returns on investment (Bebak and Wagner, 2012).

To achieve maximum efficacy, vaccines should be delivered to older, immunocompetent fish. Oral administrations of vaccines are non-stressing and can be delivered to older fish after they have been stocked into ponds. Although oral vaccination using killed vaccines has demonstrated efficacy (Plumb and Vinitnantharat, 1993; Plumb *et al.*, 1994; Thune *et al.*, 1994), the delivery of a live, attenuated vaccine shows the greatest potential. An attenuated *E. ictaluri* Δ*fur* mutant (lacking the ferric uptake regulator, Fur, protein) was effective in protecting channel catfish in laboratory challenges (Santander *et al.*, 2012). Similarly, an orally delivered live, attenuated rifampin-resistant *E. ictaluri* strain (340X2) offered a high level of protection in channel catfish against wild type *E. ictaluri* challenge in both laboratory and field trials (Wise *et al.*, 2015). In addition, cryopreserved vaccine applied after 2 years of cold storage (–74°C) still offered exceptional protection (Greenway *et al.*, 2017). In experimental pond trials, oral vaccination with this live, attenuated vaccine resulted in significant improvements in survival, feed fed, feed conversion and total harvested weight (Wise *et al.*, 2015). Oral vaccination facilitates the delivery of the vaccine to older fish, circumventing many of the limitations associated with bath immersions of fry. While vaccine coverage is limited by the variable feeding rates within the population, it is thought non-vaccinated fish that fail to ingest sufficient amounts of the vaccine still benefit from herd immunity within the pond population. At the population level, oral delivery may prove to be a more effective, practical method of immunization against *E. ictaluri* than previous immersion or injection techniques.

15.6 Conclusions and Suggestions for Future Research

An assessment of the literature published since the description of *E. piscicida* in 2013 reveals a number of *E. tarda* studies that fail to address the possibility of misclassification, adding further discord to the scientific record. The segregation of *E. tarda* into multiple taxa means that a multitude of valid historical studies on the pathogenic effects of

Edwardsiella in fish require a reassessment. Recent findings, in conjunction with archived sequence data, suggest many reports of *E. tarda* in fish are in fact *E. piscicida* or *E. anguillarum*. Moving forward, it is important for researchers and diagnosticians to employ consistent nomenclature and methods of identification in compliance with recently restructured systematics.

In addition, the development of adequate vaccination protocols to protect fish against *E. tarda* has been complicated by taxonomic ambiguity. The diverse antigenicity of the *Edwardsiella* spp. has been cited as a major obstacle to the development of an effective *E. tarda* vaccine. However, the challenges associated with the development of such a vaccine may lie in the fact that until recently *E. tarda* represented three genetically distinct species, all with varying degrees of pathogenicity in different fish hosts. That said, as is evident by the work that has been summarized here, the absence of viable vaccines against the different *Edwardsiella* spp. is not the issue. There are a substantial number of efficacious vaccine candidates encompassing a broad spectrum of strategies, including recombinant proteins, DNA vaccines, formalin-killed cells and bacterins, attenuated organisms and bacterial ghosts. Rather, the issue lays in efficient and efficacious delivery methods of these vaccines to fish. Future research should focus on developing practical vaccination protocols that logistically conform to industry standards and facilitate broader producer acceptance.

References

Abayneh, T., Colquhoun, D. and Sørum, H. (2012) Multilocus sequence analysis (MLSA) of *Edwardsiella tarda* isolates from fish. *Veterinary Microbiology* 158, 367–375.

Abayneh, T., Colquhoun, D. and Sørum, H. (2013) *Edwardsiella piscicida* sp. nov., a novel species pathogenic to fish. *Journal of Applied Microbiology* 114, 644–654.

Abdelhamed, H., Lu, J., Shaheen, A., Abbass, A., Lawrence, M. and Karsi, A. (2013) Construction and evaluation of an *Edwardsiella ictaluri fhuC* mutant. *Veterinary Microbiology* 162, 858–865.

Abraham, T., Mallick, P., Adikesavalu, H. and Banerjee, S. (2015) Pathology of *Edwardsiella tarda* infection in African catfish, *Clarias gariepinus* (Burchell 1822), fingerlings. *Archives of Polish Fisheries* 23, 141–148.

Acharya, M., Maiti, N.K., Mohanty, S., Mishra, P. and Samanta, M. (2007) Genotyping of *Edwardsiella tarda* isolated from freshwater fish culture system.

Comparative Immunology, Microbiology and Infectious Diseases 30, 33–40.

Akinbowale, O., Peng, H. and Barton, M. (2006) Antimicrobial resistance in bacteria isolated from aquaculture sources in Australia. *Journal of Applied Microbiology* 100, 1103–1113.

Alcaide, E., Herraiz, S. and Esteve, C. (2006) Occurrence of *Edwardsiella tarda* in wild European eels *Anguilla anguilla* from Mediterranean Spain. *Diseases of Aquatic Organisms* 73, 77–81.

Amandi, A., Hiu, S., Rohovec, J. and Fryer, J. (1982) Isolation and characterization of *Edwardsiella tarda* from fall chinook salmon (*Oncorhynchus tshawytscha*). *Applied and Environmental Microbiology* 43, 1380–1384.

Aoki, T. and Takahashi, A. (1987) Class D tetracycline resistance determinants of R plasmids from the fish pathogens *Aeromonas hydrophila*, *Edwardsiella tarda*, and *Pasteurella piscicida*. *Antimicrobial Agents and Chemotherapy* 31, 1278–1280.

Aoki, T., Kitao, T. and Fukudome, M. (1989) Chemotherapy against infection with multiple drug resistant strains of *Edwardsiella tarda* in cultured eels. *Fish Pathology* 24, 161–168.

Ashida, T., Okimasu, E., Ui, M., Heguri, M., Oyama, Y. and Amemura, A. (1999) Protection of Japanese flounder *Paralichthys olivaceus* against experimental edwardsiellosis by formalin-killed *Edwardsiella tarda* in combination with oral administration of immunostimulants. *Fisheries Science* 65, 527–530.

Austin, B. and Austin, D. (1993) *Bacterial Pathogens in Farmed Wild Fish*. Ellis Horwood, New York, pp. 188–226.

Bader, J., Shoemaker, C., Klesius, P., Connolly, M. and Barbaree, J. (1998) Genomic subtyping of *Edwardsiella ictaluri* isolated from diseased channel catfish by arbitrarily primed polymerase chain reaction. *Journal of Aquatic Animal Health* 10, 22–27.

Baldwin, T. and Newton, J. (1993) Pathogenesis of enteric septicemia of channel catfish, caused by *Edwardsiella ictaluri*: bacteriologic and light and electron microscopic findings. *Journal of Aquatic Animal Health* 5, 189–198.

Bartie, K., Austin, F., Diab, A., Dickson, C., Dung, T. *et al.* (2012) Intraspecific diversity of *Edwardsiella ictaluri* isolates from diseased freshwater catfish, *Pangasianodon hypophthalmus* (Sauvage), cultured in the Mekong Delta, Vietnam. *Journal of Fish Diseases* 35, 671–682.

Bebak, J. and Wagner, B. (2012) Use of vaccination against enteric septicemia of catfish and columnaris disease by the U.S. catfish industry. *Journal of Aquatic Animal Health* 24, 30–36.

Bilodeau, A., Waldbieser, G., Terhune, J., Wise, D. and Wolters, W. (2003) A real-time polymerase chain reaction assay of the bacterium *Edwardsiella ictaluri* in channel catfish. *Journal of Aquatic Animal Health* 15, 80–86.

Blanch, A., Pintó, R. and Jofre, J. (1990) Isolation and characterization of an *Edwardsiella* sp. strain, causative agent of mortalities in sea bass (*Dicentrarchus labrax*). *Aquaculture* 88, 213–222.

Camus, A., Dill, J., McDermott, A., Hatcher, N. and Griffin, M. (2016) *Edwardsiella piscicida*-associated septicaemia in a blotched fantail stingray *Taeniura meyeni* (Müeller & Henle). *Journal of Fish Diseases* 39, 1125–1131.

Castro, N., Toranzo, A., Nunez, S. and Magarinos, B. (2008) Development of an effective *Edwardsiella tarda* vaccine for cultured turbot (*Scophthalmus maximus*). *Fish and Shellfish Immunology* 25, 208–212.

Cheng, S., Hu, Y., Zhang, M. and Sun, L. (2010) Analysis of the vaccine potential of a natural avirulent *Edwardsiella tarda* isolate. *Vaccine* 28, 2716–2721.

Choi, S. and Kim, K. (2011) Generation of two auxotrophic genes knock-out *Edwardsiella tarda* and assessment of its potential as a combined vaccine in olive flounder (*Paralichthys olivaceus*). *Fish and Shellfish Immunology* 31, 58–65.

Ciembor, P., Blazer, V., Dawe, D. and Shotts, E. (1995) Susceptibility of channel catfish to infection with *Edwardsiella ictaluri*: effect of exposure method. *Journal of Aquatic Animal Health* 7, 132–140.

Clarridge, J., Musher, D., Fainstein, V. and Wallace, R. (1980) Extraintestinal human infection caused by *Edwardsiella tarda*. *Journal of Clinical Microbiology* 11, 511–514.

Clavijo, A., Conroy, G., Conroy, D., Santander, J. and Aponte, F. (2002) First report of *Edwardsiella tarda* from tilapias in Venezuela. *Bulletin of the European Association of Fish Pathologists* 22, 280–282.

Costa, A., Kanai, K. and Yoshikoshi, K. (1998) Serological characterization of atypical strains of *Edwardsiella tarda* isolated from sea breams. *Fish Pathology* 33, 265–274.

Crumlish, M., Dung, T., Turnbull, J., Ngoc, N. and Ferguson, H. (2002) Identification of *Edwardsiella ictaluri* from diseased freshwater catfish, *Pangasius hypophthalmus* (Sauvage), cultured in the Mekong Delta, Vietnam. *Journal of Fish Diseases* 25, 733–736.

Darwish, A., Plumb, J. and Newton, J. (2000) Histopathology and pathogenesis of experimental infection with *Edwardsiella tarda* in channel catfish. *Journal of Aquatic Animal Health* 12, 255–266.

Ewing, W., McWhorter, A., Escobar, M. and Lubin, A. (1965) *Edwardsiella*, a new genus of *Enterobacteriaceae* based on a new species, *E. tarda*. *International Bulletin of Bacteriological Nomenclature and Taxonomy* 15, 33–38.

Fernandez, D., Pittman-Cooley, L. and Thune, R. (2001) Sequencing and analysis of the *Edwardsiella ictaluri* plasmids. *Plasmid* 45, 52–56.

Fogelson, S., Petty, B., Reichley, S., Ware, C., Bowser, P. et al. (2016) Histologic and molecular characterization of *Edwardsiella piscicida* infection in largemouth bass (*Micropterus salmoides*). *Journal of Veterinary Diagnostic Investigation* 28, 338–344.

Francis-Floyd, R., Beleau, M., Waterstrat, P. and Bowser, P. (1987) Effect of water temperature on the clinical outcome of infection with *Edwardsiella ictaluri* in channel catfish. *Journal of the American Veterinary Medical Association* 191, 1413–1416.

Francis-Floyd, R., Reed, P., Bolon, B., Estes, J. and McKinney, S. (1993) An epizootic of *Edwardsiella tarda* in largemouth bass (*Micropterus salmoides*). *Journal of Wildlife Diseases* 29, 334–336.

Gaikowski, M., Wolf, J., Endris, R. and Gingerich, W. (2003) Safety of Aquaflor (florfenicol, 50% type A medicated article), administered in feed to channel catfish, *Ictalurus punctatus*. *Toxicologic Pathology* 31, 689–697.

Gaunt, P., Endris, R., Khoo, L., Howard, R., McGinnis, A. et al. (2004) Determination of dose rate of florfenicol in feed for control of mortality in channel catfish *Ictalurus punctatus* (Rafinesque) infected with *Edwardsiella ictaluri*, etiological agent of enteric septicemia. *Journal of the World Aquaculture Society* 35, 257–267.

Gaunt, P., McGinnis, A., Santucci, T., Cao, J., Waeger, P. and Endris, R. (2006) Field efficacy of florfenicol for control of mortality in channel catfish, *Ictalurus punctatus* (Rafinesque), caused by infection with *Edwardsiella ictaluri*. *Journal of the World Aquaculture Society*, 37, 1–11.

Gilman, R., Madasamy, M., Gan, E., Mariappan, M., Davis, C. and Kyser, K. (1971) *Edwardsiella tarda* in jungle diarrhea and a possible association with *Entamoeba histolytica*. *Southeast Asian Journal of Tropical Medicine and Public Health* 2, 186–189.

Greenway, T., Byars, T., Elliot, R., Jin, X., Griffin, M. and Wise, D. (2017) Validation of fermentation and processing procedures for the commercial scale production of a live attenuated *Edwardsiella ictaluri* vaccine for use in catfish aquaculture. *Journal of Aquatic Animal Health* (in press).

Griffin, M., Mauel, M., Greenway, T., Khoo, L. and Wise, D. (2011) A real-time polymerase chain reaction assay for quantification of *Edwardsiella ictaluri* in catfish pond water and genetic homogeneity of diagnostic case isolates from Mississippi. *Journal of Aquatic Animal Health* 23, 178–188.

Griffin, M., Quiniou, S., Cody, T., Tabuchi, M., Ware, C. et al. (2013) Comparative analysis of *Edwardsiella* isolates from fish in the eastern United States identifies two distinct genetic taxa amongst organisms phenotypically classified as *E. tarda*. *Veterinary Microbiology* 165, 358–372.

Griffin, M., Ware, C., Quiniou, S., Steadman, J., Gaunt, P. et al. (2014) *Edwardsiella piscicida* identified in the southeastern USA by *gyrB* sequence, species-specific and repetitive sequence-mediated PCR. *Diseases of Aquatic Organisms*, 108, 23–35.

Griffin, M., Reichley, S., Greenway, T., Quiniou, S., Ware, C. et al. (2015) Comparison of *Edwardsiella ictaluri* isolates from different hosts and geographic origins. *Journal of Fish Diseases* 39, 947–969.

Grimont, P., Grimont, F., Richard, C. and Sakazaki, R. (1980) *Edwardsiella hoshinae*, a new species of *Enterobacteriaceae*. *Current Microbiology* 4, 347–351.

Haenen, O., Evans, J. and Berthe, F. (2013) Bacterial infections from aquatic species: potential for and prevention of contact zoonoses. *Revue Scientifique et Technique* 32, 497–507.

Hawke, J. and Khoo, L. (2004) Infectious diseases. In: Tucker, C. and Hargreaves, J. (ed.) *Biology and Culture of the Channel Catfish*, 1st edn. Elsevier, Amsterdam, pp. 347–443.

Hawke, J., McWhorter, A., Steigerwalt, A. and Brenner, D. (1981) *Edwardsiella ictaluri* sp. nov., the causative agent of enteric septicemia of catfish. *International Journal of Systematic Bacteriology* 31, 396–400.

Hawke, J., Kent, M., Rogge, M., Baumgartner, W., Wiles, J. et al. (2013) Edwardsiellosis caused by *Edwardsiella ictaluri* in laboratory populations of zebrafish *Danio rerio. Journal of Aquatic Animal Health* 25, 171–183.

Herman, R. and Bullock, G. (1986) Pathology caused by the bacterium *Edwardsiella tarda* in striped bass. *Transactions of the American Fisheries Society* 115, 232–235.

Hilton, L. and Wilson, J. (1980) Terramycin-resistant *Edwardsiella tarda* in channel catfish. *The Progressive Fish-Culturist* 42, 159.

Hirai, Y., Ashata-Tago, S., Ainoda, Y., Fujita, T. and Kikuchi, K. (2015) *Edwardsiella tarda* bacteremia. A rare but fatal water- and foodborne infection: review of the literature and clinical cases from a single centre. *Canadian Journal of Infectious Disease and Medical Microbiology* 26, 313–318.

Hirono, I., Tange, N. and Aoki, T. (1997) Iron-regulated haemolysin gene from *Edwardsiella tarda. Molecular Microbiology* 24, 851–856.

Humphrey, J. and Langdon, J. (1986) Pathological anatomy and diseases of barramundi (*Lates calcarifer*). In: Copland, J.W. and Grey, D.L. (eds) *Management of Wild Cultured Sea Bass/Barramundi (*Lates calcarifer*): Proceedings of an International Workshop held at Darwin, N.T., Australia, 24–30 September 1986. ACIAR Proceedings No. 20, Australian Centre for International Agricultural Research, Canberra, pp. 198–203.

Humphrey, J., Lancaster, C., Gudkovs, N. and McDonald, W. (1986) Exotic bacterial pathogens *Edwardsiella tarda* and *Edwardsiella ictaluri* from imported ornamental fish *Betta splendens* and *Puntius conchonius*, respectively: isolation and quarantine significance. *Australian Veterinary Journal* 63, 369–371.

Iwanowicz, L., Griffin, A., Cartwright, D. and Blazer, V. (2006) Mortality and pathology in brown bullheads *Amieurus nebulosus* associated with a spontaneous *Edwardsiella ictaluri* outbreak under tank culture conditions. *Diseases of Aquatic Organisms* 70, 219–225.

Janda, J. and Abbott, S. (1993) Infections associated with the genus *Edwardsiella*: the role of *Edwardsiella tarda* in human disease. *Clinical Infectious Diseases* 17, 742–748.

Janda, J., Abbott, S., Kroske-Bystrom, S., Cheung, W., Powers, C. et al. (1991) Pathogenic properties of *Edwardsiella* species. *Journal of Clinical Microbiology* 29, 1997–2001.

Joh, S., Kim, M., Kwon, H., Ahn, E., Jang, H. and Kwon, J. (2011) Characterization of *Edwardsiella tarda* isolated from farm-cultured eels, *Anguilla japonica*, in the Republic of Korea. *Journal of Veterinary Medical Science* 73, 7–11.

Kasornchandra, J., Rogers, W. and Plumb, J. (1987) *Edwardsiella ictaluri* from walking catfish, *Clarias batrachus* L., in Thailand. *Journal of Fish Diseases* 10, 137–138.

Kawai, K., Liu, Y., Ohnishi, K. and Oshima, S. (2004) A conserved 37 kDa outer membrane protein of *Edwardsiella tarda* is an effective vaccine candidate. *Vaccine* 22, 3411–3418.

Kent, M. and Lyons, J. (1982) *Edwardsiella ictaluri* in the green knife fish, *Eigemannia virescens. Fish Health News* 2, 2. [Eastern Fish Disease Laboratory, US Department of the Interior, Fish and Wildlife Service, Kearneysville, West Virginia.]

Keskin, O., Seçer, S., İzgür, M., Türkyilmaz, S. and Mkakosya, R.S. (2004) *Edwardsiella ictaluri* infection in *rainbow trout* (*Oncorhynchus mykiss*). *Turkish Journal of Veterinary and Animal Sciences* 28, 649–653.

Klesius, P. (1994) Transmission of *Edwardsiella ictaluri* from infected, dead to noninfected channel catfish. *Journal of Aquatic Animal Health* 6, 180–182.

Klesius, P., Lovy, J., Evans, J., Washuta, E. and Arias, C. (2003) Isolation of *Edwardsiella ictaluri* from tadpole madtom in a southwestern New Jersey River. *Journal of Aquatic Animal Health* 15, 295–301.

Kusuda, R. and Kawai, K. (1998) Bacterial diseases of cultured marine fish in Japan. *Fish Pathology* 33, 221–227.

Kusuda, R., Toyoshima, T., Iwamura, Y. and Sako, H. (1976) *Edwardsiella tarda* from an epizootic of mullets (*Mugil cephalus*) in Okitsu Bay. *Nippon Suisan Gakkaishi* 42, 271–275.

Kusuda, R., Itami, T., Munekiyo, M. and Nakajima, H. (1977) Characteristics of a *Edwardsiella* sp. from an epizootic of cultured sea breams. *Nippon Susan Gakkaishi* 43, 129–134.

Kwon, S., Nam, Y., Kim, S. and Kim, K. (2006) Protection of tilapia (*Oreochromis mosambicus*) from edwardsiellosis by vaccination with *Edwardsiella tarda* ghosts. *Fish and Shellfish Immunology* 20, 621–626.

Kwon, S., Lee, E., Nam, Y., Kim, S. and Kim, K. (2007) Efficacy of oral immunization with *Edwardsiella tarda* ghosts against edwardsiellosis in olive flounder (*Paralichthys olivaceus*). *Aquaculture* 269, 84–88.

Lan, J., Zhang, X., Wang, Y., Chen, J. and Han, Y. (2008) Isolation of an unusual strain of *Edwardsiella tarda* from turbot and establish a PCR detection technique with the *gyrB* gene. *Journal of Applied Microbiology* 105, 644–651.

Lawrence, M., Cooper, R. and Thune, R. (1997) Attenuation, persistence, and vaccine potential of an *Edwardsiella ictaluri purA* mutant. *Infection and Immunity* 65, 4642–4651.

Leotta, G., Piñeyro, P., Serena, S. and Vigo, G. (2009) Prevalence of *Edwardsiella tarda* in Antarctic wildlife. *Polar Biology* 32, 809–812.

Liang, S., Wu, H., Liu, B., Xiao, J., Wang, Q. and Zhang, Y. (2012) Immune response of turbot (*Scophthalmus maximus* L.) to a broad spectrum vaccine candidate, recombinant glyceraldehyde-3-phosphate dehydrogenase of *Edwardsiella tarda*. *Veterinary Immunology and Immunopathology* 150, 198–205.

Lo, D., Lee, Y., Wang, J. and Kuo, H. (2014) Antimicrobial susceptibility and genetic characterisation of oxytetracycline-resistant *Edwardsiella tarda* isolated from diseased eels. *Veterinary Record* 175, 203–203.

Lv, Y., Zheng, J., Yang, M., Wang, Q. and Zhang, Y. (2012) An *Edwardsiella tarda* mutant lacking UDP-glucose dehydrogenase shows pleiotropic phenotypes, attenuated virulence, and potential as a vaccine candidate. *Veterinary Microbiology* 160, 506–512.

Mainous, M., Smith, S. and Kuhn, D. (2010) Effect of common aquaculture chemicals against *Edwardsiella ictaluri* and *E. tarda*. *Journal of Aquatic Animal Health* 22, 224–228.

Matshushima, S., Yajima, S., Taguchi, T., Takahashi, A., Shiseki, M. *et al.* (1996) A fulminating case of *Edwardsiella tarda* septicemia with necrotizing fasciitis. *Kansenshogaku Zasshi* 70, 631–636.

Matsuyama, T., Kamaishi, T., Ooseko, N., Kurohara, K. and Iida, T. (2005) Pathogenicity of motile and non-motile *Edwardsiella tarda* to some marine fish. *Fish Pathology* 40, 133–135.

Menanteau-Ledouble, S., Karsi, A. and Lawrence, M. (2011) Importance of skin abrasion as a primary site of adhesion for *Edwardsiella ictaluri* and impact on invasion and systematic infection in channel catfish *Ictalurus punctatus*. *Veterinary Microbiology* 148, 425–430.

Meyer, F. and Bullock., G. (1973) *Edwardsiella tarda*, a new pathogen of channel catfish (*Ictalurus punctatus*). *Applied Microbiology* 25, 155–156.

Miyazaki, T. and Plumb, J. (1985) Histopathology of *Edwardsiella ictaluri* in channel catfish, *Ictalurus punctatus* (Rafinesque). *Journal of Fish Diseases* 8, 389–392.

Mohanty, B. and Sahoo, P. (2007) Edwardsiellosis in fish: a brief review. *Journal of Biosciences* 32, 1331–1344.

Mqolomba, T. and Plumb, J. (1992) Effect of temperature and dissolved oxygen concentration on *Edwardsiella ictaluri* in experimentally infected channel catfish. *Journal of Aquatic Animal Health* 4, 215–217.

Nagai, T. and Nakai, T. (2014) Water temperature effect on *Edwardsiella ictaluri* infection of ayu *Plecoglossus altivelis*. *Fish Pathology* 49, 61–63.

Nagai, T., Iwamoto, E., Sakai, T., Arima, T., Tensha, K. *et al.* (2008) Characterization of *Edwardsiella ictaluri* isolated from wild ayu *Plecoglossus altivelis* in Japan. *Fish Pathology* 43, 158–163.

Nakhro, K., Devi, T. and Kamilya, D. (2013) *In vitro* immunopathogenesis of *Edwardsiella tarda* in catla *Catla catla* (Hamilton). *Fish and Shellfish Immunology* 35, 175–179.

Newton, J.C., Bird, R.C., Blevins, W.T., Wilt, G.R. and Wolfe, L.G. (1988) Isolation, characterization, and molecular cloning of cryptic plasmids isolated from *Edwardsiella ictaluri*. *American Journal of Veterinary Research* 49, 1856–1860.

Nite.go.jp (2016) Genome Projects: Whole Genome Shotgun (WGS). (National) Biological Resource Center (NBRC), NITE (National Institute of Technology and Evaluation), Shibuya-ku, Tokyo. Available at: http://www.nite.go.jp/en/nbrc/genome/project/wgs/project_wgs.html (accessed 31 May 2016).

Oguro, K., Tamura, K., Yamane, J., Shimizu, M., Yamamoto, T. *et al.* (2014) Draft genome sequences of two genetic variant strains of *Edwardsiella piscicida*, JF1305 and RSB1309, isolated from olive flounder (*Paralichythys olivaceus*) and red sea bream (*Pagrus major*) cultured in Japan, respectively. *Genome Announcements* 2(3): e00546–14.

Padrós, F., Zarza, C., Dopazo, L., Cuadrado, M. and Crespo, S. (2006) Pathology of *Edwardsiella tarda* infection in turbot, *Scophthalmus maximus* (L.). *Journal of Fish Diseases* 29, 87–94.

Pan, X., Hao, G., Yao, J., Xu, Y., Shen, J. and Yin, W. (2010) Identification and pathogenic facts studying for *Edwarsdiella tarda* from edwardsiellosis of *Trionyx sinensis* [J.]. *Freshwater Fisheries* 6, 40–45.

Panangala, V., Shoemaker, C., van Santen, V., Dybvig, K. and Klesius, P. (2007) Multiplex-PCR for simultaneous detection of three bacterial fish pathogens, *Flavobacterium columnare*, *Edwardsiella ictaluri*, and *Aeromonas hydrophila*. *Diseases of Aquatic Organisms* 74, 199–208.

Park, S., Aoki, T. and Jung, T. (2012) Pathogenesis of and strategies for preventing *Edwardsiella tarda* infection in fish. *Veterinary Research* 43, 67.

Petrie-Hanson, L. and Ainsworth, J.A. (1999) Humoral immune responses of channel catfish (*Ictalurus punctatus*) fry and fingerlings exposed to *Edwardsiella ictaluri*. *Fish and Shellfish Immunology* 9, 579–589.

Petrie-Hanson, L., Romano, C., Mackey, R., Khosravi, P., Hohn, C. and Boyle, C. (2007) Evaluation of zebrafish *Danio rerio* as a model for enteric septicemia of catfish (ESC). *Journal of Aquatic Animal Health* 19, 151–158.

Plumb, J. and Quinlan, E. (1986) Survival of *Edwardsiella ictaluri* in pond water and bottom mud. *The Progressive Fish-Culturist* 48, 212–214.

Plumb, J. and Shoemaker, C. (1995) Effects of temperature and salt concentration on latent *Edwardsiella ictaluri* infections in channel catfish. *Diseases of Aquatic Organisms* 21, 171–175.

Plumb, J. and Vinitnantharat, S. (1989) Biochemical, biophysical, and serological homogeneity of *Edwardsiella ictaluri*. *Journal of Aquatic Animal Health* 1, 51–56.

Plumb, J. and Vinitnantharat, S. (1993) Vaccination of channel catfish, *Ictalurus punctatus* (Rafinesque), by immersion and oral booster against *Edwardsiella ictaluri*. *Journal of Fish Diseases* 16, 65–71.

Plumb, J., Maestrone, G. and Quinlan, E. (1987) Use of a potentiated sulfonamide to control *Edwardsiella ictaluri* infection in channel catfish (*Ictalurus punctatus*). *Aquaculture* 62, 187–194.

Plumb, J., Vinitnantharat, S. and Paterson, W. (1994) Optimum concentration of *Edwardsiella ictaluri* vaccine in feed for oral vaccination of channel catfish. *Journal of Aquatic Animal Health* 6, 118–121.

Řehulka, J., Marejková, M. and Petráš, P. (2012) Edwardsiellosis in farmed rainbow trout (*Oncorhynchus mykiss*). *Aquaculture Research* 43, 1628–1634.

Reichley, S., Waldbieser, G., Tekedar, H., Lawrence, M. and Griffin, M. (2015a) Complete genome sequence of *Edwardsiella tarda* isolate FL95-01, recovered from channel catfish. *Genome Announcements* 3(3): e00682–15.

Reichley, S., Waldbieser, G., Ucko, M., Colorni, A., Dubytska, L. *et al*. (2015b) Complete genome sequence of an *Edwardsiella piscicida*-like species isolated from diseased grouper in Israel. *Genome Announcements* 3(4): e00829–15.

Reichley, S., Waldbieser, G., Lawrence, M. and Griffin, M. (2015c) Complete genome sequence of an *Edwardsiella piscicida*-like species, recovered from tilapia in the United States. *Genome Announcements* 3(5): e01004–15.

Reichley, S., Ware, C., Greenway, T., Wise, D. and Griffin, M. (2015d) Real-time polymerase chain reaction assays for the detection and quantification of *Edwardsiella tarda*, *Edwardsiella piscicida*, and *Edwardsiella piscicida*-like species in catfish tissues and pond water. *Journal of Veterinary Diagnostic Investigation* 27, 130–139.

Rogge, M., Dubytska, L., Jung, T., Wiles, J., Elkamel, A. *et al*. (2013) Comparison of Vietnamese and US isolates of *Edwardsiella ictaluri*. *Diseases of Aquatic Organisms* 106, 17–29.

Saeed, M.O. and Plumb, J.A. (1986) Immune response of channel catfish to lipopolysaccharide and whole cell *Edwardsiella ictaluri* vaccines. *Diseases of Aquatic Organisms* 2, 21–26.

Sae-oui, D., Muroga, K. and Nakai, T. (1984) A case of *Edwardsiella tarda* infection in cultured colored carp *Cyprinus carpio*. *Fish Pathology* 19, 197–199.

Sakai, T., Iida, T., Osatomi, K. and Kanai, K. (2007) Detection of Type 1 fimbrial genes in fish pathogenic and non-pathogenic *Edwardsiella tarda* strains by PCR. *Fish Pathology* 42, 115–117.

Sakai, T., Kamaishi, T., Sano, M., Tensha, K., Arima, T. *et al*. (2008) Outbreaks of *Edwardsiella ictaluri* infection in ayu *Plecoglossus altivelis* in Japanese rivers. *Fish Pathology* 43, 152–157.

Sakai, T., Matsuyama, T., Sano, M. and Iida, T. (2009a) Identification of novel putative virulence factors, adhesin AIDA and type VI secretion system, in atypical strains of fish pathogenic *Edwardsiella tarda* by genomic subtractive hybridization. *Microbiology and Immunology* 53, 131–139.

Sakai, T., Yuasa, K., Ozaki, A., Sano, M., Okuda, R. *et al*. (2009b) Genotyping of *Edwardsiella ictaluri* isolates in Japan using amplified-fragment length polymorphism analysis. *Letters in Applied Microbiology* 49, 443–449.

Sakai, T., Yuasa, K., Sano, M. and Iida, T. (2009c) Identification of *Edwardsiella ictaluri* and *E. tarda* by species-specific polymerase chain reaction targeted to the upstream region of the fimbrial gene. *Journal of Aquatic Animal Health* 21, 124–132.

Salati, F. and Kusuda, R. (1985) Vaccine preparations used for immunization of eel *Anguilla japonica* against *Edwardsiella tarda* infection. *Nippon Suisan Gakkaishi* 51, 1233–1237.

Santander, J., Golden, G., Wanda, S. and Curtiss, R. (2012) Fur-regulated iron uptake system of *Edwardsiella ictaluri* and its influence on pathogenesis and immunogenicity in the catfish host. *Infection and Immunity* 80, 2689–2703.

Savan, R., Igarashi, A., Matsuoka, S. and Sakai, M. (2004) Sensitive and rapid detection of edwardsiellosis in fish by a loop-mediated isothermal amplification method. *Applied and Environmental Microbiology* 70, 621–624.

Shafiei, S., Viljamaa-Dirks, S., Sundell, K., Heinikainen, S., Abayneh, T. and Wiklund, T. (2016) Recovery of *Edwardsiella piscicida* from farmed whitefish, *Coregonus lavaretus* (L.), in Finland. *Aquaculture*, 454, 19–26.

Shao, S., Lai, Q., Liu, Q., Wu, H., Xiao, J. *et al*. (2015) Phylogenomics characterization of a highly virulent *Edwardsiella* strain ET080813T encoding two distinct T3SS and three T6SS gene clusters: propose a novel species as *Edwardsiella anguillarum* sp. nov. *Systematic and Applied Microbiology* 38, 36–47.

Shao, J., Yuan, J., Shen, Y., Hu, R. and Gu, Z. (2016) First isolation and characterization of *Edwardsiella tarda* from diseased Asian swamp eel, *Monopterus albus* (Zuiew). *Aquaculture Research* 47, 3684–3688.

Shetty, M., Maiti, B., Venugopal, M., Karunasagar, I. and Karunasagar, I. (2014) First isolation and characterization of *Edwardsiella tarda* from diseased striped catfish, *Pangasianodon hypophthalmus* (Sauvage). *Journal of Fish Diseases* 37, 265–271.

Shoemaker, C. and Klesius, P. (1997) Protective immunity against enteric septicaemia in channel catfish, *Ictalurus punctatus* (Rafinesque), following controlled exposure to *Edwardsiella ictaluri*. *Journal of Fish Diseases* 20, 361–368.

Shoemaker, C., Klesius, P. and Bricker, J. (1999) Efficacy of a modified live *Edwardsiella ictaluri* vaccine in channel catfish as young as seven days post hatch. *Aquaculture* 176, 189–193.

Shoemaker, C., Klesius, P. and Evans, J. (2002) *In ovo* methods for utilizing the modified live *Edwardsiella ictaluri* vaccine against enteric septicemia in channel catfish. *Aquaculture* 203, 221–227.

Shoemaker, C., Klesius, P. and Evans, J. (2007) Immunization of eyed channel catfish, *Ictalurus punctatus*, eggs with monovalent *Flavobacterium columnare* vaccine and bivalent *F. columnare* and *Edwardsiella ictaluri* vaccine. *Vaccine* 25, 1126–1131.

Shotts, E., Blazer, V. and Waltman, W. (1986) Pathogenesis of experimental *Edwardsiella ictaluri* infections in channel catfish (*Ictalurus punctatus*). *Canadian Journal of Fisheries and Aquatic Science* 43, 36–42.

Soto, E., Griffin, M., Arauz, M., Riofrio, A., Martinez, A. and Cabrejos, M. (2012) *Edwardsiella ictaluri* as the causative agent of mortality in cultured Nile tilapia. *Journal of Aquatic Animal Health* 24, 81–90.

Starliper, C., Cooper, R., Shotts, E. and Taylor, P. (1993) Plasmid-mediated Romet resistance of *Edwardsiella ictaluri*. *Journal of Aquatic Animal Health* 5, 1–8.

Sun, Y., Liu, C.-S. and Sun, L. (2010) Identification of an *Edwardsiella tarda* surface antigen and analysis of its immunoprotective potential as a purified recombinant subunit vaccine and a surface-anchored subunit vaccine expressed by a fish commensal strain. *Vaccine* 28, 6603–6608.

Sun, Y., Liu, C.-S. and Sun, L. (2011a) Construction and analysis of the immune effect of an *Edwardsiella tarda* DNA vaccine encoding a D15-like surface antigen. *Fish and Shellfish Immunology* 30, 273–279.

Sun, Y., Liu, C.-S. and Sun, L. (2011b) Comparative study of the immune effect of an *Edwardsiella tarda* antigen in two forms: subunit vaccine vs DNA vaccine. *Vaccine* 29, 2051–2057.

Swain, P., Nayak, S., Sahu, A., Mohapatra, B. and Meher, P. (2002) Bath immunisation of spawn, fry and fingerlings of Indian major carps using a particulate bacterial antigen. *Fish and Shellfish Immunology* 13, 133–140.

Takano, T., Matsuyama, T., Oseko, N., Sakai, T., Kamaishi, T. *et al.* (2010) The efficacy of five avirulent *Edwardsiella tarda* strains in a live vaccine against edwardsiellosis in Japanese flounder, *Paralichthys olivaceus*. *Fish and Shellfish Immunology* 29, 687–693.

Takano, T., Matsuyama, T., Sakai, T. and Nakayasu, C. (2011) Protective efficacy of a formalin-killed vaccine against atypical *Edwardsiella tarda* infection in red sea bream *Pagrus major*. *Fish Pathology* 46, 120–122.

Taylor, P. (1992) Fish-eating birds as potential vectors of *Edwardsiella ictaluri*. *Journal of Aquatic Animal Health* 4, 240–243.

Tekedar, H., Karsi, A., Williams, M., Vamenta, S., Banes, M. *et al.* (2013) Complete genome sequence of channel catfish gastrointestinal septicemia isolate *Edwardsiella tarda* C07-087. *Genome Announcements* 1(6): e00959–13.

Thune, R., Stanley, L. and Cooper, R. (1993) Pathogenesis of Gram-negative bacterial infections in warmwater fish. *Annual Review of Fish Diseases* 3, 37–68.

Thune, R., Hawke, J. and Johnson, M. (1994) Studies on vaccination of channel catfish, *Ictalurus punctatus*, against *Edwardsiella ictaluri*. *Journal of Applied Aquaculture* 3, 11–24.

Thune, R., Fernandez, D. and Battista, J. (1999) An *aroA* mutant of *Edwardsiella ictaluri* is safe and efficacious as a live, attenuated vaccine. *Journal of Aquatic Animal Health* 11, 358–372.

Trung Cao, T., Tsai, M., Yang, C., Wang, P., Kuo, T. *et al.* (2014) Vaccine efficacy of glyceraldehyde-3-phosphate dehydrogenase (GAPDH) from *Edwardsiella ictaluri* against *E. tarda* in tilapia. *The Journal of General and Applied Microbiology* 60, 241–250.

Turnbaugh, P., Ley, R., Hamady, M., Fraser-Liggett, C., Knight, R. and Gordon, J. (2007) The human microbiome project: exploring the microbial part of ourselves in a changing world. *Nature* 449, 804.

Ucko, M., Colorni, A., Dubytska, L. and Thune, R. (2016) *Edwardsiella piscicida*-like pathogen in cultured grouper. *Diseases of Aquatic Organisms* 121, 141–148.

Uhland, F., Hélie, P. and Higgins, R. (2000) Infections of *Edwardsiella tarda* among brook trout in Quebec. *Journal of Aquatic Animal Health* 12, 74–77.

USDA/APHIS/NAHMS (2010) *Catfish 2010. Part I: Reference of Catfish Health and Production Practices in the United States, 2009*. US Department of Agriculture/Animal and Plant Health Inspection Service, National Animal Health Monitoring System, Fort Collins, Colorado.

Van Damme, L. and Vandepitte, J. (1980) Frequent isolation of *Edwardsiella tarda* and *Pleisiomonas shigelloides* from healthy Zairese freshwater fish: a possible source of sporadic diarrhea in the tropics. *Applied and Environmental Microbiology* 39, 475–479.

Vandepitte, J., Lemmens, P. and De Swert, L. (1983) Human edwardsiellosis traced to ornamental fish. *Journal of Clinical Microbiology* 17, 165–167.

van Soest, J., Stockhammer, O., Ordas, A., Bloemberg, G., Spaink, H. and Meijer, A. (2011) Comparison of static immersion and intravenous injection systems for exposure of zebrafish embryos to the natural pathogen *Edwardsiella tarda*. *BMC Immunology* 12:58.

Waltman, W. and Shotts, E. (1986a) Antimicrobial susceptibility of *Edwardsiella tarda* from the United States and Taiwan. *Veterinary Microbiology* 12, 277–282.

Waltman, W. and Shotts, E. (1986b) Antimicrobial susceptibility of *Edwardsiella ictaluri*. *Journal of Wildlife Diseases* 22, 173–177.

Waltman, W., Shotts, E. and Blazer, V. (1985) Recovery of *Edwardsiella ictaluri* from danio (*Danio devario*). *Aquaculture* 46, 63–66.

Waltman II, W., Shotts, E. and Wooley, R. (1989) Development and transfer of plasmid-mediated antimicrobial resistance in *Edwardsiella ictaluri*. *Canadian Journal of Fisheries and Aquatic Sciences* 46, 1114–1117.

Wang, Q., Yang, M., Xiao, J., Wu, H., Wang, X. *et al.* (2009) Genome sequence of the versatile fish pathogen *Edwardsiella tarda* provides insights into its adaptation to broad host ranges and intracellular niches. *PLoS ONE* 4(10): e7646.

Wang, R., Tekedar, H.C., Lawrence, M.L., Chouljenko, V.N., Kim, J. *et al.* (2015) Draft genome sequences of *Edwardsiella ictaluri* strains LADL11-100 and LADL11-194 isolated from zebrafish *Danio rerio*. *Genome Announcements* 3(6): e01449–15.

Wang, Y., Wang, Q., Xiao, J., Liu, Q., Wu, H. and Zhang, Y. (2011) Genetic relationships of *Edwardsiella* strains isolated in China aquaculture revealed by rep-PCR genomic fingerprinting and investigation of *Edwardsiella* virulence genes. *Journal of Applied Microbiology* 111, 1337–1348.

Waterstrat, P., Dorr, B., Glahn, J. and Tobin, M. (1999) Recovery and viability of *Edwardsiella ictaluri* from great blue herons *Ardea herodias* fed *E. ictaluri*-infected channel catfish *Ictalurus punctatus* fingerlings. *Journal of the World Aquaculture Society* 30, 115–122.

Welch, T., Evenhuis, J., White, D., McDermott, P., Harbottle, H. *et al.* (2008) IncA/C plasmid-mediated florfenicol resistance in the catfish pathogen *Edwardsiella ictaluri*. *Antimicrobial Agents and Chemotherapy* 53, 845–846.

White, F., Simpson, C. and Williams, L. (1973) Isolation of *Edwardsiella tarda* from aquatic animal species and surface waters in Florida. *Journal of Wildlife Diseases* 9, 204–208.

Williams, M., Gillaspy, A., Dyer, D., Thune, R., Waldbieser, G. *et al.* (2012) Genome sequence of *Edwardsiella ictaluri* 93-146, a strain associated with a natural channel catfish outbreak of enteric septicemia of catfish *Journal of Bacteriology* 194, 740–741.

Wise, D. and Johnson, M. (1998) Effect of feeding frequency and Romet-medicated feed on survival, antibody response, and weight gain of fingerling channel catfish *Ictalurus punctatus* after natural exposure to *Edwardsiella ictaluri*. *Journal of the World Aquaculture Society* 29, 169–175.

Wise, D., Schwedler, T. and Otis, D. (1993) Effects of stress on susceptibility of naive channel catfish in immersion challenge with *Edwardsiella ictaluri*. *Journal of Aquatic Animal Health* 5, 92–97.

Wise, D., Camus, A., Schwedler, T. and Terhune, J. (2004) Health management. In: Tucker, C. and Hargreaves, J. (eds) *Biology and Culture of the Channel Catfish*, 1st edn. Elsevier, Amsterdam, pp. 444–503.

Wise, D., Greenway, T., Li, M., Camus, A. and Robinson, E. (2008) Effects of variable periods of food deprivation on the development of enteric septicemia in channel catfish. *Journal of Aquatic Animal Health* 20, 39–44.

Wise, D., Greenway, T., Byars, T., Griffin, M. and Khoo, L. (2015) Oral vaccination of channel catfish against enteric septicemia of catfish using a live attenuated *Edwardsiella ictaluri* isolate. *Journal of Aquatic Animal Health* 27, 135–143.

Wyatt, L., Nickelson, R. and Vanderzant, C. (1979) *Edwardsiella tarda* in freshwater catfish and their environment. *Applied and Environmental Microbiology* 38, 710–714.

Xiao, J., Wang, Q., Liu, Q., Wang, X., Liu, H. and Zhang, Y. (2008) Isolation and identification of fish pathogen *Edwardsiella tarda* from mariculture in China. *Aquaculture Research* 40, 13–17.

Xu, T. and Zhang, X. (2014) *Edwardsiella tarda*: an intriguing problem in aquaculture. *Aquaculture* 431, 129–135.

Yamada, Y. and Wakabayashi, H. (1998) Enzyme electrophoresis, catalase test and PCR-RFLP analysis for the typing of *Edwardsiella tarda*. *Fish Pathology* 33, 1–5.

Yamada, Y. and Wakabayashi, H. (1999) Identification of fish-pathogenic strains belonging to the genus *Edwardsiella* by sequence analysis of *sodB*. *Fish Pathology* 34, 145–150.

Yang, M., Lv, Y., Xiao, J., Wu, H., Zheng, H. *et al.* (2012) *Edwardsiella* comparative phylogenomics reveal the new intra/inter-species taxonomic relationships, virulence evolution and niche adaptation mechanisms. *PLoS ONE* 7(5): e36987.

Yang, W., Wang, L., Zhang, L., Qu, J., Wang, Q. and Zhang, Y. (2015) An invasive and low virulent *Edwardsiella tarda esrB* mutant promising as live attenuated vaccine in aquaculture. *Applied Microbiology and Biotechnology* 99, 1765–1777.

Ye, S., Li, H., Qiao, G. and Li, Z. (2009) First case of *Edwardsiella ictaluri* infection in China farmed yellow catfish *Pelteobagrus fulvidraco*. *Aquaculture* 292, 6–10.

Ye, X., Lin, X. and Wang, Y. (2010) Identification and detection of virulence gene of the pathogenic bacteria *Edwardsiella tarda* in cultured *Scortum barcoo*. *Freshwater Fisheries* 40, 50–54.

Yeh, H., Shoemaker, C. and Klesius, P. (2005) Evaluation of a loop-mediated isothermal amplification method for rapid detection of channel catfish *Ictalurus punctatus* important bacterial pathogen *Edwardsiella ictaluri*. *Journal of Microbiological Methods* 63, 36–44.

Yu, J., Han, J., Park, K., Park, K. and Park, S. (2009) *Edwardsiella tarda* infection in Korean catfish, *Silurus asotus*, in a Korean fish farm. *Aquaculture Research* 41, 19–26.

16 *Flavobacterium* spp.

THOMAS P. LOCH* AND MOHAMED FAISAL

Department of Pathobiology and Diagnostic Investigation, College of Veterinary Medicine, Michigan State University, East Lansing, Michigan, USA

16.1 Introduction

Flavobacterial diseases in fish are mainly attributed to three Gram-negative, yellow-pigmented bacteria: *Flavobacterium psychrophilum*, the cause of bacterial cold water disease (BCWD) and rainbow trout fry syndrome (RTFS; Davis, 1946; Borg, 1948; Holt, 1987; Bernardet and Grimont, 1989); *F. columnare*, the cause of Columnaris disease (CD; Davis, 1922; Ordal and Rucker, 1944; Bernardet and Grimont, 1989); and *F. branchiophilum*, the putative agent of bacterial gill disease (BGD; Wakabayashi *et al.*, 1989).

16.1.1 *Flavobacterium psychrophilum*

Description of the microorganism

Borg (1948) described epizootics at water temperatures of 6–10°C in farmed coho salmon (*Oncorhynchus kisutch*) fry and fingerlings that were caused by masses of bacterial rods, now known as *F. psychrophilum*. The bacterium underwent multiple taxonomic reappraisals and was eventually placed within the genus *Flavobacterium* (Phylum *Bacteroidetes*; Class *Flavobacteriia*; Order *Flavobacteriales*; Family *Flavobacteriaceae*; Bernardet, 2011a) as *F. psychrophilum* (Bernardet *et al.*, 1996).

F. psychrophilum grows optimally from 15 to 18°C, but as low as 4°C (Bernardet and Bowman, 2011). Low nutrient media for the culture of *F. psychrophilum* have been described (Bernardet and Bowman, 2006; Austin and Austin, 2007; Starliper and Schill, 2011). The most common media include Cytophaga medium (Anacker and Ordal, 1959), tryptone-enriched Cytophaga medium (Bernardet and Kerouault, 1989) and tryptone yeast extract salts medium (TYES; Holt, 1987). The incorporation of antibiotics increases the selective capabilities of these media (Starliper and Schill, 2011) and assists in the recovery of *F. psychrophilum* from lesions when other microorganisms are present. Primary cultures are incubated aerobically at 15–18°C (Holt *et al.*, 1993) for ≤7 days. The bacterium produces yellow-pigmented colonies with raised centres and thin somewhat irregular margins, although variations in colony morphology do occur. The bacterium is a slender rod 1–5 µm long by 0.3–0.5 µm wide (Bernardet and Kerouault, 1989; Fig. 16.1A), displays gliding motility (Bernardet and Nakagawa, 2006) and tolerates <1% NaCl. Additional morphological, physiological and biochemical characteristics are described elsewhere (Bernardet *et al.*, 1996; Bernardet and Bowman, 2011).

At least seven *F. psychrophilum* serotypes exist (Madsen and Dalsgaard, 2000; Mata *et al.*, 2002), and intraspecific genetic heterogeneity has been described using phage typing (Castillo *et al.*, 2014), random amplified polymorphic DNA (RAPD) analysis (Chakroun *et al.*, 1997), pulsed field gel electrophoresis (Arai *et al.*, 2007), restriction digestion/ sequence analysis of ribosomal and housekeeping genes (Madsen and Dalsgaard, 2000), and multilocus sequence typing (Nicolas *et al.*, 2008; Van Vliet *et al.*, 2016). Complete genome sequences are available (Duchaud *et al.*, 2007; Wiens *et al.*, 2014).

*Corresponding author e-mail: lochthom@cvm.msu.edu

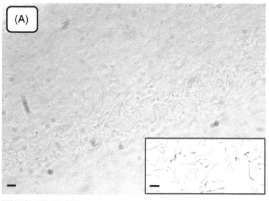

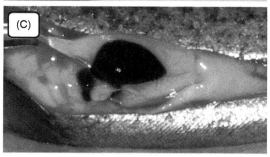

Fig. 16.1. (A) Wet-mount of a *Flavobacterium psychrophilum* isolate that was recovered from the brain of a juvenile steelhead trout (*Oncorhynchus mykiss*) that showed large numbers of the slender bacilli (scale bars, 5 µm; inset: Gram stain of the same isolate). (B) Deep ulceration at the caudal peduncle of a *F. psychrophilum*-infected Atlantic salmon (*Salmo salar*) that exposes the vertebral column. (C) Severely swollen spleen of a *F. psychrophilum*-infected steelhead trout.

Modes of transmission

Horizontal transmission is via waterborne and contact exposure (Holt, 1987). Moribund and dead fish are sources of the bacterium within fish farms (Wiklund *et al.*, 2000; Madetoja *et al.*, 2002). Vatsos *et al.* (2003) isolated *F. psychrophilum* for up to 9 months from sterile stream water, but Madetoja

et al. (2002) found that sediments enhanced survival. The bacterium also forms biofilms (Alvarez *et al.*, 2006) and occurs in benthic diatoms (Izumi *et al.*, 2005), algae (Amita *et al.*, 2000), amphibians (Brown *et al.*, 1997) and the piscicolid leech (*Myzobdella lugubris*) (Schulz and Faisal, 2010).

F. psychrophilum is isolated from the reproductive fluids of salmonids (Holt, 1987; Van Vliet *et al.*, 2015) and can be transmitted from parents to offspring via gametes/eggs (Brown *et al.*, 1997; Cipriano, 2015). Transmission can also occur if unfertilized *F. psychrophilum*-free eggs are hardened in water containing the bacterium (Kumagai *et al.*, 1998, 2000). Whether originating from water or contaminated sex products, the ability of *F. psychrophilum* to reside within eggs and/or survive iodophor disinfection enhances its dissemination (Avendaño-Herrera *et al.*, 2014).

Geographical distribution and host range

F. psychrophilum has been detected in North and South America, Europe, Asia and Oceania (Starliper, 2011; http://pubmlst.org/fpsychrophilum). The bacterium causes disease in farmed salmonids (Starliper, 2011) and rainbow trout (*O. mykiss*) and coho salmon (*O. kisutch*) are particularly susceptible (Holt *et al.*, 1993). In Europe, the disease first emerged in rainbow trout farms in France (Bernardet *et al.*, 1988) and soon after was detected among rainbow trout cultured in other European countries (Lorenzen *et al.*, 1991; Evensen and Lorenzen, 1996). Because the disease was primarily noted in rainbow trout fry (~0.2–2.0 g), it was named rainbow trout fry syndrome (RTFS; Lorenzen *et al.*, 1991). The bacterium has also been isolated from feral salmonids (Loch *et al.*, 2013; Van Vliet *et al.*, 2015) and can infect ayu (*Plecoglossus altivelus*), Japanese (*Anguilla japonica*) and European eels (*A. anguilla*), perch (*Perca fluviatilis*), gobies (*Chaenogobius urotaenia, Rhinogobius brunneus, Tridentiger brevispinis, Gymnogobius* sp.), white sturgeon (*Acipenser transmontanus*), Indian catfish (*Clarias batrachus*), sea lamprey (*Petromyzon marinus*), Japanese smelt (*Hypomesus nipponensis*), Asiatic brook lamprey (*Lethenteron reissneri*), three-spined stickleback (*Gasteroosteus aculeatus*) and European flounder (*Platichthys flesus*; Wakabayashi *et al.*, 1994; Elsayed *et al.*, 2006; Starliper, 2011; Verma and Prasad, 2014; http://pubmlst.org/fpsychrophilum).

16.1.2 *Flavobacterium columnare*

Description of the microorganism

Davis (1922) observed multiple epizootics in fishes during the summers of 1917–1919. Affected fish had 'dirty-white or yellowish areas' on the body, lesions on fins and/or gills, and eventually died. Their fins were frequently eroded to 'mere stubs' and there was necrosis of the gills. Microscopic examinations of wet mounts of affected tissues revealed large numbers of long, slender, flexible rods that formed 'column-like masses'; thus, Davis (1922) named the bacterium *Bacillus columnaris*. Ordal and Rucker (1944) made the initial isolation of a yellow-pigmented bacterium from similarly affected fish. After numerous taxonomic changes, the bacterium was classified as *Flavobacterium columnare* (Bernardet *et al.*, 1996).

Media commonly employed for the bacterium, which may incorporate antibiotic(s), include Shieh's and modified Shieh's media (Song *et al.*, 1988), Hsu–Shotts medium and Cytophaga medium (Song *et al.*, 1988; Bernardet and Bowman, 2006). Primary cultures are incubated aerobically at 20–30°C for 1–3 days; however, *F. columnare* also grows at 15°C in the authors' laboratory, albeit more slowly. Colonies of *F. columnare* are yellow pigmented, rhizoid, somewhat flat, and strongly adhere to or embed in agar. *F. columnare* is a gliding, Gram-negative bacterium (see Fig. 16.2A) that is 3–10 µm long by 0.3–0.5 µm wide. It grows in 0–0.5% NaCl and at 15–37°, with optimal growth at 20–25°C, although the bacterium can be recovered from fish at 10–12°C (Fujihara and Nakatani, 1971; Loch and Faisal, unpublished results). Additional characteristics are reported elsewhere (Bernardet *et al.*, 1996; Bernardet and Bowman, 2011).

Intraspecific serological heterogeneity is present (Anacker and Ordal, 1959) and the bacterium has genetically distinct groups, of which genomovars I, II and III were defined by Triyanto and Wakabayashi (1999). Further heterogeneity and additional genomovars have since been described (e.g. I, II, II-B, III and I/II) and linked to virulence and host specificity (LaFrentz *et al.*, 2014). The complete annotated genome (strain ATCC 49512) is available (Tekedar *et al.*, 2012).

Modes of transmission

F. columnare is transmitted horizontally and fish that survive infection become carriers that cyclically

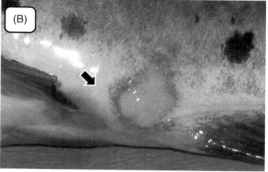

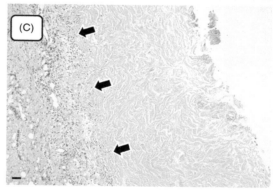

Fig. 16.2. (A) Gram-stain of the 'columns' of rod-shaped bacteria that are characteristic of *Flavobacterium columnare* (scale bar, 10 µm). (B) Dermal ulceration surrounded by diffuse haemorrhage on the trunk of a farmed channel catfish (*Ictalurus punctatus*), which is overlaid by mats of yellow-pigmented *F. columnare* (arrow). (C) Section of skin from a Chinook salmon (*Oncorhynchus tshawytscha*) infected with *F. columnare* showing ulceration of the epidermis and dermis, leaving the underlying subcutaneous connective tissue exposed (scale bar, 50 µm; haematoxylin and eosin stained section); the arrows depict the resultant inflammatory response.

shed the bacterium (Declercq et al., 2013). There is no evidence that *F. columnare* is transmitted vertically, but the bacterium can be recovered from the reproductive fluids of salmonids (Loch and Faisal, 2016a). The bacterium can survive for >5 months in distilled water and >2 years in lake water under laboratory conditions (Kunttu, 2010; Kunttu et al., 2012). However, multiple factors, such as water hardness/alkalinity, temperature, organic loads and quality affect its survival (Farmer et al., 2011; Declercq et al., 2013). *F. columnare* forms biofilms, even on inert surfaces (Cai et al., 2013).

Geographical distribution and host range

F. columnare is distributed worldwide in fresh water and nearly all freshwater (and some brackish) cool- and warm-water fishes are susceptible (Starliper and Schill, 2011), including ornamental fishes (Decostere et al., 1998). The bacterium is a significant pathogen of channel catfish (*Ictalurus punctatus*); Wagner et al., 2002) farmed within the USA. Some cold-water species, particularly trout and the freshwater stage of anadromous salmonids, are also susceptible (Avendaño-Herrera et al., 2011; LaFrentz et al., 2012a). Although many outbreaks are in cultured fishes, wild/feral fish are also affected (Loch et al., 2013).

16.1.3   *Flavobacterium branchiophilum*

Description of the microorganism

Davis (1926) documented multiple disease outbreaks in cultured fingerling brook trout (*Salvelinus fontinalis*) and steelhead trout (*O. mykiss*) suffering gill damage caused by a bacterium that formed 'luxuriant growth' over the gills. Such colonization increased mucus production and the proliferation of gill epithelia, resulting in the fusion of secondary lamellae and clubbing. Kimura et al. (1978) recovered a yellow-pigmented bacterium from similarly diseased gills that fulfilled Koch's Postulates (Wakabayashi et al., 1980). The bacterium was named *Flavobacterium branchiophilum* (Fb; Wakabayashi et al., 1989; Bernardet et al., 1996) and is considered to be the aetiological agent of bacterial gill disease (BGD) (Ostland et al., 1995; Starliper and Schill, 2011); however, environmental parameters and other bacteria (Snieszko, 1981), including other flavobacteria (Good et al., 2015; Loch and Faisal, 2016b), can produce similar gill pathology.

F. branchiophilum can be recovered from infected gills on Cytophaga medium, TYES and, in particular, casitone yeast extract agar (NCIMB medium no. 218; Bernardet and Bowman, 2006). Primary cultures should be aerobically incubated at 18–25 °C for up to 12 days (Wakabayashi et al., 1989). These Gram-negative rods produce light yellow colonies that are smooth, convex and have entire margins. The bacterium is 5–8 µm long (Fig. 16.3A) by 0.5 µm wide (Wakabayashi et al., 1989) and grows between 5 and 30°C (optimal range 18–25°C) in 0–0.2% NaCl. It is non-motile (Wakabayashi et al., 1989; Bernardet and Bowman, 2011). Additional characteristics are described elsewhere (Wakabayashi et al., 1989; Bernardet and Bowman, 2011). *F. branchiophilum* shows some antigenic heterogeneity (Huh and Wakabayashi, 1989). The annotated complete genome of strain FL-15 is available (Touchon et al., 2011).

Modes of transmission

F. branchiophilum is transmitted horizontally, but its infectivity varies among strains (Ostland et al., 1995) and environmental parameters affect gill colonization (Wakabayashi et al., 1980). The bacterium is ubiquitous in freshwater environments; therefore, water supplies containing fish, as well as the rearing of fish in a facility with a history of BGD, are important risk factors for disease (Bebak et al., 1997). High feeding, low water exchange rates and high rearing densities also favour BGD outbreaks (Good et al., 2009, 2010).

Geographical distribution and host range

F. branchiophilum has a worldwide distribution (Snieszko, 1981) and is prevalent where fish are intensively cultured. In fact, BGD is an exclusive problem for farmed fish (Starliper and Schill, 2011). While most BGD outbreaks occur in cool and cold-water fishes, warm-water fishes are susceptible (Bullock, 1990), though their susceptibility varies. For example, cultured brook trout (*Salvelinus fontinalis*), splake (*S. namaycush* × *S. fontinalis*), and rainbow trout are at higher risk of BGD than other salmonids (Good et al., 2008, 2009; Starliper and Schill, 2011). In general, juvenile salmonids are more susceptible than adults (Snieszko, 1981).

16.1.4   Impacts of the three *Flavobacterium* spp. on fish production

F. psychrophilum is the second leading pathogen-associated cause of economic losses in Chilean

---

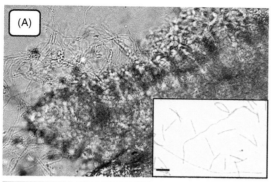

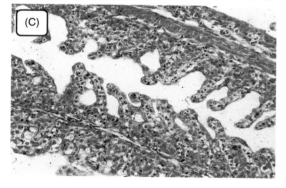

Fig. 16.3. (A) Whole mount preparation of a gill infested with masses of *Flavobacterium branchiophilum* (courtesy of the National Fish Health Research Laboratory photo archive; Leetown, West Virginia); inset, Gram stain of *F. branchiophilum*; scale bar, 10 µm. (B) Flared opercula and gill pallor that is characteristic of *F. branchiophilum* infections in a rainbow trout (*Oncorhynchus mykiss*); note also that the fish is elevated in the water column near the water–air interface. (C) Significant gill epithelial hyperplasia in a fish infected with *F. branchiophilum* that has resulted in the fusion of numerous secondary lamellae (haematoxylin and eosin stained section; courtesy of the National Fish Health Research Laboratory photo archive; Leetown, West Virginia).

aquaculture (Avendaño-Herrera *et al.*, 2014). It also causes losses in the fish farms of Europe (Garcia *et al.*, 2000), Asia (Nakayama *et al.*, 2015) and North America (Hesami *et al.*, 2008), and is among the most common pathogen-related cause of mortality among hatchery-reared salmonids in Michigan (Van Vliet *et al.*, 2015). Although it commonly causes losses among fry and fingerlings, poor eye-up has been noted among eggs infected with *F. psychrophilum* (Cipriano, 2015). *F. psychrophilum* also causes economic hardship owing to the poor growth rates in some survivors (Nematollahi *et al.*, 2003). Columnaris-associated losses were estimated at US$30 million annually in the US channel catfish industry (Wagner et al., 2002; Shoemaker *et al.*, 2011). In Finland, the frequency of CD has increased since the 1990s and is currently a significant threat to Finnish salmon/trout production (Pulkkinen *et al.*, 2010). It has also emerged as a major problem for Chilean salmonid aquaculture (Avendaño-Herrera *et al.*, 2011), farmed tilapia (*Oreochromis niloticus*; Figueiredo *et al.*, 2005), US trout aquaculture (LaFrentz *et al.*, 2012a) and ornamental fishes (Decostere *et al.*, 1998). Possibly owing to the multifactorial nature of BGD, cost estimates for losses are lacking. Nevertheless, in Ontario, Canada, BGD is a perennial problem (Daoust and Ferguson, 1983) and has persistently limited fish farm production (Good *et al.*, 2008).

16.2 Pathology

The behavioural and gross pathological changes associated with the three *Flavobacterium* spp. are important for differential diagnosis. However, some signs are not pathognomonic (Loch and Faisal, 2016b) and more than one *Flavobacterium* species may infect the same fish (Loch and Faisal, 2016a).

16.2.1 *Flavobacterium psychrophilum*

Flavobacterium psychrophilum-infected coho salmon developed lesions anterior to the tail, which resulted in 'sloughing and almost complete loss of tissue until the tail was useless and was held on by only a few strands' (Borg, 1948). Melanosis and lesions on the isthmus and dorsal aspect of the fish were also noted. Disease signs frequently include lethargy, inappetence and whitish areas on the fins (Starliper, 2011). External pathology may be limited to fin erosion, but more typically includes pale/haemorrhagic gills, ulcerations of the head/mouth, body and vent, exophthalmia (with or without corneal opacity and ocular haemorrhage), abdominal distension, and characteristic ulcerations on the dorsum and/or caudal peduncle (Fig. 16.1B). Commonly, *F. psychrophilum*-induced ulcerations at the adipose/dorsal fins, as well as exophthalmia/corneal opacity and fin erosion, will be evident, even in dim light. Spleens are swollen and enlarged (Fig. 16.1C). Renal swelling, visceral haemorrhages and ascites may be evident. In farmed coho salmon fingerlings of the Great Lakes basin, pectoral fin erosion, unilateral exophthalmia and concurrent corneal opacity and ocular haemorrhage, which is reminiscent of the *F. psychrophilum*-associated bilateral necrotic scleritis reported by Ostland *et al.* (1997), are regularly observed. In chronic BCWD, ataxia, erratic swimming and spinal abnormalities (e.g. scoliosis, kyphosis and/or lordosis) may occur. Additionally, such fish may become compressed lengthwise as 'crinklebacks' or 'pumpkinseeds' (Wood, 1974). When fry are affected, the skin covering the yolk sac is frequently eroded (Nematollahi *et al.*, 2003).

Epizootics of RTFS frequently begin 6–7 weeks after fry feed exogenously, resulting in ≤60 to 90% mortality if left untreated. Affected fish swim at the surface and show lethargy, inappetence, melanosis, exophthalmia and abdominal distension due to ascites. Fin erosion, ulcers on the body and head, and pallor of the gills, kidney, liver and intestine may be present. Splenomegaly is the most consistent internal lesion. Indeed, bacteria within the splenic tissue in squash preparations, along with some characteristic histopathology, is considered pathognomonic (Rangdale *et al.*, 1999; Bernardet and Bowman, 2006). Additionally, Evensen and Lorenzen (1996) detected *F. psychrophilum* in the retinas and choroid glands of infected rainbow trout fry using immunohistochemistry, and also observed an inflammatory response, which could explain the blindness in 5–10% of RTFS survivors. In rainbow trout fingerlings, signs of RTFS are frequently comparable with those of BCWD. In ayu, *F. psychrophilum* induces gill pallor and peduncle/perioral ulcerations; however, body ulcers and fin erosion are uncommon (Miwa and Nakayasu, 2005).

Wood and Yasutake (1956) noted 'swarms' of *F. psychrophilum* on the head, palate, lower jaw and opercula that eroded epithelial tissues and penetrated the subcutaneous tissues and musculature in coho and Chinook (*O. tshawytscha*) fingerlings. Bacteria were within the gill capillaries, but not on the epithelium. In coho salmon, renal lesions were noted, which had bacteria within degenerated glomeruli. Numerous bacteria were in the heart, with concomitant mild inflammation and necrosis in individual trabeculae. In salmonids that swam spirally, Kent *et al.* (1989) found consistent subacute to chronic periostitis, osteitis, meningitis and ganglioneurititis, as well as periosteal proliferation and inflammation of the anterior vertebrae where the vertebral column joined the cranium. They also found that the adjacent nervous tissue, and in particular the medulla, was compressed by proliferative lesions, possibly causing the abnormal swimming.

In RTFS, hyaline droplet degeneration within renal tubules, and focal necrosis of the liver, kidney and heart were observed, as were bacteria in most visceral organs (Lorenzen *et al.*, 1991). Bruno (1992) found similar lesions in older rainbow trout, but also hepatic vacuolar degeneration, pyknosis and scattered necrotic hepatocytes, along with eosinophilia and pyknosis in the renal tubular epithelium. He also noted a necrotizing lymphocytic dermatitis and myositis, with many bacteria associated with the epithelium of the olfactory capsule lumen. In rainbow trout fry, Rangdale *et al.* (1999) described splenic hypertrophy, a loss of definition of the splenic border that was replaced by a loosely structured eosinophilic layer, and degrees of peritonitis that extended from the serosal surface of the spleen. They also noted fibrinous inflammation, oedema and many filamentous bacteria in the spleen, in conjunction with splenic hypertrophy, that were pathognomonic for RTFS.

16.2.2 *Flavobacterium columnare*

Clinical signs of CD are considered by some to be nearly pathognomonic (Plumb and Hanson, 2010), although the lesion location/severity and disease course can vary. For example, highly virulent *F. columnare* causes fulminant infections accompanied by high acute mortality but little gross pathology, whereas isolates of lesser virulence progress slowly and induce chronic mortality accompanied by significant pathology (Declercq *et al.*, 2013 and references therein). *F. columnare* infections begin externally on the fins, gills and skin. Behavioural changes include inappetence, lethargy and 'hanging' at the surface. The destruction of gill lamellae results in the development of white to brown foci, which may or may not be overlaid by a yellow 'film'. *F. columnare*-induced skin lesions, which can involve regions of the head and body, often become depigmented with or without haemorrhage, and may be covered with the same yellow film and/or necrotic debris (Fig. 16.2b). Skin lesions can ulcerate into the underlying muscle. Infected fins fray and become necrotic, and fin margins will have a white/grey appearance (Plumb and Hanson, 2010). The bacterium may proliferate on the skin at the base of the dorsal and/or adipose fins and radiate outward, producing a 'saddle-back' lesion (Declercq et al., 2013). In aquarium fish, CD often manifests in the oral mucosa. Although Columnaris is frequently an external infection, systemic disease can occur (Hawke and Thune, 1992), but little internal pathology is observed (Plumb and Hanson, 2010). In the Great Lakes, F. columnare in salmonids is associated with external and systemic infections, and the bacterium is frequently recovered from the kidneys (Loch et al., 2013; Loch and Faisal, 2016a).

F. columnare causes epithelial and connective tissue necrosis of the gills (Speare and Ferguson, 1989). Declercq et al. (2013) noted that gill infections often began with epithelial proliferation and mucous cell hyperplasia, which occluded the spaces between adjacent lamellae. As the disease progressed, congestion of gill lamellae, inflammation (i.e. branchitis) and oedema resulted in epithelial lifting. Eventually, lamellar fusion can occur, which causes extensive haemorrhage and concomitant circulatory failure, as well as necrosis (Declercq et al., 2013). More recently, Declercq et al. (2015) infected common carp (Cyprinus carpio) and rainbow trout fingerlings with low and high virulence F. columnare isolates. They found that highly virulent isolates induced lesions in carp consisting of desquamation of the gill epithelium, multifocal lamellar fusion, loss of structure at the middle and distal ends of lamellae, mild inflammation and haemorrhage, and diffuse fusion of gill filaments with the lost gill lamellae replaced by necrotic debris.

In rainbow trout infected with highly virulent F. columnare, gill lesions were focally extensive and characterized by lamellar epithelial necrosis that was subsequently replaced by a mat of filamentous bacteria in an eosinophilic matrix, with branchitis and oedema in adjacent lamellae. In some farmed channel catfish, the predominant histopathological characteristic in both dermal and, to a lesser extent, branchial lesions is necrosis with little or no inflammatory response (Dr Lester Khoo, Mississippi State University, personal communication). Microscopic skin/muscle lesions due to F. columnare include epithelial ulceration (Fig. 16.2C) and replacement by mats of bacteria (particularly around the dorsal fin) that eventually invade the dermis and underlying muscle, resulting in necrosis (Morrison et al., 1981). While Declercq et al. (2015) detected no microscopic lesions in the skin of common carp infected with F. columnare, skin lesions occurred in rainbow trout which consisted of epidermal ulceration and necrosis, loss of scales, haemorrhage, bacterial infiltration and inflammation. They also noted bacterial infiltration into the deeper layers of the skin and underlying muscle, resulting in myositis.

16.2.3 Flavobacterium branchiophilum

In BGD, fish become inappetent, swim at the surface, and orient at and towards the flow of incoming water (Snieszko, 1981). Lethargy, lack of escape reflex, gasping and bilaterally flared opercula (Fig. 16.3B) are common (Speare and Ferguson, 1989). Depending upon fish age/size, stocking density and the quality of the environmental conditions, up to 80% morbidity and 10–50% mortality have been reported within 24–48 h (Snieszko, 1981; Speare et al., 1991). Because F. branchiophilum is non-invasive, pathology is frequently limited to the gills. Lamellar congestion, excess mucus/debris, and ragged or flared gill arches are evident (Speare and Ferguson, 1989). In contrast to F. columnare and F. psychrophilum, F. branchiophilum elicits epithelial hyperplasia, lamellar fusion (i.e. 'clubbing'), and even fusion of the primary lamellae (Fig. 16.3C) in severe cases (Wood and Yasutake, 1957), which cause substantial reduction in the respiratory surface area so that gaseous and ionic exchanges are inhibited, as well as osmotic homeostasis. Wood and Yasutake (1957) found that BGD-induced hyperplasia frequently began at the epithelium on the distal tips of the lamellae and eventually resulted in 'small islands of hyperplasia' scattered irregularly along gill filaments (i.e. multifocal epithelial hyperplasia).

16.3 Diagnosis of Flavobacterial Infections

Flavobacterial infections are diagnosed based on facility/case histories, behavioural changes and gross pathology/histopathology, with subsequent isolation and phenotypic characterization of the pathogen. Confirmation is best achieved using molecular and serological assays. However, different countries may follow similar, yet distinct, diagnostic guidelines, particularly for pathogens such as flavobacteria that are not notifiable to the World Organisation for Animal Health (OIE).

16.3.1 Facility inspection and history of previous epizootics

Aquaculture operations that utilize surface or spring water that contain fish are more likely to experience flavobacteria-related disease outbreaks if adequate biosecurity measures are not employed. The health of the brood stock is also important because *F. psychrophilum* can be transmitted via gametes. Also in this context, records on egg disinfections should be reviewed, as this influences the transmission of *F. psychrophilum* and *F. columnare*. Likewise, information on facility history, water quality, hatchery infrastructure and the fallowing process is useful.

16.3.2 Clinical and post-mortem examination

When skin/gill lesions suggestive of flavobacterial disease are observed (see Section 16.2), scrapings and/or whole mount preparations from the affected areas should be examined using light microscopy. Masses of slender bacilli (Fig. 16.1A), aggregates of bacilli into 'columns' or 'haystacks' (Fig. 16.2A), or masses of filamentous bacteria associated with the gills (Fig. 16.3A) suggest *F. psychrophilum*, *F. columnare* and *F. branchiophilum*, respectively. Whole-mount preparations can be Gram stained for further examination. Squash preparations of splenic tissue and examination for bacteria may also be helpful.

16.3.3 Primary isolation and presumptive identification

Tissue samples for flavobacterial culture should be collected from the leading edge of external lesions to reduce overgrowth by secondary invaders. Excised tissues from the kidney, spleen and/or brain, as well as the liver and any ascites are also used to detect systemic infections. Although *F. psychrophilum* is renowned for systemic disease and infects neurological tissues (Nematollahi *et al.*, 2003), *F. columnare* may also be recovered from brain tissue (Loch and Faisal, 2016a). Tissues can be inoculated directly on to culture

media or homogenized to enhance bacterial recovery (Starliper and Schill, 2011). Primary isolation of *F. branchiophilum* and its subsequent identification are performed less frequently because the bacterium is so fastidious (Bernardet and Bowman, 2006); the combination of facility history, clinical signs and bacterial morphology (Fig. 16.3A) are historically adequate for identification (Starliper and Schill, 2011).

Bacteria that are Gram-negative, non-spore-forming rods, produce yellow/orange colonies (due to the production of flexirubin and/or carotenoid-type pigments) and are non-motile or motile via the gliding (Bernardet, 2011a) are tentatively identified as members of the *Flavobacteriaceae*. Differential characteristics of the *Flavobacteriaceae* are reviewed by Bernardet (2011b). In the authors' experience, the genus *Chryseobacterium* is the most common 'non-*Flavobacterium*' within the *Flavobacteriaceae* that is recovered from freshwater fish and attention should be paid to differentiating these genera. Multiple phenotypic characteristics are used to differentiate *F. psychrophilum*, *F. columnare* and *F. branchiophilum* (Bernardet *et al.*, 2002; Bernardet and Bowman, 2011); (see Table 16.1).

16.3.4 Serological assays

Multiple enzyme-linked immunosorbent assays (ELISAs) and fluorescent antibody tests (FATs) based upon polyclonal antibodies that react with whole-cell *F. psychrophilum* antibodies have been developed (reviewed in Lindstrom *et al.*, 2009); however, cross-reactions with other flavobacteria limit their use.

Likewise, polyclonal sera-based capture ELISAs have been developed (Rangdale and Way, 1995; Mata *et al.*, 2002), but the serological heterogeneity of *F. psychrophilum* complicates diagnosis. More recently, Lindstrom *et al.* (2009) developed a monoclonal antibody (commercially available Mab FL-43; ImmunoPrecise Antibodies) that is specific to a protein constituent of the outer surface of *F. psychrophilum*. In an ELISA format, the monoclonal antibody detected as few as 1.6×10^3 colony forming units (cfu)/ml experimentally. The assay was validated (Long *et al.*, 2012) and proved suitable for screening salmonid kidneys, but not ovarian fluids.

Mab FL-43 was also used in a filtration-based FAT (Lindstrom *et al.*, 2009), which detected *F. psychrophilum* in ovarian fluids. Immunohistochemistry

T.P. Loch and M. Faisal

Table 16.1. Characteristics useful for presumptively differentiating *Flavobacterium psychrophilum*, *F. columnare* and *F. branchiophilum*.[a]

Characteristic	F. psychrophilum	F. columnare	F. branchiophilum
Colony morphology[b]	Low convex, circular, entire margins; 'fried egg' appearance	Flat or with slightly raised centres, rhizoid irregular margins	Low convex, circular, entire margins
Congo red adsorption	–	+	–
Flexirubin-type pigments	+	+	–
Gliding motility	(+)	+	–
Production of:			
Acid from carbohydrates (e.g. glucose, sucrose, maltose, trehalose)	–	–	+
β-Galactosidase	–	–	+
Hydrogen sulfide	–	+	–

[a]Data from Bernardet and Bowman (2011, and references therein) and Loch and Faisal, unpublished results.
[b]Morphology varies according to culture conditions and strain; –, negative reaction; +, >90% isolates give positive reaction; (+), ≤90% of isolates give a delayed or weak positive result.

can also detect *F. psychrophilum* and assess tissue localization (Madetoja *et al.*, 2000). Serodiagnostic assays for *F. columnare* include a monoclonal antibody-based indirect fluorescent antibody test (IFAT) that can visualize the bacterium on gill surfaces and within the gastrointestinal tract (Speare *et al.*, 1995). Panangalal *et al.* (2006) developed an IFAT that had a sensitivity similar to that of bacterial culture, but yielded rapid results. *F. branchiophilum* gill infections can be detected via IFAT using polyclonal antisera (Bullock *et al.*, 1994) and monoclonal antibodies (Speare *et al.*, 1995).

An ELISA was developed by MacPhee *et al.* (1995) that detected whole *F. branchiophilum* preparations at ~500 cfu /ml and ~1000 cfu/100 µg spiked gill tissue. In keeping with the fastidious nature of *F. branchiophilum*, these authors were able to isolate the bacterium in only 19/54 known infected gill samples, whereas 53/54 were positive by ELISA. Additionally, there are agglutination and immunodiffusion assays for *F. branchiophilum* (Huh and Wakabayashi, 1989; Ko and Heo, 1997).

16.3.5 Molecular assays

Polymerase chain reaction (PCR) assays, e.g. conventional PCR, nested PCR (nPCR), multiplex PCR, reverse-transcriptase PCR (RT-PCR), quantitative (real-time) PCR (qPCR), have been developed for *F. psychrophilum*, and these can detect the bacterium in fresh/frozen tissues, formalin-fixed tissues and/or environmental samples. They can also be used for confirmation. Indeed, many conventional

and nested PCR assays that target 16S rRNA and housekeeping genes are useful in this way (reviewed in Starliper and Schill, 2011). In brief, some of these PCR/nPCR assays detected as few as 17 cfu/mg spiked rainbow trout brain tissue (Wiklund *et al.*, 2000), 10 cfu/mg spleen and 5 cfu/ml ovarian fluid from farmed coho salmon and rainbow trout (Baliarda *et al.*, 2002), 110 cfu/ml spiked fresh water (Wiklund *et al.*, 2000) and 3 cfu/ml spiked non-sterile well water (Madetoja and Wiklund, 2002). Suzuki *et al.* (2008) found that the sensitivity of some *F. psychrophilum* PCR assays targeting the 16S rRNA genes was greater than those targeting the *gyrA*, *gyrB* (DNA gyrase subunits A, B) and peptidyl-prolyl *cis-trans* isomerase (*ppiC*) genes, but this increased false positive results. Del Cerro *et al.* (2002a) designed a multiplex PCR (targeting 16S rRNA genes) for the detection of *F. psychrophilum*, *Aeromonas salmonicida* and *Yersinia ruckeri*.

The use of a Taqman probe-based PCR assay that targeted the 16S rRNA gene of *F. psychrophilum* was initially reported by Del Cerro *et al.* (2002b). At least four other qPCR assays have been developed (Orieux *et al.*, 2011; Marancik and Wiens, 2013; Strepparava *et al.*, 2014; Long *et al.*, 2015). Marancik and Wiens (2013) described some pitfalls of qPCR assays that target 16S rRNA genes, such as the presence of six copies of the 16S rRNA gene (Duchaud *et al.*, 2007) that can vary in sequence and non-specifically amplify other flavobacteria. Thus, they designed their qPCR assay to target a single copy gene (RFPS00910) that detected 3.1 genome equivalents per reaction. Even so, subclinically infected fish were sometimes culture positive, but qPCR negative. Other molecular detection methods for *F. psychrophilum* include *in situ* hybridization (Vatsos *et al.*, 2002) and fluorescent *in situ* hybridization, which detect the bacterium in tissues and water (Strepparava *et al.*, 2012). Fujiwara-Nagata and Eguchi (2009) developed a quantitative LAMP (loop-mediated isothermal amplification) assay that detected ≥20 cfu equivalents of *F. psychrophilum* per reaction within 70 min. Immunomagnetic separation, in conjunction with PCR or flow cytometry, has also been utilized to detect *F. psychrophilum* and can quantify as few as 0.4 cfu/ml (Ryumae *et al.*, 2012; Hibi *et al.*, 2012).

Multiple PCR assays to detect and identify *F. columnare* are available (Triyanto *et al.*, 1999; Darwish *et al.*, 2004; Welker *et al.*, 2005). Panangala *et al.* (2007) developed a qPCR that targets the chondroitin AC lyase gene and this detects ≥3 cfu of *F. columnare* in pure culture and 3.4 cfu/ml in channel catfish. Yeh *et al.* (2006) described an *F. columnare*-specific LAMP assay that detects the bacterium in channel catfish tissues. Suebsing *et al.* (2015) developed a calcein-based LAMP for farmed tilapia that detected the bacterium in gonad/gill/blood samples at ~10-fold higher sensitivity than conventional PCR (detection limit of 2.2×10^2 cfu/reaction). *In situ* hybridization detects and visualizes *F. columnare* within infected fish (Tripathi *et al.*, 2005). *F. branchiophilum* can also be detected via conventional PCR (Toyama *et al.*, 1996) and next-generation sequencing (Good *et al.*, 2015). The DNA array-based multiplex PCR of Lievens *et al.* (2011) detects *F. psychrophilum*, *F. columnare* and *F. branchiophilum*.

16.4 Pathophysiology

The pathophysiological effects of flavobacterial diseases are complex. The skin/gill epithelium and the overlaying mucus are important defences against pathogens and are invaluable for osmoregulation. Insults to these tissues by *F. psychrophilum* and *F. columnare* compromise the homeostasis of the host. Columnaris disease is frequently accompanied by 'waterlogging', a term for skin/muscular tissues that have become oedematous (Declercq *et al.*, 2013). Likewise, BGD-induced gill epithelial hyperplasia impairs ionic exchange and interferes with osmotic homeostasis (Speare and Ferguson, 1989; Byrne *et al.*, 1995). During systemic *F. psychrophilum* and *F. columnare* infections, the host may undergo vasodilation to increase blood supply to the infection sites, and also increase vascular permeability so that leucocytes, antibodies, complement and pro-inflammatory cytokines can control the infection. The host immune response to flavobacterial infections is substantial (reviewed in Starliper and Schill, 2011; Declercq *et al.*, 2013), but this response may compromise growth, sexual maturation and reproduction in survivors. Similarly, farmed fish that recover from systemic *F. psychrophilum* infections frequently suffer from ocular and/or spinal abnormalities (LaFrentz and Cain, 2004).

Marancik *et al.* (2014) infected rainbow trout with *F. psychrophilum* and showed that their packed cell volumes (PCVs), total protein, albumin, glucose, cholesterol, chloride and calcium concentrations were altered, leading to the changes associated with BCWD. They also noted that PCV, total protein, cholesterol and calcium levels were negatively correlated with splenic loads of *F. psychrophilum*, and that the earliest observed pathophysiological changes were declines in total protein and albumin, as well as electrolyte imbalance. Reductions in PCVs were also reported in *F. psychrophilum*-infected ayu (Miwa and Nakayasu, 2005).

Tripathi *et al.* (2005) found that *F. columnare*-infected koi carp (*C. carpio*) had significantly reduced PCVs, erythrocyte counts, haemoglobin concentrations, mean corpuscular volumes and absolute lymphocyte counts, which led to a non-regenerative anaemia (microcytic and normochromic) and leucopenia. Nevertheless, there was a regeneration in the numbers of erythrocytes in some individuals. There were severe haematological changes in fish with extensive skin ulcers, a marked hyponatraemia and hypochloraemia in sera with loss of sodium and chloride ions due to passive perfusion through the gills and through disruption of the skin barrier because of ulceration.

To study the haematological changes and acid–base balance of *F. branchiophilum*-infected rainbow trout, Byrne *et al.* (1995) cannulated fish with dorsal aortic catheters, sutured in nasogastric feeding tubes and maintained individuals within Plexiglas boxes. After infection with bacteria, some fish were fed and others were not. Within 24 h, *F. branchiophilum*-infected fish that were fed exhibited severe hypoxaemia, hypercapnia, hypo-osmolality, hyponatraemia, hypochloraemia, elevated blood ammonia, increased blood ammonia, tachybranchia and more frequent rates of 'coughing' compared with similar but unfed controls.

Interestingly, unfed *F. branchiophilum*-infected fish showed similar clinical signs and blood gas alterations, albeit to lesser degrees, but electrolyte imbalances were not detected. The investigators concluded that *F. branchiophilum*-induced hypoxaemia is worsened by feeding behaviours that increase oxygen demands.

16.5 Protective and Control strategies

Optimization of husbandry conditions by minimizing stress, reducing rearing densities, maintaining hygiene and water quality, and achieving adequate water flows can help reduce flavobacteria-induced mortality. For example, elevated nitrite and increased organic matter enhance the adhesion of *F. columnare* and *F. psychrophilum* to gills (Decostere *et al.*, 1999; Nematollahi *et al.*, 2003), and poor water quality affected cumulative mortality in rainbow trout infected with *F. psychrophilum* (Garcia *et al.*, 2000). Low dissolved oxygen and elevated ammonia exacerbate and/or predispose fish to BGD (Bullock, 1990). In contrast, elevated total ammonia nitrogen interfered with *F. columnare* infections of channel catfish and actually improved survival (Farmer *et al.*, 2011). Suboptimal water exchanges are a risk factor for BGD outbreaks (Good *et al.*, 2009, 2010). In water reuse systems, species that are more resistant to flavobacteria should be positioned below more susceptible fishes.

The use of fish-free water sources (i.e. deep wells) can reduce the load of flavobacteria entering hatcheries (Madsen and Dalsgaard, 2008). Ultraviolet (UV) treatment of incoming water also reduces or eliminates flavobacterial influx. Hedrick *et al.* (2000) found that 126 mWs/cm^2 of UV irradiation killed *F. psychrophilum*, whereas many other Gram-negative bacteria are killed at ≤42 mWs/cm^2. The use of separate nets/tools for rearing-units and the disinfection of common equipment minimizes contagion, which is important because flavobacteria can survive extended periods of desiccation (e.g. >96 h for *F. psychrophilum* on plastic strips; Oplinger and Wagner, 2010).

Oplinger and Wagner (2010) showed that benzalkonium chloride (600 mg/l solution) eradicated *F. psychrophilum* on plastic after soaking for 10 min and they achieved the same result by dipping *F. psychrophilum*-laced strips into benzalkonium chloride for 10 s, followed with 1 h of drying. Oplinger and Wagner (2013) also found that *F. psychrophilum* was killed when exposed to

≥55°C, but remained viable after changes in pH (from 3.0 to 11.0 for 15 min) and osmotic pressure (0–12% NaCl for 15 min). Mainous *et al.* (2012) found that Virkon® Aquatic (0.1%; Western Chemical), Halamid® (15 ppm; Western Chemical), glutaraldehyde (2%), iodine, Lysol® (1%), Roccal® (1:256; Zoetis Services), Clorox® (50 ppm), and 50 and 70% ethanol killed *F. columnare* and *F. psychrophilum* within 1 min of contact, but formalin took 10–20 min depending upon the concentration. Interestingly, potassium permanganate eradicated *F. columnare* within 1 min, but did not substantially reduce *F. psychrophilum* even after 1 h. However, Mainous *et al.* (2012) did not study how organic matter affected disinfection efficacies, nor how biofilm affected survival, which is a matter of importance as the biofilm-forming capacity of *F. psychrophilum* enhances its survival when it is exposed to chemotherapeutants (Sundell and Wiklund, 2011). Only Virkon® Aquatic is licensed for use in aquaculture facilities by the US Environmental Protection Agency (EPA), whereas other chemicals (e.g. iodine) have low regulatory priority as determined by the US Food and Drug Administration (FDA).

The affinity of *F. psychrophilum*, and to a lesser extent of *F. columnare*, for gametes and reproductive fluids (Brown *et al.*, 1997; Loch and Faisal, 2016a), along with evidence that *F. psychrophilum* has been disseminated via egg shipments (Avendaño-Herrera *et al.*, 2014), illustrate the importance of egg disinfections. Although povidone–iodine (50 mg/l active I$_2$ for 30 min and/or 100 mg/l for 10 min) is most commonly used to disinfect eggs, hydrogen peroxide (100 mg/l for 10 min) and glutaraldehyde (200/mg l for 20 min) are also effective (Cipriano and Holt, 2005). Even though surface disinfection with iodine reduces egg-associated pathogen transmission, it does not always eliminate flavobacteria (Holt *et al.*, 1993). In fact, Kumagai *et al.* (1998) showed that *F. psychrophilum* could be recovered from eggs treated with povidone–iodine at 1000 ppm for 15 min and 200 ppm for 2 h . We (Loch and Faisal, unpublished data) have also occasionally isolated *F. columnare* from eggs that were disinfected with povidone–iodine at 50 mg/l for 30 min, followed by 100 mg/l for 10 min, and hypothesize that large bacterial quantities may affect the efficacy of disinfection.

Nutrition is an important factor in determining the risk/severity of flavobacterial diseases. For example, *F. columnare*-challenged channel catfish that were fasted for 7–10 days experienced significantly higher mortality and mounted a reduced antibody response compared with fed controls (Klesius *et al.*, 1999). Beck *et al.* (2012) found that channel catfish that were fasted for 7 days had a significant upregulation of rhamnose-binding lectin (RBL), which they previously determined was also significantly upregulated in a lineage of channel catfish that was susceptible to CD, but not in the lineage that was completely resistant. Interestingly, when channel catfish were exposed to the RBL putative ligands (e.g. L-rhamnose and D-galactose), they became protected from *F. columnare* challenge, further demonstrating the importance of RBL in Columnaris susceptibility. Liu *et al.* (2013) found that short-term fasting in *F. columnare*-challenged channel catfish caused significantly lower immune fitness and marked the dysregulation of genes involved in energy metabolism and cell cycling/proliferation. The starved fish had downregulated iNOS2b (inducible nitric oxide synthase 2b), peptidoglycan recognition protein 6 and lysozyme-C, which are crucial components of innate immunity. Paradoxically, overfeeding is a risk factor for BGD outbreaks (Good *et al.*, 2009, 2010), which may be related to increased oxygen demands that accompany feed consumption (Byrne *et al.*, 1995).

Immunostimulants for the prevention of flavobacterial disease in fish have produced mixed results. Suomalainen *et al.* (2009) evaluated the protective effects of supplementing feed with Ergosan® (Schering-Plough Aquaculture) and Alkosel® (Lallemand Animal Nutrition), which are alginic acid extracts and inactivated yeast (*Saccharomyces cerevisiae*) containing bioavailable L(+)-selenomethionine, respectively. Rainbow trout (1.2 g) that were fed Alkosel®-treated feed for 7 days and then challenged with *F. columnare* experienced significantly lower mortality than infected fish that were fed Ergosan® or standard feed. However, no significant differences in mortality were noted in 5 g fish that were fed the supplements for 14 days. Kunttu *et al.* (2009) found that treatment with yeast β-glucan and β-hydroxy-β-methylbutyrate did not protect the early life stages of rainbow trout against CD. Black cumin (*Nigella sativa*) oil strongly inhibited *F. columnare* growth *in vitro* and protected channel catfish and zebrafish (*Danio rerio*) that were fed the oil or seeds and subsequently challenged with the bacterium (Mohammed and Arias, 2016). The administration of humus extract with feed lowered BCWD mortality and skin lesion development in ayu (Nakagawa *et al.*,

2009). The incorporation of dried pink salmon (*O. gorbuscha*) testes meal into the feed of farmed rainbow trout fingerlings shows promise as an immunostimulant, as has a β-glucan-laced diet (Macrogard®; Biorigin), but Fehringer *et al.* (2014) found that providing testes meal or glucan-based diets to farmed fish for extended periods may negatively affect specific antibody development against *F. psychrophilum*.

Some pseudomonads inhibit flavobacterial growth under *in vitro* conditions (Strom-Bestor and Wiklund, 2011; Fuente Mde *et al.*, 2015) and provide protection in fish. Korkea-aho *et al.* (2011) challenged rainbow trout that that had previously been fed *Pseudomonas* M174 with *F. psychrophilum*, and this resulted in enhanced respiratory burst activity and lower mortality relative to controls. Another pseudomonad from trout eggs protected fish against RTFS, albeit through different mechanisms than *Pseudomonas* strain M174 (Korkea-aho *et al.*, 2012). Burbank *et al.* (2012) screened bacteria that were recovered from rainbow trout gastrointestinal tracts for *in vitro* antagonism of *F. psychrophilum* and eventually focused on 16 probiotic candidates within the genera *Citrobacter*, *Lysinibacillus*, *Staphylococcus*, *Aerococcus*, *Enterobacter* and *Aeromonas*. LaPatra *et al.* (2014) and Ghosh *et al.* (2016) injected one of these strains (*Enterobacter* C6-6) intraperitoneally into rainbow trout and improved their survival against BCWD. Boutin *et al.* (2012) recovered multiple probiotic strains from brook trout skin/mucus and created a 'cocktail' of strains that inhibited *F. columnare in vitro* and protected juvenile brook trout against Columnaris. Boutin *et al.* (2013) attained a probiotic effect without disrupting the natural microbiome of brook trout (i.e. dysbiosis) using a *Rhodococcus* sp. that did not colonize the fish, but rather formed biofilms in aquaria, and reduced mortality in *F. psychrophilum*-infected fish that were housed therein. Strom-Bestor and Wiklund (2011) found that the degree of *in vitro* inhibition by a probiotic pseudomonad varied according to the *F. psychrophilum* serotype.

The laboratory of Dr Greg Wiens (National Center for Cool and Cold Water Aquaculture, US Department of Agriculture-Agricultural Research Service, Leetown, West Virginia) has developed a rainbow trout lineage that has marked resistance to BCWD without any effect on growth rates (Silverstein *et al.*, 2009; Leeds *et al.*, 2010; Wiens

T.P. Loch and M. Faisal

et al., 2013). These studies have begun to uncover the mechanisms behind this resistance (Marancik et al., 2014, 2015; Wiens et al., 2015) and also identified single nucleotide polymorphisms (SNPs) that can map candidate resistance genes (Liu et al., 2015). Differential genetic resistance to CD has also been reported in channel catfish (Beck et al., 2012; LaFrentz et al., 2012b), hybrid catfish (Arias et al., 2012) and grass carp (Cternopharyngodon idellus; Yu et al., 2013). Evenhuis et al. (2015) found that resistance to BCWD in their BCWD-resistant rainbow trout lineage was correlated with CD resistance, so demonstrating that resistance to multiple diseases can be obtained in the selection process.

Harnessing the lytic activity of bacteriophages to control flavobacterial diseases is also promising. Prasad et al. (2011) recovered F. columnare-specific phages that lysed many F. columnare isolates. They infected walking catfish (C. batrachus) and subsequently exposed them to F. columnare-specific bacteriophages via multiple routes (injection, immersion and oral), which led to 100% survival in phage-treated fish with no recovery of F. columnare by culture. Comparable results were obtained by Castillo et al. (2012), who decreased experimentally induced BCWD mortality through phage treatment; however, resistance to phages can develop (Castillo et al., 2014).

The development of efficacious vaccines against flavobacteria has proven problematic. Early attempts to protect against BCWD/RTFS using whole formalin-killed bacteria under laboratory conditions were promising (Bernardet, 1997), although field evaluations of these are needed (Nematollahi et al., 2003). Since then, numerous attempts to use passive immunization or develop various bacterins, subunit vaccines, wild-type vaccines and live attenuated vaccines have been made, but none of these vaccines are commercially available in the USA (Gómez et al., 2014 and references therein). A live attenuated vaccine (CSF259-93B.17) was developed by LaFrentz et al. (2008) that generated 45% relative survival in experimentally challenged vaccinates. A later report by these authors (LaFrentz et al., 2009) suggested that several immunoreactive F. psychrophilum proteins are overexpressed under iron-limited conditions, which led Long et al. (2013) to culture CSF259-93B.17 under similar conditions. Long et al. found that immersion-vaccinated coho salmon had significantly lower mortality than mock-immunized individuals, whereas the difference in mortality between mock-vaccinated fish and those vaccinated with CSF259-93B.17 grown under iron-replete conditions were not significant. Field trials are in process at several state hatcheries (Dr Kenneth Cain, Department of Fish and Wildlife Sciences, University of Idaho; personal communication). Solís et al. (2015) reported that a formalin-killed bacterin against F. psychrophilum is commercially available for use in Chile (Flavomune®, Veterquímica).

Despite relatively promising results for vaccinating fish against F. columnare (reviewed in Declercq et al., 2013), only recently have vaccines approved for use in farmed fish become available. An attenuated live F. columnare vaccine (AQUAVAC-COL™; Merck Animal Health), is registered in the USA and is available for use in channel catfish and largemouth bass (Micropterus salmoides; Shoemaker et al., 2011). This vaccine produced 57–94% relative percent survival (RPS) in immersion-vaccinated F. columnare-challenged channel catfish fry (Shoemaker et al., 2011). In contrast, Kirkland (2010) found that under field conditions, vaccinated hybrid catfish did not have significantly improved survival rates compared with unvaccinated cohorts. Other researchers have also questioned the efficacy of AQUAVAC-COL™ outside the laboratory and suggested that its reduced efficacy in such situations may relate to the intraspecific heterogeneity of F. columnare (Mohammed et al., 2013). Nevertheless, AQUAVAC-COL™-vaccinated largemouth bass fry were significantly protected against natural CD in field trials (Bebak et al., 2009). Another F. columnare immersion bacterin (FryVacc1; Novartis Animal Health) is commercially available and licensed for salmonids (≥3 g) in the USA (Bowker et al., 2015).

There are numerous treatment options for flavobacterial diseases (Bernardet and Bowman, 2006; Austin and Austin, 2007; Starliper and Schill, 2011; Declercq et al., 2013), but the use of chemicals depends on the regulations that govern specific regions. For example, Aquaflor® (florfenicol; Merck Animal Health; 10–15 mg florfenicol/kg fish daily for 10 days) and Terramycin® (oxytetracycline dihydrate; 3.75 g 100 lb/fish daily for 10 days) are approved in the USA to treat F. psychrophilum infections in freshwater-reared salmonids (Bowker et al., 2015). Hydrogen peroxide (50–75 mg/l for 60 min, once daily, three total treatments for fingerlings/adults; and 50 mg/l for 60 min, once daily, three total treatments for fry) is approved to treat external *F. columnare* infections in freshwater-reared cool-water finfish and channel catfish. Chloramine-T is approved to treat external *F. columnare* infections in walleye (10–20 mg/l for 60 min, once daily, three total treatments) and freshwater-reared warm-water finfish (20 mg/l for 60 min, once daily, three total treatments). Aquaflor® (10–15 mg/kg fish daily for 10 days) is approved to treat systemic and/or external *F. columnare* infections in freshwater-reared finfish, and Terramycin® is approved for freshwater-reared rainbow trout at the dose given above (Bowker *et al.*, 2015). Hydrogen peroxide (100 mg/l for 30 min or 50–100 mg/l for 60 min, once daily, three total treatments) and Chloramine-T (at the dose given above) are approved to treat *F. branchiophilum* in freshwater-reared salmonids (Bowker *et al.*, 2015). Bath treatments in 5% NaCl for 2 min are effective against *F. branchiophilum* infections in rainbow trout fingerlings (Kudo and Kimura, 1983). Also, penicillin and streptomycin reduce *F. psychrophilum* transmission in seminal fluids (Oplinger and Wagner, 2015), and during the water hardening process, leading to significantly reduced prevalence of the bacterium in the resultant fertilized eggs (Oplinger *et al.*, 2015).

Increased antibiotic resistance in *F. psychrophilum* has been reported (Smith *et al.*, 2016; Van Vliet *et al.*, unpublished data), and there is resistance in *F. columnare* as well (Declercq *et al.*, 2013). The development of widespread acquired resistance to oxytetracycline (Smith *et al.*, 2016), which is frequently used to treat BCWD/RTFS, is of particular concern, as are reports of increasing minimum inhibitory concentrations for florfenicol in Chile (Henríquez-Núñez *et al.*, 2012). As a result, the recovery and identification of the causative agent from the disease event under investigation and antimicrobial susceptibility testing prior to treatment are paramount for reducing further development of antibiotic resistance.

16.6 Conclusions and Suggestions for Future Research

Flavobacterium psychrophilum, *F. columnare*, and *F. branchiophilum* limit aquaculture productivity worldwide and result in substantial economic losses. Decades of research have produced sensitive and specific methods to detect and identify these pathogens, as well as to assist in an understanding of their pathobiology. Although multiple strategies exist for their prevention and control, much remains unknown.

For example, a better understanding of the global serological and genetic population structure of these bacteria is needed to ensure that future diagnostic and immunologically based controls perform satisfactorily. Likewise, further elucidation of flavobacterial epizootiology would facilitate the development of improved and targeted control, as would continued research directed at the further development of alternative control strategies (e.g. immunostimulation/nutrition, probiotics, phage therapy). Gaps in the knowledge of flavobacterial pathogenesis and its concomitant pathophysiological changes in the host need to be filled in order to interfere with bacterial adherence and infection (possibly through ligand interference, and the use of prebiotics and probiotics), and to mitigate damage to host tissues (possibly through enhanced nutrition).

The importance of the host microbiome in preventing or augmenting flavobacterial disease must not be underestimated. In fact, we suspect that host microbiome variability plays an important role in the difficulties surrounding the reproducibility of flavobacterial experimental challenge models among and between fish species and laboratories through differential bacterial antagonism and/or synergism. Thus, the application of next-generation sequencing technologies for studying the interactions of constituents of the fish microbiome with fish-pathogenic flavobacteria will no doubt give clarity to this hypothesis, as well as to the role that some microbes have in exacerbating and/or preventing flavobacterial infections under different environmental conditions.

Yet another important area of future research is the role that biofilm plays in flavobacterial virulence, persistence in aquaculture facilities and the efficacy of chemotherapeutics. There is also a dire need to elucidate the role that uncharacterized flavobacteria play in disease outbreaks, particularly because some mimic the disease signs of better known flavobacteria. Snieszko (1981) stated 'it seems obvious that bacterial gill disease is a complex disease in which more than one type of bacterium may be involved'. This seems a fitting paradigm for many fish diseases, including those caused by flavobacteria, which result from complex interactions between the host, the environment, obligate/facultative pathogen(s) and commensalistic/mutualistic microbes.

References

Alvarez, B., Secades, P., Prieto, M., Mcbride, M.J. and Guijarro, J.A. (2006) A mutation in *Flavobacterium psychrophilum tlpB* inhibits gliding motility and induces biofilm formation. *Applied and Environmental Microbiology* 72, 4044–4053.

Amita, K., Hoshino, M., Honma, T. and Wakabayashi, H. (2000) An investigation on the distribution of *Flavobacterium psychrophilum* in the Umikawa River. *Fish Pathology* 35, 193–197.

Anacker, R.L. and Ordal, E.J. (1959) Studies on the myxobacterium *Chondrococcus columnaris*: I. Serological typing. *Journal of Bacteriology* 78, 25–32.

Arai, H., Morita, Y., Izumi, S., Katagiri, T. and Kimura, H. (2007) Molecular typing by pulsed-field gel electrophoresis of *Flavobacterium psychrophilum* isolates derived from Japanese fish. *Journal of Fish Diseases* 30, 345–355.

Arias, C.R., Cai, W., Peatman, E. and Bullard, S.A. (2012) Catfish hybrid *Ictalurus punctatus* × *I. furcatus* exhibits higher resistance to Columnaris disease than the parental species. *Diseases of Aquatic Organisms* 100, 77–81.

Austin, B. and Austin, D.A. (2007) *Bacterial Fish Pathogens: Diseases of Farmed and Wild Fish*, 4th edn. Springer/Praxis Publishing, Chichester, UK.

Avendaño-Herrera, R., Gherardelli, V., Olmos, P., Godoy, M., Heisinger, A. *et al.* (2011) *Flavobacterium columnare* associated with mortality of salmonids farmed in Chile: a case report of two outbreaks. *Bulletin of the European Association of Fish Pathologists* 31, 36–44.

Avendaño-Herrera, R., Houel, A., Irgang, R., Bernardet, J.-F., Godoy, M. *et al.* (2014) Introduction, expansion and coexistence of epidemic *Flavobacterium psychrophilum* lineages in Chilean fish farms. *Veterinary Microbiology* 170, 298–306.

Baliarda, A., Faure, D. and Urdaci, M.C. (2002) Development and application of a nested PCR to monitor brood stock salmonid ovarian fluid and spleen for detection of the fish pathogen *Flavobacterium psychrophilum*. *Journal of Applied Microbiology* 92, 510–516.

Bebak, J., Baumgarten, M. and Smith, G. (1997) Risk factors for bacterial gill disease in young rainbow trout (*Oncorhynchus mykiss*) in North America. *Preventive Veterinary Medicine* 32, 23–34.

Bebak, J., Matthews, M. and Shoemaker, C. (2009) Survival of vaccinated, feed-trained largemouth bass fry (*Micropterus salmoides floridanus*) during natural exposure to *Flavobacterium columnare*. *Vaccine* 27, 4297–4301.

Beck, B.H., Farmer, B.D., Straus, D.L., Li, C. and Peatman, E. (2012) Putative roles for a rhamnose binding lectin in *Flavobacterium columnare* pathogenesis in channel catfish *Ictalurus punctatus*. *Fish and Shellfish Immunology* 33, 1008–1015.

Bernardet, J.F. (1997) Immunization with bacterial antigens: *Flavobacterium* and *Flexibacter* infections. *Developments in Biological Standardization* 90, 179–188.

Bernardet, J.-F. (2011a) Class II. Flavobacteriia class. nov. In: Krieg, N., Staley, J., Brown, D., Hedlund, B., Paster, B., Ward, N., Ludwig, W. and Whitman, W. (eds) *Bergey's Manual of Systematic Bacteriology. Volume 4:* The Bacteroidetes, Spirochaetes, Tenericutes (Mollicutes), Acidobacteria, Fibrobacteres, 2nd edn. Springer, New York/Dordrecht, The Netherlands/Heidelberg, Germany/London, p. 105.

Bernardet, J.-F. (2011b) Family I Flavobacteriaceae Reichenbach 1992b, 327[VP]. In: Krieg, N., Staley, J., Brown, D., Hedlund, B., Paster, B., Ward, N., Ludwig, W. and Whitman, W. (eds) *Bergey's Manual of Systematic Bacteriology. Volume 4: The Bacteroidetes, Spirochaetes, Tenericutes (Mollicutes), Acidobacteria, Fibrobacteres*, 2nd edn. Springer, New York/Dordrecht, The Netherlands/Heidelberg, Germany/London, pp. 106–111.

Bernardet, J.-F. and Bowman, J.P. (2006) The genus *Flavobacterium*. In: Dworkin, M., Falkow, S., Rosenberg, E., Schleifer, K. and Stackebrandt, E. (eds) *The Prokaryotes: A Handbook on the Biology of Bacteria. Volume 7: Proteobacteria: Delta and Epsilon Subclasses. Deeply Rooting Bacteria*, 3rd edn. Springer-Verlag, New York, pp. 481–531.

Bernardet, J. and Bowman, J. (2011) Genus 1. *Flavobacterium*. In: Krieg, N., Staley, J., Brown, D., Hedlund, B., Paster, B., Ward, N., Ludwig, W. and Whitman, W. (eds) *Bergey's Manual of Systematic Bacteriology. Volume 4:* The Bacteroidetes, Spirochaetes, Tenericutes (Mollicutes), Acidobacteria, Fibrobacteres, 2nd edn. Springer, New York/Dordrecht, The Netherlands/Heidelberg, Germany/London, pp. 112–154.

Bernardet, J.-F. and Grimont, P.A. (1989) Deoxyribonucleic acid relatedness and phenotypic characterization of *Flexibacter columnaris* sp. nov., nom. rev., *Flexibacter psychrophilus* sp. nov., nom. rev., and *Flexibacter maritimus* Wakabayashi, Hikida, and Masumura 1986. *International Journal of Systematic Bacteriology* 39, 346–354.

Bernardet, J. and Kerouault, B. (1989) Phenotypic and genomic studies of *Cytophaga psychrophila* isolated from diseased rainbow trout (*Oncorhynchus mykiss*) in France. *Applied and Environmental Microbiology* 55, 1796–1800.

Bernardet, J.-F. and Nakagawa, Y. (2006) An introduction to the family Flavobacteriaceae. In: Dworkin, M., Falkow, S., Rosenberg, E., Schleifer, K. and Stackebrandt, E. (eds) *The Prokaryotes: A Handbook on the Biology of Bacteria. Volume 7: Proteobacteria: Delta and Epsilon Subclasses. Deeply Rooting Bacteria*, 3rd edn. Springer-Verlag, New York, pp. 455–480.

Bernardet, J., Baudin-Laurencin, F. and Tixerant, G. (1988) First identification of *Cytophaga psychrophila* in France. *Bulletin of the European Association of Fish Pathologists* 8, 104–105.

Bernardet, J.-F., Segers, P., Vancanneyt, M., Berthe, F., Kersters, K. et al. (1996) Cutting a Gordian knot: emended classification and description of the genus *Flavobacterium*, emended description of the family Flavobacteriaceae, and proposal of *Flavobacterium hydatis* nom. nov.(basonym, *Cytophaga aquatilis* Strohl and Tait 1978). *International Journal of Systematic Bacteriology* 46, 128–148.

Bernardet, J.-F., Nakagawa, Y. and Holmes, B. (2002) Proposed minimal standards for describing new taxa of the family Flavobacteriaceae and emended description of the family. *International Journal of Systematic and Evolutionary Microbiology* 52, 1049–1070.

Borg, A.F. (1948) *Studies on Myxobacteria Associated with Diseases in Salmonid Fishes*. University of Washington, Seattle, Washington.

Boutin, S., Bernatchez, L., Audet, C. and Derome, N. (2012) Antagonistic effect of indigenous skin bacteria of brook charr (*Salvelinus fontinalis*) against *Flavobacterium columnare* and *F. psychrophilum*. *Veterinary Microbiology* 155, 355–361.

Boutin, S., Audet, C. and Derome, N. (2013) Probiotic treatment by indigenous bacteria decreases mortality without disturbing the natural microbiota of *Salvelinus fontinalis*. *Canadian Journal of Microbiology* 59, 662–670.

Bowker, J., Trushenski, J., Gaikowski, M. and Straus, D. (2015) *Guide to Using Drugs, Biologics, and Other Chemicals in Aquaculture*. Fish Culture Section, American Fisheries Society. Revised July 2016 version available at: https://drive.google.com/file/d/0B43dblZlJqD3MTZGNHBzR21mS2c/view (accessed 11 November 2016).

Brown, L., Cox, W. and Levine, R. (1997) Evidence that the causal agent of bacterial cold-water disease *Flavobacterium psychrophilum* is transmitted within salmonid eggs. *Diseases of Aquatic Organisms* 29, 213–218.

Bruno, D. (1992) *Cytophaga psychrophila* (='*Flexibacter psychrophilus*') (Borg), histopathology associated with mortalities among farmed rainbow trout, *Oncorhynchus mykiss* (Walbaum) in the UK. *Bulletin of the European Association of Fish Pathologists* 12, 215–216.

Bullock, G. (1990) *Bacterial Gill Disease of Freshwater Fishes*. Fish Disease Leaflet No. 84, US Fish and Wildlife Service, US Department of the Interior, Washington, DC. Available at: http://citeseerx.ist.psu.edu/viewdoc/download?doi=10.1.1.650.656&rep=rep1&type=pdf (accessed 11 November 2016).

Bullock, G., Herman, R., Heinen, J., Noble, A., Weber, A. et al. (1994) Observations on the occurrence of bacterial gill disease and amoeba gill infestation in rainbow trout cultured in a water recirculation system. *Journal of Aquatic Animal Health* 6, 310–317.

Burbank, D., Lapatra, S., Fornshell, G. and Cain, K. (2012) Isolation of bacterial probiotic candidates from the gastrointestinal tract of rainbow trout, *Oncorhynchus mykiss* (Walbaum), and screening for inhibitory activity against *Flavobacterium psychrophilum*. *Journal of Fish Diseases* 35, 809–816.

Byrne, P.J., Ostland, V.E., Lumsden, J.S., Macphee, D.D. and Ferguson, H.W. (1995) Blood chemistry and acid-base balance in rainbow trout *Oncorhynchus mykiss* with experimentally-induced acute bacterial gill disease. *Fish Physiology and Biochemistry* 14, 509–518.

Cai, W., De La Fuente, L. and Arias, C.R. (2013) Biofilm formation by the fish pathogen *Flavobacterium columnare*: development and parameters affecting surface attachment. *Applied and Environmental Microbiology* 79, 5633–5642.

Castillo, D., Higuera, G., Villa, M., Middelboe, M., Dalsgaard, I. et al. (2012) Diversity of *Flavobacterium psychrophilum* and the potential use of its phages for protection against bacterial cold water disease in salmonids. *Journal of Fish Diseases* 35, 193–201.

Castillo, D., Christiansen, R.H., Espejo, R. and Middelboe, M. (2014) Diversity and geographical distribution of *Flavobacterium psychrophilum* isolates and their phages: patterns of susceptibility to phage infection and phage host range. *Microbial Ecology* 67, 748–757.

Chakroun, C., Urdaci, M.C., Faure, D., Grimont, F. and Bernardet, J.-F. (1997) Random amplified polymorphic DNA analysis provides rapid differentiation among isolates of the fish pathogen *Flavobacterium psychrophilum* and among *Flavobacterium* species. *Diseases of Aquatic Organisms* 31, 187–196.

Cipriano, R.C. (2015) Bacterial analysis of fertilized eggs of Atlantic salmon from the Penobscot, Naraguagus, and Machias rivers, Maine. *Journal of Aquatic Animal Health* 27, 172–177.

Cipriano, R.C. and Holt, R.A. (2005) *Flavobacterium Psychrophilum, Cause of Bacterial Cold-water Disease and Rainbow Trout Fry Syndrome*. Fish Disease Leaflet No. 86, National Fish Health Research Laboratory, US Geological Survey, US Department of the Interior, Kearneysville, West Virginia. Available at: https://articles.extension.org/sites/default/files/w/8/80/USFWS_Coldwater_disease.pdf (accessed 11 November 2016).

Daoust, P.Y. and Ferguson, H. (1983) Gill diseases of cultured salmonids in Ontario. *Canadian Journal of Comparative Medicine* 47, 358–362.

Darwish, A.M., Ismaiel, A.A., Newton, J.C. and Tang, J. (2004) Identification of *Flavobacterium columnare* by a species-specific polymerase chain reaction and renaming of ATCC43622 strain to *Flavobacterium johnsoniae*. *Molecular and Cellular Probes* 18, 421–427.

Davis, H.S. (1922) A new bacterial disease of fresh-water fishes. *Bulletin of the United States Bureau of Fisheries* 38, 261–280.

Davis, H.[S.] (1926) A new gill disease of trout. *Transactions of the American Fisheries Society* 56, 156–160.

Davis, H.S. (1946) *Care and Diseases of Trout*. U.S. Fish and Wildlife Service, Washington, D.C.

Declercq, A.M., Haesebrouck, F., Van den Broeck, W., Bossier, P. and Decostere, A. (2013) Columnaris disease in fish: a review with emphasis on bacterium–host interactions. *Veterinary Research* 44:27.

Declercq, A.M., Chiers, K., Haesebrouck, F., Van Den Broeck, W., Dewulf, J. *et al.* (2015) Gill infection model for columnaris disease in common carp and rainbow trout. *Journal of Aquatic Animal Health* 27, 1–11.

Decostere, A., Haesebrouck, F. and Devriese, L. (1998) Characterization of four *Flavobacterium columnare* (*Flexibacter columnaris*) strains isolated from tropical fish. *Veterinary Microbiology* 62, 35–45.

Decostere, A., Haesebrouck, F., Turnbull, J. F. and Charlier, G. (1999) Influence of water quality and temperature on adhesion of high and low virulence *Flavobacterium columnare* strains to isolated gill arches. *Journal of Fish Diseases* 22, 1–11.

Del Cerro, A., Marquez, I. and Guijarro, J.A. (2002a) Simultaneous detection of *Aeromonas salmonicida*, *Flavobacterium psychrophilum*, and *Yersinia ruckeri*, three major fish pathogens, by multiplex PCR. *Applied and Environmental Microbiology* 68, 5177–5180.

Del Cerro, A., Mendoza, M.C. and Guijarro, J.A. (2002b) Usefulness of a TaqMan-based polymerase chain reaction assay for the detection of the fish pathogen *Flavobacterium psychrophilum*. *Journal of Applied Microbiology* 93, 149–156.

Duchaud, E., Boussaha, M., Loux, V., Bernardet, J.-F., Michel, C. *et al.* (2007) Complete genome sequence of the fish pathogen *Flavobacterium psychrophilum*. *Nature Biotechnology* 25, 763–769.

Elsayed, E.E., Eissa, A.E. and Faisal, M. (2006) Isolation of *Flavobacterium psychrophilum* from sea lamprey, *Petromyzon marinus* L., with skin lesions in Lake Ontario. *Journal of Fish Diseases* 29, 629–632.

Evensen, O. and Lorenzen, E. (1996) An immunohistochemical study of *Flexibacter psychrophilus* infection in experimentally and naturally infected rainbow trout (*Oncorhynchus mykiss*) fry. *Diseases of Aquatic Organisms* 25, 53–61.

Evenhuis, J.P., Leeds, T.D., Marancik, D.P., Lapatra, S.E. and Wiens, G.D. (2015) Rainbow trout (*Oncorhynchus mykiss*) resistance to columnaris disease is heritable and favorably correlated with bacterial cold water disease resistance. *Journal of Animal Science* 93, 1546–1554.

Farmer, B.D., Mitchell, A.J. and Straus, D.L. (2011) The effect of high total ammonia concentration on the survival of channel catfish experimentally infected with *Flavobacterium columnare*. *Journal of Aquatic Animal Health* 23, 162–168.

Fehringer, T.R., Hardy, R.W. and Cain, K.D. (2014) Dietary inclusion of salmon testes meal from Alaskan seafood processing byproducts: effects on growth and immune function of rainbow trout, *Oncorhynchus mykiss* (Walbaum). *Aquaculture* 433, 34–39.

Figueiredo, H.C.P., Klesius, P.H., Arias, C.R., Evans, J., Shoemaker, C.A. *et al.* (2005) Isolation and characterization of strains of *Flavobacterium columnare* from Brazil. *Journal of Fish Diseases* 28, 199–204.

Fuente Mde, L., Miranda, C.D., Jopia, P., Gonzalez-Rocha, G., Guiliani, N. *et al.* (2015) Growth inhibition of bacterial fish pathogens and quorum-sensing blocking by bacteria recovered from Chilean salmonid farms. *Journal of Aquatic Animal Health* 27, 112–122.

Fujihara, M.P. and Nakatani, R.E. (1971) Antibody production and immune responses of rainbow trout and coho salmon to *Chondrococcus columnaris*. *Journal of the Fisheries Research Board of Canada* 28, 1253–1258.

Fujiwara-Nagata, E. and Eguchi, M. (2009) Development and evaluation of a loop-mediated isothermal amplification assay for rapid and simple detection of *Flavobacterium psychrophilum*. *Journal of Fish Diseases* 32, 873–881.

Garcia, C., Pozet, F. and Michel, C. (2000) Standardization of experimental infection with *Flavobacterium psychrophilum*, the agent of rainbow trout *Oncorhynchus mykiss* fry syndrome. *Diseases of Aquatic Organisms* 42, 191–197.

Ghosh, B., Cain, K.D., Nowak, B.F. and Bridle, A.R. (2016) Microencapsulation of a putative probiotic *Enterobacter* species, C6-6, to protect rainbow trout, *Oncorhynchus mykiss* (Walbaum), against bacterial coldwater disease. *Journal of Fish Diseases* 39, 1–11.

Gómez, E., Méndez, J., Cascales, D. and Guijarro, J.A. (2014) *Flavobacterium psychrophilum* vaccine development: a difficult task. *Microbial Biotechnology* 7, 414–423.

Good, C.M., Thorburn, M.A. and Stevenson, R.M. (2008) Factors associated with the incidence of bacterial gill disease in salmonid lots reared in Ontario, Canada government hatcheries. *Preventive Veterinary Medicine* 83, 297–307.

Good, C.M., Thorburn, M.A., Ribble, C.S. and Stevenson, R.M. (2009) Rearing unit-level factors associated with bacterial gill disease treatment in two Ontario, Canada government salmonid hatcheries. *Preventive Veterinary Medicine* 91, 254–260.

Good, C.M., Thorburn, M.A., Ribble, C.S. and Stevenson, R.M. (2010) A prospective matched nested case-control study of bacterial gill disease outbreaks in Ontario, Canada government salmonid hatcheries. *Preventive Veterinary Medicine* 95, 152–157.

Good, C.[M.], Davidson, J., Wiens, G.D., Welch, T.J. and Summerfelt, S. (2015) *Flavobacterium branchiophilum* and *F. succinicans* associated with bacterial gill disease in rainbow trout *Oncorhynchus mykiss* (Walbaum) in water recirculation aquaculture systems. *Journal of Fish Diseases* 38, 409–413.

Hawke, J.P. and Thune, R.L. (1992) Systemic isolation and antimicrobial susceptibility of *Cytophaga columnaris* from commercially reared channel catfish. *Journal of Aquatic Animal Health* 4, 109–113.

Hedrick, R.P., Mcdowell, T.S., Marty, G.D., Mukkatira, K., Antonio, D.B. *et al.* (2000) Ultraviolet irradiation inactivates the waterborne infective stages of *Myxobolus cerebralis*: a treatment for hatchery water supplies. *Diseases of Aquatic Organisms* 42, 53–59.

Henríquez-Núñez, H., Oscar, E., Goran, K. and Avendaño-Herrera, R. (2012) Antimicrobial susceptibility and plasmid profiles of *Flavobacterium psychrophilum* strains isolated in Chile. *Aquaculture* 354/355, 38–44.

Hesami, S., Allen, K.J., Metcalf, D., Ostland, V.E., Macinnes, J.I. *et al.* (2008) Phenotypic and genotypic analysis of *Flavobacterium psychrophilum* isolates from Ontario salmonids with bacterial coldwater disease. *Canadian Journal of Microbiology* 54, 619–629.

Hibi, K., Yoshiura, Y., Ushio, H., Ren, H. and Endo, H. (2012) Rapid detection of *Flavobacterium psychrophilum* using fluorescent magnetic beads and flow cytometry. *Sensors and Materials* 24, 311–322.

Holt, R.A. (1987) *Cytophaga psychrophila*, the causative agent of bacterial cold-water disease in salmonid fish. PhD thesis, Oregon State University, Corvallis, Oregon.

Holt, R.A., Rohovec, J.S. and Fryer, J.L. (1993) Bacterial cold-water disease. In: Inglis, V., Roberts, R.J. and Bromage, N.R. (eds) *Bacterial Diseases of Fish*. Blackwell Scientific, Oxford, UK.

Huh, G.-J. and Wakabayashi, H. (1989) Serological characteristics of *Flavobacterium branchiophila* isolated from gill diseases of freshwater fishes in Japan, USA, and Hungary. *Journal of Aquatic Animal Health* 1, 142–147.

Izumi, S., Fujii, H. and Aranishi, F. (2005) Detection and identification of *Flavobacterium psychrophilum* from gill washings and benthic diatoms by PCR-based sequencing analysis. *Journal of Fish Diseases* 28, 559–564.

Kent, L., Groff, J., Morrison, J., Yasutake, W. and Holt, R. (1989) Spiral swimming behavior due to cranial and vertebral lesions associated with *Cytophaga psychrophila* infections in salmonid fishes. *Diseases of Aquatic Organisms* 6, 11–16.

Kimura, N., Wakabayashi, H. and Kudo, S. (1978) Studies on bacterial gill disease in salmonids 1: selection of bacterium transmitting gill disease. *Fish Pathology* 12, 233–242.

Kirkland, S. (2010) Evaluation of the live-attenuated vaccine AquaVac-COL® on hybrid channel × blue catfish fingerlings in earthen ponds. Masters thesis, Auburn University, Auburn, Alabama.

Klesius, P., Lim, C. and Shoemaker, C. (1999) Effect of feed deprivation on innate resistance and antibody response to *Flavobacterium columnare* in channel catfish, *Ictalurus punctatus*. *Bulletin of the European Association of Fish Pathologists* 19, 156–158.

Ko, Y.-M. and Heo, G.-J. (1997) Characteristics of *Flavobacterium branchiophilum* isolated from rainbow trout in Korea. *Fish Pathology* 32, 97–102.

Korkea-Aho, T., Heikkinen, J., Thompson, K., Von Wright, A. and Austin, B. (2011) *Pseudomonas* sp. M174 inhibits the fish pathogen *Flavobacterium psychrophilum*. *Journal of Applied Microbiology* 111, 266–277.

Korkea-Aho, T.L., Papadopoulou, A., Heikkinen, J., Von Wright, A., Adams, A. *et al.* (2012) *Pseudomonas* M162 confers protection against rainbow trout fry syndrome by stimulating immunity. *Journal of Applied Microbiology* 113, 24–35.

Kudo, S. and Kimura, N. (1983) The recovery from hyperplasia in a natural infection. *Bulletin of the Japanese Society of Scientific Fisheries* 49, 1627–1633.

Kumagai, A., Takahashi, K., Yamaoka, S. and Wakabayashi, H. (1998) Ineffectiveness of iodophore treatment in disinfecting salmonid eggs carrying *Cytophaga psychrophila*. *Fish Pathology* 33, 123–128.

Kumagai, A., Yamaoka, S., Takahashi, K., Fukuda, H. and Wakabayashi, H. (2000) Waterborne transmission of *Flavobacterium psychrophilum* in coho salmon eggs. *Fish Pathology* 35, 25–28.

Kunttu, H. (2010) *Characterizing the Bacterial Fish Pathogen Flavobacterium columnare, and Some Factors Affecting Its Pathogenicity*. Jyväskylä Studies in Biology and Environmental Science 206, University of Jyväskylä, Jyväskylä, Finland. Available at: https://jyx.jyu.fi/dspace/bitstream/handle/123456789/23323/9789513938673.pdf?sequence=1 (accessed 11 November 2016).

Kunttu, H.M., Valtonen, E.T., Suomalainen, L.R., Vielma, J. and Jokinen, I.E. (2009) The efficacy of two immunostimulants against *Flavobacterium columnare* infection in juvenile rainbow trout (*Oncorhynchus mykiss*). *Fish and Shellfish Immunology* 26, 850–857.

Kunttu, H.M., Sundberg, L.R., Pulkkinen, K. and Valtonen, E.T. (2012) Environment may be the source of *Flavobacterium columnare* outbreaks at fish farms. *Environmental Microbiology Reports* 4, 398–402.

LaFrentz, B. and Cain, K. (2004) *Coldwater Disease*. Extension Bulletin, Western Regional Aquaculture Consortium (WRAC), Washington, DC.

LaFrentz, B.R., Lapatra, S.E., Call, D.R. and Cain, K.D. (2008) Isolation of rifampicin resistant *Flavobacterium psychrophilum* strains and their potential as live attenuated vaccine candidates. *Vaccine* 26, 5582–5589.

LaFrentz, B.R., Lapatra, S.E., Call, D.R., Wiens, G.D. and Cain, K.D. (2009) Proteomic analysis of *Flavobacterium psychrophilum* cultured *in vivo* and in iron-limited media. *Diseases of Aquatic Organisms* 87, 171–182.

LaFrentz, B.R., Lapatra, S.E., Shoemaker, C.A. and Klesius, P.H. (2012a) Reproducible challenge model to investigate the virulence of *Flavobacterium columnare* genomovars in rainbow trout *Oncorhynchus mykiss*. *Diseases of Aquatic Organisms* 101, 115–122.

LaFrentz, B.R., Shoemaker, C.A., Booth, N.J., Peterson, B.C. and Ourth, D.D. (2012b) Spleen index and mannose-binding lectin levels in four channel catfish families exhibiting different susceptibilities to *Flavobacterium columnare* and *Edwardsiella ictaluri*. *Journal of Aquatic Animal Health* 24, 141–147.

LaFrentz, B., Waldbieser, G., Welch, T. and Shoemaker, C. (2014) Intragenomic heterogeneity in the 16S rRNA genes of *Flavobacterium columnare* and standard protocol for genomovar assignment. *Journal of Fish Diseases* 37, 657–69.

Lapatra, S.E., Fehringer, T.R. and Cain, K.D. (2014) A probiotic *Enterobacter* sp. provides significant protection against *Flavobacterium psychrophilum* in rainbow trout (*Oncorhynchus mykiss*) after injection by two different routes. *Aquaculture* 433, 361–366.

Leeds, T., Silverstein, J., Weber, G., Vallejo, R., Palti, Y. *et al.* (2010) Response to selection for bacterial cold water disease resistance in rainbow trout. *Journal of Animal Science* 88, 1936–1946.

Lievens, B., Frans, I., Heusdens, C., Justé, A., Jonstrup, S.P. *et al.* (2011) Rapid detection and identification of viral and bacterial fish pathogens using a DNA array-based multiplex assay. *Journal of Fish Diseases* 34, 861–875.

Lindstrom, N.M., Call, D.R., House, M.L., Moffitt, C.M. and Cain, K.D. (2009) A quantitative enzyme-linked immunosorbent assay and filtration-based fluorescent antibody test as potential tools to screen broodstock for infection with *Flavobacterium psychrophilum*. *Journal of Aquatic Animal Health* 21, 43–56.

Liu, L., Li, C., Su, B., Beck, B.H. and Peatman, E. (2013) Short-term feed deprivation alters immune status of surface mucosa in channel catfish (*Ictalurus punctatus*). *PLoS ONE* 8 (9): e74581.

Liu, S., Vallejo, R.L., Palti, Y., Gao, G., Marancik, D.P. *et al.* (2015) Identification of single nucleotide polymorphism markers associated with bacterial cold water disease resistance and spleen size in rainbow trout. *Frontiers in Genetics* 6: 298.

Loch, T.P. and Faisal, M. (2016a) Flavobacteria isolated from the milt of feral Chinook salmon of the Great Lakes. *North American Journal of Aquaculture* 78, 25–33.

Loch, T.P. and Faisal, M. (2016b) *Flavobacterium spartansii* induces pathological changes and mortality in experimentally challenged Chinook salmon *Oncorhynchus tshawytscha* (Walbaum). *Journal of Fish Diseases* 39, 483–488.

Loch, T.P., Fujimoto, M., Woodiga, S.A., Walker, E.D., Marsh, T.L. *et al.* (2013) Diversity of fish-associated flavobacteria of Michigan. *Journal of Aquatic Animal Health* 25, 149–164.

Long, A., Polinski, M.P., Call, D.R. and Cain, K.D. (2012) Validation of diagnostic assays to screen broodstock for *Flavobacterium psychrophilum* infections. *Journal of Fish Diseases* 35, 407–419.

Long, A., Fehringer, T.R., Swain, M.A., LaFrentz, B.R., Call, D.R. *et al.* (2013) Enhanced efficacy of an attenuated *Flavobacterium psychrophilum* strain cultured under iron-limited conditions. *Fish and Shellfish Immunology* 35, 1477–1482.

Long, A., Call, D.R. and Cain, K.D. (2015) Comparison of quantitative PCR and ELISA for detection and quantification of *Flavobacterium psychrophilum* in salmonid broodstock. *Diseases of Aquatic Organisms* 115, 139–146.

Lorenzen, E., Dalsgaard, I., From, J., Hansen, E., Horlyck, V. *et al.* (1991) Preliminary investigations of fry mortality syndrome in rainbow trout. *Bulletin of the European Association of Fish Pathologists* 11, 77–79.

Macphee, D., Ostland, V., Lumsden, J. and Ferguson, H. (1995) Development of an enzyme-linked immunosorbent assay (ELISA) to estimate the quantity of *Flavobacterium branchiophilum* on the gills of rainbow trout *Oncorhynchus mykiss*. *Diseases of Aquatic Organisms* 21, 13–23.

Madetoja, J. and Wiklund, T. (2002) Detection of the fish pathogen *Flavobacterium psychrophilum* in water from fish farms. *Systematic and Applied Microbiology* 25, 259–266.

Madetoja, J., Nyman, P. and Wiklund, T. (2000) *Flavobacterium psychrophilum*, invasion into and shedding by rainbow trout *Oncorhynchus mykiss*. *Diseases of Aquatic Organisms* 43, 27–38.

Madetoja, J., Dalsgaard, I. and Wiklund, T. (2002) Occurrence of *Flavobacterium psychrophilum* in fish-farming environments. *Diseases of Aquatic Organisms* 52, 109–118.

Madsen, L. and Dalsgaard, I. (2000) Comparative studies of Danish *Flavobacterium psychrophilum* isolates: ribotypes, plasmid profiles, serotypes and virulence. *Journal of Fish Diseases* 23, 211–218.

Madsen, L. and Dalsgaard, I. (2008) Water recirculation and good management: potential methods to avoid disease outbreaks with *Flavobacterium psychrophilum*. *Journal of Fish Diseases* 31, 799–810.

Mainous, M.E., Kuhn, D.D. and Smith, S.A. (2012) Efficacy of common aquaculture compounds for disinfection of *Flavobacterium columnare* and *F. psychrophilum*. *Journal of Applied Aquaculture* 24, 262–270.

Marancik, D.P. and Wiens, G.D. (2013) A real-time polymerase chain reaction assay for identification and quantification of *Flavobacterium psychrophilum* and application to disease resistance studies in selectively bred rainbow trout *Oncorhynchus mykiss*. *FEMS Microbiology Letters* 339, 122–129.

Marancik, D.P., Camus, M.S., Camus, A.C., Leeds, T.D., Weber, G.M. *et al.* (2014) Biochemical reference intervals and pathophysiological changes in *Flavobacterium psychrophilum*-resistant and -susceptible rainbow trout lines. *Diseases of Aquatic Organisms* 111, 239–248.

Marancik, D., Gao, G., Paneru, B., Ma, H., Hernandez, A.G. *et al.* (2015) Whole-body transcriptome of selectively bred, resistant-, control-, and susceptible-line rainbow trout following experimental challenge with *Flavobacterium psychrophilum*. *Frontiers in Genetics* 5: 453.

Mata, M., Skarmeta, A. and Santos, Y. (2002) A proposed serotyping system for *Flavobacterium psychrophilum*. *Letters in Applied Microbiology* 35, 166–170.

Miwa, S. and Nakayasu, C. (2005) Pathogenesis of experimentally induced bacterial cold water disease in ayu *Plecoglossus altivelis*. *Diseases of Aquatic Organisms* 67, 93–104.

Mohammed, H.H. and Arias, C.R. (2016) Protective efficacy of *Nigella sativa* seeds and oil against Columnaris disease in fishes. *Journal of Fish Diseases* 39, 693–703.

Mohammed, H., Olivares-Fuster, O., LaFrentz, S. and Arias, C.R. (2013) New attenuated vaccine against Columnaris disease in fish: choosing the right parental strain is critical for vaccine efficacy. *Vaccine* 31, 5276–5280.

Morrison, C., Cornick, J., Shum, G. and Zwicker, B. (1981) Microbiology and histopathology of 'saddleback' disease of underyearling Atlantic salmon, *Salmo salar* L. *Journal of Fish Diseases* 4, 243–258.

Nakagawa, J., Iwasaki, T. and Kodama, H. (2009) Protection against *Flavobacterium psychrophilum* infection (cold water disease) in ayu fish (*Plecoglossus altivelis*) by oral administration of humus extract. *The Journal of Veterinary Medical Science* 71, 1487–1491.

Nakayama, H., Tanaka, K., Teramura, N. and Hattori, S. (2015) Expression of collagenase in *Flavobacterium psychrophilum* isolated from cold-water disease-affected ayu (*Plecoglossus altivelis*). *Bioscience, Biotechnology, and Biochemistry* 80, 135–144.

Nematollahi, A., Decostere, A., Pasmans, F. and Haesebrouck, F. (2003) *Flavobacterium psychrophilum* infections in salmonid fish. *Journal of Fish Diseases* 26, 563–574.

Nicolas, P., Mondot, S., Achaz, G., Bouchenot, C., Bernardet, J.-F. *et al.* (2008) Population structure of the fish-pathogenic bacterium *Flavobacterium psychrophilum*. *Applied and Environmental Microbiology* 74, 3702–3709.

Oplinger, R.W. and Wagner, E. (2010) Disinfection of contaminated equipment: evaluation of benzalkonium chloride exposure time and solution age and the ability of air-drying to eliminate *Flavobacterium psychrophilum*. *Journal of Aquatic Animal Health* 22, 248–253.

Oplinger, R.W. and Wagner, E.J. (2013) Control of *Flavobacterium psychrophilum*: tests of erythromycin, streptomycin, osmotic and thermal shocks, and rapid pH change. *Journal of Aquatic Animal Health* 25, 1–8.

Oplinger, R.W. and Wagner, E.J. (2015) Use of penicillin and streptomycin to reduce spread of bacterial cold-water disease I: antibiotics in sperm extenders. *Journal of Aquatic Animal Health* 27, 25–31.

Oplinger, R.W., Wagner, E.J. and Cavender, W. (2015) Use of penicillin and streptomycin to reduce spread of bacterial coldwater disease II: efficacy of using antibiotics in diluents and during water hardening. *Journal of Aquatic Animal Health* 27, 32–37.

Ordal, E.J. and Rucker, R.R. (1944) Pathogenic myxobacteria. *Proceedings of the Society for Experimental Biology and Medicine* 56, 15-18.

Orieux, N., Bourdineaud, J.P., Douet, D.G., Daniel, P. and Le Henaff, M. (2011) Quantification of *Flavobacterium psychrophilum* in rainbow trout, *Oncorhynchus mykiss* (Walbaum), tissues by qPCR. *Journal of Fish Diseases* 34, 811–821.

Ostland, V.E., Macphee, D.D., Lumsden, J.S. and Ferguson, H.W. (1995) Virulence of *Flavobacterium branchiophilum* in experimentally infected salmonids. *Journal of Fish Diseases* 18, 249–262.

Ostland, V.E., Mcgrogan, D.G. and Ferguson, H.W. (1997) Cephalic osteochondritis and necrotic scleritis in intensively reared salmonids associated with *Flexibacter psychrophilus*. *Journal of Fish Diseases* 20, 443–451.

Panangala, V.S., Shelby, R.A., Shoemaker, C.A., Klesius, P.H., Mitra, A. *et al.* (2006) Immunofluorescent test for simultaneous detection of *Edwardsiella ictaluri* and *Flavobacterium columnare*. *Diseases of Aquatic Organisms* 68, 197–207.

Panangala, V.S., Shoemaker, C.A. and Klesius, P.H. (2007) TaqMan real-time polymerase chain reaction assay for rapid detection of *Flavobacterium columnare*. *Aquaculture Research* 38, 508–517.

Plumb, J.A. and Hanson, L.A. (2010) Catfish bacterial diseases. In: Plumb, J.A. and Hanson, L.A. *Health Maintenance and Principal Microbial Diseases of Cultured Fishes*, 3rd edn. Wiley-Blackwell, Oxford, UK, pp. 275–313.

Prasad, Y., Arpana, Kumar, D. and Sharma, A.K. (2011) Lytic bacteriophages specific to *Flavobacterium columnare* rescue catfish, *Clarias batrachus* (Linn.) from columnaris disease. *Journal of Environmental Biology* 32, 161–168.

Pulkkinen, K., Suomalainen, L.-R., Read, A.F., Ebert, D., Rintamäki, P. *et al.* (2010) Intensive fish farming and the evolution of pathogen virulence: the case of Columnaris disease in Finland. *Proceedings of the Royal Society of London B: Biological Sciences* 277, 593–600.

Rangdale, R. and Way, K. (1995) Rapid identification of *C. psychrophila* from infected spleen tissue using an enzyme-linked immunosorbent assay (ELISA). *Bulletin of the European Association of Fish Pathologists* 15, 213–216.

Rangdale, R., Richards, R. and Alderman, D. (1999) Histopathological and electron microscopical observations on rainbow trout fry syndrome. *Veterinary Record* 144, 251–254.

Ryumae, U., Hibi, K., Yoshiura, Y., Ren, H. and Endo, H. (2012) Ultra highly sensitive method for detecting *Flavobacterium psychrophilum* using high-gradient immunomagnetic separation with a polymerase chain reaction. *Aquaculture Research* 43, 929–939.

Schulz, C. and Faisal, M. (2010) The bacterial community associated with the leech *Myzobdella lugubris* Leidy 1851 (Hirudinea: Piscicolidae) from Lake Erie, Michigan, USA. *Parasite* 17, 113–121.

Shoemaker, C.A., Klesius, P.H., Drennan, J.D. and Evans, J.J. (2011) Efficacy of a modified live *Flavobacterium columnare* vaccine in fish. *Fish and Shellfish Immunology* 30, 304–308.

Silverstein, J.T., Vallejo, R.L., Palti, Y., Leeds, T.D., Rexroad, C.E. *et al.* (2009) Rainbow trout resistance to bacterial cold-water disease is moderately heritable and is not adversely correlated with growth. *Journal of Animal Science* 87, 860–867.

Smith, P., Endris, R., Kronvall, G., Thomas, V., Verner-Jeffreys, D., Wilhelm, C. and Dalsgaard, I. (2016) Epidemiological cut-off values for *Flavobacterium psychrophilum* MIC data generated by a standard test protocol. *Journal of Fish Diseases* 39, 143–154.

Snieszko, S.F. (1981) *Bacterial Gill Disease of Freshwater Fishes*. Fish Disease Leaflet No. 62, US Fish Wildlife Service, US Department of the Interior, Washington, DC. Available at: http://digitalcommons.unl.edu/cgi/viewcontent.cgi?article=1146&context=usfwspubs (accessed 11 November 2016).

Solís, C.J., Poblete-Morales, M., Cabral, S., Valdés, J.A., Reyes, A.E. *et al.* (2015) Neutrophil migration in the activation of the innate immune response to different *Flavobacterium psychrophilum* vaccines in zebrafish (*Danio rerio*). *Journal of Immunology Research* 2015: 515187.

Song, Y.L., Fryer, J.L. and Rohovec, J.S. (1988) Comparison of six media for the cultivation of *Flexibacter columnaris*. *Fish Pathology* 23, 91–94.

Speare, D.J. and Ferguson, H.W. (1989) Clinical and pathological features of common gill diseases of cultured salmonids in Ontario. *The Canadian Veterinary Journal* 30, 882–887.

Speare, D.J., Ferguson, H.W., Beamish, F.W.M., Yager, J.A. and Yamashiro, S. (1991) Pathology of bacterial gill disease: sequential development of lesions during natural outbreaks of disease. *Journal of Fish Diseases* 14, 21–32.

Speare, D.J., Markham, R.J., Despres, B., Whitman, K. and Macnair, N. (1995) Examination of gills from salmonids with bacterial gill disease using monoclonal antibody probes for *Flavobacterium branchiophilum* and *Cytophaga columnaris*. *Journal of Veterinary Diagnostic Investigation* 7, 500–505.

Starliper, C.E. (2011) Bacterial coldwater disease of fishes caused by *Flavobacterium psychrophilum*. *Journal of Advanced Research* 2, 97–108.

Starliper, C. and Schill, W. (2011) Flavobacterial diseases: Columnaris disease, coldwater disease, and bacterial gill disease. In: Woo, P.T.K. and Bruno, D.B. (eds) *Fish Diseases and Disorders. Volume 3: Viral, Bacterial and Fungal Infections*, 2nd edn. CAB International, Wallingford, UK, pp. 606–631.

Strepparava, N., Wahli, T., Segner, H., Polli, B. and Petrini, O. (2012) Fluorescent *in situ* hybridization: a new tool for the direct identification and detection of *F. psychrophilum*. *PLoS ONE* 7 (11): e49280.

Strepparava, N., Wahli, T., Segner, H. and Petrini, O. (2014) Detection and quantification of *Flavobacterium psychrophilum* in water and fish tissue samples by quantitative real time PCR. *BMC Microbiology* 14:105.

Strom-Bestor, M. and Wiklund, T. (2011) Inhibitory activity of *Pseudomonas* sp. on *Flavobacterium psychrophilum, in vitro*. *Journal of Fish Diseases* 34, 255–264.

Suebsing, R., Kampeera, J., Sirithammajak, S., Withyachumnarnkul, B., Turner, W. *et al.* (2015) Colorimetric method of loop-mediated isothermal amplification with the pre-addition of calcein for detecting *Flavobacterium columnare* and its assessment in tilapia farms. *Journal of Aquatic Animal Health* 27, 38–44.

Sundell, K. and Wiklund, T. (2011) Effect of biofilm formation on antimicrobial tolerance of *Flavobacterium psychrophilum*. *Journal of Fish Diseases* 34, 373–383.

Suomalainen, L.R., Bandilla, M. and Valtonen, E.T. (2009) Immunostimulants in prevention of Columnaris disease of rainbow trout, *Oncorhynchus mykiss* (Walbaum). *Journal of Fish Diseases* 32, 723–726.

Suzuki, K., Arai, H., Kuge, T., Katagiri, T. and Izumi, S. (2008) Reliability of PCR methods for the detection of *Flavobacterium psychrophilum*. *Fish Pathology* 43, 124–127.

Tekedar, H.C., Karsi, A., Gillaspy, A.F., Dyer, D.W., Benton, N.R. *et al.* (2012) Genome sequence of the fish pathogen *Flavobacterium columnare* ATCC 49512. *Journal of Bacteriology* 194, 2763–2764.

Touchon, M., Barbier, P., Bernardet, J.-F., Loux, V., Vacherie, B. *et al.* (2011) Complete genome sequence of the fish pathogen *Flavobacterium branchiophilum*. *Applied and Environmental Microbiology* 77, 7656–7662.

Toyama, T., Kita-Tsukamoto, K. and Wakabayashi, H. (1996) Identification of *Flexibacter maritimus*, *Flavobacterium branchiophilum* and *Cytophaga columnaris* by PCR targeted 16S ribosomal DNA. *Fish Pathology* 31, 25–31.

Tripathi, N.K., Latimer, K.S., Gregory, C.R., Ritchie, B.W., Wooley, R.E. *et al.* (2005) Development and evaluation of an experimental model of cutaneous columnaris disease in koi *Cyprinus carpio*. *Journal of Veterinary Diagnostic Investigation* 17, 45–54.

Triyanto, A. and Wakabayashi, H. (1999) Genotypic diversity of strains of *Flavobacterium columnare* from diseased fishes. *Fish Pathology* 34, 65–71.

Triyanto, A., Kumamaru, A. and Wakabayashi, H. (1999) The use of PCR targeted 16S rDNA for identification of genomovars of *Flavobacterium columnare*. *Fish Pathology* 34, 217–218.

Van Vliet, D., Loch, T.P. and Faisal, M. (2015) *Flavobacterium psychrophilum* Infections in salmonid broodstock and hatchery-propagated stocks of the Great Lakes basin. *Journal of Aquatic Animal Health* 27, 192–202.

Van Vliet, D., Wiens, G., Loch, T., Nicolas, P. and Faisal, M. (2016) Genetic diversity of *Flavobacterium psychrophilum* isolated from three *Oncorhynchus* spp. in the U.S.A. revealed by multilocus sequence typing. *Applied and Environmental Microbiology*. doi:10.1128/AEM.00411–16.

Vatsos, I.N., Thompson, K.D. and Adams, A. (2002) Development of an immunofluorescent antibody technique (IFAT) and *in situ* hybridization to detect

Flavobacterium psychrophilum in water samples. *Aquaculture Research* 33, 1087–1090.

Vatsos, I.N., Thompson, K.D. and Adams, A. (2003) Starvation of *Flavobacterium psychrophilum* in broth, stream water and distilled water. *Diseases of Aquatic Organisms* 56, 115–126.

Verma, V. and Prasad, Y. (2014) Isolation and immunohistochemical identification of *Flavobacterium psychrophilum* from the tissue of catfish, *Clarias batrachus*. *Journal of Environmental Biology* 35, 389–393.

Wagner, B.A., Wise, D.J., Khoo, L.H. and Terhune, J.S. (2002) The epidemiology of bacterial diseases in food-size channel catfish. *Journal of Aquatic Animal Health* 14, 263–272.

Wakabayashi, H., Egusa, S. and Fryer, J. (1980) Characteristics of filamentous bacteria isolated from a gill disease of salmonids. *Canadian Journal of Fisheries and Aquatic Sciences* 37, 1499–1504.

Wakabayashi, H., Huh, G.J. and Kimura, N. (1989) *Flavobacterium branchiophila* sp. nov., a causative agent of bacterial gill disease of freshwater fishes. *International Journal of Systematic Bacteriology* 39, 213–216.

Wakabayashi, H., Toyama, T. and Iida, T. (1994) A study on serotyping of *Cytophaga psychrophila* isolated from fishes in Japan. *Fish Pathology* 29, 101–104.

Welker, T.L., Shoemaker, C.A., Arias, C.R. and Klesius, P.H. (2005) Transmission and detection of *Flavobacterium columnare* in channel catfish *Ictalurus punctatus*. *Diseases of Aquatic Organisms* 63, 129–138.

Wiens, G.D., Lapatra, S.E., Welch, T.J., Evenhuis, J.P., Rexroad, C.E. III *et al.* (2013) On-farm performance of rainbow trout (*Oncorhynchus mykiss*) selectively bred for resistance to bacterial cold water disease: effect of rearing environment on survival phenotype. *Aquaculture* 388, 128–136.

Wiens, G.D., Lapatra, S.E., Welch, T.J., Rexroad, C. 3rd, Call, D.R. *et al.* (2014) Complete genome sequence of *Flavobacterium psychrophilum* strain CSF259-93, used to select rainbow trout for increased genetic resistance against bacterial cold water disease. *Genome Announcements* 2(5): e00889-14.

Wiens, G.D., Marancik, D.P., Zwollo, P. and Kaattari, S.L. (2015) Reduction of rainbow trout spleen size by splenectomy does not alter resistance against bacterial cold water disease. *Developmental and Comparative Immunology* 49, 31–37.

Wiklund, T., Madsen, L., Bruun, M.S. and Dalsgaard, I. (2000) Detection of *Flavobacterium psychrophilum* from fish tissue and water samples by PCR amplification. *Journal of Applied Microbiology* 88, 299–307.

Wood, J. (1974) *Diseases of Pacific Salmon, Their Prevention and Treatment*. Washington Department of Fisheries, Olympia, Washington, DC.

Wood, E. and Yasutake, W. (1956) Histopathology of fish: III. Peduncle (cold-water) disease. *The Progressive Fish-Culturist* 18, 58–61.

Wood, E. and Yasutake, W. (1957) Histopathology of fish. V. Gill disease. *The Progressive Fish-Culturist* 19, 7–13.

Yeh, H.Y., Shoemaker, C. and Klesius, P. (2006) Sensitive and rapid detection of *Flavobacterium columnare* in channel catfish *Ictalurus punctatus* by a loop-mediated isothermal amplification method. *Journal of Applied Microbiology* 100, 919–925.

Yu, H., Tan, S., Zhao, H. and Li, H. (2013) MH-DAB gene polymorphism and disease resistance to *Flavobacterium columnare* in grass carp (*Ctenopharyngodon idellus*). *Gene* 526, 217–222.

17 Francisella noatunensis

ESTEBAN SOTO[1]* AND JOHN P. HAWKE[2]

[1]Department of Medicine and Epidemiology, School of Veterinary Medicine, University of California Davis, Davis, California, USA; [2]Department of Pathobiological Sciences, School of Veterinary Medicine, Louisiana State University, Baton Rouge, Louisiana, USA

17.1 Introduction

After *Piscirickettsia salmonis*, which was the first *Rickettsia*-like organism that was demonstrated to cause disease in fish (Fryer *et al.*, 1992), other *Rickettsia*-like organisms were reported in Nile tilapia (*Oreochromis niloticus*) (Chern and Chao, 1993; Chen *et al.*, 1994), blue-eyed plecostomus (*Panaque suttoni*) (Eigenmann & Eigenmann) and the grouper *Epinephelus melanostigma* (Khoo *et al.*, 1995; Chen *et al.*, 2000; Mauel *et al.*, 2003). Kamaishi *et al.* (2005) amplified and sequenced the 16S rDNA from an agent in frozen kidneys from diseased three-line grunt (*Parapristipoma trilineatum*) from aquaculture farms. The sequence aligned with other eubacterial 16S rDNA sequences with >97% similarity to *Francisella* species. Bacteria were isolated from the internal organs in cysteine heart agar (Difco, USA) supplemented with 1% haemoglobin, and Koch's postulates were fulfilled (Kamaishi *et al.*, 2005). This was the first report of *Francisella* causing disease in cultured fish. Subsequently, *Francisella noatunensis* was identified worldwide as a pathogen in cultured marine and freshwater fishes, as shown in Tables 17.1 and 17.2. In some cases, mortalities exceeded 90%.

Francisella is the only genus within the family *Francisellaceae*, order *Thiotrichales*. These bacteria are small Gram-negative, pleomorphic, non-motile, obligatorily aerobic and asporogenic (Foley and Nieto, 2010). Historically the three recognized species in the genus included *F. tularensis*, *F. novicida* and *F. philomiragia*. All of them can infect humans, with *F. tularensis* being the most pathogenic and the aetiological agent of tularaemia, a potentially fatal, multisystemic zoonotic disease (Petersen *et al.*, 2009; Foley and Nieto, 2010).

Genetic, phenotypic and biochemical differences have been found in *Francisella* isolates from temperate/warm-water fishes (e.g. cultured tilapia; the three-line grunt; and the hybrid striped bass, *Morone saxatilis* × *M. chrysops*), and from cod (*Gadus morhua*) and Atlantic salmon (*Salmo salar*) in the cold waters of Norway and Chile (Mikalsen *et al.*, 2007; Mikalsen and Colquhoun, 2009; Ottem *et al.*, 2009). Because of the degree of difference in the composition between the lipopolysaccharide (LPS) of *Francisella* sp. isolates from tilapia and those from the species *F. tularensis*, Kay *et al.* (2006) described the tilapia isolate as a new species, *F. victoria*. Ottem *et al.* (2007a,b) characterized isolates from Norwegian cod using molecular methods, fatty acid analysis, and biochemical and phenotypic characteristics. Because of genetic differences, they proposed a new species and named it *F. piscicida*. At the same time, other researchers in Norway phenotypically and molecularly characterized isolates from farmed Atlantic cod displaying chronic granulomatous disease (Olsen *et al.*, 2006). These bacteria were also confirmed as a *Francisella*, but the species was distinct from, although closely related to, *F. philomiragia* (Olsen *et al.*, 2006). The isolates from Atlantic cod were described as a new subspecies, *F. philomiragia* subsp. *noatunensis* subsp. nov. (Mikalsen *et al.*, 2007).

Later, Mikalsen and Colquhoun (2009) characterized isolates from diseased farmed tilapia (*Oreochromis* sp.) from Costa Rica, Atlantic salmon from Chile and the three-line grunt from

---

*Corresponding author e-mail: sotomartinez@ucdavis.edu

Table 17.1. *Francisella noatunensis* subsp. *orientalis* in fishes.

Common name	Scientific name	Location	Reference(s)
Tilapia	*Oreochromis* spp.	Brazil	Leal *et al.*, 2014
		Costa Rica	Mikalsen and Colquhoun, 2009; Soto *et al.*, 2009a, 2012c
		Indonesia	Ottem *et al.*, 2009
		Thailand	Nguyen *et al.*, 2016
		UK (England)	Jeffery *et al.*, 2010
		USA	Soto *et al.*, 2011a
		USA (Hawaii)	Soto *et al.*, 2012c, 2013c
Malawi cichlids	*Nimbochromis venustus, Nimbochromis linni, Aulonocara stuartgranti, Placidochromis* sp., *Protomelas* sp., *Naevochromis chryosogaster, Copadichromis mloto, Otopharynx tetrastigma*	Austria	Lewisch, 2014
Three-line grunt	*Parapristipoma trilinineatum*	Japan	Fukuda *et al.*, 2002
Blue-white labido	*Labidochromis caeruleus*	Taiwan	Hsieh *et al.*, 2007
Brown discus	*Symphysodon aequifasciatus*		
Deep-water hap	*Haplochromis electra*		
Electric blue hap	*Sciaenochromis fryeri*		
Elegans	*Pseudotropheus elegans*		
Firebird	*Aulonocara rubescens*		
Frontosa cichlid	*Cyphotilapia frontosa*		
Malawi eyebiter	*Dimidiochromis compressiceps*		
Maylandia zebra	*Pseudotropheus zebra*		
Rhodes's chilo	*Chilotilapia rhoadesii*		
Blue-green damselfish	*Chromis viridis*	USA	Camus *et al.* (2013)
Caesar grunt	*Haemulon carbonarium*		Soto *et al.*, 2014b
Fairy wrasses	*Cirrhilabrus*		Camus *et al.*, 2013
French grunt	*Haemulon flavolineatum*		Soto *et al.*, 2014b
Hybrid striped bass	*Morone saxatilis* × *M. chrysops*	USA	Ostland *et al.*, 2006
Zebrafish	*Danio rerio*	USA (experimental)	Vojtech *et al.*, 2009

Table 17.2. *Francisella noatunensis* subsp. *noatunensis* in fishes and other aquatic organisms.

Common name	Scientific name	Location	Reference
Atlantic cod, wild cod	*Gadus morhua*	Celtic Sea, Denmark, Ireland, Norway, Sweden, UK	Mikalsen *et al.*, 2007; Ruane *et al.*, 2015
Atlantic salmon	*Salmo salar*	Chile, Norway	Birkbeck *et al.*, 2007; Bohle *et al.*, 2009
Atlantic mackerel	*Scomber scombrus*	Norway	Ottem *et al.*, 2008; Wangen *et al.*, 2012
Blue mussel	*Mytilus edulis*		
Crab	*Cancer paguru*		
European plaice	*Pleuronectes platessa*		
Megrim	*Lepidorhombus whiffiagonis*		
Pollock, saithe	*Pollachius virens*		
Zebrafish	*Danio rerio*	Norway (experimental)	Brudal *et al.*, 2014

Japan using phenotypic and molecular taxonomic methods. On the basis of molecular genetics, they proposed that the isolates from Costa Rica and Japan be recognized as *F. asiatica* sp. nov. Because there were no differences between *F. piscicida* and *F. philomiragia* subsp. *noatunensis*, and there was increased evidence of both ecological differentiation within the *F. philomiragia* group and the existence of a specific fish pathogenic clade, they proposed that *F. philomiragia* subsp. *noatunensis* be elevated to species level as *F. noatunensis* comb. nov. sp. Ottem *et al.* (2009) used several housekeeping gene sequences and the phenotypic characterization of isolates from the three-line grunt from Japan and cod from Norway to elevate *F. philomiragia* subsp. *noatunensis* to species rank as *F. noatunensis* comb. nov. *F. piscicida* was considered a heterotypic synonym of *F. noatunensis* comb. nov. These authors also proposed *Francisella* sp. Ehime-1 (from Japan) to represent a novel subspecies, *F. noatunensis* subsp. *orientalis* subsp. nov.

The pathogenic piscine *Francisella* are now divided into two subspecies that correlate with their occurrence in cold-water or warm-water fishes. The cold-water *F. noatunensis* subsp. *noatunensis* (*Fnn*), and the temperate/warm-water, *F. noatunensis* subsp. *orientalis* (*Fno*) are pathogens in cultured and wild fishes. They are Gram-negative, non-motile and highly pleomorphic coccobacilli that range from 0.2 to 0.4 µm in width and 0.4 to 1.9 µm in length. The bacteria are catalase positive and oxidase negative aerobes that require cysteine for growth (Soto *et al.*, 2009a; Birkbeck *et al.*, 2011; Colquhoun and Duodu, 2011).

F. noatunensis is transmitted horizontally (Kamaishi *et al.*, 2005; Nylund *et al.*, 2006; Ostland *et al.*, 2006). However, the bacterium has been observed in fish gonads (Soto *et al.*, 2009a), and outbreaks in fry have been reported (Soto *et al.*, 2009a; Duodu and Colquhoun, 2010), so vertical transmission remains a possibility. Additionally, it is not known whether *Fnn* can multiply outside a host. It has been suggested that *Fnn* and *Fno* enter viable but non-culturable states after weeks in either fresh water or salt water (Duodu and Colquhoun, 2010; Duodu *et al.*, 2012a; Soto and Revan, 2012). Viable but non-culturable organisms do not grow on media, but are metabolically active. Additionally, *Fno* can produce biofilms within 24 h, and the bacteria are viable in biofilms for more than 5 days in fresh, brackish and marine waters below 30°C. Moreover, the biofilm renders *Fno* greater resistance

to hydrogen peroxide, bleach and Virkon®; thus demonstrating its survival importance in the environment (Soto *et al.*, 2015b).

F. philomiragia and *F. tularensis* can infect and replicate in *Acanthamoeba* (Abd *et al.*, 2003; Verhoeven *et al.*, 2010) and *F. noatunensis* subsp. *endociliophora* is an endosymbiont of the ciliate *Euplotes* (Schrallhammer *et al.*, 2011). The survival of *F. noatunensis* in other aquatic organisms is unknown, but *Fnn* DNA has been detected in mussels and crabs near contaminated aquaculture sites (Ottem *et al.*, 2008; Duodu and Colquhoun, 2010). These invertebrates could play a role in trophic epizootiology. Blue mussels (*Mytilus edulis*) are filter feeders and could have acquired the bacteria from the water or from fish carcasses or from epithelial cells, mucus, etc. produced by infected fish. Edible crabs, *Cancer pagurus* could have become infected in water or after feeding on infected mussels (Ottem *et al.*, 2008). However, the viability of the bacteria in such organisms is unknown because the researchers only tested for *Fnn* DNA. It was suggested that the farming of non-clinical carriers and ornamental fishes (see Tables 17.1 and 17.2) could spread the bacterium (Camus *et al.*, 2013); similarly, the importation and transportation of infected commercial fishes could spread the bacteria to new areas (Ottem *et al.*, 2008; Birkbeck *et al.*, 2011).

17.2 Diagnosis of the Infection

Clinical signs of francisellosis are non-specific and vary. Besides a wide-range of mortality rates (from 10 to >90%) in different sizes and ages of fish, abnormal swimming and anorexia are frequently observed. The great variance in mortality rates (5–90%) on tilapia farms is associated with water temperatures, water quality, existing co-infections and handling stress (Soto *et al.*, 2009a, 2012b). Most outbreaks in warm/temperate water fishes occur between 20 and 25° C (Soto *et al.*, 2009a, 2011a, 2014b). Mortality rates in farmed cod range from 5 to 40%, and infection in wild Norwegian cod can approach 13% (Ottem *et al.*, 2008). The water temperatures observed during outbreaks in Norwegian cod farms were 10–15° C (Nylund *et al.*, 2006; Olsen *et al.*, 2006) and in Chilean salmon farms they were 8–10° C (Birkbeck *et al.*, 2007).

The preliminary diagnosis of francisellosis is based on clinical history, mortality patterns, gross and histological findings, usually along with non-specific Giemsa and Gram stains (see Figs 17.1–17.3). Differential diagnoses include mycobacteriosis, nocardiosis, piscirickettsiosis, renibacteriosis, pasteurellosis and edwardsiellosis. Definitive diagnosis can be made by isolation of the organism, molecular diagnosis, serological methods and, ideally, using a combination of at least two diagnostic methods.

Media developed to culture fastidious organisms have been used to isolate *F. noatunensis* (Birkbeck *et al.*, 2011; Colquhoun and Duodu, 2011). Additionally, cell lines that successfully recover *Fnn* include: salmon head kidney, Atlantic salmon kidney cells and the rainbow trout gonad cell line, while *Fno* has been isolated using Chinook salmon embryo cells (Hsieh *et al.*, 2006; Nylund *et al.*, 2006).

Molecular diagnostic methods that identify and characterize *F. noatunensis* include genus-specific conventional PCR (Forsman *et al.*, 1994), multiplex real-time PCR for *Fnn* (Ottem *et al.*, 2008), quantitative (real-time) PCR for *Fno* and *Fnn* (Soto *et al.*, 2010a; Duodu *et al.*, 2012b), and loop-mediated isothermal amplification (LAMP) (Caipang *et al.*, 2010).

17.3 Pathology

Clinical signs include pale or dark skin, haemorrhages (from petechiae to ecchymosis), ulcers with loss of scales, frayed fins, exophthalmia and pale gills with white nodules. There may be granulomatous or pyogranulomatous inflammation, abdominal enlargement with coelomitis, ascites and widespread, multifocal pale tan or cream-coloured nodules dispersed in the heart, liver, gonads, anterior and posterior kidney and spleen, with marked splenomegaly and renomegaly (Fig. 17.1). White nodules also occur in the gills, liver (especially in cod), eye and, sporadically, in the gastrointestinal walls and mesentery fat.

Tissues can have up to 90% parenchymal replacement by coalescing granulomatous inflammation (Soto *et al.*, 2009a; Birkbeck *et al.*, 2011). Well-formed granulomas are observed in severe cases during the microscopic examination of tissue wet-mounts (Fig. 17.2A, B), particularly of the gills, spleen and kidney. Granulomas are characterized by necrotic centres surrounded with a layer of macrophages, often containing numerous 0.5–1.0 μm coccoid bacteria, and an outer layer of fibroblasts, vacuolated cells, lymphocytes and small blood vessels (Fig. 17.2C, D; 17.3). In chronically infected fish, well-formed granulomas containing melanomacrophages are particularly evident in the spleen and head kidney. Livers have small foci of necrosis and bacteria-laden macrophages. Granulomatous inflammation in the trunk kidney is relatively mild, largely sparing the nephrons. The gills have moderate epithelial hyperplasia and lamellar fusion with a few small granulomas and macrophages filled with bacteria. Atrial endothelial hypertrophy is seen in the heart. In severe cases, a widespread cellular infiltrate and the presence of granulomas are observed in the pericardium and myocardium. Granuloma is not common in the brain, and a massive mononuclear inflammatory infiltrate has been reported in severely infected fish (Soto *et al.*, 2009a). Small, pleomorphic coccobacilli are detected inside and outside cells, especially in acute and subacute infections (Fig. 17.3).

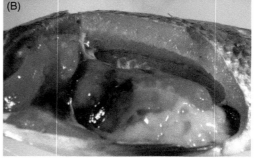

Fig. 17.1. Gross pathology in piscine francisellosis. (A) Splenomegaly and renomegaly with multifocal white nodulations in Atlantic cod (*Gadus morhua*) infected with *Francisella noatunensis* subsp. *noatunensis*. Courtesy of Dr Duncan Colquhoun, National Veterinary Institute, Norway. (B) Nile tilapia (*Oreochromis niloticus*) with *F.n.* subsp. *orientalis*. Courtesy of Dr Juan A. Morales, Universidad Nacional of Costa Rica.

E. Soto and J.P. Hawke

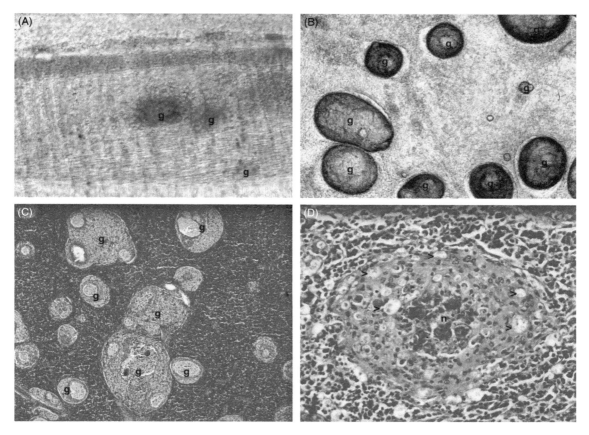

Fig. 17.2. Microscopic findings in *Francisella noatunensis* subsp. *orientalis* infected Nile tilapia (*Oreochromis niloticus*). (A) Gills granulomas in wet mount; 'g' denotes granulomas (haematoxylin and eosin stain, 40× magnification). Courtesy of Dr Juan A. Morales, Universidad Nacional of Costa Rica. (B) Spleen granulomas in wet mount; 'g' denotes granulomas (haematoxylin and eosin stain, 40× magnification). Courtesy of Dr Juan A. Morales, Universidad Nacional of Costa Rica. (C) Widespread multifocal granulomatous lesions with mixed inflammatory infiltrates in the spleen; 'g' denotes granulomas (haematoxylin and eosin stain, 10× magnification). (D) Typical granuloma in the anterior (head) kidney, with a necrotic hyperchromatic core (n) surrounded by epithelioid macrophages, lymphocytes and necrotic cells; the arrowheads denote intracellular coccoid bacteria in head kidney phagocytes (haematoxylin and eosin stain, 40× magnification).

17.4 Pathophysiology, Pathogenesis and Virulence

Soto *et al.* (2013a) studied francisellosis in Nile tilapia after an immersion infection with wild type *Fno*. Surface mucus collected 3 h postexposure contained the highest number of *Fno* genome equivalents. After 96 h fish were septic and presented marked increases of *Fno* genome equivalents in the spleen, anterior kidney, posterior kidney, gills, heart, liver, brain, gonads and the gastrointestinal tract. Also associated with the appearance of the genome equivalents were the granulomas typical of

francisellosis, which were found predominantly in the spleen and anterior and posterior kidney (Soto *et al.*, 2013a). Gjessing *et al.* (2011) studied the tissue and cell dynamics in the spleen of *Fnn* laboratory-infected cod. They found macrophage-like cells and granulocyte-like cells staining for peroxidase and lysozyme within infectious foci in the spleen. Thus, the authors suggested that, at least in cod, francisellosis causes a pyogranulomatous inflammatory response.

F. noatunensis resides and replicates not only in extracellular environments but also inside eukaryotic

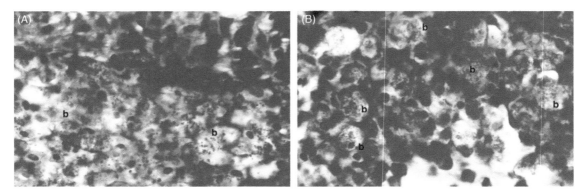

Fig. 17.3. Photomicrographs of granulomatous inflammation in (A) the gills and (B) the spleen of Nile tilapia (*Oreochromis niloticus*) naturally infected with *Francisella noatunensis* subsp. *orientalis*. Note the presence of widespread positively-stained bacteria denoted by 'b' intra- and extracellularly (Giemsa stain, 100× magnification). Courtesy of Dr Juan A. Morales, Universidad Nacional of Costa Rica.

cells, especially those of the monocyte/macrophage lineage (Soto *et al.*, 2010b; Bakkemo *et al.*, 2011; Furevik *et al.*, 2011; Vestvik *et al.*, 2013; Brudal *et al.*, 2014). In tilapia head kidney-derived macrophages, *Fno* replicates effectively inside phagocytes over 72 h and is ultimately cytotoxic. In cod, *Fnn* adheres to, penetrates and replicates inside cells of the monocyte/macrophage lineage and to a lesser extent in neutrophils and B cells (Furevik *et al.*, 2011). Brudal *et al.* (2014) demonstrated that both neutrophils and macrophages take up *F. noatunensis* in zebrafish (*Danio rerio*) embryos, in which macrophages are the main site of replication.

The genes in *Francisella* pathogenicity islands (FPIs; a cluster of 16–19 genes that is duplicated in most of the *Francisella* genomes that have been sequenced) appear to encode virulence factors that mediate the intracellular survival of the bacteria (Sridhar *et al.*, 2012; Sjödin *et al.*, 2012). The *iglABCD* operon is important for virulence determination in *Fno* and necessary for the induction of disease and intra-macrophage survival in tilapia and zebrafish (Soto *et al.*, 2009b, 2010b; Hansen personal communication). Little is known of the mutations associated with virulence within the FPI of *F. noatunensis* (Soto *et al.*, 2009b, 2010b; Hansen *et al.*, 2013). Soto *et al.* (2009b, 2010b) described the attenuation of *Fno* by mutation of the *iglC* gene, and in the earlier study of the two listed, reported 100% mortality in tilapia intraperitoneally injected with ~1.5 × 10⁸ colony forming units (cfu)/ml wild type *Fno*. Fish infected with the wild type strain (LADL 07-285A) died after only

48 h, while only one fish died in the 30 days post challenge via intraperitoneal injection with an *iglC* mutant. When immersion challenged with the wild type strain (1.8–3.7 × 10⁷ cfu/ml), ~50% of the fish survived, whereas 100% survived when challenged with the attenuated mutant. Soto *et al.* (2010b) found that mutation of the *iglC* gene had no impact on the susceptibility of *Fno* to complement lysis. However, the *iglC* gene was identified as important for intra-macrophage and intra-endothelial cell replication (Soto *et al.*, unpublished). While the macrophages and epithelial cells internalized both the wild type and the mutant bacteria, the mutant did not replicate. Mutation of the *iglC* gene also reduced cytotoxicity and apoptosis to macrophages (Soto *et al.*, 2010b). The *iglC* gene is one of the most extensively studied genes within the FPI owing to its high expression during intracellular growth, so demonstrating its importance for pathogenicity and virulence (Nano and Schmerk, 2007).

Hansen *et al.* (2013) demonstrated that mutation of the *pdpA* gene (for the pathogenicity determinant protein A, PdpA) in *Fno* resulted in significant attenuation in the zebrafish model. The gene is approximately 2.4k bp in length and is one of the largest open reading frames in the FPI (Nano and Schmerk, 2007). It is located at the beginning of a putative operon containing the *pdpB*, *vgrG* and *dotU* genes. The role and function of *pdpA* is poorly understood and research continues on this. Much of our understanding of the gene comes from studies on *F. novicida* and *F. tularensis* (Nano *et al.*, 2004; Schmerk *et al.*, 2009a,b; Chou *et al.*, 2013).

E. Soto and J.P. Hawke

The gene is conserved throughout the latter species, with more than 97% similarity (Schmerk et al., 2009a,b). Although it is not suspected to be an integral assembly protein in the type 6-secretion system, PdpA may function as a chaperone component or may be an effector protein (Schmerk et al., 2009b; Bröms et al., 2012a,b). Nano et al. (2004) demonstrated that mice did not die after intradermal injections with up to 10^7 cfu/ml of the F. novicida pdpA mutant, while the LD_{50} of the wild type strain was ~2–5 × 10^2 cfu/ml. Additionally, Schmerk et al. (2009b) showed that the pdpA mutant in F. novicida did not escape the phagosomes of the infected macrophages and exhibited highly attenuated virulence in chicken embryos and mice.

The Fno pdpA gene encodes the F. tularensis pathogenicity determinant protein A-homologue. In F. tularensis, PdpA is necessary for intracellular growth and virulence; however, the role of the Fno pdpA gene in the pathogenesis of piscine francisellosis is unknown. The role of the type VI secretion system in the pathogenicity of bacteria has been well established as an important mechanism for virulence. The type VI secretion system delivers macromolecules from the bacterial cytoplasm into a host cell, traversing both bacterial and eukaryotic membranes through conserved substrates that resemble the needle complex of other secretion systems (de Bruin et al., 2007; Nano and Schmerk, 2007). The Francisella spp. type VI secretion system is encoded within the FPI, and deletion of the core components of the system results in the inability of the bacterium to escape the phagosome and replicate intracellularly (Bröms et al., 2012a,b). Further, even though the core components of the Francisella type VI secretion system do not share a structural similarity with typical type VI secretion proteins, their genetic location and secretion patterns indicate that they may share similar functions (Ludu et al., 2008; Bröms et al., 2012a,b).

Recently, Farrell (2015) studied the virulence of two different marker-based Fno pdpA mutants (ΔpdpA-1 and ΔpdpA-2) generated in opposing polarity, along with the wild type parent strain, by immersion infections in hybrid red tilapia (Oreochromis sp.). Both mutant strains were highly attenuated and their LD_{50} values were >2 × 10^6 cfu/ml water, whereas the LD_{50} of the wild strain was 891 cfu/ml water. Compared with the ΔpdpA mutants, the wild type strain was more resistant to oxidative killing by hydrogen peroxide, but all strains had similar susceptibilities to sodium hypochlorite and serum-mediated lysis. This study further substantiates the role of the pdpA gene product as an important virulence factor in Fno and confirms the hypothesis that mutations in this gene result in attenuation.

Little is known of the immunopathological response in fish. The Fnn LPS appears to stimulate a modest immune response upon challenge to cod phagocytes (Bakkemo et al., 2011). Additionally, a high IL-10 response in cod indicated that Fnn could drive a stronger immune response towards a non-completely effective Th2 (T helper cell type 2) response (Bakkemo et al., 2011). A weak (5–10 fold) increase in the expression of the interleukins IL-1β and IL-8 was measured after cod macrophage infection with Fnn (Bakkemo et al., 2011). Moreover, Fnn inhibits the respiratory burst in cod leucocytes, a trait used by intracellular organisms to survive in the phagolysosomes of phagocytic cells such as macrophages (Vestvik et al., 2013).

These studies show that F. noatunensis can live intracellularly by modulating host immunity, although this response may vary in different species. In vivo analysis of immune-related gene expression in zebrafish infected with Fno indicated that the zebrafish mount a significant tissue-specific pro-inflammatory response by upregulating IL-1β, gamma interferon (IFNγ) and tumour necrosis factor alpha (TNFα) mRNA within 6 h postinfection that persists for 7 days (Vojtech et al., 2009).

17.5 Protective and Control Strategies

The development of vaccines, treatments and other strategies to prevent and control piscine francisellosis must take into account the epizootiology of F. noatunensis and the pathogenesis of piscine francisellosis. The facultative intracellular nature of the bacteria and their capacity to form biofilm and viable non-culturable forms need to be considered in the context of prophylaxis and therapeutic success. Currently, there are no commercially available vaccines and few treatments have been reported against piscine francisellosis, though a protective immune response has been reported in laboratory-immunized and challenged fish. The production of antibodies to F. noatunensis has been achieved in cod and tilapia (Schröder et al., 2009; Soto et al., 2011b), but Fnn LPS stimulated a modest immune response upon challenge to cod phagocytes, which might subrogate the immune response (Bakkemo et al., 2011). Recent findings suggest that mucosal and serum antibodies play important roles in protection (Soto *et al.*, 2015a) by binding to specific adhesins on pathogens, agglutinating them and thereby preventing attachment to target cells. They opsonize the pathogens for phagocytes, and also activate the classical pathway of complement and potentiate antibody-mediated cell killing. Antibodies from vaccinated adult tilapia provided significant protection ($P < 0.001$) in passively immunized fish challenged with 10^4 and 10^5 cfu/fish of wild type *Fno*; but not when challenged with 10^6 cfu/fish (Soto *et al.*, 2011b).

A strong cell-mediated immune response prevents *Francisella* spp. infection in vertebrates (Kirimanjeswara *et al.*, 2008; Chou *et al.*, 2013). Consequently, adjuvants will probably be needed in killed or subunit vaccines to stimulate cell-mediated immunity (Soto *et al.*, 2011b; Brudal *et al.*, 2015). Another possibility is to use live-attenuated vaccines that persist in tissues and stimulate protective immunity (Soto *et al.*, 2011b, 2014a; Brudal *et al.*, 2015). Soto *et al.* (2009b, 2011b, 2014a) demonstrated that an insertional mutation in the *iglC* gene of *Fno* attenuated the bacteria, and that significant protection was detected in vaccinated tilapia fingerlings when challenged with wild type *Fno* via immersion. The Δ*iglC* protected tilapia against challenge with high doses (LD_{80}) of wild type *Fno* strain. Naive tilapia vaccinated via immersion with Δ*iglC* and subsequently challenged with *Fno* were protected (90% mean percentage survival) from the lethal challenges (Soto *et al.*, 2011b). Even though this was a laboratory study, it demonstrated that live attenuated vaccines have potential future applications. More recently, Brudal *et al.* (2014) developed a vaccine based on 'outer membrane vesicles' that protected zebrafish embryos against pathogenic *Fnn*.

Increasing the water temperature to 30°C prevented the development of clinical signs and mortality. (Soto *et al.*, 2012b) in laboratory infected Nile tilapia and has been successfully used to prevent diseases in ornamental fish (Soto *et al.*, 2014b). Although temperature significantly influenced the development of francisellosis in tilapia, salinity had no effect. This could be applied in indoor cultured facilities where the increasing temperature is feasible or practical.

Antimicrobial susceptibility tests for *F. noatunensis* are based on methods devised for *F. tularensis* (Baker *et al.*, 1985). A broth microdilution test is performed in a medium consisting of a modified Mueller-Hinton II cation adjusted broth (Becton, Dickinson and Company (BD), Sparks, Maryland), supplemented with 2% IsoVitalex (BD BBL medium) and 0.1% glucose. Both subspecies have low minimal inhibitory concentrations (MIC) to florfenicol and oxolinic acid; and high MICs to erythromycin and trimethoprim/sulfamethoxazole. Additionally, *Fno* presents low MICs to enrofloxacin, gentamicin, neomycin, oxytetracycline, tetracycline, streptomycin and nitrofurantoin, but high MICs to penicillin, ampicillin, amoxicillin, ceftiofur, sulfadimethoxine, sulfathiazole, novobiocin, tylosin tartrate and clindamycin (Soto *et al.*, 2012a). Isachsen *et al.* (2012) found that *Fnn* had high MICs to oxytetracycline, ciprofloxacin and streptomycin, but were susceptible to flumequine and rifampin (rifampicin).

Only three antimicrobials are currently approved for use in food fish in the USA. Of these, florfenicol and oxytetracycline are potential therapeutants against francisellosis. Tilapia mortalities (in both natural and experimental infections) were reduced when fish were treated with oxytetracycline and florfenicol in the feed (Mauel *et al.*, 2003; Soto *et al.*, 2010c, 2013b); however rapid treatment is an absolute requirement because infected fish quickly become inappetent. Antimicrobial therapy is particularly successful during acute disease, but it is not as effective for subacute or chronic infections (Soto *et al.*, 2010c, 2013b). We hypothesize that antimicrobials are unable to penetrate the granulomas that form during subacute and chronic infections. Additionally, the ability of drugs to penetrate inside eukaryotic cells must be determined before recommending treatments.

Florfenicol administered at doses of 15 and 20 mg florfenicol/kg body weight daily for 10 days significantly reduced the mortality associated with francisellosis in Nile tilapia (Soto *et al.*, 2013b). Mauel *et al.* (2003) reported a survival of 83% in fish challenged by cohabitation and fed Rangen™ feed fortified with 4.0 g of oxytetracycline per pound (~2.2 kg). In comparison, only 17% of fish in the exposed and non-medicated treatment survived.

Recently, a two-pronged approach that involves temperature manipulation and antimicrobial treatment was recommended (Soto *et al.*, 2014b). The water temperature was gradually increased to 30°C over a 2 week period and maintained for 3 months. Once the water temperature reached 30°C, oxytetracycline-medicated feed was provided at a total dosage of 3 g oxytetracycline/100 lb fish daily for 10 days (Soto *et al.*, 2014b).

It is wise to depopulate fish and to disinfect the facility once an outbreak occurs. Three frequently used disinfectants in aquatic facilities are effective against both the planktonic and biofilm forms of *Fno*. These include 1% Virkon for 10 min, 10% household bleach (~6000 ppm available chlorine) for 10 min or 3% hydrogen peroxide for 10 min *in vitro* (Soto *et al.*, 2015b).

17.6 Suggestions for Future Research

Piscine francisellosis is recognized as one of the most important emergent diseases of cultured fish. Its wide host range, fastidious nature, capacity to survive in multiple environments and global presence highlight the significance of this intriguing pathogen, yet we know very little about it. The environmental persistence and pathogenesis of piscine francisellosis needs further investigation, and prophylactic, therapeutic and eradication protocols are needed with efficacy against planktonic, biofilm and viable non-culturable states of the bacteria.

Vaccines against piscine francisellosis should be based on selecting antigens that could potentially stimulate both mucosal and systemic immune responses. Recent research suggests that effector proteins released via secretion systems show promise; however, inactivated vaccines should stimulate strong cellular immunity. This is also a challenge with other pathogens where adjuvants and injectable methods are required. If live attenuated mutants are to be used to vaccinate tilapia or cod, evidence of attenuation in other fish species should be investigated. Although countries and regions have different regulations concerning safety, the researchers and pharmaceutical companies providing vaccines should utilize stringent policies in order to provide the safest product for the fish, humans and the environment. Vaccines should be highly effective and easy to administer, e.g. via immersion or by oral methods. Antimicrobials should ideally be bactericidal and capable of penetrating granulomas and *in vivo* biofilms. Research on the use of 'organic' products for treatment and prevention, particularly for industries like aquaponics where 'organic' practices are all important, is necessary.

Better practices for biosecurity and the early diagnosis of diseases are also desirable. The detection of the pathogen in subclinically infected, carrier or latently infected fish before transportation should aid in the prevention of new outbreaks and environmental contamination. Similarly, the development of non-lethal diagnostic methods is desirable for testing brood stocks as the vertical transmission of francisellosis remains a possibility. Disinfection and eradication protocols should consider the biology of the bacteria. The selection of resistant stocks of fish, particularly for important cultured fish species, is warranted. The host range of *Francisella*, including reptiles and amphibians, should be determined. The susceptibility of chronically infected, vaccinated and naive fish to co-infection with opportunistic and primary fish pathogens such as species of *Streptococcus*, *Vibrio* and *Aeromonas*, should be investigated because reports of co-infection are common. Finally, a better understanding of the overall cost of piscine francisellosis in the different life stages and species affected should provide strong evidence of the need for future research.

References

Abd, H., Johansson, T., Golovliov, I., Sandström, G. and Forsman, M. (2003) Survival and growth of *Francisella tularensis* in *Acanthamoeba castellanii*. *Applied and Environmental Microbiology* 69, 600–606.

Baker, C.N., Hollis, D.G. and Thornsberry, C. (1985) Antimicrobial susceptibility testing of *Francisella tularensis* with a modified Mueller-Hinton Broth. *Journal of Clinical Microbiology* 22, 212–215.

Bakkemo, K.R., Mikkelsen, H., Bordevik, M., Torgersen, J., Winther-Larsen, H.C. *et al.* (2011) Intracellular localisation and innate immune responses following *Francisella noatunensis* infection of Atlantic cod (*Gadus morhua*) macrophages. *Fish and Shellfish Immunology* 31, 993–1004.

Birkbeck, T.H., Bordevik, M., Frøystad, M.K. and Baklien, A. (2007) Identification of *Francisella* sp. from Atlantic salmon, *Salmo salar* L., in Chile. *Journal of Fish Diseases* 30, 505–507.

Birkbeck, T.H, Feist, S.W. and Verner-Jeffreys, D.W. (2011) *Francisella* infections in fish and shellfish. *Journal of Fish Diseases* 34, 173–187.

Bohle, H., Tapia, E., Martínez, A., Rozas, M. Figueroa, A. *et al.* (2009) *Francisella philomiragia*, bacteria asociada con altas mortalidades en salmones del Atlántico (*Salmo salar*) cultivados en balsas-jaulas en el Lago Llanquihue. *Archivos de Medicina Veterinaria* 41, 237–244.

Bröms, J.E., Meyer, L., Lavander, M., Larsson, P., Sjöstedt, A. *et al.* (2012a) DotU and VgrG, core components of type VI secretion systems, are essential for *Francisella* LVS pathogenicity. *PLoS One* 7(4): e34639.

Bröms, J.E., Meyer, L., Sun, K., Lavander, M. and Sjöstedt, A. (2012b) Unique substrates secreted by the type VI secretion system of *Francisella tularensis* during intramacrophage infection. *PLoS One* 7(11): e50473.

Brudal, E., Ulanova, L.S., O Lampe, E., Rishovd, A.L., Griffiths, G. et al. (2014) Establishment of three Francisella infections in zebrafish embryos at different temperatures. Infection and Immunity 82, 2180–2194.

Brudal, E., Lampe, E.O., Reubsaet, L., Roos, N., Hegna, I.K. et al. (2015) Vaccination with outer membrane vesicles from Francisella noatunensis reduces development of francisellosis in a zebrafish model. Fish and Shellfish Immunology 42, 50–57.

Caipang, C.M., Kulkarni, A., Brinchmann, M.F., Korsnes, K. and Kiron, V. (2010) Detection of Francisella piscicida in Atlantic cod (Gadus morhua L.) by the loop-mediated isothermal amplification (LAMP) reaction. Veterinary Journal 184, 357–361.

Camus, A.C., Dill, J.A., McDermott, A.J., Clauss, T.M., Berliner, A.L. et al. (2013) Francisella noatunensis subsp. orientalis infection in Indo-Pacific reef fish entering the United States through the ornamental fish trade. Journal of Fish Diseases 36, 681–684.

Chen, S.-C., Tung, M.-C., Chen, S.-P., Tsai, J.-F. and Wang, P.-C. (1994) Systematic granulomas caused by a rickettsia-like organism in Nile tilapia, Oreochromis niloticus (L.), from southern Taiwan. Journal of Fish Diseases 17, 591–599.

Chen, S.-C., Wang, P.-C., Tung, M.-C., Thompson, K.D. and Adams, A. (2000) A Piscirickettsia salmonis-like organism in grouper, Epinephelus melanostigma in Taiwan. Journal of Fish Diseases 23, 415–418.

Chern, R.S. and Chao, C.B. (1993) Outbreaks of a disease caused by rickettsia-like organism in cultured tilapias in Taiwan. Fish Pathology 29, 61–71.

Chou, A.Y., Kennett, N.J., Nix, E.B., Schmerk, C.L., Nano, F.E. et al. (2013) Generation of protection against Francisella novicida in mice depends on the pathogenicity protein PdpA, but not PdpC or PdpD. Microbes and Infection 15, 816–827.

Colquhoun, D.J. and Duodu, S. (2011) Francisella infections in farmed and wild aquatic organisms. Veterinary Research 42:47.

de Bruin, O.M., Ludu, J.S. and Nano, F.E. (2007) The Francisella pathogenicity island protein IglA localizes to the bacterial cytoplasm and is needed for intracellular growth. BMC Microbiology 7:1.

Duodu, S. and Colquhoun, D. (2010) Monitoring the survival of fish-pathogenic Francisella in water microcosms. FEMS Microbiology Ecology 74, 534–541.

Duodu, S., Larsson, P., Sjödin, A., Forsman, M. and Colquhoun, D.J. (2012a) The distribution of Francisella-like bacteria associated with coastal waters in Norway. Microbial Ecology 64, 370–377.

Duodu, S., Larsson, P., Sjödin, A., Soto, E., Forsman, M. et al. (2012b) Real-time PCR assays targeting unique DNA sequences of fish-pathogenic Francisella noatunensis subspecies noatunensis and orientalis. Diseases of Aquatic Organisms 101, 225–234.

Farrell, F. (2015) Disruption of the pathogenicity determinant protein A gene (pdpA) in Francisella noatunensis subsp. orientalis results in attenuation and greater susceptibility to oxidative stress. MSc thesis, Ross University School of Veterinary Medicine, Bassaterre, Saint Kitts and Nevis, West Indies.

Foley, J.E. and Nieto, N.C. (2010) Tularemia. Veterinary Microbiology 140, 332–338.

Forsman, M., Sandstrom, G. and Sjostedt, A. (1994) Analysis of 16S ribosomal DNA sequences of Francisella strains and utilization for determination of the phylogeny of the genus and for identification of strains by PCR. International Journal of Systematic Bacteriology 44, 38–46.

Fryer, J.L., Lannan, C.N., Giovannoni, S.J. and Wood, N.D. (1992) Piscirickettsia salmonis gen. nov., sp. nov., the causative agent of an epizootic disease in salmonid fishes. International Journal of Systematic Bacteriology 42, 120–126.

Fukuda, Y., Okamura, A., Nishiyama, M., Kawakami, H., Kamaishi, T. et al. (2002) Granulomatosis of cultured three-line grunt Parapristipoma trilineatum caused by an intracellular bacterium. Fish Pathology 37, 119–124.

Furevik, A., Pettersen, E.F., Colquhoun, D. and Wergeland, H.I. (2011) The intracellular lifestyle of Francisella noatunensis in Atlantic cod (Gadus morhua L.) leucocytes. Fish and Shellfish Immunology 30, 488–494.

Gjessing, M.C., Inami, M., Weli, S.C., Ellingsen, T., Falk, K. et al. (2011) Presence and interaction of inflammatory cells in the spleen of Atlantic cod, Gadus morhua L., infected with Francisella noatunensis. Journal of Fish Diseases 34, 687–699.

Hansen, J.D., Ray, K., Woodson, J.C., Soto, E. and Welch, T.J. (2013) Disruption of the Francisella noatunensis pdpA gene results in virulence attenuation. Fish and Shellfish Immunology 34, 1655.

Hsieh, C.Y., Tung, M.C., Tu, C., Chang, C.D. and Tsai, S.S. (2006) Enzootics of visceral granulomas associated with Francisella-like organism infection in tilapia (Oreochromis sp.) Aquaculture 254, 129–138.

Hsieh, C.Y., Wu, Z.B., Tung, M.C. and Tsai, S.S. (2007) PCR and in situ hybridization for the detection and localization of a new pathogen Francisella-like bacterium (FLB) in ornamental cichlids. Diseases of Aquatic Organisms 75, 29–36.

Isachsen, C.H., Vågnes, O., Jakobsen, R.A. and Samuelsen, O.B. (2012) Antimicrobial susceptibility of Francisella noatunensis subsp. noatunensis strains isolated from Atlantic cod Gadus morhua in Norway. Diseases of Aquatic Organisms 98, 57–62.

Jeffery, K.R., Stone, D., Feist, S.W. and Verner-Jeffreys, D.W. (2010) An outbreak of disease caused by Francisella sp. in Nile tilapia Oreochromis niloticus at a recirculation fish farm in the UK. Diseases of Aquatic Organisms 91, 161–165.

Kamaishi, T., Fukuda, Y. and Nishiyama, M. (2005) Identification and pathogenicity of intracellular Francisella bacterium in three-line grunt Parapristipoma trilineatum. Fish Pathology 40, 67–71.

Kay, W., Petersen, B.O., Duus, J.Ø., Perry, M.B. and Vinogradov, E. (2006) Characterization of the lipopolysaccharide and beta-glucan of the fish pathogen *Francisella victoria*. *FEBS Journal* 273, 3002–3013.

Khoo, L., Dennis, P.M. and Lewbart, G.A. (1995) Rickettsia-like organisms in the blue-eyed plecostomus, *Panaque suttoni* (Eigenmann & Eigenmann). *Journal of Fish Diseases* 18, 157–164.

Kirimanjeswara, G.S., Olmos, S., Bakshi, C.S. and Metzger, D.W. (2008) Humoral and cell-mediated immunity to the intracellular pathogen *Francisella tularensis*. *Immunological Reviews* 225, 244–255.

Leal, C.A.G., Tavares, G.C. and Figueiredo, H.C.P. (2014) Outbreaks and genetic diversity of *Francisella noatunensis* subsp *orientalis* isolated from farm-raised Nile tilapia (*Oreochromis niloticus*) in Brazil. *Genetics and Molecular Research* 13, 5704–5712.

Lewisch, E. (2014) Francisellosis in ornamental African cichlids in Austria. *Bulletin of the European Association of Fish Pathologists* 34, 63–70.

Ludu, J.S., de Bruin, O.M., Duplantis, B.N., Schmerk, C.L., Chou, A.Y. *et al.* (2008) The *Francisella* pathogenicity island protein PdpD is required for full virulence and associates with homologues of the type VI secretion system. *Journal of Bacteriology* 190, 4584–4595.

Mauel, M.J., Miller, D.L. and Frazier, K. (2003) Characterization of a Piscirickettsiosis-like disease in Hawaiian tilapia. *Diseases of Aquatic Organisms* 53, 249–255.

Mikalsen, J. and Colquhoun, D.J. (2009) *Francisella asiatica* sp. nov. isolated from farmed tilapia (*Oreochromis* sp.) and elevation of *Francisella philomiragia* subsp. *noatunensis* to species rank as *Francisella noatunensis* comb. nov., sp. nov. *International Journal of Systematic and Evolutionary Microbiology*. Available as abstract only at: doi:10.1099/ijs.0.002139-0 (accessed 14 November 2016).

Mikalsen, J., Olsen, A.B., Tengs, T. and Colquhoun, D.J. (2007) *Francisella philomiragia* subsp. *noatunensis* subsp. nov., isolated from farmed Atlantic cod (*Gadus morhua* L.). *International Journal of Systematic and Evolutionary Microbiology* 57, 1960–1965.

Nano, F.E. and Schmerk, C. (2007) The *Francisella* pathogenicity island. *Annals of the New York Academy of Sciences* 1105, 122–137.

Nano, F.E., Zhang, N., Cowley, S.C., Klose, K.E., Cheung, K.K. *et al.* (2004) A *Francisella tularensis* pathogenicity island required for intramacrophage growth. *Journal of Bacteriology* 186, 6430–6436.

Nguyen, V.V., Dong, H.T., Senapin, S., Pirarat, N. and Rodkhum, C. (2016) *Francisella noatunensis* subsp. *orientalis*, an emerging bacterial pathogen affecting cultured red tilapia (*Oreochromis* sp.) in Thailand. *Aquaculture Research* 47, 3697–3702.

Nylund, A., Ottem, K.F, Watanabe, K., Karlsbakk, E. and Krossøy, B. (2006) *Francisella* sp. (Family *Francisellaceae*) causing mortality in Norwegian cod (*Gadus morhua*) farming. *Archives of Microbiology* 185, 383–392.

Olsen, A.B., Mikalsen, J., Rode, M., Alfjorden, A., Hoel, E. *et al.* (2006) A novel systemic granulomatous inflammatory disease in farmed Atlantic cod, *Gadus morhua* L., associated with a bacterium belonging to the genus *Francisella*. *Journal of Fish Diseases* 29, 307–311.

Ostland, V.E., Stannard, J.A., Creek, J.J., Hedrick, R.P. and Ferguson, H.W. (2006) Aquatic *Francisella*-like bacterium associated with mortality of intensively cultured hybrid striped bass *Morone chrysops* × *M. saxatilis*. *Diseases of Aquatic Organisms* 72, 135–145.

Ottem, K.F., Nylund, A., Karlsbakk, E., Friis-Møller, A. and Krossøy, B. (2007a) Characterization of *Francisella* sp., GM2212, the first *Francisella* isolate from marine fish, Atlantic cod (*Gadus morhua*). *Archives of Microbiology* 187, 343–350.

Ottem, K.F., Nylund, A., Karlsbakk, E., Friis-Møller, A., Krossøy, B. *et al.* (2007b), New species in the genus *Francisella* (*Gammaproteobacteria*; *Francisellaceae*); *Francisella piscicida* sp. nov. isolated from cod (*Gadus morhua*). *Archives of Microbiology* 188, 547–550.

Ottem, K.F., Nylund, A., Isaksen, T.E., Karlsbakk, E. and Bergh, Ø. (2008) Occurrence of *Francisella piscicida* in farmed and wild Atlantic cod, *Gadus morhua* L., in Norway. *Journal of Fish Diseases* 31, 525–534.

Ottem, K.F., Nylund, A., Karlsbakk, E., Friis-Møller, A. and Kamaishi, T. (2009) Elevation of *Francisella philomiragia* subsp. *noatunensis* Mikalsen *et al.* (2007) to *Francisella noatunensis* comb. nov. (syn. *Francisella piscicida* Ottem *et al.* (2008) syn. nov.) and characterization of *Francisella noatunensis* subsp. *orientalis* subsp. nov., two important fish pathogens. *Journal of Applied Microbiology* 106, 1231–1243.

Petersen, J.M., Carlson, J., Yockey, B., Pillai, S., Kuske, C. *et al.* (2009) Direct isolation of *Francisella* spp. from environmental samples. *Letters in Applied Microbiology* 48, 663–667.

Ruane, N.M., Bolton-Warberg, M., Rodger, H.D., Colquhoun, D.J., Geary, M. *et al.* (2015) An outbreak of francisellosis in wild-caught Celtic Sea Atlantic cod, *Gadus morhua* L., juveniles reared in captivity. *Journal of Fish Diseases* 38, 97–102.

Schmerk, C.L., Duplantis, B.N., Wang, D., Burke, R.D., Chou, A.Y. *et al.* (2009a) Characterization of the pathogenicity island protein PdpA and its role in the virulence of *Francisella novicida*. *Microbiology* 155, 1489–1497.

Schmerk, C.L., Duplantis, B.N., Howard, P.L. and Nano, F.E. (2009b) A *Francisella novicida pdpA* mutant exhibits limited intracellular replication and remains associated with the lysosomal marker LAMP-1. *Microbiology* 155, 1498–1504.

Schrallhammer, M., Schweikert, M., Vallesi, A., Verni, F. and Petroni, G. (2011) Detection of a novel subspecies of *Francisella noatunensis* as an endosymbiont of the ciliate *Euplotes raikovi*. *Microbial Ecology* 61, 455–464.

Schrøder, M.B., Ellingsen, T., Mikkelsen, H., Norderhus, E.A. and Lund, V. (2009) Comparison of antibody responses in Atlantic cod (*Gadus morhua* L.) to *Vibrio anguillarum*, *Aeromonas salmonicida* and *Francisella* sp. *Fish and Shellfish Immunology* 27, 112–119.

Sjödin, A., Svensson, K., Ohrman, C., Ahlinder, J., Lindgren, P. et al. (2012) Genome characterisation of the genus *Francisella* reveals insight into similar evolutionary paths in pathogens of mammals and fish. *BMC Genomics* 13:268.

Soto, E. and Revan, F. (2012) Culturability and persistence of *Francisella noatunensis* subsp. *orientalis* (syn. *Francisella asiatica*) in sea- and freshwater microcosms. *Microbial Ecology* 63, 398–404.

Soto, E., Hawke, J.P., Fernandez, D. and Morales, J.A. (2009a) *Francisella* sp., an emerging pathogen of tilapia, *Oreochromis niloticus* (L.) in Costa Rica. *Journal of Fish Diseases* 32, 713–722.

Soto, E., Fernandez, D. and Hawke, J.P. (2009b) Attenuation of the fish pathogen *Francisella* sp. by mutation of the *iglC* gene. *Journal of Aquatic Animal Health* 21, 140–149.

Soto, E., Bowles, K., Fernandez, D. and Hawke, J.P. (2010a) Development of a real-time PCR assay for identification and quantification of the fish pathogen *Francisella noatunensis* subsp. *orientalis*. *Diseases of Aquatic Organisms* 89, 199–207.

Soto, E., Fernandez, D., Thune, R. and Hawke, J.P. (2010b) Interaction of *Francisella asiatica* with tilapia (*Oreochromis niloticus*) innate immunity. *Infection and Immunity* 78, 2070–2078.

Soto, E., Endris, R.G. and Hawke, J.P. (2010c) *In vitro* and *in vivo* efficacy of florfenicol for treatment of *Francisella asiatica* infection in tilapia. *Antimicrobial Agents and Chemotherapy* 54, 4664–4670.

Soto, E., Baumgartner, W., Wiles, J. and Hawke, J.P. (2011a) *Francisella asiatica* as the causative agent of piscine francisellosis in cultured tilapia (*Oreochromis* sp.) in the United States. *Journal of Veterinary Diagnostic Investigation* 23, 821–825.

Soto, E., Wiles, J., Elzer, P., Macaluso, K. and Hawke, J.P. (2011b) Attenuated *Francisella asiatica iglC* mutant induces protective immunity to francisellosis in tilapia. *Vaccine* 29, 593–598.

Soto, E., Griffin, M., Wiles, J. and Hawke, J.P. (2012a) Genetic analysis and antimicrobial susceptibility of *Francisella noatunensis* subsp. *orientalis* (syn. *F. asiatica*) isolates from fish. *Veterinary Microbiology* 154, 407–412.

Soto, E., Abrams, S.B. and Revan, F. (2012b) Effects of temperature and salt concentration on *Francisella noatunensis* subsp. *orientalis* infections in Nile tilapia *Oreochromis niloticus*. *Diseases of Aquatic Organisms* 101, 217–223.

Soto, E., Illanes, O., Hilchie, D., Morales, J.A., Sunyakumthorn, P. et al. (2012c) Molecular and immunohistochemical diagnosis of *Francisella noatunensis* subsp. *orientalis* from formalin-fixed, paraffin-embedded tissues. *Journal of Veterinary Diagnostic Investigation* 24, 840–845.

Soto, E., Kidd, S., Mendez, S., Marancik, D., Revan, F. et al. (2013a) *Francisella noatunensis* subsp. *orientalis* pathogenesis analyzed by experimental immersion challenge in Nile tilapia, *Oreochromis niloticus* (L.). *Veterinary Microbiology* 164, 77–84.

Soto, E., Kidd, S., Gaunt, P.S. and Endris, R. (2013b) Efficacy of florfenicol for control of mortality associated with *Francisella noatunensis* subsp. *orientalis* in Nile tilapia, *Oreochromis niloticus* (L.). *Journal of Fish Diseases* 36, 411–418.

Soto, E., McGovern-Hopkins, K., Klinger-Bowen, R., Fox, B.K., Brock, J. et al. (2013c) Prevalence of *Francisella noatunensis* subsp. *orientalis* in cultured tilapia on the island of Oahu, Hawaii. *Journal of Aquatic Animal Health* 25, 104–109.

Soto, E., Brown, N., Gardenfors, Z.O., Yount, S., Revan, F. et al. (2014a) Effect of size and temperature at vaccination on immunization and protection conferred by a live attenuated *Francisella noatunensis* immersion vaccine in red hybrid tilapia. *Fish and Shellfish Immunology* 41, 593–599.

Soto, E., Primus, A.E., Pouder, D.B., George, R.H., Gerlach, T.J. et al. (2014b) Identification of *Francisella noatunensis* in novel host species French grunt (*Haemulon flavolineatum*) and Caesar grunt (*Haemulon carbonarium*). *Journal of Zoo and Wildlife Medicine* 45, 727–731.

Soto, E., Tobar, J. and Griffin, M. (2015a). Mucosal vaccines. In: Beck, B.H. and Peatman, E. (eds) *Mucosal Health in Aquaculture*. Academic Press/Elsevier, London/San Diego, California/Waltham, Massachusetts/Kidlington, UK, pp. 297–323.

Soto, E., Halliday-Simmonds, I., Francis, S., Kearney, M.T. and Hansen, J.D. (2015b) Biofilm formation of *Francisella noatunensis* subsp. *orientalis*. *Veterinary Microbiology* 181, 313–317.

Sridhar, S., Sharma, A., Kongshaug, H., Nilsen, F. and Jonassen, I. (2012) Whole genome sequencing of the fish pathogen *Francisella noatunensis* subsp. *orientalis* Toba04 gives novel insights into *Francisella* evolution and pathogenicity. *BMC Genomics* 13:598.

Verhoeven, A.B., Durham-Colleran, M.W., Pierson, T., Boswell, W.T. and Van Hoek, M.L. (2010) *Francisella philomiragia* biofilm formation and interaction with the aquatic rotest *Acanthamoeba castellanii*. *The Biological Bulletin* 219, 178–188.

Vestvik, N., Rønneseth, A., Kalgraff, C.A., Winther-Larsen, H.C., Wergeland, H.I. et al. (2013) *Francisella noatunensis* subsp. *noatunensis* replicates within Atlantic cod (*Gadus morhua* L.) leucocytes and inhibits respiratory burst activity. *Fish and Shellfish Immunology* 35, 725–733.

Vojtech, L.N., Sanders, G.E., Conway, C., Ostland, V. and Hansen, J.D. (2009) Host immune response and acute disease in a zebrafish model of *Francisella* pathogenesis. *Infection and Immunity* 77, 914–925.

Wangen, I.H., Karlsbakk, E., Einen, A.C., Ottem, K.F., Nylund, A. et al. (2012) Fate of *Francisella noatunensis*, a pathogen of Atlantic cod *Gadus morhua*, in blue mussels *Mytilus edulis*. *Diseases of Aquatic Organisms* 98, 63–72.

18 *Mycobacterium* spp.

DAVID T. GAUTHIER[1]* AND MARTHA W. RHODES[2]

[1]*Department of Biological Sciences, Old Dominion University, Norfolk, Virginia, USA;* [2]*Department of Aquatic Health Sciences, Virginia Institute of Marine Science, The College of William and Mary, Gloucester Point, Virginia, USA*

18.1 Introduction

18.1.1 *Mycobacterium* spp.

Members of the genus *Mycobacterium* (Order *Actinomycetales*, Family *Mycobacteriaceae*) are aerobic to microaerophilic, non-motile, rod-shaped bacteria that stain Gram positive and acid fast. Excepting non-culturable species (e.g. *M. leprae*), mycobacteria are frequently grouped by the phenotypic characters of growth rate and pigmentation (Runyon, 1959). Runyon Groups I–III are fastidious and take more than 5 days to produce colonies on solid media. Group I mycobacteria are photochromogenic, producing yellow–orange pigment (Fig. 18.1) on exposure to light, and include species such as *M. marinum*, *M. pseudoshottsii* and *M. kansasii*. Group II mycobacteria are scotochromogenic, producing pigment regardless of light exposure, and include species such as *M. gordonae* and *M. scrofulaceum*. Group III mycobacteria are non-pigmented, and include many of the notable pathogens in the genus, including *M. tuberculosis*, *M. avium*, *M. ulcerans* and the fish pathogen *M. shottsii*. Runyon Group IV mycobacteria are 'fast-growing', taking <5 days to produce colonies on agar. This group is usually non-pigmented (Fig. 18.1) or 'late-pigmenting', and includes pathogens such as *M. fortuitum*, *M. chelonae*, *M. abscessus* and *M. peregrinum*.

Although the Runyon characterization is still used, molecular taxonomy has revealed its shortcomings. For example, *M. szulgai* may be photochromogenic or scotochromogenic depending on the growth temperature, and *M. ulcerans* is non-pigmented, while the closely related *M. pseudoshottsii*, a proposed ecovar of *M. ulcerans* (Doig *et al.*, 2012), is pigmented. Also, growth rate is more closely correlated with phylogeny; Group I–III mycobacteria and Group IV mycobacteria form distinct clades, and a distinctive deletion in the 16S rRNA gene, as well as the presence of two rRNA operons, is characteristic of the latter (Bercovier *et al.*, 1986; Rogall *et al.*, 1990; Menendez *et al.*, 2002). Multilocus sequencing, especially of 16S rRNA and of the genes *hsp*65 (heat-shock protein, 65 kDa), *rpo*B (RNA polymerase β) and *erp* (external repeated protein), is an effective methodology for the identification of known species (Devulder *et al.*, 2005; Gauthier *et al.*, 2011).

The mycobacteria include several significant pathogens of humans: *M. tuberculosis* and *M. bovis*, which cause tuberculosis; *M. leprae*, the cause of Hansen's disease or leprosy; and *M. ulcerans*, the cause of Buruli ulcer, an emerging disease in West Africa and Australia (van der Werf *et al.*, 2005). Mycobacteria also infect other animal hosts and occupy a variety of environmental niches. Species such as *M. vanbaalenii* and *M. hodleri* are associated with hydrocarbon-contaminated soils and degrade polycyclic aromatic hydrocarbons (Kleespies *et al.*, 1996; Khan *et al.*, 2002). Other environmental mycobacteria, such as *M. terrae* and *M. gordonae*, are occasionally implicated in disease among immunocompromised animals and humans (Bonnet *et al.*, 1996; Carbonara *et al.*, 2000). Members of the *M. avium–intracellulare–scrofulaceum* (MAIS) group of mycobacteria, in particular, are significant opportunistic pathogens of persons with AIDS and other forms of immunocompromise (Kirschner *et al.*, 1992).

18.1.2 Mycobacteria in fish: aetiological agents

Mycobacteriosis is a major cause of morbidity and mortality in fish worldwide. *M. marinum*, *M. fortuitum*

---

*Corresponding author e-mail: dgauthie@odu.edu

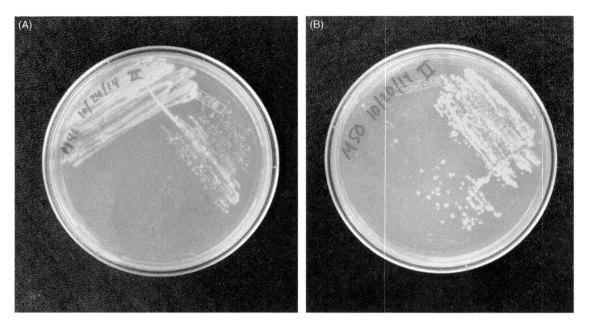

Fig. 18.1. (A) Fast-growing non-pigmented *Mycobacterium chelonae*; (B) slow-growing photochromogenic *Mycobacterium marinum*.

and *M. chelonae* are the most commonly reported aetiological agents, although recent research has demonstrated a growing diversity of *Mycobacterium* spp. that cause diseases in fish (e.g. Gauthier and Rhodes, 2009; Jacobs *et al.*, 2009a).

M. marinum, the most frequently described piscine mycobacterium, was isolated in 1926 from several marine aquarium fish (Aronson, 1926). Infections have been described worldwide from fresh and salt water, as well as warm- and cold-water fish. *M. marinum* is a fastidious photochromogen (Runyon Type I) that is restricted to growth at <35°C on primary isolation, but may adapt to higher temperatures in culture. It produces rough to smooth colonies on solid agar and is differentiated biochemically from other photochromogenic slow-growing mycobacteria by negative nitrate reduction, positive urease production and Tween-20 hydrolysis (Leão *et al.*, 2004).

M. fortuitum, first isolated from a freshwater aquarium fish (*Hyphessobrycon innesi*) in 1953 (Ross and Brancato, 1959), has a wide host, temperature and salinity range, similar to *M. marinum*. *M. fortuitum* is a non-chromogenic fast-growing mycobacterium, which is differentiated from other Runyon Group IV mycobacteria by a positive arylsulfatase reaction, nitrate reduction, NaCl tolerance and

negative mannitol utilization (Leão *et al.*, 2004). *M. peregrinum*, a closely related mycobacterium at times considered as a subspecies of *M. fortuitum*, is a pathogen in cultured zebrafish (*Danio rerio*) (Kent *et al.*, 2004).

M. chelonae (syn. *M. chelonei*), originally isolated in 1903 from a turtle (*Chelona corticata*, nom. obl.; now *Caretta caretta*) is a pathogen of salmonids worldwide (Ashburner, 1977; Grange, 1981; Bruno *et al.*, 1998). Infections have also been reported in other fish, including laboratory-reared zebrafish (Whipps *et al.*, 2008). *M. chelonae* is a Runyon Type IV mycobacterium and is differentiated from others in this group by a positive arylsulfatase reaction, negative nitrate reduction and a lack of tolerance to NaCl (Leão *et al.*, 2004). *M. abscessus* was considered a subspecies of *M. chelonae* until it was elevated to a species (Kusunoki and Ezaki, 1992), and its host range is similar to that of *M. chelonae*. *M. salmoniphilum* (ex Ross, 1960) was recently re-elevated to species status after being a subspecies of *M. chelonae* (Whipps *et al.*, 2007a). Primarily a pathogen of salmonids, *M. salmoniphilum* also infects burbot (*Lota lota*) and sturgeon (*Acipenser gueldenstaedtii*) (Berg *et al.*, 2013; Righetti *et al.*, 2014).

Two other *Mycobacterium* spp., *M. shottsii* and *M. pseudoshottsii*, were described from Chesapeake

Bay, an estuary lying inland on the mid-Atlantic east coast of the USA. Dermal and internal granulomatous lesions containing acid-fast bacteria, consistent with piscine mycobacteriosis, were first observed in striped bass (*Morone saxatilis*) from Chesapeake Bay in 1997 (Vogelbein *et al.*, 1998; Pieper, 2006). Initial attempts to culture *Mycobacterium* spp. from striped bass with acid-fast bacteria were unsuccessful, but subsequently long-term culture was performed on splenic homogenates. A novel very slow-growing isolate was described by Rhodes *et al.* (2001) and was later named *M. shottsii* (Rhodes *et al.*, 2003); an additional isolate, also from Chesapeake Bay striped bass, and provisionally named '*M. chesapeaki*', was described at about the same time (Heckert *et al.*, 2001); it is likely that these two species are synonymous (Gauthier and Rhodes, 2009). *M. pseudoshottsii*, a second, related, very slow-growing isolate from Chesapeake Bay striped bass, was described in 2005 (Rhodes *et al.*, 2005). Both *M. pseudoshottsii* and *M. shottsii* have high culture- and PCR-based prevalence in Chesapeake Bay striped bass, and appear to be the major aetiological agents of mycobacteriosis in this system (Rhodes *et al.*, 2004; Gauthier *et al.*, 2008b).

M. shottsii and *M. pseudoshottsii* have also been detected in striped bass at other US east coast locations, as well as in the related white perch (*Morone americana*) (Ottinger *et al.*, 2007; Stine *et al.*, 2009). Isolates similar to *M. pseudoshottsii* have also been reported from marine fish in the Red Sea (Colorni, 1992; Ranger *et al.*, 2006) and from the Japanese yellowtail (*Seriola quinqueradiata*) (Kurokawa *et al.*, 2013). PCR-based studies have also detected this bacterium in menhaden (*Brevoortia tyrannus*) and Bay anchovy (*Anchoa mitchilli*) from Chesapeake Bay (Gauthier *et al.*, 2010).

M. shottsii and *M. pseudoshottsii* are similar (>99%) to *M. marinum* in their 16S rRNA gene, and are biochemically similar to each other with the exception of lower heat tolerance (<30°) in *M. shottsii* and positive niacin production in both species (Rhodes *et al.*, 2005). *M. shottsii* is nonpigmented (Type III), whereas *M. pseudoshottsii* is a photochromogen (Type I). Genomic analysis indicates that *M. pseudoshottsii* is basal to the *M. ulcerans* clade, and shares genomic features of *M. ulcerans*, including a high copy number of the insertion sequence IS2404 and the presence of a megaplasmid of >170 kb, which encodes the mycolactone toxin. These similarities suggest that *M. pseudoshottsii* should be considered an ecovar

of *M. ulcerans* (Doig *et al.*, 2012), although *M. pseudoshottsii* retains its species standing in the current nomenclature. *M. ulcerans* has been reported to be widely distributed in south-eastern Louisiana (Hennigan *et al.*, 2013), though this study used IS2404 and a limited set of additional loci that were unable to discriminate the human *M. ulcerans* pathogen and proposed ecovars (such as *M. pseudoshottsii*) that are primarily isolated from poikilotherms and have uncertain virulence to humans. Indeed, the IS2404 sequence variants are present in organisms that are biochemically and genetically identified as *M. marinum* (Gauthier *et al.*, 2010), so its detection should not be taken as definitive evidence of human-pathogenic *M. ulcerans*. *M. ulcerans* and *M. pseudoshottsii* produce different forms of mycolactone toxin, with the mycolactone F from *M. pseudoshottsii* producing significantly fewer cytopathic effects in mouse tissue culture than the mycolactone A/B from *M. ulcerans* (Mve-Obiang *et al.*, 2003; Mve-Obiang *et al.*, 2005). Further, there is no evidence that *M. pseudoshottsii* infects humans, and *M. ulcerans* isolates from humans appear to have little virulence in fish (Mosi *et al.*, 2012).

Preliminary genomic analysis of *M. shottsii* (type strain M175) indicates that it is a highly derived variant of *M. marinum*, which bears little resemblance to *M. pseudoshottsii* in its differences from generalist *M. marinum* strains. *M. shottsii* does not possess IS2404, a mycolactone plasmid or other *M. ulcerans*-clade genomic changes, and contains several unique insertion sequences that distinguish it from *M. marinum*. *M. shottsii* and *M. pseudoshottsii* also appear to be ecologically different in that *M. pseudoshottsii* is widely distributed in the water and sediments of Chesapeake Bay, whereas *M. shottsii* has only been isolated from *Morone* spp. (Gauthier *et al.*, 2010).

18.1.3 Transmission

Nigrelli and Vogel (1963) cited 151 fish species susceptible to *Mycobacterium* spp., which suggests that most teleosts are susceptible to infection with these bacteria, if not to disease. However, the transmission of mycobacteria is still poorly understood. Early anecdotal observations on hatchery salmonids implicated oral transmission as juvenile fish fed with offal of infected adults became diseased (Wood and Ordal, 1958; Ross *et al.*, 1959). Additionally, contaminated *Tubifex* worms were implicated in infections of aquarium fish (Nenoff and Uhlemann, 2006). The aqueous (or particle-borne) transmission of *M. marinum* to striped bass was suggested by Gauthier *et al.* (2003), and zebrafish embryos may be infected with *M. marinum* via aqueous exposure (Davis *et al.*, 2002). Presumptive waterborne transmission between systems containing hybrid striped bass (*Morone saxatilis* × *M. chrysops*) infected and uninfected with *M. marinum* was also reported (Li and Gatlin, 2005).

The potential for both aqueous and dietary transmission of *M. marinum* and *M. peregrinum* to zebrafish was demonstrated by Hariff *et al.* (2007). Oral transmission has been confirmed for *M. marinum* and *M. chelonae* in zebrafish, and was enhanced by feeding them with infected *Paramecium caudatum* (Peterson *et al.*, 2013). The vertical transmission of mycobacteria has not yet been clearly demonstrated, with the possible exception of transovarian transmission in live-bearing fish (Conroy, 1966). Ross and Johnson (1962) were not able to demonstrate vertical transmission in Chinook salmon (*Onchorhynchus tshawytscha*), but transmission from parent to offspring of the same species was indicated by Ashburner (1977). Mycobacteria were not evident within gametes, and the possibility of waterborne infection from surface-contaminated milt and/or ovarian fluid could not be ruled out. Mycobacteria have also been cultured from the gametes and gametic fluid of striped bass, but it was unclear whether the bacteria were actually present within gametes (Stine *et al.*, 2006).

18.1.4   Impacts

Mortality and morbidity associated with mycobacteriosis is frequently observed in aquaculture (Hedrick *et al.*, 1987; Bruno *et al.*, 1998), especially in conjunction with stress and/or poor water quality. The mortality is usually chronic, with occasional episodes of higher mortality (e.g. Bruno *et al.*, 1998). Acute, high-level mortality occurs particularly in warm-water aquaria stocked at high fish densities (Whipps *et al.*, 2007b). Significant disease in wild fish is less common, and the aetiological agents concerned are rarely confirmed. Lund and Abernethy (1978) reported an 8% prevalence of granulomatous lesions with acid-fast bacilli in mountain whitefish (*Prosopium williamsoni*) from the Yakima River in Washington state. Presumptive mycobacteriosis was observed among north-east Atlantic mackerel (*Scomber scombrus*) (MacKenzie,

1988), where prevalence increased slowly with age, attaining >90% in older fish. A high prevalence of visceral mycobacteriosis (25–80%) was reported in Pacific coast striped bass (Sakanari *et al.*, 1983). Outbreaks of disease associated with *M. chelonae* have also been observed in wild yellow perch (*Perca flavescens*) in Alberta, Canada (Daoust *et al.*, 1989) that had >80% visceral prevalence and 5–25% prevalence of dermal lesions.

A high prevalence (>50%) of mycobacteriosis has been consistently detected in striped bass from the Chesapeake Bay (Cardinal, 2001; Overton *et al.*, 2003; Gauthier *et al.*, 2008a). In striped bass from the mainstem of Chesapeake Bay, visceral mycobacteriosis increased with age in male and female fish until 5–6 years of age. Prevalence remained high (>70%) for males aged 7+ years, but decreased considerably in females (to ~20–30%) (Gauthier *et al.*, 2008a). The analysis of tag-recapture data indicated that natural (non-fishing) mortality of striped bass in the Bay increased after 1999 (Jiang *et al.*, 2007), and modelling of the apparent prevalence data supports disease-associated mortality (Gauthier *et al.*, 2008a). A tag-recapture study examining the survival of fish with and without dermal lesions had similar results (Sadler *et al.*, 2012), indicating that mycobacteriosis may have an impact on survival in this wild stock.

18.2   Diagnosis

Piscine mycobacteriosis ranges from an acute, fulminant disease with significant bacteraemia (e.g. Wolf and Smith, 1999; Whipps *et al.*, 2007b), to the more frequently observed chronic forms characterized by low bacterial loads and host granulomatous inflammation. External clinical signs, when present, are non-specific and include emaciation, ascites, exophthalmia, scoliosis and scale loss. Dermal ulceration may occur in the advanced stages, and is characteristic of the mycobacterial infections in wild striped bass in Chesapeake Bay (Fig. 18.2). Behavioural changes include lethargy, loss of equilibrium or buoyancy control, and anorexia. Granulomatous inflammation that appears as grey to off-white nodules may occur in the visceral organs and muscles (Fig. 18.3). The anterior kidney, spleen and liver are targeted, but all tissues may be affected.

A number of strategies are used to detect and differentiate mycobacteria in fish tissues, including histology, culture and molecular methods (see Kaattari *et al.*, 2006; and Gauthier and Rhodes,

Fig. 18.2. Dermal mycobacteriosis in Chesapeake Bay striped bass (*Morone saxatilis*): (A) early dermal lesions appear as pigmented foci (arrows) with localized scale erosion; (B) advanced multifocal ulceration and pigmented foci with localized ulceration also present (arrow).

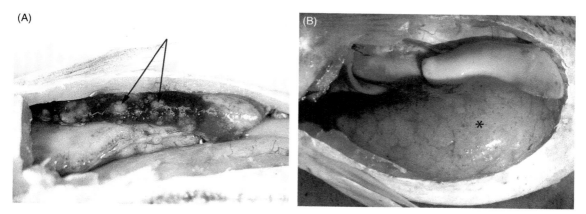

Fig. 18.3. Splenic mycobacteriosis in Chesapeake Bay striped bass (*Morone saxatilis*): (A) moderate granulomatous inflammation and multifocal grey nodules (arrows) within the spleen; (B) a severe granulomatous inflammation in which the majority of the splenic parenchyma (asterisk) has been replaced by granulomatous tissue. Photo: Wolfgang Vogelbein.

2009, for a review). Histology is the most traditional method, generally with the detection of acid-fast bacilli via the Ziehl–Neelsen staining method or modifications thereof. Acid-fast, non-branching bacilli within granulomas are considered indicative of *Mycobacterium* infections, although granulomas containing no visible acid-fast bacilli are frequently reported in experimental infections (Colorni *et al.*, 1998; Gauthier *et al.*, 2003; Watral and Kent, 2007).

Mycobacteria are typically cultured on Middlebrook 7H9 broth or 7H10 agar. Tween-20 is used as a carbon source and detergent in the former to prevent clumping. Löwenstein–Jensen agar is also frequently employed for mycobacterial culture (Kent

and Kubica, 1985). Because many piscine mycobacteria are inhibited above 35°C, and in some cases, at 30°C, culture at 22–24°C and/or the environmental temperature of the fish being studied is recommended. Some isolates are also extremely dysgonic (slow and relatively poorly growing) on artificial medium (e.g. *M. shottsii*), so extended culture periods (>60 days) should be employed for maximum sensitivity. Diagnostic keys using biochemical characters are available for piscine mycobacteria, generally focusing on *M. marinum*, *M. fortuitum* and *M. chelonae* (Frerichs, 1993; Chinabut, 1999). Other commonly used keys (Lévy-Frébault and Portaels, 1992; Tortoli, 2003; Leão *et al.*, 2004) may also be employed, though isolates from fish frequently do not precisely fit the biochemical definitions for named species. It is, therefore, prudent to describe novel isolates from fish or other poikilotherms as '-like' described species (e.g. 'M. triplex-like'), unless a preponderance of biochemical and multilocus genetic information is presented, especially if the existence of human pathogen is claimed from a fish. This has precedent in the literature, including description of a moray eel (Gymnothorax funebris) pathogen as M. montefiorense, a new species (Levi et al., 2003), following initial characterization of this isolate as 'M. triplex-like' based on 16S rRNA gene similarities (Herbst et al., 2001).

The culture of mycobacteria from fish is best performed from internal sites using aseptic techniques. However, techniques for 'decontaminating' external samples are available. The acid-fast cell wall makes mycobacteria highly hydrophobic and comparatively resistant to biocides and antibiotics. Consequently, compounds such as hypochlorite and benzalkonium chloride (N-alkyldimethylbenzylammonium chloride) can limit the overgrowth of other environmental bacteria and allow better isolation of mycobacteria when an aseptic technique is not possible (Brooks et al., 1984; Rhodes et al., 2004).

Molecular techniques are increasingly being used to detect mycobacteria in fish, including varieties of PCR, PCR/RFLP (restriction fragment length polymorphism), qPCR (quantitative PCR), loop-mediated amplification (LAMP) and other methods (see Kaattari et al., 2006 for review). The majority of these methods target the genes for 16S rRNA (the small ribosomal subunit), heat-shock protein, 65 kDa (hsp65), RNA polymerase β (rpoB) and the external repeated protein (erp). Although insertion sequences (bacterial transposable elements), which can be highly species- or strain-specific, are used extensively to detect and type human-pathogenic mycobacteria (e.g. IS6110 in M. tuberculosis), they are less frequently used for fish isolates, with the exception of IS2404, which is used to detect M. ulcerans clade members, including M. pseudoshottsii. IS901/IS1245 has also been used to identify M. avium-related bacteria in a tropical fish (Lescenko et al., 2003). As with biochemical characterization, many mycobacterial amplicons from fish may not exactly match the published sequences for species with nomenclatural standing, and the definitive identification of a named species based on a sequence of one or a limited number of amplicons can be misleading.

The detection of mycobacterial exposure in humans and animals is usually performed by a skin test, whereby a purified protein derivative (PPD) of mycobacteria is injected subcutaneously and the presence and size of induration from the Type IV hypersensitivity reaction is subsequently measured. This test has been supplemented with interferon capture assays such as QuantiFERON-Gold®, which detect T cell sensitization in vitro using peripheral blood leucocytes. Contrary to the well-developed state of serological tests for mycobacterial exposure in humans and animals of veterinary importance, serological assays for Mycobacterium spp. in fish are rarely used. Fish mount antibody responses against Mycobacterium extracellular products (Chen et al., 1996), whole-cell sonicates (Chen et al., 1997) and live infections (Colorni et al., 1998). Furthermore, rainbow trout (Oncorhynchus mykiss) develop delayed type-hypersensitivity to M. tuberculosis antigen (Bartos and Sommer, 1981), and present a classic induration upon injection of killed M. tuberculosis or M. salmoniphilum cells subsequent to immunization with Freund's Complete adjuvant. Therefore, the development of serological diagnostics for mycobacterial exposure in fish appears tractable, but has not yet progressed.

18.3 Pathology

The visceral granulomatous inflammation that occurs in piscine mycobacteriosis superficially resembles that in human tuberculosis. The term 'fish/piscine tuberculosis' was used in the past, but the disease is more appropriately referred to as 'fish/piscine mycobacteriosis', given the difference in aetiology. Piscine granulomas are typical in organization, and are composed of epithelioid macrophages surrounded by inflammatory leucocytes (Fig. 18.4). The granulomas may be cellular or necrotic and may display caseation. Frequently, a layer of highly compressed spindle-shaped epithelioid cells surrounds the necrotic cores. Giant multinucleate cells are present in the mycobacterial granulomas of some fish, but are absent in others. Granulomatous inflammation may range from minor, being detectable only using histology, to the presence of large multifocal or confluent accumulations of granulomatous inflammation with considerable displacement and destruction of the normal tissue architecture. Grey to off-white nodules may be obvious in such cases, but care should be taken to differentiate these from other granulomatous

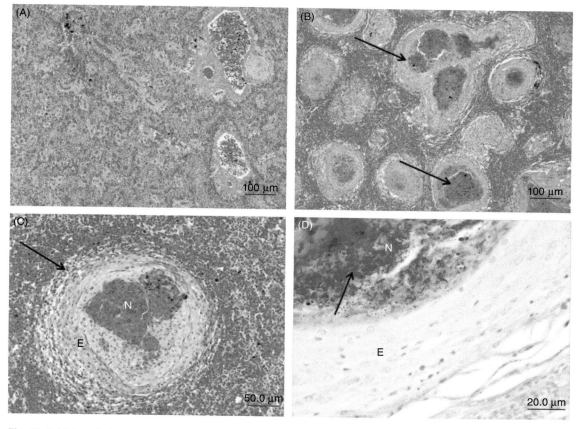

Fig. 18.4. Histopathology of mycobacteriosis in the spleen of fishes: (A) healthy spleen (haematoxylin and eosin (H&E) stain; (B) heavily affected spleen exhibiting multiple granulomas (arrows) (H&E stain); (C) single splenic granuloma exhibiting central necrosis (N), a surrounding cellular layer composed of epithelioid cells (E) and an indistinct outer capsule of leucocytes (arrow) (H&E stain); (D) single splenic granuloma exhibiting central necrosis (N), acid-fast mycobacteria (arrow) and encapsulating epithelioid cells (E) (Ziehl–Neelsen stain).

conditions (e.g. infection with *Francisella*, *Photobacterium* or *Nocardia*) or encapsulated helminths. Visceral adhesions are also observed in severe cases. Acute mycobacteriosis is characterized by incomplete or absent granuloma formation, high numbers of intra- and extracellular acid-fast bacteria, and widespread tissue necrosis (Talaat *et al.*, 1998).

Dermal lesions are also granulomatous, and they may extend into the hypodermis. The associated inflammation leads to the loss of epithelium and scales, and coalescing shallow ulcers can result. Early dermal mycobacteriosis in striped bass manifests as a focal erosion of a single scale, with an underlying pigmented focus (Fig.18.2). Granulomatous inflammation may underlie these focal erosions.

18.4 Pathophysiology

Swanson *et al.* (2002) reported reduced swimming performance (critical swimming velocity; Brett, 1964) in infected delta smelt (*Hypomesus transpacificus*). Lapointe *et al.* (2014) examined the metabolic parameters in diseased and healthy striped bass under varying oxygen concentrations and temperatures, and concluded that high temperature, hypoxia and disease interact synergistically to markedly reduce aerobic scope in diseased fish. Heavy splenic disease reduced hypoxia tolerance in striped bass, especially under elevated water temperature. However, severe splenic disease did not influence maximum metabolic rate or aerobic scope under conditions of optimal temperature and normoxia. Furthermore, neither the haematocrit nor haemoglobin concentrations were affected. Similarly, neither gill nor intestinal Na+-K+-ATPase activities were affected in fish with dermal disease, which were hypothesized to have compromised osmoregulation. These findings suggest that wild striped bass tolerate significant pathology with low impacts on oxygen transport and osmoregulation, although disease does reduce metabolic resilience in suboptimal environmental conditions (such as elevated temperature, hypoxia).

Reduced growth and stunting are common in mycobacterial infections, most likely due to the metabolic costs associated with a long-term immune response to chronic disease. Emaciation and a reduced condition factor are observed in severely diseased striped bass (Lapointe *et al.*, 2014), and Latour *et al.* (2012) found that striped bass with visceral mycobacteriosis had compromised growth (length-at-age) relative to non-diseased fish.

18.5 Protective and Control Strategies

The treatment of mycobacterial infections in fish has been attempted with antibiotics, including tetracycline (Boos *et al.*, 1995), erythromycin (Kawakami and Kusuda, 1990) and streptomycin (Colorni *et al.*, 1998), none of which successfully cleared infections. Some success was reported in treating guppies (*Lebistes reticulatus*) with aqueous kanamycin sulfate (50ppm), but it was not confirmed that the antibiotic successfully cleared the infection (Conroy and Conroy, 1999). Rifampicin and ethambutol have also been used in fish with limited success (Colorni, 1992; Chinabut, 1999). The other first-line tuberculosis drugs isoniazid and pyrazinamide are not indicated for the treatment of piscine mycobacteriosis, as most non-tuberculous strains are inherently resistant (Rastogi *et al.*, 1992).

Commercial vaccines against piscine *Mycobacterium* spp. are not available. Recombinant and DNA vaccines based on the fibronectin binding protein Ag85A elicit specific and non-specific immune responses and, in some cases, confer short-term protection in hybrid striped bass (Pasnik *et al.*, 2003; Pasnik and Smith, 2005, 2006), but long-term protection has not been achieved.

Because of the limited treatment options, the control of mycobacteriosis in aquaculture is primarily based on strict quarantine and screening practices (Astrofsky *et al.*, 2000). Li and Gatlin (2005) reported marginal improvement of survival in

M. marinum-infected hybrid striped bass receiving probiotic supplements (GroBiotic®-A), in feed. The growth rate and feed efficiency in fish receiving probiotics (GroBiotic®-A, autolysed brewer's yeast) also increased relative to controls, so increased survival in infected fish may have been related to improved fish condition, which is consistent with the studies of Jacobs *et al.* (2009a), who demonstrated exacerbated mycobacteriosis in striped bass receiving suboptimal nutrition.

The only option for the eradication of mycobacteriosis in aquaculture involves the destruction of affected stock and decontamination of the facilities (Noga, 2000; Roberts, 2001). The decontamination of aquaria is a challenge because commonly used disinfectants such as N-alkyldimethylbenzylammonium chloride (Roccal-D, Micronex) and potassium peroxymonosulfate (Virkon-S) are ineffective against *M. marinum*. Ethanol and benzyl-4-chlorophenol/phenylphenol (Lysol) are effective, as is sodium hypochlorite (bleach), albeit with extended contact times (Mainous and Smith, 2005).

18.6 Conclusions and Suggestions for Further Research

The paradigm of the 'big-three' mycobacteria (i.e. *M. marinum*, *M. fortuitum* and *M. chelonae*) in fish has changed, and it is likely that the full diversity of mycobacteria infecting fish remains underrecognized at present. Within Chesapeake Bay alone, Stine *et al.* (2010), using gas chromatography of fatty acid and methyl esters (FAME), reported 29 groups of fish-infecting mycobacteria, and Rhodes *et al.* (2004) presented at least eight phenotypically distinct groups, most of which probably contain distinct species or strains. The DNA sequences of mycobacteria amplified from formalin-preserved tissues of migratory striped bass (Matsche *et al.*, 2010) have indicated further diversity. The phylogeny of these mycobacteria, as well as their pathogenicity to fish under different conditions, are virtually unexplored. Because of this diversity, it is important to accurately identify and focus efforts on the numerically dominant isolates that are associated with disease. Moreover, because mycobacteria range from non-pathogenic saprophytes to obligate pathogens, it is necessary to identify cultured and/or PCR-detected *Mycobacterium* spp. to the lowest taxonomic level possible, including the differentiation of strains. Without such differentiation, there is high potential for the incorrect conclusions to be drawn about the pathogenicity of individual mycobacteria.

While the impacts of mycobacteriosis in aquaculture and aquaria are serious and easily measurable, the effects on wild fish populations are less clear. Given the chronic nature of the disease and the difficulties in detecting cryptic mortality in the field, determining population-level impacts is challenging. Mathematical modelling of apparent prevalence data and tag-recapture studies can help; however, the inclusion of these effects in management models is necessary to understand the impacts of mycobacteriosis on biomass and population health. In addition, it is important to examine the sublethal effects of mycobacteriosis, such as metabolic impairment and growth suppression, to obtain a more complete picture of disease impacts. As mycobacteriosis is most frequently observed in aquaculture facilities under suboptimal conditions, endemic disease in wild populations may indicate that environmental stressors are important. Epidemiological identification of these stressors in wild populations is a necessary area of future research that can increase our understanding of disease ecology.

Culture-based, histological and molecular diagnostics for mycobacterial infections in fish are fairly well established, and there are ample molecular and biochemical reference resources for determining the relatedness of novel isolates. These are currently all lethal methods, which can limit their practical use among valuable fish, or among large numbers of individuals. Non-lethal blood-based diagnostics, whether serological or cellular, remain virtually undescribed, and their development could facilitate screenings of brood stock and other fish before they are imported into facilities.

Clearly, with the inherently antibiotic-refractory nature of mycobacteria and general lack of success in employing these agents, vaccine development will be important. The development of efficacious vaccines may be challenging, if the difficulties in developing vaccines to *M. tuberculosis* in humans are taken as an indicator.

References

Aronson, J.D. (1926) Spontaneous tuberculosis in saltwater fish. *Journal of Infectious Diseases* 39, 315–320.

Ashburner, L.D. (1977) Mycobacteriosis in hatchery-confined chinook salmon (*Oncorhynchus tshawytscua* Walbaum) in Australia. *Journal of Fish Biology* 10, 523–528.

Astrofsky, K.M., Schrenzel, M.D., Bullis, R.A., Smolowitz, R.M. and Fox, J.G. (2000) Diagnosis and management of atypical *Mycobacterium* spp. infections in established laboratory zebrafish (*Brachydanio rerio*) facilities. *Comparative Medicine* 50, 666–672.

Bartos, J.M. and Sommer, C.V. (1981) *In vivo* cell-mediated immune response to *M. tuberculosis* and *M. salmoniphilum* in rainbow trout (*Salmo gairdneri*). *Developmental and Comparative Immunology* 5, 75–83.

Bercovier, H., Kafri, O. and Sela, S. (1986) Mycobacteria possess a surprisingly small number of ribosomal RNA genes in relation to the size of their genome. *Biochemical and Biophysical Research Communications* 136, 1136–1141.

Berg, V., Zerihun, M.A., Jørgensen, A., Lie, E., Dale, O.B. et al. (2013) High prevalence of infections and pathological changes in burbot (*Lota lota*) from a polluted lake (Lake Mjøsa, Norway). *Chemosphere* 90, 1711–1718.

Bonnet, E., Massip, P., Bauriaud, R., Alric, L. and Auvergnat, J.-C. (1996) Disseminated *Mycobacterium gordonae* infection in a patient infected with human immunodeficiency virus. *Clinical Infectious Diseases* 23, 644–645.

Boos, S., Schmidt, H., Ritter, G. and Manz, D. (1995) Untersuchungen zur oralen Wirksamkeit von Rifampicin gegen die Mykobakteriose der Zierfische [Effectiveness of oral rifampicin against mycobacteriosis in tropical fish]. *Berliner und Münchener* Tierärztliche Wochenschrift 108, 253–255.

Brett, J.R. (1964) The respiratory metabolism and swimming performance of young sockeye salmon. *Journal of the Fisheries Research Board of Canada* 21, 1183–1226.

Brooks, R.W., George, K.L., Parker, B.C. and Falkinham, J.O. III (1984) Recovery and survival of nontuberculous mycobacteria under various growth and decontamination conditions. *Canadian Journal of Microbiology* 30, 1112–1117.

Bruno, D.W., Griffiths, J., Mitchell, C.G., Wood, B.P., Fletcher, Z.J. et al. (1998) Pathology attributed to *Mycobacterium chelonae* infection among farmed and laboratory-infected Atlantic salmon *Salmo salar*. *Diseases of Aquatic Organisms* 33, 101–109.

Carbonara, S., Tortoli, E., Costa, D., Monno, L., Fiorentino, G. et al. (2000) Disseminated *Mycobacterium terrae* infection in a patient with advanced human immunodeficiency virus disease. *Clinical Infectious Diseases* 30, 831–835.

Cardinal, J.L. (2001) Mycobacteriosis in striped bass, *Morone saxatilis*, from Virginia waters of Chesapeake Bay. Master's thesis MSc., College of William and Mary, Virginia Institute of Marine Science, Gloucester Point, Virginia.

Chen, S.-C., Yoshida, T., Adams, A., Thompson, K.D. and Richards, R.H. (1996) Immune response of rainbow trout to extracellular products of *Mycobacterium* spp. *Journal of Aquatic Animal Health* 8, 216–222.

Chen, S.-C., Adams, A., Thompson, K.D. and Richards, R.H. (1997) A comparison of the antigenicity of the extracellular products and whole-cell sonicates from *Mycobacterium* spp. in rabbits, mice and fish by immunoblotting and enzyme-linked immunosorbent assay. *Journal of Fish Diseases* 20, 427–442.

Chinabut, S. (1999) Mycobacteriosis and nocardiosis. In: Woo, P.T.K. and Bruno, D.W. (eds) *Fish Diseases and Disorders. Volume 3: Viral, Bacterial and Fungal Infections.* CAB International, Wallingford, UK, pp. 319–340.

Colorni, A. (1992) A systemic mycobacteriosis in the European sea bass *Dicentrarchus labrax* cultured in Eilat (Red Sea). *The Israeli Journal of Aquaculture – Bamidgeh* 44, 75–81.

Colorni, A., Avtalion, R., Knibb, W., Berger, E., Colorni, B. and Timan, B. (1998) Histopathology of sea bass (*Dicentrarchus labrax*) experimentally infected with *Mycobacterium marinum* and treated with streptomycin and garlic (*Allium sativum*) extract. *Aquaculture* 160, 1–17.

Conroy, D. (1966) A report on the problems of bacterial fish diseases in the Argentine Republic. *Bulletin, Office International des Epizooties* 65, 755–768.

Conroy, G. and Conroy, D.A. (1999) Acid-fast bacterial infection and its control in guppies (*Lebistes reticulatus*) reared on an ornamental fish farm in Venezuela. *Veterinary Record* 144, 177–178.

Daoust, P.-Y., Larson, B.E. and Johnson, G.R. (1989) Mycobacteriosis in yellow perch (*Perca flavescens*) from two lakes in Alberta. *Journal of Wildlife Diseases* 25, 31–37.

Davis, J.M., Clay, H., Lewis, J.L., Ghori, N., Herbomel, P. and Ramakrishnan, L. (2002) Real-time visualization of *Mycobacterium*–macrophage interactions leading to initiation of granuloma formation in zebrafish embryos. *Immunity* 17, 693–702.

Devulder, G., De Montclos, M.P. and Flandrois, J.P. (2005) A multigene approach to phylogenetic analysis using the genus *Mycobacterium* as a model. *International Journal of Systematic and Evolutionary Microbiology* 55, 293–302.

Doig, K.D., Holt, K.E., Fyfe, J.A.M., Lavender, C.J., Eddyani, M. *et al.* (2012) On the origin of *Mycobacterium ulcerans*, the causative agent of Buruli ulcer. *BMC Genomics* 13:258.

Frerichs, G.N. (1993) Mycobacteriosis: nocardiosis. In: Inglis, V., Roberts, R.J. and Bromage, N.R. (eds) *Bacterial Diseases of Fish.* Blackwell Scientific, Oxford, UK, pp. 219–235.

Gauthier, D.T. and Rhodes, M.W. (2009) Mycobacteriosis in fishes: a review. *The Veterinary Journal* 180, 33–47.

Gauthier, D.T., Rhodes, M.W., Vogelbein, W.K., Kator, H. and Ottinger, C.A. (2003) Experimental mycobacteriosis in striped bass (*Morone saxatilis*). *Diseases of Aquatic Organisms* 54, 105–117.

Gauthier, D.T., Latour, R.J., Heisey, D.M., Bonzek, C.F., Gartland, J. *et al.* (2008a) Mycobacteriosis-associated mortality in wild striped bass (*Morone saxatilis*) from Chesapeake Bay, USA. *Ecological Applications* 18, 1718–1727.

Gauthier, D.T., Vogelbein, W.K., Rhodes, M.W. and Reece, K. (2008b) Nested PCR assay for detection of *Mycobacterium shottsii* and *Mycobacterium pseudoshottsii* in striped bass (*Morone saxatilis*). *Journal of Aquatic Animal Health* 20, 192–201.

Gauthier, D.T., Reece, K.S., Xiao, J., Rhodes, M.W., Kator, H.I. *et al.* (2010) Quantitative PCR assay for *Mycobacterium pseudoshottsii* and *Mycobacterium shottsii* and application to environmental samples and fishes from the Chesepeake Bay. *Applied and Environmental Microbiology* 76, 6171–6179.

Gauthier, D.T., Helenthal, A.M., Rhodes, M.W., Vogelbein, W.K. and Kator, H.I. (2011) Characterization of photochromogenic *Mycobacterium* spp. from Chesapeake Bay striped bass (*Morone saxatilis*). *Diseases of Aquatic Organisms* 95, 113–124.

Grange, J.M. (1981) *Mycobacterium chelonae. Tubercle* 62, 273–276.

Hariff, M.J., Bermudez, L.E. and Kent, M.L. (2007) Experimental exposure of zebrafish, *Danio rerio* (Hamilton) to *Mycobacterium marinum* and *Mycobacterium peregrinum* reveals the gastrointestinal tract as the primary route of infection: a potential model for environmental mycobacterial infection. *Journal of Fish Diseases* 30, 587–600.

Heckert, R.A., Elankumaran, S., Milani, A. and Baya, A. (2001) Detection of a new *Mycobacterium* species in wild striped bass in the Chesapeake Bay. *Journal of Clinical Microbiology* 39, 710–715.

Hedrick, R.P., Mcdowell, T. and Groff, J. (1987) Mycobacteriosis in cultured striped bass from California. *Journal of Wildlife Diseases* 23, 391–395.

Hennigan, C.E., Myers, L. and Ferris, M.J. (2013) Environmental distribution and seasonal prevalence of *Mycobacterium ulcerans* in southern Louisiana. *Antimicrobial Agents and Chemotherapy* 79, 2648–2656.

Herbst, L.H., Costa, S.F., Weiss, L.M., Johnson, L.K., Bartell, J. *et al.* (2001) Granulomatous skin lesions in moray eels caused by a novel *Mycobacterium* species related to *Mycobacterium triplex. Infection and Immunity* 69, 4639–4646.

Jacobs, J.M., Rhodes, M.R., Baya, A., Reimschuessel, R., Townsend, H. and Harrell, R.M. (2009a) Influence of nutritional state on the progression and severity of mycobacteriosis in striped bass *Morone saxatilis. Diseases of Aquatic Organisms* 87, 183–197.

Jacobs, J.M., Stine, C.B., Baya, A.M. and Kent, M.L. (2009b) A review of mycobacteriosis in marine fish. *Journal of Fish Diseases* 32, 119–130.

Jiang, H., Pollock, J.M., Brownie, C., Hoenig, J.M., Latour, R.J. *et al.* (2007) Tag return models allowing for harvest and catch and release: evidence of environmental and management impacts on striped bass fishing and natural mortality rates. *North American Journal of Fisheries Management* 27, 387–396.

Kaattari, I., Rhodes, M.W., Kaattari, S.L. and Shotts, E.B. (2006) The evolving story of *Mycobacterium tuberculosis* clade members detected in fish. *Journal of Fish Diseases* 29, 509–520.

Kawakami, K. and Kusuda, R. (1990) Efficacy of rifampicin, streptomycin, and erythromycin against experimental *Mycobacterium* infection in cultured yellowtail. *Nippon Suisan Gakkaishi* 56, 51–53.

Kent, M.L., Whipps, C.M., Matthews, J.L., Florio, D., Watral, V. et al. (2004) Mycobacteriosis in zebrafish (*Danio rerio*) research facilities. *Comparative Biochemistry and Physiology – Part C: Toxicology and Pharmacology* 138, 383–390.

Kent, P.T. and Kubica, G.P. (1985) *Public Health Mycobacteriology. A Guide for the Level III Laboratory*. US Department of Health and Human Services Publication No. (CDC) 86-8230, Centers for Disease Control, Atlanta, Georgia.

Khan, A.A., Kim, S.-J., Paine, D.D. and Cerniglia, C.E. (2002) Classification of a polycyclic aromatic hydrocarbon-metabolizing bacterium, *Mycobacterium* sp. strain PYR-1, as *Mycobacterium vanbaalenii* sp. nov. *International Journal of Systematic and Evolutionary Microbiology* 52, 1997–2002.

Kirschner, R.A., Parker, B.C. and Falkinham, J.O. III (1992) Epidemiology of infection by nontuberculous mycobacteria: *Mycobacterium avium, Mycobacterium intracellulare,* and *Mycobacterium scrofulaceum* in acid, brown-water swamps of the southeastern United States and their association with environmental variables. *American Review of Respiratory Disease* 145, 271–275.

Kleespies, M., Kroppenstedt, R.M., Rainey, F.A., Webb, L.E. and Stackebrandt, E. (1996) *Mycobacterium hodleri* sp. nov., a new member of the fast-growing mycobacteria capable of degrading polycyclic aromatic hydrocarbons. *International Journal of Systematic Bacteriology* 46, 683–687.

Kurokawa, S., Kabayama, J., Fukuyasu, T., Hwang, S.D., Park, C.I. et al. (2013) Bacterial classification of fish-pathogenic *Mycobacterium* species by multigene phylogenetic analyses and MALDI Biotyper identification system. *Marine Biotechnology* 15, 340–348.

Kusunoki, S. and Ezaki, T. (1992) Proposal of *Mycobacterium peregrinum* sp. nov., nom. rev., and elevation of *Mycobacterium chelonae* subsp. *abscessus* (Kubica *et al.*) to species status: *Mycobacterium abscessus* comb. nov. *International Journal of Systematic Bacteriology* 42, 240–245.

Lapointe, D., Vogelbein, W.K., Fabrizio, M.C., Gauthier, D.T. and Brill, R.W. (2014) Temperature, hypoxia, and mycobacteriosis: effects on adult striped bass *Morone saxatilis* metabolic performance. *Diseases of Aquatic Organisms* 108, 113–127.

Latour, R.J., Gauthier, D.T., Gartland, J., Bonzek, C.F., Mcnamee, K. and Vogelbein, W.K. (2012) Impacts of mycobacteriosis on the growth of striped bass (*Morone saxatilis*) in Chesapeake Bay. *Canadian Journal of Fisheries and Aquatic Sciences* 69, 247–258.

Leão, S.C., Martin, A., Mejia, G.I., Palomino, J.C., Robledo, J. et al. (2004) *Practical Handbook for the Phenotypic and Genotypic Identification of Mycobacteria*. Vanden Broelle, Brugges, Belgium.

Lescenko, P., Matlova, L., Dvorska, L., Bartos, M., Vavra, O. et al. (2003) Mycobacterial infection in aquarium fish. *Veterinární Medicína* 48, 71–78.

Levi, M.H., Bartell, J., Gandolfo, L., Smole, S.C., Costa, S.F. et al. (2003) Characterization of *Mycobacterium montefiorense* sp. nov., a novel pathogenic mycobacterium from moray eels that is related to *Mycobacterium triplex*. *Journal of Clinical Microbiology* 41, 2147–2152.

Lévy-Frébault, V.V. and Portaels, F. (1992) Proposed minimal standards for the genus *Mycobacterium* and for description of new slowly growing *Mycobacterium* species. *International Journal of Systematic Bacteriology* 42, 315–323.

Li, P. and Gatlin, D.M.I. (2005) Evaluation of the prebiotic GroBiotic®-A and brewers yeast as dietary supplements for sub-adult hybrid striped bass (*Morone chrysops* × *M. saxatilis*) challenged *in situ* with *Mycobacterium marinum*. *Aquaculture* 248, 197–205.

Lund, J.E. and Abernethy, C.S. (1978) Lesions of tuberculosis in mountain whitefish (*Prosopium williamsoni*). *Journal of Wildlife Diseases* 14, 222–228.

Mackenzie, K. (1988) Presumptive mycobacteriosis in north-east Atlantic mackerel, *Scomber scombrus* L. *Journal of Fish Biology* 32, 263–275.

Mainous, M.E. and Smith, S.A. (2005) Efficacy of common disinfectants against *Mycobacterium marinum*. *Journal of Aquatic Animal Health* 17, 284–288.

Matsche, M., Overton, A., Jacobs, J., Rhodes, M.R. and Rosemary, K.M. (2010) Low prevalence of splenic mycobacteriosis in migratory striped bass *Morone saxatilis* from North Carolina and Chesapeake Bay, USA. *Diseases of Aquatic Organisms* 90, 181–189.

Menendez, M.C., Garcia, M.J., Navarro, M.C., Gonzalez-Y-Merchand, J.A., Rivera-Gutierrez, S. et al. (2002) Characterization of an rRNA operon (rrnB) of *Mycobacterium fortuitum* and other mycobacterial species: implications for the classification of mycobacteria. *Journal of Bacteriology* 184, 1078–1088.

Mosi, L., Mutoji, N.K., Basile, F.A., Donnell, R., Jackson, K.L. et al. (2012) *Mycobacterium ulcerans* causes minimal pathogenesis and colonization in medaka (*Oryzias latipes*): an experimental fish model of disease transmission. *Microbes and Infection* 14, 719–729.

Mve-Obiang, A., Lee, R.E., Portaels, F. and Small, P.L.C. (2003) Heterogeneity of mycolactones produced by clinical isolates of *Mycobacterium ulcerans*:

implications for virulence. *Infection and Immunity* 71, 774–783.

Mve-Obiang, A., Lee, R.E., Umstot, E.S., Trott, K.A., Grammer, T.C. *et al.* (2005) A newly discovered mycobacterial pathogen isolated from laboratory colonies of *Xenopus* species with lethal infections produces a novel form of mycolactone, the *Mycobacterium ulcerans* macrolide toxin. *Infection and Immunity* 73, 3307–3312.

Nenoff, P. and Uhlemann, R. (2006) Mycobacteriosis in mangrove killifish (*Rivulus magdalanae*) caused by living fish food (*Tubifex tubifex*) infected with *Mycobacterium marinum*. *Deutsche Tierärztliche Wochenschrift* 113, 209–248.

Nigrelli, R.F. and Vogel, H. (1963) Spontaneous tuberculosis in fishes and in other cold-blooded vertebrates with special reference to *Mycobacterium fortuitum* Cruz from fish and human lesions. *Zoologica: Scientific Contributions of the New York Zoological Society* 48, 131–144.

Noga, E.J. (2000) *Fish Disease: Diagnosis and Treatment*. Iowa State University Press, Ames, Iowa.

Ottinger, C.A., Brown, J.J., Densmore, C., Starliper, C.E., Blazer, V. *et al.* (2007) Mycobacterial infections in striped bass from Delaware Bay. *Journal of Aquatic Animal Health* 19, 99–108.

Overton, A.S., Margraf, F.J., Weedon, C.A., Pieper, L.H. and May, E.B. (2003) The prevalence of mycobacterial infections in striped bass in Chesapeake Bay. *Fisheries Management and Ecology* 10, 301–308.

Pasnik, D.J. and Smith, S.A. (2005) Immunogenic and protective effects of a DNA vaccine for *Mycobacterium marinum* in fish. *Veterinary Immunology and Immunopathology* 103, 195–206.

Pasnik, D.J. and Smith, S.A. (2006) Immune and histopathologic responses of DNA-vaccinated hybrid striped bass *Morone saxatilis* × *M. chrysops* after acute *Mycobacterium marinum* infection. *Diseases of Aquatic Organisms* 73, 33–41.

Pasnik, D.J., Vemulapalli, R., Smith, S.A. and Schurig, G.G. (2003) A recombinant vaccine expressing a mammalian *Mycobacterium* sp. antigen is immunostimulatory but not protective in striped bass. *Veterinary Immunology and Immunopathology* 95, 43–52.

Peterson, T.S., Ferguson, J.A., Watral, V.G., Mutoji, K.N., Ennis, D.G. and Kent, M.L. (2013) *Paramecium caudatum* enhances transmission and infectivity of *Mycobacterium marinum* and *Mycobacterium chelonae* in zebrafish (*Danio rerio*). *Diseases of Aquatic Organisms* 106, 229–239.

Pieper, L. (2006) Striped bass disease overview for the past ten years plus. In: Ottinger, C.A. and Jacobs, J. (eds) *USGS/NOAA Workshop on Mycobacteriosis in Striped Bass, May 7–10, 2006, Annapolis, Maryland*. USGS Scientific Investigations Report 2006-5214/NOAA Technical Memorandum NOS NCCOS 41, US Geological Survey/National Oceanic and Atmospheric Administration, Reston, Virginia, pp. 10–11. Available at:

http://aquaticcommons.org/2233/1/SIR2006-5214-complete.pdf (accessed 15 November 2016).

Ranger, B.S., Mahrous, E.A., Mosi, L., Adusumilli, S., Lee, R.E. *et al.* (2006) Globally distributed mycobacterial fish pathogens produce a novel plasmid-encoded toxic macrolide, mycolactone F. *Infection and Immunity* 74, 6037–6045.

Rastogi, N., Goh, K.S., Guillou, N. and Labrousse, V. (1992) Spectrum of drugs against atypical mycobacteria: how valid is the current practice of drug susceptibility testing and the choice of drugs? *Zentralblatt für Bakteriologie* 277, 474–84.

Rhodes, M.W., Kator, H., Kotob, S., Van Berkum, P., Kaattari, I. *et al.* (2001) A unique *Mycobacterium* species isolated from an epizootic of striped bass (*Morone saxatilis*). *Emerging Infectious Diseases* 7, 1–3.

Rhodes, M.W., Kator, H., Kotob, S., Van Berkum, P., Kaattari, I. *et al.* (2003) *Mycobacterium shottsii* sp. nov., a slowly growing species isolated from Chesapeake Bay striped bass. *International Journal of Systematic and Evolutionary Microbiology* 53, 421–424.

Rhodes, M.W., Kator, H., Kaattari, I., Gauthier, D., Vogelbein, W.K. and Ottinger, C.A. (2004) Isolation and characterization of mycobacteria from striped bass *Morone saxatilis* from the Chesapeake Bay. *Diseases of Aquatic Organisms* 61, 41–51.

Rhodes, M.W., Kator, H., McNabb, A., Deshayes, C., Reyrat, J.-M. *et al.* (2005) *Mycobacterium pseudoshottsii* sp. nov., a slowly growing chromogenic species isolated from Chesapeake Bay striped bass (*Morone saxatilis*). *International Journal of Systematic and Evolutionary Microbiology* 55, 1139–1147.

Righetti, M., Favaro, L., Antuofermo, E., Caffara, M., Nuvoli, S. *et al.* (2014) *Mycobacterium salmoniphilum* infection in a farmed Russian sturgeon, *Acipenser gueldenstaedtii* (Brandt & Ratzeburg). *Journal of Fish Diseases* 37, 671–674.

Roberts, R.J. (2001) *Fish Pathology*, 3rd edn. W.B. Saunders, London.

Rogall, T., Wolters, J., Flohr, T. and Böttger, E.C. (1990) Towards a phylogeny and definition of species at the molecular level within the genus *Mycobacterium*. *International Journal of Systematic Bacteriology* 40, 323–330.

Ross, A.J. (1960) *Mycobacterium salmoniphilum* sp. nov. from salmonid fishes. *American Review of Respiratory Disease* 81, 241–250.

Ross, A.J. and Brancato, F.P. (1959) *Mycobacterium fortuitum* Cruz from the tropical fish, *Hyphessobrycon innessi*. *Journal of Bacteriology* 78, 392–395.

Ross, A.J. and Johnson, H.E. (1962) Studies of transmission of mycobacterial infections in Chinook salmon. *The Progressive Fish-Culturist* 24, 147–149.

Ross, A.J., Earp, B.J. and Wood, J.W. (1959) *Mycobacterial Infections in Adult Salmon and Steelhead Trout Returning to the Columbia River basin and Other Areas in 1957*. Special Scientific Report – Fisheries

D.T. Gauthier and M.W. Rhodes

No. 332, US Department of the Interior. Available at: http://spo.nmfs.noaa.gov/SSRF/SSRF332.pdf (accessed 15 November 2016).

Runyon, E.H. (1959) Anonymous mycobacteria in pulmonary disease. *The Medical Clinics of North America* 43, 273–290.

Sadler, P., Smith, M.W., Sullivan, S.E., Hoenig, J.M., Harris, R.E. Jr and Goins, L. (2012) *Evaluation of Striped Bass Stocks in Virginia: Monitoring and Tagging Studies, 2010–2014*. Progress Report F-77-R-24 to Virginia Marine Resources Commission, Newport News, Virginia, 26 January 2012. Department of Fisheries Science, School of Marine Science, Virginia Institute of Marine Science, The College of William and Mary, Gloucester Point, Virginia. Available at: http://fluke.vims.edu/hoenig/pdfs/Final_report_2011.pdf (accessed 15 November 2016).

Sakanari, J.A., Reilly, C.A. and Moser, M. (1983) Tubercular lesions in Pacific coast populations of striped bass. *Transactions of the American Fisheries Society* 112, 565–566.

Stine, C.B., Kane, A.S., Matsche, M., Pieper, L., Rosemary, K.M. *et al.* (2006) Microbiology of gametes and age 0–3 striped bass (*Morone saxatilis*). In: Ottinger, C.A. and Jacobs, J. (eds) *USGS/NOAA Workshop on Mycobacteriosis in Striped Bass, May 7–10, 2006, Annapolis, Maryland*. USGS Scientific Investigations Report 2006-5214/NOAA Technical Memorandum NOS NCCOS 41, US Geological Survey/National Oceanic and Atmospheric Administration, Reston, Virginia, pp. 13–14. Available at: http://aquaticcommons.org/2233/1/SIR2006-5214-complete.pdf (accessed 15 November 2016).

Stine, C.B., Jacobs, J.M., Rhodes, M.R., Overton, A., Fast, M. and Baya, A.M. (2009) Expanded range and new host species of *Mycobacterium shottsii* and *M. pseudoshottsii*. *Journal of Aquatic Animal Health* 21, 179–183.

Stine, C.B., Kane, A.S. and Baya, A.M. (2010) Mycobacteria isolated from Chesapeake Bay fish. *Journal of Fish Diseases* 33, 39–46.

Swanson, C., Baxa, D.V., Young, P.S., Cech, J.J.J. and Hedrick, R.P. (2002) Reduced swimming performance in delta smelt infected with *Mycobacterium* spp. *Journal of Fish Biology* 61, 1012–1020.

Talaat, A.M., Reimschuessel, R., Wasserman, S.S. and Trucksis, M. (1998) Goldfish, *Carassius auratus*, a novel animal model for the study of *Mycobacterium marinum* pathogenesis. *Infection and Immunity* 66, 2938–2942.

Tortoli, E. (2003) Impact of genotypic studies on mycobacterial taxonomy: the new mycobacteria of the 1990s. *Clinical Microbiology Reviews* 16, 319–354.

Van Der Werf, T.S., Steinstra, Y., Johnson, R.C., Phillips, R., Adjei, O. *et al.* (2005) *Mycobacterium ulcerans* disease. *Bulletin of the World Health Organization* 83, 785–791.

Vogelbein, W.K., Zwerner, D., Kator, H., Rhodes, M.W., Kotob, S.I. and Faisal, M. (1998) Mycobacteriosis in the striped bass, *Morone saxatilis*, from Chesapeake Bay. In: Kane, A.S. and Poynton, S.L. (eds) *3rd Symposium on Aquatic Animal Health*. APC Press, Baltimore, Maryland.

Watral, V. and Kent, M.L. (2007) Pathogenesis of *Mycobacterium* spp. in zebrafish (*Danio rerio*) from research facilities. *Comparative Biochemistry and Physiology – Part C: Toxicology and Pharmacology* 145, 55–60.

Whipps, C.M., Butler, W.R., Pourahmad, F., Watral, V. and Kent, M.L. (2007a) Molecular systematics support the revival of *Mycobacterium salmoniphilum* (ex Ross 1960) sp. nov., nom. rev., a species closely related to *Mycobacterium chelonae*. *International Journal of Systematic and Evolutionary Microbiology* 57, 2525–2531.

Whipps, C.M., Dougan, S.T. and Kent, M.L. (2007b) *Mycobacterium haemophilum* infections of zebrafish (*Danio rerio*) in research facilities. *FEMS Microbiology Letters* 270, 21–26.

Whipps, C.M., Matthews, J.L. and Kent, M.L. (2008) Distribution and genetic characterization of *Mycobacterium chelonae* in laboratory zebrafish *Danio rerio*. *Diseases of Aquatic Organisms* 82, 45–54.

Wolf, J.C. and Smith, S.A. (1999) Comparative severity of experimentally induced mycobacteriosis in striped bass *Morone saxatilis* and hybrid tilapia *Oreochromis* spp. *Diseases of Aquatic Organisms* 38, 191–200.

Wood, J.W. and Ordal, E.J. (1958) *Tuberculosis in Pacific Salmon and Steelhead Trout*. Contribution No. 25, Fish Commission of Oregon, Portland, Oregon. Available at: http://library.state.or.us/repository/2013/201310171436594/index.pdf (accessed 15 November 2016).

19 Photobacterium damselae

JOHN P. HAWKE*

*Department of Pathobiological Sciences, School of Veterinary Medicine,
Louisiana State University, Baton Rouge, Louisiana, USA*

19.1 Introduction

Photobacterium damselae is the causative agent of
the disease of photobacteriosis. It is a Gram-
negative halophilic bacterium that belongs to the
Class *Gammaproteobacteria*, Order *Vibrionales*,
Family *Vibrionaceae*. The genus includes bioluminescent species that colonize the light organs of
deep-sea fishes, hence the name *Photobacterium*. The
two pathogenic subspecies in fishes are *P. damselae*
subsp. *piscicida* and *P. damselae* subsp. *damselae*,
neither of which exhibit bioluminescence. The disease syndromes caused by these two subspecies, as
well as their clinical signs, are quite different, as are
the phenotypic characteristics of the bacteria themselves (see Table 19.2).

19.2 Photobacterium damselae subsp. piscicida

P. damselae subsp. *piscicida* is the causative agent
of a disease that has been called 'fish pasteurellosis'
and 'pseudotuberculosis'. 'Fish pasteurellosis' was
described in 1963 following a massive disease outbreak in Chesapeake Bay, Maryland, which killed
approximately 50% of the native white perch
(*Morone americana*) and striped bass (*M. saxatilis*)
(Snieszko *et al.*, 1964). An isolate from this outbreak
was deposited in the American Type Culture
Collection, Manassas, Virginia (ATCC 17911).
Based on its physical and biochemical characteristics,
the bacterium was tentatively placed in the genus
Pasteurella. Janssen and Surgalla (1968) concluded
that the bacterium was a new species on the basis
of morphological, physiological and serological
studies; they proposed the name *Pasteurella piscicida*.

However, this name was not validated by bacterial
taxonomists owing to physiological inconsistencies
with the characteristics that had been described for
Pasteurella. These included the lack of nitrate
reductase, tolerance of pH values outside the normal range, halophilia, a lower optimum growth
temperature and an unusual host range.
Consequently, the bacterium was not listed in
Bergey's Manual of Systematic Bacteriology
(Mannheim, 1984), nor in the *Approved List of
Bacterial Names* (Skerman *et al.*, 1989).
Nevertheless, the name was used until 1995, when
the organism was renamed *Photobacterium damsela*
subsp. *piscicida* based on 16S ribosomal RNA
sequences (Gauthier *et al.*, 1995); this name was
later corrected to the present taxonomic designation of *P. damselae* subsp. *piscicida* (Truper and
De'Clari, 1997).

There are several accounts of acute septicaemia
caused by *P. damselae* subsp. *piscicida* in wild and
cultured fishes in the USA. Subsequent to the original outbreak in Chesapeake Bay, the bacterium was
responsible for smaller fish kills of striped bass in
Chesapeake Bay (Paperna and Zwerner, 1976) and
also in western Long Island Sound, New York state
(Robohm, 1983). The disease was further documented from cultured fish in the USA among striped
bass fingerlings reared for stock enhancement in
earthen brackish water ponds on the Alabama Gulf
Coast at the Claude Peteet Mariculture Center,
Alabama Marine Resources Division (Hawke *et al.*,
1987). Significant mortality was attributed to this
pathogen in commercially reared hybrid striped
bass (HSB; *Morone saxatilis* × *M. chrysops*) cultured in net pens located in Louisiana brackish
water lakes (Hawke *et al.*, 2003). Table 19.1

*E-mail: jhawke1@lsu.edu

Table 19.1. Hosts, localities and references for outbreaks of *Photobacterium damselae* subsp. *piscicida* in wild and cultured fishes.

Fish host	Location	Reference(s)
Ayu (*Plecoglossus altevelis*)	Japan	Kusuda and Miura, 1972
Black seabream (*Mylio macrocephalus*)	Japan	Muroga *et al.*, 1977
Cobia (*Rachycentron canadum*)	Taiwan, Brazil	Lopez *et al.*, 2002; Moraes *et al.*, 2015
Gilthead sea bream (*Sparus aurata*)	Spain, Portugal, Malta, Italy	Toranzo *et al.*, 1991; Magariños *et al.*, 1992; Baptista *et al.*, 1996; Bakopoulos *et al.*, 1997
Golden pompano (*Trachinotus ovatus*)	China	Wang *et al.*, 2013
Hybrid striped bass (*Morone saxatilis* × *M. chrysops*)	Gulf Coast, Alabama and Louisiana; Israel	Hawke *et al.*, 1987, 2003; Nitzan *et al.*, 2001
Oval file fish (*Navodan modestus*)	Japan	Yasunaga *et al.*, 1984
Paradise fish (*Macropodus opercularis*)	Taiwan	Liu *et al.*, 2011
Red seabream (*Pagrus major*)	Japan	Yasunaga *et al.*, 1983
Redspotted grouper (*Epinephalus okarra*)	Japan	Ueki *et al.*, 1990
Sea bass (*Dicentrarchus labrax*)	France, Turkey, Greece	Magariños *et al.*, 1992; Bakopoulos *et al.*, 1995; Candan *et al.*, 1996
Snakehead (*Channa maculata*)	Taiwan	Tung *et al.*, 1985
Striped bass (*Morone saxatilis*)	Chesapeake Bay, Maryland; Long Island Sound, New York	Paperna and Zwerner, 1976; Robohm, 1983
Striped jack (*Pseudocaranx dentex*)	Japan	Nakai *et al.*, 1992
White perch (*Morone americana*)	Chesapeake Bay, Maryland	Snieszko *et al.*, 1964
Yatabe blenny (*Pictiblennius yatabei*)	Japan	Hamaguchi *et al.*, 1991
Yellowtail (*Seriola quinqueradiata*)	Japan	Kubota *et al.*, 1970

summarizes data on the hosts and localities of outbreaks of *P. damselae* subsp. *piscicida* in wild and cultured fishes.

The importance of *P. damselae* subsp. *piscicida* as a pathogen of wild and farmed fishes worldwide has now been established, as has been highlighted in several reviews (Austin and Austin, 1993; Magariños *et al.*, 1996; Romalde, 2002; Plumb and Hanson, 2007; Andreoni and Magnani, 2014).

19.2.1 Description of the bacterium

P. damselae subsp. *piscicida* is a Gram-negative, pleomorphic, rod-shaped bacterium (0.5–0.8 by 0.7–2.6 µm). The bacterium is non-motile, non-flagellated, and usually exhibits bipolar staining with Gram and Geimsa stains. It is moderately halophilic, oxidase positive (it takes 60 s for colour development), catalase positive and is sensitive to the vibriostatic agent 0/129. It grows slowly on tryptic soy agar with 5% sheep blood (TSAB) or brain heart infusion agar (BHIA) supplemented with 2% salt (Hawke *et al.*, 2003). The bacterium fails to grow on thiosulfate citrate bile sucrose agar (TCBS), but on TSAB it produces non-haemolytic, 1 mm wide greyish white colonies after 48 h incubation at 28°C. Optimal growth occurs between 22.5 and 30°C, in 1.0–2.5% NaCl and at pH 6.47–7.24 (Hashimoto *et al.*, 1985). The bacterium is a facultative anaerobe that produces acid but not gas from glucose, mannose, maltose, fructose and galactose. The organism generates a unique code number (2005004) in the API 20E system bioMé-rieux (Hawke *et al.*, 1987). All strains – from diverse hosts and geographic regions – have similar biochemical and physiological characteristics (Magarinõs *et al.*, 1992).

19.2.2 Epizootiology

Outbreaks of photobacteriosis caused by *P. damselae* subsp. *piscicida* occur within water temperatures of 14–29°C and at salinities of 3–21 ppt, although the optimum ranges for acute disease are 18–25°C and 5–15 ppt. The Chesapeake Bay isolate survived for 3 days in sterile brackish water (Janssen and Surgalla, 1968). However, there is evidence that viable but non-culturable forms may exist for extended periods in both seawater and sediments (Magariños *et al.*, 1994a). It has been suggested that another species of fish or invertebrate, in which susceptible species are cultured, may serve as a carrier and/or reservoir of infection (Robohm, 1983). The Louisiana Gulf Coast isolates are almost identical in biochemical phenotype and enzyme activity to isolates from Chesapeake Bay, Greece, Japan and Israel, but differ in their plasmid profiles and antimicrobial susceptibilities. The Louisiana isolates possess a unique plasmid banding profile and they typically produce two large plasmid bands of >30 kb and two smaller bands of 8.0 kb and 5.0 kb. The isolates from Israel and Greece produce 10 and 8 kb bands while the Japanese isolates possess plasmids of 5 and 3.5 kb (Hawke, 1996). The resistance to Romet® and/or Terramycin® demonstrated by some Louisiana strains was the result of acquisition of an R-plasmid (Hawke *et al.*, 2003; Kim *et al.*, 2008). Louisiana Gulf Coast strains analysed using random amplified polymorphic DNA (RAPD) analysis belonged to clonal lineage group 2, which was similar to the Japanese strains (Magariños *et al.*, 2000; Hawke *et al.*, 2003).

19.2.3 Virulence factors

A multitude of virulence mechanisms is utilized by *P. damselae* subsp. *piscicida* to overcome fish immune defence mechanisms. The normal antibacterial activity of the skin mucus of sea bream (*Sparus aurata*) and sea bass (*Dicentrarchus labrax*) is ineffective against the pathogen (Magariños *et al.*, 1995). The bacterium enters the host via the gills by an undetermined process and migrates via

the bloodstream to target cells and organs in HSB (Hawke, 1996). In addition to the gills, adherence to the intestinal mucosae is important for infection (Magariños *et al.*, 1996). This is mediated by a protein or glycoprotein receptor and internalization of the bacterium occurs via an actin microfilament-dependent mechanism. A polysaccharide capsule surrounding the outer bacterial membrane provides resistance to complement and serum-induced killing. Serum resistance enables the persistence of the bacteria in the bloodstream, and their migration to target cells and tissues; it also allows bacterial numbers to reach extremely high levels – approximately 10^6 cfu (colony forming units)/ml – in the blood of HSB (Hawke, 1996). The bacterium is a facultative intracellular pathogen. Once phagocytosed by macrophages, it replicates within the phagosome, ultimately killing the macrophage and releasing numerous bacterial cells into the surrounding tissues and the blood. In HSB, the intracellular replication involves inhibition of phagosome–lysosome fusion within the macrophage (Elkamel *et al.*, 2003); the studies by these authors were performed both *in vivo* in HSB and *in vitro* in HSB head kidney-derived macrophages.

Striped bass and HSB seem to be highly susceptible to *P. damselae* subsp *piscicida*, regardless of age, though other species of fish show increased resistance to infection and disease when they exceed a weight of 50 g (Andreoni and Magnani, 2014). Noya *et al.* (1995) found that the bacterium replicated uninhibitedly in peritoneal exudate (PE) cells of juvenile (0.5 g) gilthead sea bream (*S. aurata*), but were inhibited in the PE cells of larger (20–30 g) sea bream. Mortality was 100% in infected juvenile sea bream, whereas there was no mortality in larger fish. A key pathogenicity factor is a plasmid-encoded 56 kDa apoptosis-inducing protein (AIP56), which induces apoptosis in sea bass macrophages and neutrophils (do Vale *et al.*, 2005).

Finally, one of the most important virulence mechanisms of *P. damselae* subsp. *piscicida* is the acquisition of iron from the host by siderophores. Iron is extremely limited in the host, and siderophores secreted by the bacterium scavenge iron bound to host proteins such as transferrin. This allows the iron to be acquired by the bacterium via specific outer membrane receptors. Strains of the bacterium with a mutation in a siderophore biosynthesis gene were avirulent in HBS (Hawke, 1996). A siderophore biosynthesis gene cluster was identified in

P. damselae subsp. *piscicida* strain DI21 (from sea bream) that was similar to the *Yersinia* high-pathogenicity island (HPI) (Osorio *et al.*, 2006). This suggested that the siderophore from DI21 may be functionally and structurally related to yersiniabactin, an important siderophore of the human pathogen *Y. enterocolitica*. Mutation of the *irp1* gene of DI21 resulted in impaired growth under iron-limited conditions, failure to produce siderophore and 100-fold decrease in virulence in juvenile turbot (*Scophthalmus maximus*) (Osorio *et al.*, 2006).

19.2.4 Diagnostic procedures

Presumptive diagnosis

For presumptive diagnosis, the bacterium should be isolated from the host in pure culture and identified using the procedures outlined in Section 19.2.1. The phenotypic characteristics of *P. damselae* subsp. *piscicida* are outlined in Table 19.2.

Table 19.2. Phenotypic comparison of typical isolates of *Photobacterium damselae* subsp. *piscicida* and *P. damselae* subsp. *damselae*.

Test[a]	*P. damselae* subsp. *piscicida*[b]	*P. damselae* subsp. *damselae*[b]
0/129	S	S
API 20E code	2005004	2015004
		6015004
		2011004
Bipolar staining	+	−
Catalase	+	+
Gram stain	−	−
Growth at 35°C	−	+
Growth on TCBS	−	+G
Haemolysis	−	+
Lysine decarboxylase	−	75%+
Motility	−	+
Nitrate reduction	−	+
Oxidase	W+	+
Pleomorphism	+	−
Urease	−	+
Voges–Proskauer	+	10%+

[a]0/129, a vibriostatic agent; API 20E code in the system bioMérieux; TCBS, thiosulfate citrate bile sucrose agar; Voges–Proskauer, a test for acetoin production.
[b]Key: +, positive result; −, negative result; +G = positive growth of a green colony; S = sensitive; W+ = weak or delayed positive result; 10%+, only 10% of strains are positive; 75%+, 75% of strains are positive.

Confirmatory diagnosis

MOLECULAR METHODS

16S rRNA sequencing. Universal primers to 16S rRNA can be utilized to generate PCR products. The products (complete or partial sequences) may be confirmed as *P. damselae* using sequenced PCR products and comparing them using NCBI BLAST (the US National Center for Biotechnology Information Basic Local Alignment Search Tool; https://blast.ncbi.nlm.nih.gov/Blast.cgi) analysis with the sequences in GenBank (the US National Institutes of Health genetic sequence database; https://www.ncbi.nlm.nih.gov/genbank/). Unfortunately, this procedure cannot be used to differentiate the two subspecies.

Species-specific PCR. The first species-specific PCR method was developed by Aoki *et al.* (1997) using sequences in a species-specific plasmid, pZP1. However, this method proved inadequate because European strains did not possess the same plasmid profile. A rapid method of identification and differentiation between the subspecies of *P. damselae* was later developed by using primers designed to amplify the capsular polysaccharide gene. A particular primer pair is used to amplify a 410 bp fragment of the capsular polysaccharide gene derived from *P. damselae* subsp. *piscicida* (GenBank accession number AB074290). The forward primer, CPSF, is 5′-AGGGGATCCGATTATTACTG-3′, which corresponds to positions 531–550 of the *P. damselae* subsp. *piscicida* gene for capsular polysaccharide; the reverse primer, CPSR, is 5′-TCCCATTGAGAAGATTTGAT-3′, corresponding to positions 921–940. As both subspecies of *P. damselae* are detected using this PCR, the subspecies must be differentiated by growth on TCBS agar. *P. damselae* subsp. *damselae* grows on TCBS producing a green colony, while *P. damselae* subsp. *piscicida* fails to grow (Rajan *et al.*, 2003).

Subspecies-specific multiplex PCR. A multiplex-PCR approach, employing two primer pairs directed to the internal regions of the *16S rRNA* and *ureC* genes was developed to identify and discriminate between the subspecies (Osorio *et al.*, 2000a). With this procedure, *P. damselae* subsp. *damselae* strains yield two amplification products, one of 267 bp and the other of 448 bp, corresponding to internal fragments of the *16S rRNA* and *ureC* genes, respectively. In contrast, *P. damselae* subsp. *piscicida* only shows the PCR product of 267 bp from a *16S rRNA* fragment, indicating the absence of the urease gene. A partial sequence of the *ureC* gene from *P. damselae* subsp. *damselae* was retrieved from the GenBank database with accession number U40071. A forward primer, Ure-5′ (20-mer 5′-TCCGGAATAGGTAAAGCGGG-3′), and a reverse primer, Ure-3′ (22-mer 5′-CTTGAATATCCATCTCATCTGC-3′), were designed flanking a 448 bp long stretch of the *ureC* gene. A forward primer (118-mer 5′-GCTTGAAGAGATTCGAGT-3′; positions 1016–1033 in the *Escherichia coli 16s rRNA* gene), and a reverse primer (18-mer 5′-CACCTCGCGGTCTTGCTG-3′; positions 1266–1283), flanking a 267 bp fragment of the *16S rRNA* gene of strain ATCC 29690 of *P. damselae* subsp. *piscicida* (GenBank accession number Y18496) are used in conjunction with Ure-5′ and Ure-3′ in a multiplex PCR reaction (Osorio *et al.*, 1999). More recently, an additional multiplex PCR protocol was developed targeting the gene coding for penicillin-binding protein 1A and the *ureC* gene (Amagliani *et al.*, 2009). The optimized multiplex PCR identifies and discriminates both subspecies of *P. damselae* with a detection limit of 500 fg DNA (100 genomic equivalent units).

SEROLOGICAL METHODS. Acute infections with minimal clinical signs or subclinical infections may be detected using the BIONOR™ AQUARAPID test kit (BIONOR, Skein, Norway). This kit utilizes a specific antibody against *P. damselae* subsp. *piscicida* antigen in an enzyme-linked immunosorbent assay (ELISA) on tissue homogenates (Romalde *et al.*, 1995a). To identify bacteria from diseased fish that are suspected of being *P. damselae* subsp. *piscicida*, the BIONOR™ Mono Aqua kit may be used (Romalde *et al.*, 1995b). This procedure relies on a particle agglutination procedure applied to suspensions of pure cultures of the bacterium.

19.2.5 Clinical signs of disease, gross pathology

Infected fish may show different clinical signs and pathology based on the species of fish, environmental conditions and drug intervention. In general, fish affected by photobacteriosis become lethargic, swim slowly near the surface, and ultimately sink and rest on the bottom prior to death. Increased ventilation rates and loss of equilibrium may be evident. Acute

and chronic forms of photobacteriosis have been described from various fishes (Thune *et al.*, 1993). In acute disease, very little gross pathology is observed (Bullock, 1978; Toranzo *et al.*, 1991). In chronic disease, small whitish lesions resembling granulomas are visible in the spleen and kidney, hence the use of the name 'pseudotuberculosis' for the disease.

Diseased cultured striped bass and HBS may be lethargic, and display pallor of the gills, petechiae in the opercular region and darker than normal pigmentation (Hawke *et al.*, 1987, 2003). Internally, an enlarged and friable spleen is the only obvious gross clinical sign (Fig. 19.1) but the liver may be slightly mottled and the kidney may be haemorrhagic. White perch show only slight haemorrhage of the operculum and base of fins. In wild striped bass and white perch, small white 'miliary' lesions may be seen in the swollen spleen and kidney (Wolke, 1975; Bullock, 1978). The chronic form of the disease may vary depending on the fish species affected and whether or not antibiotic feeds were administered. Similar small white granuloma-like lesions (Fig. 19.2) may be visible in the spleen and kidney of cultured HBS following antibiotic therapy (Hawke *et al.*, 2003). It is suggested that antibiotics slow the progress of infection and thus allow granulomatous inflammation to be grossly visible.

Cultured yellowtail and amberjack [*Seriola* spp.] present with oedema and failure to regulate pigmentation. In yellowtail, chronic lesions are typified by 1–2 mm granuloma-like lesions in the spleen, kidney and heart that are composed of masses of the causal bacterium, epithelial cells and fibroblasts.

Fig. 19.1. Hybrid striped bass collected in a moribund state with acute photobacteriosis caused by *Photobacterium damselae* subsp. *piscicida*. The spleen is swollen and friable; otherwise, the clinical signs are minimal. Photo by Dr Joe Newton.

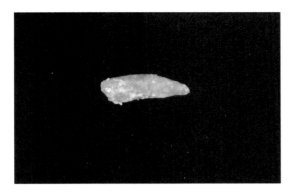

Fig. 19.2. Chronic photobacterial infection caused by *Photobacterium damselae* subsp. *piscicida* in hybrid striped bass following antibiotic therapy. The spleen is characterized by necrotic foci, bacterial colonies and granuloma-like lesions in the splenic parenchyma visible as white 'miliary' lesions. Photo by Dr Al Camus.

Fig. 19.3. Focal necrosis in the spleen of hybrid striped bass associated with numerous colonies of *Photobacterium damselae* subsp. *piscicida*. Inflammatory cell accumulations are lacking. Photo by Dr John Hawke.

The lesions grossly resemble granulomas (Kubota *et al.*, 1970). Diseased gilthead sea bream exhibit no apparent external clinical signs except in rare individuals that haemorrhage around the head and operculum. Cultured cobia (*Rachycentron canadum*) have whitish granuloma-like lesions in the kidney, liver and spleen. Moribund cobia are lethargic and are darker than normal (Liu *et al.*, 2003).

19.2.6 Histopathology

The histopathology of naturally infected white perch and striped bass from Chesapeake Bay was first reported by Wolke (1975). In what was apparently a chronic form of the disease, collections of necrotic lymphoid and peripheral blood cells were observed in the spleen. Focal areas of hepatocytes undergoing coagulation necrosis and a conspicuous lack of inflammatory cell responses were evident in the liver. In the acute disease in striped bass and HBS, there was acute multifocal necrosis of the lymphoid tissue of the spleen characterized by loss of cells, coagulation necrosis, karyorrhexis and large colonies of the bacterium (Fig. 19.3). In the liver, acute multifocal necrosis with prominent karyorrhexis is common and inflammatory cellular accumulations are absent (Hawke *et al.*, 1987). Similar microscopic lesions were reported in gilthead sea bream (Toranzo *et al.*, 1991). In acute disease, bacterial numbers approach 1 million bacteria/ml blood (Fig. 19.4; Hawke, 1996). The response in yellowtail and cobia is different, with

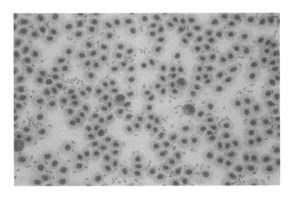

Fig. 19.4. Blood smear from a moribund hybrid striped bass with approximately 1 × 10⁶ cfu/ml of *Photobacterium damselae* subsp. *piscicida* in the blood. Photo by Dr John Hawke.

bacterial colonies in the spleen forming granuloma-like structures – 'pseudotubercles' – containing eosinophils and surrounded by epithelial cells. In all species, masses of bacteria (bacterial emboli) form in the capillaries and block blood flow in the interstitial spaces of the internal organs. In experimentally infected HBS, death was attributed to bacterial emboli in the gill capillaries and asphyxiation due to inability to achieve blood flow and gas exchange (Hawke, 1996).

19.2.7 Treatment of the disease

Early detection is paramount for effective treatment and control. The most common approach is the

---

administration of antibiotic-medicated feeds early in the infection (Andreoni and Magnani, 2014). As discussed earlier, the progress of infection and mortality is influenced by the size, age and species of fish. Decisions on the time and method to administer medicated feeds are influenced by culture type (pond, sea cage, net pen, indoor system) and factors such as water temperature, mortality rate and season. In the USA, there are very few approved antibiotics and a veterinary prescription or investigational new animal drug (INAD) permit may be required. The use of Romet®- and oxytetracycline-medicated feeds to treat infected HBS in cage and net pens was initially successful in reducing and even stopping mortality (Hawke *et al.*, 2003). Unfortunately disease recurrence was common and isolates from secondary outbreaks often had multidrug resistance because of acquired R-plasmids (Hawke *et al.*, 2003). R-plasmids from Japanese strains of the pathogen conferred resistance to kanamycin, chloramphenicol, tetracycline and sulphonamides, whereas US strains carried resistance to tetracycline, trimethoprim and sulfonamides (Kim *et al.*, 2008).

19.2.8 Prevention of photobacteriosis

The lack of consistent efficacy of antibiotic feeds and the emergence of antibiotic-resistant strains of the *P. damselae* subsp. *piscicida* have led to the development of vaccines, probiotics and prophylactic dietary supplements. One alternative to the use of antibiotics is the incorporation of plant extracts in the diet (e.g. the addition of garlic (*Allium sativum*) or the active ingredient, allicin, to the diet). Garlic has a long history of medicinal applications and has a wide spectrum of antibacterial properties (Guo *et al.*, 2015). In experimental trials with cobia, garlic powder incorporated into the feed and offered at 1.2 g/kg daily for 21 days conferred significant resistance to challenge with *P. damselae* subsp. *piscicida*. *Vibrio* strain NM 10 from the intestine of marine fish produced a heat-labile proteinaceous substance (<5 kDa), with inhibitory activity against the bacterium. This strain could have potential as a probiotic against photobacteriosis (Sugita *et al.*, 1997).

Early vaccines – either formalin-killed bacterins or extracellular products (ECPs – lipopolysaccharides (LPS) and ribosomal fractions) – against photobacteriosis in Japanese yellowtail (*S. quinqueradiata*) were ineffective (Fukuda and Kusuda, 1981). Bacterins enriched with a 97kDa outer membrane protein and a 52 kDa protein found in culture

supernatants elicited antibody responses in the gills and mucosae and were protective (Barnes *et al.*, 2005). An ECP-enriched bacterin commercialized for use in Europe has shown favourable results in sea bass and yellowtail by immersion (Magariños *et al.*, 1994b).

Two live attenuated strains of *P. damselae* subsp. *piscicida*, LSU-P1 and LSU-P2, were produced from mutants in the parent strain LADL 91-197 isolated from HSB. Strain LSU-P1, a siderophore biosynthesis mutant, was produced by transformation of LADL91-197 by electroporation and inactivation of a siderophore biosynthesis gene with random insertion of a mini *tn10* transposon carrying a kanamycin resistance gene on a suicide plasmid pEIS. Selection for mutants was carried out on media containing kanamycin and chrome azurol S to detect siderophore production (Hawke, 1996). Strain LSU-P2 has a mutation in an aromatic amino acid biosynthesis gene (*aroA*). The mutation was made by creating a frameshift mutation in the *aroA* gene by homologous recombination and insertion of a kanamycin resistance gene (Thune *et al.*, 2003). In laboratory trials, both strains provided significant protection in HBS following immersion vaccination and challenge with virulent LADL91-197. However, because of the decline and eventual demise of HBS aquaculture on the US Gulf Coast, the vaccines were never commercialized.

Subunit vaccines have been explored in Taiwan to prevent photobacteriosis in cobia (Ho *et al.*, 2011). In these studies, immunoproteomics was used to identify antigens of *P. damselae* subsp. *piscicida* that can be cloned and produced as recombinant proteins for the development of potential vaccines. An approach referred to as 'reverse vaccinology' utilizes bioinformatics to identify the bacterial proteins that are important vaccine candidates. Antigens have been identified that are involved in the adherence and internalization of the bacterium *in vitro*. Challenge experiments following the immunization of sea bass with a recombinant antigen identified by this process induced specific antibodies and protection (Andreoni *et al.*, 2013).

19.3 *Photobacterium damselae* subsp. *damselae*

P. damselae subsp. *damselae*, which was originally named *Vibrio damsela*, causes ulcer disease and haemorrhagic septicaemia in marine fish. Love *et al.* (1981) reported *V. damsela* as the aetiological

J.P. Hawke

agent that caused skin ulcers in damselfish (*Chromis punctipinnis*) inhabiting the coastal waters of southern California, and also noted that the bacterium had been isolated from water and from two human wounds and could be a cause of wound infections and disease in humans. The organism shares many features and phenotypic characteristics with *Vibrio*. It has been renamed *Listonella damsela* (MacDonell and Colwell, 1985), *Photobacterium damsela* (Smith *et al.*, 1991), *P. damsela* subsp. *damsela* (Gauthier *et al.*, 1995) and, finally, its present name of *P. damselae* subsp. *damselae* (Truper and De'Clari, 1997).

The bacterium is pathogenic to many marine fishes, crustaceans, molluscs and cetaceans. In humans, it causes opportunistic infections that can progress into necrotizing fasciitis with resulting loss of limbs or even fatalities (Rivas *et al.*, 2013). Specific fish hosts are the blacksmith damselfish (*Chromis punctipinnis*) (Love *et al.*, 1981), various species of shark (Grimes *et al.*, 1984, 1985; Han *et al.*, 2009), yellowtail (*S. quinqueradiata*) (Sakata *et al.*, 1989), gilthead sea bream (Vera *et al.*, 1991), European sea bass (*D. labrax*) (Botella *et al.*, 2002), turbot (Fouz *et al.*, 1992), barramundi (*Lates calcarifer*) in Australia (Renault *et al.*, 1994) and rainbow trout (*Oncorynchus mykiss*) (Pedersen *et al.*, 2009). The bacterium was also isolated from newly cultured marine fishes (*Pagrus auriga*, *P. pagrus*, *Diplodus sargus* and *Argyrosomus regius*) in Spain (Labella *et al.*, 2011). In addition to its isolation from fish, *P. damselae* subsp. *damselae* has been isolated from diseased shrimp (*Penaeus monodon*) (Wang and Chen, 2006), octopus (*Octopus joubini*) (Hanlon *et al.*, 1984), turtles (*Dermochelys coriacea*) (Obendorf *et al.*, 1987), dolphins (*Tursiops truncatus*) (Fujioka *et al.*, 1988) and wound infections in humans (Love *et al.*, 1981; Clarridge and Zighelboim-Daum, 1985).

19.3.1 Description of the bacterium

P. damselae subsp. *damselae* is a Gram-negative rod or coccobacillus, 0.5–0.8 by 0.7–2.0 µm in size. The bacterium is motile by single or multiple unsheathed polar flagella and typically is oxidase, catalase and urease positive (Smith *et al.*, 1991). It is halophilic, growing over a range of 1–6% NaCl, and grows on BHIA with 2% salt from 20 to 37°C, with optimum growth at 35°C. On TSAB with 5% sheep blood colonies are typically haemolytic, but this characteristic is variable depending on the

strain and the haemolysins produced. The organism produces a green colony on TCBS agar and whitish grey colonies 1–2 mm wide on TSAB or BHI with 2% salt after 24 h at 35°C. The bacterium is a facultative anaerobe that produces acid and gas from glucose, and acid from mannose, melibiose, maltose, cellobiose and D-galactose. It is sensitive to the vibriostat 0/129 (Fouz *et al.*, 1992). Typical fish pathogenic strains of the bacterium produce a unique code number (2015004) in the API 20E system (bioMérieux); however, variable results for the Voges–Proskauer test (for acetoin production) and lysine decarboxylase test occur in some strains, resulting in different codes (see Table 19.2). Several studies have demonstrated additional biochemical heterogeneity among strains from different hosts, with variable results seen in tests for gas production from glucose, urease production and the utilization of certain sugars (Labella *et al.*, 2011).

19.3.2 Epizootiology

The bacterium is a normal inhabitant of seawater and marine sediments and prefers warm waters (20–30°C). Virulent strains can survive in seawater microcosms at 14–22°C as culturable bacteria for extended periods, which maintain their infectivity for fish (Fouz *et al.*, 1998). Disease outbreaks have occurred on earthen pond fish farms and in net pens and sea cages, as well as in indoor tanks and aquaria. The organism is transmitted through the water to highly susceptible host organisms. The infections are often opportunistic, and less susceptible fish require some stress or injury as predisposing factors. In Denmark, where rainbow trout are cultured in seawater, the bacterium caused disease at higher water temperatures. In laboratory trials, it was 1000 times more virulent in trout at 20°C than at 13°C (Pedersen *et al.*, 2009). In turbot, spikes in mortality occur when the water temperature is increased from 18 to 25°C (Fouz *et al.*, 1992). In southern Spain, the bacterium has caused numerous disease outbreaks in cultured marine species, i.e. redbanded sea bream (*P. auriga*), common sea bream (*P. pagrus*) and white sea bream (*D. sargus*) (Labella *et al.*, 2011). Mortality rates on Spanish fish farms ranged from 5% in December to 94% in August. Because of its predilection for higher water temperatures, some authors anticipate disease problems caused by this organism to increase in light of global climate changes (Pedersen *et al.*, 2009).

19.3.3 Virulence factors

ECPs from *P. damselae* subsp. *damselae* display cytotoxic activity for different fish and mammalian cell lines. Only virulent strains produce toxic ECPs and the cytotoxic components are thermolabile (Labella *et al.*, 2010). The primary virulence factor, which has traditionally been listed as damselysin, is a phospholipase D. Other phospholipases may also participate in virulence because some virulent strains lack the *dly* gene for phospholipase D, but still have phospholipase activity in their ECPs (Osorio *et al.*, 2000b). Recently, a 150 kb plasmid, pPHDD1, was identified that carries the gene for Dly as well as an Hly toxin of the pore-forming toxin family (Rivas *et al.*, 2011). Only a small percentage of strains harbour pPHDD1, and these strains exhibit a wider zone of beta-haemolysis than strains without this plasmid (Rivas *et al.*, 2013). Both haemolysins are important and mutation of the genes causes avirulence in fish and reduced virulence in mice (Rivas *et al.*, 2013). The *hlyA* gene can also be chromosome encoded in some strains. In virulence studies, it was demonstrated that Dly acts in synergy with HlyA, whether it is plasmid or chromosomally encoded (Rivas *et al.*, 2013). In addition, enzymatic activities detected in the ECPs could be related with toxicity *in vitro* or *in vivo* (Labella *et al.*, 2010). Most strains do not have detectable proteases (Fouz *et al.*, 1992), which suggests that other uncharacterized molecules play a role in toxicity. It is anticipated that complete genome sequencing of *P. damselae* subsp. *damselae* will help to answer these deficiencies in the state of our knowledge.

19.3.4 Diagnostic procedures

Presumptive diagnosis

For presumptive diagnosis, the bacterium should be isolated from the host in pure culture and identified using the procedures outlined in Section 19.3.1. The phenotypic characteristics of *P. damselae* subsp. *damselae* are outlined in Table 19.2 along with those of *P. damselae* subsp. *piscicida*.

Confirmatory diagnosis

A confirmatory diagnosis may be achieved using the molecular methods described for *P. damselae* subsp. *piscicida* in Section 19.2.4. The different strains of *P. damselae* subsp. *damselae* can also be investigated based on their serological differences and genetic variation, as briefly described below.

SEROLOGY AND STRAIN HETEROGENEITY. Strains of *P. damselae* subsp. *damselae* are heterogeneous serologically (Smith *et al.*, 1991; Fouz *et al.*, 1992; Pedersen *et al.*, 2009). Four recognizable LPS-based serogroups (A–D) were identified from turbot (Fouz *et al.*, 1992). Genetic variation as determined by amplified fragment length polymorphism (AFLP) was significant: 24 of 33 strains from European sea bass and gilthead sea bream showed different banding patterns in which a relationship existed with the geographic origin of the isolate or the host fish (Botella *et al.*, 2002).

19.3.5 Clinical signs of disease and gross pathology

Clinical signs may vary slightly according to the host species and behavioural signs have not been described in most hosts. In the Australian snapper (*P. auratus*), moribund fish float on the surface in lateral recumbency for several hours prior to death (Stephens *et al.*, 2006). In damselfish, ulcerative lesions form near the pectoral fin and caudal peduncle and may increase in size to 20 mm prior to death (Love *et al.*, 1981). In rainbow trout, extensive haemorrhages are observed in the skin around the vent, in the parietal peritoneum and in the intestinal tract. Skin haemorrhages are most pronounced at the fin bases and along the ventral midline of the abdomen. Petechiae are often observed in the liver, along with haemorrhagic ascites in the abdomen (Pedersen *et al.*, 2009). In sea bass and sea bream, the clinical signs mimic vibriosis. The main external signs are exophthalmia, dark skin pigmentation, pale gills and eroded fins. Haemorrhagic areas and ulcers are common on the body surface. Internally, a pale liver, splenomegaly (swollen spleen) and bloody ascites are noted (Labella *et al.*, 2011). In diseased turbot, the most remarkable clinical signs are abdominal distention, haemorrhagic areas in the eyes and mouth (including the palate and jaws), and around the anus (Fouz *et al.*, 1992). Wild caught red snapper (*Lutjanus campechanus*) from the Gulf of Mexico have shallow skin ulcerations (Hawke and Baumgartner, 2011).

19.3.6 Histopathology

Very little information exists on the histopathology and microscopic lesions of this disease. Infected Australian snapper had granulocytic enteritis, with

---

oedema and infiltrations of eosinophilic granular cells. This same cell type was prominent in the gills. There were increased numbers of melanomacrophage centres and haemosiderin deposits in the spleen, kidney and liver of infected fish (Stephens *et al.*, 2006). Red snapper with skin lesions had ulcerative dermatitis with moderate dermal fibrosis and inflammatory cell infiltrates (Hawke and Baumgartner, 2011).

19.3.7 Treatment of the disease

Outbreaks of photobacteriosis are typically treated with medicated feeds and the antibiotic susceptibility of typical isolates from fish farms has been reported (Pedersen *et al.*, 2009; Labella *et al.*, 2011). Typical strains from Spanish fish farms are susceptible to trimethoprim–sulfamethoxazole, chloramphenicol, enrofloxacin, flumequine, nalidixic acid, oxolinic acid and nitrofurantoin. Antibiotic resistance has been observed to streptomycin, erythromycin, tetracycline, oxytetracycline, amoxicillin and ampicillin (Labella *et al.*, 2011). Because of the antigenic heterogeneity among isolates, vaccines have not been employed (Botella *et al.*, 2002).

19.4 Summary and Conclusions

Both subspecies of *P. damselae* are important disease-causing agents, but they differ phenotypically and epidemiologically. *P. damselae* subsp. *piscicida* is an obligate fish pathogen that does not survive long in the environment, whereas *P. damselae* subsp. *damselae* remains infective in the environment (water and sediments) for extended periods, and is an opportunistic pathogen of fish. *P. damselae* subsp. *damselae* is also an opportunistic pathogen of homeothermic animals, including humans. *P. damselae* subsp. *piscicida* is a fish pathogen that occurs in temperate climates at water temperatures in the range of 18–28°C and causes disease with few accompanying clinical signs, while *P. damselae* subsp. *damselae* prefers warmer climates and causes disease with obvious clinical signs, including multifocal haemorrhages and ulcers in the skin. Killed bacterins have shown little efficacy in the prevention of photobacteriosis; however, live attenuated vaccines to protect against *P. damselae* subsp. *piscicida* infection show promise for the future management and control of this disease in aquaculture. By contrast, *P. damselae* subsp. *damselae* is

serologically diverse, which complicates immunotherapy. The management of this subspecies will most likely rely on improved husbandry practices and effective antibiotic therapy.

References

Amagliani, G., Omiccioli, E., Andreoni, F., Boiani, R., Bianconi, I. *et al.* (2009) Development of a multiplex PCR assay for *Photobacterium damselae* subsp. *piscicida* identification in fish samples. *Journal of Fish Diseases* 32, 645–653.

Andreoni, F. and Magnani, M. (2014) Photobacteriosis: prevention and diagnosis. *Journal of Immunology Research* 2014: Article ID 793817.

Andreoni, F., Boiani, R., Serafini, G., Amagliani, G., Dominici, S. *et al.* (2013) Isolation of a novel gene from *Photobacterium damselae* subsp. *piscicida* and analysis of the recombinant antigen as a promising vaccine candidate. *Vaccine* 31, 820–826.

Aoki, T., Ikerda, D., Katagiri, T. and Hirono, I. (1997) Rapid detection of the fish pathogenic bacterium *Pasteurella piscicida* by polymerase chain reaction targeting nucleotide sequences of the species specific plasmid pZP1. *Fish Pathology* 32, 143–151.

Austin, B. and Austin, D.A. (1993) *Bacterial Fish Pathogens. Disease of Farmed and Wild Fish*, 2nd edn. Ellis Horwood, Chichester, UK, pp. 23–42.

Bakopoulos, V., Adams, A. and Richards, R.H. (1995) Some biochemical properties and antibiotic sensitivities of *Pasteurella piscicida* isolated in Greece and comparison with strains from Japan, France and Italy. *Journal of Fish Diseases* 18, 1–7.

Bakopoulos, V., Peric, Z., Rodger, H., Adams, A. and Richards, R. (1997) First report of fish pasteurellosis from Malta. *Journal of Aquatic Animal Health* 9, 26–33.

Baptista, T., Romalde, J.L. and Toranzo, A.E. (1996) First occurrence of pasteurellosis in Portugal affecting cultured gilthead seabream (*Sparus aurata*). *Bulletin of the European Association of Fish Pathologists* 16, 92–95.

Barnes, A.C., dos Santos, N.M.S. and Ellis, A.E. (2005) Update on bacterial vaccines: *Photobacterium damselae* subsp. *piscicida*. *Developments in Biologicals* 121, 75–84.

Botella, S., Pujalta, M.-J., Macian, M.-C., Ferrus, M.-A., Hernandez, J. *et al.* (2002) Amplified fragment length polymorphism (AFLP) and biochemical typing of *Photobacterium damselae* subsp. *damselae*. *Journal of Applied Microbiology* 93, 681–688.

Bullock, G.L. (1978) *Pasteurellosis of Fishes. Fish Disease Leaflet* No. 54, US Department of the Interior, Fish and Wildlife Service, Washington, DC.

Candan, A., Kucuker, M.A. and Karatas, S. (1996) Pasteurellosis in cultured sea bass (*Dicentrarchus* labrax) in Turkey. *Bulletin of the European Association of Fish Pathologists* 16, 150–153.

Clarridge, J.E. and Zighelboim-Daum, S. (1985) Isolation and characterization of two haemolytic phenotypes of *Vibrio damsela* associated with a fatal wound infection. *Journal of Clinical Microbiology* 21, 302–306.

doVale, A., Silva, M.T., dosSantos, N.M.S., Nascimento, D.S., Reis-Rodrigues, P. *et al.* (2005) AIP56, a novel plasmid-encoded virulence factor of *Photobacterium damselae* subsp. *piscicida* with apoptotic activity against sea bass macrophages and neutrophils. *Molecular Microbiology* 58, 1025–1038.

Elkamel, A., Hawke, J.P., Henk, W.G. and Thune, R.L. (2003) *Photobacterium damselae* subsp. *piscicida* is capable of replicating in hybrid striped bass macrophages. *Journal of Aquatic Animal Health* 15, 175–183.

Fujioka, R.S., Greco, S.B., Cates, M.B. and Schroeder, J.P. (1988) *Vibrio damsela* from wounds in bottlenose dolphins *Tursiops truncatus*. *Diseases of Aquatic Organisms* 4, 1–8.

Fouz, B., Larsen, J.L., Nielsen, B, Barja, J. and Toranzo, A.E. (1992) Characterization of *Vibrio damsela* strains isolated from turbot *Scophthalmus maximus* in Spain. *Diseases of Aquatic Organisms* 12, 155–166.

Fouz, B., Toranzo, A.E., Marco-Noales, E. and Amaro, C. (1998) Survival of fish-virulent strains of *Photobacterium damselae* subsp. *damselae* in seawater under starvation conditions. *FEMS Microbiology Letters* 168, 181–186.

Fukuda, Y. and Kusuda, R. (1981) Efficacy of vaccination for pseudotuberculosis in cultured yellowtail by various routes of administration. *Bulletin of the Japanese Society for Scientific Fisheries* 47, 147–150.

Gauthier, G., LaFay, B., Ruimy, R., Breittmayer, V., Nicolas, J.L. *et al.* (1995) Small-subunit rRNA sequences and whole DNA relatedness concur for the reassignment of *Pasteurella piscicida* (Snieszko *et al.*) (Janssen and Surgalla) to the genus *Photobacterium* as *Photobacterium damsela* subsp. *piscicida* comb. nov. *International Journal of Systematic Bacteriology* 45, 139–144.

Grimes, D.J., Colwell, R.R., Stemmler, J., Hada, H., Maneval, D. *et al.* (1984) *Vibrio* species as agents of elasmobranch disease. *Helgoländer Meeresunters* 37, 309–315.

Grimes, D.J., Brayton, P., Colwell, R.R. and Gruber, S.H. (1985) Vibrios as authocthonous flora of neritic sharks. *Systematic and Applied Microbiology* 6, 221–226.

Guo, J.J, Kuo, C.M., Hong, J.W., Chou, R.L., Lee, Y.H. *et al.* (2015) The effects of garlic supplemented diets on antibacterial activities against *Photobacterium damselae* subsp. *piscicida* and *Streptococcus iniae* and on growth in cobia, *Rachycentron canadum Aquaculture* 435, 111–115.

Hamaguchi, M., Usui, H. and Kusuda, R. (1991) *Pasteurella piscicida* infection in yatabe blenny (*Pictiblennius yatabei*). *Gyobio Kenkyu* 26, 93–94.

Han, J.E., Gomez, D.K., Kim, J.H., Choresca, C.H., Shin, S.P. *et al.* (2009) Isolation of *Photobacterium damselae* subsp. *damselae* from zebra shark *Stegostoma fasciatum*. *Korean Journal of Veterinary Research* 49, 35–38.

Hanlon, R.T., Forsythe, J.W., Cooper, K.M., Dinuzzo, A.R., Folse, D.S. *et al.* (1984) Fatal penetrating skin ulcers in laboratory reared octopuses. *Journal of Invertebrate Pathology* 44, 67–83.

Hashimoto, S., Muraoka, A., Mihara, S. and Kusuda, R. (1985) Effects of cultivation temperature, NaCl concentration, and pH on the growth of *Pasteurella piscicida*. *Bulletin of the Japanese Society for Scientific Fisheries* 51, 63–67.

Hawke, J.P. (1996) Importance of a siderophore in the pathogenesis and virulence of *Photobacterium damsela* subsp. *piscicida* in hybrid striped bass (*Morone saxatilis* × *Morone chrysops*). Ph.D. dissertation, Louisiana State University, Baton Rouge, Louisiana.

Hawke, J.P. and Baumgartner, W.A. (2011) *Unpublished Case Reports*. Louisiana Aquatic Diagnostic Laboratory, Louisiana State University, Baton Rouge, Louisiana.

Hawke, J.P., Plakas, S.M., Minton, R.V., McPhearson, R.M., Snider, T.G. *et al.* (1987) Fish pasteurellosis of cultured striped bass (*Morone saxatilis*) in coastal Alabama. *Aquaculture* 65, 193–204.

Hawke, J.P., Thune, R.L., Cooper, R.K., Judice, E. and Kelly-Smith, M. (2003) Molecular and phenotypic characterization of strains of *Photobacterium damselae* subsp. *piscicida* from hybrid striped bass cultured in Louisiana, USA. *Journal of Aquatic Animal Health* 15, 189–201.

Ho, L.-P., Lin, J.H.-Y., Liu, H.-C., Chen, H.-E., Chen, T.-Y. *et al.* (2011) Identification of antigens for the development of a subunit vaccine against *Photobacterium damselae* subsp. *piscicida*. *Fish and Shellfish Immunology* 30, 412–419.

Janssen, W.A. and Surgalla, M.J. (1968) Morphology, physiology, and serology of a *Pasteurella* species pathogenic for white perch (*Roccus americanus*). *Journal of Bacteriology* 96, 1606–1610.

Kim, M., Hirono, I., Kurokawa, K., Maki, T., Hawke, J. *et al.* (2008) Complete DNA sequence and analysis of the transferable multiple drug resistance plasmids (R-plasmids) from *Photobacterium damselae* subsp. *piscicida* isolated in Japan and USA. *Antimicrobial Agents and Chemotherapeutics* 52, 606–611.

Kubota, S.S., Kimura, M. and Egusa, S. (1970) Studies of a bacterial tuberculoidosis of the yellowtail. I. Symptomatology and histopathology. *Fish Pathology* 4, 111–118.

Kusuda, R. and Miura, W. (1972) Characteristics of a *Pasteurella* sp. pathogenic for pond cultured ayu. *Fish Pathology* 7, 51–57.

Labella, A.C., Sanchez-Montes, N., Berbel, C., Aparicio, M. Castro, D. *et al.* (2010) Toxicity of *Photobacterium damselae* subsp. *damselae* isolated from new cultured marine fish. *Diseases of Aquatic Organisms* 92, 31–40.

Labella, A., Berbel, C., Manchado, M., Castro, D. and Borrego, J.J. (2011) Chapter 9. *Photobacterium damselae* subsp. *damselae*, an emerging pathogen affecting new cultured marine fish species in southern Spain. In: Aral, F. and Doğu, Z. (eds) *Recent Advances in Fish Farms*. InTech Open Science, Rijeka, Croatia. Available at: http://www.intechopen.com/articles/show/title/photobacterium-damselae-subsp-damselae-an-emerging-pathogen-affecting-new-cultured-marine-fish-speci (accessed 16 November 2016).

Liu, P.-C., Lin, J.-Y. and Lee, K.-K. (2003) Virulence of *Photobacterium damselae* subsp. *piscicida* in cultured cobia *Rachycentron canadum*. *Journal of Basic Microbiology* 43, 499–507.

Liu, P.-C., Cheng, C.-F., Chang, C.-H., Lin, S.-L., Wang, W.-S. *et al.* (2011) Highly virulent *Photobacterium damselae* subsp. *piscicida* isolated from Taiwan paradise fish, *Macropodus opercularis* (L.) in Taiwan. *African Journal of Microbiology Research* 5, 2107–2113.

Lopez, C., Rajan, P.R., Lin, J.H.-Y. and Yang, H.-L. (2002) Disease outbreak in sea farmed cobia, *Rachycentron canadum* associated with *Vibrio* spp., *Photobacterium damselae* subsp. *piscicida*, monogenean, and myxosporean parasites. *Bulletin of the European Association of Fish Pathologists* 23, 206–211.

Love, M., Teebkin-Fisher, D., Hose, J.E., Farmer, J.J., Hickman, F.W. *et al.* (1981) *Vibrio damsela*, a marine bacterium, causes skin ulcers on the damselfish *Chromis punctipinnis*. *Science* 214, 1139–1140.

MacDonell, M.T. and Colwell, R.R. (1985) Phylogeny of the *Vibrionaceae*, and recommendation of two new genera, *Listonella* and *Shewanella*. *Systematic and Applied Microbiology* 6, 171–182.

Magariños, B., Romalde, J.L., Bandin, I., Fouz, B. and Toranzo, A.E. (1992) Phenotypic, antigenic, and molecular characterization of *Pasteurella piscicida* strains isolated from fish. *Applied and Environmental Microbiology* 58, 3316–3322.

Magariños, B., Romalde, J.L., Barja, J.L. and Toranzo, A.E. (1994a) Evidence of a dormant but infective state of the fish pathogen *Pasteurella piscicida* in seawater and sediment. *Applied and Environmental Bacteriology* 60, 180–186.

Magariños, B., Noya, M., Romalde, J.L., Perez, G. and Toranzo, A.E. (1994b) Influence of fish size and vaccine formulation on the protection of gilthead seabream against *Pasteurella piscicida*. *Bulletin of the European Association of Fish Pathologists* 14, 120–122.

Magariños, B., Pazos, F., Santos, Y., Romalde, J.L. and Toranzo, A.E. (1995) Response of *Pasteurella piscicida* and *Flexibacter maritimus* to skin mucus of

marine fish. *Diseases of Aquatic Organisms* 21, 103–108.

Magariños, B, Toranzo, A.E. and Romalde, J.L. (1996) Phenotypic and pathobiological characteristics of *Pasteurella piscicida*. *Annual Review of Fish Diseases* 6, 41–64.

Magariños, B., Toranzo, A.E., Barja, J.L. and Romalde, J.L. (2000) Existence of two geographically-linked clonal lineages in the bacterial fish pathogen *Photobacterium damselae* subsp. *piscicida* evidenced by random amplified polymorphic DNA analysis. *Epidemiology and Infection* 125, 213–219.

Mannheim, W. (1984) Family III. Pasteurellaceae Pohl 1981a, 382. In: Krieg, N.R. and Holt, J.G. (eds) *Bergey's Manual of Systematic Bacteriology*, 9th edn, Vol. 1. Williams and Wilkins, Baltimore, Maryland, pp. 550–552.

Moraes, J.R.E., Shimada, M.T., Yunis, J., Claudiano, G.S., Filho, J.R.E. *et al.* (2015) Photobacteriosis outbreak in cage reared *Rachycentron canadum*: predisposing conditions. In: *Abstracts of the Annual Meeting of the European Aquaculture Society, Aquaculture Europe 15, Rotterdam, Netherlands*. World Aquaculture Society, Ostend, Belgium, Session abstracts 119. Available at: https://www.was.org/easOnline/AbstractDetail.aspx?i=4421 (accessed 16 November 2016).

Muroga, K., Sugiyama, T. and Ueki, N. (1977) Pasteurellosis in cultured black sea bream *Mylio macrocephalus*. *Journal of the Faculty of Fisheries and Animal Husbandry, Hiroshima University* 16, 17–21.

Nakai, T., Fujiie, N., Muroga, K., Arimoto, M., Mizuta, Y. *et al.* (1992) *Pasteurella piscicida* in hatchery-reared striped jack. *Gyobio Kenkyu* 27, 103–108.

Nitzan, S., Shwartsburd, B., Vaiman, R. and Heller, E.D. (2001) Some characteristics of *Photobacterium damselae* subsp. *piscicida* isolated in Israel during outbreaks of pasteurellosis in hybrid striped bass (*Morone saxatilis* × *M. chrysops*). *Bulletin of the European Association of Fish Pathologists* 21, 77–80.

Noya, M., Magariños, B., Toranzo, A.E. and Lamas, J, (1995) Sequential pathology of experimental pasteurellosis in gilthead seabream *Sparus aurata*. A light- and electron-microscopic study. *Diseases of Aquatic Organisms* 21, 177–186.

Obendorf, D.L., Carson, J. and McManus, T.J. (1987) *Vibrio damsela* infection in a stranded leatherback turtle (*Dermochelys coriacea*). *Journal of Wildlife Diseases* 23, 666–668.

Osorio, C.R., Collins, M.D., Toranzo, A.E., Barja, J.L. and Romalde, J.L. (1999) 16S rRNA gene sequence analysis of *Photobacterium damselae* and nested PCR method for rapid detection of the causative agent of fish pasteurellosis. *Applied and Environmental Microbiology* 65, 2942–2946.

Osorio, C.R., Toranzo, A.E., Romalde, J.L. and Barja, J.L. (2000a) Multiplex PCR assay for *ureC* and 16s rRNA genes clearly discriminates between both subspecies of *Photobacterium damselae*. *Diseases of Aquatic Organisms* 40, 177–183.

Osorio, C.R., Romalde, J.L., Barja, J.L. and Toranzo, A.E. (2000b) Presence of phospholipase-D (*dly*) gene encoding for damselysin production is not a pre-requisite for pathogenicity in *Photobacterium damselae* subsp. *damselae*. *Microbial Pathogenesis* 28, 119–126.

Osorio, C.R., Juiz-Río, S. and Lemos, M.L. (2006) A siderophore biosynthesis gene cluster from the fish pathogen *Photobacterium damselae* subsp. *piscicida* is structurally and functionally related to the *Yersinia* high-pathogenicity island. *Microbiology* 152, 3327–3341.

Paperna, I. and Zwerner, D.E. (1976) Parasites and diseases of striped bass, *Morone saxatilis* (Walbaum) from the lower Chesapeake Bay. *Journal of Fish Biology* 9, 267–287.

Pedersen, K., Skall, H.F., Lassen-Nielsen, H.M., Bjerrum, L. and Olesen, N.J. (2009) *Photobacterium damselae* subsp. *Damselae*, an emerging pathogen in Danish rainbow trout *Oncorhynchus mykiss* (Walbaum) mariculture. *Journal of Fish Diseases* 32, 465–472.

Plumb, J.A. and Hanson, L.A. (2007) Striped bass bacterial diseases. In: Plumb, J.A. and Hanson, L.A. *Health Maintenance and Principal Microbial Diseases of Cultured Fishes*, 3rd edn. Wiley-Blackwell, Ames, Iowa, pp. 429–433.

Rajan, P.R., Lin, J.H.-Y., Ho, M.-S. and Yang, H.-L. (2003) Simple and rapid detection of *Photobacterium damselae* ssp. *piscicida* by a PCR technique and plating method. *Journal of Applied Microbiology* 95, 1375–1380.

Renault, T., Haffner, P., Malfondet, C. and Weppe, M. (1994) *Vibrio damsela* as a pathogenic agent causing mortalities in cultured sea bass (*Lates calcarifer*). *Bulletin of the European Association of Fish Pathologists* 14, 117–119.

Rivas, A.J., Balado, M., Lemos, M.L. and Osorio, C.R. (2011) The *Photobacterium damselae* subsp. *damselae* hemolysins damselyn and HlyA are encoded within a new virulence plasmid. *Infection and Immunity* 79, 4617–4627.

Rivas, A.J., Lemos, M.L. and Osorio, C.R. (2013) *Photobacterium damselae* subsp. *damselae*, a bacterium pathogenic for marine animals and humans. *Frontiers in Microbiology* 4:283.

Robohm, R.A. (1983) *Pasteurella piscicida*. In: Anderson, D.P., Dorson, M. and Dubourget, P. (eds) *Antigens of Fish Pathogens: Development and Production for Vaccines and Serodiagnostics*. Collection Foundation Marcel Merieux, Association Corporative des Etudiants en Médecine de Lyon, Lyon, France, pp. 161–175.

Romalde, J.L. (2002) *Photobacterium damselae* subsp. *piscicida*: an integrated view of a bacterial fish pathogen. *International Microbiology* 5, 3–9.

Romalde, J.L., LeBreton, A., Magariños, B. and Toranzo, A.E. (1995a) Use of BIONOR Aquarapid-Pp kit for the diagnosis of *Pasteurella piscicida* infections. *Bulletin of the European Association of Fish Pathologists* 15, 64–66.

Romalde, J.L., Magariños, B., Fouz, D.B., Bandin, I., Nunez, S. *et al.* (1995b) Evaluation of BIONOR Mono-kits for rapid detection of bacterial fish pathogens. *Diseases of Aquatic Organisms* 21, 25–34.

Sakata, T., Matsuura, M. and Shimkawwa, Y. (1989) Characteristics of *Vibrio damsela* isolated from diseased yellowtail *Seriola quinqueradiata*. *Nippon Suisan Gakkaishi* 55, 135–141.

Skerman, V.B.D., McGowan, V. and Sneath, P.H.A. (eds) (1989) *Approved Lists of Bacterial Names,* amended edn. American Society for Microbiology, Washington, DC, p. 72.

Smith, S.K., Sutton, D.C., Fuerst, J.A. and Reichelt, J.L. (1991) Evaluation of the genus *Listonella* and reassignment of *Listonella damsela* (Love *et al.*) MacDonell and Colwell to the genus *Photobacterium* as *Photobacterium damsela* comb. nov. with an emended description. *International Journal of Systematic Bacteriology* 41, 529–534.

Snieszko, S.F., Bullock, G.L., Hollis, E. and Boone, J.G. (1964) *Pasteurella sp.* from an epizootic of white perch (*Roccus americanus*) in Chesapeake Bay tidewater areas. *Journal of Bacteriology* 88, 1814–1815.

Stephens, F.J., Raidal, S.R., Buller, N. and Jones, B. (2006) Infection with *Photobacterium damselae* subsp. *damselae* and *Vibrio harveyi* in snapper, *Pagrus auratus* with bloat. *Australian Veterinary Journal* 84, 173–177.

Sugita, H., Matsuo, N., Hirose, Y., Iwato, M. and Deguchi, Y. (1997) *Vibrio* strain NM 10, isolated from the intestine of a Japanese coastal fish, has an inhibitory effect against *Pasteurella piscicida*. *Applied and Environmental Microbiology* 63, 4986–4989.

Thune, R.L., Stanley, L.A. and Cooper, R.K. (1993) Pathogenesis of Gram negative bacterial infections in warmwater fish. *Annual Review of Fish Diseases* 3, 37–68.

Thune, R.L., Fernandez, D.H., Hawke, J.P. and Miller, R. (2003) Construction of a safe, stable, efficacious vaccine against *Photobacterium damselae* ssp. *piscicida*. *Diseases of Aquatic Organisms* 57, 51–58.

Toranzo, A.E., Barreiro, S., Casal, J.F., Figueras, A., Magariños, B. *et al.* (1991) Pasteurellosis in cultured gilthead seabream (*Sparus aurata*): first report in Spain. *Aquaculture* 99, 1–15.

Truper, H.G. and De'Clari, L. (1997) Taxonomic note: necessary correction of epithets formed as substantives (nouns) "in apposition". *International Journal of Systematic Bacteriology* 47, 908–909.

Tung, M.C., Tsai, S.S., Ho, L.F., Huang, S.T. and Chen, S.C. (1985) An acute septicemic infection of *Pasteurella* organism in pond cultured Formosa snake-head fish

(*Channa maculata*) in Taiwan. *Fish Pathology* 20, 143–148.

Ueki, N., Kayano, Y. and Muroga, K. (1990) *Pasteurella piscicida* in juvenile red grouper. *Fish Pathology* 25, 43–44.

Vera, P., Navas, J.I. and Fouz, B. (1991) First isolation of *Vibrio damsela* from sea bream *Sparus aurata*. *Bulletin of the European Association of Fish Pathologists* 11, 112–113.

Wang, F.-I. and Chen, J.-C. (2006) Effect of salinity on the immune response of tiger shrimp *Penaeus monodon* and its susceptibility to *Photobacterium damselae* subsp. *damselae. Fish and Shellfish Immunity* 20, 671–681.

Wang, R., Feng, J., Su, Y., Ye, L. and Wang, J. (2013) Studies on the isolation of *Photobacterium damselae*

subsp. *piscicida* from diseased golden pompano (*Trachinotus ovatus*) and antibacterial agents sensitivity. *Veterinary Microbiology* 162, 957–963.

Wolke, R.E. (1975) Pathology of bacterial and fungal diseases affecting fish. In: Ribelin, W.E. and Migaki, G. (eds) *The Pathology of Fishes*. The University of Wisconsin Press, Madison, Wisconsin, pp. 33–116.

Yasunaga, N., Hatai, K and Tsukahara, J. (1983) *Pasteurella piscicida* from an epizootic of cultured red seabream. *Fish Pathology* 18, 107–110.

Yasunaga, N., Yasumoto, S., Hirakawa, E. and Tsukahara, J. (1984) On a massive mortality of oval file fish (*Navodan modestus*) caused by *Pasteurella piscicida. Fish Pathology* 19, 51–55.

20 *Piscirickettsia salmonis*

JERRI BARTHOLOMEW,[1]* KRISTEN D. ARKUSH[2]
AND ESTEBAN SOTO[3]

[1]*Department of Microbiology, Oregon State University, Corvallis, Oregon, USA;*
[2]*formerly of Bodega Marine Laboratory, University of California-Davis, Bodega
Bay, California, USA;* [3]*Department of Medicine and Epidemiology, School of
Veterinary Medicine, University of California, Davis, California, USA*

20.1   Introduction

20.1.1   Description

Piscirickettsia salmonis (Fryer *et al.*, 1992) is a
Gram-negative, non-motile, facultative intracellular bacterium. The type strain, LF-89 (ATCC VR
1361), was recovered from an epizootic among
coho salmon (*Oncorhynchus kisutch*) in seawater
net pens in Chile (Fryer *et al.*, 1990). The bacterium
has since been recovered from other marine and
freshwater fishes (see Table 20.1). Although isolates
recovered from non-salmonid fishes are morphologically similar, only some of their identities have
been confirmed (Lannan *et al.*, 1991; Alday-Sanz
et al., 1994).

P. *salmonis* is predominantly coccoid and may
occur in pairs of curved rods or rings (Fryer *et al.*,
1990). The diameters of individual cells are 0.5–
1.5 µm. Replication occurs within membrane-
bound cytoplasmic vacuoles in host cells
(Fig. 20.1). The bacterium can be cultured in cell
lines (Fryer *et al.*, 1990) or on enriched agar and
broth (Mikalsen *et al.*, 2008). The bacterium stains
dark blue with Giemsa reagents, is Giménez-stain
negative (Giménez, 1964) and retains basic fuchsin
when stained using Pinkerton's method for rickettsia
and chlamydia (US Surgeon-General's Office, 1994);
see Fryer *et al.* (1990).

Comparison of 16S, internal transcribed spacer
(ITS) and 23S ribosomal DNA sequences indicated
that *P. salmonis* from Chile, Norway, Ireland,
Scotland and Canada are similar and cluster as a
monophyletic group, but the Chilean isolate EM-90

is different (Mauel *et al.*, 1999; Heath *et al.*, 2000).
Variations in antibiotic sensitivity and cytopathic
effect (CPE), in tissue culture (Smith *et al.*, 1996b),
in host range (Cvitanich *et al.*, 1991) and in virulence
(Smith *et al.*, 1996a; House *et al.*, 1999) have been
reported within strains.

20.1.2   Transmission

In terrestrial environments, an intermediate arthro-
pod host or vector transmits rickettsiae (Weiss and
Moulder, 1984), and the detection of *P. salmonis* in
the parasitic isopod *Ceratothoa gaudichaudii*
(Garcés *et al.*, 1994) suggests that ectoparasites may
facilitate the transmission of *P. salmonis*.

However, experimental studies have shown that
horizontal transmission can occur in fresh water
and seawater without a vector (Cvitanich *et al.*,
1991; Almendras *et al.*, 1997a). Marine fishes are
also potential reservoirs of the pathogen, and infected
white seabass (*Atractoscion nobilis*) transmit the
infection to Chinook (*O. tshawytscha*) and coho
(Arkush *et al.*, 2005) salmon via cohabitation.
Experimentally, *P. salmonis* can invade via the skin
and gills as well as via intestinal and gastric intubation (Smith *et al.*, 1999, 2004, 2015), indicating
that transmission could also occur by the consumption of infected prey. The detection of bacteria in
the intestines and kidneys of infected fish, along
with necrosis and sloughing of the mucosal epithelium and renal tubules, indicate the potential routes
of shed material (Cvitanich *et al.*, 1991; Arkush *et al.*,
2005; Smith *et al.*, 2015). The shedding of infected

---

*Corresponding author e-mail: bartholj@science.oregonstate.edu

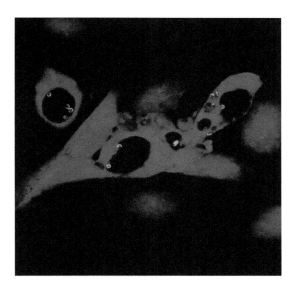

Fig. 20.1. Confocal image of *Piscirickettsia salmonis* replicating in cytoplasmic vacuoles in CHSE-214 (Chinook salmon, *Oncorhynchus tshawytscha*, embryo) cells. Scale bar = 5 μm. Image reprinted with permission.

cells from gill filaments, and even cannibalism, may cause further spread of the pathogen (Arkush *et al.*, 2005). It has been demonstrated experimentally that *P. salmonis* can survive 14 days in salt water but is almost immediately inactivated in fresh water (Lannan and Fryer, 1994). Longevity in seawater was also demonstrated in a field study in which *P. salmonis* was detected in the waters surrounding commercial aquaculture operations for 40 days after an outbreak of the disease occurred (Olivares and Marshall, 2010). Although sporogonic stages have not been observed, the discovery of a small infective variant of *P. salmonis* from ageing tissue cultures and from naturally infected fish suggests that the bacterium survives when conditions limit multiplication (Rojas *et al.*, 2008).

The vertical transmission of *P. salmonis* was demonstrated by intraperitoneally inoculating rainbow trout (*O. mykiss*) brood stock with the bacteria (Larenas *et al.*, 1996, 2003); these were later detected in the gonads, coelomic and seminal fluids, and the ova of inoculated fish. Larenas *et al.* (2003) detected bacteria in fry when one or both rainbow trout parents were inoculated, but even though the bacterium was detected in a proportion of the offspring as they grew (16–24% in alevins; 12–16% in fingerlings), they did not develop piscirickettsiosis. Progeny were also infected via the incubation of

eggs in a medium containing *P. salmonis* during fertilization. Using scanning electron microscopy, Larenas *et al.* (2003) demonstrated that *P. salmonis* penetrated the ovum through the chorion as early as 1 min after contact with the egg. The low number of naturally occurring piscirickettsiosis cases reported in juvenile fish in fresh water (Bravo, 1994; Gaggero *et al.*, 1995) suggests that vertical transmission is rare. Nevertheless, because *P. salmonis* is rapidly deactivated in fresh water (Lannan and Fryer, 1994), horizontal transmission is unlikely unless it is protected by a host cell membrane or other biological material.

20.1.3   Geographic and host distribution

Bacteria identified as *P. salmonis*, *Piscirickettsia*-like organisms (PLOs) or *Rickettsia*-like organisms (RLOs) occur globally in freshwater and marine fishes and invertebrates (see Fryer and Lannan, 1994; Mauel and Miller, 2002; Rozas and Enríquez, 2014). Many of these PLOs or RLOs were subsequently shown to have an affinity to *Francisella* spp.

P. salmonis associated with piscirickettsiosis of salmon has been reported from Chile, Ireland, Norway, Scotland and Canada (Table 20.1) and it is likely that the host range and distribution of *P. salmonis* will expand with further studies. This disease was probably first recognized during the early 1970s in British Columbia, Canada in pink salmon (*O. gorbuscha*) held for experimental purposes in seawater at the Pacific Biological Station (Evelyn *et al.*, 1998). Despite intermittent observations in pink, Chinook and coho salmon, it was only in 1992 that *P. salmonis* was isolated from diseased Atlantic and Chinook salmon in seawater net pens (Brocklebank *et al.*, 1993), when culture methods for the bacterium (Lannan and Fryer, 1994) and fluorescent antibody-based techniques (Lannan *et al.*, 1991) had become available.

Disease outbreaks in cultured salmonids occurred near Puerto Montt in southern Chile in 1989, but there was evidence to suggest that the disease had been present since the early 1980s (Bravo and Campos, 1989). Infections were detected in coho salmon, Atlantic salmon (*Salmo salar*), Chinook salmon, rainbow trout (Cvitanich *et al.*, 1991) and masu salmon (*O. masou*) (Bravo, 1994). Although the infections were mainly in sea and brackish water fish, the disease also occurs in freshwater fish (Cvitanich *et al.*, 1991). During the 1990s, *P. salmonis* was recovered from saltwater net-pen reared

Table 20.1. *Piscirickettsia salmonis* geographic and host range.

Location	Fish	Organism observed (O) Organism isolated (I) DNA detected (DNA)	Reference(s)
Australia, Tasmania	*Salmo salar*	O/DNA	Corbeil *et al.*, 2005
Canada, British Columbia	*Oncorhynchus gorbuscha, O. kisutch, O. tshawytscha, Salmo salar*	O/I/DNA	Brocklebank *et al.*, 1993; Evelyn *et al.*, 1998
Canada, Nova Scotia	*Salmo salar*	O/I/DNA	Jones *et al.*, 1998; Cusack *et al.*, 2002
Chile	*Oncorhynchus gorbuscha, O. kisutch, O. masou, O. mykiss, O. tshawytscha, Salmo salar*	O/I/DNA	Bravo and Campos, 1989; Fryer *et al.*, 1990; Branson and Nieto Díaz-Munoz, 1991; Cvitanich *et al.*, 1991; Garcés *et al.*, 1991; Bravo, 1994; Gaggero *et al.*, 1995; Smith *et al.*, (1995)
Chile	*Eleginops maclovinus, Odontesthes regia, Salilota australis, Sebastes capensis*	DNA	Contreras-Lynch *et al.*, 2015
Europe	*Dicentrarchus labrax*	O/I/DNA	Comps *et al.*, 1996; Steiropoulos *et al.*, 2002; Athanassopoulou *et al.*, 2004
Ireland	*Salmo salar*	O/I/DNA	Rodger and Drinan, 1993
Norway	*Salmo salar*	O/I/DNA	Olsen *et al.*, 1997
USA, California	*Atractoscion nobilis*	O/I/DNA	Chen *et al.*, 2000
USA, Oregon	Coastal water sample	DNA	Mauel and Fryer (2001)

Atlantic salmon in Norway (Olsen *et al.*, 1997), Ireland (Rodger and Drinan, 1993), Scotland (Grant *et al.*, 1996) and Canada (Jones *et al.*, 1998). A genetically distinct strain of *P. salmonis*, most closely related to strain EM-90 from Atlantic salmon in Chile was subsequently detected in farmed Atlantic salmon from south-east Tasmania (Corbeil *et al.*, 2005).

P. salmonis also occurs in non-salmonids. An isolate obtained from hatchery-reared white seabass in California, USA (WSB-98; Chen MF *et al.*, 2000) was 99% homologous with the type strain LF-89 (Arkush *et al.*, 2005), and this isolate induced disease and mortality in experimentally infected juvenile white seabass (Arkush *et al.*, 2005). An RLO from European sea bass (*Dicentrarchus labrax*) from the Mediterranean Sea that had severe encephalitis was also related to LF-89 (McCarthy *et al.*, 2005). Additional detections of the bacterium in Chile made using polymerase chain reaction (PCR) assays were from the Patagonian blenny (*Eleginops maclovinus*), Cape redfish (*Sebastes capensis*) and tadpole codling (*Salilota australis*) (Contreras-Lynch *et al.*, 2015).

20.1.4 Impact

Annual losses due to piscirickettsiosis have affected the Chilean economy, a country that is second to Norway as the largest exporter of salmon and trout (Wilhelm *et al.*, 2006). In 1989, >1.5 million coho salmon died and the economic loss was US$10 million (Larenas *et al.*, 2000). In 1995, >10 million coho salmon died during the seawater rearing phase, mostly from piscirickettsiosis, and the loss was US$49 million (Smith *et al.*, 1997). Improved prevention and control methods, as well as a shift from rearing coho salmon to the more resistant Atlantic salmon, reduced the severe effects of the disease in the early 2000s (Fryer and Hedrick, 2003). Mortality from the infectious salmon anaemia crisis in the mid-2000s eclipsed that attributed to piscirickettsiosis. However, piscirickettsiosis has re-emerged as a primary fish health challenge in Chile (Rozas and Enríquez, 2014), and Ibieta *et al.* (2011) reported that it represents 90% of the infectious disease mortality in seawater. In 2014, piscirickettsiosis caused 74.1, 36.7 and 73.5% of mortality in cultured Atlantic and coho salmon and rainbow trout, respectively (Sernapesca, 2014).

Between 2010 and 2014, >1700 tonnes of antimicrobials were used within the salmonid industry in Chile. Most of these antibiotics were used in marine aquaculture (96% of cases), with 90% used against piscirickettsiosis (Sernapesca, 2015).

The impacts of piscirickettsiosis on salmonid culture in other areas of the world are not as well documented. In British Columbia, infections are usually coincidental to other infectious diseases, though in some cases *P. salmonis* is the primary cause of mortality (Evelyn *et al.*, 1998). In Scotland, even though high mortality has been reported from Atlantic salmon farms, the incidence and overall impacts of the disease have been considered low (Rozas and Enríquez, 2014). In Norway, mortality was associated with poor environmental conditions (particularly algal blooms), overstocked net pens and poor smolt condition (Olsen *et al.*, 1997). Improvement of these conditions reduced the impact of the disease.

20.2 Diagnosis of the Infection

20.2.1 Clinical disease signs

Infections of *P. salmonis* in salmonids have been referred to as 'coho salmon syndrome' or 'Hitra disease' (Branson and Nieto Díaz-Munoz, 1991), 'salmonid rickettsial septicemia' (Cvitanich *et al.*, 1991) and 'piscirickettsiosis' (Fryer *et al.*, 1992).

The external signs vary, but they include inappetance and darkened body coloration. Infected fish may be lethargic and swim near the water surface or at the perimeter of enclosures. Erratic swimming was noted in infected coho salmon, and the bacterium was isolated from the brains of fish with clinical signs (Skarmeta *et al.*, 2000). Similarly, abnormal swimming and whirling behaviours were exhibited by infected juvenile seabass (*D. labrax*) (McCarthy *et al.*, 2005). Infected fish may have skin lesions that range from small areas of raised scales to shallow haemorrhagic ulcers (Branson and Nieto Díaz-Munoz, 1991). Anaemia, evidenced as pale gills, is consistently present (Cvitanich *et al.*, 1991), although mortality may occur without signs of disease (Arkush *et al.*, 2005).

The disease in salmonids generally occurs 6–12 weeks after transfer to seawater (Bravo and Campos, 1989; Fryer *et al.*, 1990; Cvitanich *et al.*, 1991). Even though natural infections in fresh water are rare, Smith *et al.* (2015) reported >90% mortality in laboratory infections of rainbow trout fry in fresh water, and the clinical signs were consistent with those in naturally infected salmonids during acute piscirickettsiosis.

20.2.2 Diagnosis

Presumptive diagnosis is based on microscopic observations of the bacterium in stained tissue imprints or smears, or isolation of the bacterium using cell culture (Fryer *et al.*, 1990) or agar media (Mauel *et al.*, 2008; Mikalsen *et al.*, 2008; Yañez *et al.*, 2013). Confirmation of *P. salmonis* is made using an immunofluorescence antibody test (IFAT; Lannan *et al.*, 1991; OIE, 2006), immunohistochemistry (Alday-Sanz *et al.*, 1994; OIE, 2006) or PCR assays (House and Fryer, 2002; OIE, 2006).

Infected tissues (e.g. kidney, liver, brain and blood) are co-cultured with a susceptible cell line. In addition to replicating in most salmonid cell lines (Fryer *et al.*, 1990), the bacterium has been isolated using cell lines from fathead minnow (*Pimephales promelas*) (Fryer *et al.*, 1990), brown bullhead catfish (*Ameiurus nebulosus*) (Almendras *et al.*, 1997b) and the fall armyworm (*Spodoptera frugiperda*, the Sf21 cell line) (Birkbeck *et al.*, 2004a). Because the organism is sensitive to most of the antibiotics routinely used in cell culture (Lannan and Fryer, 1991), all media, including tissue transport or processing buffers, should be antibiotic free. The bacterium is also sensitive to elevated temperatures and freezing, so aseptically removed tissues should be kept at 4°C. Inoculated cultures are incubated for optimal growth at 15–18°C and examined for 28 days for CPEs (Fryer *et al.*, 1990), which appear as plaque-like clusters of rounded cells with large vacuoles. At primary isolation, any CPE (Fig 20.2) may require 21 or more days to appear. In subsequent passages, the CPE becomes apparent 4–7 days after inoculation. In salmonid cell lines, replication is inhibited above 20°C and below 10°C. Titres of *P. salmonis* may decrease by 99% after a single freeze–thaw cycle at −70°C; however, the addition of 10% dimethylsulfoxide as a cryoprotectant improves survival tenfold (Fryer *et al.*, 1990). The organism can produce a titre of 10^6 to 10^7 50% tissue culture infective doses/ml in cultured fish cells, and approximately a 100 times greater yield in Sf21 cells (Birkbeck *et al.*, 2004a).

The bacterium can also be cultured on agar and in broth supplemented with sheep's blood and cysteine (Mauel *et al.*, 2008; Mikalsen *et al.*, 2008). Colonies appear convex, greyish white, shiny, centrally opaque and with translucent

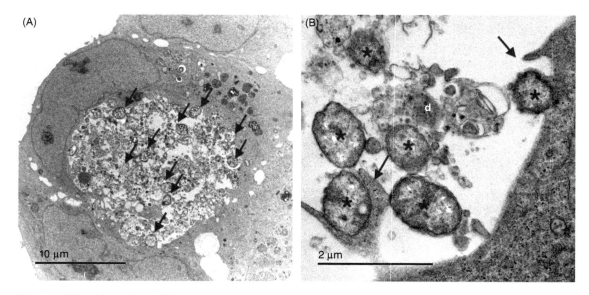

Fig. 20.2. Ultra-thin section of *Piscirickettsia salmonis* infecting CHSE-214 (Chinook salmon, *Oncorhynchus tshawytscha*, embryo) cells: (A) coccoid bacteria (arrows) and associated necrotic cellular debris within a ~20 um diameter vacuole; (B) internalization (arrows) of recently released bacteria (asterisk) by CHSE-214 cells, with debris from recently lysed cell noted by 'd'.

slightly undulating margins (Fig. 20.3). Two blood-free agar media have recently been described (Yañez *et al.*, 2013).

A less sensitive but more rapid method for presumptive diagnosis is via the microscopic examination of smears, impressions or histological sections of kidney, liver or spleen tissue stained with Gram, Giemsa or methylene blue solutions (Lannan and Fryer, 1991). In Giemsa-stained preparations, the pleomorphic bacteria appear within host cell cytoplasmic vacuoles darkly stained in coccoid or ring forms, frequently in pairs, with a diameter of 0.5–1.5 μm (Fig. 20.4).

Following its initial detection, the identity of *P. salmonis* can be confirmed serologically using immunofluorescence (Lannan *et al.*, 1991; Jamett *et al.*, 2001), immunohistochemistry (Alday-Sanz *et al.*, 1994; Steiropoulos *et al.*, 2002) or enzyme-linked immunosorbent assay (ELISA) (Aguayo *et al.*, 2002). Alternatively, the presence of *P. salmonis* can be established using molecular techniques. PCR assays amplify target sequences in the small subunit ribosomal (16S) gene (Mauel *et al.*, 1996) or the ITS region (Marshall *et al.*, 1998). A dot blot DNA hybridization test may detect *P. salmonis* from fish tissues and cell culture supernatants, while *in situ* hybridization can visualize infected cells in tissues

Fig. 20.3. *Piscirickettsia salmonis* colonies on modified Thayer–Martin (MTMII) agar (Becton Dickenson BBL, Sparks, Maryland) for 10 days at 20°C.

(Venegas *et al.*, 2004). A DNA microarray, by means of which PCR products or genomic DNA samples can be interrogated for the presence of *P. salmonis* and 14 other fish pathogens, has been described by Warsen *et al.* (2004). More recently, a multiplex PCR was designed to detect *P. salmonis*, *Streptococcus phocae*, *Aeromonas salmonicida* and *Vibrio anguillarum* (Tapia-Cammas *et al.*, 2011).

J. Bartholomew *et al.*

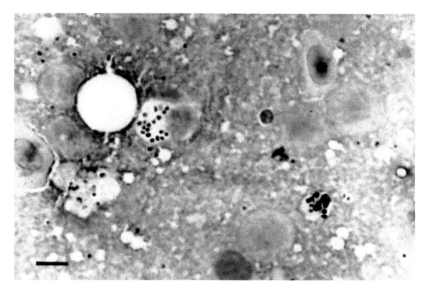

Fig. 20.4. Kidney impression from moribund coho salmon (*Oncorhynchus kisutch*) showing *Piscirickettsia salmonis* organisms within and among host cells. Giemsa stain. Bar = 10 µm. Photograph courtesy of Sandra Bravo, Puerto Montt, Chile. Image reprinted with permission.

20.3 Pathology

Infected fish often have skin lesions that range from small areas of raised scales to shallow ulcers (Fig. 20.5) (Cvitanich *et al.*, 1990; Fryer *et al.*, 1990) and may evolve into diffuse skin ulcers over the body, including the operculum and peduncle (Rozas and Enríquez, 2014).

Internally, the piscirickettsiae spread systemically, resulting in a swollen kidney and spleen. Although they are present only in 5–10% of infected fish, the most diagnostic lesions occur in the liver where they present as 5–6 mm diameter white or cream-coloured circular opaque nodules or as ring-shaped foci (Lannan and Fryer, 1993). Characteristic of septicaemic infections, ascites is often present, as are petechial haemorrhages in the musculature, in organs including the stomach, intestines, pyloric caeca, swim bladder and adipose tissue (Fig. 20.5) (Cvitanich *et al.*, 1991; Brocklebank *et al.*, 1993; Rodger and Drinan, 1993). Caverns with serosanguinous exudate in the skeletal muscle are increasingly reported, especially in trout. These unique lesions are generally associated with a small variant of *P. salmonis* (Rojas *et al.*, 2008).

The histopathology associated with *P. salmonis* has been reviewed (see Almendras and Fuentealba, 1997; Fryer and Hedrick, 2003; Rozas and Enríquez, 2014). Normal haematopoietic and lymphoid tissues

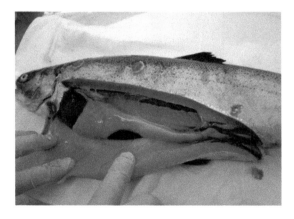

Fig. 20.5. Rainbow trout (*Oncorhynchus mykiss*) infected with *Piscirickettsia salmonis* showing shallow haemorrhagic ulcers in the skin and petechial haemorrhages in the liver and coelom. Photo courtesy of Jaime Tobar, Centrovet Ltda, Santiago, Chile.

in the kidney and spleen are replaced with chronic inflammatory cells and host cell debris. Liver lesions (Fig. 20.6) are severe and the piscirickettsiae are often seen in the cytoplasm of degenerating hepatocytes and in macrophages, or free in the blood. Granulomas have been reported in the liver tissue, and in acute infections this presents as a mottled

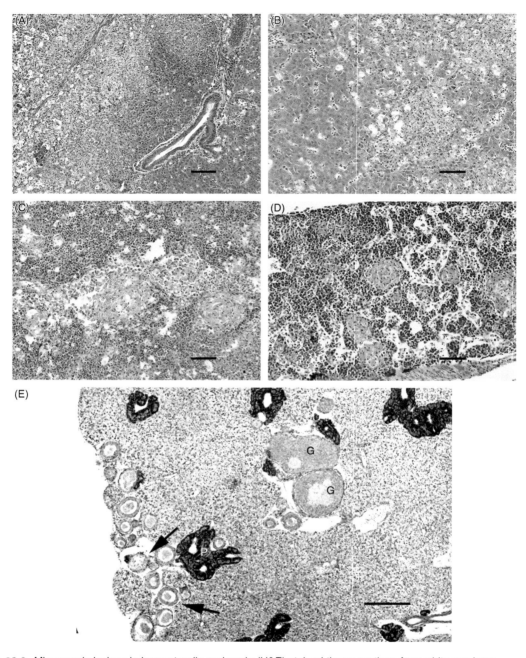

Fig. 20.6. Microscopic lesions in haematoxylin and eosin (H&E) stained tissue sections from white sea bass (*Atractoscion nobilis*) following experimental infections with *Piscirickettsia salmonis*: (A) focal hepatic lesions characterized by necrosis and infiltration with mononuclear cells, scale bar = 100 µm; (B) foci of necrosis in the liver, scale bar = 50 µm; (C) foci of necrosis in the spleen, scale bar = 100 µm; (D) multiple and focal aggregates of macrophages in the interstitium of the anterior kidney, scale bar = 100 µm; (E) large and well-developed granulomas (G) and smaller and numerous granulomas (arrows) near islet of exocrine pancreas (P) in the liver 123 days after intraperitoneal injection with the bacterium, scale bar = 200 µm.

J. Bartholomew *et al.*

appearance to the organ (Arkush *et al.*, 2005). Vascular and perivascular necrosis may be present in the liver, kidney and spleen, with intravascular coagulation resulting in fibrin thrombi within small blood vessels and infiltration by inflammatory cells (Branson and Nieto Díaz-Munoz, 1991; Cvitanich *et al.*, 1991). Granulomatous inflammation of the large intestine is common and often results in necrosis and sloughing of the mucosal epithelium. Granulomatous inflammation of the gill epithelia results in fusion of the lamellae (Branson and Nieto Díaz-Munoz, 1991). In the dermis and epidermis, granulomatous inflammation and necrosis, with degeneration of the subdermal musculature, may be observed (Brocklebank *et al.*, 1993; Chen *et al.*, 1994). Meningitis, endocarditis and pancreatitis may also be present.

Almendras *et al.* (1997a) infected Atlantic salmon via the intraperitoneal, oral and gill routes, and by cohabitation with infected fish. In intraperitoneally infected fish, a capsular (serosal) pattern occurred in the liver and spleen and spread to the leucocytes and other parenchymal cells of the spleen, liver and kidney. In fish exposed via the oral and gill routes, and fish indirectly infected via cohabitation, infections were characterized by a haematogenous pattern in which bacteria invaded organs from the blood vessels to the surrounding tissues. Vascular lesions appeared to result from the tropism of *P. salmonis* for endothelial cells, whereas lesions in the liver were the result of ischaemic necrosis and direct injury by the intracytoplasmic bacteria (Almendras *et al.*, 2000).

Using immunohistochemical methods, Smith *et al.* (2015) detected *P. salmonis* attached to the surface of the external epithelial cell layer of the skin and gills of rainbow trout fry within 5 min post immersion, in the oesophagus after 15 min and in the blood after 18 h. The skin and gills were also entry portals in experimentally exposed coho salmon (Smith *et al.*, 2004). Whether by initial serosal or haematogenous dissemination of the bacteria, during the later stages of infection, the internal gross and microscopic pathological changes in these species are similar, probably because of septicaemia.

In white seabass, a histiocytic inflammatory response and organization of macrophages into discrete aggregates facilitate invasion of the blood vessels and spread to other tissues by infected macrophages (Arkush *et al.*, 2005). The survival and persistence of intracellular *P. salmonis*, especially within granulomatous lesions, enables evasion of the host's immunological surveillance and allows the bacteria to persist in the host. Reactivation of these lesions has not been experimentally demonstrated, and thus a true carrier state has not been established.

The specific antibody response of experimentally infected coho salmon and rainbow trout was weak, and most likely directed against lipooligosaccharide components of the lipopolysaccharide (Kuzyk *et al.*, 1996). In white seabass that survived experimental exposures to the bacterium, a high percentage of the fish produced serum anti-*P. salmonis* antibodies (81.4%), but the strength of those responses varied (Arkush *et al.*, 2006). Such variations in antibody response are not unusual in fish, and the role of cell-mediated immunity against *P. salmonis* requires further investigation.

20.3.1 Virulence factors

Few virulence genes have been described in *P. salmonis*, and the expression and functionality of the predicted proteins is not well studied. A number of studies have linked phenotypic differences between *P. salmonis* strains with genetic diversity. Polymorphisms in lipopolysaccharide (LPS) genes were linked with mucoid phenotypic diversity (Bohle *et al.*, 2014); this study also identified homologues to type IV secretion systems (T4SSs) in *P. salmonis* that showed diversity in the number of T4SS-related genes. Interestingly, the overexpression of some of the T4SS homologues in *P. salmonis* occurred at pH <5, indicative of their potential role in bacterial survival after phagocytosis (Gómez *et al.*, 2013). Utilizing the SHK-1 (Atlantic salmon head kidney) cell line infected with *P. salmonis*, Isla *et al.* (2014) found significant increases in the expression of *P. salmonis* proteins ClpB and BipA. These proteins are important virulence factors for mammalian facultative intracellular pathogens like *Legionella* and *Francisella*, and probably play a similar role in the intracellular survival of *P. salmonis* (Meibom *et al.*, 2008; Isla *et al.*, 2014). Whole genome sequencing also revealed putative Fur boxes, to which Fur (ferric uptake repressor) proteins bind with high affinity, and the expression of these genes under iron-depleted conditions may allow the bacterium to cope with iron-deprivation mechanisms in the host (Pulgar *et al.*, 2015).

Finally, differences in gyrA (DNA gyrase subunit A) sequences were associated with different susceptibilities to quionolones (Henríquez *et al.*, 2015). Unfortunately, the lack of molecular methods for the genetic manipulation of *P. salmonis* has hampered the functional analysis of these virulence factor candidates.

20.4 Protective and Control Strategies

20.4.1 Husbandry

Simple husbandry, including the fallowing of farms, reduced rearing densities and holding single-year classes at a site to reduce horizontal transmission may facilitate avoidance of piscirickettsiosis. Although vertical transmission has only been demonstrated experimentally, the screening of adult brood stock may further reduce the risk of infection. In Chile, a surveillance programme was adopted to decrease the impact of disease through early detection and appropriate controls (Sernapesca, 2013).

20.4.2 Antimicrobial therapy

P. salmonis is sensitive under *in vitro* conditions to many antibiotics, including gentamicin, streptomycin, erythromycin, tetracycline and chloramphenicol (Fryer *et al.*, 1990; Cvitanich *et al.*, 1991). Oxolinic acid, amoxicillin, erythromycin, florfenicol, flumequine and oxytetracycline are used to treat piscirickettsiosis in Chile, although use of oxolinic acid and flumequine has declined because of their importance in human medicine (Rozas and Enríquez, 2014). Moreover, antibiotic treatments have proven inconsistent in the field (Smith *et al.*, 1996b, 1997). For example, feeds medicated with oxolinic acid (10–30 mg/kg daily for 10–15 days) and oxytetracycline (55–100 mg/kg daily for 10–15 days) are most effective, but infected fish are slow to respond and require repeated applications (Palmer *et al.*, 1997). Resistance to certain antibiotics has been reported (Rodger and Drinan, 1993), and variable antimicrobial sensitivity has increased the number of treatments and dosages (Rozas and Enríquez, 2014).

20.4.3 Vaccines

There is a large number (33 as reported by Rozas and Enríquez, 2014) of commercially available vaccines in Chile, of which 29 are injectable inactivated bacterins and four are subunit vaccines that may be administered in combination with other bacterial or viral vaccines. Vaccines based on *P. salmonis* bacterins have yielded mixed results. In coho salmon, a formalin-inactivated bacterin elicited a dose-dependent response, which in some instances exacerbated the disease (Kuzyk *et al.*, 2001). Birkbeck *et al.* (2004b) showed that vaccines prepared from either heat- or formalin-inactivated *P. salmonis* gave significant protection. By using inactivated bacterins encapsulated in a proprietary MicroMatrix® technology, oral administration of vaccine during 10 consecutive days induced protection either as a primary vaccination or as a booster to a primary IP vaccination (Tobar *et al.*, 2011).

Protein and carbohydrate antigens of *P. salmonis* detected using rabbit polyclonal antisera were proposed as potential constituents within subunit vaccines (Kuzyk *et al.*, 1996; Barnes *et al.*, 1998), though when convalescent coho salmon and rainbow trout sera were tested using some of the same purified antigens, only weak reactions were detected (Kuzyk *et al.*, 1996). Because *P. salmonis* is an intracellular pathogen, the stimulation of cell-mediated immunity, including enhanced phagocytosis and intracellular killing, may be critical for successful vaccination (Barnes *et al.*, 1998).

Several recombinant vaccines have been tested. An outer surface lipoprotein of *P. salmonis*, OspA, recombinantly produced in *Escherichia coli*, conferred 59% relative percent survival (RPS) in vaccinated coho salmon (Kuzyk *et al.*, 2001). When T cell epitopes from tetanus toxin and measles virus fusion protein (both of which are immunogenic in mammals) were incorporated into an OspA fusion protein, vaccine efficacy was nearly three-fold (83.0% RPS) that of OspA protein alone (30.2% RPS) (Kuzyk *et al.*, 2001).

The use of more than one recombinant antigen may potentiate the immune response. A recombinant vaccine formulated with heat shock protein (Hsp) 60 and Hsp70 plus the flagellar protein of *P. salmonis*, had a 95% RPS in Atlantic salmon (Wilhelm *et al.*, 2006). Reactive antibodies were detected in the sera of fish at least 8 months postvaccination. A highly immunogenic protein, ChaPs, from naturally infected coho salmon in southern Chile is another potential candidate for vaccine development (Marshall *et al.*, 2007). Sequence analysis of ChaPs demonstrated that it is also a heat shock protein, and these are molecules that have previously been exploited in recombinant vaccines (Wilhelm *et al.*, 2006).

In aquaculture, the efficacy of commercially available inactivated and subunit vaccines is generally high through the first winter but decreases in both vaccinated and unvaccinated groups in early spring (Leal and Woywood, 2007). Injectable vaccines are relatively effective in preventing outbreaks that occur after initial transfer to seawater, but not to subsequent outbreaks (Tobar *et al.*, 2011). These later outbreaks affect larger fish and result in greater economic loss. A booster vaccination is neither technically nor economically feasible. However, an oral booster vaccination could be an attractive alternative. A solution of injectable vaccine delivered per os produced antibodies against *P. salmonis* by 300 degree days after feeding and was protective when administered either as a primary or booster vaccine (Rozas and Enríquez, 2014). Despite these advances in experimental vaccination, there has been no significant reduction of mortality on aquaculture farms (Rozas and Enríquez, 2014).

Treatment approaches are also being revolutionized through the identification of molecular biomarkers. Microarray-based experiments have been used to identify differentially expressed genes in Atlantic salmon macrophages and anterior kidney cells in response to infection (Rise *et al.*, 2004). Transcripts were selected for potential utility as biomarkers in the molecular pathogenesis of *P. salmonis* infection and in the evaluation of vaccines and therapeutants. For example, changes were detected in the redox status of infected macrophages that may enable these cells to tolerate *P. salmonis* infection. It might be possible, then, to reduce haematopoietic tissue damage using an antioxidant.

20.5   Conclusions and Suggestions for Future Research

In recent years, significant advances have been made in the development of recombinant vaccines to control piscirickettsiosis. Presumably, field trials of these control measures will be conducted with commercialization to follow. However, key elements of the biology of the pathogen, such as the possibility of a carrier state in recovered fish, transmission among and between anadromous salmonids and marine fish species, and environmental or host factors that may precipitate disease outbreaks have to be more fully explored. Broader geographic and host comparisons of *P. salmonis* and PLO agents

via molecular analysis are needed to delineate phylogenetic relationships.

With improvements in *P. salmonis* culture methods, early detection of the bacterium should facilitate rapid diagnosis and mitigate losses caused by this pathogen. The greatest challenges include elucidation of *P. salmonis* pathogenesis, the characterization of virulence and vaccine development against this fastidious pathogen.

References

Aguayo, J., Miquel, A., Aranki, N., Jarnett, A., Valenzuela, P.D.T. et al. (2002) Detection of *Piscirickettsia salmonis* in fish tissue by an enzyme-linked immunosorbent assay using specific monoclonal antibodies. *Diseases of Aquatic Organisms* 49, 33–38.

Alday-Sanz, V., Rodger, H., Turnbull, T., Adams, A. and Richards, R.H. (1994) An immunohistochemical diagnostic test for rickettsial disease. *Journal of Fish Diseases* 17, 189–191.

Almendras, F.E. and Fuentealba, I.C. (1997) Salmonid rickettsial septicaemia caused by *Piscirickettsia salmonis*: a review. *Diseases of Aquatic Organisms* 29, 137–144.

Almendras, F.E., Fuentealba, I.C., Jones, S.R.M., Markham, F. and Spangler, E. (1997a) Experimental infection and horizontal transmission of *Piscirickettsia salmonis* in freshwater-raised Atlantic salmon, *Salmo salar* L. *Journal of Fish Diseases* 20, 409–418.

Almendras, F.E., Jones, S.R.M., Fuentealba, C. and Wright, G.M. (1997b) *In vitro* infection of a cell line from *Ictalurus nebulosus* with *Piscirickettsia salmonis*. *Canadian Journal of Veterinary Research* 61, 66–68.

Almendras, F.E., Fuentealba, I.C., Frederic Markham, R.F. and Speare, D.J. (2000) Pathogenesis of liver lesions caused by experimental infection with *Piscirickettsia salmonis* in juvenile Atlantic salmon, *Salmo salar* L. *Journal of Diagnostic Investigation* 12, 552–557.

Arkush, K.D., McBride, A.M., Mendonca, H.L., Okihiro, M.S., Andree, K.B. et al. (2005) Genetic characterization and experimental pathogenesis of *Piscirickettsia salmonis* isolated from white seabass *Atractoscion nobilis*. *Diseases of Aquatic Organisms* 73, 131–139.

Arkush, K.D., Edes, H.L., McBride, A.M., Adkinson, M.A. and Hedrick, R.P. (2006) Persistence of *Piscirickettsia salmonis* and the detection of serum antibodies to the bacterium in white seabass *Atractoscion nobilis* following experimental exposure. *Diseases of Aquatic Organisms* 73, 139–149.

Athanassopoulou, F., Groman, D., Prapas, T. and Sabatakou, O. (2004) Pathological and epidemiological observations on rickettsiosis in cultured sea bass (*Dicentrarchus labrax* L.) from Greece. *Journal of Applied Ichthyology* 20, 525–529.

Barnes, M.N., Landolt, M.L., Powell, D.B. and Winston, J.R. (1998) Purification of *Piscirickettsia salmonis* and partial characterization of antigens. *Disease of Aquatic Organisms* 33, 33–41.

Birkbeck, T.H., Griffen, A.A., Reid, H.I., Laidler, L.A. and Wadsworth, S. (2004a) Growth of *Piscirickettsia salmonis* to high titers in insect tissue culture cells. *Infection and Immunity* 72, 3693–3694.

Birkbeck, T.H., Rennie, S., Hunter, D., Laidler, L.A. and Wadsworth, S. (2004b) Infectivity 400 of a Scottish isolate of *Piscirickettsia salmonis* for Atlantic salmon *Salmo salar* and 401 immune response of salmon to this agent. *Diseases of Aquatic Organisms* 60, 97–103.

Bohle, H., Henríquez, P., Grothusen, H., Navas, E., Sandoval, A. *et al.* (2014) Comparative genome analysis of two isolates of the fish pathogen *Piscirickettsia salmonis* from different hosts reveals major differences in virulence-associated secretion systems. *Genome Announcements* 2(6):e01219–14.

Branson, E.J. and Nieto Díaz-Munoz, D. (1991) Description of a new disease condition occurring in farmed coho salmon, *Oncorhynchus kisutch* (Walbaum), in South America. *Journal of Fish Diseases* 14, 147–156.

Bravo, S. (1994) Piscirickettsiosis in freshwater. *Bulletin of the European Association of Fish Pathologists* 14, 137–138.

Bravo, S. and Campos, M. (1989) Coho salmon syndrome in Chile. *American Fisheries Society/Fish Health Section Newsletter* 17(3), 3.

Brocklebank, J.R., Evelyn, T.P., Speare, D.J. and Armstrong, R.D. (1993) Rickettsial septicemia in farmed Atlantic and Chinook salmon in British Columbia: clinical presentation and experimental transmission. *The Canadian Veterinary Journal* 34, 745–748.

Chen, M.F., Yun, S., Marty, G.D., McDowell, T.S., House, M.L. *et al.* (2000) A *Piscirickettsia salmonis*-like bacterium associated with mortality of white seabass *Atractoscion nobilis*. *Diseases of Aquatic Organisms* 43, 117–126.

Chen, S.C., Tung, M.C., Chen, S.P., Tsai, J.F., Wang, P.C. *et al.* (1994) System granulomas caused by a rickettsia-like organism in Nile tilapia, *Oreochromis niloticus* (L.), from southern Taiwan. *Journal of Fish Diseases* 17, 591–599.

Comps, M., Raymond, J.C. and Plassiart, G.N. (1996) Rickettsia-like organism infecting juvenile sea-bass *Dicentrarchus labrax*. *Bulletin of the European Association of Fish Pathologists* 16, 30–33.

Contreras-Lynch, S., Olmos, P., Vargas, A., Figueroa, J., González-Stegmaier, R. *et al.* (2015) Identification and genetic characterization of *Piscirickettsia salmonis* in native fish from southern Chile. *Diseases of Aquatic Organisms* 115, 233–244.

Corbeil, S., Hyatt, A.D. and Crane, M.S.J. (2005) Characterisation of an emerging rickettsia-like organism in Tasmanian farmed Atlantic salmon *Salmo salar*. *Diseases of Aquatic Organisms* 64, 37–44.

Cusack, R.R., Groman, D.B. and Jones, S.R.M. (2002) The first case of rickettsial infections in farmed Atlantic salmon in eastern North America. *The Canadian Veterinary Journal* 43, 435–440.

Cvitanich, J.D., Garate, N.O. and Smith, C.E. (1990) Etiological agent in Chilean Coho disease isolated and confirmed by Koch's postulates. *American Fisheries Society/Fish Health Section Newsletter* 18(1), 1–2.

Cvitanich, J.D., Garate, N.O. and Smith, C.E. (1991) The isolation of a rickettsia-like organism causing disease and mortality in Chilean salmonids and its confirmation by Koch's postulate [sic]. *Journal of Fish Diseases* 14, 121–145.

Evelyn, T.P.T., Kent, M.L., Poppe, T.T. and Bustos, P. (1998) Bacterial diseases. In: Kent, M.L. and Poppe, T.T (eds) *Diseases of Seawater Netpen-reared Salmonid Fishes*. Pacific Biological Station, Nanaimo, British Columbia, Canada, pp. 17–35.

Fryer, J.L. and Hedrick, R.P. (2003) *Piscirickettsia salmonis*: a Gram-negative intracellular bacterial pathogen of fish. *Journal of Fish Diseases* 26, 251–262.

Fryer, J.L. and Lannan, C.N. (1994) Rickettsial and chlamydial infections of freshwater and marine fishes, bivalves, and crustaceans. *Zoological Studies* 33, 95–107.

Fryer, J.L., Lannan, C.N., Garcés, L.H., Larenas, J.J. and Smith, P.A. (1990) Isolation of a *Rickettsiales*-like organism from diseased coho salmon *Oncorhynchus kisutch* in Chile. *Fish Pathology* 25, 107–114.

Fryer, J.L., Lannan, C.N., Giovannoni, S.J. and Wood, N.D. (1992) *Piscirickettsia salmonis* gen. nov., sp. nov., the causative agent of an epizootic disease in salmonid fishes. *International Journal of Systematic Bacteriology* 42, 120–126.

Gaggero, A., Castro, H. and Sandino, A.M. (1995) First isolation of *Piscirickettsia salmonis* from coho salmon, *Oncorhynchus kisutch* (Walbaum), and rainbow trout, *Oncorhynchus mykiss* (Walbaum), during the freshwater stage of their life cycle. *Journal of Fish Diseases* 18, 277–279.

Garcés, L.H., Larenas, J.J., Smith, P.A., Sandino, S., Lannan, C.N. and Fryer, J.L. (1991) Infectivity of a rickettsia isolated from coho salmon *Oncorhynchus kisutch*. *Diseases of Aquatic Organisms* 11, 93–97.

Garcés, L.H., Correal, P., Larenas, J.J., Contreras, J., Oyandel, S. *et al.* (1994) Finding *Piscirickettsia salmonis* on *Cerathothoa gaudichaudii*. In: Hedrick, R.P. and Winton, J.R. (organizers) *International Symposium on Aquatic Fish Health*: Program and Abstracts, Sheraton Hotel, Seattle, Washington, September 4–8, 1994, abstract, p. 109.

Giménez, D.F. (1964) Staining rickettsiae in yolk-sac cultures. *Stain Technology* [now *Biotechnic and Histochemistry*] 39, 139–140.

Gómez, F., Tobar, J.A., Henríquez, V., Sola, M., Altamirano, C. and Marshall, S. (2013) Evidence of the presence of a functional Dot/Icm type IV-B secretion system in the fish bacterial pathogen *Piscirickettsia salmonis*. *PLoS ONE* 8(1):e54934.

Grant, A.N., Brown, A.C., Cox, D.I., Birkbeck, T.H. and Griffen, A.A. (1996) Rickettsia-like organism in farmed salmon. *Veterinary Record* 138, 423.

Heath, S., Pak, S., Marshall, S., Prager, E.M. and Orrego, C. (2000) Monitoring *Piscirickettsia salmonis* by denaturant gel electrophoresis and competitive PCR. *Diseases of Aquatic Organisms* 41, 19–29.

Henríquez, P., Bohle, H., Bustamante, F., Bustos, P. and Mancilla, M. (2015) Polymorphism in gyrA is associated to quinolones resistance in Chilean *Piscirickettsia salmonis* field isolates. *Journal of Fish Diseases* 38, 415–418.

House, M.L. and Fryer, J.L. (2002) The biology and molecular detection of *Piscirickettsia salmonis*. In: Cunningham, C.O. (ed.) *Molecular Diagnosis of Salmonid Diseases*. Kluwer, Dordrecht, The Netherlands, pp. 141–155.

House, M.L., Bartholomew, J.L., Winton, J.R. and Fryer, J.L. (1999) Relative virulence of three isolates of *Piscirickettsia salmonis* for coho salmon *Oncorhynchus kisutch*. *Diseases of Aquatic Organisms* 35, 107–113.

Ibieta, P., Venegas, C., Takle, H., Hausdorf, M. and Tapia, V. (2011) *Chilean Salmon Farming on the Horizon of Sustainability: Review of the Development of a Highly Intensive Production, the ISA Crisis and Implemented Actions to Reconstruct a more Sustainable Aquaculture Industry*. InTech Open Science, Rijeka, Croatia.

Isla, A., Haussmann, D., Vera, T., Kausel, G. and Figueroa, J. (2014) Identification of the *clpB* and *bipA* genes and an evaluation of their expression as related to intracellular survival for the bacterial pathogen *Piscirickettsia salmonis*. *Veterinary Microbiology* 173, 390–394.

Jamett, A., Aguayo, J., Miquel, A., Muller, I., Arriagada, R. *et al.* (2001) Characteristics of monoclonal antibodies against *Piscirickettsia salmonis*. *Journal of Fish Diseases* 24, 205–215.

Jones, S.R.M., Markham, R.J.F., Groman, D.B. and Cusack, R.R. (1998) Virulence and antigenic characteristics of a cultured *Rickettsiales*-like organism isolated from farmed Atlantic Salmon *Salmo salar* in eastern Canada. *Diseases of Aquatic Organisms* 33, 25–31.

Kuzyk, M.A., Thornton, J.C. and Kay, W.W. (1996) Antigenic characterization of the salmonid pathogen *Piscirickettsia salmonis*. *Infection and Immunity* 64, 5205–5210.

Kuzyk, M.A., Burian, J., Machander, D., Dolhaine, D., Cameron, S. *et al.* (2001) An efficacious recombinant subunit vaccine against the salmonid rickettsial pathogen *Piscirickettsia salmonis*. *Vaccine* 19, 2337–2344.

Lannan, C.N. and Fryer, J.L. (1991) Recommended methods for inspection of fish for the salmonid rickettsia. *Bulletin of the European Association of Fish Pathologists* 11, 135–136.

Lannan, C.N. and Fryer, J.L. (1993) *Piscirickettsia salmonis*, a major pathogen of salmonid fish in Chile. *Fisheries Research* 17, 115–121.

Lannan, C.N. and Fryer, J.L. (1994) Extracellular survival of *Piscirickettsia salmonis*. *Journal of Fish Diseases* 17, 545–548.

Lannan, C.N., Ewing, S.A. and Fryer, J.L. (1991) A fluorescent antibody test for detection of the rickettsia causing disease in Chilean salmonids. *Journal of Aquatic Animal Health* 3, 229–234.

Larenas, J.J., Astorga, C., Contreras, J. and Smith, P. (1996) *Piscirickettsia salmonis* in ova obtained from rainbow trout (*Oncorhynchus mykiss*) experimentally inoculated. *Archivos de Medicina Veterinaria* 28, 161–166.

Larenas, J., Contreras, J. and Smith, P. (2000) *Piscirickettsiosis*: uno de los principales problemas en cultivos de salmones en Chile. *Revista de Extension Tecno Vet* 6, 28–30.

Larenas, J.J., Bartholomew, J.L., Troncoso, O., Fernandez, S., Ledezma, H. *et al.* (2003) Experimental vertical transmission of *Piscirickettsia salmonis* and *in vitro* study of attachment and mode of entrance into the fish ovum. *Diseases of Aquatic Organisms* 56, 25–30.

Leal, J. and Woywood, D. (2007) *Piscirickettsiosis* en Chile: avances y perspectivas para su control. *Salmociencia* 2, 34–42.

Marshall, S., Heath, S., Henríquez, V. and Orrego, C. (1998) Minimally invasive detection of *Piscirickettsia salmonis* in cultivated salmonids via the PCR. *Applied and Environmental Microbiology* 64, 3066–3069.

Marshall, S.H., Conejeros, P., Zahr, M., Olivares, J., Gómez, F. *et al.* (2007) Immunological characterization of a bacterial protein isolated from salmonid fish naturally infected with *Piscirickettsia salmonis*. *Vaccine* 25, 2095–2102.

Mauel, M.J. and Fryer, J.L. (2001) Amplification of a *Piscirickettsia salmonis*-like 16S rDNA product from bacterioplankton DNA collected from the coastal waters of Oregon, USA. *Journal of Aquatic Animal Health* 13, 280–284.

Mauel, M.J. and Miller, D.L. (2002) Piscirickettsiosis and piscirickettsiosis-like infections in fish: a review. *Veterinary Microbiology* 87, 279–289.

Mauel, M.J., Giovannoni, S.J. and Fryer, J.L. (1996) Development of polymerase chain reaction assays for detection, identification, and differentiation of *Piscirickettsia salmonis*. *Diseases of Aquatic Organisms* 26, 189–195.

Mauel, M.J., Giovannoni, S.J. and Fryer, J.L. (1999) Phylogenetic analysis of *Piscirickettsia salmonis* by 16S, internal transcribed spacer (ITS) and 23S ribosomal DNA sequencing. *Diseases of Aquatic Organisms* 35, 115–123.

Mauel, M.J., Ware, C. and Smith, P.A. (2008) Culture of *Piscirickettsia salmonis* on enriched blood agar. *Journal of Veterinary Diagnostic Investigation* 20, 213–214.

McCarthy, U., Steiropoulos, N.A., Thompson, K.D., Adams, A., Ellis, A.E. *et al.* (2005) Confirmation of *Piscirickettsia salmonis* as a pathogen in European sea bass *Dicentrarchus labrax* and phylogenetic comparison with salmonid strains. *Diseases of Aquatic Organisms* 64, 107–119.

Meibom, K.L., Dubail, I., Dupuis, M., Barel, M., Lenco, J., Stulik, J. *et al.* (2008) The heat-shock protein ClpB of *Francisella tularensis* is involved in stress tolerance and is required for multiplication in target organs of infected mice. *Molecular Microbiology* 67, 1384–1401.

Mikalsen, J., Skjaervik, O., Wiik-Nielsen, J., Wasmuth, M.A. and Colquhoun, D.J. (2008) Agar culture of *Piscirickettsia salmonis*, a serious pathogen of farmed salmonid and marine fish. *FEMS Microbiology Letters* 278, 43–47.

OIE (2006) Chapter 2.1.13. Piscirickettsiosis (*Piscirickettsia salmonis*). In: *Manual of Diagnostic Tests for Aquatic Animals 2006*, 5th edn. World Organisation for Animal Health, Paris, pp. 236–241. Available at: http://www.oie.int/doc/ged/D6510.PDF (accessed 17 November 2016).

Olivares, J. and Marshall, S.H. (2010) Determination of minimal concentration of *Piscirickettsia salmonis* in water columns to establish a fallowing period in salmon farms. *Journal of Fish Diseases* 33, 261–266.

Olsen, A.B., Melby, H.P., Speilberg, L., Evensen, Ø. and Håstein, T. (1997) *Piscirickettsia salmonis* infection in Atlantic salmon *Salmo salar* in Norway – epidemiological, pathological and microbiological findings. *Diseases of Aquatic Organisms* 31, 35–48.

Palmer, R., Rutledge, M., Callanan, K. and Drinan, E. (1997) A piscirickettsiosis-like disease in farmed Atlantic salmon in Ireland – isolation of the agent. *Bulletin of the European Association* of *Fish Pathologists* 17, 68–72.

Pulgar, R., Hödar, C., Travisany, D., Zuñiga, A., Domínguez, C. *et al.* (2015) Transcriptional response of Atlantic salmon families to *Piscirickettsia salmonis* infection highlights the relevance of the iron-deprivation defense system. *BMC Genomics* 16:495.

Rodger, H.D. and Drinan, E.M. (1993) Observation of a rickettsia-like organism in Atlantic salmon, *Salmo salar* L., in Ireland. *Journal of Fish Diseases* 16, 361–369.

Rojas, V., Olivares, J., del Río, R. and Marshall, S.H. (2008) Characterization of a novel and genetically different small infective variant of *Piscirickettsia salmonis*. *Microbial Pathogenesis* 44, 370–378.

Rozas, M. and Enríquez, R. (2014) *Piscirickettsiosis* and *Piscirickettsia salmonis* in fish: a review. *Journal of Fish Diseases* 37, 163–188.

Rise, M.L., Jones, S.R.M., Brown, G.D., von Schalburg, K.R., Davidson, W.S. *et al.* (2004) Microarray analyses identify molecular biomarkers of Atlantic salmon macrophage and hematopoietic kidney response to *Piscirickettsia salmonis* infection. *Physiological Genomics* 20, 21–35.

Sernapesca (2013) *Establece Programa Sanitario Específico de Vigilancia y Control de Piscirickettsiosis (PSEVC-PISCIRICKETTSIOSIS)*. Servicio Nacional

de Pesca y Acuicultura, Valparaíso, Chile. Available at: http://www.sernapesca.cl/index.php?option=com_ remository&Itemid=246&func=startdown&id=6726 (accessed 1 February 2016).

Sernapesca (2014) *Informe Sanitario de Salmonicultura en Centros Marinos – Año 2014*. Servicio Nacional de Pesca y Acuicultura, Valparaíso, Chile. Available at: http://www.sernapesca.cl/index.php?option=com_ remository&Itemid=246&func=fileinfo&id=11083 (accessed 17 November 2016).

Sernapesca (2015) *Informe Sobre el Uso de Antimicrobianos en Salmonicultura Nacional 2015*. Servicio Nacional de Pesca y Acuicultura, Valparaíso, Chile. Available at: http://www.sernapesca.cl/presentaciones/Comunicaciones/Informe_Sobre_Uso_de_ Antimicrobianos_2015.pdf (accessed 15 February 2016).

Skarmeta, A.M., Henriquez, V. Zahr, M., Orrego, C. and Marshall, S.H. (2000) Isolation of a virulent *Piscirickettsia salmonis* from the brain of naturally infected coho salmon (*Oncorhynchus kisutch*). *Bulletin of the European Association of Fish Pathologists* 20, 261–264.

Smith, P.A., Lannan, C.N., Garcés, L.H., Jarpa, M., Larenas, J. *et al.* (1995) Piscirickettsiosis: a bacterin field trial in coho salmon (*Oncorhynchus kisutch*). *Bulletin of the European Association of Fish Pathologists* 15, 137–141.

Smith, P.A., Contreras, J.R., Garcés, L.H., Larenas, J., Oyanedel, S. *et al.* (1996a) Experimental challenge of coho salmon and rainbow trout with *Piscirickettsia salmonis*. *Journal of Aquatic Animal Health* 8, 130–134.

Smith, P.A., Vecchiolla, I.M., Oyanedel, S., Garcés, L.H., Larenas, J. *et al.* (1996b) Antimicrobial sensitivity of four isolates of *Piscirickettsia salmonis*. *Bulletin of the European Association of Fish Pathologists* 16, 164–168.

Smith, P.A., Contreras, J.R., Larenas, J., Aguillon, J.C., Garcés *et al.* (1997) Immunization with bacterial antigens: piscirickettsiosis. In: Gudding, R., Lillehaug, A., Midtlyng, P.J. and Brown, F. (eds) *Fish Vaccinology*. Development of Biological Standards, Vol. 90, Karger, Basel, Switzerland, pp. 161–166.

Smith, P.A., Pizarro, P., Ojeda, P., Contreras, J.R., Oyanedel, S. *et al.* (1999) Routes of entry of *Piscirickettsia salmonis* in rainbow trout *Oncorhynchus mykiss*. *Diseases of Aquatic Organisms* 37, 165–172.

Smith, P.A., Pizarro, P., Ojeda, P., Contreras, J.R., Oyanedel, S. *et al.* (2004) Experimental infection of coho salmon *Oncorhynchus mykiss* by exposure of skin, gills and intestine with *Piscirickettsia salmonis*. *Diseases of Aquatic Organisms* 61, 53–57.

Smith, P.A., Díaz, F.E., Rojas, M.E., Díaz, S., Galleguillos, M. *et al.* (2015) Effect of *Piscirickettsia salmonis* inoculation on the ASK continuous cell line. *Journal of Fish Diseases* 38, 321–324.

Steiropoulos, N.A., Yuksel, S.A., Thompson, K.D., Adams, A. and Ferguson, H.W. (2002) Detection of *Rickettsia*-like organisms (RLOs) in European sea bass

(*Dicentrarchus labrax*) by immunohistochemistry. *Bulletin of the European Association of Fish Pathologists* 22, 260–267.

Tapia-Cammas, D., Yañez, A., Arancibia, G., Toranzo, A.E. and Avendaño-Herrera, R. (2011) Multiplex PCR for the detection of *Piscirickettsia salmonis*, *Vibrio anguillarum*, *Aeromonas salmonicida* and *Streptococcus phocae* in Chilean marine farms. *Diseases of Aquatic Organisms* 97, 135–142.

Tobar, J.A., Jerez, S., Caruffo, M., Bravo, C., Contreras, F. *et al.* (2011) Oral vaccination of Atlantic salmon (Salmo salar) against salmonid rickettsial septicaemia. *Vaccine* 29, 2336–2340.

US Surgeon-General's Office (1994) Laboratory methods of the United States Army. In: Simmons, J.S. and Gentzkow, C.J. (eds) *Laboratory Methods in the United States Army*, 5th edn. Lea and Febiger, Philadelphia, Pennsylvania, p. 572.

Venegas, C.A., Contreras, J.R., Larenas, J.J. and Smith, P.A. (2004) DNA hybridization assay for the detection of *Piscirickettsia salmonis* in salmonid fish. *Journal of Fish Diseases* 27, 431–433.

Warsen, A.E., Krug, M.J., LaFrentz, S., Stanek, D.R., Loge, F.J. *et al.* (2004) Simultaneous discrimination between 15 fish pathogens by using 16S ribosomal DNA PCR and DNA microarrays. *Applied and Environmental Microbiology* 70, 4216–4221.

Weiss, E. and Moulder, J.W. (1984) Order I. Rickettsiales Gieszczkiewicz 1939, 25[AL]. In: Kreig, N.R. (ed.) *Bergey's Manual of Systematic Bacteriology, Vol.* 1. Williams and Wilkins, Baltimore/London, pp. 687–729.

Wilhelm, V., Miquel, A., Burzio, L.O., Rosemblatt, M., Engel, E., Valenzuela, S. *et al.* (2006) A vaccine against the salmonid pathogen *Piscirickettsia salmonis* based on recombinant proteins. *Vaccine* 24, 5083–5091.

Yañez, A.J., Silva, H., Valenzuela, K., Pontigo, J.P., Godoy, M. *et al.* (2013) Two novel blood-free solid media for the culture of the salmonid pathogen *Piscirickettsia salmonis*. *Journal of Fish Diseases* 36, 587–591.

21   *Renibacterium salmoninarum*

DIANE G. ELLIOTT*

US Geological Survey, Western Fisheries Research Center, Seattle, Washington, USA

21.1   Introduction

Since the initial description of bacterial kidney disease (BKD) in wild Atlantic salmon (*Salmo salar*) in Scotland in the early 1930s (Smith, 1964), considerable effort has been expended to better understand and control this serious disease, and several comprehensive reviews have been written about BKD and its causative agent, *Renibacterium salmoninarum* (Fryer and Sanders, 1981; Austin and Austin, 1987; Elliott *et al.*, 1989; Evelyn, 1993; Evenden *et al.*, 1993; Fryer and Lannan, 1993; Pascho *et al.*, 2002; Wiens, 2011).

The geographic range of BKD encompasses both freshwater and marine habitats nearly worldwide where wild or cultured salmonids occur, including countries in North America, South America, Europe and Asia (Wiens, 2011; Kristmundsson *et al.*, 2016). Although all salmonids are susceptible to BKD, Pacific salmon (genus *Oncorhynchus*) are generally the most susceptible (Elliott *et al.*, 2014). Natural outbreaks have been restricted to the family Salmonidae (Fryer and Lannan, 1993; Pascho *et al.*, 2002), but *R. salmoninarum* has also been detected from non-salmonid fishes and bivalve molluscs collected within or near culture facilities or watersheds harbouring salmonids (US Fish and Wildlife Service, 2011; Elliott *et al.*, 2014). Several non-salmonids have also been experimentally infected (Wiens, 2011), raising the possibility that other species may serve as reservoirs or vectors for this pathogen.

The transmission of *R. salmoninarum* by both horizontal and vertical routes enhances its persistence within fish populations (Evelyn, 1993). Horizontal transmission occurs in both freshwater and seawater environments (Pascho *et al.*, 2002). The bacterium is shed in the faeces of infected fish

(Balfry *et al.*, 1996) and in cells exfoliating from fish surface tissues (Elliott *et al.*, 2015). Survival in the environment may be restricted by competition with normal aquatic microflora, and relatively short survival times (4–21 days at 10–18°C) have been reported in environmental samples (Pascho *et al.*, 2002). Therefore, transmission through water probably occurs over short distances (Murray *et al.*, 2012). Sites of entry for horizontal transmission are believed to include the gastrointestinal tract following the ingestion of infected fish carcasses or faeces, or surface tissues through injury sites (see Elliott *et al.*, 2015).

R. salmoninarum is transmitted vertically to progeny in association with the eggs (Pascho *et al.*, 2002); some bacteria are carried intra ovum within the yolk. Eggs may become infected via ovarian (coelomic) fluid containing high concentrations of bacteria, with post-ovulation infections presumably occurring through the micropyle. Egg infection may also occur directly from ovarian tissue early in oocyte development (Elliott *et al.*, 2014).

The bacterium is a small (0.3–1.0 μm by 1.0–1.5 μm), non-motile, non-spore-forming, non-acid-fast, slowly replicating Gram-positive rod and often occurs in pairs (Sanders and Fryer, 1980). The closest relatives of *R. salmoninarum* in the family *Micrococcaceae* are *Arthrobacter* spp., which are non-pathogenic soil organisms (Wiens *et al.*, 2008). The genome of *R. salmoninarum* comprises a single circular 3.15 Mb pair chromosome without plasmids or phage elements (Wiens *et al.*, 2008). While *R. salmoninarum* is geographically widespread in salmonids, different isolates of the bacterium show limited variation in their biochemical and serological profiles (Pascho *et al.*, 2002; Elliott *et al.*, 2014)

---

*E-mail: dgelliott@usgs.gov

and low genetic diversity (Pascho *et al.*, 2002; Brynildsrud *et al.*, 2014). Recently, however, next-generation sequencing technology used to generate genome-wide single-nucleotide polymorphism (SNP) data from diverse *R. salmoninarum* isolates has distinguished sublineages of the bacterium for epizootiological studies (Brynildsrud *et al.*, 2014).

BKD is a prevalent salmonid disease with mortality as high as 70 and 80% in some populations of Atlantic salmon and Pacific salmon, respectively (see Evenden *et al.*, 1993; Wiens, 2011), though accurate estimates of the mortality and overall impacts of *R. salmoninarum* on fish populations are difficult to determine due to the chronic nature of infections and frequent presence of co-infections (Evenden *et al.*, 1993). Mortality estimates are especially problematic for free-ranging populations because diseased fish may be vulnerable to predation (Mesa *et al.*, 1998). Nevertheless, clinical BKD has been observed in feral salmonids, including wild spawning populations with no history of supplementation with hatchery fish (Pascho *et al.*, 2002; Elliott *et al.*, 2014). Outbreaks are most frequently reported in cultured fish and the spread of *R. salmoninarum* has been facilitated by the expansion of salmonid aquaculture and the associated transfers of infected fish and eggs (Evenden *et al.*, 1993; Murray *et al.*, 2012). In addition, the immunosuppressive effects of *R. salmoninarum* infections (see Section 21.4) may contribute to deaths that are ascribed to secondary pathogens (Munson *et al.*, 2010).

21.2 Clinical Signs and Diagnosis

21.2.1 Clinical signs

R. salmoninarum infections usually progress slowly, with overt disease uncommon until fish are 6–12 months old (Evelyn, 1993). Fish severely infected with *R. salmoninarum* may exhibit no external signs or may show one or more of the following: lethargy, darkened or mottled skin, abdominal distension caused by ascites, pale gills associated with anaemia, exophthalmos, skin vesicles filled with clear, bloody or turbid fluid, petechial haemorrhages on the body, haemorrhages at the fin bases and near the vent, shallow skin ulcers and large cystic cavities or abscesses extending into the skeletal muscle (Pascho *et al.*, 2002; Bruno *et al.*, 2013). In some mature salmonids, particularly rainbow trout (*O. mykiss*), a superficial infection known as 'spawning rash' may manifest as a pustulous dermatitis

covering large areas of the skin, with many small blisters or haemorrhagic nodules in the epidermis (Bruno *et al.*, 2013). None of these signs are pathognomonic for BKD, and confirmation of *R. salmoninarum* requires accurate diagnosis. *R. salmoninarum* may also persist as subclinical infections in fish populations for prolonged periods (Pascho *et al.*, 2002; Elliott *et al.*, 2013).

21.2.2 Diagnosis

A number of different methods have been developed for the detection of *R. salmoninarum*, although a single ideal diagnostic test has not yet been identified (Pascho *et al.*, 2002; Elliott *et al.*, 2013). The use of standardized, validated assays with known limits of sensitivity, specificity and reliability is recommended (Purcell *et al.*, 2011; Elliott *et al.*, 2013).

Clinical BKD is often quickly diagnosed via the characteristic signs and lesions of the disease, followed by Gram or fluorescent antibody staining of tissue smears from lesions. The reliability of Gram staining for detecting subclinical infections is limited by its low sensitivity ($1 \times 10^{7-9}$ cells/g of tissue; see Wiens, 2011). The presence of melanin granules in tissues also can obscure low numbers of bacteria (Pascho *et al.*, 2002), and Gram stains do not provide specific identification.

Culture is the benchmark assay to verify the viability of *R. salmoninarum*, and several media have been developed for this (Pascho *et al.*, 2002; Wiens, 2011). Most of these are modifications of the KDM2 (kidney disease medium-2) medium, which contains 1% (w/v) peptone, 0.05% (w/v) yeast extract, 0.1% (w/v) cysteine and 7–20% (v/v) serum. The growth of *R. salmoninarum* in culture is aerobic and is optimal at 15–18°C (Sanders and Fryer, 1980). Visible colonies may appear within 5–7 days of inoculation (Faisal *et al.*, 2010a), but incubation times of up to 19 weeks may be required in subclinical cases (Pascho *et al.*, 2002). Reported *R. salmoninarum* detection limits for cultures of kidney homogenates are between 8×10^1 (80) and 5×10^2 cells/g (see Elliott *et al.*, 2013). However, contamination of cultures by faster growing heterotrophic bacteria can occur even when selective media containing antibiotics are used (see Pascho *et al.*, 2002), which can hinder the detection of *R. salmoninarum*. Identity is confirmed using biochemical tests (Austin and Austin, 1987) or serological and/or molecular methods (OIE, 2003).

Renibacterium salmoninarum

Because of the slow replication and fastidious requirements of *R. salmoninarum*, numerous immunological and molecular methods have been developed for its detection (Pascho *et al.*, 2002). Immunological methods include the fluorescent antibody technique (FAT) and enzyme-linked immunosorbent assay (ELISA), which are widely used and recommended (OIE, 2003; AFS-FHS, 2014). Both direct (DFAT) and indirect (IFAT) FAT methods using either polyclonal or monoclonal antibodies can detect *R. salmoninarum* in tissue smears or histological sections (Pascho *et al.*, 2002). Membrane-filtration-FAT (MF-FAT) assays have been employed for ovarian fluid and for diluted homogenates of kidney tissue (Pascho *et al.*, 2002). The detection limit for *R. salmoninarum* by FAT in tissue smears is 10^3–10^7 cells/g tissue (Pascho *et al.*, 2002; Elliott *et al.*, 2013), whereas the detection limit in filtered samples by MF-FAT is less than 10^2 cells/ml (Wiens, 2011; Elliott *et al.*, 2013). FAT identifies *R. salmoninarum* by both morphological and specific staining characteristics, but it does not determine viability, and the assays are labour intensive.

The ELISAs developed to detect specific *R. salmoninarum* antigens in tissues use either polyclonal or monoclonal antibodies (Pascho *et al.*, 2002; Wiens, 2011). Limited studies indicate the higher sensitivity of polyclonal ELISAs compared with monoclonal ELISAs (Jansson *et al.*, 1996; Elliott, unpublished data), owing to their recognition of additional antigenic epitopes (Pascho *et al.*, 2002). Affinity purification of antisera is recommended to ensure that the specificity of polyclonal ELISAs is adequate (Wiens, 2011). The minimum numbers of *R. salmoninarum* cells that are needed to produce soluble antigen levels detectable by ELISAs have been difficult to determine (Pascho *et al.*, 2002; Elliott *et al.*, 2013), but the estimated detection limit of a polyclonal ELISA is about 10^3 cells/g kidney (Jansson *et al.*, 1996). ELISAs are semi-quantitative (Pascho *et al.*, 2002; Elliott *et al.*, 2013) and are frequently used to screen brood stocks (see Section 21.5) or monitor infection in populations where BKD is endemic (e.g. Maule *et al.*, 1996; Elliott *et al.*, 1997; Meyers *et al.*, 2003; Faisal *et al.*, 2010b, 2012; Kristmundsson *et al.*, 2016). One advantage of ELISA is its ability to detect soluble *R. salmoninarum* antigens that circulate from localized tissue infections (Elliott *et al.*, 2015). As the soluble antigens can persist in the absence of live bacteria, a positive ELISA result does not necessarily indicate an active infection,

and it may require confirmation by another test (Wiens, 2011). ELISA is, therefore, more suited to screening rather than use as a confirmatory assay (Elliott *et al.*, 2013).

Molecular tests, particularly polymerase chain reaction (PCR) assays, have gained popularity for *R. salmoninarum* detection because of their specificity (Pascho *et al.*, 2002; Wiens, 2011), and a few PCRs are recommended as confirmatory tests (OIE, 2003; AFS-FHS, 2014). Conventional (cPCR), nested (nPCR) and real-time quantitative (qPCR) assays have been developed to detect *R. salmoninarum* DNA sequences (Pascho *et al.*, 2002; Wiens, 2011). The relative analytical sensitivities of these PCRs are assay specific and depend on factors such as primer sequences and reaction conditions (see Purcell *et al.*, 2011). One strength of PCR is its ability to amplify the target nucleic acid sequence from low numbers of organisms present in samples. In practice, PCR sensitivity is often reduced due to the focal or non-uniform distribution of *R. salmoninarum* in tissues of subclinically infected fish, combined with the small proportion of tissue that is actually tested in each PCR reaction following DNA extraction (Purcell *et al.*, 2011; Elliott *et al.*, 2013). The limit of detection for *R. salmoninarum* in kidney tissue by nPCR or qPCR is about $\geq10^3$ cells/g tissue (Elliott *et al.*, 2013). The PCR assays designed to detect DNA sequences do not determine viability. However, reverse transcription (RT) PCR assays have been developed to detect messenger RNA (mRNA), which has a short half-life and is indicative of the presence of viable bacteria (Pascho *et al.*, 2002; Elliott *et al.*, 2013). Nevertheless, mRNA expression may be affected by the differential expression of bacterial genes during active versus latent infections, and so the analytical sensitivity of RT-PCR for mRNA may be lower than that of PCR for the detection of *R. salmoninarum* DNA (Elliott *et al.*, 2013).

21.3 Pathology

21.3.1 Macroscopic internal lesions

At necropsy, the internal examination of diseased fish with BKD often reveals focal to multifocal nodular lesions in the kidney (Fig. 21.1) and in other organs, such as the heart, liver or spleen. Enlargement of the kidney and spleen is frequent (Fig. 21.1). Other macroscopic observations may include turbid or serosanguineous fluid in the

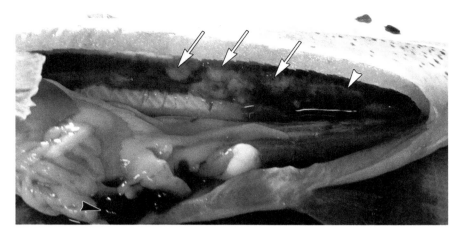

Fig. 21.1. Adult male (305 mm length) hatchery-reared cutthroat trout (*Oncorhynchus clarkii*) naturally infected with *Renibacterium salmoninarum*. Characteristic macroscopic greyish white granulomatous lesions are visible in the kidney (arrows), and the spleen (black arrowhead) and posterior kidney (white arrowhead) are markedly swollen. Photograph courtesy Dr John Drennan, Aquatic Animal Health Laboratory, Colorado Parks and Wildlife, Brush, Colorado.

abdominal and pericardial cavities, the presence of viscous yellow or blood-tinged fluid in in the intestine, haemorrhages in the visceral organs and the abdominal wall, and a diffuse white membranous layer (pseudomembrane) on the surface of one or more internal organs (Fryer and Sanders, 1981; Evelyn, 1993; Evenden *et al.*, 1993).

21.3.2 Histopathology

BKD is often described as a bacteraemia characterized by diffuse systemic chronic granulomatous inflammation (Bruno, 1986). Although the kidney interstitial tissue may be predominantly affected (Ferguson, 1989), granulomatous inflammation is generally present in most infected tissues. Granulomas in Pacific salmon are often diffuse with poorly defined borders (Fig. 21.2a), whereas those in Atlantic salmon are more encapsulated and contain abundant epithelioid macrophages (Evelyn, 1993). Older granulomas may have a central zone of caseous necrosis surrounded by epithelioid cells, fibrosis and infiltrating lymphoid cells (Fig. 21.2b). Pseudomembranes comprising thin layers of fibrin and collagen with trapped phagocytes and bacteria may form over the capsules of organs such as the kidney, liver and spleen.

In severe cases of BKD, fibrin exudation is especially common in areas with vascular lesions, which may be characterized by necrosis of the tunica media or endothelium, with limited haemorrhage or

thrombosis (Ferguson, 1989). The heart is a frequent site of BKD lesions that present as diphtheritic epicarditis as well as granulomatous myocarditis at the interface of the compact and spongy myocardium. In the renal kidney, membranous glomerulopathy/glomerulonephritis is common (Fig. 21.3) and is linked to subendothelial deposition of antigen–antibody complexes in the glomeruli (Elliott *et al.*, 2014), resulting in thickening of the glomerular basement membrane.

While BKD often occurs as a systemic disease, localized infections with lesions restricted to tissues such as the central nervous system (CNS), eye or postorbital tissues, or skin and fins have been reported (Pascho *et al.*, 2002). Similar to systemic infections, *R. salmoninarum* is associated with granulomatous or pyogranulomatous lesions in affected tissues. For example, the bacterium may cause meningitis and encephalitis in the absence of systemic infections (Speare *et al.*, 1993). CNS infections may result from antibiotic treatments that clear the infection from other organs but not the brain (Speare, 1997), or through retrograde extension from the posterior uvea (choroid) of the eye to the floor of the diencephalon via the connective tissue sheaths (epineurium and perineurium) of the optic nerve (Speare *et al.*, 1993). In fish affected by *R. salmoninarum*-associated spawning rash, granulomatous inflammation may invade adjacent scale pockets and extend laterally along the fibrous layer of the dermis. The bacterium may not be detected

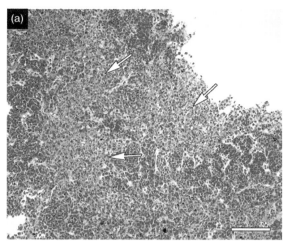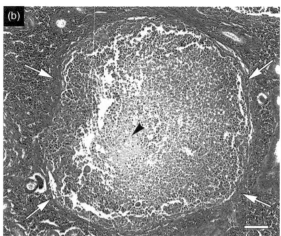

Fig. 21.2. (a) Histological section of anterior kidney from a naturally *Renibacterium salmoninarum*-infected juvenile coho salmon (*Oncorhynchus kisutch*) from a public aquarium. Diffuse granulomatous inflammation (arrows), with poorly defined borders, is evident. (b) Histological section of posterior kidney from a hatchery-reared juvenile Chinook salmon (*O. tshawytscha*) naturally infected with *R. salmoninarum*. The established bacterial kidney disease granuloma shows a central area of necrosis (arrowhead) and is encapsulated by fibrosis (arrows). Haematoxylin and eosin stain, scale bars = 100 µm. Photographs courtesy Carla Conway, US Geological Survey, Western Fisheries Research Center, Seattle, Washington.

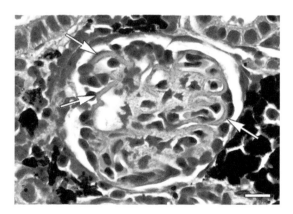

Fig. 21.3. Histological section of posterior kidney tissue showing membranous glomerulopathy in a juvenile Chinook salmon (*Oncorhynchus tshawytscha*) experimentally infected with *Renibacterium salmoninarum*. Hyaline thickening of the glomerular basement membrane (arrows) is linked to subendothelial deposition of antigen–antibody complexes in glomerular capillaries. Haematoxylin and eosin stain, scale bar = 10 µm. Photograph courtesy Carla Conway, US Geological Survey, Western Fisheries Research Center, Seattle, Washington.

elsewhere in the body and the skin lesions resolve after spawning (Ferguson, 1989). Some localized infections probably represent the proliferation of *R. salmoninarum* after introduction of the bacterium

through breaks in surface tissues such as the skin or eyes (Elliott *et al.*, 2015).

R. salmoninarum is a facultative intracellular pathogen that occurs both extracellularly and intracellularly within the phagocytic cells of infected fish. Intracellular *R. salmoninarum* are present within the cytoplasm of epithelial or endothelial cells or neutrophils (Bruno, 1986; Ferguson, 1989), as well as in reticular cells, sinusoidal cells or fibroblastic barrier cells of kidney and spleen haematopoietic tissue (Flaño *et al.*, 1996). The largest numbers of bacteria are usually observed within macrophages (Fig. 21.4). *R. salmoninarum* can survive and perhaps replicate within macrophages (Pascho *et al.*, 2002), a strategy to evade host defences. Epithelioid macrophages may show limited phagocytic activity, whereas melanomacrophages are actively phagocytic (Flaño *et al.*, 1996); the dispersal of pigment in tissues – which is a histopathological characteristic of BKD (see Fig. 21.5) – may result from the disruption and lysis of melanomacrophages (Bruno, 1986; Flaño *et al.*, 1996).

21.4 Pathophysiology

The haematological parameters associated with natural or experimental *R. salmoninarum* infections

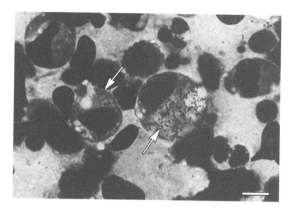

Fig. 21.4. Spleen imprint from a hatchery-reared adult cutthroat trout (*Oncorhynchus clarkii*) naturally infected with *Renibacterium salmoninarum*. Bacteria are visible within the cytoplasm of macrophages (arrows). Dip Quick[1] (Jorgensen Laboratories) proprietary Romanowski stain, scale bar = 10 µm. Photograph courtesy Dr John Drennan, Aquatic Animal Health Laboratory, Colorado Parks and Wildlife, Brush, Colorado.

have been investigated (see Wiens, 2011). Erythrocyte-associated changes include reductions in haematocrit and circulating erythrocyte counts, as well as decreases in haemoglobin, erythrocyte diameter and the ratio of mature to immature erythrocytes, and increases in the erythrocyte sedimentation rate. A reduction in circulating erythrocytes is related to the retention of these cells in the spleen, which causes splenomegaly in *R. salmoninarum*-infected fish. Experimentally infected fish also show increases in monocytes and transitory increases in thrombocytes and neutrophils, but no changes in the numbers of circulating small and large lymphocytes. The progression of BKD correlates with decreased total serum protein (especially the electrophoretically faster migrating fractions), cholesterol and sodium, and increased serum levels of bilirubin, blood urea nitrogen and potassium.

Certain physiological changes may reduce the performance and survival of fish. For example, Mesa *et al.* (1998, 2000) demonstrated significant increases in plasma cortisol and lactate levels, and

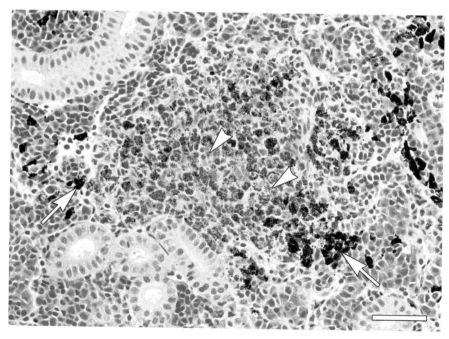

Fig. 21.5. Histological section of a focal granulomatous lesion in posterior kidney tissue of a juvenile Chinook salmon (*Oncorhynchus tshawytscha*) experimentally infected with *Renibacterium salmoninarum*. The lesion has poorly defined borders, and shows characteristic dispersal of pigment (arrowheads) from the disruption of melanomacrophages. Intact melanomacrophages with dense cytoplasmic pigment (arrows) are visible in tissue adjacent to the lesion. Haematoxylin and eosin stain, scale bar = 50 µm. Photograph courtesy Carla Conway, US Geological Survey, Western Fisheries Research Center, Seattle, Washington.

significant decreases in glucose, as BKD progressed in juvenile Chinook salmon (*O. tshawytscha*) challenged with *R. salmoninarum*. The elevated cortisol indicated that *R. salmoninarum* infections are stressful, particularly in the later stages, and the increased lactate suggested hypoxaemia, resulting in poor oxygen transport, which is consistent with the severe haematological changes. Mesa *et al.* (1998) hypothesized that decreased plasma glucose resulted from excessive use of this energy substrate to combat *R. salmoninarum* infection. The authors also determined that juvenile Chinook salmon with moderate-to-high *R. salmoninarum* infection were significantly more vulnerable to predation, although no overt behavioural differences were observed between infected and control fish. The authors postulated that the bioenergetic demands of chronic infections reduced the amount of energy available to perform activities such as predator avoidance. The decreased metabolic scope of Chinook salmon with more severe infections also may be an underlying mechanism of the greater susceptibility of these fish to stressors such as dissolved gas supersaturation (Wiens, 2011). However, Mesa *et al.* (2000) showed no differences in mortality or the progression of *R. salmoninarum* infection in Chinook salmon subjected to multiple bouts of handling, hypoxia and mild agitation, in comparison with infected fish that were not handled. Nevertheless, unlike the uninfected fish subjected to multiple stressors, the *R. salmoninarum*-infected fish were unable to consistently produce significant hyperglycaemia in response to stressors.

Interactions between infections by *R. salmoninarum* and the physiological processes in the host may have additional consequences. Limited research suggests that chronic infections in salmonids adversely affect growth (Sandell *et al.*, 2015). Also, smoltification can dramatically increase the severity of BKD in fish (Mesa *et al.*, 1999), and may contribute to mortality shortly after seawater entry (see Pascho *et al.*, 2002; Wiens, 2011). Glomerulonephritis associated with BKD in returning adult Atlantic salmon may be severe and cause morbidity from osmoregulatory failure (Bruno *et al.*, 2013).

Finally, immunosuppression by *R. salmoninarum* may contribute to decreased resistance to secondary pathogens. The *in vitro* and *in vivo* immunosuppressive ability of *R. salmoninarum* has been well demonstrated. The abundant 57 kDa extracellular major soluble antigen (p57 or MSA) of *R. salmoninarum* has been identified as a dominant virulence factor and is a key mediator of immunosuppression, as exemplified by a reduction of antibody production and phagocyte bactericidal activity (see Pascho *et al.*, 2002; Wiens, 2011). A 22 kDa *R. salmoninarum* surface protein, p22, has also been implicated in the suppression of antibody production (Fredriksen *et al.*, 1997).

21.5 Prevention and Control

Bacterial kidney disease is one of the more difficult bacterial fish diseases to control (Elliott *et al.*, 1989). Management strategies for BKD have been reviewed (e.g. Elliott *et al.*, 1989, 2014; Pascho *et al.*, 2002; Wiens, 2011), and have included the application of chemotherapeutants and vaccines, the interruption of vertical transmission, and changes in hygiene, husbandry and biosecurity practices.

Chemotherapeutants with antimicrobial activity against *R. salmoninarum* have been applied with partial success (Pascho *et al.*, 2002; Elliott *et al.*, 2014). The antimicrobial agent that has been most widely tested is the macrolide antibiotic erythromycin, which in the USA is available only as an investigational new animal drug (INAD) through the US Food and Drug Administration (FDA) (see Wiens, 2011). The oral administration of erythromycin can result in tissue concentrations of the drug that exceed the minimum inhibitory concentration for *R. salmoninarum* and these can reduce BKD mortality (Pascho *et al.*, 2002). A dose of 100 mg/kg body weight of erythromycin thiocyanate in medicated feed for 28 days is the most effective treatment against a severe intraperitoneal *R. salmoninarum* infection (Wiens, 2011). However, the antibiotic does not completely eliminate the pathogen (Pascho *et al.*, 2002), and there is potential for the emergence of antibiotic resistance (Elliott *et al.*, 2014).

The injection of erythromycin phosphate (11 mg/kg body weight) into adult Chinook salmon at 21- to 30-day intervals until spawning can reduce prespawning mortality, but higher doses can cause jaundice and death (Wiens, 2011). The injection of erythromycin into female fish between 9 and 60 days before spawning can result in therapeutic antibiotic levels within mature eggs (Haukenes and Moffitt, 2002; Pascho *et al.*, 2002), and can prevent the vertical transmission of *R. salmoninarum* under experimental conditions (Pascho *et al.*, 2002); its effectiveness in preventing vertical transmission on a fish-production scale has not been determined. Water hardening of eggs in erythromycin has not consistently reduced *R. salmoninarum* prevalence in fry hatched from the treated eggs (Pascho *et al.*, 2002); in this case, the rapid leaching of the antibiotic from eggs may prevent sufficient contact time with the bacterium.

Vaccination was recently reviewed by Elliott *et al.* (2014). Only one commercially licensed BKD vaccine is currently available. The vaccine, sold under the trade name of Renogen®[1] (Elanco Animal Health), is a lyophilized preparation containing live cells of the non-pathogenic environmental bacterium *Arthrobacter davidanieli* (proposed nomenclature), and is licensed for sale in the USA, Canada and Chile, but not in Europe nor Japan. The vaccine culture is re-suspended in sterile saline prior to the intraperitoneal injection of pre-smolts, and the manufacturer recommends a minimum of 400 degree days between vaccination and exposure to *R. salmoninarum*. The close phylogenetic relationship between *Arthrobacter* and *R. salmoninarum* is most likely the basis for the cross-species protection exhibited by Renogen® (Wiens *et al.*, 2008). Laboratory and field trials of the vaccine demonstrated significant protection against BKD in juvenile Atlantic salmon vaccinated with the live *Arthrobacter* vaccine, but there was limited or no protection in juvenile Chinook salmon experimentally challenged with *R. salmoninarum* (Elliott *et al.*, 2014). The duration of protection against BKD following Renogen® vaccination is unknown, but significantly higher survival was reported among Renogen®-vaccinated Atlantic salmon relative to control fish at about 23 and 27 months after vaccination in two commercial facilities where natural BKD outbreaks had occurred. Nevertheless, these fish had probably been exposed to *R. salmoninarum* long before BKD mortality began at about 21 and 19 months after vaccination, respectively (see Elliott *et al.*, 2014).

Most additional BKD vaccine development has focused on heat- or formalin-inactivated whole cell or lysed bacterins prepared from *R. salmoninarum* cultures, or live vaccines containing attenuated *R. salmoninarum* isolates (Elliott *et al.*, 2014). Among the more promising experimental bacterins are those incorporating bacterial cells with reduced surface MSA, achieved by the heat treatment of *R. salmoninarum* cells or by the selection of bacterial strains expressing low levels of MSA. One reduced-MSA bacterin demonstrated a low but detectable therapeutic effect, resulting in higher survival of vaccinated Chinook salmon that were already infected with *R. salmoninarum* relative to infected fish that were not vaccinated. In contrast, live attenuated *R. salmoninarum* strains with normal cell-associated MSA provided greater (but incomplete) protection than attenuated strains with reduced MSA in Atlantic salmon vaccinated by intraperitoneal injection.

Because of a lack of completely efficacious vaccines and chemotherapeutants, other management approaches for BKD have been implemented. In endemic areas of the disease, the culling or segregation of eggs from infected fish is used to reduce the impact of vertical transmission. Successful culling/segregation programmes have generally employed culture (Gudmundsdóttir *et al.*, 2000) or, more recently, a polyclonal antibody ELISA (Pascho *et al.*, 1991; Gudmundsdóttir *et al.*, 2000; Meyers *et al.*, 2003; Munson *et al.*, 2010; Faisal *et al.*, 2012) that is used to screen the kidney tissues of spawning females. Eggs from females testing positive for *R. salmoninarum* are usually discarded. However, in populations with high *R. salmoninarum* prevalence or low numbers of spawning fish, egg lots from positive females (or from females showing moderate-to-high antigen levels by ELISA) may be segregated for rearing rather than culled. Alternatively, the threshold ELISA absorbance (optical density) value for culling egg lots may be raised to include positive females judged to be at lower risk of vertical transmission (Munson *et al.*, 2010).

Consistent brood stock screening and culling programmes have contributed to significant decreases in BKD prevalence and severity, and in the overall mortality of fish during hatchery rearing, and to reductions in the prevalence and levels of *R. salmoninarum* infections in subsequent generations of returning adults (Gudmundsdóttir *et al.*, 2000; Meyers *et al.*, 2003; Munson *et al.*, 2010; Faisal *et al.*, 2012). The evidence also suggests that there is higher survival during downriver migration and seawater entry of offspring of parents with low or undetectable *R. salmoninarum* levels, as well as higher adult returns, in comparison with the progeny of adults with moderate-to-high *R. salmoninarum* levels (see Wiens, 2011). Additionally, such programmes can decrease the need for antibiotic chemotherapy and reduce the numbers of spawning adults and juveniles that are necessary to compensate for the anticipated losses (Munson *et al.*, 2010).

Despite the success of brood stock culling programmes, the selective culling of infected females could result in undesirable genetic changes such as progeny that are more susceptible to BKD (Munson *et al.*, 2010). Hard *et al.* (2006) reported that the resistance of progeny Chinook salmon to *R. salmoninarum* infection was not genetically correlated with *R. salmoninarum* antigen levels in the parent fish, suggesting that antigen levels in adults at spawning reflected environmental rather than genetic sources of variation. Nevertheless, it is uncertain whether the results of Hard *et al.* (2006) can be extended to other salmonid species and environmental settings in aquaculture.

Other strategies, including improved biosecurity, hygiene and husbandry practices, have been applied to reduce vertical and horizontal transmission and to alleviate stressful rearing conditions that augment the development of clinical BKD. Commonly implemented measures include the following: antibiotic injection of pre-spawning adults; surface disinfection of green eggs; use of separate nets and brushes for each rearing unit; use of disinfectant footbaths or mats between rearing units or facilities; regular cleaning and disinfection of rearing units; judicious prophylactic oral application of antibiotics during stressful events; separation of year classes and fallowing of rearing units; use of single-pass rather than recirculating water; and rearing of fish in indoor or covered units (Maule *et al.*, 1996; Gudmundsdóttir *et al.*, 2000; Meyers *et al.*, 2003; Munson *et al.*, 2010; Faisal *et al.*, 2012; Murray *et al.*, 2012).

The eradication of *R. salmoninarum* from fish culture facilities is difficult because of the dual modes of pathogen transmission, the presence of infection in wild or escaped salmonids, and difficulties in the consistent detection of subclinical infections (Murray *et al.*, 2012; Peeler and Otte, 2016). The removal of feral fish from water supplies is worth consideration (Meyers *et al.*, 2003; Murray *et al.*, 2012). Because of the possibility of *R. salmoninarum* transmission between ponds or cages, fallowing at farms between production cycles is a minimum recommendation (Murray *et al.*, 2012). An exception is a highly biosecure land-based farm in which fish are reared in indoor tanks, and individual tanks can be depopulated and disinfected (Murray *et al.*, 2012). None the less, even farm-level fallowing may be ineffective if *R. salmoninarum* is reintroduced with the ova or fish that are used to repopulate a farm or if wild or feral fish reservoirs remain (Murray *et al.*, 2012). In the UK, epizootiological and cost–benefit analyses have resulted in modifying the goals of BKD management programmes from pathogen eradication to controlling pathogen spread (Murray *et al.*, 2012; Hall *et al.*, 2014; Peeler and Otte, 2016). Murray *et al.* (2012) concluded that compartmentalization of BKD management strategies by fish species and geographically defined areas could result in more effective targeting of BKD controls. Hall *et al.* (2014) showed that a BKD control policy to limit spread via the detection of BKD-affected farms, and the restriction of fish movement from those farms unless *R. salmoninarum* is successfully eradicated, is more cost-effective in Atlantic salmon aquaculture than alternative policies of greater or lesser stringency.

21.6 Conclusions and Suggestions for Future Research

BKD remains an important cause of morbidity and mortality of salmonid fishes. Prophylactics and therapeutics that are fully effective for multiple fish species are not yet commercially available. A further improvement of BKD management strategies would benefit from better understanding of the mechanisms of *R. salmoninarum* transmission, pathogenesis and the host response.

Recent advances in genome sequencing and analysis have provided new insights into pathogenic mechanisms that could be useful in vaccine development (Elliott *et al.*, 2014). For example, *R. salmoninarum* has two iron acquisition systems, one of which involves the production of siderophores (Bethke *et al.*, 2016). Commercial vaccines have been developed to reduce the colonization of pathogenic bacteria in mammals by limiting iron acquisition via siderophore-based immunization (e.g. Cull *et al.*, 2012); similar strategies could be considered for vaccines against *R. salmoninarum*.

Additional research has identified *R. salmoninarum* genes that putatively encode a sortase and sortase substrates (Wiens *et al.*, 2008). In Gram-positive bacteria, sortases and their surface protein substrates are vital in adhesion to host cells, colonization and evasion of the host immune response. Protection against infections by virulent *Streptococcus* and *Staphylococcus* species has been demonstrated in mice vaccinated with recombinant sortase or sortase substrate proteins produced *in vitro* (see Elliott *et al.*, 2014). Drug discovery efforts suggest that *R. salmoninarum* sortase could also be a therapeutic target for sortase inhibitors, which may prove useful for the prevention of bacterial adherence and invasion of host cells *in vivo* (see Wiens, 2011).

Results from next-generation genome sequencing could be utilized in compartmentalized BKD management (Murray *et al.*, 2012) to limit disease spread. Data from the SNP analysis of different *R. salmoninarum* isolates has allowed inferences of the patterns of global *R. salmoninarum* dissemination as well as intraspecific and interspecific transmission of isolates within and between farms (Brynildsrud *et al.*, 2014). These data, and data from future studies with larger numbers of isolates from more diverse geographic locations, could be used to assess the potential sources and risks of *R. salmoninarum* transmission under different BKD management scenarios.

Another potential application of next-generation sequencing is for investigations of the host response to aid in the development of improved vaccines and other strategies for BKD management. Whole-transcriptome shotgun sequencing (RNA-Seq) is used to detect and quantify the complete set of gene transcripts in a cell at a given time, and often reveals changes in gene expression profiles under different conditions (Wang *et al.*, 2009). Previous studies have identified changes in the expression of certain host genes that may be involved in immunity to *R. salmoninarum* infection (see Wiens, 2011). However, RNA-Seq evaluations could provide a more complete identification of the host genes that are important for BKD resistance in studies involving different BKD vaccines, virulent and attenuated *R. salmoninarum* isolates, or salmonid stocks with differing BKD susceptibility.

Finally, improvements in diagnostic methods are desirable for a more accurate monitoring of subclinical *R. salmoninarum* infections in fish populations so as to better inform the timing of treatments or other BKD management strategies, and to assess the success of those strategies. Non-lethal sampling methods are preferable for monitoring endangered salmonid populations or valuable brood stock. For example, Elliott *et al.* (2015) determined that the qPCR analysis of surface mucus samples was the most suitable alternative to the testing of lethal kidney samples from juvenile Chinook salmon for *R. salmoninarum*. In another laboratory investigation, Shulz (2014) reported that samples of urine/faeces showed good potential for the non-lethal evaluation of *R. salmoninarum* prevalence and intensity when samples were tested using a combination of nPCR

and polyclonal ELISA. Although the results of these studies are promising, additional research is needed with fish in naturally infected populations and with other salmonid species.

Note

[1] The mention of trade, firm, or corporation names in this publication is for the information and convenience of the reader and does not constitute an official endorsement or approval by the U.S. Government of any product or service to the exclusion of others that may be suitable.

References

AFS-FHS (2014) FHS *Blue Book: Suggested Procedures for the Detection and Identification of Certain Finfish and Shellfish Pathogens*, 2014 edn. American Fisheries Society-Fish Health Section, Bethesda, Maryland. Available at: http://www.afs-fhs.org/bluebook/bluebook-index.php (accessed 5 January 2016).

Austin, B. and Austin, D.A. (1987) *Bacterial Fish Pathogens: Disease in Farmed and Wild Fish.* Ellis Horwood, Chichester, UK.

Balfry, S.K., Albright, L.J. and Evelyn, T.P.T. (1996) Horizontal transfer of *Renibacterium salmoninarum* among farmed salmonids via the fecal–oral route. *Diseases of Aquatic Organisms* 25, 63–69.

Bethke, M., Poblete-Morales, M., Irgang, R., Yáñez, A. and Avendaño-Herrera, R. (2016) Iron acquisition and siderophore production in the fish pathogen *Renibacterium salmoninarum. Journal of Fish Diseases* 39, 1275–1283.

Bruno, D.W. (1986) Histopathology of bacterial kidney disease in laboratory infected rainbow trout, *Salmo gairdneri* Richardson, and Atlantic salmon, *Salmo salar* L., with reference to naturally infected fish. *Journal of Fish Diseases* 9, 523–537.

Bruno, D.W., Noguera, P.A. and Poppe, T.T. (2013) *A Colour Atlas of Salmonid Diseases*, 2nd edn. Springer, London, UK.

Brynildsrud, O., Feil, E.J., Bohlin, J., Castillo-Ramirez, S., Colquhoun, D. *et al.* (2014) Microevolution of *Renibacterium salmoninarum*: evidence for intercontinental dissemination associated with fish movements. *The ISME Journal* 8, 746–756.

Cull, C.A., Paddock, Z.D., Nagaraja, T.G., Bello, N.M., Babcock, A.H. *et al.* (2012) Efficacy of a vaccine and a direct-fed microbial against fecal shedding of *Escherichia coli* O157:H7 in a randomized pen-level field trial of commercial feedlot cattle. *Vaccine* 30, 6210–6215.

Elliott, D.G., Pascho, R.J. and Bullock, G.L. (1989) Developments in the control of bacterial kidney disease of salmonid fishes. *Diseases of Aquatic Organisms* 6, 201–215.

Elliott, D.G., Pascho, R.J., Jackson, L.M., Matthews, G.M. and Harmon, J.R. (1997) *Renibacterium salmoninarum* in spring–summer Chinook salmon smolts at dams on the Columbia and Snake Rivers. *Journal of Aquatic Animal Health* 9, 114–126.

Elliott, D.G., Applegate, L.J., Murray, A.L., Purcell, M.K. and McKibben, C.L. (2013) Bench-top validation testing of selected immunological and molecular *Renibacterium salmoninarum* diagnostic assays by comparison with quantitative bacteriological culture. *Journal of Fish Diseases* 36, 779–809.

Elliott, D.G., Wiens, G.D., Hammell, K.L. and Rhodes, L.D. (2014) Vaccination against bacterial kidney disease. In: Gudding, R., Lillehaug, A. and Evensen, Ø. (eds) *Fish Vaccination*. Wiley Blackwell, Chichester/Oxford, UK, pp. 255–272.

Elliott, D.G., McKibben, C.L., Conway, C.M., Purcell, M.K., Chase, D.M. *et al.* (2015) Testing of candidate non-lethal sampling methods for detection of *Renibacterium salmoninarum* in juvenile Chinook salmon *Oncorhynchus tshawytscha*. *Diseases of Aquatic Organisms* 114, 21–43.

Evelyn, T.P.T. (1993) Bacterial kidney disease – BKD. In: Inglis, V., Roberts, R.J. and Bromage, N.R. (eds) *Bacterial Diseases of Fish*. Halsted Press, New York, pp. 177–195.

Evenden, A.J., Grayson, T.H., Gilpin, M.L. and Munn, C.B. (1993) *Renibacterium salmoninarum* and bacterial kidney disease – the unfinished jigsaw. *Annual Review of Fish Diseases* 3, 87–104.

Faisal, M., Eissa, A.E. and Starliper, C.E. (2010a) Recovery of *Renibacterium salmoninarum* from naturally infected salmonine stocks in Michigan using a modified culture protocol. *Journal of Advanced Research* 1, 95–102.

Faisal, M., Loch, T.P., Brenden, T.O., Eissa, A.E., Ebener, M.P. *et al.* (2010b) Assessment of *Renibacterium salmoninarum* infections in four lake whitefish (*Coregonus clupeaformis*) stocks from northern Lakes Huron and Michigan. *Journal of Great Lakes Research* 36, 29–37.

Faisal, M., Schulz, C., Eissa, A., Brenden, T., Winters, A. *et al.* (2012) Epidemiological investigation of *Renibacterium salmoninarum* in three *Oncorhynchus* spp. in Michigan from 2001 to 2010. *Preventive Veterinary Medicine* 107, 260–274.

Ferguson, H.W. (1989) *Systemic Pathology of Fish*. Iowa State University Press, Ames, Iowa.

Flaño, E., López-Fierro, P., Razquin, B., Kaattari, S.L. and Villena, A. (1996) Histopathology of the renal and splenic haemopoietic tissues of coho salmon *Oncorhynchus kisutch* experimentally infected with *Renibacterium salmoninarum*. *Diseases of Aquatic Organisms* 24, 107–115.

Fredriksen, A., Endresen, C. and Wergeland, H.I. (1997) Immunosuppressive effect of a low molecular weight surface protein from *Renibacterium salmoninarum* on lymphocytes from Atlantic salmon (*Salmo salar* L). *Fish and Shellfish Immunology* 7, 273–282.

Fryer, J.L. and Lannan, C.N. (1993) The history and current status of *Renibacterium salmoninarum*, the causative agent of bacterial kidney disease in Pacific salmon. *Fisheries Research* 17, 15–33.

Fryer, J.L. and Sanders, J.E. (1981) Bacterial kidney disease of salmonid fish. *Annual Review of Microbiology* 35, 273–298.

Gudmundsdóttir, S., Helgason, S., Sigurjónsdóttir, H., Matthíasdóttir, S., Jónsdóttir, H., Laxdal, B. and Benediktsdóttir, E. (2000) Measures applied to control *Renibacterium salmoninarum* infection in Atlantic salmon: a retrospective study of two sea ranches in Iceland. *Aquaculture* 186, 193–203.

Hall, M., Soje, J., Kilburn, R., Maguire, S. and Murray, A.G. (2014) Cost-effectiveness of alternative management policies for bacterial kidney disease in Atlantic salmon aquaculture. *Aquaculture* 434, 88–92.

Hard, J.J., Elliott, D.G., Pascho, R.J., Chase, D.M., Park, L.K. *et al.* (2006) Genetic effects of ELISA-based segregation for control of bacterial kidney disease in Chinook salmon (*Oncorhynchus tshawytscha*). *Canadian Journal of Fisheries and Aquatic Sciences* 63, 2793–2808.

Haukenes, A.H. and Moffitt, C.M. (2002) Hatchery evaluation of erythromycin phosphate injections in pre-spawning Chinook salmon. *North American Journal of Aquaculture* 64, 167–174.

Jansson, E., Hongslo, T., Höglund, J. and Ljungberg, O. (1996) Comparative evaluation of bacterial culture and two ELISA techniques for the detection of *Renibacterium salmoninarum* antigens in salmonid kidney tissues. *Diseases of Aquatic Organisms* 27, 197–206.

Kristmundsson, Á., Árnason, F., Gudmundsdóttir, S. and Antonsson, T. (2016) Levels of *Renibacterium salmoninarum* antigens in resident and anadromous salmonids in the River Ellidaár system in Iceland. *Journal of Fish Diseases* 39, 681–692.

Maule, A.G., Rondorf, D.W., Beeman, J. and Haner, P. (1996) Incidence of *Renibacterium salmoninarum* infections in juvenile hatchery spring Chinook salmon in the Columbia and Snake rivers. *Journal of Aquatic Animal Health* 8, 37–46.

Mesa, M.G., Poe, T.P., Maule, A.G. and Schreck, C.B. (1998) Vulnerability to predation and physiological stress responses in juvenile Chinook salmon (*Oncorhynchus tshawytscha*) experimentally infected with *Renibacterium salmoninarum*. *Canadian Journal of Fisheries and Aquatic Sciences* 55, 1599–1606.

Mesa, M.G., Maule, A.G., Poe, T.P. and Schreck, C.B. (1999) Influence of bacterial kidney disease on smoltification in salmonids: is it a case of double jeopardy? *Aquaculture* 174, 25–41.

Mesa, M.G., Maule, A.G. and Schreck, C.B. (2000) Interaction of infection with *Renibacterium salmoninarum* and physical stress in juvenile Chinook salmon:

physiological responses, disease progression, and mortality. *Transactions of the American Fisheries Society* 129, 158–173.

Meyers, T.R., Korn, D., Glass, K., Burton, T., Short, S. *et al.* (2003) Retrospective analysis of antigen prevalences of *Renibacterium salmoninarum* (Rs) detected by enzyme-linked immunosorbent assay in Alaskan Pacific salmon and trout from 1988 to 2000 and management of Rs in hatchery Chinook and coho salmon. *Journal of Aquatic Animal Health* 15, 101–110.

Munson, A.D., Elliott, D.G. and Johnson, K. (2010) Management of bacterial kidney disease in Chinook salmon hatcheries based on broodstock testing by enzyme-linked immunosorbent assay: a multiyear study. *North American Journal of Fisheries Management* 30, 940–955.

Murray, A.G., Munro, L.A., Wallace, I., Allan, C.E.T., Peeler, E.J. *et al.* (2012) Epidemiology of *Renibacterium salmoninarum* in Scotland and the potential for compartmentalised management of salmon and trout farming areas. *Aquaculture* 324–325, 1–13.

OIE (2003) Chapter 2.1.11. Bacterial kidney disease (*Renibacterium salmoninarum*). In: *Manual of Diagnostic Tests for Aquatic Animals*, 4th edn. World Organisation for Animal Health, Paris, France, pp. 167–184. Available at: http://www.oie.int/doc/ged/D6505.PDF (accessed 17 November 2016).

Pascho, R.J., Elliott, D.G. and Streufert, J.M. (1991) Brood stock segregation of spring Chinook salmon *Oncorhynchus tshawytscha* by use of the enzyme-linked immunosorbent assay (ELISA) and the fluorescent antibody technique (FAT) affects the prevalence and levels of *Renibacterium salmoninarum* infection in progeny. *Diseases of Aquatic Organisms* 12, 25–40.

Pascho, R.J., Elliott, D.G. and Chase, D.M. (2002) Comparison of traditional and molecular methods for detection of *Renibacterium salmoninarum*. In: Cunningham, C.O. (ed.) *Molecular Diagnosis of Salmonid Diseases.* Kluwer, Dordrecht, The Netherlands, pp. 157–209.

Peeler, E.J. and Otte, M.J. (2016) Epidemiology and economics support decisions about freedom from aquatic animal diseases. *Transboundary and Emerging Diseases* 63, 266–277.

Purcell, M.K., Getchell, R.G., McClure, C.A. and Garver, K.A. (2011) Quantitative polymerase chain reaction (PCR) for detection of aquatic animal pathogens in a diagnostic laboratory setting. *Journal of Aquatic Animal Health* 23, 148–161.

Sandell, T.A., Teel, D.J., Fisher, J., Beckman, B. and Jacobson, K.C. (2015) Infections by *Renibacterium salmoninarum* and *Nanophyetus salmincola* Chapin are associated with reduced growth in juvenile Chinook salmon, *Oncorhynchus tshawytscha* (Walbaum), in the Northeast Pacific Ocean. *Journal of Fish Diseases* 38, 365–378.

Sanders, J.E. and Fryer, J.L. (1980) *Renibacterium salmoninarum* gen. nov., sp. nov., the causative agent of bacterial kidney disease in salmonid fishes. *International Journal of Systematic Bacteriology* 30, 496–502.

Shulz, C.A. (2014) Factors and pitfalls influencing the detection of bacterial kidney disease. Ph.D. dissertation, Michigan State University, East Lansing, Michigan.

Smith, I.W. (1964) The occurrence and pathology of Dee disease. *Freshwater and Salmon Fisheries Research* 34, 1–12.

Speare, D.J. (1997) Differences in patterns of meningoencephalitis due to bacterial kidney disease in farmed Atlantic and Chinook salmon. *Research in Veterinary Science* 62, 79–80.

Speare, D.J., Ostland, V.E. and Ferguson, H.W. (1993) Pathology associated with meningoencephalitis during bacterial kidney disease in salmonids. *Research in Veterinary Science* 54, 25–31.

US Fish and Wildlife Service (2011) National Wild Fish Health Survey Database, US Fish and Wildlife Service, Washington, DC. Available at: http://ecos.fws.gov/wildfishsurvey/database/nwfhs/ (accessed 19 November 2015).

Wang, Z., Gerstein, M. and Snyder, M. (2009) RNA-Seq: a revolutionary tool for transcriptomics. *Nature Reviews Genetics* 10, 57–63.

Wiens, G.D. (2011) Bacterial kidney disease (*Renibacterium salmoninarum*). In: Woo, P.T.K. and Bruno, D.W. (eds) *Fish Diseases and Disorders, Volume 3: Viral, Bacterial, and Fungal Pathogens*, 2nd edn. CABI, Wallingford, UK, pp. 338–374.

Wiens, G.D., Rockey, D.D., Wu, Z., Chang, J., Levy, R. *et al.* (2008) Genome sequence of the fish pathogen *Renibacterium salmoninarum* suggests reductive evolution away from an environmental *Arthrobacter* ancestor. *Journal of Bacteriology* 190, 6970–6982.

22 *Streptococcus iniae* and *S. agalactiae*

CRAIG A. SHOEMAKER,[1]* DE-HAI XU[1] AND ESTEBAN SOTO[2]

[1]US Department of Agriculture-Agricultural Research Service, Aquatic Animal Health Research Unit, Auburn, Alabama, USA; [2]Department of Medicine and Epidemiology, School of Veterinary Medicine, University of California Davis, Davis, California, USA

22.1 Introduction

Streptococcus iniae and *S. agalactiae* are Gram-positive bacterial pathogens of cultured and wild fish, which have been the subject of several previous reviews (e.g. Agnew and Barnes, 2007; Klesius *et al.*, 2008; Plumb and Hanson, 2010; Salati, 2011). Both bacteria may present zoonotic concerns (Weinstein *et al.*, 1997; Delannoy *et al.*, 2013). *S. iniae* infects immunocompromised patients who have handled live fish (Gauthier, 2015). Comparative genomic analysis of piscine *S. agalactiae* isolates suggests that human strains of *S. agalactiae* are present in fish, frogs and aquatic mammals, thus posing a potential risk for human disease (Delannoy *et al.*, 2013; Liu *et al.*, 2013). *S. iniae* was the main pathogen of farmed fish in the late 1990s–2000s (Shoemaker *et al.*, 2001; Agnew and Barnes, 2007). Presently, *S. agalactiae* has emerged as the major pathogen in cultured tilapia (*Oreochromis* spp.) in Asia and in Latin and South America (Suanyuk *et al.*, 2008; Mian *et al.*, 2009; Sheehan *et al.*, 2009; Chen *et al.*, 2012; Zamri-Saad *et al.*, 2014). The annual worldwide monetary loss due to these pathogens was originally underestimated at US$100 million (Shoemaker *et al.*, 2001). China alone accounts for about 40% of global tilapia production (~US$3 billion), and Chinese producers have reported losses of 30–80% due to *S. agalactiae* (Ye *et al.*, 2011; Chen *et al.*, 2012). Assuming an annual average loss of 40%, that equals about US$1 billion lost revenue in China alone.

22.1.1 Descriptions of bacteria

S. iniae is a Lancefield non-groupable *Streptococcus* that was described by Pier and Madin (1976) from Amazon dolphin (*Inia geoffrensis*). *S. agalactiae* is a Lancefield group B *Streptococcus* that was first reported from wild fish (Robinson and Meyer, 1966) and was subsequently described as *S. difficile* (Eldar *et al.*, 1994). Vandamme *et al.* (1997) reclassified *S. difficile* as non-haemolytic *S. agalactiae* Ib. Two biotypes of *S. agalactiae* were described based on phenotypic characterization and capsular typing (Sheehan *et al.*, 2009). Based on multilocus sequence analysis (Evans *et al.*, 2008; Delannoy *et al.*, 2013) and on genome sequencing, which has defined fish-specific and fish-associated genes (Rosinski-Chupin *et al.*, 2013; Delannoy *et al.*, 2016), *S. agalactiae* comprises two clonal complexes. Using primers described by Imperi *et al.* (2010), Shoemaker *et al.* ('unpublished data') has demonstrated that *S. agalactiae* comprises three molecular capsular types (Fig. 22.1). The classical biochemical characteristics, capsular types and common multilocus sequence types for *S. iniae* and the two *S. agalactiae* clonal complexes are given in Table 22.1.

22.1.2 Transmission

Streptococcus spp. are transmitted horizontally through the water with newly introduced carrier fish being the source of infection (Eldar *et al.*,

---

*Corresponding author e-mail: craig.shoemaker@ars.usda.gov

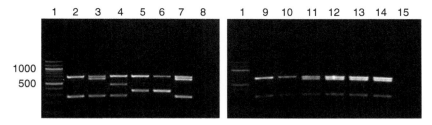

Fig. 22.1. A gel image obtained using a modification of a multiplex PCR assay of Imperi *et al.* (2010) for the capsular typing of *Streptococcus agalactiae* pathogenic to fish. American type culture collection (ATCC) isolates were used to validate the assay. Lane numbers are shown across the top, with numbers of base pairs of the various *cps* (capsular polysaccharide) genes indicated to the left. Lane 1, 100 bp molecular marker; Lane 2, ATCC 12400, Ia (688bp *cpsL* and 272bp *cpsG*); Lane 3, ATCC 51487, Ib (688 bp *cpsL*, 621 bp *cpsJ* and 272 bp *cpsG*); Lane 4, ATCC 13813, II (688bp *cpsL*, 465 bp *cpsJ* and 272 bp *cpsG*); Lane 5, ATCC 31475, III (688 bp *cpsL* and 352 bp *cpsG*); Lane 6, Malaysia isolate (III); Lane 7, Mexico isolate (John Plumb, Ib); Lane 8, no template control; Lane 9, Kuwait isolate (KU-MU-11B, Ia); Lane 10, Brazil isolate (04-ARS-BZ-TN-002, Ia); Lane 11, Honduras isolate (LADL-05-108A, Ib); Lane 12,–Brazil isolate (ARS-BZ-TN-004, Ib); Lane 13, Honduras isolate (TN-HON-08A, Ib); Lane 14, Ecuador isolate (8 Br, Ib); and Lane 15, *S. iniae* (ARS-98-60). The 688 bp band based on the *cpsL* gene is diagnostic for *S. agalactiae*.

Table 22.1. Characteristics of *Streptococcus iniae* and *S. agalactiae* isolated from fish.

Characteristic	*S. iniae*	Species/strain	
		S. agalactiae (Ib)	*S. agalactiae* (Ia or III)
Arginine dihydrolase (ADH)	+ (variable)	+	+
Capsular serotype	I, II[a]	Ib[b]	Ia,[b] III[b]
Clonal complex	NA[c]	552[d]	7[d,e,f]
Growth at 37°C	+ (variable)	+ (weak)	+
Haemolysis	beta[g]	none	beta
Hippurate hydrolysis	–	+	+
Lancefield group	none	B	B
Multilocus sequence type	NA	257,[h] 258,[h] 259,[h] 260,[h,i] 261[h,i]	7,[e,h,i] 283,[i] 500,[i] 491[i]
Pyrrolidonyl arylamidase (PYR)	+	–	–
Starch hydrolysis	+	–	–

[a]Two serotypes have been described (Barnes *et al.*, 2003).
[b]Based on molecular capsular typing using primers by Imperi *et al.* (2010); and capsular typing antiserum Denka Seiken Co., Ltd., Japan (Shoemaker and Xu, unpublished).
[c]NA = Not applicable.
[d]From Delannoy *et al.* (2016).
[e]From Kayansamruaj *et al.* (2015).
[f]From Rosinski-Chupin *et al.* (2013).
[g]Type of blood agar may influence the haemolysis, alpha, gamma and beta haemolysis have been noted on tryptic soy agar supplemented with 5% sheep blood (Chou *et al.*, 2014).
[h]From Evans *et al.* (2008).
[i]From Delannoy *et al.* (2013).

1994; Shoemaker *et al.*, 2001; Tavares *et al.*, 2016). The pathogens can persist in water and sediments near fish farms for over a year (Nguyen *et al.*, 2002). Faecal–oral transmission can occur when infected dead fish are fed to fish (Robinson and Meyer, 1966; Kim *et al.*, 2007). Iregui *et al.* (2016) demonstrated that oral gavaged *S. agalactiae* entered red tilapia (*Oreochromis* sp.) through the gastrointestinal epithelium, causing septicaemia. Interestingly, grouper (*Epinephelus lanceolatus*) fed *S. agalactiae*-spiked feed became anorexic and lethargic, and *S. agalactiae* was isolated from the internal organs of the fish, but no mortality occurred (Delamare-Deboutteville *et al.*, 2015). Shoemaker *et al.* (2000) observed that healthy fish cannibalized the internal organs and eyes of infected fish with subsequent transmission of *S. iniae* and resultant mortality. However, an alternative route via the nares, skin and gills (McNulty *et al.*, 2003) cannot be ruled out. Regardless, the removal of dead and moribund fish should be a priority because these fish shed the pathogens.

Experimental infection with both *S. iniae* and *S. agalactiae* has been induced via waterborne exposure, but the results are sometimes inconsistent. Mortality is often low (e.g. 6–20%) and the abrasion of fish kept at high stocking density is needed to establish infection (Rasheed and Plumb, 1984; Chang and Plumb, 1996; Shoemaker *et al.*, 2000; Mian *et al.*, 2009; Soto *et al.*, 2015). Most studies have used intraperitoneal (IP) injections to fulfil Koch's postulates and reproduce disease under laboratory conditions. Infection via injection is reproducible and results in clinical disease similar to that which is observed in the field. Further, the virulence of *S. agalactiae* Ib isolates for tilapia (Mian *et al.*, 2009; Evans *et al.*, 2015; Delannoy *et al.*, 2016) and grouper (Delamare-Deboutteville *et al.*, 2015) via IP injection has seen extremely low median lethal dose (LD_{50}) values ($<1 \times 10^2$ cfu, colony forming units). Injection (either IP or intramuscular, IM) bypasses the innate immune system, especially the skin and mucosal compartments. LaFrentz *et al.* (2016) recently cohabited naive Nile tilapia (*O. niloticus*) with tilapia injected IP with *S. iniae*. The trial resulted in 8–10% mortality among the naive fish versus 60–70% mortality in the IP-injected fish; this result was repeated in a large single tank trial involving more than 2300 (176 ± 50 g) fish, which gave 6% mortality in the cohabitant's versus 60% in the injected fish). For *S. agalactiae* Ib (Shoemaker *et al.*, unpublished data), three different doses of bacteria (2×10^3, 2×10^4, 2×10^5) were injected IM into Nile tilapia (25 fin clipped fish) and these fish were cohabited with 25 naive fish (one tank per dose, 50 tilapia in total). Similar results were obtained in this trial to those in the *S. iniae* trial, with 12–20% mortality among the cohabitants versus 76–88% mortality in the IM-injected fish (Fig. 22.2).

The vertical transmission of both *S. iniae* and *S. agalactiae* was suggested in tilapia because bacteria were detected both in fertilized eggs and in the resultant progeny using the loop-mediated isothermal amplification (LAMP) assay (Suebseing *et al.*, 2013). Pradeep *et al.* (2016) artificially spawned tilapia brood stock and found both pathogens in unfertilized eggs (70%), milt (90%) and larvae (40%) tested with the LAMP assay. The potential for vertical transmission makes the control of *S. iniae* and *S. agalactiae* problematic.

22.1.3 Geographical distribution

S. iniae and *S. agalactiae* are distributed worldwide and infect more than 27 species of fish, including tilapia (Klesius *et al.*, 2008; Sheehan *et al.*, 2009). Both of the pathogens affect wild and farmed species in fresh, brackish and marine waters (Bowater *et al.*, 2012; Chou *et al.*, 2014; Keirstead *et al.*, 2014).

22.2 Diagnosis

22.2.1 Clinical signs of disease

Clinical signs (Fig. 22.3) include loss of appetite, darkening of the skin, grouping at the bottom of tanks, lethargy, body curvature, and haemorrhage at the bases of fins and opercula (Shoemaker *et al.*, 2006b). Some infected fish swim erratically by either spiralling or spinning just below the water surface. The most prominent signs are uni- or bi-lateral exophthalmia (Fig. 22.3A, B), abdominal distension (Amal and Zamri-Saad, 2011) and coelomitis (Fig. 22.3C); however, these signs are not pathognomonic. Most often, exophthalmia and the associated eye opacity are the last signs observed, suggesting an association with chronic disease (Shoemaker *et al.*, 2006b). LaFrentz *et al.* (2016) noted jaw and caudal pustules in dead and surviving *S. iniae*-infected Nile tilapia (Fig. 22.3D). Similar lesions are also associated with *S. agalactiae* infection, along with buccal paralysis (Tavares *et al.*, 2016) and faecal casts or strings extruding from the anus (Pasnik *et al.*, 2009). Post-mortem examinations of the *S. agalactiae*-infected fish revealed the presence of bloody fluid in the body cavity, an enlarged and reddened spleen and a pale enlarged liver, as well as inflammation of the heart and kidney (Amal and Zamri-Saad, 2011). In some cases, infected fish do not show obvious clinical signs before dying, and death is attributed to septicaemia with infection of the brain and nervous system (Barham *et al.*, 1979).

C.A. Shoemaker *et al.*

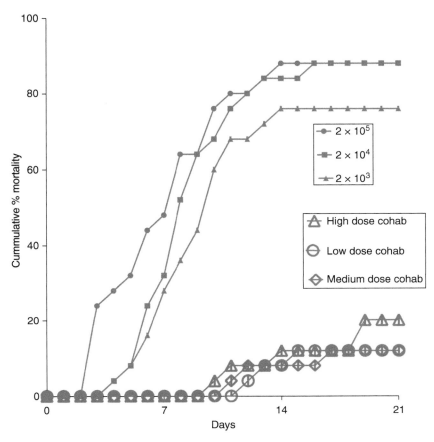

Fig. 22.2. Experimental infection of Nile tilapia (*Oreochromis niloticus*) via intramuscular injection of three doses of *Streptococcus agalactiae* Ib and subsequent cohabitation with naive fish. There was mortality of cohabitants (cohab), but it was low, suggesting that bacteria shed from the injected fish transmitted infection among naive cohabitants either by waterborne and/or faecal–oral routes.

22.2.2   Diagnosis of infection

Diagnosis relies on culture of the bacteria on 5% sheep blood agar plates (Remel, Lenexa, Kansas). Fresh fish are the best source for culturing the bacteria, typically from the kidney and brain. Researchers have also used the dry swab technique (Shoemaker *et al.*, 2001) and commercially available swabs (e.g. Remel BactiSwab) (Keirstead *et al.*, 2014). Presumptive diagnosis is made following the isolation of Gram-positive, catalase-negative bacteria in pure culture. Biochemical characterization (Pier and Madin, 1976; Vandamme *et al.*, 1997) and capsular typing (e.g. *S. agalactiae*) with commercially available antiserum (Denka Seiken, Japan) can provide a definitive diagnosis if molecular technology is unavailable. Miniaturized rapid

systems are useful and *S. agalactiae* is readily identified by the API 20 Strep and API rapid ID 32 STREP test kits (bioMérieux, Durham, North Carolina). Various authors have reported typical API 20 Strep profiles identified for *S. iniae* (e.g. Chou *et al.*, 2014). Molecular biology can further aid with characterization by using primers that have been developed based on the amplification of repetitive genetic regions (Chou *et al.*, 2014); in the case of *S. agalactiae*, molecular capsular typing is accomplished by using a multiplex scheme (Imperi *et al.*, 2010 and Fig. 22.1). A more thorough description of recent polymerase chain reaction (PCR) technologies is provided in Section 22.2.3.

Selective media are advantageous for isolation of streptococci. Columbia blood agar consisting of

Fig. 22.3. Gross pathological findings in naturally (A,B and C) and laboratory (D) infected sexually mature (>180 g) tilapia (*Oreochromis* sp.) with *Streptococcus* spp.: (A, B) exophthalmia and eye opacity; (C) severe coelomitis with tissue adhesion; (D) haemorrhagic skin pustules in the base of the mouth. Photos A, B, C courtesy of Dr Juan A. Morales.

animal derived peptone, enzymatic digests of casein, and defibrinated sheep blood was first described by Ellner *et al.* (1966). The medium was further improved by adding nalidixic acid and colistin (Columbia CNA agar) to suppress growth of most Gram-negative bacteria and enhance recovery of streptococci. Two other selective media for *S. iniae* are thallium acetate-oxolinic acid-blood agar (Nguyen *et al.*, 2002) and Todd-Hewitt broth supplemented with thallium acetate, colostin and oxolinic acid. Nguyen *et al.* (2002) demonstrated the effectiveness of both media to isolate *S. iniae* from water and sediment.

22.2.3 Molecular diagnostics

The molecular identification of *S. iniae* and *S. agalactiae* has relied on PCR amplification of the 16S rRNA gene or the 16S–23S intergenic spacer regions (IGS), with sequence analysis for confirmation (Berridge *et al.*, 1998; Chou *et al.*, 2014; Soto *et al.*, 2015). Mata *et al.* (2004) developed a species-specific PCR with diagnostic potential based on the lactate oxidase (*lctO*) gene that detected 32-61 *S. iniae* cells. Biochemical tests, conventional PCR and the associated 16S rRNA gene sequencing, which are dependent on bacterial isolation, are time-consuming processes. Real-time quantitative polymerase chain reaction (qPCR) can quantify *S. agalactiae* in fish samples. Sebastião *et al.* (2015) described a qPCR based on the *cfb* gene, which encodes the CAMP factor (an extracellular protein) that is specific for *S. agalactiae*; however, both *S. iniae* and *S. agalactiae* were amplified; thus, the potential for both pathogens to occur within the same sample makes the use of this assay problematic. Su *et al.* (2016) developed a qPCR assay for *S. agalactiae* using three different primer pairs based on the *cpsE* (capsular polysaccharide) gene, the *sip* (surface immunogenic protein) gene and the 16S-23S rRNA IGS. The primers based on *cpsE* and *sip* cross-reacted with other bacteria or fish DNA, but the primers targeting the 16S–23S IGS provided specific detection of approximately nine copies of *S. agalactiae* DNA. Su *et al.* (2016) quantified the bacteria from infected fish tissues and assessed bacterial persistence in convalescent tilapia. Tavares *et al.* (2016) used a non-lethal sampling technique on tilapia to validate the assay developed by Su *et al.* (2016); aspirated kidney samples and venipuncture in combination with qPCR proved highly efficient for the detection of *S. agalactiae*.

The LAMP assay has value in low-tech settings. In this method, the target sequence is amplified at a constant temperature (60–65°C) using three sets of primers in the case of *S. iniae* and *S. agalactiae* and a *Bst* (*Bacillus stearothermophilus*) DNA polymerase. The addition of a colour reagent (e.g. calcein) to the reaction tube allows easy visualization of a positive reaction if electrophoresis equipment is not available. Suebsing *et al.* (2013) demonstrated the detection of both *S. iniae* and *S. agalactiae* using a species-specific LAMP assay based on primers designed to the superoxide dismutase (*sodA*) gene. The technology has assessed the carrier status of brood stock, gametes and fry for both *S. iniae* and *S. agalactiae* (Pradeep *et al.*, 2016).

22.3 Pathology

Figures 22.4 (A–D) and 22.5 (A–D) show the histopathological changes associated with infection by *S. iniae* in the yellowtail snapper (*Ocyurus chrysurus*) and white sturgeon (*Acipenser transmontanus*), and by *S. agalactiae* Ib in Nile tilapia and the gulf killifish (*Fundulus grandis*). Tilapia exposed to *Streptococcus* sp. had meningitis, polyserositis, splenitis, ovaritis and myocarditis (Chang and Plumb, 1996). Pericarditis of the heart is characterized by the infiltration of macrophages, lymphocytes and erythrocytes. Macrophages also infiltrate the bulbous arteriosus and ventricles. Bacteria were present in the heart epicardium and the tunica adventitia of the bulbus arteriosus (Miyazaki *et al.*, 1984). Tissue sections showed the infiltration of bacteria-laden necrotic macrophages and precipitation of fibrin (Miyazaki *et al.*, 1984).

The histology of infected liver tissue showed loss of hepatic parenchyma structure, hepatic necrosis, eosinophilic granular inflammatory cell infiltration and severe blood congestion (Abuseliana *et al.*, 2011). The spleen of red tilapia infected by *S. agalactiae* had granulomas and mononuclear cell infiltration, red pulp degeneration and haemosiderosis (Abuseliana *et al.*, 2011). The splenic blood vessels were markedly congested. Using transmission electron microscopy (TEM), Iregui *et al.* (2016) found dividing bacteria attached to the apical surface of intestinal epithelial cells of tilapia orally exposed to *S. agalactiae*. The bacteria multiplied and reached massive numbers within the cytoplasm of the epithelial cells.

Blood vessels in the brain of infected fish were congested and exhibited dilatation and detachment of blood vessel walls. Macrophages intermingled with lymphocytes and fibroblasts in haemorrhagic areas of the meninges of the telencephalon and cerebellum. The infection of the eye caused lenticular fibre disruption and separation, vacuolar formation, or fibroblast proliferation in the lenticular cortex (Chang and Plumb, 1996). Miyazaki *et al.* (1984) also reported that there were bacteria disseminated in the cornea, orbital adipose tissue and oculomotor musculature that resulted in either exudative or granulomatous inflammation, necrosis and haemorrhage.

Infected testis and ovary exhibited either exudative or granulomatous inflammations as the infection progressed (Miyazaki *et al.*, 1984). The testis with exudative inflammation showed extreme bacterial multiplication, necrosis, congestion of the capillaries, oedema and the infiltration of bacteria-laden macrophages and neutrophils.

Suanyuk *et al.* (2010) noted that the histopathological changes found in the liver, pancreas, heart, eye and brain of Asian sea bass (*Lates calcarifer*) were more severe than those in red tilapia. Liver tissue exhibited dilation of hepatic sinusoids and lymphocytic infiltration. The hepatocytes had high levels of vacuolization, degeneration and focal necrosis. The pancreatic tissues showed the degeneration of acinar cells and a loss of zymogen granules. Brains showed lymphocytic infiltration around the infected areas.

S. iniae also causes disease in cultured rainbow trout (*Oncorhynchus mykiss*) (Akhlaghi and Mahjor, 2004; Lahav *et al.*, 2004). Lahav *et al.* (2004) collected fish from trout farms in northern Israel that had experienced heavy fish mortalities caused by *S. iniae* type II. The histopathology

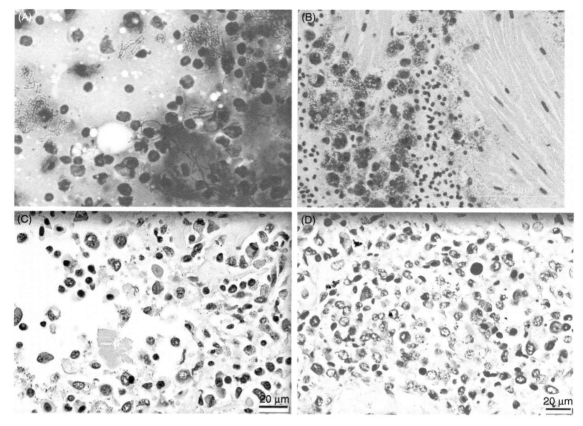

Fig. 22.4. Histopathological findings in fish tissues infected with *Streptococcus iniae*: (A) hepatic tissue of yellowtail snapper (*Ocyurus chrysurus*) with bacteria associated with lysed leucocytes, suggesting that they have been phagocytosed, Wright's stain, 100× magnification; (B) the myocardial fibres in the heart of infected yellowtail snapper are separated and distorted by an inflammatory infiltrate that also contains abundant Gram-positive cocci and diplococci, either within macrophages or forming extracellular colonies, Gram Brown and Brenn stain for differential staining of bacteria in tissue sections; (C, D) infected kidney from white sturgeon (*Acipenser transmontanus*) with changes limited primarily to interstitial haematopoietic areas consisting of necrosis of scattered individual-to-small groups of cells and the presence of (rarely) larger foci of necrosis, often in periarterial locations, and including deposition of amorphous, pale eosinophilic, fibrin-like material, with a few Gram-positive cocci in pairs and short chains widely scattered throughout interstitial areas, C, Gram stain and D, Giemsa stain. Photos courtesy Dr MaryAnna Thrall (A), Dr Natalie Keirstead (B) and Dr Al Camus (C, D).

associated with *S. iniae* type II differed from that regularly observed with *S. iniae* type I. Although no injuries were noted in the skeletal muscle of fish infected by *S. iniae* type I, fish infected with *S. iniae* type II showed multifocal to diffuse areas of degeneration and necrosis, accompanied by heterophilic histiocytic infiltration.

An outbreak of *S. iniae* in channel catfish (*Ictalurus punctatus*) occurred at fish farms in Guangxi, China resulting in acute septicaemia and death (Chen *et al.*, 2011). Histological examination of the viscera revealed severe hepatitis and necrotic hepatocytes around congested blood vessels. Other pathological changes in the catfish were similar to those in tilapia.

The pathological changes in tilapia infected by either *S. iniae* or *S. agalactiae* include pericarditis, epicarditis, myocarditis, endocarditis and meningitis (Chen *et al.*, 2007). There was proliferation of lymphoid tissue in the kidneys, with accumulations of eosinophilic material in the cytoplasm of the tubular cells, where the nuclei were displaced to

C.A. Shoemaker *et al*.

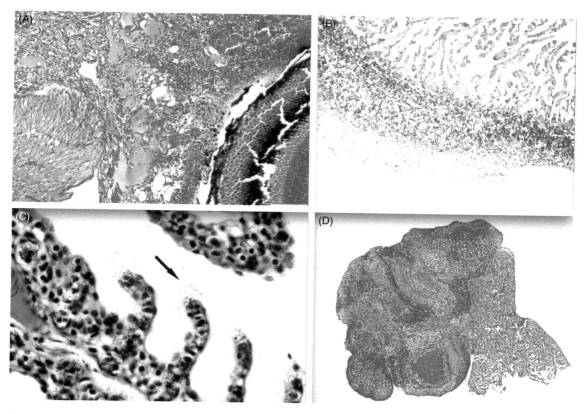

Fig. 22.5. Histopathological findings in fish tissues infected with *Streptococcus agalactiae*: (A) severe granulomatous choroiditis in laboratory-infected Nile tilapia (*Oreochromis niloticus*) challenged by intracoelomic injection (haematoxilin and eosin stain, 20×); (B) severe pericarditis in naturally infected Nile tilapia (haematoxylin and eosin stain, 20×); (C) coccoid bacteria in lamellar capillaries and the filament interstitium, causing necrosis of the lamellar tips and the release of bacteria (arrow) in gulf killifish (*Fundulus grandis*) challenged by intraperitoneal injection of *S. agalactiae* (haematoxylin and eosin stain, 100×); (D) granulomatous epicarditis in the bulbus arteriosus of Nile tilapia fingerlings challenged with *S. agalactiae* by intracoelomic injection (haematoxylin and eosin stain, 10×). Photos courtesy Dr Michele Dennis (A), Dr Juan A. Morales (B), Dr Wes Baumgartner (C) and Dr Oscar Illanes (D).

the side. In *S. agalactiae* infections, there was dissolution of the renal tubules, with cocci surrounding the tubules and contained within the interstitial cells and glomeruli. In addition, bacteria were distributed throughout the spleen and head kidney. During *S. agalactiae* infection, large numbers of cocci occurred in the tissues and within the systemic circulation of infected fish, but this was not a prominent feature of an *S. iniae* infection.

22.4 Pathophysiology

Environmental stressors, such as heavy stocking density, poor husbandry, extremes of water temperature, high ammonia, low dissolved oxygen,

pH (>8), salinity (Zamri-Saad *et al.*, 2014) and external parasitism (e.g. *Trichodina*, *Gyrodactylus* and *Ichthyophthirius* infestations) (Xu *et al.*, 2007, 2009) can result in pathophysiology and enhance streptococcosis.

The effect of stocking density on the mortality caused by *S. iniae* was evaluated in tilapia stocked at low (5.6 g/l), medium (11.2 g/l) and high (22.4 g/l) densities (Shoemaker *et al.*, 2000). A stocking density of ≥11.2 g/l significantly increased *S. iniae* mortality. The mortality rate in the low-density treatment was 4.8% whereas that in the medium- and high-density groups was 28.4 and 25.6%, respectively.

Rodkhum *et al.* (2011) infected Nile tilapia with *S. agalactiae* Ia via immersion in 10^8 cfu/ml and maintained fish at 25, 30 or 33°C for 1 week to assess the effect of temperature on mortality. Mortality (60%) was highest in fish maintained at 33°C, while fish at 30°C exhibited 40% mortality and those held at 25°C survived without the development of clinical disease. These results suggested that tilapia reared at high temperatures (≥30°C) are more susceptible to *S. agalactiae* Ia. Reports from fish farms also suggest that temperatures above 30°C lead to severe *S. agalactiae*-associated mortality among market-sized tilapia (Mian *et al.*, 2009; Noraini *et al.*, 2013).

Ndong *et al.* (2007) evaluated the susceptibility of Mozambique tilapia (*O. mossambicus*) to *S. iniae* at different temperatures. The cumulative mortality of *S. iniae*-injected fish moved from 27 to 19°C (57%) and to 35°C (50%) was significantly higher than that of fish transferred and/or held at 23°C (20%), 27°C (10%) and 31°C (17%), respectively. Additionally, fish transferred from 27°C to either 19 or 35°C showed a decreased leucocyte count and respiratory burst, as well as phagocytic activity and index. The results suggest that a temperature change of 8°C above or below 27°C is immunosuppressive and leads to increased mortality from *S. iniae*.

Evans *et al.* (2003) compared the blood glucose and mortality of tilapia exposed to sublethal dissolved oxygen (DO ≤ 1 mg/l) and under normoxic conditions (≥ 4 mg/l). The blood glucose level (>100 mg/dl) and mortality (80%) were significantly higher in tilapia exposed to sublethal DO after IP injection with *S. agalactiae*. In cultured or wild fish, a sublethal DO may be the result of harmful algal blooms, algal bloom die-offs, eutrophication or physical stratification (Evans *et al.*, 2003; Amal *et al.*, 2013).

Amal *et al.* (2013) showed that the ammonia concentration of the water positively correlated with the presence of *S. agalactiae* in red tilapia cultured in rivers and lakes. Massive mortality by *S. agalactiae* in wild mullet (*Liza klunzingeri*) was associated with ammonium concentrations elevated from 0.11 to 0.33 mg/l (Glibert *et al.*, 2002). Evans *et al.* (2006) noted that exposure to unionized ammonia concentrations of 0.37 mg/l for 24 h caused acute and transient stress responses in Nile tilapia as assessed by elevated blood glucose levels, but did not increase their susceptibility to *S. agalactiae*.

Perera *et al.* (1997) demonstrated that hybrid tilapia (*O. nilotica* × *O. aurea*) in alkaline water (pH > 8) had a higher mortality than fish kept in acidic or neutral water (pH 6–7) after challenge with *S. iniae*. Milud *et al.* (2013) found that a water salinity of 15 ppt increased the susceptibility of red tilapia to *S. agalactiae* compared with fish at salinities of 5 and 10 ppt.

External parasites also increase the susceptibility of fish to *S. iniae* and *S. agalactiae*. Damaged skin, lesions and ulcers are putative routes for bacterial invasion and subsequent manifestation of disease (Xu *et al.*, 2007). Xu *et al.* (2007) demonstrated that mortality was significantly higher (42.2%) in *S. iniae*-challenged tilapia parasitized with *Gyrodactylus niloticus* than in non-parasitized challenged fish (6.7%). Xu *et al.* (2009) also reported enhanced mortality in Nile tilapia following co-infections with *Ichthyophthirius multifiliis* (Ich) and *S. iniae*. Around 20% mortality was noted in single infections (i.e. *S. iniae* or Ich), with up to 88% mortality reported in co-infected fish.

22.5 Prevention and Control Strategies

The control and/or prevention of *S. iniae* or *S. agalactiae* is best incorporated into fish health management plans (see Plumb and Hanson, 2010) that rely on sound fish husbandry, including biosecurity, maintaining the water quality, proper nutrition, etc.

High productivity in tilapia farming is achieved by balancing stocking density with survival and performance (Zamri-Saad *et al.*, 2014). When mortality increases, decreasing the stock density can reduce fish stress and pathogen load, and farmers must balance stocking rates to maximize production while limiting the risk of disease due to poor water quality and enhanced disease transmission (Shoemaker *et al.*, 2000).

Extreme precautions should be exercised when introducing brood stock or eggs into a new or existing brood facility. The disinfection of fish eggs infected with *S. iniae* or *S. agalactiae* is difficult. Chemicals that are approved for use on the eggs of food fish are surface disinfectants that can reduce pathogen presence on the egg's shell but have limited efficacy on bacteria within the fish eggs (Salvesen and Vadstein, 1995). Fish eggs and fry should, therefore, be obtained from pathogen-free sources.

22.5.1 Probiotics, prebiotics and symbiotics

Shelby *et al.* (2006) found that commercial probiotics (Biomate SF-20; Bioplus 2B; Bactocell PA10 MD; Levucell SB 20) did not enhance innate immunity or survival to *S. iniae*. Aly *et al.* (2008) fed tilapia with the probiotics *Bacillus subtilis* and *Lacotobacillus acidophilus*, alone and in combination with each other. Innate immune parameters increased after feeding the probiotics, but resistance to *S. iniae* was only seen after 2 months of feeding *L. acidophilus*. Iwashita *et al.* (2015) fed Nile tilapia a combination of probiotics (*B. subtilis*, *Aspergillus oryzae* and *Saccharomyces cerevisiae*) at 10 g/kg diet, which improved survival from 4 to 34%) after challenge with *S. iniae*. Feeding *Lactococcus lactis* (BFE920), *Bacillus* sp. (IS-2) and *Bacillus* spp. improved the innate immunity and resistance of olive flounder (*Paralichthys olivaceus*) to *S. iniae* (Cha *et al.*, 2013; Jang *et al.*, 2013; Kim *et al.*, 2013).

Widanarni and Tanbiyaskur (2015) fed *Bacillus* sp. NP5, sweet potato oligosaccharides or the combination of these to Nile tilapia. Survival was 74, 74 and 85%, respectively, compared with the control (18.5%) following *S. agalactiae* challenge. Agung *et al.* (2015) microencapsulated *Bacillus* NP5 and mannan oligosaccharide and the combination thereof. Tilapia were fed for 40 days and treatment efficacies were evaluated after subsequent *S. agalactiae* challenge. Survival (55–60%) was significantly better in the probiotic-fed groups than in the controls (33%) (Agung *et al.*, 2015). Other probiotics include *Lactobacillus rhamnosus* GG (LGG), *B. subtilis*, *B. licheniformis* and *Bacillus* sp. combined with *Pediococcus* sp. (Ng *et al.*, 2014; Pirarat *et al.*, 2015). Work on LGG showed that an alginate matrix improved the *in vitro* survival of the lactobacilli under stimulated gastric conditions, as well as the *in vivo* survival of Nile tilapia (50% with encapsulated lactobacilli versus 12% in the controls) (Pirarat *et al.*, 2015). Ng *et al.* (2014) conducted an 8 week feeding study using commercial/prototype probiotics, including *B. subtilis*, *B. licheniformis* and *Bacillus* sp. combined with *Pediococcus* sp. With the exception of *B. licheniformis*, survival against *S. agalactiae* was improved and Ng *et al.* (2014) suggested that *B. subtilis* worked best.

22.5.2 Vaccines and vaccination

Vaccination strategies for *S. iniae* and *S. agalactiae* have relied on killed vaccines (Klesius *et al.*, 2006; Sheehan *et al.*, 2009). Two formalin-killed vaccines for *S. iniae* are available from Merck Animal Health (www.merck-animal-health.com). Norvax® Strep Si is a formalin-killed bacterin that has been tested by

Merck Animal Health for safety and efficacy in the Asian sea bass. In the Merck laboratory trial, fifty 5 g fish were vaccinated by immersion with the recommended dose of Norvax® Strep Si, and challenged by immersion with *S. iniae* 3 weeks later; the trial resulted in a relative percentage survival (RPS) of 68% in the vaccinated fish. In the field, the vaccination by injection of 20 g fish resulted in 100% RPS when they were challenged with *S. iniae* 6 weeks later. Aquavac-Garvetil™ is a killed vaccine containing *S. iniae* and *L. garviae* antigens administered by immersion and by an oral booster vaccination. Field trials by Merck in tilapia gave RPS of 15 (immersion only) and 50% (immersion plus oral) following a natural outbreak of streptococcal disease. The vaccines are not licensed for use in all countries, so producers should consult the appropriate regulatory authorities for their regions. Protection is attributed to the production of antibody (LaFrentz *et al.*, 2011), but other immune mechanisms are also involved (Allen and Neely, 2012; Aviles *et al.*, 2013). The emergence of new serotypes of *S. iniae* in Israel (Eyngor *et al.*, 2008) and Australia (Millard *et al.*, 2012) has decreased vaccine efficacy, suggesting the need for polyvalent vaccines that include the different serotypes.

Modified extracellular product (ECP) bacterins against *S. iniae* have been developed in the USA (Klesius *et al.*, 2000). Shoemaker *et al.* (2010) noted that protection occurred among sex-reversed tilapia IP vaccinated with the ECP bacterin and challenged with heterologous isolates (RPS, 79–100%). Serological response as evidenced by Western blot analysis showed that antigens were conserved among the isolates. Furthermore, the protective nature of antibodies evoked by vaccination was demonstrated by passive immunization and immunoproteomics (LaFrentz *et al.*, 2011). Co-infections with other bacteria (Shoemaker *et al.*, 2012) and parasites (Martins *et al.*, 2011) diminished vaccine performance in the field.

Two licensed IP injectable vaccines for *S. agalactiae* are available in some regions. AQUAVAC® Strep Sa, from Merck, is a water-in-oil emulsion *S. agalactiae* biotype 2 inactivated vaccine that is applied as a single dose (0.05 ml) in fish ≥15 g. The RPS was 70–90% in laboratory studies, and field trials showed 13 and 1% improved survival on two farms, respectively (www.merck-animal-health.com). AlphaJect® micro1 TiLa, a similar formulation from Pharmaq, produced a RPS ≥ 60% as reported on the product's fact sheet; its availability is limited to

Costa Rica, Panama and Honduras (www.pharmaq. no). Killed vaccines against *S. agalactiae* biotype 1 (beta-haemolytic) and biotype 2 (non-haemolytic) have been formulated by Intervet/Schering-Plough, and protective immunity shown to be biotype specific (Sheehan *et al.*, 2009). Evans *et al.* (2004) developed an ECP *S. agalactiae* bacterin using a beta-haemolytic *S. agalactiae* Ia isolate (Evans *et al.*, 2008), which protected tilapia following IP immunization. Chen *et al.* (2012) developed killed *S. agalactiae* vaccines based on different combinations of ten isolates exhibiting distinct pulse-field gel electrophoresis (PFGE) genotypes. They demonstrated the potential of a polyvalent vaccine that may provide broad protection to field isolates.

A few studies have reported the efficacy of killed vaccines administered via immersion and/or oral delivery for *S. iniae* (Shoemaker *et al.*, 2006a) and *S. agalactiae* (Firdaus-Nawi *et al.*, 2014; Nur-Nazifah *et al.*, 2014), but injection is the most effective route of administration. Live vaccines developed for other fish pathogens have been successfully applied using immersion and oral delivery (Shoemaker and Klesius, 2014). Live attenuated *S. iniae* vaccines were developed using transposon mutagenesis or allelic exchange to delete the genes responsible for virulence (Buchanan *et al.*, 2005; Lowe *et al.*, 2007). Locke *et al.* (2010) tested the efficacy of the attenuated strains in hybrid striped bass (*Morone chrysops* × *M. saxatilis*) via injection and immersion delivery in comparison with a whole-cell *S. iniae* bacterin. The results demonstrated that the best efficacy (100% protection but some vaccine-related mortality) was produced by the Δ*simA* (surface M-like protein) mutant after bath immersion, though the attenuated isolate retained some virulence. The other vaccine candidates tested via immersion were completely attenuated but their RPS values were reduced – 26% with the Δ*pgmA* (phosphoglucomutase) mutant and 51% with the Δ*cpsD* mutant (no vaccine-related mortality with either) – but they were still better than the bacterin RPS (19%).

A live attenuated *S. agalactiae* strain was developed using continuous *in vitro* passage in tryptic soy broth (Li *et al.*, 2015). The starting material was *S. agalactiae* capsular type Ia isolate HN016 (Chen *et al.*, 2012), and Li *et al.* (2015) designated the strain YM001, which exhibited longer chain length (more than five cells), a non-haemolytic phenotype on blood agar and an altered PFGE profile compared with the parent isolate. The passaged strain was attenuated and its efficacy following IP, immersion and oral vaccination (RPS, 53–96%) was demonstrated against challenge by the parent strain (Li *et al.*, 2015).

Recently, an oral DNA vaccine encoding the surface immunogenic protein (sip) of *S. agalactiae* was delivered by an attenuated *Salmonella typhimurium* (Huang *et al.*, 2014). The authors designed a recombinant plasmid, pVAX1-sip, and transferred this to the live attenuated *Salmonella* carrier to orally vaccinate and protect tilapia. Although the plasmid did not integrate into the tilapia chromosomes (Huang *et al.*, 2014), the vaccine is a genetically modified organism (GMO), and the public acceptance of such products is low in Europe and the US.

Other vaccine development strategies include recombinant protein vaccines delivered via injection and/or orally for both *S. iniae* (Aviles *et al.*, 2013) and *S. agalactiae* (Nur-Nazifah *et al.*, 2014). The *S. iniae* vaccine used by Aviles *et al.* (2013) was based on a conserved surface M-protein of *S. iniae*. The recombinant protein induced specific antibody but did not protect Asian sea bass (Aviles *et al.*, 2013). The feed-based recombinant vaccine described by Nur-Nazifah *et al.* (2014) used a cell wall surface anchor protein from *S. agalactiae* expressed in *Escherichia coli*. Tilapia fed the recombinant vaccine exhibited 70% survival, but all of the fish in the other treatments died (Nur-Nazifah *et al.*, 2014).

22.5.3 Antimicrobial therapy

Prudent antimicrobial therapy is an essential tool for aquaculturists when other strategies fail to maintain fish health. The only approved antibiotic for controlling mortality due to *S. iniae* in the USA is AQUAFLOR® (florfenicol). The dose is 15 mg florfenicol/kg fish daily for 10 consecutive days with a 15 day withdrawal time (Gaunt *et al.*, 2010; Gaikowski *et al.*, 2013). AQUAFLOR® must be used with a veterinary feed directive. Zamri-Saad *et al.* (2014) reported that farmers in Malaysia use erythromycin and oxytetracycline as a treatment and/or prophylactic against streptococcosis. The prophylactic use of antimicrobials is prohibited in many countries (e.g. the USA), and the judicious use of antimicrobials is encouraged worldwide. Little information is available on the use of antibiotics to control mortality induced by *S. agalactiae*.

22.5.4 Selective breeding

The selective breeding of fish for improved growth and disease resistance is an important component of effective fish health management. LaFrentz *et al.* (2016) conducted the first large-scale trial of resistance in Nile tilapia against *S. iniae*, in which full- and paternal half-sib families were challenged to determine the potential for additive genetic variation in resistance to *S. iniae*. A moderately high heritability estimate (0.42 ± 0.07) was found, which suggested that *S. iniae* resistance was heritable and that family selection could result in genetic improvement in tilapia.

22.6 Conclusions with Suggestions for Future Research

S. iniae and *S. agalactiae* are economically important pathogens of wild and cultured fish worldwide. Prudent reporting of the capsular type, biotype and/or genotype of *S. agalactiae* should be encouraged in scientific publications. Many recent reports describe studies on *S. agalactiae* but fail to define the strain that was used; such a definition would allow better comparison between studies and may speed the scientific discovery of immune mechanisms and potential biocontrol strategies.

Comparative studies on histopathology and/or pathogenesis between *S. iniae*, *S. agalactiae* type Ia, III and Ib (i.e. different biotypes or genotypes of *S. agalactiae*) are lacking. As fish-specific and fish-associated genotypes of *S. agalactiae* (Delannoy *et al.*, 2016) have recently been defined, this information should be included in such comparisons. Additionally, recent studies from Malaysia and Thailand have reported the occurrence of concurrent *S. iniae* and *S. agalactiae* infections on fish farms. Therefore, laboratory studies examining host–pathogen interactions in the presence of multiple streptococci are needed. These experiments may have direct implications for the design of effective control and vaccination strategies for streptococcosis in fish.

Acknowledgements

The work was supported by USDA-ARS (CRIS Project No. 6010-32000-026-00D) and Spring Genetics, Homestead, Florida, USA (MTA-CRADA No. 58-6010-6-005). Mention of trade names or commercial products in this publication is solely for the purpose of providing specific information and does not imply recommendation or endorsement by the US Department of Agriculture.

References

Abuseliana, A.F., Daud, H.H.M., Aziz, S.A., Bejo, S.K. and Alsaid, M. (2011) Pathogenicity of *Streptococcus agalactiae* isolated from a fish farm in Selangor to juvenile red tilapia (*Oreochromis* sp.). *Journal of Animal and Veterinary Advances* 10, 914–919.

Agnew, W. and Barnes, A.C. (2007) *Streptococcus iniae*: an aquatic pathogen of global veterinary significance and a challenging candidate for reliable vaccination. *Veterinary Microbiology* 122, 1–15.

Agung, L.A., Widanarni and Yuhana, M. (2015) Application of micro-encapsulated probiotic *Bacillus* NP5 and prebiotic mannan oligosaccharide (MOS) to prevent streptococcosis on tilapia *Oreochromis niloticus*. *Research Journal of Microbiology* 10, 571–581.

Akhlaghi, M. and Mahjor, A.A. (2004) Some histopathological aspects of streptococcosis in cultured rainbow trout (*Oncorhynchus mykiss*). *Bulletin of the European Association of Fish Pathologists* 24, 132–136.

Allen, J.P. and Neely, M.N. (2012) CpsY influences *Streptococcus iniae* cell wall adaptations important for neutrophil intracellular survival. *Infection and Immunity* 80, 1707–1715.

Aly, S.M., Abdel-Galil Ahmed, Y., Abdel-Aziz Ghareeb, A. and Mohamed, M.F. (2008) Studies on *Bacillus subtilis* and *Lactobacillus acidophilus*, as potential probiotics, on the immune response and resistance of *Tilapia nilotica* (*Oreochromis niloticus*) to challenge infections. *Fish and Shellfish Immunology* 25, 128–136.

Amal, M.N.A. and Zamri-Saad, M. (2011) Streptococcosis in tilapia (*Oreochromis niloticus*): a review. *Pertanika Journal of Tropical Agricultural Science* 34, 195–206.

Amal, M.N.A., Zamri-Saad, M., Siti-Zahrah, A. and Zulkafli, A.R. (2013) Water quality influences the presence of *Streptococcus agalactiae* in cage cultured red hybrid tilapia, *Oreochromis niloticus* × *Oreochromis mossambicus*. *Aquaculture Research* 46, 313–323.

Aviles, T., Zhang, M.M., Chan, J., Delamare-Deboutteville, J., Green, T.J. *et al.* (2013) The conserved surface M-protein SiMA of *Streptococcus iniae* is not effective as a cross-protective vaccine against differing capsular serotypes in farmed fish. *Veterinary Microbiology* 162, 151–159.

Barham, W.T., Schoobee, H. and Smit, G.L. (1979) The occurrence of *Aeromonas* sp. and *Streptococcus* spp. in rainbow trout (*Salmo gairdneri*). *Journal of Fish Biology* 15, 457–460.

Barnes, A.C., Young, F.M., Horne, M.T. and Ellis, A.E. (2003) *Streptococcus iniae*: serological differences, presence of capsule and resistance to immune serum killing. *Diseases of Aquatic Organisms* 53, 241–247.

Berridge, B.R., Fuller, J.D., de Azavedo, J., Low, D.E., Bercovier, H. *et al.* (1998) Development of a specific nested oligonucleotide PCR primer for *Streptococcus iniae* 16S–23S ribosomal DNA intergenic spacer. *Journal of Clinical Microbiology* 36, 2778–2781.

Bowater, R.O., Forbes-Faulkner, J., Anderson, I.G., Condon, K., Robinson, B. *et al.* (2012) Natural outbreak of *Streptococcus agalactiae* (GBS) infection in wild giant Queensland grouper, *Epinephelus lanceolatus* (Bloch), and other wild fish in northern Queensland, Australia. *Journal of Fish Diseases* 35, 173–186.

Buchanan, J.T., Stannard, J.A., Lauth, X., Ostland, V.E., Powell, H.C. *et al.* (2005) *Streptococcus iniae* phosphoglucomutase is a virulence factor and target for vaccine development. *Infection and Immunity* 73, 6935–6944.

Cha, J.-H., Rahimnejad, S., Yang, S.-Y., Kim, K.-W. and Lee, K.-J. (2013) Evaluations of *Bacillus* spp. as dietary additive on growth performance, innate immunity and disease resistance of olive flounder (*Paralichthys olivaceus*) against *Streptococcus iniae* and as water additives. *Aquaculture* 402–403, 50–57.

Chang, P.H. and Plumb, J.A. (1996) Histopathology of experimental *Streptococcus* sp. infection in tilapia, *Oreochromis niloticus* (L.) and channel catfish, *Ictalurus punctatus* (Rafinesque). *Journal of Fish Diseases* 19, 235–241.

Chen, C.Y., Chao, C.B. and Bowser, P.R. (2007) Comparative histopathology of *Streptococcus iniae* and *Streptococcus agalactiae*-infected tilapia. *Bulletin of the European Association of Fish Pathologists* 27, 2–9.

Chen, D.F., Wang, K.Y., Geng, Y., Wang, J., Huang, X.L. *et al.* (2011) Pathological changes in cultured channel catfish *Ictalurus punctatus* spontaneously infected with *Streptococcus iniae*. *Diseases of Aquatic Organisms* 95, 203–208.

Chen, M., Wang, R., Li, LP, Liang, WW, Li, J. *et al.* (2012) Screening vaccine candidate strains against *Streptococcus agalactiae* of tilapia based on PFGE genotype. *Vaccine* 30, 6088–6092.

Chou, L., Griffin, M.J., Fraites, T., Ware, C., Ferguson, H. *et al.* (2014) Phenotypic and genotypic heterogeneity among *Streptococcus iniae* isolates recovered from cultured and wild fish in North America, Central America and Caribbean islands. *Journal of Aquatic Animal Health* 26, 263–271.

Delamare-Deboutteville, J., Bowater, R., Condon, K., Reynolds, A., Fisk, A. *et al.* (2015) Infection and pathology in Queensland grouper, *Epinephelus lanceolatus*, (Blosh), caused by exposure to *S. agalactiae* via different routes. *Journal of Fish Diseases* 38, 1021–1035.

Delannoy, C., Crumlish, M., Fontaine, M.C., Pollock, J., Foster, G. *et al.* (2013) Human *Streptococcus agalactiae* strains in aquatic mammals and fish. *BMC Microbiology* 13:41.

Delannoy, C., Zadoks, R.N., Crumlish, M., Rodgers, D., Lainson, F.A. *et al.* (2016) Genomic comparison of virulent and non-virulent *Streptococcus agalactiae* in fish. *Journal of Fish Diseases* 39, 13–29.

Eldar, A., Bejerano, Y. and Bercovier, H. (1994) *Streptococcus shiloi* and *Streptococcus difficile*: two new streptococcal species causing meningoencephalitis in fish. *Current Microbiology* 28, 139–143.

Ellner, P.D., Stoessel, C.J., Drakeford, E. and Vasi, F. (1966) A new culture medium for medical bacteriology. *American Journal of Clinical Pathology* 45, 502–504.

Evans, J.J., Shoemaker, C.A. and Klesius, P.H. (2003) Effects of sublethal dissolved oxygen stress on blood glucose and susceptibility to *Streptococcus agalactiae* in Nile tilapia, *Oreochromis niloticus*. *Journal of Aquatic Animal Health* 15, 202–208.

Evans, J.J., Shoemaker, C.A. and Klesius, P.H. (2004) Efficacy of *Streptococcus agalactiae* (Group B) vaccine in tilapia (*Oreochromis niloticus*) by intraperitoneal and bath immersion administration. *Vaccine* 22, 3769–3773.

Evans, J.J., Pasnik, D.J., Bril, G.C. and Klesius, P.H. (2006) Un-ionized ammonia exposure in Nile tilapia: toxicity, stress response and susceptibility to *Streptococcus agalactiae*. *North American Journal of Aquaculture* 68, 23–33.

Evans, J.J., Bohnsack, J.F., Klesius, P.H., Whiting, A.A., Garcia, J.G. *et al.* (2008) Phylogenetic relationships among *Streptococcus agalactiae* isolated from piscine, dolphin, bovine and human sources: a dolphin and piscine lineage associated with a fish epidemic in Kuwait is also associated with human neonatal infections in Japan. *Journal of Medical Microbiology* 57, 1369–1376.

Evans, J.J., Pasnik, D.J. and Klesius, P.H. (2015) Differential pathogenicity of five *Streptococcus agalactiae* isolates of diverse geographic origin in Nile tilapia (*Oreochromis niloticus* L.). *Aquaculture Research* 46, 2374–2381.

Eyngor, M., Tekoah, Y., Shapira, R., Hurvitz, A., Zlotkin, A. *et al.* (2008) Emergence of novel *Streptococcus iniae* exopolysaccharide-producing strains following vaccination with nonproducing strains. *Applied and Environmental Microbiology* 74, 6892–6897.

Firdaus-Nawi, M., Yusoff, S.M., Yusof, H., Abdulla, S.-Z. and Zamri-Saad, M. (2014) Efficacy of feed-based adjuvant vaccine against *Streptococcus agalactiae* in *Oreochromis* spp. in Malaysia. *Aquaculture Research* 45, 87–96.

Gaikowski, M.P., Wolf, J.C., Schleis, S.M., Tuomari, D. and Endris, R.G. (2013) Safety of florfenicol administered in feed to tilapia (*Oreochromis* sp.). *Toxicological Pathology* 41, 639–652.

Gaunt, P.S., Endris, R., McGinnis, A., Baumgartner, W., Camus, A. *et al.* (2010) Determination of florfenicol dose rate in feed for control of mortality in Nile tilapia infected with *Streptococcus iniae*. *Journal of Aquatic Animal Health* 22, 158–166.

Gauthier, D.T. (2015) Bacterial zoonoses of fishes: a review and appraisal of evidence for likages between fish and human infections. *The Veterinary Journal* 203, 27–35.

Glibert, P.M., Landsberg, J., Evans, J.J., Al Sarawi, M.A., Faraj, M. *et al.* (2002) A fish kill of massive proportion in Kuwait Bay, Arabian Gulf, 2001: the roles of disease, harmful algae, and eutrophication. *Harmful Algae* 12, 1–17.

Huang, L.Y., Wang, K.Y., Xiao, D., Chen, D.F., Geng, Y. *et al.* (2014) Safety and immunogenicity of an oral DNA vaccine encoding Sip of *Streptococcus agalactiae* from Nile tilapia *Oreochromis niloticus* delivered by live attenuated *Salmonella typhimurium*. *Fish and Shellfish Immunology* 38, 34–41.

Imperi, M., Pataracchia, M., Alfarone, G., Baldassarri, L., Orefici, G. *et al.* (2010) A multiplex PCR assay for the direct identification of the capsular type (Ia to IX) of *Streptococcus agalactiae*. *Journal of Microbiological Methods* 80, 212–214.

Iregui, C.A., Comas, J., Vasquez, G.M. and Verjan, N. (2016) Experimental early pathogenesis of *Streptococcus agalactiae* infection in red tilapia (*Oreochromis* sp.) *Journal of Fish Diseases* 39, 205–215.

Iwashita, M.K.P., Nakandakare, I.B., Terhune, J.S., Wood, T. and Ranzini-Paiva, M.J.T. (2015) Dietary supplementation with *Bacillus subtilis*, *Saccharomyces cerevisiae* and *Aspergillus oryzae* enhance immunity and disease resistance against *Aeromonas hydrophila* and *Streptococcus iniae* infection in juvenile tilapia *Oreochromis niloticus*. *Fish and Shellfish Immunology* 43, 60–66.

Jang, I.-S., Kim, D.-H. and Heo, M.-S. (2013) Dietary administration of probiotics, *Bacillus* sp. IS-2, enhance the innate immune response and disease resistance of *Paralichthys olivaceus* against *Streptococcus iniae*. *Korean Journal of Microbiology* 49, 172–178.

Kayansamruaj, P., Pirarat, N., Kondo, H., Hirono, I. and Rodkhum, C. (2015) Genomic comparison between pathogenic *Streptococcus agalactiae* isolated from Nile tilapia in Thailand and fish-derived ST7 strains. *Infection, Genetics and Evolution* 36, 307–314.

Keirstead, N.D., Brake, J.W., Griffin, M.J., Halliday-Simmonds, I., Thrall, M.A. *et al.* (2014) Fatal septicemia caused by the zoonotic bacterium *Streptococcus iniae* during and outbreak in Caribbean reef fish. *Veterinary Pathology* 51, 1035–1041.

Kim, D., Beck, B.R., Heo, S.-B, Kim, H.D., Lee, S.-M. *et al.* (2013) *Lactococcus lactis* BFE920 activates the innate immune system of olive flounder (*Paralichthyes olivaceus*), resulting in protection against *Streptococccus iniae* infection and enhancing feed efficiency and weight gain in large-scale field studies. *Fish and Shellfish Immunology* 35, 1585–1590.

Kim, J.H., Gomez, D.K., Choresca, C.H. and Park, S.C. (2007) Detection of major bacterial and viral pathogens in trash fish used to feed cultured flounder in Korea. *Aquaculture* 272, 105–110.

Klesius, P.H., Shoemaker, C.A. and Evans, J.J. (2000) Efficacy of a single and combined *Streptococcus iniae* isolate vaccine administered by intraperitoneal and intramuscular routes in tilapia (*Oreochromis niloticus*). *Aquaculture* 188, 237–246.

Klesius, P.H., Evans, J.J., Shoemaker, C.A. and Pasnik, D.J. (2006) Streptococcal vaccinology in aquaculture. In: Lim, C. and Webster, C. (eds) *Tilapia: Biology, Culture and Nutrition*. Food Products Press, imprint of Haworth Press, Binghampton, New York, pp. 583–606.

Klesius, P.H., Shoemaker, C.A. and Evans, J.J. (2008) *Streptococcus*: a worldwide fish health problem. In: Elghobashy, H., Fitzsimmons, K. and Diab, A.S. (eds) *From the Pharaohs to the Future: Proceedings of the 8th International Symposium on Tilapia in Aquaculture (ISTA8), Cairo, Egypt, Volume 1*. AQUAFISH Collaborative Research Support Program, Corvallis, Oregon, pp. 83–107.

LaFrentz, B.R., Shoemaker, C.A. and Klesius, P.H. (2011) Immunoproteomic analysis of the antibody response obtained in Nile tilapia following vaccination with a *Streptococcus iniae* vaccine. *Veterinary Microbiology* 152, 346–352.

LaFrentz, B.R., Lozano, C.A., Shoemaker, C.A., García, J.C., Xu, D.-H. *et al.* (2016) Controlled challenge experiment demonstrates substantial additive genetic variation in resistance of Nile tilapia (*Oreochromis niloticus*) to *Streptococcus iniae*. *Aquaculture* 458, 134–139.

Lahav, D., Eyngor, M., Hurvitz, A., Ghittino, C., Lublin, A. *et al.* (2004) *Streptococcus iniae* type II infections in rainbow trout *Oncorhynchus mykiss*. *Diseases of Aquatic Organisms* 62, 177–180.

Li, L.P., Wang, R., Liang, W.W., Huang, T., Huang, Y. *et al.* (2015) Development of live attenuated *Streptococcus agalactiae* vaccine for tilapia via continuous passage in vitro. *Fish and Shellfish Immunology* 45, 955–963.

Liu, G., Zhang, W. and Lu, C. (2013) Comparative genomics analysis of *Streptococcus agalactiae* reveals that isolates from cultured tilapia in China are closely related to the human strain A909. *BMC Genomics* 17:775.

Locke, J.B., Vicknair, M.R., Ostland, V.E., Nizet, V. and Buchanan, J.T. (2010) Evaluation of *Streptococcus iniae* killed bacterin and live attenuated vaccines in hybrid striped bass through injection and bath immersion. *Diseases of Aquatic Organisms* 89, 117–123.

Lowe, B.A., Miller, J.D. and Neely, M.N. (2007) Analysis of the polysaccharide capsule of the systemic pathogen *Streptococcus iniae* and its implications in virulence. *Infection and Immunity* 75, 1255–1264.

Martins, M.L., Shoemaker, C.A., Xu, D. and Klesius, P.H. (2011) Effect of parasitism on vaccine efficacy against

Streptococcus iniae in Nile tilapia. *Aquaculture* 314, 18–23.

Mata, A.I., Blanco, M.M., Domínguez, L., Fernandez-Garayzábal, J.F. and Gibello, A. (2004) Development of a PCR assay for *Streptococcus iniae* based on the lactate oxidase (*lctO*) gene with potential diagnostic value. *Veterinary Microbiology* 101, 109–116.

McNulty, S.T., Klesius, P.H., Shoemaker, C.A. and Evans, J.J. (2003) *Streptococcus iniae* infection and tissue distribution in hybrid striped bass (*Morone chrysops* × *Morone saxitilis*). *Aquaculture* 220, 165–173.

Mian, G.F., Godoy, D.T., Leal, C.A.G., Yuhara, T.Y., Costa, G.M. *et al.* (2009) Aspects of the natural history and virulence of *S. agalactiae* infection in Nile tilapia. *Veterinary Microbiology* 136, 180–183.

Millard, C.M., Baiano, J.C.F., Chan, C., Yuen, B., Aviles, F. *et al.* (2012) Evolution of the capsular operon of *Streptococcus iniae* in response to vaccination. *Applied and Environmental Microbiology* 78, 8219–8226.

Milud, A., Hassan, D., Noordin, M., Khairani, S.B., Yasser, M. *et al.* (2013) Environmental factors influencing the susceptibility of red hybrid tilapia (*Orechromis* sp.) to *Streptococcus agalactiae* infection. *Advanced Science Letters* 19, 3600–3604.

Miyazaki, T., Kubota, S.S., Kaige, N. and Miyashita, T. (1984) A histopathological study of streptococcal disease in tilapia. *Fish Pathology* 19, 167–172.

Ndong, D., Chen, Y.Y., Lin, Y.H., Vaseeharan, B. and Chen, J.C. (2007) The immune response of tilapia *Oreochromis mossambicus* and its susceptibility to *Streptococcus iniae* under stress in low and high temperatures. *Fish and Shellfish Immunology* 22, 686–94.

Ng, W.-K., Kim, Y.-C., Romano, N., Koh, C.-B. and Yang, S.-Y. (2014) Effects of dietary probiotic on the growth and feeding efficiency of red hybrid tilapia, *Oreochromis* sp., and subsequent resistance to *Streptococcus agalactiae*. *Journal of Applied Aquaculture* 26, 22–31.

Nguyen, H.T., Kanai, K. and Yoshikoshi, K. (2002) Ecological investigation of *Streptococcus iniae* isolated in cultured Japanese flounder (*Paralichthys olivaceus*) using selective isolation procedures. *Aquaculture* 205, 7–17.

Noraini, O., Jahwarhar, N.A., Sabri, M.Y., Emikpe, B.O., Tanko, P.N. *et al.* (2013) The effect of heat stress on clinicopathological changes and immunolocalization of antigens in experimental *Streptococcus agalactiae* infection in red hybrid tilapia. *Veterinary World* 6, 997–1003.

Nur-Nazifah, M., Sabri, M.Y. and Siti-Zahrah, A. (2014) Development and efficacy of feed-based recombinant vaccine encoding the cell wall surface anchor family protein of *Streptococcus agalactiae* against streptococcosis in *Oreochromis* sp. *Fish and Shellfish Immunology* 37, 193–200.

Pasnik, D.J., Evans, J.J. and Klesius, P.H. (2009) Fecal strings associated with *Streptococcus agalactiae*

infection in Nile tilapia (*Oreochromis niloticus*). *Open Veterinary Science Journal* 3, 6–8.

Perera, R.P., Sterling, K.J. and Lewis, D.H. (1997) Epizootiological aspects of *Streptococcus iniae* affecting tilapia in Texas. *Aquaculture* 152, 25–33.

Pier, G.B. and Madin, S.H. (1976) *Streptococcus iniae* sp. nov., a beta hemolytic streptococcus isolated from an Amazon freshwater dolphin, *Inia geoffrensis*. *International Journal of Systematic Bacteriology* 26, 545–553.

Pirarat, N., Pinpimai, K., Rodkhum, C., Chansue, N., Ooi, E.L. *et al.* (2015) Viability and morphological evaluation of alginate-encapsulated *Lactobacillus rhamnosus* GG under simulated tilapia gastrointestinal conditions and its effects on growth performance, intestinal morphology and protection against *Streptococus agalactiae*. *Animal Feed Science and Technology* 207, 93–103.

Plumb, J.A. and Hanson, L.A. (2010) Tilapia bacterial diseases. In: Plumb, J.A. and Hanson, L.A. *Health Maintenance and Principal Microbial Diseases of Cultured Fishes*, 3rd edn. Wiley-Blackwell, Ames, Iowa, pp. 445–463.

Pradeep, P.J., Suebsing, R., Sirthammajak, S., Kampeera, J., Jitrakorn, S. *et al.* (2016) Evidence of vertical transmission and tissue tropism of streptococcosis from naturally infected red tilapia (*Oreochromis* spp.). *Aquaculture Reports* 3, 58–66.

Rasheed, V. and Plumb, J.A. (1984) Pathogeneicity of non-hemolytic group B *Streptococcus* sp. in gulf killifish, *Fundulus grandis* Baird and Girard. *Aquaculture* 37, 97–105.

Robinson, J.A. and Meyer, F.P. (1966) Streptococcal fish pathogen. *Journal of Bacteriology* 95, 512.

Rodkhum, C., Kayansamruaj, P. and Pirarat, N. (2011) Effect of water temperature on susceptibility to *Streptococcus agalactiae* serotype Ia infection in Nile tilapia (*Oreochromis niloticus*). *Thai Journal of Veterinary Medicine* 41, 309–314.

Rosinski-Chupin, I., Sauvage, E., Mairey, B., Mangenot, S., Ma, L. *et al.* (2013) Reductive evolution in *Streptococcus agalactiae* and emergence of a host adapted lineage. *BMC Genomics* 14:252.

Salati, F. (2011) *Enterococcus seriolicida* and *Streptococcus* spp. (*S. iniae, S. agalactiae and S. dysgalactiae*). *Fish Diseases and Disorders* 3, 375–396.

Salvesen, I. and Vadstein, O. (1995) Surface disinfection of eggs from marine fish: evaluation of four chemicals. *Aquaculture International* 3, 155–171.

Sebastião, F. de A., Lemos, E.G.M. and Pilarski, F. (2015) Validation of absolute quantitative real-time PCR for the diagnosis of *Streptococcus agalactiae* in fish. *Journal of Microbiological Methods* 119, 168–175.

Sheehan, B., Labrie, L., Lee, Y.S., Wong, F.S., Chan, J. *et al.* (2009) Streptococcosis in tilapia: vacccination effective against main strep species. *Global Aquaculture Advocate*, July/August, 72–74. Available at:

http://pdf.gaalliance.org/pdf/GAA-Sheehan-July09.pdf (accessed 18 November 2016).

Shelby, R.A., Lim, C., Yildirim-Aksoy, M. and Delaney, M.A. (2006) Effects of probiotic diet supplements on disease resistance and immune response of young Nile tilapia, *Oreochromis niloticus*. *Journal of Applied Aquaculture* 18, 23–34.

Shoemaker, C.A. and Klesius, P.H. (2014) Replicating vaccines. In: Gudding, R., Lillehaug, A. and Evensen, Ø. (eds) *Fish Vaccination*. Wiley-Blackwell, Chichester/Oxford, UK, pp. 33–46.

Shoemaker, C.A., Evans, J.J. and Klesius, P.H. (2000) Density and dose: factors affecting mortality of *Streptococcus iniae* infected tilapia (*Oreochromis niloticus*). *Aquaculture* 188, 229–235.

Shoemaker, C.A., Klesius, P.H. and Evans, J.J. (2001) Prevalence of *Streptococcus iniae* in tilapia, hybrid striped bass, and channel catfish on commercial fish farms in the United States. *American Journal of Veterinary Research* 62, 174–177.

Shoemaker, C.A., Vandenberg, G.W., Desormeaux, A., Klesius, P.H. and Evans, J.J. (2006a) Efficacy of a *Streptococcus iniae* modified bacterin delivered using Oralject™ technology in Nile tilapia (*Oreochromis niloticus*). *Aquaculture* 255, 151–156.

Shoemaker, C.A., Xu, D.H., Evans, J.J. and Klesius, P.H. (2006b) Parasites and diseases. In: Lim, C. and Webster, C. (eds) *Tilapia: Biology, Culture and Nutrition*. Food Products Press, imprint of Haworth Press, Binghampton, New York, pp. 561–582.

Shoemaker, C.A., LaFrentz, B.R., Klesius, P.H. and Evans, J.J. (2010) Protection against heterologous *Streptococcus iniae* isolates using a modified bacterin vaccine in Nile tilapia, *Oreochromis niloticus* (L.). *Journal of Fish Diseases* 33, 537–544.

Shoemaker, C.A., LaFrentz, B.R. and Klesius, P.H. (2012) Bivalent vaccination of sex reversed hybrid tilapia against *Streptococcus iniae* and *Vibrio vulnificus*. *Aquaculture* 354–355, 45–49.

Soto, E., Wang, R., Wiles, J., Baumgartner, W., Green, C. *et al.* (2015) Characterization of isolates of *Streptococcus agalactiae* from diseased farmed and wild marine fish from the U.S. Gulf Coast, Latin America and Thailand. *Journal of Aquatic Animal Health* 27, 123–134.

Su, Y.-L., Feng, J., Li, Y.-W., Bai, J.-S. and Li, A.-X. (2016) Development of a quantitative PCR assay for monitoring *Streptococcus agalactiae* colonization and tissue tropism in experimentally infected tilapia. *Journal of Fish Diseases* 39, 229–238.

Suanyuk, N., Kong, F., Ko, D., Gilvert, G.L. and Supamattaya, K. (2008) Occurrence of rare genotypes of *Streptococcus agalactiae* in cultured red tilapia (*Oreochromis* sp.) and Nile tilapia (*O. niloticus*) in Thailand – relationship to human isolates? *Aquaculture* 284, 35–40.

Suanyuk, N., Sukkasame, N., Tanmark, N., Yoshida, T. and Itami, T. (2010) *Streptococcus iniae* infection in cultured Asian seabass (*Lates calcarifer*) and red tilapia (*Oreochromis* sp.) in southern Thailand. *Songklanakarin Journal of Science and Technology* 32, 341–348.

Suebsing, R., Kampeera, J., Tookdee, B., Withyachumnarnkul, B., Turner, W. *et al.* (2013) Evaluation of colorimetric loop-mediated isothermal amplification assay for visual detection of *Streptococcus agalactiae* and *Streptococcus iniae* in tilapia. *Letters in Applied Microbiology* 57, 317–324.

Tavares, G.C., de Alcântara Costa, F.A., Santos, R.R.D., Barony, G.M., Leal, C.A.G. *et al.* (2016) Nonlethal sampling methods for diagnosis of *Streptococcus agalactiae* infection in Nile tilapia, *Oreochromis niloticus* (L.). *Aquaculture* 454, 237–242.

Vandamme, P., Devriese, L.A., Pot, B., Kersters, K. and Melin, P. (1997) *Streptococcus difficile* is a nonhemolytic Group B, type Ib *Streptococcus*. *International Journal of Systematic Bacteriology* 47, 81–85.

Weinstein, M.R., Litt, M., Kertesz, D.A., Wyper, P., Rose, D. *et al.* (1997) Invasive infections due to a fish pathogen, *Streptococcus iniae*. *New England Journal of Medicine* 337, 589–594.

Widanarni and Tanbiyaskur (2015) Application of probiotic, prebiotic and synbiotic for the control of streptococcosis in tilapia *Oreochromis niloticus*. *Pakistan Journal of Biological Sciences* 18, 59–66.

Xu, D.-H., Shoemaker, C.A. and Klesius, P.H. (2007) Evaluation of the link between gyrodactylosis and streptococcosis of Nile tilapia, *Oreochromis niloticus*. *Journal of Fish Diseases* 30, 233–238.

Xu, D.H., Shoemaker, C. and Klesius, P. (2009) Enhanced mortality in Nile tilapia, *Oreochromis niloticus* (L.) following coinfections with ichthyophthiriasis and streptococcosis. *Diseases of Aquatic Organisms* 85, 187–192.

Ye, X., Li, J., Lu, M., Deng, G., Jiang, X. *et al.* (2011) Identification and molecular typing of *Streptococcus agalactiae* isolated from pond-cultured tilapia in China. *Fisheries Science* 77, 623–632.

Zamri-Saad, M., Amal, M.N.A., Siti-Zahrah, A. and Zulkafli, A.R. (2014) Control and prevention of streptococcosis in cultured tilapia in Malaysia: a review. *Pertanika Journal of Tropical Agricultural Science* 37, 389–410.

23 Vibriosis: *Vibrio anguillarum, V. ordalii* and *Aliivibrio salmonicida*

ALICIA E. TORANZO,[1]* BEATRIZ MAGARIÑOS[1] AND RUBEN AVENDAÑO-HERRERA[2]

[1]*Departamento de Microbiología y Parasitología, Universidade de Santiago de Compostela, Santiago de Compostela, Spain;* [2]*Laboratorio de Patología de Organismos Acuáticos y Biotecnología Acuícola, Universidad Andrés Bello, Viña del Mar, Chile*

23.1 Introduction

Vibriosis is a group of diseases caused by different bacterial species formerly classified as belonging to the *Vibrio* genus, although some of them have since been transferred to new genera within the family *Vibrionaceae*. These include *Moritella viscosa* (formerly *Vibrio viscosus*) (Benediktsdóttir *et al.*, 2000), *Aliivibrio salmonicida* (formerly *Vibrio salmonicida*) (Urbanczyk *et al.*, 2007) and *Photobacterium damselae* (formerly *Vibrio damsela*) (Smith *et al.*, 1991).

This chapter focuses on the detection/diagnosis, antigenic/genetic characterization, disease mechanism and control/prevention of three species – *Vibrio anguillarum*, *V. ordalii* and *Aliivibrio salmonicida* – which are responsible for serious haemorrhagic septicaemia worldwide in marine and brackish water fishes. All are Gram-negative motile rods, oxidase and catalase positive, halophilic, and facultative anaerobes. Whereas *V. anguillarum* has a wide geographical distribution and host range, *V. ordalii* and *A. salmonicida* are mainly restricted to salmonids in particular geographical regions. The main differential characteristics among them include the capacity to form colonies in cultures on a medium specific for vibrios, a broad temperature range for growth, the presence of the enzyme arginine dihydrolase and the capacity to produce acid from several sugars, such as sucrose, maltose, mannose, trehalose, amygdaline and mannitol.

23.2 *Vibrio anguillarum*

23.2.1 Description of the microorganism

V. anguillarum, the aetiological agent of classical vibriosis, causes typical haemorrhagic septicaemia in warm- and cold-water fishes of economic importance, including Pacific and Atlantic salmon (*Oncorhynchus* spp. and *Salmo salar*), rainbow trout (*O. mykiss*), turbot (*Scophthalmus maximus*), sea bass (*Dicentrarchus labrax*), gilthead sea bream (*Sparus aurata*), striped bass (*Morone saxatilis*), cod (*Gadus morhua*), Japanese and European eels (*Anguilla japonica* and *A. anguilla*) and ayu (*Plecoglosus altivelis*) (Toranzo and Barja, 1990, 1993; Actis *et al.*, 1999; Toranzo *et al.*, 2005; Samuelsen *et al.*, 2006; Austin and Austin, 2007; Angelidis, 2014). It is a Gram-negative bacillus of the family *Vibrionaceae*, motile, halophilic and facultatively anaerobic, which grows rapidly from 20 to 30°C in media containing 1–2% NaCl. The genome of *V. anguillarum* is about 4.2–4.3 Mbp with a guanine–cytosine (GC) content of 43–46% (Naka *et al.*, 2011; Li *et al.*, 2013; Holm *et al.*, 2015), and it consists of two circular chromosomes, which is a common feature among *Vibrio* species. Even though a total of 23 O serotypes occur among *V. anguillarum* isolates (Sørensen and Larsen, 1986; Pedersen *et al.*, 1999), only serotypes O1, O2 and, to a lesser

*Corresponding author e-mail: alicia.estevez.toranzo@usc.es

---

extent, O3, cause mortalities in fish (Tajima *et al.*, 1985; Toranzo and Barja, 1990, 1993; Larsen *et al.*, 1994; Toranzo *et al.*, 1997, 2005). While serotypes O1 and O2 have a wide distribution of host species and geographical region, serotype O3 affects mainly eel and ayu cultured in Japan and Atlantic salmon reared in Chile (Silva-Rubio *et al.*, 2008a).

In contrast to serotype O1, which is antigenically homogeneous, serotypes O2 and O3 are heterogeneous, and different subgroups within each serotype are named O2a, O2b, O2c and O3A and O3B, (Olsen and Larsen, 1993; Santos *et al.*, 1995; Mikkelsen *et al.*, 2007; Silva-Rubio *et al.*, 2008a). Whereas serotypes O1 and O2a occur in salmonid and non-salmonid fishes, serotypes O2b and O2c have only been detected in marine fishes. Serotype O3A is recovered from diseased fishes, but subgroup O3B is only found in the environment.

Genetic studies have been done on the intraspecific variability within the major pathogenic serotypes of *V. anguillarum* (O1, O2 and O3) (Pedersen and Larsen, 1993; Skov *et al.*, 1995; Tiainen *et al.*, 1995; Toranzo *et al.*, 1997; Mikkelsen *et al.*, 2007). Regardless of the method employed – ribotyping, amplified fragment length polymorphism (AFLP) and pulsed-field gel electrophoresis (PFGE) – O1 strains were the most genetically homogeneous serotype. However, using PFGE, different clonal lineages with epidemiological value were detected within the three serotypes (Skov *et al.*, 1995; Toranzo *et al.*, 2011).

The mode of transmission of *V. anguillarum* is controversial. In most cases, infection is initiated by bacterial penetration of the skin through areas in which the mucus has been lost (Muroga and de la Cruz, 1987; Kanno *et al.*, 1989; Svendsen and Bogwald, 1997; Spanggaard *et al.*, 2000; O'Toole *et al.*, 2004; Croxatto *et al.*, 2007; Weber *et al.*, 2010). In fact, the mucus has high antibacterial activity against this microorganism (Harrel *et al.*, 1976a; Fouz *et al.*, 1990). Oral intake of *V. anguillarum* through water or food can also initiate vibriosis, mainly in larval fish because their gastrointestinal tracts are not acidic enough to inactivate the bacterium (Grisez *et al.*, 1996; Olsson *et al.*, 1998; O'Toole *et al.*, 1999, 2004; Engelsen *et al.*, 2008). *V. anguillarum* can survive in a culturable form for more than a year in saline waters (Hoff, 1989a), which favour broad dissemination in the water column.

23.2.2   Diagnosis of the infection

Clinical signs of the disease

Fish with classical vibriosis have a generalized septicaemia with haemorrhaging on the bases of the fins, the ventral and lateral areas, mouth, operculum and eyes (Fig. 23.1). Exophthalmia and corneal opacity are often seen. Moribund fish are frequently anorexic with pale gills, which reflects a severe anaemia. Oedematous lesions, predominantly centred on the hypodermis, are often seen. Internally, splenomegaly is observed, the kidney is swollen, and the intestines may be distended and filled with a clear viscous liquid. There are haemorrhages in the renal parenchyma and the cells of the epithelium of the kidney tubules are vacuolated. In the melanomacrophage centres of the kidney, accumulations of iron deposits are observed. In the spleen, liver and kidneys, there are haemorrhagic and necrotic areas, and large necrotic lesions may be present in the muscles (Ransom *et al.*, 1984; Lamas *et al.*, 1994; Toranzo *et al.*, 2005; Austin and Austin, 2007; Angelidis, 2014).

Detection of V. anguillarum

V. anguillarum is a halophilic bacterium that grows easily in general culture media supplemented with 1% NaCl. Moreover, the simultaneous use of thiosulfate-citrate-bile-salts-sucrose (TCBS) (Oxoid) medium facilitates the differential recognition of typical yellow (sucrose positive) colonies.

V. anguillarum is presumptively diagnosed using standard biochemical tests (see Table 23.1). It is a Gram-negative, motile rod-shaped bacterium that is oxidase, catalase, Voges-Proskauer (acetoin), β-galactosidase and arginine dihydrolase positive, but is negative for the decarboxylation of lysine and ornithine. The bacterium hydrolyses gelatin and starch, and ferments glucose, sucrose, maltose, mannose, trehalose, amygdaline and mannitol (Toranzo and Barja, 1990). A serological confirmation using serotype-specific antisera is necessary (Toranzo *et al.*, 1987). Although commercial diagnostic kits based on slide agglutination and ELISA are available, they do not discriminate between serotypes (Romalde *et al.*, 1995) and, therefore, have limited epidemiological value.

Sensitive polymerase chain reaction (PCR) protocols have been developed for the rapid and specific identification of *V. anguillarum* from pure cultures and fish tissues. The *rpoN* gene, encoding

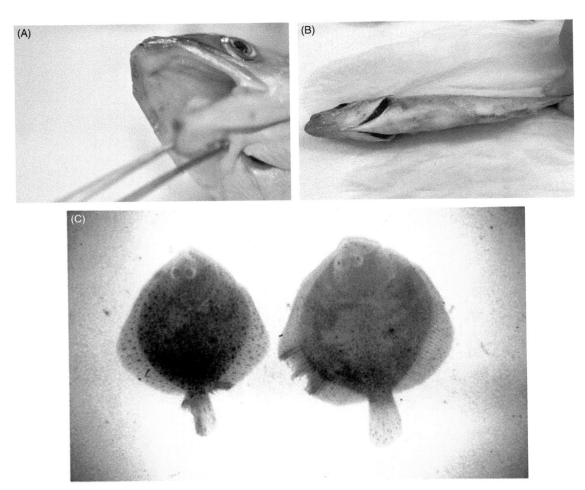

Fig. 23.1. Clinical signs of vibriosis caused by *Vibrio anguillarum*: (A) haemorrhages in the mouth and around the eyes of Atlantic salmon (500 g weight); (B) ventral and lateral haemorrhagic areas in sea bass (25 g weight); and (C) haemorrhages on the fins of turbot (10 g weight).

the sigma factor 54, the *rpoS* gene, encoding the sigma factor 38, the *empA* gene, encoding an EmpA extracellular zinc metalloprotease, the *amiB* gene, encoding a peptidoglycan hydrolase, or the *groEL* gene, encoding bacterial chaperonins, were employed as targets (Osorio and Toranzo, 2002; Hong *et al.*, 2007; Xiao *et al.*, 2009; Kim *et al.*, 2010). These protocols are especially useful for discrimination between *V. anguillarum* and *V. ordalii*, which share nearly 99% 16S rDNA sequence identity. In addition, the *rpoN* gene can be targeted in a multiplex PCR for the simultaneous detection of *V. anguillarum*, *Aeromonas salmonicida*, *Piscirickettsia salmonis* and *Streptococcus phocae*

(Tapia-Cammas *et al.*, 2011). Loop-mediated isothermal amplification (LAMP) based on the *amiB* gene or *empA* gene is a cost-effective alternative to PCR, and provides rapid, specific and sensitive detection of this pathogen (Kulkarni *et al.*, 2009; Gao *et al.*, 2010).

23.2.3 Disease mechanism

Although the disease mechanism of *V. anguillarum* is not completely understood, its virulence factors include adhesion, colonization and invasion factors, exotoxin production, cell surface components and systems for iron uptake (Frans *et al.*, 2011).

A.E. Toranzo *et al.*

Table 23.1. Phenotypic characteristics of *Vibrio anguillarum, V. ordalii* and *Aliivibrio salmonicida*.[a]

Characteristic	*V. anguillarum*	*V. ordalii*	*A. salmonicida*
Acid production from:			
amygdaline	+	–	–
glucose	+	+	+
inositol	–	–	–
maltose	+	–	+
mannitol	+	V	+
mannose	+	–	–
sucrose	+	–	–
trehalose	+	V	+
Arginine dihydrolase	+	–	–
Catalase	+	+	+
Citrate utilization	+	–	–
Cytochrome oxidase	+	+	+
β-Galactosidase	+	–	–
Gelatine hydrolysis	+	V	–
Gram stain	–	–	–
Growth:			
at 25°C	+	+	–
at 37°C	V	–	–
in 0% NaCl	–	–	–
in 3% NaCl	+	+	+
in 5% NaCl	–	–	–
on TCBS	+ (Y)	–	–
Indole production	+	–	–
Lysine decarboxylase	–	–	–
Motility	+	+	+
Nitrate reduction	+	V	–
O/F (oxidative–fermentative) Leifson test	F	F	F
Ornithine decarboxylase	–	–	–
Starch hydrolysis	+	–	–
Voges–Proskauer acetoin test	+	–	–

[a]Key: F, fermentative; TCBS, thiosulfate-citrate-bile-salts-sucrose medium; V, variable reaction among strains; Y, yellow colonies.

Moreover, some of these factors are regulated by quorum sensing systems.

Factors of adhesion, colonization and invasion

During entry to the host, the bacterial flagellum facilitates attachment and penetration of the pathogen to the host mucus (O'Toole *et al.*, 1996). *V. anguillarum* shows a chemotactic response to the mucus of the gills, skin and intestine of fish (Bordas *et al.*, 1998; O'Toole *et al.*, 1999; Larsen *et al.*, 2001) and adheres to and colonizes the intestinal tract (Bordas *et al.*, 1998). Furthermore, when *V. anguillarum* is cultured in mucus from the intestinal tract of salmon, various proteins are specifically expressed, including the metalloprotease EmpA, whose role in virulence was suggested by Milton *et al.* (1992) and further confirmed by Denkin and Nelson (2004). In addition, three genes related to proteolytic enzymes were identified and are thought to be required, together with the metalloprotease, in the invasion of host tissues (Rodkhum *et al.*, 2006).

Exotoxins

V. anguillarum produces extracellular enzymes (haemolysins and RTX (toxins) that could contribute to tissue damage and facilitate bacterial proliferation. Five different genes (*vah1*, Hirono *et al.*, 1996; *vah2, vah3, vah4, vah5*, Rodkhum *et al.*, 2005) that encode haemolysins in *V. anguillarum* have been described. Individual mutants defective in *vah2* to *vah5* are less virulent for rainbow trout than the parent strain, suggesting that each of these haemolysins contributes to the virulence of the pathogen.

The pathogen also produces RTX toxin (Rodkhum *et al.*, 2006), a pore-forming toxin, but the role of this toxin in its virulence is not clear.

Cell-surface components

V. anguillarum can resist the bactericidal effect of fish serum and such resistance is mediated by the synthesis of long chains of O-antigen polysaccharides (Boesen *et al.*, 1999). What is more, the O-antigen of *V. anguillarum* appears to be involved in siderophore-mediated iron uptake because a mutation in the O-antigen biosynthesis operon resulted in ferri-siderophore uptake deficiency (Welch and Crosa, 2005).

Additionally, *V. anguillarum* can efficiently osmoregulate through five salt-responsive outer membrane proteins that help the bacterium to adapt to the different salinities of the host and seawater (Kao *et al.*, 2009). Two other major outer membrane proteins regulate resistance to bile, which enhances bacterial survival and colonization of fish intestines (Wang *et al.*, 2009).

Iron-uptake systems

V. anguillarum scavenges the iron needed for its metabolism from the host tissues or from the environment using siderophores. Two different siderophore-mediated systems have been described in *V. anguillarum* O1 and O2 (Lemos *et al.*, 2010). In most pathogenic strains of serotype O1, the 65 kb pJM1 plasmid harbours genes for the siderophore anguibactin (Crosa, 1980; Stork *et al.*, 2002), though its biosynthesis also needs chromosomally encoded enzymes as well as those encoded by the plasmid genes (Alice *et al.*, 2005).

Serotype O2 strains and some plasmid-less serotype O1 strains, produce a catecholate-type siderophore that is unrelated to the pJM1-mediated system (Lemos *et al.*, 1988; Conchas *et al.*, 1991). This chromosome-encoded system produces the siderophore vanchrobactin (Balado *et al.*, 2006, 2008; Soengas *et al.*, 2006).

The importance of siderophores in the virulence of *V. anguillarum* has been demonstrated in plasmid-less cured strains as well as in mutants impaired in the genes involved in the siderophore biosynthesis or in the genes needed for siderophore transport to the bacterial cell. Such deficient strains/mutants of *V. anguillarum* suffered a severe reduction in their infective capability for fish (Wolf and Crosa, 1986; Stork *et al.*, 2002, 2004).

Other iron assimilation systems, such as the utilization of haemin, haemoglobin and haemoglobin–haptoglobin as iron sources by haem-transport specific mechanisms are present in *V. anguillarum* (Mouriño *et al.*, 2004). The incidence of the haem-transport system in the biology of the bacterium is uncertain, but it is a conserved system among vibrios (Mazoy *et al.*, 2003; Lemos and Osorio, 2007). While the system is not essential for virulence, it does favour the ability of the bacterium to colonize the host when iron is available as haem groups (Mazoy *et al.*, 2003).

Regulation of virulence factors

Quorum sensing (QS) systems are involved in the regulation of virulence genes in several Gram-negative fish pathogens, including *V. anguillarum* (Bruhn *et al.*, 2005). The most intensively studied QS systems rely on the production of *N*-acylhomoserine lactones (AHLs), which are signalling molecules in a cell-to-cell communication system that enables monitoring of the bacterial population density in order to coordinate gene expression. Three QS systems have been identified in *V. anguillarum* serotype O1, and they are similar to those present in the bioluminescent species *V. fischeri* and *V. harveyi* (Milton *et al.*, 1997, 2001; Croxatto *et al.*, 2004; Milton, 2006; Weber *et al.*, 2008). It was demonstrated that a *V. anguillarum* mutant in the QS-regulated transcriptional activator VanT was deficient in the metalloprotease EmpA and in biofilm formation, which are required for the invasion and survival of the pathogen (Croxatto *et al.*, 2002).

23.2.4 Prevention and control

Most commercial *V. anguillarum* vaccines are inactivated whole-cell bacterins delivered via immersion or intraperitoneal (IP) injection as part of multicomponent non-adjuvanted or oil-adjuvanted vaccines (Toranzo *et al.*, 1997, 2005; Håstein *et al.*, 2005; Angelidis, 2014; Colquhoun and Lillehaug, 2014). Marine fishes, such as turbot or sea bass, are immersion vaccinated in aqueous *V. anguillarum* bacterins when the fish are 1–2 g (average weight), followed by a booster immersion vaccination 1 month later. In the case of sea bass, another booster via IP injection is usually employed before transfer to cages (at 20–25 g). For salmonid species, polyvalent oil-based vaccines are administered IP at 1 month before smoltification, which provides protection until harvest (at about 24 months). The use of non-mineral oil-adjuvants has reduced the retardation of growth and development of intra-abdominal lesions that were often reported with mineral oil-adjuvanted vaccines (Midtlyng and Lillehaug, 1998; Poppe and Koppang, 2014).

The development of live attenuated *V. anguillarum* strains that can be used as immersion and oral vaccines represented an important trend in research (Yu *et al.*, 2012; Zhang *et al.*, 2013). However, the immunoprotection conferred by these live vaccines was evaluated only 1 month after vaccination and, therefore, there are no data on the duration of the protection conferred.

Although *V. anguillarum* vaccines are effective, antimicrobial therapy is still required to contain some outbreaks. Oxytetracycline, oxolinic acid, flumequine, potentiated sulfonamides, florfenicol and fluoroquinolones are used to treat *V. anguillarum* infections (Lillehaug *et al.*, 2003; Samuelsen and Bergh, 2004; Samuelsen *et al.*, 2006; Angelidis, 2014), though resistance has developed to oxolinic acid (Colquhoun *et al.*, 2007). At present, the drugs of choice and the dosing regimens are oxytetracycline (75–100 mg/kg body weight daily for 10 days), florfenicol (10 mg/ kg body weight daily for 7 days) and enrofloxacine (10 mg/ kg body weight daily for 7 days) (Rigos and Troisi, 2005).

Probiotics have also been investigated for the control of vibriosis. *Pseudomonas fluorescens*, *Aeromonas sobria* and *Bacillus* sp. inhibited *V. anguillarum* infection in rainbow trout that had been previously immersed in a solution containing each antibacterial strain (Gram *et al.*, 1999; Brunt *et al.*, 2007). *Lactobacillus delbrueckii* subsp. *delbrueckii* and *Vagococcus fluviales* protected sea bass against challenge with *V. anguillarum* when the probiotic bacteria were administered in the water (Carnevali *et al.*, 2006) or through the food (Sorroza *et al.*, 2012). Chabrillon *et al.* (2006) isolated bacteria belonging to the *Vibrionaceae* family, which when administered mixed with their commercial diet, significantly protected sea bream challenged with *V. anguillarum*. Bacteria belonging to the genera *Phaeobacter*, *Roseobacter* and *Ruegeria* are also promising probiotics against *V. anguillarum* infections in turbot larvae when they are bioencapsulated in rotifers (Planas *et al.*, 2006; Porsby *et al.*, 2008).

Because of the prevalence of QS systems in *V. anguillarum*, the inhibition of AHL-mediated QS processes has also been proposed as a method of control (Defoirdt *et al.*, 2004). In fact, it was found that the use of specific QS inhibitors, such as furarone C-30, caused a reduced mortality in rainbow trout infected with *V. anguillarum* (Rasch *et al.*, 2004), and Brackman *et al.* (2009) demonstrated that a nucleoside analogue, LMC-21 (an adenosine derivative), affected biofilm formation and protease production in *V. anguillarum*. In addition, the isolation and characterization of marine bacterial strains that inhibit QS, either enzymatically or by the production of inhibitors or antagonists, may help in the development of new biotechnological tools in aquaculture. In order to use such tools, practical approaches, such as how to deliver the QS disrupting agent or organism to the specific site of infection, as well as the cost of such treatment, need to be considered.

23.3 *Vibrio ordalii*

23.3.1 Description of the microorganism

V. ordalii was initially referred to as atypical *Vibrio* sp. 1669 (Harrell *et al.*, 1976b), *Vibrio* sp. RT (Ohnishi and Muroga, 1976), *V. anguillarum* biotype II (Schiewe *et al.*, 1977) and *V. anguillarum* phenon II (Ezura *et al.*, 1980). Schiewe *et al.* (1981) clarified the taxonomic status of this pathogen and named it *V. ordalii*.

The species was initially isolated as the aetiological agent of vibriosis in wild and cultured coho salmon (*O. kisutch*) from the coastal waters of the American Pacific Northwest (Harrell *et al.*, 1976b). It has since been reported in Japan, Australia and New Zealand, with infections mainly in cultured salmonids (Ransom *et al.*, 1984; Toranzo *et al.*, 1997). In 2004, the pathogen was reported in southern Chile inducing mortality in cultured Atlantic salmon, Pacific salmon and rainbow trout (Colquhoun *et al.*, 2004; Silva-Rubio *et al.*, 2008b). The pathogen has also been reported in other fishes, such as ayu and rockfish (*Sebastes schlegeli*) in Japan (Muroga *et al.*, 1986) and gilthead sea bream in Turkey (Akayli *et al.*, 2010).

V. ordalii is a Gram-negative, motile, rod-shaped bacterium that ferments glucose. It is catalase and oxidase positive, and sensitive to the vibriostatic agent O/129 (Farmer *et al.*, 2005). The bacterium grows in conventional media supplemented with 1–2% NaCl.

The total genomic size of *V. ordalii* is roughly 3.4 Mbp, with 43–44% GC content (Schiewe *et al.*, 1981;

Naka et al., 2011). The genome consists of two circular chromosomes, which is a common feature among *Vibrio* species. However, *V. ordalii* presents a large chromosomal deletion (≈600 kb) in comparison with *V. anguillarum*. Additionally, another common feature among all *V. ordalii* is the presence of a 30 kb cryptic plasmid designated pMJ101 (Schiewe and Crosa, 1981). Interestingly, this extra-chromosomal element is unrelated to the pJM1 virulence plasmid that is present in most *V. anguillarum* serotype O1 strains (Crosa, 1980).

V. ordalii is antigenically homogeneous and shares antigens with the *V. anguillarum* sero-subgroup O2a (Schiewe et al., 1981). Both microorganisms have O-antigen polysaccharides and these differ only in their degree of polymerization (Sadovskaya et al., 1998). *V. ordalii* and *V. anguillarum* O2a contain specific, differentiating antigenic determinants (Chart and Trust, 1984; Mutharia et al., 1993), but there are also antigenic differences within *V. ordalii* (Bohle et al., 2007; Silva-Rubio et al., 2008b). In fact, Silva-Rubio et al. (2008b) described the electrophoretic patterns of Chilean *V. ordalii* lipopolysaccharides as similar, with a ladder of low molecular weight O-antigen bands, but different from the type strain isolated from coho salmon in the USA, which showed antigenic reactions in both the low and high molecular weight regions. This suggests that the Chilean isolates may constitute a new serological subgroup.

To determine the degree of intraspecific genetic diversity that exists within *V. ordalii*, analyses were conducted by ribotyping, PFGE, random amplified polymorphic DNA (RAPD), repetitive element palindromic PCR (REP-PCR), and enterobacterial repetitive intergenic consensus sequence-based PCR (ERIC-PCR). The results indicate homology among isolates, independent of host and geographic origin of the isolate, thereby demonstrating the bacterial clonality of this species (Wards and Patel, 1991; Tiainen et al., 1995; Silva-Rubio et al., 2008b). However, Silva-Rubio et al. (2008b) detected minor genetic differences between the coho salmon type strain and Chilean isolates, thus suggesting that the *V. ordalii* obtained from Atlantic salmon in Chile could constitute a distinct clonal group.

There is limited information on the nature of *V. ordalii* in aquatic environments. Therefore, the possible role of water as a natural reservoir and route of transmission for the disease is unclear. Ruiz et al. (2016b) demonstrated that *V. ordalii* strains survived for a year in sterile seawater without the addition of nutrients, and long-term maintenance did not affect its biochemical or genetic properties. Additionally, *V. ordalii* remained infective for rainbow trout for 60 days in sterile seawater, although under non-sterile conditions the number of culturable *V. ordalii* declined very rapidly, with no *V. ordalii* colonies detectable by biochemical tests just 2 days after the start of the experiment.

This observation was supported by results obtained from quantitative PCR (qPCR) analysis for *V. ordalii* DNA in microcosms, in which qPCR gave a positive reaction during only the first 3 days of the experiment. Antagonism assays showed that native microbiota decreased the survival of *V. ordalii* in non-sterile microcosms, indicating that the autochthonous microbiota affect its survival (Ruiz et al., 2016b). Other alternative strategies for surviving during adverse environmental conditions may also be employed by *V. ordalii* in the aquatic environment, such as biofilm formation, as described by Naka et al. (2011); here, the *V. ordalii* genome carries an entire biofilm formation cluster (Syp system) with the exception of the negative regulator SypE. Interestingly, the Syp system is essential for another species, *V. fischeri*, to colonize squid organs (Yip et al., 2006). *V. ordalii* forms biofilm, adheres to glass surfaces coated with fish mucus and possesses hydrophobic characteristics (Ruiz et al., 2015) that can facilitate its colonization on fish.

23.3.2 Diagnosis of the infection

Clinical signs of the disease

Vibriosis caused by *V. ordalii* is associated with necrosis and haemorrhagic lesions in the tissues surrounding the infection sites, including the ventral fin and anal pore (Fig. 23.2a). Some cases present abscesses with haemorrhaging of the skin (Ransom et al., 1984). Bohle et al. (2007) described 14 clinical cases of Atlantic salmon in Chile with recurrent boils on the lateral and laterodorsal sides, extensive lateral haemorrhagic ulcers, haemorrhagic oedema of the peduncle, ulcers and haemorrhaging at the base of the pectoral fins, and petechiae on the abdomen. Internally, infected fish had whitish spots, primarily on the liver and spleen, splenomegaly and, in some cases, a pale liver with ecchymosis (Fig. 23.2b).

The histopathology associated with experimental infections by *V. ordalii* suggests that the bacteria concentrate in aggregates or colonies that are more

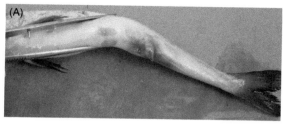

Fig. 23.2. Clinical signs of vibriosis caused by *Vibrio ordalii* in Atlantic salmon (50 g average weight): (A) haemorrhagic lesions in tissues surrounding the area around the anal pore; and (B) splenomegaly and a pale liver with ecchymosis.

abundant in muscle tissues (skeletal, cardiac and smooth) than in the blood (Ransom *et al.*, 1984). Also, the mucosal pathology indicates that overt infections began in the descending intestine and rectum, with the most severe changes in these regions, while pathology was less pronounced in the anterior gastrointestinal tract. Alternatively, the presence of *V. ordalii* on the skin suggested that bacteria enter the host by invasion of the integument (Ransom, 1978), with sloughing of the tissues of the lower gastrointestinal tract indicating a likely mode of excretion and transmission (Ransom *et al.*, 1984).

Detection of V. ordalii

V. ordalii is a halophilic, motile, rod-shaped, Gram-negative bacterium that does not grow on TCBS or MacConkey agar (Bohle *et al.*, 2007). The bacterium can be presumptively diagnosed based on the isolation of bacterial colonies on appropriate media supplemented with 1–2% NaCl, followed by biochemical tests using conventional tube and plate test procedures (see Table 23.1). *V. ordalii* produces translucent, circular, convex, 1–2 mm wide colonies after 2–3 days incubation at 18°C (Poblete-Morales *et al.*, 2013). The bacterium can easily be differentiated from *V. anguillarum* based on its negative reactions for arginine dihydrolase, β-galactosidase and indole production (Schiewe *et al.*, 1981). In addition to this, BIOLOG-GN (Biolog Inc., California) fingerprinting and API-20E (BioMérieux) profiling further reveals differences among *V. ordalii* strains (Austin *et al.*, 1997). Such intraspecific differences are related to the ability of isolates to produce acid from trehalose and mannitol, and to hydrolyse gelatin (Colquhoun *et al.*, 2004; Bohle *et al.*, 2007; Silva-Rubio *et al.*, 2008b). Although *V. ordalii* has only one serotype, serological confirmation using

commercial diagnostic kits provides poor discrimination of the species (Bohle *et al.*, 2007) owing to cross reactions with *V. anguillarum* O2 (Chart and Trust, 1984).

Chromosomal DNA–DNA pairing revealed that the overall DNA sequences of *V. ordalii* and *V. anguillarum* are highly similar (Schiewe *et al.*, 1981). Additionally, DNA comparisons of the 5S rRNA gene sequences reveal a close relationship between *V. anguillarum* and *V. ordalii*, as they differ in only seven out of 120 bp (MacDonell and Colwell, 1984; Pillidge and Colwell, 1988; Ito *et al.*, 1995). Similarly, the 16S rRNA gene sequence of *V. ordalii* is indistinguishable from that of *V. anguillarum* (Wiik *et al.*, 1995). In contrast, the *vohB* gene, which codes for haemolysin production, is a suitable target gene for detecting and distinguishing *V. ordalii* in PCR and qPCR assays (Avendaño-Herrera *et al.*, 2014). This protocol is fast, specific and sensitive enough to detect and quantify *V. ordalii* in infected tissues.

23.3.3 Disease mechanisms

V. ordalii has numerous genes that are potential virulence genes for adhesion, colonization and invasion factors; exotoxin production, cell surface components and systems for iron uptake (Naka *et al.*, 2011).

Factors of adhesion, colonization and invasion

A prerequisite for successful colonization is the ability to adhere to the host tissue, because this promotes the delivery of bacterial toxins and precedes bacterial penetration into target cells (Montgomery and Kirchman, 1994). Ruiz *et al.* (2015) suggested that the ability to grow on the surface of a fish could facilitate colonization and subsequent invasion. *In vitro* studies have demonstrated that the pathogenicity of *V. ordalii* is not correlated with the haemagglutination of Atlantic salmon red blood cells or with biofilm characteristics, but it is associated with hydrophobic properties (Ruiz *et al.*, 2015). Hydrophobicity and biofilm production are major factors in the adhesion process and the survival of pathogens in cells (Daly and Stevenson, 1987). Specific adherence is mediated through compounds on the surface of the bacteria, which bind via rigid stereochemical bonds to particular molecules on the support to which they are adhering. In contrast, non-specific adherence depends on hydrophobic or ionic interactions between certain structures on the surface of the bacterium and on the support (Ofek and Doyle, 1994). Naka *et al.* (2011) identified the metalloprotease gene *empA* in *V. ordalii*, a gene that has already been discussed as an element in the virulence of *V. anguillarum* (Milton *et al.*, 1992; Denkin and Nelson, 2004). However, it is not known whether this gene has a similar function in *V. ordalii*.

Exotoxins

In *V. ordalii*, the production of a leucocytolytic factor may be important for pathogenesis because circulating leucocyte counts are decreased by 80–95% in the blood of diseased fish (Ransom *et al.*, 1984). Bacterial haemolysins are cytolytic toxins generally considered to be virulence factors due to their ability to affect erythrocytes and other cells (Rowe and Welch, 1994). Nevertheless, the relationship between haemolysin production and pathogenicity in *V. ordalii* is still controversial. Kodama *et al.* (1984) reported that the bacterium did not produce haemolysin, but analysis of the complete genome sequences of *V. ordalii* ATCC 33509[T] by Naka *et al.* (2011) found nearly all of the genes coding for the biosynthesis of haemolysins (a putative virulence factor) to be present except for *vah*4.

Cell-surface components

V. ordalii resists the bactericidal actions of non-immune rainbow trout serum (Trust *et al.*, 1981) and agglutinates different eukaryotic cells. These studies indicate that the pathogen attaches to and interacts with host cells (Larsen and Mellergaard, 1984). Ruiz *et al.* (2015) found that Atlantic salmon mucus favoured the growth of Chilean *V. ordalii* isolates, but not of the coho salmon type strain.

This suggests the presence in the Chilean *V. ordalii* isolates of a cell structure (e.g. capsular material) that enhances bacterial resistance. Sadovskaya *et al.* (1998) demonstrated that *V. ordalii* produces capsular polysaccharides, but their significance to pathogenesis is unknown.

Iron-uptake systems

Another putative virulence factor of *V. ordalii* is its capacity to express high-affinity iron-uptake mechanisms which compete with host iron-binding proteins. Initially, Pybus *et al.* (1994) suggested that the siderophores of *V. anguillarum* inhibit *V. ordalii*. This, together with the inability of *V. ordalii* to use anguibactin ferric complexes, suggests that there are different iron-uptake systems in these two microorganisms. More recently, Naka *et al.* (2011) described the genes involved in iron-uptake mechanisms in the type strain of *V. ordalii*, including the genes essential for ferric anguibactin biosynthesis and the iron-transport genes, which include *vabA–vabE*, *tonB2*, *exbB*, *exbD1* and *fur*.

Ruiz *et al.* (2016a) demonstrated that *V. ordalii* possesses different systems of iron acquisition, one involving siderophore synthesis and the other directly binding haem. Using strains of *V. ordalii* from Atlantic salmon and the *V. ordalii* type strain from coho salmon, it was found that all strains grew in presence of the chelating agent 2,2′-dipyridyl and produced siderophores in solid and liquid media.

Cross-feeding assays among the *V. ordalii* strains showed that all were able to be cross-fed by the type strain and one Chilean strain, but the remaining Chilean isolates were unable to promote the growth of any other *V. ordalii* strain, evidencing the existence of an intraspecific variability in the siderophores produced (Ruiz *et al.*, 2016a). According to *in silico* searches and to the genomic data reported by Naka *et al.* (2011), *V. ordalii* ATCC 33509[T] contains the *vab* cluster genes, which encode the synthesis of vanchrobactin, a typical catechol siderophore identified in *V. anguillarum* (Balado *et al.*, 2006). In addition, the type strain harbours a gene cluster that is homologous to the piscibactin (a mixed-type siderophore) biosynthesis genes described in *P. damselae* subsp. *piscicida* (Osorio *et al.*, 2006; Souto *et al.*, 2012). The expression of the genes (assessed by reverse transcription (RT)-PCR) involved in the synthesis (*irp1* and *irp2*), transport (*frpA*) and regulation (*araC1* and *araC2*) of piscibactin was strongly upregulated when cells were grown under low iron availability. It should be noted that even though such PCR results are obtained, it does not mean that all of the necessary genes for the synthesis and transport of vanchrobactin and piscibactin are present in *V. ordalii*, or that all of them are expressed. However, Ruiz *et al.* (2016a), using a bioassay with pure siderophores, demonstrated that none of the *V. ordalii* strains could utilize vanchrobactin, but all of them could use piscibactin as a siderophore.

Furthermore, all *V. ordalii* strains use haemin and haemoglobin as a sole source of iron. Additionally, *V. ordalii* cells display iron-regulated haem-binding activity, and no difference in haemoglobin acquisition was observed among the strains tested, indicating the possible existence of constitutive binding molecules located at the *V. ordalii* cell surface (Ruiz *et al.*, 2016a). Finally, virulence tests using rainbow trout as a model of infection revealed a clear relationship between iron uptake ability and pathogenicity in *V. ordalii*. In fact, the results from challenge studies showed that the type strain and a representative Chilean Atlantic salmon strain increased in their degree of virulence when cultured under iron-restricted conditions.

23.3.4 Prevention and control

Vibriosis caused by *V. ordalii* is currently a major health concern in Chilean salmonid farming. There are nine registered commercial vaccines for *V. ordalii* in Chile. Seven are administered as an injectable emulsion, one orally and one by immersion. Three of these commercial products are produced as monovalent vaccines, one is bivalent, three others are trivalent and two more are quadrivalent; as such, these preparations contain one to four antigens, including antigens from other fish pathogens, such as atypical *Aeromonas salmonicida*, *Piscirickettsia salmonis* and the viruses for infectious pancreatic necrosis and infectious salmon anaemia.

The most common antibiotics used in the treatment of infections by *V. ordalii* in Chile are oxytetracycline (55–100 mg/kg fish daily for 10–15 days) and florfenicol (10 mg/kg fish daily for 7–10 days) (Poblete-Morales *et al.*, 2013); both of these are wide-spectrum bacteriostatic agents. However, several fish farmers in Chile have reported that these antimicrobials have largely been replaced by oxolinic acid (10–30 mg/kg fish daily for 10–15 days) (San Martín *et al.*, 2010), though there is a general lack of knowledge about its effectiveness.

For the monitoring of the correct application of antibiotics and to prevent resistance, it is essential to standardize and validate the protocols for evaluating susceptibility to antimicrobials. It is important to note that the concept of resistance covers epidemiological, clinical and pharmacological aspects, and requires an antibiotic concentration value or break point to distinguish between strain categories (Turnidge and Paterson, 2007). The clinical break point refers to a threshold value that divides strains with a high probability of being treated from those in which the treatment is likely to fail. The epidemiological break point addresses the issue of resistance mechanisms, and separates wild type (WT) bacteria from those non-wild types (NWTs) that have acquired or selected a resistance mechanism; this requires the study of a large number of isolates from diverse geographical and host origins, collected over an extended period of time (Henríquez *et al.*, 2016). *In vitro* studies using the guidelines published by the Clinical Laboratory Standards Institute (CLSI, 2006) on the susceptibility of *V. ordalii* to oxytetracycline, florfenicol and oxolinic acid indicate that the majority of Chilean *V. ordalii* isolates are NWT (Poblete-Morales *et al.*, 2013).

23.4 *Aliivibrio salmonicida*

23.4.1 Description of the microorganism

A. salmonicida, formerly classified as *V. salmonicida* (Urbanczyk *et al.*, 2007), is the aetiological agent of 'Hitra disease' or 'cold water vibriosis', which affects salmonids and cod cultured in Canada, Norway and the UK (Egidius *et al.*, 1981, 1986; Bruno *et al.*, 1986; Sørum *et al.*, 1990). The disease occurs mainly in the late autumn, winter and early spring when seawater temperatures are below 15°C.

The bacterium is a Gram-negative curved and motile rod (0.5 by 2–3 µm), which has up to nine polar sheathed flagella but no lateral flagellum. The colonies are small, grey and smooth, with entire edges. Growth occurs from 1 to 22°C, but optimum growth is at 15°C. *A. salmonicida* is halophilic and grows in 0.5–4% NaCl, with an optimum concentration of 1.5% NaCl.

The genome of *A. salmonicida* is 4.6 Mb with a GC content of 39.6%. It consists of two circular chromosomes of 3.3 and 1.2 Mb, four circular plasmids and 111 protein-coding sequences (Hjerde *et al.*, 2008).

Serologically, *A. salmonicida* is heterogeneous. Although a hydrophobic protein called VS-P1, present in the surface layer, is the dominant antigen in all strains (Espelid *et al.*, 1987; Hjelmeland *et al.*, 1988), two distinct serotypes have been described based on their lipopolysaccharide structure: the first serotype groups all of the isolates from salmon and the majority from cod; the second serotype includes some isolates from cod farms in northern Norway (Schrøder *et al.*, 1992).

Hitra disease appeared in 1977, but occurred for the first time on a large scale in 1979 within Norwegian salmonid farms around the island of Hitra, south of Trondheim (Egidius *et al.*, 1981, 1986). This outbreak of the disease persisted at Hitra and northward until the late autumn of 1983, when it was then reported further south near Stavanger. The Bergen region, which has the highest density of fish farms, was badly affected. Outbreaks have also been reported in Scotland (Bruno *et al.*, 1985, 1986), the Faroe Islands (Dalsgaard *et al.*, 1988) and New Brunswick and Nova Scotia in Canada (Sørum *et al.*, 1992).

The mode of transmission and route of infection of *A. salmonicida* are unknown. The microorganism shows a high potential to enter into a viable but non-culturable stage in the ocean environment (Enger *et al.*, 1990). In addition, it was suggested that it can be transmitted through seawater either as bacterioplankton or on the surface of particulate matter (Holm *et al.*, 1985; Sørum *et al.*, 1990; Nordmo *et al.*, 1997).

A. salmonicida can be isolated from sediments obtained from farms in which fish have sustained epizootics of cold-water vibriosis (Hoff, 1989a,b). The microorganism has been also detected in the faeces of experimentally infected fish. Thus, *A. salmonicida* could persist in marine sediments through contaminated faeces (Enger *et al.*, 1989). The bacterium has also been detected in fish farm sediments where there was no previous history of Hitra disease (Enger *et al.*, 1990, 1991). Based on these observations, asymptomatic carriers may exist and contribute to the spread of Hitra disease in the marine environment (Actis *et al.*, 1999).

23.4.2 Diagnosis of the infection

Clinical signs of the disease

The clinical signs of Hitra disease are non-specific, but usually include lethargy and cessation of feeding.

Affected fish turn dark, exhibit exophthalmos, a swollen vent and pinpoint haemorrhaging along the abdomen and at the bases of the pectoral, pelvic and anal fins. The gills are usually pale. Internally, the disease is characterized by anaemia and haemorrhages with a generalized septicaemia, presenting as large numbers of bacteria in the blood of moribund or recently dead fish. The haemorrhages are mainly found in the integument surrounding internal organs including the caeca, abdominal fat and kidney (Poppe *et al.*, 1985; Egidius *et al.*, 1986; Kent and Poppe, 2002). Experimental studies performed in cod fry showed that cataracts, cranial haemorrhage and splenomegaly may occur (Sørum *et al.*, 1990).

Histopathologically, a close relationship was found in Atlantic salmon between the numbers of bacteria detected in the tissues and the degree of morphological damage; the most severe cell damage occurred where there was a rich supply of blood (Totland *et al.*, 1988). The earliest signs of damage were detected in the cell coats and cell membranes on the luminal side of the endothelial cells of capillaries. Later, bacteria were also found intracellularly within endothelial cells and leucocytes. The endothelial cells may be completely disintegrated, and actively proliferating bacteria can be detected in the extravascular spaces of surrounding tissues.

Detection of A. salmonicida

Presumptive diagnosis of Hitra disease is based on culture and subsequent identification of the causative bacterium using standard biochemical tests (see Table 23.1). *A. salmonicida* is a Gram-negative, catalase and cytochrome oxidase positive bacterium that does not grow on TCBS medium. The microorganism is facultatively anaerobic and strongly susceptible to the vibriostatic agent O129. It is negative for indole and hydrogen sulfide production; citrate cannot be used as the sole source of carbon and nitrates are not reduced. The bacterium produces acid from fructose, galactose, glucose, maltose, ribose, trehalose and mannitol. It does not haemolyse human or sheep erythrocytes. Serological confirmation of *A. salmonicida* is based on a slide agglutination test using specific commercial polyclonal antiserum, and fluorescent antibody tests are usually employed for routine purposes (Toranzo *et al.*, 2005). To our knowledge, no PCR-based approaches have been developed for the identification of this pathogen.

23.4.3 Disease mechanism

Although *A. salmonicida* has been known in Norwegian aquaculture for more than 25 years, only several studies have identified potential virulence factors for the microorganism. Motility is linked to colonization and virulence in several bacteria. *A. salmonicida* depends on its motility to gain access to a fish (Bjelland *et al.*, 2012). Karlsen *et al.* (2008) showed that *A. salmonicida* motility is regulated by environmental factors such as osmolarity and temperature, while Raeder *et al.* (2007) showed that the presence of fish skin mucus in the culture synthetic media increased the levels of proteins involved in motility (the flagellar proteins FlaC, FlaD and FlaE). These three flagellin molecules polymerize to form the filaments of flagella and though alone they are not essential for flagellar synthesis and motility, in *V. anguillarum*, the deletion of FlaD and FlaE significantly reduces virulence. Therefore, the observed induction of *A. salmonicida* flagellar proteins makes them possible virulence factors.

A. salmonicida is described as cryptically bioluminescent (Fidopiastis *et al.*, 1999). Its cells produce light in culture, but only when they are exposed to either an aliphatic aldehyde and/or the major *V. fischeri* autoinducer N-(3-oxo-hexanoyl)-L-homoserine lactone, a transcriptional activator of the luminescence (*lux*) genes. The cloning and sequencing of the *A. salmonicida lux* operon revealed that homologues of all of the genes required for luminescence are present: *luxAB* (luciferase) and *luxCDE* (aliphatic-aldehyde synthesis). In addition, *A. salmonicida* strains have a novel arrangement and number of homologues of the *luxR* and *luxI* quorum-sensing regulatory genes that reduce bioluminescence. Mutation of the *luxA* gene resulted in a marked delay in mortality among Atlantic salmon (Nelson *et al.*, 2007).

The tissue damage observed in fish with Hitra disease suggests that *A. salmonicida* secretes proteins during infection. Several protein secretion systems have been identified in its genome, including three type I secretion systems (T1SS), one type II secretion system (T2SS), two type VI secretion systems (T6SSI and T6SS2) and one Flp-type pilus system. The presence of these systems demonstrates that the bacterium can secrete extracellular toxins and/or enzymes (Hjerde *et al.*, 2008; Bjelland *et al.*, 2013). It has been hypothesized that bacterial extracellular toxins are responsible for the extended petechial haemorrhages observed during a Hitra infection (Holm *et al.*, 1985; Totland *et al.*, 1988). Although genomic analysis has identified three putative haemolysins similar to those of *V. anguillarum* and *V. vulnificus*, there is low expression of these haemolysins during active infection. Consequently, the extensive haemolysis that occurs during the disease may be a result of the activity of the immune system of the fish (Bjelland *et al.*, 2013). In fact, these authors described how the complement system in Atlantic salmon has the potential to be extremely damaging to self-tissues and that the C3 factor could address the pathological signs observed during Hitra disease.

Colquhoun and Sørum (2001) demonstrated the presence of a siderophore-based iron sequestration mechanism in *A. salmonicida* comprising a single iron-chelating molecule belonging to the hydroxamate family of siderophores and an array of several iron-regulated outer membrane proteins. They also reported that the production of this siderophore was highest at temperatures around and below 10°C, which correlates well with the temperatures when Hitra disease occurs. Winkelmann *et al.* (2002) further demonstrated that this molecule is bisucaberin, a cyclic dihydroxamate previously isolated from the marine bacterium *Alteromonas haloplanktis*. Additionally, analysis of the genome sequence of *A. salmonicida* revealed the presence of a haem uptake system with high sequence similarity and synteny to that of other *Vibrionaceae*. A functional TonB system which contributes to transport of the haem complex and ferric siderophores across the outer membrane was also described (Hjerde *et al.*, 2008).

A. salmonicida can proliferate in fish blood and resists the bactericidal activity of serum (Bjelland *et al.*, 2013). Studies performed in salmon indicate that the bacterium is recognized by the host and induces a rapid and strong, but short-lasting, immune response. This shows that the microorganism is resistant to the important salmon cytokines and may also possess mechanisms that inhibit the host immune response during infection. It has also been demonstrated that this pathogen has developed resistance mechanisms against the bactericidal effect of complement.

23.4.4 Prevention and control

Outbreaks of cold-water vibriosis are associated with stressful events such as grading, transport, poor husbandry and environmental conditions, high stocking densities and poor water quality. When outbreaks of the disease are confirmed, chemo-therapeutic agents such as oxytetracycline (75–100 mg/kg body weight daily for 10 days), oxolinic acid (10–30 mg/kg body weight daily for 10 days) and sulfonamides (25 mg/kg body weight daily, for 5–10 days) (Rigos and Troisi, 2005) are employed.

A number of studies analysing the effectiveness of immunization as a preventive method against cold-water vibriosis have been performed. The majority of vaccines tested consist of formalin-killed cells which were evaluated by different routes of admin-istration (Holm and Jorgensen, 1987; Hjeltnes *et al.*, 1989; Lillehaug *et al.*, 1990; Schrøder *et al.*, 1992). The results showed that these formulations show promise (giving 90–100% protection) and can be used to prevent disease caused by *A. salmonicida* in Atlantic salmon and cod. Moreover, oil-adjuvanted bacterins containing at least two pathogenic vibrios – *V. anguillarum* and *A. salmonicida* – are systematically utilized to vaccinate sal-monids in Nordic countries (Toranzo *et al.*, 1997).

23.5 Conclusions and Suggestions for Future Research

The vibriosis that is caused by the three distinct species of the *Vibrionaceae* that have been described in this chapter is an economically important disease in marine aquaculture. Great efforts have been made in the diagnosis, prevention and control of the aetiological agents of these vibriosis to protect fish against them, as well as to characterize the virulence factors associated with these pathogens.

Although the studies so far done on the interac-tions between the species and their hosts have produced important advances, the mechanisms of their pathogenesis are not completely elucidated. The development of new research strategies such as 'dual RNA-seq', which profiles RNA expression simultaneously in the pathogen and host during infection, should unravel the gene functions that are occurring, and could further effective disease prevention.

Finally, the increase of water temperature and ocean acidification caused by climate change can alter the physiology of the fish being cultured as well as the abundance and diversity of the popula-tions of the bacteria causing vibriosis. Therefore, the impact of climate change on the prevalence of vibriosis in marine fish needs more evaluation.

Acknowledgements

A.E. Toranzo and B. Magariños thank the Xunta de Galicia (Spain) project No. GRC-2014/007 for financial support. This work was supported in part by FONDAP N°15110027 and FONDECYT N°1150695 awarded by the Comisión Nacional de Investigación Científica y Tecnológica (CONICYT, Chile) as well as by the Universidad Andrés Bello of Chile.

References

Actis, L.A., Tolmasky, M.E. and Crosa, J.H. (1999) Vibriosis. In: Woo, P.T.K. and Bruno, D.W. (eds) *Fish Diseases and Disorders, Volume. 3: Viral, Bacterial and Fungal Infections*. CAB International, Wallingford, UK, pp. 523–558.

Akayli, T., Timur, G., Albayrak, G. and Aydemir, B. (2010) Identification and genotyping of *Vibrio ordalii*: a com-parison of different methods. *Israeli Journal of Aquaculture Bamidgeh* 62, 9–18.

Alice, A.F., López, C.S. and Crosa, J.H. (2005) Plasmid- and chromosome-encoded redundant and specific functions are involved in biosynthesis of the sidero-phore anguibactin in *Vibrio anguillarum* 775: a case of chance and necessity? *Journal of Bacteriology* 187, 2209–2214.

Angelidis, P. (2014) *Vibrio anguillarum*-associated vibrio-sis in the Mediterranean aquaculture. In: Angelidis, P. (ed.) Aspects of Mediterranean Marine Aquaculture. Blue Crab PC, Thessaloniki, Greece, pp. 243–264.

Austin, B. and Austin, D.A. (2007) *Bacterial Fish Pathogens: Diseases of Farmed and Wild Fish*, 4th edn. Springer and Praxis Publishing, Dordrecht, The Netherlands, Berlin/Heidelberg, New York and Chichester, UK

Austin, B., Austin, D.A., Blanch, A.R., Cerda, M., Grimont, F. *et al.* (1997) A comparison of methods for the typing of fish-pathogenic *Vibrio* spp. *Systematic and Applied Microbiology* 20, 89–101.

Avendaño-Herrera, R., Maldonado, J.P., Tapia-Cammas, D., Feijoó, C.G., Calleja, F. *et al.* (2014) PCR protocol for the detection of *Vibrio ordalii* by amplification of the *vohB* (hemolysin) gene. *Diseases of Aquatic Organisms* 107, 223–234.

Balado, M., Osorio, C.R. and Lemos, M.L. (2006) A gene cluster involved in the biosynthesis of vanchrobactin, a chromosome-encoded siderophore produced by *Vibrio anguillarum*. *Microbiology* 152, 3517–3528.

Balado, M., Osorio, C.R. and Lemos, M.L. (2008) Biosynthetic and regulatory elements involved in the production of the siderophore vanchrobactin in *Vibrio anguillarum*. *Microbiology* 154, 1400–1413.

Benediktsdóttir, E., Verdonk, L., Spröer, C., Helgason, S. and Swings, J. (2000) Characterization of *Vibrio* viscosus and Vibrio wodanis isolated at different geographical locations: a proposal for reclassification of Vibrio viscosus as Moritella viscosa comb. nov. International Journal of Systematic and Evolutionary Microbiology 50, 479–488.

Bjelland, A.M., Johansen, R., Brudal, E., Hansen, H., Winther-Larsen, H. et al. (2012) Vibrio salmonicida pathogenesis analyzed by experimental challenge of Atlantic salmon (Salmo salar). Microbial Pathogenesis 52, 77–84.

Bjelland, A.M., Fauske, A.K., Nguyen, A., Orlien, I.E., Østgaard, I.M. et al. (2013) Expression of Vibrio salmonicida virulence genes and immune response parameters in experimentally challenged Atlantic salmon (Salmo salar L.). Frontiers in Microbiology 4: 401.

Boesen, H.T., Pedersen, K., Larsen, J.L., Koch, C. and Ellis, A.E. (1999) Vibrio anguillarum resistance to rainbow trout (Oncorhynchus mykiss) serum: role of O-antigen structure of lipopolysaccharide. Infection and Immunity 67, 294–301.

Bohle, H., Kjetil, F., Bustos, P., Riofrío, A. and Peters, C. (2007) Fenotipo atípico de Vibrio ordalii, bacteria altamente patogénica aislada desde salmón del Atlántico cultivado en las costas marinas del sur de Chile. Archivos de Medicina Veterinaria 39, 43–52.

Bordas, M.A., Balebona, M.C., Rodríguez-Maroto, J.M., Borrego, J.J. and Moriñigo, M.A. (1998) Chemotaxis of pathogenic Vibrio strains towards mucus surfaces of gilt-head sea bream (Sparus aurata L.). Applied and Environmental Microbiology 64, 1573–1575.

Brackman, G., Celen, S., Baruah, K., Bossier, P., Van Calenbergh, S. et al. (2009) AI-2 quorum-sensing inhibitors affect the starvation response and reduce virulence in several Vibrio species, most likely by interfering with Lux-PQ. Microbiology 155, 4114–4122.

Bruhn, J.B., Dalsgaard, I., Nielsen, K.F., Buchholtz, C., Larsen, J.L. et al. (2005) Quorum sensing signal molecules (acylated homoserine lactones) in Gram-negative fish pathogenic bacteria. Diseases of Aquatic Organisms 65, 43–52.

Bruno, D.W., Hasting, T.S., Ellis, A.E. and Wotten, R. (1985) Outbreak of a cold water vibriosis in Atlantic salmon in Scotland. Bulletin of the European Association of Fish Pathologists 5, 62–63.

Bruno, D.W., Hastings, T.S. and Ellis, A.E. (1986) Histopathology, bacteriology and experimental transmission of a cold water vibriosis in Atlantic salmon Salmo salar. Diseases of Aquatic Organisms 1, 163–168.

Brunt, J., Newaj-Fyzul, A. and Austin, B. (2007) The development of probiotics for the control of multiple bacterial diseases of rainbow trout, Oncorhynchus mykiss (Walbaum). Journal of Fish Diseases 30, 573–579.

Carnevali, O., de Vivo, L., Sulpizio, R., Gioacchini, G., Olivotto, I. et al. (2006) Growth improvement by probiotic in European sea bass juveniles (Dicentrarchus labrax, L.), with particular attention to IGF-1, myostatin and cortisol gene expression. Aquaculture 258, 430–438.

Chabrillon, M., Arijo, S., Díaz-Rosales, P., Balebona, M.C. and Moriñigo, M.A. (2006) Interference of Listonella anguillarum with potential probiotic microorganisms isolated from farmed gilthead sea bream (Sparus aurata L.). Aquaculture Research 37, 78–86.

Chart, H. and Trust, T.J. (1984) Characterization of surface antigens of the marine fish pathogens Vibrio anguillarum and Vibrio ordalii. Canadian Journal of Microbiology 30, 703–710.

CLSI (2006) Methods for Antimicrobial Disk Susceptibility Testing of Bacteria Isolated from Aquatic Animals. Approved Guidelines. CLSI Document M42-A, Clinical and Laboratory Standards Institute, Wayne, Pennsylvania.

Colquhoun, D.J. and Lillehaug, A. (2014) Vaccination against vibriosis. In: Gudding, R., Lillehaug, A. and Evensen, Ø. (eds) Fish Vaccination. Wiley-Blackwell, Chichester/Oxford, UK, pp. 172–184.

Colquhoun, D.J. and Sørum, H. (2001) Temperature dependent siderophore production in Vibrio salmonicida. Microbial Pathogenesis 31, 213–219.

Colquhoun, D.J., Aase, I.L., Wallace, C., Baklien, Å. and Gravningen, K. (2004) First description of Vibrio ordalii from Chile. Bulletin of the European Association of Fish Pathologists 24, 185–188.

Colquhoun, D.J., Aarflot, L. and Melvold, C.F. (2007) gyrA and parC mutations and associated quinolone resistance in Vibrio anguillarum serotype O2b strains isolated from farmed Atlantic cod (Gadus morhua) in Norway. Antimicrobial Agents and Chemotherapy 51, 2597–2599.

Conchas, R.F., Lemos, M.L., Barja, J.L. and Toranzo, A.E. (1991) Distribution of plasmid- and chromosome-mediated iron uptake systems in Vibrio anguillarum strains of different origins. Applied and Environmental Microbiology 57, 2956–2962.

Crosa, J.H. (1980) A plasmid associated with virulence in the marine fish pathogen Vibrio anguillarum specifies an iron-sequestering system. Nature 284, 566–568.

Croxatto, A., Chalker, V.J., Lauritz, J., Jass, J., Hardman, A. et al. (2002) Van T, a homologue of Vibrio harveyi LuxR, regulates serine, metalloprotease, pigment, and biofilm production in Vibrio anguillarum. Journal of Bacteriology 184, 1616–1629.

Croxatto, A., Pride, J., Hardman, A., Williams, P., Camara, M. et al. (2004) A distinctive dual-channel quorum-sensing system operates in Vibrio anguillarum. Molecular Microbiology 52, 1677–1689.

Croxatto, A., Lauritz, J., Chen, C. and Milton, D.L. (2007) Vibrio anguillarum colonization of rainbow trout integument requires a DNA locus involved in exopolysaccharide transport and biosynthesis. Environmental Microbiology 9, 370–383.

Dalsgaard, I., Jürgens, O. and Mortensen, A. (1988) Vibrio salmonicida isolated from farmed Atlantic salmon in Faroe Islands. *Bulletin of the European Association of Fish Pathologists* 8, 53–54.

Daly, J. and Stevenson, R. (1987) Hydrophobic and haemagglutinating properties of *Renibacterium salmoninarum*. *Journal of General Microbiology* 133, 3575–3580.

Defoirdt, T., Boon, N., Bossier, P. and Verstraete, W. (2004) Disruption of bacterial quorum sensing: an unexplored strategy to fight infections in aquaculture. *Aquaculture* 240, 69–88.

Denkin, S.M. and Nelson, D.R. (2004) Regulation of *Vibrio anguillarum empA* metalloprotease expression and its role in virulence. *Applied and Environmental Microbiology* 70, 4193–4204.

Egidius, E., Andersen, K., Claussen, E. and Raa, J. (1981) Cold-water vibriosis or Hitra-disease in Norwegian salmonid farming. *Journal of Fish Diseases* 4, 353–354.

Egidius, E., Wiik, R., Andersen, K., Holff, K.A. and Hjeltness, B. (1986) *Vibrio salmonicida* sp. nov., a new fish pathogen. *International Journal of Systematic Bacteriology* 36, 518–520.

Engelsen, A.R., Sandlund, N., Fiksdal, I.U. and Bergh, Ø. (2008) Immunohistochemistry of Atlantic cod larvae *Gadus morhua* experimentally challenged with *Vibrio anguillarum*. *Diseases of Aquatic Organisms* 80, 13–20.

Enger, Ø., Husevag, B. and Goksøyr, J. (1989) Presence of *Vibrio salmonicida* in fish farm sediments. *Applied and Environmental Microbiology* 55, 2815–2818.

Enger, Ø., Hoff, K.A., Schei, G.H. and Dundas, I. (1990) Starvation survival of the fish pathogenic bacteria *Vibrio anguillarum* and *Vibrio salmonicida* in marine environments. *FEMS Microbiology Letters* 74, 215–220.

Enger, Ø., Husevag, B. and Goksøyr, J. (1991) Seasonal variation in presence of *Vibrio salmonicida* and total bacterial counts in Norwegian fish-farm water. *Canadian Journal of Microbiology* 37, 618–623.

Espelid, S., Hjelmeland, K. and Jørgensen, T. (1987) The specificity of Atlantic salmon antibodies made against the fish pathogen *Vibrio salmonicida*, establishing the surface protein VS-P1 as the predominant antigen. *Developmental and Comparative Immunology* 11, 529–537.

Ezura, Y., Tajima, K., Yoshimizu, M. and Kimura, T. (1980) Studies on the taxonomy and serology of causative organisms of fish vibriosis. *Fish Pathology* 14, 167–179.

Farmer, J.J., Janda, M., Brenner, F.W., Cameron, D.N. and Birkhead, K.M. (2005) Genus I. *Vibrio* Pacini 1854. In: Brenner, D.J., Krieg, N.R. and Staley, J.T. (eds) *Bergeys's Manual of Systematic Bacteriology, Volume 2: The Proteobacteria, Part B: The Gammaproteobacteria*, 2nd edn. Springer, New York, pp. 495–545.

Fidopiastis, P.M., Sørum, H. and Ruby, E.G. (1999) Cryptic luminescence in the cold-water fish pathogen *Vibrio salmonicida*. *Archives in Microbiology* 171, 205–209.

Fouz, B., Devesa, S., Gravningen, K., Barja, J.L. and Toranzo, A.E. (1990) Antibacterial action of the mucus of turbot. *Bulletin of the European Association of Fish Pathologists* 10, 56–59.

Frans, I., Michiels, C.W., Bossier, P., Willems, K.A., Lievens, B. et al. (2011) *Vibrio anguillarum* as a fish pathogen: virulence factors, diagnosis and prevention. *Journal of Fish Diseases* 34, 643–661.

Gao, H., Li, F., Zhang, X., Wang, B. and Xiang, J. (2010) Rapid, sensitive detection of *Vibrio anguillarum* using loop mediated isothermal amplification. *Chinese Journal of Oceanology and Limnology* 28, 62–66.

Gram, L., Melchiorsen, J., Spanggard, B., Huber, I. and Nielsen, T.F. (1999) Inhibition of *Vibrio anguillarum* by *Pseudomonas fuorescens* AH2, a possible treatment of fish. *Applied and Environmental Microbiology* 65, 963–973.

Grisez, L., Chair, M., Zorruelos, P. and Ollevier, F. (1996) Mode of infection and spread of *Vibrio anguillarum* in turbot *Scophthalmus maximus* larvae after oral challenge through live feed. *Diseases of Aquatic Organisms* 26, 181–187.

Harrel, L.W., Etlinger, H.M. and Hodgins, H.O. (1976a) Humoral factors important in resistance of salmonid fish to bacterial disease. II. Anti-*Vibrio anguillarum* activity in mucus and observations on complement. *Aquaculture* 7, 363–370.

Harrell, L.W., Novotny, A.J., Schiewe, M.J. and Hodgins, H.O. (1976b) Isolation and description of two vibrios pathogenic to Pacific salmon in Puget Sound, Washington. *Fishery Bulletin* 74, 447–449.

Håstein, T., Gudding, R. and Evensen, Ø. (2005) Bacterial vaccines for fish – an update of the current situation worldwide. *Developmental Biology* 121, 55–74.

Henríquez, P., Kaiser, M., Bohle, H., Bustos, P. and Mancilla, M. (2016) Comprehensive antibiotic susceptibility profiling of Chilean *Piscirickettsia salmonis* field isolates. *Journal of Fish Diseases* 39, 441–448.

Hirono, I., Masuda, T. and Aoki, T. (1996) Cloning and detection of the hemolysin gene of *Vibrio anguillarum*. *Microbial Pathogenesis* 21, 173–182.

Hjelmeland, K., Stensvåg, K., Jørgensen, T. and Espelid, S. (1988) Isolation and characterization of a surface layer antigen from *Vibrio salmonicida*. *Journal of Fish Diseases* 11, 197–205.

Hjeltnes, B., Andersen, K. and Ellingsen, H. (1989) Vaccination against *Vibrio salmonicida*. The effect of different routes of administration and of revaccination. *Aquaculture* 83, 1–2.

Hjerde, E., Lorentzen, M.S., Holden, M.T.G., Seeger, K., Paulsen, S. et al. (2008) The genome sequence of the fish pathogen *Aliivibrio salmonicida* strain LFI1238 shows extensive evidence of gene decay. *BMC Genomics* 9:616.

Hoff, K.A (1989a) Survival of *Vibrio anguillarum* and *Vibrio salmonicida* at different salinities. *Applied and Environmental Microbiology* 55, 1775–1786.

A.E. Toranzo *et al.*

Hoff, K.A. (1989b) Occurrence of the fish pathogen *Vibrio salmonicida* in faeces from Atlantic salmon, *Salmo salar* L., after experimental infection. *Journal of Fish Diseases* 12, 595–597.

Holm, K.O. and Jørgensen, T. (1987) A successful vaccination of Atlantic salmon, *Salmo salar* L. against "Hitra disease" or coldwater vibriosis. *Journal of Fish Diseases* 10, 85–90.

Holm, K.O., Strøm, E., Stensvaag, K., Raa, J. and Jørgense, T.Ø. (1985) Characteristics of a *Vibrio* sp. associated with the "Hitra disease" of Atlantic salmon in Norwegian fish farms. *Fish Pathology* 20, 125–129.

Holm, K.O., Nilsson, K., Hjerde, E., Willassen, N.-P. and Milton, D. (2015) Complete genome sequence of *Vibrio anguillarum* NB10, a virulent isolate from the Gulf of Bothnia. *Standards in Genomic Sciences* 10: 60.

Hong, G.-E., Kim, D.-G., Bae, J.-Y., Ahn, S.-H., Bai, S.C. et al. (2007) Species-specific PCR detection of the fish pathogen, *Vibrio anguillarum*, using the *amiB* gene, which encodes *N*-acetylmuramoyl-L-alanine amidase. *FEMS Microbiology Letters* 269, 201–206.

Ito, H., Uchida, I., Sekizaki, T. and Terakado, N. (1995) A specific oligonucleotide probe based on 5S rRNA sequences for identification of *Vibrio anguillarum* and *Vibrio ordalii*. *Veterinary Microbiology* 43, 167–171.

Kanno, T., Nakai, T and Muroga, K. (1989) Mode of transmission of vibriosis among ayu, *Plecoglossus altivelis*. *Journal of Aquatic Animal Health* 1, 2–6.

Kao, D.Y., Cheng, Y.C., Kuo, T.Y., Lin, S.B., Lin, C.C. et al. (2009) Salt-responsive outer membrane proteins of *Vibrio anguillarum* serotype O1 as revealed by comparative proteome. *Journal of Applied Microbiology* 106, 2079–2085.

Karlsen, C., Paulsen, S.M., Tunsjø, H.S., Krinner, S., Sørum, H. et al. (2008) Motility and flagellin gene expression in the fish pathogen *Vibrio salmonicida*: effects of salinity and temperature. *Microbial Pathogenesis* 45, 258–264.

Kent, M.L. and Poppe, T.T. (2002) Infectious diseases coldwater fish in marine and brackish water. In: Woo, P.T.K., Bruno, D.W. and Lim, H.S. (eds) *Diseases and Disorders of Finfish in Cage Culture*, 1st edn. CAB International, Wallingford, UK, pp. 61–105.

Kim, D.-G., Kim, Y.-R., Kim, E.-Y., Cho, H.M., Ahn, S.-H. et al. (2010) Isolation of the *groESL* cluster from *Vibrio anguillarum* and PCR detection targeting *groEL* gene. *Fisheries Science* 76, 803–810.

Kodama, H., Moustafa, M., Ishiguro, S., Mikami, T. and Izawa, H. (1984) Extracellular virulence factors of fish *Vibrio*: relationships between toxic material, hemolysin, and proteolytic enzyme. *American Journal of Veterinary Research* 45, 2203–2207.

Kulkarni, A., Caipang, C.M.A., Brinchmann, M.F., Korsnes, K. and Kiron, V. (2009) Use of loop-mediated isothermal amplification (LAMP) assay for the detection of *Vibrio anguillarum* O2b, the causative agent of vibriosis in Atlantic cod, *Gadus morhua*. *Journal of Rapid Methods and Automation in Microbiology* 17, 503–518.

Lamas, J., Santos, Y., Bruno, D., Toranzo, A.E. and Anadon, R. (1994) A comparison of pathological changes caused by *Vibrio anguillarum* and its extracellular products in rainbow trout (*Oncorhynchus mykiss*). *Fish Pathology* 29, 79–89.

Larsen, J.L. and Mellergaard, S. (1984) Agglutination typing of *Vibrio anguillarum* isolates from diseased fish and from the environment. *Applied and Environmental Microbiology* 47, 1261–1265.

Larsen, J.L., Pedersen, K. and Dalsgaard, I. (1994) *Vibrio anguillarum* serovars associated with vibriosis in fish. *Journal of Fish Diseases* 17, 259–267.

Larsen, M.H., Larsen, J.L. and Olsen, J.E. (2001) Chemotaxis of *Vibrio anguillarum* to fish mucus: role of the origin of the fish mucus, the fish species and the serogroup of the pathogen. *FEMS Microbiology Ecology* 38, 77–80.

Lemos, M.L. and Osorio, C.R. (2007) Heme, an iron supply for vibrios pathogenic for fish. *BioMetals* 20, 615–626.

Lemos, M.L., Salinas, P., Toranzo, A.E., Barja, J.L. and Crosa, J.H. (1988) Chromosome-mediated iron uptake systems in pathogenic strains of *Vibrio anguillarum*. *Journal of Bacteriology* 170, 1920–1925.

Lemos, M.L., Balado, M. and Osorio, C.R. (2010) Anguibactin versus vanchrobactin-mediated iron uptake in *Vibrio anguillarum*: evolution and ecology of a fish pathogen. *Environmental Microbiology Reports* 2, 19–26.

Li, G., Mo, Z., Li, J., Xiao, P. and Hao, B. (2013) Complete genome sequence of *Vibrio anguillarum* M3, a serotype O1 strain isolated from Japanese flounder in china. *Genome Announcement* 1(5): e00769-13.

Lillehaug, A., Sørum, R.H. and Ramstad, A. (1990) Cross-protection after immunization of Atlantic salmon *Salmo salar* L., against different strains of *Vibrio salmonicida*. *Journal of Fish Diseases* 13, 519–523.

Lillehaug, A., Lunestad, B.T. and Grave, K. (2003) Epidemiology of bacterial diseases in Norwegian aquaculture – a description based on antibiotic prescription data for the ten-year period 1991 to 2000. *Diseases of Aquatic Organisms* 53, 115–125.

MacDonell, M.T. and Colwell, R.R. (1984) Nucleotide base sequence of *Vibrionaceae* 5S rRNA. *FEBS Letters* 175, 183–188.

Mazoy, R., Osorio, C.R., Toranzo, A.E. and Lemos, M.L. (2003) Isolation of mutants of *Vibrio anguillarum* defective in haeme utilization and cloning of *huvA*, a gene coding for an outer membrane protein involved in the use of haeme as iron source. *Archives of Microbiology* 179, 329–338.

Midtlyng, P.J. and Lillehaug, A. (1998) Growth of Atlantic salmon *Salmo salar* after intraperitoneal administration of vaccines containing adjuvants. *Diseases of Aquatic Organisms* 32, 91–97.

Mikkelsen, H., Lund, V., Martinsen, L.C., Gravningen, K. and Schrøeder, M.B. (2007) Variability among *Vibrio anguillarum* O2 isolates from Atlantic cod (*Gadus morhua* L.): characterization and vaccination studies. *Aquaculture* 266, 16–25.

Milton, D.L. (2006) Quorum sensing in vibrios: complexity for diversification. *International Journal of Medical Microbiology* 296, 61–71.

Milton, D.L., Norqvist, A. and Wolfwatz, H. (1992) Cloning of a metalloprotease gene involved in the virulence mechanism of *Vibrio anguillarum*. *Journal of Bacteriology* 174, 7235–7244.

Milton, D.L., Hardman, A., Camara, M., Chhabra, S.R., Bycroft, B.W. *et al.* (1997) Quorum sensing in *Vibrio anguillarum*: characterization of the *vanI/vanR* locus and identification of the autoinducer *N*-(3-oxodecanoyl)-L-homoserine lactone. *Journal of Bacteriology* 179, 3004–3012.

Milton, D.L., Chalker, V.J., Kirke, D., Hardman, A., Camara, M. and Williams, P. (2001) The LuxM homologue VanM from *Vibrio anguillarum* directs the synthesis of *N*-(3-hydroxyhexanoyl) homoserine lactone and *N*-hexanoylhomoserine lactone. *Journal of Bacteriology* 183, 3537–3547.

Montgomery, M.T. and Kirchman, D.L. (1994) Induction of chitin-binding proteins during the specific attachment of the marine bacterium *Vibrio harveyi* to chitin. *Applied and Environmental Microbiology* 60, 4284–4288.

Mouriño, S., Osorio, C.R. and Lemos, M.L. (2004) Characterization of heme uptake cluster genes in the fish pathogen *Vibrio anguillarum*. *Journal of Bacteriology* 186, 6159–6167.

Muroga, K. and de la Cruz, M.C. (1987) Fate and localization of *Vibrio anguillarum* in tissues of artificially infected ayu (*Plecoglossus altivelis*). *Fish Pathology* 22, 99–103.

Muroga, K., Yasuhiko, J. and Masumura, K. (1986) *Vibrio ordalii* isolated from diseased ayu (*Plecoglossus altivelis*) and rockfish (*Sebastes schlegeli*). *Fish Pathology* 21, 239–243.

Mutharia, L.W., Raymond, B.T., Dekievit, T.R. and Stevenson, R.M.W. (1993) Antibody specificities of polyclonal rabbit and rainbow trout antisera against *Vibrio ordalii* and serotype O2 strains of *Vibrio anguillarum*. *Canadian Journal of Microbiology* 39, 492–499.

Naka, H., Dias, G.M., Thompson, C.C., Dubay, C., Thompson, F.L. *et al.* (2011) Complete genome sequence of the marine fish pathogen *Vibrio anguillarum* harboring the pJM1 virulence plasmid and genomic comparison with other virulent strains of *V. anguillarum* and *V. ordalii*. *Infection and Immunity* 79, 2889–2900.

Nelson, E.J., Tunsjø, H.S., Fidopiastis, P.M., Sørum, H. and Ruby, E.G. (2007) A novel *lux* operon in the cryptically bioluminescent fish pathogen *Vibrio salmonicida* is associated with virulence. *Applied and Environmental Microbiology* 73, 1825–1833.

Nordmo, R., Sevatdal, S. and Ramstad, A. (1997) Experimental infection with *Vibrio salmonicida* in Atlantic salmon (*Salmo salar* L.): an evaluation of three different challenge methods. *Aquaculture* 158, 23–32.

Ofek, I. and Doyle, R.J. (1994) *Bacterial Adhesion to Cells and Tissues*. Chapman and Hall, New York.

Ohnishi, K. and Muroga, K. (1976) *Vibrio* sp. as a cause of disease in rainbow trout cultured in Japan. I. Biochemical characteristics. *Fish Pathology* 11, 159–165.

Olsen, J.E. and Larsen, J.L. (1993) Ribotypes and plasmid contents of *Vibrio anguillarum* strains in relation to serovars. *Applied and Environmental Microbiology* 59, 3863–3870.

Olsson, J.C., Joborn, A., Westerdahl, A., Blomberg, L., Kjelleberg, S. *et al.* (1998) Survival, persistence and proliferation of *Vibrio anguillarum* in juvenile turbot, *Scophthalmus maximus* (L.), in intestine and faeces. *Journal of Fish Diseases* 21, 1–9.

Osorio, C. and Toranzo, A.E. (2002) DNA-based diagnostics in sea farming. In: Fingerman, M. and Nagabhushanam, R. (eds) *Recent Advances in Marine Biotechnology, Volume 7: Seafood Safety and Human Health*. Science Publishers, Plymouth, UK, pp. 253–310.

Osorio, C.R., Juiz-Río, S. and Lemos, M.L. (2006) A siderophore biosynthesis gene cluster from the fish pathogen *Photobacterium damselae* subsp. *piscicida* is structurally and functionally related to the *Yersinia* high-pathogenicity island. *Microbiology* 152, 3327–3341.

O'Toole, R., Milton, D.L. and Wolf-Watz, H. (1996) Chemotactic motility is required for invasion of the host by the fish pathogen *Vibrio anguillarum*. *Molecular Microbiology* 19, 625–637.

O'Toole, R., Lundberg, S., Fredricksson, S.A., Jansson, A., Nilsson, B. *et al.* (1999) The chemotactic response of *Vibrio anguillarum* to fish intestinal mucus is mediated by a combination of multiple mucus components. *Journal of Bacteriology* 181, 4308–4317.

O'Toole, R., von Hofsten, J., Rosqvist, R., Olsson, P.E. and Wolf-Watz, H. (2004) Visualisation of zebrafish infection by GFP labelled *Vibrio anguillarum*. *Microbial Pathogenesis* 37, 41–46.

Pedersen, K. and Larsen, J.L. (1993) rRNA gene restriction patterns of *Vibrio anguillarum* serogroup O1. *Diseases of Aquatic Organisms* 16, 121–126.

Pedersen, K., Grisez, L., van Houdt, R., Tiainen, T., Ollevier, F. *et al.* (1999) Extended serotyping scheme for *Vibrio anguillarum* with the definition of seven provisional O-serogroups. *Current Microbiology* 38, 183–189.

Pillidge, C.J. and Colwell, R.R. (1988) Nucleotide sequence of the 5S rRNA from *Listonella* (*Vibrio*) *ordalii* ATCC 33509 and *Listonella* (*Vibrio*) *tubiashii* ATCC 19105. *Nucleic Acids Research* 16, 3111.

A.E. Toranzo *et al.*

Planas, M., Pérez-Lorenzo, M., Hjelm, M., Gram, L., Fiksdal, I.U. *et al.* (2006) Probiotic effect *in vivo* of *Roseobacter* strain 27-4 against *Vibrio* (*Listonella*) *anguillarum* infections in turbot (*Scophthalmus maximus* L.) larvae. *Aquaculture* 255, 323–333.

Poblete-Morales, M., Irgang, R., Henríquez-Núñez, H., Toranzo, A.E., Kronvall, G. *et al.* (2013) *Vibrio ordalii* antimicrobial susceptibility testing – modified culture conditions required and laboratory-specific epidemiological cut-off values. *Veterinary Microbiology* 165, 434–442.

Poppe, T.T. and Koppang, E. (2014) Vaccination against vibriosis. In: Gudding, R., Lillehaug, A. and Evensen, Ø. (eds) *Fish Vaccination*, Wiley-Blackwell, Chichester/Oxford, UK, pp. 153–161.

Poppe, T.T., Håstein, T. and Salte, R. (1985) Hitra disease (haemorrhagic syndrome) in Norwegian salmon farming: present status. In: Ellis, A.E. (ed.) *Fish and Shellfish Pathology. First International Conference of the European Association of Fish Pathologists, Plymouth, England, Sept. 20–23, 1983.* Academic Press, London/Orlando, Florida, pp. 223–229.

Porsby, C.H., Nielsen, K.F. and Gram, L. (2008) *Phaeobacter* and *Ruegeria* species of the *Roseobacter* clade colonize separate niches in a Danish turbot (*Scophthalmus maximus*)-rearing farm and antagonize *Vibrio anguillarum* under different growth conditions. *Applied and Environmental Microbiology* 74, 7356–7364.

Pybus, V., Loutit, M.W., Lamont, I.L and Tagg, J.R. (1994) Growth inhibition of the salmon pathogen *Vibrio ordalii* by a siderophore produced by *Vibrio anguillarum* strain VL4355. *Journal of Fish Diseases* 17, 311–324.

Raeder, I.L., Paulsen, S.M., Smalås, A.O. and Willassen, N.P. (2007) Effect of skin mucus on the soluble proteome of *Vibrio salmonicida* analysed by 2-D gel electrophoresis and tandem mass spectrometry. *Microbial Pathogenesis* 42, 36–45.

Ransom, D.P. (1978) Bacteriologic, immunologic and pathologic studies of *Vibrio* sp. pathogenic to salmonids. PhD thesis, Oregon State University, Corvallis, Oregon.

Ransom, D.P., Lannan, C.N., Rohovec, J.S. and Fryer, J.L. (1984) Comparison of histopathology caused by *Vibrio anguillarum* and *Vibrio ordalii* in three species of Pacific salmon. *Journal of Fish Diseases* 7, 107–115.

Rasch, M., Buch, C., Austin, B., Slierendrecht, W.J., Ekmann, K.S. *et al.* (2004) An inhibitor of bacterial quorum sensing reduces mortalities caused by vibriosis in rainbow trout (*Oncorhynchus mykiss*, Walbaum). *Systematic and Applied Microbiology* 27, 350–359.

Rigos, G. and Troisi, G.M. (2005) Antibacterial agents in Mediterranean finfish farming: a synopsis of pharmacokinetics in important euryhaline fish species and possible environmental implications. *Reviews in Fish Biology and Fisheries* 15, 53–73.

Rodkhum, C., Hirono, I., Crosa, J.H. and Aoki, T. (2005) Four novel hemolysin genes of *Vibrio anguillarum* and their virulence to rainbow trout. *Microbial Pathogenesis* 39, 109–119.

Rodkhum, C., Hirono, I., Stork, M., DiLorenzo, M., Crosa, J.H. and Aoki, T. (2006) Putative virulence-related genes in *Vibrio anguillarum* identified by random genome sequencing. *Journal of Fish Diseases* 29, 157–166.

Romalde, J.L., Magariños, B., Fouz, B., Bandín, I., Nuñez, S. *et al.* (1995) Evaluation of Bionor mono-kits for rapid detection of bacterial fish pathogens. *Diseases of Aquatic Organisms* 21, 25–34.

Rowe, G.E. and Welch, R.A. (1994) Assays of hemolytic toxins. *Methods in Enzymology* 235, 657–667.

Ruiz, P., Poblete, M., Yáñez, A.J., Irgang, R., Toranzo, A.E. *et al.* (2015) Cell surface properties of *Vibrio ordalii* strains isolated from Atlantic salmon *Salmo salar* in Chilean farms. *Diseases of Aquatic Organisms* 113, 9–23.

Ruiz, P., Balado, M., Toranzo, A.E., Poblete-Morales, M., Lemos, M.L. *et al.* (2016a) Iron assimilation and siderophore production by *Vibrio ordalii* strains isolated from diseased Atlantic salmon (*Salmo salar*) in Chile. *Diseases of Aquatic Organisms* 118, 217–226.

Ruiz, P., Poblete-Morales, M., Irgang, R., Toranzo, A.E. and Avendaño-Herrera, R. (2016b) Survival behavior and virulence of the fish pathogen *Vibrio ordalii* in seawater microcosms. *Diseases of Aquatic Organisms* 120, 27–38.

Sadovskaya, I., Brisson, J.R., Khieu, N.H., Mutharia, L.M. and Altman, E. (1998) Structural characterization of the lipopolysaccharide O-antigen and capsular polysaccharide of *Vibrio ordalii* serotype O:2. *European Journal of Biochemistry* 253, 319–327.

Samuelsen, O.B. and Bergh, Ø. (2004) Efficacy of orally administered florfenicol and oxolinic acid for the treatment of vibriosis in cod (*Gadus morhua*). *Aquaculture* 235, 27–35.

Samuelsen, O.B., Nerland, A.H., Jørgensen, T., Schøder, M.B., Svåsand, T. *et al.* (2006) Viral and bacterial diseases of Atlantic cod *Gadus morhua*, their prophylaxis and treatment: a review. *Diseases of Aquatic Organisms* 71, 239–254.

San Martín N., B., Yatabe, T., Gallardo, A. and Medina, P. (2010) *Manual de Buenas Prácticas en el Uso de Antibióticos y Antiparasitarios en la Salmonicultura Chilena.* Universidad de Chile and Servicio Nacional de Pesca (SERNAPESCA), Santiago, Chile.

Santos, Y., Pazos, F., Bandín, I. and Toranzo, A.E. (1995) Analysis of antigens present in the extracellular products and cell surface of *Vibrio anguillarum* O1, O2 and O3. *Applied and Environmental Microbiology* 61, 2493–2498.

Schiewe, M.H. and Crosa, J.H. (1981) Molecular characterization of *Vibrio anguillarum* biotype 2. *Canadian Journal of Microbiology* 27, 1011–1018.

Schiewe, M.H., Crosa, J.H. and Ordal, E.J. (1977) Deoxyribonucleic acid relationships among marine vibrios pathogenic to fish. *Canadian Journal of Microbiology* 23, 954–958.

Schiewe, M.H., Trust, T.J. and Crosa, J.H. (1981) *Vibrio ordalii* sp. nov.: a causative agent of vibriosis in fish. *Current Microbiology* 6, 343–348.

Schrøder, M.B., Espelid, S. and Jørgensen, T.O. (1992) Two serotypes of *Vibrio salmonicida* isolated from diseased cod (*Gadus morhua* L.): virulence, immunological studies and vaccination experiments. *Fish and Shellfish Immunol*ogy 2, 211–221.

Silva-Rubio, A., Avendaño-Herrera, R., Jaureguiberry, B., Toranzo, A.E. and Magariños, B. (2008a) First description of serotype O3 in *Vibrio anguillarum* strains isolated from salmonids in Chile. *Journal of Fish Diseases* 31, 235–239.

Silva-Rubio, A., Acevedo, C, Magariños, B., Jaureguiberry, B, Toranzo, A.E. *et al.* (2008b) Antigenic and molecular characterization of *Vibrio ordalii* strains isolated from Atlantic salmon *Salmo salar* in Chile. *Diseases of Aquatic Organisms* 79, 27–35.

Skov, M.N., Pedersen, K. and Larsen, J.L. (1995) Comparison of pulse-field gel electrophoresis, ribotyping and plasmid profiling for typing of *Vibrio anguillarum* serovar O1. *Applied and Environmental Microbiology* 61, 1540–1545.

Smith, S.K., Sutton, D.C., Fuerst, J.A. and Reichelt, J.L. (1991) Evaluation of the genus *Listonella* and reassignment of *Listonella damsela* (Love *et al.*) MacDonell and Colwell to the genus *Photobacterium* as *Photobacterium damsela* comb. nov. with an emended description. *International Journal of Systematic and Evolutionary Microbiology* 41, 529–534.

Soengas, R.G., Anta, C., Espada, A., Paz, V., Ares, I.R. *et al.* (2006) Structural characterization of vanchrobactin, a new catechol siderophore produced by the fish pathogen *Vibrio anguillarum* serotype O2. *Tetrahedron Letters* 47, 7113–7116.

Sørensen, U.B.S. and Larsen, J.L. (1986) Serotyping of *Vibrio anguillarum*. *Applied and Environmental Microbiology* 51, 593–597.

Sørum, H., Hvaal, A.B., Heum, M., Daae, F.L. and Wiik, R. (1990) Plasmid profiling of *Vibrio salmonicida* for epidemiological studies of cold-water vibriosis in Atlantic salmon (*Salmo salar*) and cod (*Gadus morhua*). *Applied and Environmental Microbiology* 56, 1033–1037.

Sørum, H., Roberts, M.C. and Crosa, J.H. (1992) Identification and cloning of a tetracycline resistance gene from the fish pathogen *Vibrio salmonicida*. *Antimicrobial Agents and Chemotherapy* 36, 611–615.

Sorroza, L., Padilla, D., Acosta, F., Román, L., Grasso, V., Vega, J. and Real, F. (2012) Characterization of the probiotic strain *Vagococcus fluvialis* in the protection of European sea bass (*Dicentrarchus labrax*) against vibriosis by *Vibrio anguillarum*. *Veterinary Microbiology* 155, 369–373.

Souto, A., Montaos, M.A., Rivas, A.J., Balado, M., Osorio, C. *et al.* (2012) Structure and biosynthetic assembly of piscibactin, a siderophore from *Photobacterium damselae* subsp. *piscicida*, predicted from genome analysis. *European Journal of Organic Chemistry* 29, 5693–5700.

Spanggaard, B., Huber, I., Nielsen, J., Nielsen, T. and Gram, L. (2000) Proliferation and location of *Vibrio anguillarum* during infection of rainbow trout, *Oncorhynchus mykiss* (Walbaum). *Journal of Fish Diseases* 23, 423–427.

Stork, M., Di Lorenzo, M., Welch, T.J., Crosa, L.M. and Crosa, J.H. (2002) Plasmid-mediated iron uptake and virulence in *Vibrio anguillarum*. *Plasmid* 48, 222–228.

Stork, M., Di Lorenzo, M., Mouriño, S., Osorio, C.R., Lemos, M.L. and Crosa, J.H. (2004) Two *tonB* systems function in iron transport in *Vibrio anguillarum*, but only one is essential for virulence. *Infection and Immunity* 72, 7326–7329.

Svendsen, Y.S. and Bogwald, J. (1997) Influence of artificial wound and non-intact mucus layer on mortality of Atlantic salmon (*Salmo salar* L.) following a bath challenge with *Vibrio anguillarum* and *Aeromonas salmonicida*. *Fish and Shellfish Immunology* 7, 317–325.

Tajima, K., Ezura, Y. and Kimura, T. (1985) Studies on the taxonomy and serology of causative organisms of fish vibriosis. *Fish Pathology* 20, 131–142.

Tapia-Cammas, D., Yañez, A., Arancibia, G., Toranzo, A.E. and Avendaño-Herrera, R. (2011) Multiplex PCR for the detection of *Piscirickettsia salmonis*, *Vibrio anguillarum*, *Aeromonas salmonicida* and *Streptococcus phocae* in Chilean marine farms. *Diseases of Aquatic Organisms* 97, 135–142.

Tiainen, R., Pedersen, K. and Larsen, J.L. (1995) Ribotyping and plasmid profiling of *Vibrio anguillarum* serovar O2 and *Vibrio ordalii*. *Journal of Applied Bacteriology* 79, 384–392.

Toranzo, A.E. and Barja, J.L. (1990) A review of the taxonomy and seroepizootiology of *Vibrio anguillarum*, with special reference to aquaculture in the northwest of Spain. *Diseases of Aquatic Organisms* 9, 73–82.

Toranzo, A.E. and Barja, J.L. (1993) Virulence factors of bacteria pathogenic for cold water fish. *Annual Review of Fish Diseases* 3, 5–36.

Toranzo, A.E., Baya, A., Roberson, B.S., Barja, J.L., Grimes, D.J. *et al.* (1987) Specificity of slide agglutination test for detecting bacterial fish pathogens. *Aquaculture* 61, 81–97.

Toranzo, A.E., Santos, Y. and Barja, J.L. (1997) Immunization with bacterial antigens: *Vibrio* infections. In: Gudding, R., Lillehaug, A., Midtlyng, P.J. and Brown, F. (eds) *Fish Vaccinology*. Karger, Basel, Switzerland, pp. 93–105.

Toranzo, A.E., Magariños, B. and Romalde, J.L. (2005) A review of the main bacterial fish diseases in mariculture systems. *Aquaculture* 246, 37–61.

Toranzo, A.E., Avendaño-Herrera, R., Lemos, M.L. and Osorio, C.R. (2011) Vibriosis. In: Avendaño-Herrera, R. (ed.) *Enfermedades Infecciosas del Cultivo de Salmónidos en Chile y el Mundo*. NIVA, Puerto Varas, Chile, pp. 133–159.

Totland, G.K., Nylund, A. and Holm, K.O. (1988) An ultrastructural study of morphological changes in Atlantic salmon, *Salmo salar* L., during the development of cold water vibriosis. *Journal of Fish Diseases* 11, 1–13.

Trust, T.J., Courtice, I.D., Khouri, A.G., Crosa, J.H. and Schiewe, M.H. (1981) Serum resistance and hemagglutination ability of marine vibrios pathogenic to fish. *Infection and Immunity* 34, 702–707.

Turnidge, J. and Paterson, D.L. (2007) Setting and revising antibacterial susceptibility breakpoints. *Clinical Microbiology Reviews* 20, 391–408.

Urbanczyk, H., Ast, J.C., Higgins, M.J., Carson, J. and Dunlap, P.V. (2007) Reclassification of *Vibrio fischeri*, *Vibrio logei*, *Vibrio salmonicida* and *Vibrio wodanis* as *Aliivibrio fischeri* gen. nov., comb. nov., *Aliivibrio salmonicida* comb nov. and *Aliivibrio wodanis* comb. nov. *International Journal of Systematic and Evolutionary Microbiology* 57, 2823–2829.

Wang, Y., Xu, Z., Jia, A., Chen, J., Mo, Z. and Zhang, X. (2009) Genetic diversity between two *Vibrio anguillarum* strains exhibiting different virulence by suppression subtractive hybridization. *Wei Sheng Wu Xue Bao* (*Acta Microbiologica Sinica*) 49, 363–371.

Wards, B.J. and Patel, H.H. (1991) Characterization by restriction endonuclease analysis and plasmid profiling of *Vibrio ordalii* strains from salmon (*Oncorhynchus tshawytscha* and *Oncorhynchus nerka*) with vibriosis in New Zealand. *Journal of Marine and Freshwater Research* 25, 345–350.

Weber, B., Croxatto, A., Chen, C. and Milton, D.L. (2008) RpoS induces expression of the *Vibrio anguillarum* quorum-sensing regulator VanT. *Microbiology* 154, 767–780.

Weber, B., Chen, C. and Milton, D.L. (2010) Colonization of fish skin is vital for *Vibrio anguillarum* to cause disease. *Environmental Microbiology Reports* 21, 133–139.

Welch, T.J. and Crosa, J.H. (2005) Novel role of lipopolysaccharide O1 side chain in ferric siderophore transport and virulence of *Vibrio anguillarum*. *Infection and Immunity* 73, 5864–5872.

Wiik, R., Stackebrandt, E., Valle, O., Daae, F.L., Rødseth, O.M. *et al.* (1995) Classification of fish-pathogenic vibrios based on comparative 16S rRNA analysis. *International Journal of Systematic Bacteriology* 45, 421–428.

Winkelmann, G., Schmid, D.G., Nicholson, G., Jung, G. and Colquhoun, D.J. (2002) Bisucaberin: a dihydroxamate siderophore isolated from *Vibrio salmonicida*, an important pathogen of farmed Atlantic salmon (*Salmo salar*). *Biometals* 15, 153–160.

Wolf, M.K. and Crosa, J.H. (1986) Evidence for the role of a siderophore in promoting *Vibrio anguillarum* infections. *Journal of General Microbiology* 132, 2949–2952.

Xiao, P., Mo, Z.L., Mao, Y.X., Wang, C.L., Zou, Y.X. and Li, J. (2009) Detection of *Vibrio anguillarum* by PCR amplification of the *empA* gene. *Journal of Fish Diseases* 32, 293–296.

Yip, E.S., Geszvain, K., DeLoney-Marino, C.R. and Visick, K.L. (2006) The symbiosis regulator *rscS* controls the *syp* gene locus, biofilm formation and symbiotic aggregation by *Vibrio fischeri*. *Molecular Microbiology* 62, 1586–1600.

Yu, L.P., Hu, Y.H., Sun, B.G. and Sun, L. (2012) C312M: an attenuated *Vibrio anguillarum* strain that induces immunoprotection as an oral and immersion vaccine. *Diseases of Aquatic Organisms* 102, 33–42.

Zhang, Z., Wu, H., Xiao, J., Wang, Q., Liu, Q. *et al.* (2013) Immune response evoked by infection with *Vibrio anguillarum* in zebrafish bath-vaccinated with a live attenuated strain. *Veterinary Immunology and Immunopathology* 154, 138–144.

24 *Weissella ceti*

Timothy J. Welch,[1]* David P. Marancik[2] and Christopher M. Good[3]

[1]US Department of Agriculture Agricultural Research Service, National Center for Cool and Cold Water Aquaculture, Kearneysville, West Virginia, USA; [2]Department of Pathobiology, St. George's University of Veterinary Medicine, True Blue, St. George's, Grenada, West Indies; [3]The Conservation Fund's Freshwater Institute, Shepherdstown, West Virginia, USA

24.1   Introduction

Weissella species are usually not associated with disease; however, novel strains of *Weissella ceti* were recently recognized as pathogens for rainbow trout (*Oncorhynchus mykiss*). *W. ceti* was identified in 2007 at a commercial rainbow trout farm in China (Liu *et al.*, 2009) and later in farmed rainbow trout within Brazil (Figueiredo *et al.*, 2012; Costa *et al.*, 2015) and North Carolina (Welch and Good, 2013). Genome sequences of representative US and Brazilian strains show high genetic similarity despite the lack of epizootic linkages between the farms and countries concerned (Figueiredo *et al.*, 2015). The origins of the bacteria associated with these outbreaks are unknown, but the occurrence of *W. ceti* on three continents suggests that it is an emerging pathogen.

W. ceti is a Gram-positive, catalase and oxidase negative and non-motile rod that is 1.5 μm long by 0.30 μm in diameter (Fig. 24.1). The bacterium forms small (0.25 mm) white α-haemolytic colonies on trypticase soya agar (TSA, Remel) supplemented with 5% sheep blood (Fig. 24.1). Thus far, weissellosis has been reported only among intensively cultured rainbow trout, although the pathogen could be missed in health surveys of wild and cultured fish owing to its fastidious nature and its predisposition to infect the brain, a tissue that is typically not sampled in fish health surveys. Weissellosis appears to predominantly affect fish of 0.25–1 kg (Figueiredo *et al.*, 2012; Welch and Good, 2013), so this disease can cause severe economic losses. This propensity to affect larger fish is consistent with experimental infections, which have established a correlation between body weight and mortality (Marancik *et al.*, 2013). Elevated temperature was a critical predisposing factor for outbreaks in North Carolina (Welch and Good, 2013), where weissellosis was only detected when temperatures exceeded 16°C. The disease subsided during the autumn when temperatures cooled, and was absent during the winter. Elevated water temperature was also critical in the Chinese and Brazilian outbreaks (Figueiredo *et al.*, 2012; Liu *et al.*, 2009). The route of infection and reservoirs of infection are under investigation.

24.2   Clinical Signs and Diagnosis

Liu *et al.* (2009) made the first report of weissellosis in a commercial rainbow trout farm in China. Diseased trout were septicaemic, with symptoms that included haemorrhages in the eyes and vent; internal examination confirmed widespread tissue involvement, including haemorrhages in the intestinal tract and peritoneal wall, as well as petechiae in the liver. The outbreak lasted for more than 2 months and cumulative mortality was about 40%. Subsequently, Figueiredo *et al.* (2012) observed weissellosis outbreaks in rainbow trout at five Brazilian farms; clinical signs included anorexia, after which fish became lethargic and exhibited exophthalmia, ascites and gross haemorrhages in the eyes and mouth. Most of the fish affected were near market

---

*Corresponding author e-mail: tim.welch@ars.usda.gov

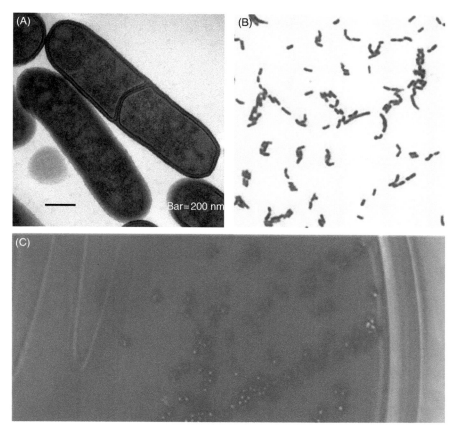

Fig. 24.1. (A) Transmission electron micrograph and (B) light micrograph of Gram-positive stained cells of *Weissella ceti* NC36; (C) shows *W. ceti* NC36 colonies displaying α-haemolytic activity on trypticase soya agar supplemented with 5% sheep blood after 18 h at 30°C.

size (250 g average weight), although in one farm the affected fish ranged from fingerlings to brood stock. Fish mortality was high (50–80%) and occurred approximately 4–5 days after clinical signs were noted. Welch and Good (2013) reported weissellosis outbreaks at two farms in North Carolina; the affected fish were of 0.5–1 kg and mortalities were high, at approximately 2000 fish a day. Clinical signs included lethargy at the water surface, darkened coloration, bilateral exophthalmia, corneal opacity, ocular haemorrhage and occasional corneal rupture (Fig. 24.2).

Laboratory studies with *W. ceti* have largely replicated the clinical signs observed during natural outbreaks, including high mortality in experimentally infected rainbow trout, Nile tilapia (*Oreochromis niloticus*) (Figueredo *et al.* 2012), and crucian carp (*Carassius auratus gibelio*) (Liu *et al.*, 2009). Rainbow

trout infected intraperitoneally displayed anorexia, lethargy, ascites, exophthalmia, unusual swimming, rectal prolapse, and haemorrhages of the eyes, fins and vent (Figueredo *et al.*, 2012). Marancik *et al.* (2013) reported similar clinical signs in experimentally infected trout. Together, these studies indicated that the clinical signs of weissellosis were consistently similar in natural and experimental infections.

Haemorrhagic septicaemia occurs in many bacterial infections, and consequently is not pathognomonic for weissellosis. The culture media used for primary isolation of the bacterium include TSA modified with 5% sheep blood and brain heart infusion agar (BHI, Remel) (Liu *et al.*, 2009; Figueiredo *et al.*, 2012; Welch and Good, 2013). *W. ceti* can be isolated from the spleen, anterior kidney and brain. Bacterial loads are often highest in the brain and, in some cases, the pathogen can only be isolated from this organ.

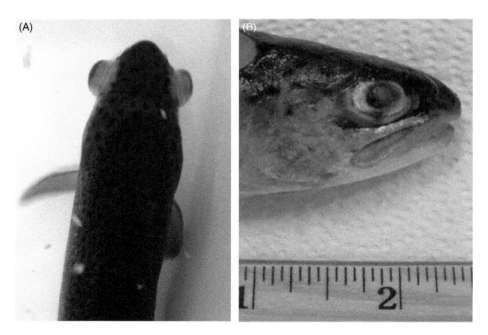

Fig. 24.2. Rainbow trout exhibiting gross ocular signs typical of weissellosis, including: (A) bilateral exophthalmia; and (B) ruptured cornea and intraocular haemorrhage.

Detectable growth and α-haemolysis develop within 15–18 h at 30°C after incubation on TSA–blood. This rapid growth is unique among salmonid bacteria. In the initial reports, it was necessary to confirm diagnosis using phenotypic characterization and 16S rRNA gene sequencing, but PCR-based methods have now been developed that allow for rapid identification of *W. ceti* in bacteriological cultures and/or directly from infected tissues (Snyder *et al.*, 2015). These recent methods (conventional and quantitative PCR) rely on PCR amplification of putative platelet and collagen adhesion genes in the genome sequence of the US trout strain NC36 (Ladner *et al.*, 2013). The usefulness of these assays was confirmed in field evaluations at farms in North Carolina where the pathogen was present. Further work is necessary to verify the specificity of these PCR-based methods for *W. ceti* isolates from other geographic locations.

24.3 Pathology and Pathophysiology

Histopathological lesions have only been described in experimentally infected rainbow trout (Figueiredo *et al.*, 2012; Marancik *et al.*, 2013). Similar to the gross signs of the disease, the microscopic changes are largely consistent with a bacte-

rial septicaemia. The widespread haemorrhaging in internal organs is associated with vasculitis (Fig. 24.3), characterized with transmural infiltration of the blood vessels by macrophages, lymphocytes and neutrophils, and degeneration of the vascular tunica and elastic membrane (Marancik *et al.*, 2013). Granulomatous inflammation and necrosis are also found within the retrobulbar region of the eye (with progression to granulomatous panophthalmitis and corneal ulceration), the epicardium and myocardium, and the brain (Figueiredo *et al.*, 2012; Marancik *et al.*, 2013).

The pathophysiological effects have not been well studied. Anecdotal observations indicate that exophthalmia and subsequent blindness affects feeding. Genome sequencing of the Brazilian and US isolates suggests that *W. ceti* possesses virulence factors that potentially influence the health of infected fish and the disease process. (Ladner *et al.*, 2013; Figueiredo *et al.*, 2014a,b). For example, homologues to haemolysins, collagen binding proteins, a platelet-associated adhesion protein, and a mucus-binding protein were detected, which might contribute to adherence to the host tissue and invasion of that tissue by the bacteria. However, the influences of these factors on the disease process,

T.J. Welch *et al.*

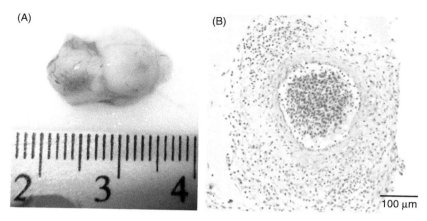

Fig. 24.3. (A) *Weissella ceti*-caused cerebral haemorrhage, (B) most likely associated with vasculitis, as demonstrated histologically by the transmural infiltration of a medium-sized artery with mononuclear cells and neutrophils (20× magnification, stained with haematoxylin and eosin).

including vasculitis and damage to vital organs such as the brain and heart, have not yet been determined.

24.4 Protective and Control Strategies

Under *in vitro* conditions, the Chinese *W. ceti* isolates were sensitive to tetracycline and chloramphenicol but resistant to sulfamethoxazole/trimethoprim and norfloxacin. Prior to that analysis, diseased fish were treated with sulfamethoxazole/trimethoprim and norfloxacin without success. When the treatment was changed to oxytetracycline, mortality abated (Liu *et al.*, 2009). The Brazilian strains were resistant to sulfonamides but most were susceptible to oxytetracycline and florfenicol under *in vitro* conditions (Figueiredo *et al.*, 2012). Antimicrobial susceptibility profiles have not been determined for the US strains and there were no attempts made to treat the disease (Welch and Good, 2013). The Brazilian and US isolates have several genes encoding putative antibiotic resistance proteins and multidrug efflux pumps (Figueiredo *et al.*, 2015). Currently, there are no drugs approved for treatment of weissellosis in food fish.

Intraperitoneal injection with formalin-killed *W. ceti* in an aqueous formulation provided significant protection against experimental infections with the North Carolina strain (Welch and Good, 2013). A relative percentage protection (RPS) of 87.5% was observed at 38 days (608 degree days) and an 85% RPS at 72 days (1152 degree days) postvaccination. This *W. ceti* bacterin was equally effective when combined with a commercially

available *Yersinia ruckeri* vaccine (Novartis Animal Health) and bivalent delivery did not alter the efficacy of the *Y. ruckeri* component under laboratory conditions (Welch and Good, 2013). This is important, because North Carolina trout farms routinely vaccinate to control *Y. ruckeri*, and delivering the *W. ceti* and *Y. ruckeri* antigens together does not increase either the production costs or the handling stress associated with multiple vaccinations. The efficacy of the bivalent vaccine was verified by challenging vaccinated fish approximately 12 months after vaccination and comparing mortality with similarly aged, non-vaccinated controls (Welch and Hinshaw, unpublished). Currently, approximately 4 million fish are vaccinated annually in North Carolina with the bivalent product, and since this vaccination programme was initiated in 2012, the pathogen has not been detected at the farms where the initial outbreaks occurred (Welch, unpublished). These vaccines are produced by a custom vaccine manufacturer and are applied by veterinary prescription. In contrast, Costa *et al.* (2015) reported that the killed whole-cell aqueous-based vaccine was ineffective (RPS 58%) against the Brazilian strain of *W. ceti*, although an oil-adjuvanted formulation conferred a high level of protection (RPS 92%).

24.5 Conclusions and Suggestions for Future Research

W. ceti is an emerging pathogen of farmed rainbow trout and much is not known about the biology of the bacterium and the disease. Weissellosis can become a major problem for the aquaculture industry. Future research should include studies of: (i) the range of fish species that can subclinically carry the pathogen and/or develop clinical weissellosis; (ii) how and where the pathogen can persist and overwinter in the environment prior to reinfecting fish during warmer periods; (iii) the origins of the pathogen and the epizootiological pathways that have facilitated its rapid dissemination to highly disparate locations; (iv) elucidation of pathogen transmission and its mechanisms of pathogenesis; (v) methods for the prevention and containment of the disease in hatcheries; and (vi) the humoral and cell-mediated responses of trout to both the pathogen and the vaccine. Because *W. ceti* is not broadly disseminated in the USA, the trout industry should restrict the movement of infected lots beyond contaminated facilities.

References

Costa, F.A., Leal, C.A., Schuenker, N.D., Leite, R.C. and Figueiredo, H.C. (2015) Characterization of *Weissella ceti* infections in Brazilian rainbow trout, *Oncorhynchus mykiss* (Walbaum), farms and development of an oil-adjuvanted vaccine. *Journal of Fish Disease* 38, 295–302.

Figueiredo, H.C., Costa, F.A., Leal, C.A., Carvalho-Castro, G.A. and Leite, R.C. (2012) *Weissella* sp. outbreaks in commercial rainbow trout (*Oncorhynchus mykiss*) farms in Brazil. *Veterinary Microbiology* 156, 359–366.

Figueiredo, H.C., Leal, G., Pereira, F.L., Soares, S.C., Dorella, F.A. *et al.* (2014a) Whole-genome sequence of *Weissella ceti* strain WS08, isolated from diseased rainbow trout in Brazil. *Genome Announcement* 2(4): e00851-14.

Figueiredo, H.C., Leal, C.A., Dorella, F.A., Carvalho, A.F., Soares, S.C. *et al.* (2014b) Complete genome sequences of fish pathogenic *Weissella ceti* strains WS74 and WS105. *Genome Announcement* 2(5): e01014-14.

Figueiredo, H.C., Soares, S.C., Pereira, F.L., Dorella, F.A., Carvalho, A.F. *et al.* (2015) Comparative genome analysis of *Weissella ceti*, an emerging pathogen of farm-raised rainbow trout. *BMC Genomics* 16:1095.

Ladner, J.T., Welch, T.J., Whitehouse, C.A. and Palacios, G.F. (2013) Genome sequence of *Weissella ceti* NC36, an emerging pathogen of farmed rainbow trout in the United States. *Genome Announcement* 1(1): e00187-12.

Liu, J.Y., Li, A.H., Ji, C. and Yang, W.M. (2009) First description of a novel *Weissella* species as an opportunistic pathogen for rainbow trout *Oncorhynchus mykiss* (Walbaum) in China. *Veterinary Microbiology* 136, 314–320.

Marancik, D.P., Welch, T.J., Leeds, T.D. and Wiens, G.D. (2013) Acute mortality, bacterial load, and pathology of select lines of adult rainbow trout challenged with *Weissella* sp. NC36. *Journal of Aquatic Animal Health* 25, 230–236.

Snyder, A.K., Hinshaw, J.M. and Welch, T.J. (2015) Diagnostic tools for rapid detection and quantification of *Weissella ceti* NC36 infections in rainbow trout. *Letters in Applied Microbiology* 60, 103–110.

Welch, T.J. and Good, C.M. (2013) Mortality associated with weissellosis (*Weissella* sp.) in USA farmed rainbow trout: potential for control by vaccination. *Aquaculture* 388, 122–127.

25 Yersinia ruckeri

MICHAEL ORMSBY AND ROBERT DAVIES*

Institute of Infection, Immunity and Inflammation, College of Medical, Veterinary and Life Sciences, University of Glasgow, Glasgow, UK

25.1 Introduction

Yersinia ruckeri is a Gram-negative member of the *Enterobacteriaceae* and causes enteric redmouth (ERM) disease or yersiniosis of salmonids. Since its isolation in the USA and Canada (Ross *et al.*, 1966; Bullock *et al.*, 1978; Busch, 1978; Stevenson and Daly, 1982), *Y. ruckeri* has also been detected in Europe, South America, Africa, Asia and Australasia (Horne and Barnes, 1999). The disease was first isolated in rainbow trout (*Oncorhynchus mykiss*) but it has since been recovered from both salmonid and non-salmonid fishes (Horne and Barnes, 1999; Carson and Wilson, 2009). *Y. ruckeri* has increasingly become an important pathogen of Atlantic salmon (*Salmo salar*) in Australia (Carson and Wilson, 2009), Chile (Bastardo *et al.*, 2011), Norway (Shah *et al.*, 2012) and Scotland (Ormsby *et al.*, 2016), where salmon-production is economically important. The pathogen also infects other farmed species such as channel catfish (*Ictalurus punctatus*) (Danley *et al.*, 1999), sturgeons (*Acipenser baeri* and *A. schrenckii*) (Vuillaume *et al.*, 1987; Shaowu *et al.*, 2013) and whitefish (*Coregonus peled*) (Rintamaki *et al.*, 1986).

25.2 Enteric Redmouth Disease

The clinical signs, pathology and histopathology of ERM in rainbow trout and Atlantic salmon have been reviewed (Tobback *et al.*, 2007; Carson and Wilson, 2009; Kumar *et al.*, 2015). Briefly, the disease is an acute infection, primarily of rainbow trout, that is typically characterized by subcutaneous haemorrhages in and around the oral cavity (Fig. 25.1A) and at the base of the pectoral and pelvic fins and along the flanks (Fig. 25.1B); haemorrhaging of the gill filaments may also occur.

A less severe form, known as yersiniosis or 'salmon blood spot disease', is recognized in Atlantic salmon (Carson and Wilson, 2009; Costa *et al.*, 2011). In addition to subcutaneous haemorrhages in and around the oral cavity and at the bases of the fins, yersiniosis is also characterized by patches ('blood spots') of haemorrhagic congestion on the iris of the eye (Fig. 25.1C) and with a marked unilateral or bilateral exophthalmos (Fig. 25.1D).

The bacterium causes a general septicaemia with an inflammatory response and bacterial colonization is common of the kidneys, spleen, heart, liver and gills, and in areas of petechial haemorrhages (Rucker, 1966). Pathological changes occur in the gills, including hyperaemia, oedema and desquamation of epithelial cells in the secondary lamellae (Horne and Barnes, 1999; Tobback *et al.*, 2007). Focal necroses in the liver, kidneys and spleen are also common. In Atlantic salmon, the histopathological features are those of a typical septicaemia; bacteria are readily detected in the blood and in circulating macrophages, and may also localize at sites of tissue haemorrhage (Carson and Wilson, 2009). Infection by *Y. ruckeri* of Atlantic salmon smolts during acclimatization stress is characterized by bacteria in the blood, with congestion, haemorrhage and tissue localization (Carson and Wilson, 2009).

The pathogen is spread via direct contact with infected fish or through water, either from infected or carrier fish. A carrier state for *Y. ruckeri* was first recognized by Rucker (1966). Intestinal shedding of the pathogen leads to recurrent infection and mortality on a cyclical basis, and periodic shedding is crucial to the spread of disease (Rucker, 1966; Busch and Lingg, 1975). The bacterium can survive for >100 days in river water, lake water, estuarine water and sediments at temperatures of 6 and 18°C

*Corresponding author e-mail: Robert.Davies@glasgow.ac.uk

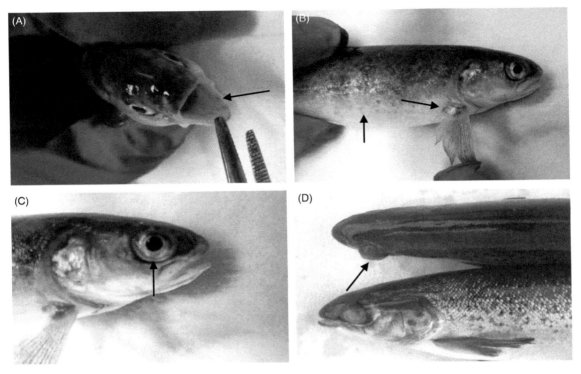

Fig. 25.1. Clinical features of enteric redmouth disease/yersiniosis in an experimentally infected Atlantic salmon showing: haemorrhaging (arrows) of (A) the oral cavity, (B) the base of the pectoral fin and flank, and (C) the eye; with (D) exophthalmos of the eye (arrow).

(Romalde *et al.*, 1994), although survival is greatly reduced at higher (35%) salinities (Thorsen *et al.*, 1992). Biofilms may be a source of recurrent infection in fish farms (Coquet *et al.*, 2002). The gills are an important portal of entry, from which the bacteria spread to other organs (Tobback *et al.*, 2009; Ohtani *et al.*, 2014). However, the gills and gut are of equal importance as entry points and in the initial interactions of *Y. ruckeri* with its host (Tobback *et al.*, 2010a).

25.3 Diagnosis

Diagnostic assays include culture-based, serological and molecular biological techniques. Bacteria are normally isolated from the head kidney, spleen or sites of haemorrhage (Carson and Wilson, 2009), but they may also be cultured from the posterior intestine to determine carrier status (Busch and Lingg, 1975; Rodgers, 1992). Selective media, including Waltman–Shotts (Waltman and Shotts, 1984) and ROD (ribose ornithine deoxycholate) agar (Rodgers, 1992) are useful for discrimination. The identification of *Y. ruckeri*

can be confirmed using biochemical characterization, and key characteristics are described in Davies and Frerichs (1989), and Austin and Austin (2012). Serological techniques for the rapid diagnosis of *Y. ruckeri* include enzyme-linked immunosorbent assay (ELISA) (Cossarini-Dunier, 1985), immunofluorescence antibody technique (IFAT) (Smith *et al.*, 1987) and commercially available agglutination kits (Romalde *et al.*, 1995). Various polymerase chain reaction (PCR)-based assays have been developed for direct detection of the bacterium from the blood or other tissues of both infected fish and subclinical carriers (Gibello *et al.*, 1999; Altinok *et al.*, 2001; Temprano *et al.*, 2001; Saleh *et al.*, 2008; Seker *et al.*, 2012). In particular, real-time PCR assays have been developed for rapid detection and confirmation (Bastardo *et al.*, 2012a; Keeling *et al.*, 2012).

25.4 Strain Differentiation

Strain differentiation of *Y. ruckeri* is important in disease surveillance and control. The strain that is concerned also has an impact on the pathobiology because different strains vary in their host specificity and virulence (Davies, 1991c; Haig *et al.*, 2011; Calvez *et al.*, 2014). For example, serotype O1 isolates are responsible for the majority of disease outbreaks in rainbow trout worldwide, while non-O1 serotypes are more commonly associated with other fishes. Phenotypic typing approaches such as biotyping, serotyping and outer membrane protein (OMP)-typing are commonly used to characterize and differentiate the strains of *Y. ruckeri* (Tobback *et al.*, 2007; Bastardo *et al.*, 2012b; Kumar *et al.*, 2015). These typing methods utilize variations in structures or molecules on the bacterial cell surface: biotyping assesses variation in the presence or absence of flagella; serotyping assesses variation in surface lipopolysaccharide (LPS); and OMP typing assesses variation in the molecular mass of specific 'major' OMPs. The bacterial surface is at the interface between the pathogen and host, and plays an important role in pathogenesis. Variation in cell-surface structures and molecules such as flagella, LPS and OMPs is driven by selective pressure, including the host immune response, and these structures and molecules are important markers for strain differentiation. Molecular methods have also been used to assess genetic variation within *Y. ruckeri* and these have allowed important insights of the population biology and epizootiology of this pathogen.

25.4.1 Biotyping

Y. ruckeri has two biotypes: biotype 1 strains are motile (they have flagella) and lipase positive (they hydrolyse Tween), whereas biotype 2 strains are negative for both of these characteristics (Fig. 25.2). Since the identification of biotype 2 isolates in the UK (Davies and Frerichs, 1989), non-motile variants have been widely detected in Europe (Horne and Barnes, 1999; Wheeler *et al.*, 2009) and they have also been identified in the USA (Arias *et al.*, 2007; Welch *et al.*, 2011) and in Australia (Carson and Wilson, 2009). Biotype 2 strains are of increased concern because they cause disease in fish vaccinated against biotype 1 strains (Tinsley *et al.*, 2011a).

25.4.2 Serotyping

Serological classification is important for strain differentiation. After the first identification of serotypes I and II (O'Leary, 1977), additional serotypes were identified subsequently (Bullock *et al.*, 1978;

Stevenson and Airdrie, 1984; Pyle and Schill, 1985; Daly *et al.*, 1986; Pyle *et al.*, 1987). Unfortunately, confusion arose because serotype designations were assigned which, in most cases, did not match those of the other schemes, as demonstrated in Table 25.1. To resolve the confusion, Davies (1990) developed a simple, rapid-slide agglutination assay that was based on the O-antigen and, crucially, used key reference strains common to the previously published schemes. A serotyping scheme was proposed that recognized five O-serotypes designated O1, O2, O5, O6 and O7 (Table 25.1) and this scheme has been widely adopted (Strom-Bestor *et al.*, 2010; Tinsley *et al.*, 2011a; Shah *et al.*, 2012; Calvez *et al.*, 2014). Subsequently, Romalde *et al.* (1993) proposed another scheme that recognized four O-serotypes. This scheme introduces further complexity because it utilizes membrane protein profiles and proteins in extracellular products (ECPs); the scheme also fails to adequately distinguish between the O2 and O7 serotypes because these are classified under the same O2 umbrella (see Table 25.1). Nevertheless, this approach identified a serotype O1 subtype (O1b) that is found in Australian (Carson and Wilson, 2009) and Chilean (Bastardo *et al.*, 2011) Atlantic salmon. More recently, a new O-serotype, designated serotype O8 (Ormsby *et al.*, 2016), was identified from Scottish Atlantic salmon. The new LPS-type has a polysaccharide core region similar to that of serotype O1 LPS, but it has a unique O-antigen side chain (Fig. 25.3).

25.4.3 OMP typing

An OMP-typing scheme, based on variations in the molecular mass of the major OMPs (OmpA, OmpC and OmpF), was developed to further discriminate *Y. ruckeri* strains (Davies, 1989, 1991a). OmpA is distinguished from OmpC and OmpF by its unique heat-modification properties, and OmpC can be distinguished from OmpF because it is expressed under anaerobic but not aerobic growth conditions (Fig. 25.4). Comparative OMP profiling has played an important role in various epizootiological studies (Davies, 1991a; Romalde *et al.*, 1993; Sousa *et al.*, 2001; Wheeler *et al.*, 2009; Bastardo *et al.*, 2011, 2012b; Tinsley *et al.*, 2011a). Bastardo *et al.* (2012b) concluded that the greatest discrimination was obtained when three or more typing methods were employed, especially biotyping with the API 20E system (an analytical profile index for *Enterobacteriaceae* from bioMérieux), LPS

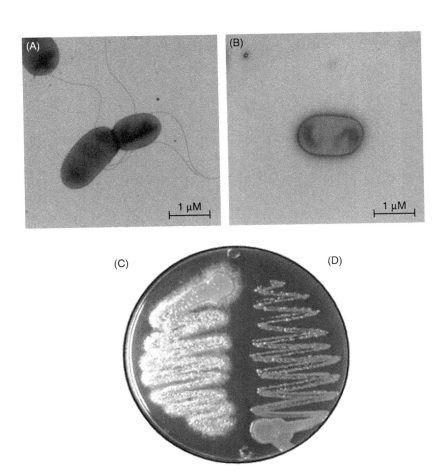

Fig. 25.2. The presence of flagella and Tween hydrolysis in representative biotype 1 and 2 isolates of *Y. ruckeri*. (A) shows the presence of flagella in a motile isolate of *Y. ruckeri* and (B) their absence in a non-motile isolate as demonstrated by transmission electron microscopy. The corresponding (C) Tween-positive and (D) Tween-negative hydrolysis results on tryptone soya agar supplemented with Tween 80 are also shown. The motile and Tween-positive isolate (A and C) represents biotype 1; the non-motile and Tween-negative isolate (B and D) represents biotype 2. Copyright © American Society for Microbiology, Applied and Environmental Microbiology 82, 5785–5794, 2016.

Table 25.1. Comparison of *Yersinia ruckeri* serotyping schemes. The serotype schemes and designations shown are based on: [a]whole-cell antigens; [b]O-antigens; [c]O-antigens and extracellular products.

Reference							
O'Leary, 1977[a]	Bullock *et al.*, 1978[a]	Stevenson and Airdrie, 1984[a]	Pyle and Schill, 1985[a]	Daly *et al.*, 1986[a]	Pyle *et al.*, 1987[a]	Davies, 1990[b]	Romalde *et al.*, 1993[c]
I	I	I	–	I	1	O1	O1a
II	II	II	2	II	2	O2	O2a, b, c
–	III	III	–	III	6	O1	O1b
–	–	IV	–	–	–	–	–
–	–	V	6	V	5	O5	O3
–	–	VI	5	VI	4	O6	O4
–	–	–	4	–	3	O7	–

M. Ormsby and R. Davies

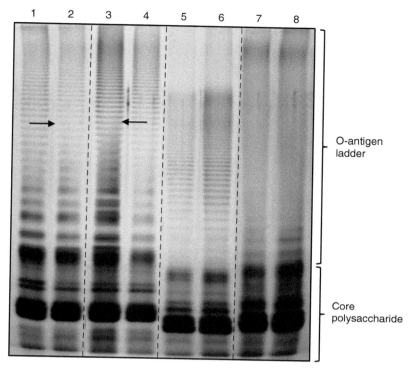

Fig. 25.3. Representative lipopolysaccharide (LPS) profiles of serotype O1, O2, O5 and O8 isolates of *Yersinia ruckeri*. Pairs of isolates representing serotypes O1 (lanes 1 and 2), O8 (lanes 3 and 4), O2 (lanes 5 and 6) and O5 (lanes 7 and 8) are shown. The core polysaccharide and O-antigen regions of the LPS molecule are labelled. The LPS of serotype O1 (lanes 1 and 2) and O8 (lanes 3 and 4) isolates share a common core polysaccharide region but have slightly different O-antigen ladder patterns. Arrows indicate the differing mobilities of O-antigen units in serotype O1 and O8 LPS. Copyright © American Society for Microbiology, Applied and Environmental Microbiology 82, 5785–5794, 2016.

serotyping and OMP profiling. Based on biotype, serotype and OMP type, distinct clonal groups were identified (Davies, 1991b) that differed in their virulence and host-specificity (Davies, 1991c). These clonal groups are considered as different 'pathotypes' and are likely to differ in their pathobiology.

25.4.4 Molecular typing

Plasmid profiling (De Grandis and Stevenson, 1982; Garcia *et al.*, 1998), ribotyping (Lucangeli *et al.*, 2000), repetitive sequence-based PCR (Huang *et al.*, 2013), pulsed field gel electrophoresis (PFGE) (Wheeler *et al.*, 2009; Strom-Bestor *et al.*, 2010; Huang *et al.*, 2013; Calvez *et al.*, 2015), multilocus enzyme electrophoresis (MLEE) (Schill *et al.*, 1984) and multilocus sequence typing (MLST) (Bastardo *et al.*, 2012c; Calvez *et al.*, 2015) have all been used

to assess genetic variation within *Y. ruckeri*. Schill *et al.* (1984) demonstrated that the pathogen is extremely homogeneous. More recently, Bastardo *et al.* (2012a), using MLST, confirmed that the pathogen has a relatively low degree of genetic diversity even though the 103 strains analysed were from diverse geographic locations and host species. Recombination rather than mutation accounted for most of this genetic variation, leading the authors to suggest that *Y. ruckeri* has an epizootic population structure. The authors also identified associations between specific sequence types (STs) and host fish species, and suggested that this might be due to adaptive niche specialisation. Calvez *et al.* (2015) confirmed the general homogeneity of *Y. ruckeri* and also demonstrated that strains from different host species are genetically distinct. They concluded that movements of farmed fish, geographic

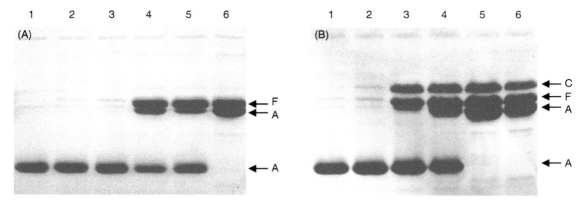

Fig. 25.4. The identification of the outer membrane proteins OmpA, OmpC and OmpF of *Yersinia ruckeri* by SDS-PAGE using different solubilization temperatures and aeration conditions. *Y. ruckeri* was grown under (A) aerobic and (B) anaerobic growth conditions. The labels on the right of each panel are: A, OmpA protein; C, OmpC protein; and F, OmpF protein. Lanes 1 to 6 represent solubilization temperatures of 50, 60, 70, 80, 90 and 100°C, respectively. The heat-modifiable protein, OmpA, can be differentiated from OmpC and OmpF because OmpA undergoes a characteristic shift in molecular mass, from approximately 30 to 40 kDa, when heated at 90 or 100°C. OmpC can be distinguished from OmpF by its known behaviour under different conditions of aeration; it is expressed under anaerobic growth conditions (B) but not under aerobic growth conditions (A). Copyright © American Society for Microbiology, Applied and Environmental Microbiology 82, 5785–5794, 2016.

origins and ecological niches are involved in the dissemination and evolution of *Y. ruckeri*.

25.5 Virulence Factors and Pathobiology

Relatively little is known about the pathobiology of infection and the roles of specific virulence factors in the disease process. Putative virulence factors of *Y. ruckeri* are related to adherence and invasion, iron uptake, and intracellular survival and immune evasion, or by virtue of being components of the ECPs.

25.5.1 Adherence and invasion

The adherence and invasion of *Y. ruckeri* have been demonstrated in cell lines (Romalde and Toranzo, 1993; Kawula *et al.*, 1996; Tobback *et al.*, 2010b), and in gill and gut tissue (Tobback *et al.*, 2010a,b). However, the specific molecules and mechanisms involved in these processes have not been identified.

25.5.2 Iron uptake

Pathogenic bacteria have evolved a range of mechanisms to acquire iron. One mechanism involves the secretion of siderophores that chelate iron with an extremely high affinity (Faraldo-Gómez and Sansom, 2003). Davies (1991d) demonstrated that growth under iron-limited conditions upregulated four high molecular mass OMPs of 66, 68, 69.5 and 72 kDa, although siderophore production was not detected. Romalde *et al.* (1991) identified three iron-regulated OMPs but, in contrast, were able to detect a siderophore in iron-limited media. Subsequently, a catechol-type siderophore, termed ruckerbactin, was identified and shown to be involved in iron acquisition *in vitro* (Fernández *et al.*, 2004); ruckerbactin was also demonstrated to have a potential role in the pathogenesis of *Y. ruckeri*.

25.5.3 Intracellular survival and immune evasion

Intracellular survival and immune evasion are not well understood but play potentially important roles in the pathogenesis of ERM (Tobback *et al.*, 2009). Ryckaert *et al.* (2010) showed that *Y. ruckeri* survives and replicates within trout macrophages *in vitro*. However, Tobback *et al.* (2010b) demonstrated that the bacterium does not multiply or survive within cultured cell lines. Also, the greater number of extracellular bacteria compared to intracellular bacteria in tissue sections led these authors to conclude that intracellular survival is of minor importance in the pathogenesis of *Y. ruckeri*.

Tobback et al. (2009) demonstrated that the gills are an important portal of entry for both virulent and avirulent strains, but the subsequent persistence of virulent strains in internal organs, and not that of avirulent strains, led to the conclusion that immune evasion is a major virulence factor of this pathogen. Davies (1991c) demonstrated a relationship between the serum resistance of *Y. ruckeri* and its virulence in rainbow trout. LPS type was a determinant in providing resistance to complement-mediated lysis, but other factors could also be involved. Tobback et al. (2010b) confirmed that serum resistance is important to the extracellular survival of the pathogen.

25.5.4 Extracellular products

Romalde and Toranzo (1993) demonstrated that injection of *Y. ruckeri* ECPs into rainbow trout elicited some of the characteristic signs of ERM and concluded that ECPs are likely to play important roles in the pathogenesis of the disease. Component molecules involved in the virulence of ECPs include the metalloprotease Yrp1 (Secades and Guijarro, 1999; Fernández et al., 2002, 2003), the haemolysin YhlA (Fernández et al., 2007), and an azocasein protease (Secades and Guijarro, 1999). Yrp1 is a 47 kDa protease produced at the end of the exponential growth phase (Secades and Guijarro, 1999). It digests various extracellular matrix and muscle proteins and is thought to contribute to the virulence of *Y. ruckeri*; for example, it may cause leakage of the blood capillaries and so give rise to the typical haemorrhages around the oral cavity (Fernández et al., 2003). The *Y. ruckeri* haemolysin YhlA is considered to play an active role in pathogenicity (Fernández et al., 2007). The production of YhlA increased when bacteria were grown under iron-limited conditions, suggesting a possible role in the acquisition of iron from the host (Fernández et al., 2007; Tobback, 2009).

25.6 Control Strategies

25.6.1 Antibiotic therapy

Antimicrobial compounds are commonly used to treat *Y. ruckeri* infections, though the antibiotics used are limited to amoxicillin, oxolinic acid, oxytetracycline and sulfadiazine, in combination with trimethroprim and florfenicol (Michel et al., 2003); this limitation may lead to the development

of antibiotic resistance. In the UK, the only fully licensed and available antibiotics are florfenicol, oxytetracycline and amoxicillin; both florfenicol and oxytetracycline are commonly used in the salmon industry. The antibiotic oxolinic acid (at 10 mg/kg body weight for 10 days) is also used in the UK (Verner-Jeffreys and Taylor, 2015) in conjunction with a Special Import Licence, as the drug is considered to be the most efficacious and cost-effective available (Richard Hopewell, Dawnfresh Seafoods, personal communication).

In general, the regulations differ in permitted antibiotic usage from country to country and the appropriate regulatory authorities need to be consulted. Most European isolates of *Y. ruckeri* are still widely responsive to antibiotic therapy (Calvez et al., 2014), but some isolates in the USA are completely resistant to therapeutic levels of sulfamerazine and oxytetracycline (Post, 1987), and resistance to tetracyclines and sulfonamides has also been reported (De Grandis and Stevenson, 1985). A β-lactamase gene, which would counter the use of β-lactam antibiotics, has been discovered in *Y. ruckeri* (Mammeri et al., 2006), although the evidence suggests that the gene is not likely to be expressed at high levels (Stock et al., 2002). Rodgers (2001) used an *in vitro* test to demonstrate the rapid development of resistance against oxolinic acid, oxytetracycline and potentiated sulfonamides in *Y. ruckeri*. This has been confirmed with isolates of *Y. ruckeri* from trout farms in Bulgaria (Orozova et al., 2014).

25.6.2 Probiotics

The emergence of antibiotic resistance has driven the search for alternative control measures and the use of probiotics is one option. Probiotics are cultured products or live microbial feed supplements that have a beneficial effect on the host by improving the intestinal microbial balance (Fuller, 1989). Their use in aquaculture has been reviewed by Irianto and Austin (2002) and Newaj-Fyzul et al. (2014). The protective mechanisms of probiotics are generally unknown, though the most likely modes of action include stimulation of the immune response, the alteration of microbial metabolism, and competitive exclusion by the production of inhibitory compounds or competition for nutrients, space or oxygen (Fuller, 1989; Newaj-Fyzul et al., 2014).

There are a number of probiotics that improve resistance to *Y. ruckeri* infection and this approach has significant potential for the control of ERM

(Raida et al., 2003; Kim and Austin, 2006; Brunt and Austin, 2007; Balcazar et al., 2008; Capkin and Altinok, 2009; Sica et al., 2012). Subcellular components of Aeromonas sobria and Bacillus subtilis, including cell wall proteins, OMPs, LPS and whole cell proteins, also provided increased fish survival against Y. ruckeri (Abbass et al., 2010). However, the use of probiotics has a potential downside (Irianto and Austin, 2002; Brunt and Austin, 2007) because certain bacteria (e.g. A. sobria) with probiotic potential may also be pathogenic to fish (Austin and Austin, 2012). The incorporation of bacteria into animal feed could also lead to a reservoir of drug resistance, which would have wider implications for human and animal health. Furthermore, the genetic exchange of virulence factors could occur between probiotic bacteria and the natural flora of the host, leading to the emergence of new pathogens.

25.6.3 Vaccination

Vaccines against ERM are widely used and vaccination is a success story. Bacterin-based vaccines have been commercially available for more than 30 years (Ross and Klontz, 1965), although the exact protective mechanisms are poorly understood. Vaccines are administered via immersion, spraying, injection or orally, and they provide good protection against disease (see Kumar et al., 2015). The current vaccination strategy involves immersion or injection (intraperitoneal or intramuscular) at the fry stage. After immersion vaccination, it is standard practice to administer a booster 6 months post vaccination to provide continued protection (Tatner and Horne, 1985). The oral administration of first-feeding Atlantic salmon fry with a microencapsulated vaccine formulation provides more efficacious protection than immersion (Ghosh et al., 2016).

Vaccines against ERM are generally monovalent, inactivated whole-cell suspensions of serotype O1, biotype 1 isolates (see Kumar et al., 2015), but a bivalent vaccine containing biotype 1 and 2 isolates (AquaVac RELERA™) is available for use in rainbow trout (Tinsley et al., 2011b; Deshmukh et al., 2012), while a monovalent serotype O1b vaccine (Yersinivac-B) is used in Atlantic salmon (Costa et al., 2011). Serotype O1-based vaccines are designed primarily to prevent ERM in rainbow trout, but they are also widely used in the salmon-farming industry. The serotype O1b-based vaccine

is aimed specifically at Atlantic salmon in Australasia (Costa et al., 2011).

Alternative approaches to whole-cell vaccines include subunit and DNA vaccines. Fernández et al. (2003) demonstrated that a Yrp1 toxoid injected intramuscularly (IM) generated a high level of protection (relative percentage survival (RPS) 79%) in rainbow trout. An ECP preparation also provided protection (74–81% RPS) in rainbow trout challenged intraperitoneally (IP) (Ispir and Dorucu, 2010). The use of non-adjuvanted flagellin as a subunit vaccine provided non-specific protection of rainbow trout against challenge with both biotype 1 (flagellated) and biotype 2 (non-flagellated) strains (Scott et al., 2013). LPS administered via IP injection and immersion provided excellent protection (77–84% RPS) after an IP challenge (Ispir and Dorucu, 2014). Purified LPS is a highly potent protective antigen (Welch and LaPatra, 2016) and it was suggested that LPS is the only component evoking protection that is elicited by whole-cell vaccines. These observations confirm the conclusions of Tinsley et al. (2011b) that the O-antigen is the dominant immunogen. Consequently, these findings suggest that serotype O1 vaccines are likely to offer limited cross-protection against other serotypes because, by definition, each serotype is highly immunospecific.

In general, live-attenuated vaccines elicit stronger cell-mediated immunity than whole-cell vaccines. A highly attenuated serotype O1 aroA (the gene for 5-enolpyruvylshikimate 3-phosphate synthase) mutant was created for use as a live vaccine against ERM (Temprano et al., 2005). The vaccination of rainbow trout conferred significant protection (90% RPS) after IP challenge with a biotype 1 Y. ruckeri strain. However, concerns about the environmental safety of genetically modified organisms are likely to restrict the commercial use of such live attenuated vaccines.

Vaccination will probably become a more important disease control strategy as antibiotic resistance increases, although its widespread and long-term use also comes at a cost. This is because vaccination exerts a selective pressure on bacterial populations that may shift the predominant serotypes circulating within the host species via serotype replacement or switching. Serotype switching has occurred in Streptococcus iniae (Bachrach et al., 2001) and it is due to specific mutations within a limited number of genes in the capsule biosynthesis operon (Millard *et al.*, 2012). So is there evidence of serotype replacement and/or switching occurring in Y. *ruckeri*? Serotype O1-based vaccines were introduced in Chile in 1995 to control yersiniosis in Atlantic salmon (Bravo and Midtlyng, 2007), but the O1b serotype emerged in 2008, and Bastardo *et al.* (2011) suggested that vaccination may have provided the selective pressure for the emergence of this serotype. Similarly, the emergence of serotype O8 in Scottish Atlantic salmon in recent years may be the result of the widespread use of serotype O1-based vaccines in the salmon-farming industry (Ormsby *et al.*, 2016). It has also been suggested that the emergence of biotype 2 strains of Y. *ruckeri* has been driven by the use of biotype 1, serotype O1 vaccines (Fouz *et al.*, 2006; Welch *et al.*, 2011; Scott *et al.*, 2013). Clearly, future surveillance will be important for monitoring potential changes in the dominant disease-causing serotypes that are likely to result from vaccination.

25.7 Conclusions and Suggestions for Future Research

Due to the increasing importance of aquaculture and the potential economic impact of Y. *ruckeri* on the fish-farming industry, advances in the rapid diagnosis, prevention and treatment of ERM should be among the main priorities in the future. Diagnostic tools, such as real-time PCR assays, that enable rapid confirmation of the species identity, biotype and serotype of clinical specimens would be particularly beneficial. Due to increasing antibiotic resistance, vaccines and probiotics are likely to take on greater significance in disease control. Ultimately, an improved understanding of host–pathogen interactions and the molecular basis of pathogenesis will allow more reliable and effective control strategies to be developed. The development of improved vaccines is likely to utilize approaches that can monitor the individual responses of specific host genes as predictive indicators of vaccine success, such as cDNA microarray (Bridle *et al.*, 2012) and 'omics' technologies. Such approaches will also allow the identification of the specific bacterial antigens involved in stimulating a protective immune response. The development of appropriate *in vitro* models will reduce the need for costly, time-consuming and unethical *in vivo* experiments.

References

Abbass, A., Sharifuzzaman, S.M. and Austin, B. (2010) Cellular components of probiotics control *Yersinia ruckeri* infection in rainbow trout, *Oncorhynchus mykiss* (Walbaum). *Journal of Fish Diseases* 33, 31–37.

Altinok, I., Grizzle, J.M. and Liu, Z. (2001) Detection of *Yersinia ruckeri* in rainbow trout blood by use of the polymerase chain reaction. *Diseases of Aquatic Organisms* 44, 29–34.

Arias, C.R., Olivares-Fuster, O., Hayden, K., Shoemaker, C.A., Grizzle, J.M. and Klesius, P.H. (2007) First report of *Yersinia ruckeri* biotype 2 in the USA. *Journal of Aquatic Animal Health* 19, 35–40.

Austin, B. and Austin, D.A. (2012) *Bacterial Fish Pathogens: Disease in Farmed and Wild Fish*, 5th edn. Springer/Praxis Publishing, Chichester, UK.

Bachrach, G., Zlotkin, A., Hurvitz, A., Evans, D.L. and Eldar, A. (2001) Recovery of *Streptococcus iniae* from diseased fish previously vaccinated with a *Streptococcus* vaccine. *Applied and Environmental Microbiology* 67, 3756–3758.

Balcazar, J.L., Vendrell, D., de Blas, I., Ruiz-Zarzuela, I., Muzquiz, J.L. and Girones, O. (2008) Characterization of probiotic properties of lactic acid bacteria isolated from intestinal microbiota of fish. *Aquaculture* 278, 188–191.

Bastardo, A., Bohle, H., Ravelo, C., Toranzo, A.E. and Romalde, J.L. (2011) Serological and molecular heterogeneity among *Yersinia ruckeri* strains isolated from farmed Atlantic salmon (*Salmo salar*) in Chile. *Diseases of Aquatic Organisms* 93, 207–214.

Bastardo, A., Ravelo, C. and Romalde, J.L. (2012a) Highly sensitive detection and quantification of the pathogen *Yersinia ruckeri* in fish tissues by using real-time PCR. *Applied Microbiology and Biotechnology* 96, 511–520.

Bastardo, A., Ravelo, C. and Romalde, J.L. (2012b) A polyphasic approach to study the intraspecific diversity of *Yersinia ruckeri* strains isolated from recent outbreaks in salmonid culture. *Veterinary Microbiology* 160, 176–182.

Bastardo, A., Ravelo, C. and Romalde, J.L. (2012c) Multilocus sequence typing reveals high genetic diversity and epidemic population structure for the fish pathogen *Yersinia ruckeri*. *Environmental Microbiology* 14, 1888–1897.

Bravo, S. and Midtlyng, P.J. (2007) The use of fish vaccines in the Chilean salmon industry 1999–2003. *Aquaculture* 270, 36–42.

Bridle, A.R., Koop, B.F. and Nowak, B.F. (2012) Identification of surrogates of protection against yersiniosis in immersion vaccinated Atlantic salmon. *PLoS One* 7(7): e40841.

Brunt, J. and Austin, B. (2007) The development of probiotics for the control of multiple bacterial diseases of rainbow trout, *Oncorhynchus mykiss* (Walbaum). *Journal of Fish Diseases* 30, 573–579.

Bullock, G.L., Stuckley, H.M. and Shotts, E.B. (1978) Enteric redmouth bacterium: comparison of isolates from different geographic areas. *Journal of Fish Diseases* 1, 351–356.

Busch, R.A. (1978) Enteric redmouth disease (Hagerman strain). *Marine Fisheries Review* 40, 42–51.

Busch, R.A. and Lingg, A.J. (1975) Establishment of an asymptomatic carrier strate infection of enteric redmouth disease (Hagerman Strain) in rainbow trout (*Salmo giardneri*). *Journal of the Fisheries Research Board of Canada* 32, 2429–2432.

Calvez, S., Gantelet, H., Blanc, G., Douet, D.G. and Daniel, P. (2014) *Yersinia ruckeri* biotypes 1 and 2 in France: presence and antibiotic susceptibility. *Diseases of Aquatic Organisms* 109, 117–126.

Calvez, S., Mangion, C., Douet, D.G. and Daniel, P. (2015) Pulsed-field gel electrophoresis and multi locus sequence typing for characterizing genotype variability of *Yersinia ruckeri* isolated from farmed fish in France. *Veterinary Research* 46, 1–13.

Capkin, E. and Altinok, I. (2009) Effects of dietary probiotic supplementations on prevention/treatment of yersiniosis disease. *Journal of Applied Microbiology* 106, 1147–1153.

Carson, J. and Wilson, T. (2009) *Yersiniosis in Fish*. Australia and New Zealand Standard Diagnostic Procedure, Committee on Animal Health Laboratory Standards (SCAHLS) [of Australia and New Zealand], Animal Health Committee, Australia. Available at: http://www.scahls.org.au/procedures/documents/aqanzsdp/yersiniosis.pdf (accessed 23 November 2016).

Coquet, L., Cosette, P., Quillet, L., Petit, F., Junter, G.A. and Jouenne, T. (2002) Occurrence and phenotypic characterization of *Yersinia ruckeri* strains with biofilm-forming capacity in a rainbow trout farm. *Applied and Environmental Microbiology* 68, 470–475.

Cossarini-Dunier, M. (1985) Indirect enzyme-linked immunosorbent assay (ELISA) to titrate rainbow trout serum antibodies against two pathogens: *Yersinia ruckeri* and Egtved virus. *Aquaculture* 49, 197–208.

Costa, A.A., Leef, M.J., Bridle, A.R., Carson, J. and Nowak, B.F. (2011) Effect of vaccination against yersiniosis on the relative percent survival, bactericidal and lysozyme response of Atlantic salmon, *Salmo salar*. *Aquaculture* 315, 201–206.

Daly, J.G., Lindvik, B. and Stevenson, R.M.W. (1986) Serological heterogeneity of recent isolates of *Yersinia ruckeri* from Ontario and British Columbia. *Diseases of Aquatic Organisms* 1, 151–153.

Danley, M.L., Goodwin, A.E. and Killian, H.S. (1999) Epizootics in farm-raised channel catfish, *Ictalurus punctatus* (Rafinesque), caused by the enteric redmouth bacterium *Yersinia ruckeri*. *Journal of Fish Diseases* 22, 451–456.

Davies, R.L. (1989) Biochemical and cell-surface characteristics of *Yersinia ruckeri* in relation to the epizootology and pathogenesis of infections in fish. PhD thesis, University of Stirling, Stirling, UK.

Davies, R.L. (1990) O-serotyping of *Yersinia ruckeri* with special emphasis on European isolates. *Veterinary Microbiology* 22, 299–307.

Davies, R.L. (1991a) Outer membrane protein profiles of *Yersinia ruckeri*. *Veterinary Microbiology* 26, 125–140.

Davies, R.L. (1991b) Clonal analysis of *Yersinia ruckeri* based on biotypes, serotypes and outer membrane protein types. *Journal of Fish Diseases* 14, 221–228.

Davies, R.L. (1991c) Virulence and serum-resistance in different clonal groups and serotypes of *Yersinia ruckeri*. *Veterinary Microbiology* 29, 289–297.

Davies, R.L. (1991d) *Yersinia ruckeri* produces four iron-regulated outer membrane proteins but does not produce detectable siderophores. *Journal of Fish Diseases* 14, 563–570.

Davies, R.L. and Frerichs, G.N. (1989) Morphological and biochemical differences among isolates of *Yersinia ruckeri* obtained from wide geographical areas. *Journal of Fish Diseases* 12, 357–365.

De Grandis, S.A. and Stevenson, R.M.W. (1982) Variations in plasmid profiles and growth characteristics of *Yersinia ruckeri* strains. *FEMS Microbiology Letters* 15, 199–202.

De Grandis, S.A. and Stevenson, R.M.W. (1985) Antimicrobial susceptibility patterns and R plasmid-mediated resistance of the fish pathogen *Yersinia ruckeri*. *Antimicrobial Agents and Chemotherapy* 27, 938–942.

Deshmukh, S., Raida, M.K., Dalsgaard, I., Chettri, J.K., Kania, P.W. and Buchmann, K. (2012) Comparative protection of two different commercial vaccines against *Yersinia ruckeri* serotype O1 and biotype 2 in rainbow trout (*Oncorhynchus mykiss*). *Veterinary Immunology and Immunopathology* 145, 379–385.

Faraldo-Gómez, J.D. and Sansom, M.S.P. (2003) Acquisition of siderophores in Gram-negative bacteria. *Nature Reviews Molecular Cell Biology* 4, 105–116.

Fernández, L., Secades, P., Lopez, J.R., Márquez, I. and Guijarro, J.A. (2002) Isolation and analysis of a protease gene with an ABC transport system in the fish pathogen *Yersinia ruckeri*: insertional mutagenesis and involvement in virulence. *Microbiology* 148, 2233–2243.

Fernández, L., Lopez, J.R., Secades, P., Menendez, A. and Guijarro, J.A. (2003) *In vitro* and *in vivo* studies of the Yrp1 protease from *Yersinia ruckeri* and its role in protective immunity against enteric redmouth disease of salmonids. *Applied and Environmental Microbiology* 69, 7328–7335.

Fernández, L., Marquez, I., Guijarro, J.A. and Ma, I. (2004) Identification of specific *in vivo*-induced (*ivi*) genes in *Yersinia ruckeri* and analysis of ruckerbactin, a catecholate siderophore iron acquisition system.

M. Ormsby and R. Davies

Applied and Environmental Microbiology 70, 5199–5207.

Fernández, L., Prieto, M. and Guijarro, J.A. (2007) The iron- and temperature-regulated haemolysin YhlA is a virulence factor of *Yersinia ruckeri*. *Microbiology* 153, 483–489.

Fouz, B., Zarza, C. and Amaro, C. (2006) First description of non-motile *Yersinia ruckeri* serovar I strains causing disease in rainbow trout, *Oncorhynchus mykiss* (Walbaum), cultured in Spain. *Journal of Fish Diseases* 29, 339–346.

Fuller, R. (1989) Probiotics in man and animals. *Journal of Applied Bacteriology* 66, 365–378.

Garcia, J.A., Dominguez, L., Larsen, J.L. and Pedersen, K. (1998) Ribotyping and plasmid profiling of *Yersinia ruckeri*. *Journal of Applied Microbiology* 85, 949–955.

Ghosh, B., Nguyen, T.D., Crosbie, P.B.B., Nowak, B.F. and Bridle, A.R. (2016) Oral vaccination of first-feeding Atlantic salmon, *Salmo salar L.*, confers greater protection against yersiniosis than immersion vaccination. *Vaccine* 34, 1–10.

Gibello, A., Blanco, M.M., Moreno, M.A., Cutuli, M.T. and Domínguez, L. (1999) Development of a PCR assay for detection of *Yersinia ruckeri* in tissues of inoculated and naturally infected trout. *Applied and Environmental Microbiology* 65, 346–350.

Haig, S.J., Davies, R.L., Welch, T.J., Reese, R.A. and Verner-Jeffreys, D.W. (2011) Comparative susceptibility of Atlantic salmon and rainbow trout to *Yersinia ruckeri*: relationship to O antigen serotype and resistance to serum killing. *Veterinary Microbiology* 147, 155–161.

Horne, M.T. and Barnes, A.C. (1999) Enteric redmouth disease (*Yersinia ruckeri*). In: Woo, P.K.T. and Bruno, D.W. (eds) *Fish Diseases and Disorders, Volume 3: Viral, Bacterial and Fungal Infections*. CAB International, Wallingford, UK, pp. 455–477.

Huang, Y., Runge, M., Michael, G.B., Schwarz, S., Jung, A. and Steinhagen, D. (2013) Biochemical and molecular heterogeneity among isolates of *Yersinia ruckeri* from rainbow trout (*Oncorhynchus mykiss*, Walbaum) in north west Germany. *BMC Veterinary Research* 9:215.

Irianto, A. and Austin, B. (2002) Probiotics in aquaculture. *Journal of Fish Diseases* 25, 633–642.

Ispir, U. and Dorucu, M. (2010) Effect of immersion booster vaccination with *Yersinia ruckeri* extracellular products (ECP) on rainbow trout *Oncorhynchus mykiss*. *International Aquatic Research* 2, 127–130.

Ispir, U. and Dorucu, M. (2014) Efficacy of lipopolysaccharide antigen of *Yersinia ruckeri* in rainbow trout by intraperitoneal and bath immersion administration. *Research in Veterinary Science* 97, 271–273.

Kawula, T.H., Lelivelt, M.J. and Orndorff, P.E. (1996) Using a new inbred fish model and cultured fish tissue cells to study *Aeromonas hydrophila* and *Yersinia*

ruckeri pathogenesis. *Microbial Pathogenesis* 20, 119–125.

Keeling, S.E., Johnston, C., Wallis, R., Brosnahan, C.L., Gudkovs, N. and McDonald, W.L. (2012) Development and validation of real-time PCR for the detection of *Yersinia ruckeri*. *Journal of Fish Diseases* 35, 119–125.

Kim, D.-H. and Austin, B. (2006) Innate immune responses in rainbow trout (*Oncorhynchus mykiss*, Walbaum) induced by probiotics. *Fish and Shellfish Immunology* 21, 513–524.

Kumar, G., Menanteau-Ledouble, S., Saleh, M. and El-Matbouli, M. (2015) *Yersinia ruckeri*, the causative agent of enteric redmouth disease in fish. *Veterinary Research* 46, 1–10.

Lucangeli, C., Morabito, S., Caprioli, A., Achene, L., Busani, L. *et al.* (2000) Molecular fingerprinting of strains of *Yersinia ruckeri* serovar O1 and *Photobacterium damsela* subsp. *piscicida* isolated in Italy. *Veterinary Microbiology* 76, 273–281.

Mammeri, H., Poirel, L., Nazik, H. and Nordmann, P. (2006) Cloning and functional characterization of the ambler class C β-lactamase of *Yersinia ruckeri*. *FEMS Microbiology Letters* 257, 57–62.

Michel, C., Kerouault, B. and Martin, C. (2003) Chloramphenicol and florfenicol susceptibility of fish-pathogenic bacteria isolated in France: comparison of minimum inhibitory concentration, using recommended provisory standards for fish bacteria. *Journal of Applied Microbiology* 95, 1008–1015.

Millard, C.M., Baiano, J.C.F., Chan, C., Yuen, B., Aviles, F. *et al.* (2012) Evolution of the capsular operon of *Streptococcus iniae* in response to vaccination. *Applied and Environmental Microbiology* 78, 8219–8226.

Newaj-Fyzul, A., Al-Harbi, A.H. and Austin, B. (2014) Review: developments in the use of probiotics for disease control in aquaculture. *Aquaculture* 431, 1–11.

O'Leary, P.J. (1977) Enteric redmouth bacterium of salmonids: a biochemical and serological comparison of selected isolates. PhD Thesis, Oregon State University, Corvallis, Oregon.

Ohtani, M., Villumsen, K.R., Strøm, H.K. and Raida, M.K. (2014) 3D Visualization of the initial *Yersinia ruckeri* infection route in rainbow trout (*Oncorhynchus mykiss*) by optical projection tomography. *PLoS One* 9(2): e89672.

Ormsby, M.J., Caws, T., Burchmore, R., Wallis, T., Verner-Jeffreys, D.W. and Davies, R.L. (2016) *Yersinia ruckeri* isolates recovered from diseased Atlantic salmon (*Salmo salar*) in Scotland are more diverse than those from rainbow trout (*Oncorhynchus mykiss*) and represent distinct sub-populations. *Applied and Environmental Microbiology* 82, 5785–5794.

Orozova, P., Chikova, V. and Sirakov, I. (2014) Diagnostics and antibiotic resistance of *Yersinia ruckeri* strains isolated from trout fish farms in Bulgaria. *International Journal of Development Research* 4, 2727–2733.

Post, G. (1987) Pathogenic *Yersinia ruckeri*, enteric red-mouth disease (yersiniosis). In: Post, G (ed.) *Textbook of Fish Health*. THF Publications, Neptune City, New Jersey, pp. 47–51.

Pyle, S.W. and Schill, W.B. (1985) Rapid serological analysis of bacterial lipopolysaccharides by electrotransfer to nitrocellulose. *Journal of Immunological Methods* 85, 371–382.

Pyle, S.W., Ruppenthal, T., Cipriano, R. and Shotts, E.B. (1987) Further characterization of biochemical and serological characteristics of *Yersinia ruckeri* from different geographic areas. *Microbios Letters* 35, 87–93.

Raida, M.K., Larsen, J.L., Nielsen, M.E. and Buchmann, K. (2003) Enhanced resistance of rainbow trout, *Oncorhynchus mykiss* (Walbaum), against *Yersinia ruckeri* challenge following oral administration of *Bacillus subtilis* and *Bacillus licheniformis* (BioPlus2B). *Journal of Fish Diseases* 26, 495–498.

Rintamaki, P., Tellerovo Valtonen, E. and Frerichs, G.N. (1986) Occurrence of *Yersinia ruckeri* infection in farmed whitefish, *Coregonus peled* Gmelin and *Coregonus muksun* Pallas, and Atlantic salmon, *Salmo salar* L., in northern Finland. *Journal of Fish Diseases* 9, 137–140.

Rodgers, C.J. (1992) Development of a selective-differential medium for the isolation of *Yersinia ruckeri* and its application in epidemiological studies. *Journal of Fish Diseases* 15, 243–254.

Rodgers, C.J. (2001) Resistance of *Yersinia ruckeri* to antimicrobial agents *in vitro*. *Aquaculture* 196, 325–345.

Romalde, J.L. and Toranzo, A.E. (1993) Pathological activities of *Yersinia ruckeri*, the enteric redmouth (ERM) bacterium. *FEMS Microbiology Letters* 112, 291–299.

Romalde, J.L., Conchas, R.F. and Toranzo, A.E. (1991) Evidence that *Yersinia ruckeri* possesses a high affinity iron uptake system. *FEMS Microbiology Letters* 15, 121–125.

Romalde, J.L., Magarinos, B., Barja, J.L. and Toranzo, A.E. (1993) Antigenic and molecular characterization of *Yersinia ruckeri*: proposal for a new intraspecies classification. *Systematic and Applied Microbiology* 16, 411–419.

Romalde, J.L., Barja, J.L., Magarinos, B. and Toranzo, A.E. (1994) Starvation-survival processes of the bacterial fish pathogen *Yersinia ruckeri*. *Systematic and Applied Microbiology* 17, 161–168.

Romalde, J.L., Magarinos, B., Fouz, B., Bandin, I., Nunez, S. and Toranzo, A.E. (1995) Evaluation of BIONOR Mono-kits for rapid detection of bacterial fish pathogens. *Diseases of Aquatic Organisms* 21, 25–34.

Ross, A.J. and Klontz, W. (1965) Oral immunization of rainbow trout (*Salmo gairdneri*) against an etiological agent of 'redmouth disease'. *Journal of the Fisheries Research Board of Canada* 22, 3–9.

Ross, A.J., Rucker, R.R. and Ewing, W.H. (1966) Description of a bacterium associated with redmouth disease of rainbow trout (*Salmo gairdneri*). *Canadian Journal of Microbiology* 12, 763–770.

Rucker, R.R. (1966) Redmouth disease of rainbow trout (*Salmo giardneri*). *Bulletin de l' Office International des Epizooties* 65, 825–830.

Ryckaert, J., Bossier, P., D'Herde, K., Diez-Fraile, A., Sorgeloos, P. *et al.* (2010) Persistence of *Yersinia ruckeri* in trout macrophages. *Fish and Shellfish Immunology* 29, 648–655.

Saleh, M., Soliman, H. and El-Matbouli, M. (2008) Loop-mediated isothermal amplification as an emerging technology for detection of *Yersinia ruckeri* the causative agent of enteric redmouth disease in fish. *BMC Veterinary Research* 4:31.

Schill, W.B., Phelps, S.R. and Pyle, S.W. (1984) Multilocus electrophoretic assessment of the genetic structure and diversity of *Yersinia ruckeri*. *Applied and Environmental Microbiology* 48, 975–979.

Scott, C.J.W., Austin, B., Austin, D.A. and Morris, P.C. (2013) Non-adjuvanted flagellin elicits a non-specific protective immune response in rainbow trout (*Oncorhynchus mykiss*, Walbaum) towards bacterial infections. *Vaccine* 31, 3262–3267.

Secades, P. and Guijarro, J.A. (1999) Purification and characterization of an extracellular protease from the fish pathogen *Yersinia ruckeri* and effect of culture conditions on production. *Applied and Environmental Microbiology* 65, 3969–3975.

Seker, E., Karahan, M., Ispir, U., Cetinkaya, B., Saglam, N. and Sarieyyupoglu, M. (2012) Investigation of *Yersinia ruckeri* infection in rainbow trout (*Oncorhynchus mykiss*, Walbaum 1792) farms by polymerase chain reaction (PCR) and bacteriological culture. *The Journal of the Faculty of Veterinary Medicine, University of Kafkas* 18, 913–916.

Shah, S.Q.A., Karatas, S., Nilsen, H., Steinum, T.M., Colquhoun, D.J. and Sørum, H. (2012) Characterization and expression of the *gyrA* gene from quinolone resistant *Yersinia ruckeri* strains isolated from Atlantic salmon (*Salmo salar L.*) in Norway. *Aquaculture* 350–353, 37–41.

Shaowu, L., Di, W., Hongbai, L. and Tongyan, L. (2013) Isolation of *Yersinia ruckeri* strain H01 from farm-raised Amur sturgeon *Acipenser schrencki* in China. *Journal of Aquatic Animal Health* 25, 9–14.

Sica, M.G., Brugnoni, L.I., Marucci, P.L. and Cubitto, M.A. (2012) Characterization of probiotic properties of lactic acid bacteria isolated from an estuarine environment for application in rainbow trout (*Oncorhynchus mykiss*, Walbaum) farming. *Antonie Van Leeuwenhoek* 101, 869–879.

Smith, A.M., Goldring, O.L. and Dear, G. (1987) The production and methods of use of polyclonal antisera to the pathogenic organisms *Aeromonas salmonicida*, *Yersinia ruckeri* and *Renibacterium salmonicida. Journal of Fish Biology* 31, 225–226.

Sousa, J.A., Magariños, B., Eiras, J.C., Toranzo, A.E. and Romalde, J.L. (2001) Molecular characterization of Portuguese strains of *Yersinia ruckeri* isolated from fish culture systems. *Journal of Fish Diseases* 24, 151–159.

Stevenson, R.M.W. and Airdrie, D.W. (1984) Isolation of *Yersinia ruckeri* bacteriophages. *Applied and Environmental Microbiology* 47, 1201–1205.

Stevenson, R.M.W. and Daly, J.G. (1982) Biochemical and serological characteristics of Ontario isolates of *Yersinia ruckeri*. *Canadian Journal of Fisheries and Aquatic Sciences* 39, 870–876.

Stock, I., Henrichfreise, B. and Wiedemann, B. (2002) Natural antibiotic susceptibility and biochemical profiles of *Yersinia enterocolitica*-like strains: *Y. bercovieri*, *Y. mollaretii*, *Y. aldovae* and '*Y. ruckeri*'. *Journal of Medical Microbiology* 51, 56–69.

Strom-Bestor, M., Mustamaki, N., Heinikainen, S., Hirvela-Koski, V., Verner-Jeffreys, D. and Wiklund, T. (2010) Introduction of *Yersinia ruckeri* biotype 2 into Finnish fish farms. *Aquaculture* 308, 1–5.

Tatner, M.F. and Horne, M. (1985) The effects of vaccine dilution, length of immersion time, and booster vaccinations on the protection levels induced by direct immersion vaccination of brown trout, *Salmo trutta*, with *Yersinia ruckeri* (ERM) vaccine. *Aquaculture* 46, 11–18.

Temprano, A., Yugueros, J., Hernanz, C., Sunchez, M., Berzal, B. and Luengo, J.M. (2001) Rapid identification of *Yersinia ruckeri* by PCR amplification of *yruI–yruR* quorum sensing. *Journal of Fish Diseases* 24, 253–261.

Temprano, A., Riaño, J., Yugueros, J., González, P., de Castro, L. *et al.* (2005) Potential use of a *Yersinia ruckeri* O1 auxotrophic *aroA* mutant as a live attenuated vaccine. *Journal of Fish Diseases* 28, 419–427.

Thorsen, B.K., Enger, O., Norland, S. and Hoff, K.A. (1992) Long-term starvation survival of *Yersinia ruckeri* at different salinities studied by microscopical and flow cytometric methods. *Applied and Environmental Microbiology* 58, 1624–1628.

Tinsley, J.W., Austin, D.A., Lyndon, A.R. and Austin, B. (2011a) Novel non-motile phenotypes of *Yersinia ruckeri* suggest expansion of the current clonal complex theory. *Journal of Fish Diseases* 34, 311–317.

Tinsley, J.W., Lyndon, A.R. and Austin, B. (2011b) Antigenic and cross-protection studies of biotype 1 and biotype 2 isolates of *Yersinia ruckeri* in rainbow trout, *Oncorhynchus mykiss* (Walbaum). *Journal of Applied Microbiology* 111, 8–16.

Tobback, E. (2009) Early pathogenesis of *Yersinia ruckeri* infections in rainbow trout (*Oncorhynchus mykiss*, Walbaum). PhD thesis, Ghent University, Ghent, Belgium.

Tobback, E., Decostere, A., Hermans, K., Haesebrouck, F. and Chiers, K. (2007) Review of *Yersinia ruckeri* infections in salmonid fish. *Journal of Fish Diseases* 30, 257–268.

Tobback, E., Decostere, A., Hermans, K., Ryckaert, J., Duchateau, L. *et al.* (2009) Route of entry and tissue distribution of *Yersinia ruckeri* in experimentally infected rainbow trout *Oncorhynchus mykiss*. *Diseases of Aquatic Organisms* 84, 219–228.

Tobback, E., Hermans, K., Decostere, A., Van Den Broeck, W., Haesebrouck, F. and Chiers, K. (2010a) Interactions of virulent and avirulent *Yersinia ruckeri* strains with isolated gill arches and intestinal explants of rainbow trout *Oncorhynchus mykiss*. *Diseases of Aquatic Organisms* 90, 175–179.

Tobback, E., Decostere, A., Hermans, K., Van den Broeck, W., Haesebrouck, F. and Chiers, K. (2010b) *In vitro* markers for virulence in *Yersinia ruckeri*. *Journal of Fish Diseases* 33, 197–209.

Verner-Jeffreys, D.W. and Taylor, N.J. (2015) *SARF100 – Review of Freshwater Treatments Used in the Scottish Freshwater Rainbow Trout Aquaculture Industry*. Cefas (Centre for Environment, Fisheries and Aquaculture Science, Weymouth Laboratory, Weymouth, UK) contract report C6175A commissioned and published by SARF. Scottish Aquaculture Research Forum, Pitlochry, UK. Available at: http://www.sarf.org.uk/cms-assets/documents/208213-793666.sarf100.pdf (accessed 23 November 2016).

Vuillaume, A., Brun, R., Chene, P., Sochon, E. and Lesel, R. (1987) First isolation of *Yersinia ruckeri* from sturgeon, *Acipenser baeri* Brandt, in south west of France. *Bulletin of the European Association of Fish Pathologists* 7, 18–19.

Waltman, W.D. and Shotts, E.B. (1984) A medium for the isolation and differentiation of *Yersinia ruckeri*. *Canadian Journal of Fisheries and Aquatic Sciences* 41, 804–806.

Welch, T.J. and LaPatra, S. (2016) *Yersinia ruckeri* lipopolysaccharide is necessary and sufficient for eliciting a protective immune response in rainbow trout (*Oncorhynchus mykiss*, Walbaum). *Fish and Shellfish Immunology* 49, 420–426.

Welch, T.J., Verner-Jeffreys, D.W., Dalsgaard, I., Wiklund, T., Evenhuis, J.P. *et al.* (2011) Independent emergence of *Yersinia ruckeri* biotype 2 in the United States and Europe. *Applied and Environmental Microbiology* 77, 3493–3499.

Wheeler, R.W., Davies, R.L., Dalsgaard, I., Garcia, J., Welch, T.J. *et al.* (2009) *Yersinia ruckeri* biotype 2 isolates from mainland Europe and the UK likely represent different clonal groups. *Diseases of Aquatic Organisms* 84, 25–33.

Index

Page numbers in **bold** type refer to figures and tables.

bacterial gill disease (BGD)
    causal agent 211, 214
    impacts and control 215, 221, 222
    pathology of infection **215**, 217, 220
bacterial kidney disease (BKD)
    causal agent 286–287
    clinical signs and diagnosis 287–289, **289**
    histopathology 289–290, **290**, **291**
    physiology of infections 290–292
    prevention and control 292–295
behavioural changes *see* swimming behaviour, abnormal
Betanodavirus (genus) 128–129, 133, **137**, 138–140
BGD *see* bacterial gill disease
biofilms, bacterial
    in aquaria, probiotic effect 222
    resistance to disinfection 110, 221, 241
    role in pathogen survival 212, 214, 235
        Vibrio spp. 318, 320, 322
    as source of recurrent infection 340
bioluminescent bacteria 258, 318, 325
biosecurity
    effect of measures on disease containment 68, 74,
        167, 294
    guidelines for disease-endemic areas 18–19
    import quarantine measures 45
    legislation and standards 85–86, 124
    measures for untreatable viral diseases 6, 139
biotypes (motile/non-motile *Y. ruckeri*) 341, **342**
birds, role in transmission 81, 201–202
Birnaviridae (family) 1, 9
BKD *see* bacterial kidney disease
black bullhead catfish *(Ameiurus melas)* 38, 41, **44**, 79, 91
blood samples
    bacteraemia in photobacteriosis 263, **263**
    histopathology of IHN infection 16
    ISAV infection on red blood cells 72, 73–74
blue catfish *(Ictalurus furcatus)* 91, 99
breeding *see* selection for resistance

capsid structure, viruses
    betanodavirus serotypes 129
    herpesviruses 61
    infectious pancreatic necrosis virus (IPNV) 1–2
    ranaviruses (EHNV, ECV) 39–40
    salmonid alphavirus 160
capsular typing, *Streptococcus* spp. 298, **299**
carp pox
    control strategies 63–64
    diagnosis and pathology **62**, 62–63, **63**
    prevalence 51, 61
    viral agent, cyprinid herpesvirus 1 (CyHV-1) 61–62
catfish
    European incidence of ECV/EHNV diseases 41
    signs of *Edwardsiella* infection 198
    see also black bullhead catfish; channel catfish;
        sheatfish

CCVD *see* channel catfish viral disease
CD *see* Columnaris disease
cell-mediated cytotoxicity (CMC) 31
cell-surface components
    bacterial, and adaptation to host 318, 322
    membrane proteins as viral receptors 138
    MSA virulence factor, *R. salmoninarum* 292, 293
    surface adhesins, *Aeromonas* spp. 178
cells, diseased appearance *see* cytopathic effects
channel catfish *(Ictalurus punctatus)* 91–99, **95**, 174, 195
    Columnaris disease **213**, 214, 215, 217, 223
    enteric septicaemia (ESC) 190, 200–203
    streptococcosis 304
channel catfish viral disease (CCVD)
    clinical signs and diagnosis 92–95, **93**, **94**, **95**
    histopathology and pathophysiology 95–98, **96**, **97**, **98**
    prevalence of virus and disease 91–92
    protective and control strategies 98–99
    viral agent, IcHV-1 (CCV) 91
chum salmon *(Oncorhynchus keta)* 54, 56–58, **59**, 60
classification *see* taxonomy
clinical signs
    bacterial diseases
        Aeromonas spp. infections 174, 175–177, **176**
        bacterial kidney disease (BKD)
            287, 288–289, **289**
        Edwardsiella spp. infections 195, 196, 198,
            201, 202
        enteric redmouth disease 339, **340**
        flavobacterial infections **212**, **213**, **215**, 215–217
        mycobacterial infections 248, **249**, 250–251
        photobacteriosis **262**, 262–263, 266
        piscine francisellosis 235–236, **236**, **237**
        piscirickettsiosis 275, **277**
        streptococcal infections 300, **302**
        vibriosis 315, **316**, 320–321, **321**, 324
        weissellosis 334–336, **336**
    viral diseases
        channel catfish viral disease (CCVD) 92,
            94–95, **95**
        infectious haematopoietic necrosis (IHN) 14, **15**
        infectious pancreatic necrosis (IPN) 3, 163
        infectious salmon anaemia (ISA) 70, **70**
        iridoviral diseases (RSIVD, WSIV) 148, 153
        koi herpesvirus disease (KHVD) 116, **117**,
            118, 119
        largemouth bass viral (LMBV) disease **104**,
            108, **109**
        lineage 3 ranavirus diseases (EHN, ECV) 42–43
        Oncorhynchus masou virus disease
            (OMVD) 56–57
        pancreas disease of salmonids (SAV) **162**,
            162–163
        spring viraemia of carp (SVC) 81–82, **82**
        viral haemorrhagic septicaemia (VHS) 27
        viral nervous necrosis (VNN) 133–135, **134**,
            136, **136**

economic impacts of disease (*continued*)
  flavobacteria in farmed fish 214–215
  industry-wide and individual 91
  in mariculture, due to iridoviruses 147
  Oncorhynchus masou virus disease 55–56
  significant setbacks due to ISA 68, 74, 274
  value of losses due to specific diseases
      estimates for streptococcosis 298
      IPN 2–3
      KHVD 115, 116
      piscirikettsiosis in Chile 274–275
      SAV in salmon industry 162
ectoparasites 116, 272, 306
ECV *see* European catfish virus
Edwardsiella (genus) 190–194, **191–192**, **193**, **194**
Edwardsiella anguillarum 190, **191**, 199–200
Edwardsiella ictaluri 190, **191**, 195, 200–203
Edwardsiella piscicida 190, **191**, 197–199
Edwardsiella tarda 190, **191–192**, **193**–197
edwardsiellosis 194, 197, 199
EHN *see* epizootic haematopoietic necrosis
electron microscopy 119
ELISA assays 288
endothelial cells, infection responses 72–73
enteric redmouth disease (ERM, yersiniosis)
  causal bacterium 339
  clinical signs, diagnosis and pathology 339–340, **340**
  control strategies 345–347
  see also Yersinia ruckeri
enteric septicaemia of catfish (ESC) 190, 200–203
Enterobacteriaceae (family) 190
epizootic haematopoietic necrosis (EHN)
  disease signs and diagnosis 41–43
  pathology **43**, 43–45
  protective and control strategies 41, 45–46
  viral agent, EHNV 38–41
epizootics
  condition of survivors 5
  developing from untreated minor outbreaks 74
  in farmed fish fry/fingerlings 211, 216
  genetic niche specialization of strains 343–344
  predisposing factors in fish 260, 265
  reported in cultured sturgeon 152
  survivors as carriers 99, 121, 176
  of tumours caused by herpesviruses 51, 55, 57, 60
ERM *see* enteric redmouth disease
ESC (enteric septicaemia of catfish) 190, 200–203
European catfish virus (ECV) disease
  clinical signs and diagnosis 41–43
  pathology 43–45, **44**
  protective and control strategies 45, 46
  viral agent, ECV 38–41
extracellular products
  exotoxins, *Vibrio* spp. 317–318, 322, 325
  Francisella type VI secretion system 239
  host antibody responses 250
  secreted virulence factors

Aeromonas spp. 178, 179
MSA, *R. salmoninarum* 292, 293
Yersinia ruckeri 345
used as vaccines
  against photobacteriosis 264
  against streptococcal infections 307, 308

faecal casts
  causing acute enteritis 5
  as sign of infection 3, 14, 300
  in water during farm outbreaks 162
feed additives
  antiviral agents 19
  pro- and prebiotics 180–181, 222, 252, 306–307
fish husbandry *see* husbandry practices
fishing, recreational 40, 81, 108–109
  disease prevention guidelines 110
flagella, bacterial 178, 317, 325, 341
flavobacterial diseases
  causal agents 211, 213, 214, 224
  diagnosis 217–220, **219**
  distribution, hosts and transmission 212, 213–214
  economic impacts 214–215
  pathology and pathophysiology **212**, **213**, **215**,
      215–217, 220–221
  protection and control strategies 221–224
Flavobacterium branchiophilum 214, **215**, 217
Flavobacterium columnare **213**, 213–214, 216–217
Flavobacterium psychrophilum 211–212, **212**, 214–216
fluorescent antibody technique (FAT) 56, 218
  indirect (IFAT) **44**, 62, 117, 148, 219
  membrane-filtration (MF-FAT) 288
Francisella noatunensis 233–235, **234**, 237–238
  see also piscine francisellosis
Francisella tularensis 233, 235, 239
functional feeds 168
furunculosis 173, 175–176, **176**, 180–181

gastrointestinal tract
  effects of CCVD 96, **98**
  histopathology of IHNV infection 16
gene transfer, horizontal 173, 177
genome components
  Aeromonas spp. virulence systems 177, 178
  betanodaviruses (NNVs) 128–129
  channel catfish disease virus (IcHV1) 97
  Flavobacterium spp., analysis of genotypes 211, 213
  infectious pancreatic necrosis virus (IPNV) 1, **2**
  infectious salmon anaemia virus (ISAV) 69
  lineage 3 ranaviruses (EHNV, ECV) 38–39
  Mycobacterium spp. 247
  Oncorhynchus masou virus (OMV) 54
  pathogenicity islands in *Francisella* spp. 238–239
  plasmid profiles, *Photobacterium* isolates 260
  Renibacterium salmoninarum 286, 294–295

oncogenic viruses 51–54, 61–62, 64
   see also tumours
Oncorhynchus masou virus disease (OMVD)
      control and protection strategies 60–61
      geographic distribution and importance 54–56
      pathology 56–60, **58**
      signs and diagnosis 56, 57, **58**, **59**
      viral agent, SalHV-2 (OMV) 52–54, 64
oral delivery of vaccines
      advantages over injection/immersion 203
      alginate microsphere encapsulation 8, 9, 20
      for booster vaccination 281
      chitosan-based particles 75
      ghosts (empty bacterial envelopes) 181, 199
      transgenic carriers in pelleted feed 85
ornamental fish as carriers 148, 152, 194, 235
Orthomyxoviridae (family) 68, 73
osmoregulation
      bacterial adaptations 318
      failure as cause of death 121
      impaired by infection 72, 84, 220, 252, 292
outbreaks, influencing factors
      age and stress in hosts 92
      extent of control measures 68, 74
      feeding restriction 202
      human activity 40, 41, 108–109
      imports of aquaculture stock 2, 287
      practical loss reduction methods 99
      strain variation 341
      water supply risk factors 30, 195, 221
      water temperature 148, 201, 265
outer membrane proteins (OMPs)
      identified in *Aeromonas* spp. 178, 179
      iron-regulated 325, 344
      typing, *Yersinia ruckeri* strains 341, 343, **344**
ozone disinfection of eggs 139

pancreas disease (PD)
      clinical signs and diagnosis **162**, 162–163
      control strategies 166–168
      histopathology and physiology 163–166, **164**, **165**
      mortality rates 3, 162, 163
      viral agent, SAV 160–162
         virulence variation 166
pancreatic tissue, disease damage
      acinar cell necrosis, CCVD 95
      atrophy due to pancreatic disease (SAV) 164, **165**
      necrosis due to IPN 4, 5, **5**
papillomas, viral 61–63, **62**
Pasteurella (genus) 258
pathogenesis (disease development)
      channel catfish viral disease 97–98
      francisellosis 237–238
      infectious haematopoietic necrosis 16, 18
      infectious salmon anaemia 72–74
      koi herpesvirus disease 120–121

piscirickettsiosis 279, 281
ranavirus infections (EHNV, ECV) 44–45
vibriosis 316–318, 321–323, 325
viral haemorrhagic septicaemia 29–30
viral nervous necrosis 138
pathogenicity *see* virulence
pathology *see under* individual diseases
pathophysiology *see* physiological effects of disease
PD *see* pancreas disease
phage therapy 223
photobacteriosis
      diagnosis and pathology **261**, 261–263, **263**, 266–267
      disease treatment 263–264, 267
      outbreaks and hosts 258–259, **259**, 260, 264–265
      pasteurellosis type (*P. d. piscicida*) 258–264, **262**
      ulcer disease type (*P. d. damselae*) 264–267
Photobacterium damselae 258, 314
      ssp. *damselae* **261**, 265, 267
         virulence factors 266
      ssp. *piscicida* 259–260, **261**, 267
         virulence factors 260–261
phylogenetic analysis
      betanodaviruses (VNNs) 128–129
      classification of iridoviruses 38–39, **39**, 153
      geographic distribution of genetic groups 14, 69, 84
      methods, *Edwardsiella* spp. 190, **192**, 192–194, **193**, **194**
      molecular methods for *Mycobacterium* spp. 245
      relatedness and cross-protection 293
      relationships among herpesviruses 54, **55**
      tracking of introduction events 160, 168
physiological effects of disease
      associated with anaemia (ISA, RSIVD) 72, 151
      causes of skin disintegration 121
      consequences of reduced circulatory capacity 166, 290–291, 292
      electrolyte/fluid imbalance from renal failure 18, 96–97, 121
      growth impacts 109, 166
      host responses to SVCV infection 84
      impacts of mucosal damage 5
      interaction with environmental stressors 109–110, 292, 305–306
      interactions with fasting 220–221, 222
      long-term effects in survivors 220
      virulence mechanisms
         bacterial 177–179, 238–239
         viral 29–30, 72–74
      wasting and anorexia 155
pigments
      bacterial
         brown (*Aeromonas* spp.) 174, **175**, 179
         photochromogenic (*Mycobacterium* spp.) 245, 246, **246**
         yellow (*Flavobacterium* spp.) 211, 213, **213**, 214, 218
      dispersal by melanomacrophages (BKD) 290, **291**

pike fry rhabdovirus (PFRV) 83
piscine francisellosis
    causal agent and hosts 233–235, **234**, 241
    diagnosis and pathology 235–236, **236**, **237**
    physiology and virulence 237–239, **238**
    protection and control strategies 239–241
Piscirickettsia salmonis
    bacterial characteristics 272, 275–276, **276**
    virulence factors 279–280
piscirickettsiosis
    causal agent 272, **273**
        transmission 272–273
    diagnosis and pathology 275–279, **276**, **277**, **278**
    distribution, hosts and impacts 273–275, **274**
    host responses to virulence 279–280
    protection and control strategies 280–281
plaque assays 15, 56, 108
prebiotics 180–181, 222, 307
prescheduled harvesting 167
probiotics 180, 222, 252, 306–307, 319
    rationale and limitations of use 345–346
protective strategies *see* control and protection strategies
proteins, viral
    as antigens in vaccines 8, 9
    effects of mutations on virulence 30
    infectious pancreatic necrosis virus (IPNV) 1–2, **2**, **3**
    infectious salmon anaemia virus (ISAV) 68–69
    involved in immune evasion 121, 128
    OMV-specific polypeptides 53–54
    receptor-binding, and tissue tropism 165
pseudotuberculosis 258, 262

qPCR *see* real-time (quantitative) PCR
quantitative trait loci (QTLs) 6, 19, 31, 168
quorum sensing
    involved in bioluminescence 325
    quenching (inhibition) 180, 319
    regulatory systems in *Aeromonas* spp.
        174, **175**, 179
    virulence regulation in *Vibrio* spp. 318

rainbow trout *(Oncorhynchus mykiss)*
    enteric redmouth disease 339, 346
    farmed
        Edwardsiella tarda infection 196
        EHNV infection 40, 41, 42–43, 46
        SAV infection and transmission 162
        streptococcal infections 303–304
        Weissella ceti as pathogen 334–335, **336**
    mixed viral infections 6
    oncogenic herpesvirus infection 53–58, **58**, 60–61
    persistence of IHNV in survivors 13
    spawning rash 287
    virulence of VHSV genotypes 29
rainbow trout fry syndrome (RTFS) 212, 216, 222, 223

Ranavirus (genus)
    distribution and host ranges 45, 105
    taxonomic relationships 38–39, **39**, 104, 108
real-time (quantitative) PCR (qPCR)
    antagonism assays 320
    confirmation of IHC diagnosis 70
    sensitivity and specificity 42, 93, 94, **94**, 135
        for different gene targets 219–220
    used for disease monitoring 108, 302–303
    virus detection 4, 16, **17**, 27
        advantages over conventional PCR 118
        laboratory service providers 163
        limitations 84
recombinant bacteria
    expressing viral protein/subunits 20
    incorporating immunogenic proteins 280, 308
    Lactobacillus casei for IPN vaccine delivery 9, **9**
    Lactobacillus plantarum expressing viral
        proteins 85, 124
    resistance gene insertion, *P. damselae* 264
    for vaccines against edwardsiellosis 197, 199
red sea bream iridoviral disease (RSIVD)
    diagnostic methods 148, 150, **150**
    pathology and infection progress 150–151
    protective and control strategies 151
    viral agent, RSIV
        hosts and related megalocytiviruses
            147–148, **149**
        transmission 148, 151
redfin perch *(Perca fluviatilis)* 40–46, **43**
Renibacterium salmoninarum 286–287
    see also bacterial kidney disease
replication, viral
    capability related to virulence 29
    followed by viraemia 165–166
    identification of sites within host 18, 120–121, 138
    inhibition 19, 53
    process typical of ranaviruses 40
reservoirs of infection
    exclusion, in aquaculture facilities 46
    introduction events from wild fish 160–161
    persistence, variation in hosts and tissues 13
    in viruses with broad host range 40, 45
resistant strains, fish *see* selection for resistance
retroviruses 51, 135
Rhabdoviridae (family) 13, 26, 79–80, 83
rickettsias 272, 273
    see also Piscirickettsia salmonis
RSIVD *see* red sea bream iridoviral disease
RTFS *see* rainbow trout fry syndrome

salmon blood spot disease 339
salmon farming industry
    economic threat posed by ISA 68
    mortality rates and losses from IPN 2–3
    patterns of disease spread 70, 160–161